Sequential
Statistical
Procedures

This is Volume 26 in
PROBABILITY AND MATHEMATICAL STATISTICS

A Series of Monographs and Textbooks

Editors: Z. W. BIRNBAUM AND E. LUKACS

A complete list of titles in this series appears at the end of this volume.

SEQUENTIAL STATISTICAL PROCEDURES

Z. *Govindarajulu*

DEPARTMENT OF STATISTICS
UNIVERSITY OF KENTUCKY
LEXINGTON, KENTUCKY

ACADEMIC PRESS New York San Francisco London 1975

A Subsidiary of Harcourt Brace Jovanovich, Publishers

ACADEMIC PRESS, INC.
111 Fifth Avenue, New York, New York 10003

United Kingdom Edition published by
ACADEMIC PRESS, INC. (LONDON) LTD.
24/28 Oval Road, London NW1

Library of Congress Cataloging in Publication Data

Govindarajulu, Z
Sequential statistical procedures.

(Probability and mathematical statistics)
Bibliography: p.
1. Sequential analysis. I. Title.
QA279.7.G68 519.5′3 73-18984
ISBN 0-12-294250-7
AMS (MOS) 1970 Subject Classifications: 60G40, 62L10, 62L12, 62L15

PRINTED IN THE UNITED STATES OF AMERICA

To my mother

The unreal has no existence, and the real never ceases to be; the reality of both has thus been perceived by the seers of truth.

Bhagavad Gita II.16

For a brief space it is granted to us, if we will, to enlighten the darkness that surrounds our path . . . , we press forward, torch in hand, along the path. Soon from behind comes the runner who will out-pace us. All our skill lies in giving into his hand the living torch, bright and unflickering, as we ourselves disappear in the darkness.

Havelock Ellis

A strict accordance between practical work and a corresponding theory is never possible and yet all our life is based on constant practical applications of inapplicable theories.

Jerzy Neyman

Contents

Preface

Sequential procedures are concerned with statistical analysis of data when the number of observations is not predetermined. My interest in sequential analysis grew out of teaching a course on the topic several times to graduate students at the University of Kentucky and also from my research interests in nonparametric sequential procedures. Since A. Wald (1947) wrote his celebrated book *Sequential Analysis*, the subject has grown considerably, especially in the aspect of sequential estimation. The purpose of the present work is to gather in one place the numerous contributions on sequential estimation that are scattered throughout the literature.

Originally, I thought of including material on sequential selection procedures and stochastic approximation. However, limitations of space have excluded such possibilities. The book is made self-contained by providing most of the proofs. Certain mathematical details, which are not required for the main thought but required for the sake of completeness, are given in appendixes. Certain new results are included in Sections 2.4, 3.14, and 4.22. This book is by no means complete as far as all sequential procedures are concerned. A reasonably long list of references is included. A few uncited references, denoted by asterisks, are also included, with an indication as to the chapter or section to which they pertain. My apologies to those whose work has been slighted inadvertently.

Useful illustrations and numerical tables are provided wherever possible, and exercises are given at the end of certain sections. Each result is properly attributed to the relevant author. I have tried to make the statements of the theorems and their proofs formal and precise and the discussions somewhat informal.

I have tried to serve the needs of the student, the research worker, and users of sequential procedures. This book is appropriate for a one-semester or two-quarter course on sequential procedures. The prerequisites are a basic course in mathematical statistics and probability and a course in advanced calculus. Although measure theoretic language is used in a few sections, knowledge of measure theory is not necessary. The sections are somewhat independent in the sense that the reader can read sections of his choice independently of the others. A word of advice to instructors: a semester's course can be based on Sections 1.1–1.4, 2.1–2.5, 2.8, 2.9, 2.11, 2.13, 3.1–3.3, 3.6 (light), 3.7, 3.10 (light), 3.11, 4.1, 4.5, 4.7, 4.6, 4.2 (light), 4.3, 4.8, 4.9, 4.12, 4.13 (light) in the specified order. The instructor can devote about four lectures to the specified sections in Chapter 1, eleven lectures to those sections in Chapter 2, ten lectures to sections in Chapter 3, and fourteen lectures to Chapter 4, each lecture lasting for 50 minutes.

Readers are invited to convey to the author their comments if any and to draw to his attention typographical and other errors, for which the author takes full responsibility.

My sincere thanks are due to Professor Jack Kiefer for his advice on many points and for his useful comments. I extend my thanks to Professor Jack Hall for his helpful comments and his drawing my attention to certain unpublished papers, and to Professor Robert Berk for his careful reading of the earlier draft and his useful suggestions. I would like to express my sincere thanks to Professors Bartoszyński and Saul Blumenthal for their help in critically reading the galley proofs. My special thanks are due to the editorial staff of Academic Press for the generous cooperation extended to me. I wish to express my deep appreciation to Mrs. Nancy Swank who has typed most of the manuscript with great diligence and accuracy. I also appreciate Mrs. Beverly Morrison's and Mrs. Tresa Humphrey's typing certain parts of the book and other help. For permission to reproduce tables, thanks are due to the editors of *Annals of Mathematical Statistics*, *Biometrika*, *Journal of the American Statistical Association*, *Journal of the Royal Statistical Society*, *Proceedings of the Cambridge Philosophical Society*, *Technometrics*, and other journals. Finally I am grateful to the University of Kentucky Research Foundation for providing me with support.

Notation

List of Symbols

Page	*Symbol*	*Meaning*
2	R_m	m-dimensional Euclidean space
2	θ	Parameter
2	H_0	Null hypothesis
2	H_1	Alternative hypothesis
2	α	Probability of type I error
2	β	Probability of type II error
3	ν	Expected sample size
3	D	A procedure
4	N	Stopping times of a sequential procedure
4	$\pi(\theta)$	Operating characteristic function at θ
7	θ or μ	Normal mean
7	σ^2	Normal variance
8	n_0	Initial sample size
15	$f(x; \theta) = B(\theta)e^{-Q(\theta)R(x)}h(x)$	Probability density function
15	Λ_n	Likelihood ratio based on a sample of size n

Page	*Symbol*	*Meaning*
15	$A, B\,(0 < A < B < \infty)$	Stopping bounds
15	K	A generic constant
15	$f(x;\,\theta) = B(\theta)e^{Q(\theta)R(x)}h(x)$	Density of the exponential family
23	Z	$\ln\{f(X;\,\theta_1)/f(X;\,\theta_0)\}$
32	$S_N = \sum_1^N Z_i$	Randomly stopped sum of random variables
32	ϕ	Null set
32	Ω	Sample space
40	$C(t)$	$E(e^{Zt})$
91	Φ	Standard normal distribution function
126	G	Group of transformations on the observation space
127	$\bar{G}$	The induced group of transformations on the parameter space
129	Θ	Parameter space
131	$\mathscr{N}$	Null subset of k-dimensional Euclidean space
132	s_n^2	Sample variance based on a sample of size n
133	$v(G)$	Right Haar measure
140	$V_n(\mathbf{X})_n$	Invariantly sufficient statistic
144	$\mathscr{I}_n$	The G-invariant subsets of the Borel field generated by $X_1, \ldots, X_n$
153	R_t	Excess over the boundary
182	$\Pi_1, \ldots, \Pi_k$	A set of k populations
209	$\mathrm{Re}(\theta) = \mathrm{Re}(\alpha, \beta, \theta)$	Efficiency of the SPRT at θ relative to the optimum fixed-sample size procedure
217	$\mathrm{Re}(\alpha, \beta, \theta, \theta^*)$	The limiting efficiency of the SPRT at θ relative to the fixed-sample size procedure having strength (α, β) and which is designed to test $H_0 : \theta = -\theta^*$ versus $H_1 : \theta = \theta^*$
239	$\mathscr{B}(c)$	Bayes continuation region
254	$F_n(x), G_n(y)$	The empirical distribution functions of X and Y, respectively, based on samples of size n

Page	*Symbol*	*Meaning*
258	$\mathscr{G}$	The class of all distribution functions closed under a certain metric
294	$2d$	Length of the confidence interval
295	$1-\alpha$	Confidence coefficient
303	$\overline{X}_{ij}$	Sample mean based on j observations from ith population
317	$r(\delta(x); \theta)$	The loss incurred when one makes a terminal decision $\delta(x)$ and θ is the true value of the parameter
317	$C(n)$	The cost of taking n observations
332	$R(\delta; \theta)$	$Er(\delta(X); \theta)$
335	$W(\delta, g)$	Expected risk with respect to prior density $g(\theta)$
407	n^0	Optimum fixed-sample size for obtaining a fixed-width interval for the normal mean when σ is known
427	$\phi(y)$	Standard normal density function
452	$\mathscr{n}(\mu, \Sigma)$	A multivariate normal population having mean vector μ and variance–covariance matrix Σ
498	$e(S, M)(e(S, W))$	Asymptotic efficiency of the sign procedure relative to that of Chow and Robbins (Wilcoxon)

List of Abbreviations

Page	*Abbreviation*	*Expansion*
2	OC	Operating characteristic
2	ASN	Average sample number
3	SPRT	Sequential probability ratio test
79	OP	Optimal property
79	M	Monotonicity property
81	ROP	Restricted optimum property
85	MLR	Monotone likelihood ratio
87	GSPRT	Generalized sequential probability ratio test
98	wp1	With probability one
115	FSST	Fixed-sample size test

Page	*Abbreviation*	*Expansion*
147	BAN	Best asymptotically normal
148	mle	Maximum likelihood estimator
209	UME test	Uniformly most-economical test
226	ARE	Asymptotic relative efficiency
235	EVSI	Expected value of sample information
235	EVPI	Expected value of perfect information
259	SCR test	Sequential configural rank test
260	MSCR test	Modified sequential configural rank test
266	ROSPRT	Rank order SPRT
279	SST	Sequential sign test
284	LMP	Locally most powerful
305	UC	Uniformly consistent
305	MM	Minimax
412	a.e.	Almost everywhere
413	a.s.	Almost surely
482	AO	Asymptotically optimal
483	APO	Asymptotically pointwise optimal
491	iid	Independent and identically distributed

CHAPTER 1

Introduction and Certain Double Sampling Procedures

1.1 Introduction to Sequential Testing

Sequential procedures differ from the usual statistical procedures in that the sample size is not fixed in advance. The experimenter has the freedom to look at a sequence of observations one (or some fixed number) at a time and to decide whether to stop sampling and make a decision or to continue sampling and take an action later. It is assumed that the order of the sequence of observations which the experimenter may take is prescribed in advance. Decision problems in which the experimenter has, besides, a choice of experiments to perform at each stage or to vary sequentially the treatments are of a higher order of difficulty. Such problems are called sequential design problems. For instance, consider the following problem.

Problem

Suppose we wish to compare several drugs or treatments (as in sequential screening of cancer drugs), then it should be possible to drop some drugs out of the trials at an early stage, if the results from these compare unfavorably with others.

Thus an essential feature of a sequential procedure is that the number of observations required to terminate the experiment is a random variable since it depends on the outcome of the observations. One is often interested in sequential procedures because they are economical in the sense that we may reach a decision earlier via a sequential procedure than via a fixed-sample-size procedure. In practical life, people seem to make decisions sequentially. In sequential experimentation, we need the following:

(i) the initial sample size,
(ii) a rule for termination of the experiment,
(iii) what additional number of observations to take if the experiment is to be continued, and
(iv) a terminal decision rule.

(ii) and (iii) can be combined into a single rule. When it is only the number of observations in an experiment which is sequentially dependent, the theory tends to be rather simpler than in sequential design problems and the theory will be of general applicability.

If the experimentation has continued until there are m observations, namely $X_1, \ldots, X_m$, a sequential test is completely defined by specifying sets $R_m{}^0$, $R_m{}^1$, and $R_m{}^c$ for all positive integral values m. Since the above sets are to be mutually exclusive and have union R_m, the m-dimensional Euclidean space, it suffices to specify any two of the three sets. If $(X_1, \ldots, X_m)$ belongs to $R_m{}^0$, we accept hypothesis H; if the sample point belongs to $R_m{}^1$, we reject H; and if it falls in the region $R_m{}^c$, we continue sampling. The sets $R_m{}^0$, $R_m{}^1$, $R_m{}^c$ $(m = 1, 2, \ldots)$ can be chosen in many ways, and a basic problem in the theory of sequential tests is that of a suitable choice for these sets. The criteria for the choice of these sets will be dictated by the operating characteristic (OC) and the average sample number (ASN) functions which will be made clear in the following.

Let the underlying distribution function be indexed by a real parameter θ and suppose the statistician is faced with the problem of testing two hypotheses H_0 and H_1. The OC (θ) function is defined to be the probability of accepting H_0 when θ is the true value of the parameter. The OC function describes what the sequential test procedure accomplishes. Thus a desirable property of a sequential procedure is that its OC function should be high for values of θ that are consistent with H_0 and low for values of θ that are consistent with H_1. For instance, $\mathrm{OC}(\theta) \geq 1 - \alpha$ for θ in H_0 and $\mathrm{OC}(\theta) \leq \beta$ for θ in

H_1 where α and β denote the error probabilities. It is desirable that any sequential test S should satisfy the above requirements on the OC function. As noted earlier, the number of observations required by a sequential procedure is random, and of particular interest is the expected value of the sample size when θ is the value of the parameter. So the expected value will be a function of θ and it is called the ASN function. Another desirable property of a sequential rule will be to have a small ASN function, in particular, for a given α and β. We want the expected sample sizes to be smaller than the corresponding sample sizes required by the fixed-sample-size procedure. Let $\nu(\theta \mid D)$ denote the expected sample size of the sequential procedure D when θ is the true value of the parameter. If an admissible procedure D_0 exists for which $\nu(\theta \mid D_0) = \min_D \nu(\theta \mid D)$ (that is, the expected number of observations for D_0 is minimized) for all θ, then D_0 is considered to be a "uniformly best" test. In general, however, no uniformly best test exists. That is, it will not be possible to minimize the ASN function for all θ. Thus, a compromise is in order for the selection of a sequential procedure. The problem of finding an optimal sequential procedure is somewhat easier when H_0 and H_1 are simple hypotheses. Wald's sequential probability ratio test (SPRT) gives the minimum ASN at both H_0 and H_1. The efficiency of a procedure D at θ is defined as the ratio of the minimum expected sample number at θ to the expected sample number of D at θ. Wald's SPRT has efficiency equal to 1 at H_0 and H_1.

1.2 Sampling Inspection

The earliest sequential procedure is the double sampling plan proposed by Dodge and Romig (1929) for sampling inspection. A lot has to be accepted or rejected. The usual single-sample plan is to take a sample of size n and if the number of defectives exceeds c, we reject the lot; and accept the lot if the number of defectives is less than c (Fig. 1.2.1). The drawback of a single-sample size procedure is that at earlier than sample size n, we might have had more than c defectives, which would lead us to reject the lot.

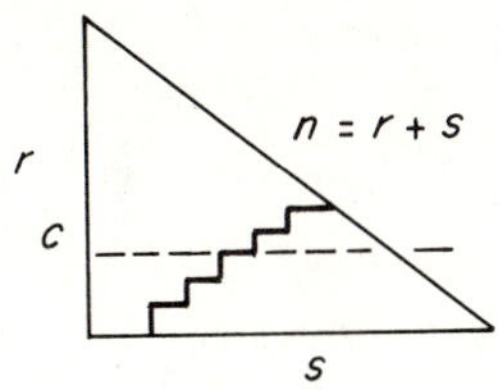

Fig. 1.2.1. Here r is the number of defective items, and s is the number of effective items.

An alternative rule is: Sample one item at a time, without replacement; at any sampling stage let r be the number of defectives and s the number of effective items in the sample; stop and reject the lot if $r \geq c$, stop and accept the lot if $s \geq n - c + 1$; otherwise continue sampling. The sample size required is as small as c (if all the items in the sample are defective), or it may be as large as n.

PROPERTIES OF CURTAILED SAMPLING Let $\pi(\theta) = P(\text{acc } H_0 \mid \theta)$ denote the operating characteristic function. Then some candidates for $\pi(\theta)$ are

$$\pi(\theta) = \begin{cases} 1, & \theta < \theta', \\ 0, & \theta > \theta', \end{cases} \quad \text{or} \quad \pi(\theta) = \begin{cases} 1, & \theta < \theta_0, \\ 0, & \theta > \theta_1, \end{cases} \quad \text{or} \quad \pi(\theta) = \begin{cases} 1 - \alpha, & \theta = \theta_0, \\ \beta, & \theta = \theta_1. \end{cases} \tag{1.2.1}$$

SAMPLE SIZE DISTRIBUTION Let N denote the random sample size required to terminate the experiment. Then,

$$P_\theta(N = c \text{ and rej } H_0) = \theta^c,$$

$$P_\theta(N = c + r \text{ and rej } H_0) = \binom{c + r - 1}{c - 1} \theta^c (1 - \theta)^r, \quad r = 0, 1, \ldots, n - c,$$

$$P_\theta(n = n - c + 1 + s \text{ and acc } H_0)$$

$$= \binom{n - c + s}{s} \theta^s (1 - \theta)^{n-c+1}, \qquad s = 0, 1, \ldots, c - 1.$$

Now, $E(N) = \sum_{m=1}^{n} m p_m$ where p_m denotes the probability that a decision is reached at the mth trial. Notice that $p_m = 0$ for $m < c$. Further,

$$\begin{aligned} p_m &= P(\text{rej at stage } m \text{ or acc at stage } m, m \geq c) \\ &= P_\theta(\text{rej}) + P(\text{acc at stage } m) \\ &= \binom{m - 1}{c - 1} \theta^c (1 - \theta)^{m-c} + \binom{m - 1}{n - c} (1 - \theta)^{n-c+1} \theta^{m-(n-c+1)}; \end{aligned} \tag{1.2.2}$$

$$\begin{aligned} E(N) &= c\theta^c \sum_{m=c}^{n} \binom{m}{c} (1 - \theta)^{m-c} \\ &\quad + (n - c + 1)(1 - \theta)^{n-c+1} \sum_{m=n-c+1}^{n} \binom{m}{n - c + 1} \theta^{m-(n-c+1)} \\ &= c\theta^c \sum_{r=0}^{n-c} \binom{r + c}{c} (1 - \theta)^r + (n - c + 1)(1 - \theta)^{n-c+1} \\ &\quad \times \sum_{r=0}^{c-1} \binom{n - c + 1 + r}{r} \theta^r. \end{aligned} \tag{1.2.3}$$

The main advantage of curtailed inspection can now be stated formally: In comparison with an equivalent single-sampling plan, curtailed inspection has the same OC curve but has ASN (average sample number) curve nowhere higher and for some ranges of θ appreciably lower than the sample size of the single-sampling plan.

One may ask what boundaries could give some optimum property of the ASN curve, while satisfying some restrictions on the OC curve [such as (1.2.1)]. Wald's (1947) boundaries, which are parallel lines, satisfy the third of (1.2.1) and give the lowest possible ASN at both θ_0 and θ_1 (but not at intermediate values of θ). Closed boundaries have a finite upper limit. Open boundaries have no finite upper limit. For Wald's SPRT, since the boundaries are parallel lines, there is a positive (but rapidly decreasing) probability of obtaining a sample size larger than any given number, however large.

The curtailed single-sample procedure for $c = 1$ has been studied in great detail by Anderson and Friedman (1960). Let us denote this plan by C; it can be described as follows: After taking m observations ($m = 1, \ldots, n-1$) stop sampling and reject the lot if $X_m = 1$ (a defective at the mth trial) and continue sampling if $X_m = 0$; after taking n observations stop sampling and reject the lot if $X_n = 1$ and accept the lot if $X_n = 0$.

The plan has one constant, namely n, to be adjusted. The probability of accepting the lot $\pi(\theta \mid C)$ is the probability of n nondefectives in a sample of size n. That is

$$\pi(\theta \mid C) = (1 - \theta)^n. \tag{1.2.4}$$

Obviously $\pi(0 \mid C) = 1$, $\pi(\theta \mid C)$ decreases, and $\pi(1 \mid C) = 0$. One way to fix n is to impose a requirement on $\pi(\theta \mid C)$ at some particular value of θ, for example, that $\pi(\theta \mid C) \leq \beta$ for some θ_1 ($0 < \theta_1 < 1$) and β ($0 < \beta < 1$), that is, that the probability of accepting the lot should not exceed β if the proportion of defectives is as high as θ_1. Then n is chosen as the smallest integer for which

$$(1 - \theta_1)^n \leq \beta. \tag{1.2.5}$$

Then we have the following result pertaining to the strong optimality of the plan C.

Theorem 1.2.1 (Anderson and Friedman, 1960) For any procedure D satisfying (i) $\pi(0 \mid D) = 1$ and (ii) $\pi(\theta_1 \mid D) \leq \beta$ where $0 < \beta < 1$, the number of observations required to reach a decision is for each sample sequence at least as large as the number for the curtailed single-sampling plan C. If the number is the same for every sample, then D is identical to C (assume that for each θ, the sampling terminates finitely with probability one).

Proof If $\theta = 0$, $X_i = 0$ for every i. Condition (i) implies that for any procedure D, the sequence of observations $0, 0, \ldots$ leads to acceptance.

Since the probability must be 1 (for $\theta = 0$) of reaching a decision after a finite number of observations, there must be a number n^* such that the lot is accepted at the nth observation when $X_1 = \cdots = X_{n*} = 0$. Then $\pi(\theta_1 \mid D)$ is at least equal to the probability of this sequence and condition (ii) implies that

$$(1 - \theta_1)^{n^*} \leq \beta, \tag{1.2.6}$$

which implies $n^* \geq n$. For any sequence with $X_1 = \cdots = X_n = 0$, the number of observations required to reach a decision is, therefore, the same for D and for C if $n^* = n$, and larger for D than for C if $n^* > n$. Hence, a sequence $X_1 = \cdots = X_m = 0$ leads to continuation for D at the mth stage, $m = 1, \ldots, n-1$. Now consider a sequence with $X_1 = \cdots = X_{l-1} = 0$, $X_l = 1$, $l = 1, \ldots, n$, which under C leads to termination at the lth observation; unless D has the same decision rule at the lth observation as C, such a sequence requires more observations to terminate under D than under C.

REMARK 1.2.1 The above strong optimum property of C states that for *every* sample sequence, C is at least as good as any other procedure D in that it leads to a decision with no more observations. This of course implies the weaker optimum property that the expected sample size under C is no greater than that under D for every value of θ. Theorem 1.2.1 may be applied directly to problems in which it is desired solely to specify the risk of acceptance or rejection at a single point. For example, it might be desired solely to specify the risk of accepting a product with a stated percentage of defectives—the consumer's risk—and not to specify the producer's risk. For such a problem, a curtailed single-sampling plan C would be appropriate.

1.3 Loss and Risk Functions

Example 1.3.1 Boxes of electronic components are available for assembly of a certain system. Each component can be inspected and classified as good or defective. Every good component rejected costs a known amount. Almost every bad item passed causes trouble.

Let $W_0(\theta)$ $[W_1(\theta)]$ denote the cost of accepting [rejecting] a box having a percentage θ of bad items. Suppose that the cost of inspecting items is independent of the number n inspected; unit cost = cost of inspecting an item. Let $P_A(\theta \mid D)$ $[P_R(\theta \mid D)]$ be the probability of accepting [rejecting] a batch of quality θ under a sampling plan D. If $E(N \mid \theta, D)$ is the expected sample size, then

$$R(\theta \mid D) = E(N \mid \theta) + P_A(\theta \mid D) W_0(\theta) + P_R(\theta \mid D) W_1(\theta).$$

We therefore need a plan that is optimal in some sense. Wald suggested using the plan D' that minimizes the maximum loss or regret. That is, consider

$$\min_D \max_\theta R(\theta \mid D) \qquad \text{or} \qquad \min_D \max_\theta \{R(\theta \mid D) - \min_{\tilde{D}} R(\theta \mid \tilde{D})\}.$$

If the a priori distribution of θ is known, then one can minimize $\mathrm{ER}(\theta \mid D)$ and obtain a Bayes procedure—which does exist if the cost functions are known precisely. Sequential procedures that are minimax are hard to find except in trivial cases; e.g., it is optimal not to inspect.

1.4 Certain Double Sampling Procedures

In this section we shall point out certain hypothesis-testing problems for which fixed-sample and meaningful procedures do not exist. However, one can obtain for these problems meaningful procedures that are based on double samples.

Given $X_1, \ldots, X_n$, a random sample from a normal population having mean θ and variance σ^2. We are interested in testing $H_0: \theta = \theta_0$ against the alternative hypothesis $H_1: \theta > : \theta_0$, which is known as Student's hypothesis. It is well known that the uniformly most powerful unbiased test of H_0 against H_1 is given by

$$\text{rej } H_0 \qquad \text{if} \quad T = (\bar{X} - \theta_0)n^{1/2}/s > t_{n-1,\,1-\alpha} \tag{1.4.1}$$

where $\bar{X}$ and s denote the mean and the standard deviation of the observed X_i and $t_{n-1,\,1-\alpha}$ denotes the $100(1-\alpha)$th percentage point on the t-distribution with $n-1$ degrees of freedom. If H_1 is of the form $\theta \neq \theta_0$, then the rule

$$\text{rej } H_0 \qquad \text{if} \quad |T| > t_{n-1,\,1-\alpha/2}$$

is the uniformly most powerful unbiased test of H_0. Let $1 - \pi(\theta, \sigma)$ denote the power of the test in (1.4.1); then $\pi(\theta_0, \sigma) = 1 - \alpha$, irrespective of the value of σ. However, when one is planning an experiment, one is interested in knowing the probability with which the statistical test will detect a difference or an effect when it actually exists. Although the form of the power function of Student's test is known and tabled, it depends on σ, of which we have only a vague notion. Hence, it is of interest to devise a test of Student's hypothesis such that its power function would be independent of σ. However, Dantzig (1940) has shown the nonexistence of fixed-sample and meaningful test procedures for the same; this is given in Theorem 1.4.1.

Theorem 1.4.1 (Dantzig, 1940) If there exists a region W such that, whatever be the value of σ,

$$(2\pi\sigma^2)^{-n/2} \int_W \cdots \int \exp\left[-\sum_{i=1}^{n} (x_i - \theta_0)^2/2\sigma^2\right] dx_1 \cdots dx_n \equiv \alpha, \tag{1.4.2}$$

$$(2\pi\sigma^2)^{-n/2} \int_W \cdots \int \exp\left[-\sum_{i=1}^{n} (x_i - \theta_1)^2/2\sigma^2\right] dx_1 \cdots dx_n \equiv 1 - \beta \tag{1.4.3}$$

where $\theta_1 \neq \theta_0$ and α, β are constants, then

$$\alpha \equiv 1 - \beta. \tag{1.4.4}$$

REMARK 1.4.1.1 We do not present the proof of Dantzig's theorem since it is a little bit involved. However we shall present a much simpler proof due to Stein (1945) for the case of linear hypothesis.

There do exist regions such that the power function is independent of both θ and σ. For example, let s_n denote the sample standard deviation based on $(X_1, \ldots, X_n)$ and s_{n-1} the sample standard deviation based on $(X_1, \ldots, X_{n-1})$. Then the region given by all points $(X_1, \ldots, X_n)$ that satisfy the inequality $s_{n-1}/s_n \geq c$ is such a region. That is, $P(s_{n-1}/s_n \geq c \mid \theta, \sigma)$ is free of θ and free of σ. Such regions are, however, unsuitable for testing Student's hypothesis because they will reject this hypothesis when it is wrong and when it is correct with equal frequency.

Stein (1945) has proposed a two-sample test with the above property, the size of the second sample depending on the result of the first. The problem of determining confidence intervals with preassigned length and confidence coefficient for the mean of a normal distribution with unknown variance is solved using the same procedure. In order to make the power of a test or the length of a confidence interval exactly independent of the variance, it appears necessary to waste a small part of the information. In the following we shall describe Stein's two-sample procedure in order to test Student's hypothesis.

Procedure A sample of n_0 observations $X_1, \ldots, X_{n_0}$ is taken, and the sample estimate s^2 of the variance computed by

$$s^2 = (n_0 - 1)^{-1}\left\{\sum_1^{n_0} X_i^2 - n_0^{-1}\left(\sum_1^{n_0} X_i\right)^2\right\}. \tag{1.4.5}$$

Then n is defined by

$$n = \max\{[s^2/z] + 1, n_0 + 1\} \tag{1.4.6}$$

where z is a previously specified constant and $[\cdot]$ denotes the largest integer less than $\cdot$. Additional observations $X_{n_0+1}, \ldots, X_n$ are drawn and according to a specified rule that depends only on s^2, real numbers a_i, $i = 1, \ldots, n$ are chosen so that

$$\sum_1^n a_i = 1, \qquad a_1 = a_2 = \cdots = a_{n_0}, \qquad s^2 \sum_1^n a_i^2 = z. \tag{1.4.7}$$

This is clearly possible since $\min \sum_1^n a_i^2 = 1/n \le z/s^2$ by (1.4.7), the minimum being taken subject to the conditions $\sum_1^n a_i = 1$, $a_1 = a_2 = \cdots = a_{n_0}$. Then define

$$\begin{aligned} T' &= \left[\sum a_i X_i - \theta_0\right]/z^{1/2} \\ &= \left[\sum a_i(X_i - \theta)\right]/z^{1/2} + (\theta - \theta_0)/z^{1/2} \\ &= U + (\theta - \theta_0)/z^{1/2} \quad \text{(say).} \end{aligned}$$

Because U, for given s, is normally distributed with mean 0 and variance $\sigma^2 \sum a_i^2/z = \sigma^2/s^2$, now one can easily verify that the unconditional distribution of U is Student's t with $n_0 - 1$ degrees of freedom. Thus, U has the t-distribution with $n_0 - 1$ degrees of freedom irrespective of the value of σ. Then in order to test for one-sided alternatives: $\theta > \theta_0$, the critical region of size α is defined by

$$\left(\sum_1^n a_i X_i - \theta_0\right)/z^{1/2} > t_{n_0-1,\,1-\alpha} \tag{1.4.8}$$

and the power function is

$$1 - \pi(\theta) = P\{t_{n_0-1} > t_{n_0-1,\,1-\alpha} + (\theta_0 - \theta)/z^{1/2}\}. \tag{1.4.9}$$

An analogous critical region and an expression for the power function which is independent of σ hold for the two-sided alternative: $\theta \ne \theta_0$.

As mentioned earlier, these tests will not be used in practice since they waste information in order to make the power of the test strictly independent of the variance. Instead of (1.4.8) we take a total of

$$n = \max\{[s^2/z] + 1, n_0\} \tag{1.4.10}$$

observations and define

$$\begin{aligned} T'' &= \left(n^{-1} \sum_1^n X_i - \theta_0\right) n^{1/2}/s \\ &= n^{-1/2} \sum_1^n (X_i - \theta)/s + (\theta - \theta_0) n^{1/2}/s \\ &= U' + (\theta - \theta_0) n^{1/2}/s. \end{aligned} \tag{1.4.11}$$

One can easily establish that U' has a t-distribution with $n_0 - 1$ degrees of freedom, since $n \geq s^2/z$ and $|(\theta - \theta_0)n^{1/2}/s| \geq |(\theta - \theta_0)/z^{1/2}|$. So, if we employ $T'' > t_{n_0-1,1-\alpha}$ instead of (1.4.8), the power of the test is always increased. Also, the expected number of observations will be reduced.

Also, Stein (1945) has extended these considerations to the case of a linear hypothesis including the nonexistence of a single-sample test whose power is independent of the variance. Let X_i, $i = 1, \ldots, n$, be independently normally distributed with means

$$EX_i = \theta_i, \quad i = 1, \ldots, p; \qquad EX_i = 0, \quad i = p+1, \ldots, n, \tag{1.4.12}$$

and variance σ^2. The θ_i and σ^2 are unknown and we wish to test

$$H_0\colon \quad \theta_i = 0, \qquad i = 1, \ldots, p.$$

Using the Neyman–Pearson lemma, we find the most powerful test of H_0 against the alternative $\theta_i = \theta_{i0}$, $i = 1, \ldots, p$, if the variance σ^2 is known can be based on $T = \sum_1^n \theta_{i0} X_i$. The critical region of size α based on $\sum_1^n \theta_{i0} X_i$ is given by

$$W_0(\sigma) = \left\{(X_1, \ldots, X_n) : \frac{\sum_1^p \theta_{i0} X_i}{\sigma(\sum_1^p \theta_{i0}^2)^{1/2}} > z\right\} \tag{1.4.13}$$

where $\Phi(z) = \alpha$, Φ denoting the standard normal distribution function, since, under H_0, T is normally distributed with mean 0 and variance $\sigma^2 \sum_1^p \theta_{i0}^2$. Under H_1, T is normally distributed with mean $\sum_1^p \theta_{i0}^2$ and variance $\sigma^2 \sum_1^p \theta_{i0}^2$. Thus the power of the test for the alternative H_1 as a function of σ^2 is

$$1 - \pi_0(\sigma) = \Phi\left\{z - \left(\sum_1^p \theta_{i0}^2\right)^{1/2} \Big/ \sigma\right\}. \tag{1.4.14}$$

Now assume that there exists a test based on the critical region W of size α whose power $1 - \beta$ is independent of σ^2. Since $W_0(\sigma)$ is the best critical region of size α for any σ, we must have

$$1 - \beta \leq 1 - \pi_0(\sigma) = \Phi\left(z - \left\{\sum_1^p \theta_{i0}^2\right\}^{1/2} \Big/ \sigma\right) \tag{1.4.15}$$

so that

$$1 - \beta \leq \liminf_{\sigma} \{1 - \pi_0(\sigma)\} = \Phi(z) = \alpha. \tag{1.4.16}$$

By interchanging H_0 and H_1 we can reverse the inequality (1.4.16), proving

$$1 - \beta = \alpha. \tag{1.4.17}$$

Thus, any single-sample test for a linear hypothesis whose power is independent of the variance has constant power equal to the size of the critical region.

REMARK 1.4.1.2 When $p = n$ and $\theta_i \equiv \theta$, we get the case considered by Dantzig (1940).

Moshman (1958) has explored the proper choice of the initial sample size in Stein's two-stage procedure and shows that an upper percentage point of the distribution of the total sample size, when used in conjunction with the expectation, is a rapidly calculable guide to an efficient choice of the size of the first sample.

From (1.4.10) we have

$$P(n = n_0) = P(s^2/z \le n_0) = P\{\chi^2(n_0 - 1) \le (n_0 - 1)n_0 z/\sigma^2\} \tag{1.4.18}$$

and for any $m > n_0$

$$\begin{aligned} P(n = m) &= P(m - 1 < s^2/z \le m) \\ &= P\{(n_0 - 1)(m - 1)/\sigma^2 \le \chi^2(n_0 - 1) \le (n_0 - 1)mz/\sigma^2\}, \end{aligned} \tag{1.4.19}$$

where $\chi^2(k)$ is a chi-square random variable with k degrees of freedom. Therefore, letting $\lambda = z/\sigma^2$ one may easily show that

$$\begin{aligned} E(n) = n_0 P\{\chi^2(n_0 - 1) < (n_0 - 1)n_0 \lambda\} \\ + \lambda^{-1} P\{\chi^2(n_0 + 1) > (n_0 - 1)n_0 \lambda\} \\ + \theta_1 P\{\chi^2(n_0 - 1) > (n_0 - 1)n_0 \lambda\} \end{aligned} \tag{1.4.20}$$

and

$$\begin{aligned} \operatorname{var}(n) = {n_0}^2 P\{\chi^2(n_0 - 1) < (n_0 - 1)n_0 \lambda\} \\ + \frac{(n_0 + 1)}{(n_0 - 1)\lambda^2} P\{\chi^2(n_0 + 3) > (n_0 - 1)n_0 \lambda\} \\ + \frac{2\theta_2}{\lambda} P\{\chi^2(n_0 + 1) > (n_0 - 1)n_0 \lambda\} \\ + \theta_3 P\{\chi^2(n_0 - 1) > (n_0 - 1)n_0 \lambda\} - (En)^2 \end{aligned} \tag{1.4.21}$$

where $0 \leq \theta_i \leq 1$ $(i = 1, 2, 3)$. Whereas (1.4.20) defines $E(n)$ within a maximum error of unity, (1.4.21) is not as useful inasmuch as the factor $1/\lambda$ may be, and frequently is, large. Furthermore, it is somewhat difficult to translate $\text{var}(n)$ into working percentage points of the distribution of n. A more useful procedure is to calculate a given percentage point n_p of the distribution such that

$$P(n \leq n_p) = \sum_{m=n_0}^{n_p} P(n = m) \geq p$$

which is equivalent, if $p_k(y)$ denotes the probability density function of $\chi^2(k)$, to setting

$$\int_0^{(n_0-1)n_p} p_{(n_0-1)}(y)\, dy \geq p$$

and choosing n_p such that

$$n_p = \max\{n_0, [\lambda^{-1}(100\,p\text{th percentage point of } \chi^2(n_0 - 1)/(n_0 - 1))]\}. \tag{1.4.22}$$

Note that the upper percentage points of $\chi^2(k)/k$ decrease monotonically as k increases. A rough, but objective rule of thumb for the choice of n_0 may be derived as follows.

Let $E(n \mid n_0^*)$ be the expectation of n if $n_0 = n_0^*$ and $n_p(n_0^*)$ the 100pth percentile of n if $n_0 = n_0^*$. Define

$$P(n_0^*) = \int_0^{(n_0-1)\lambda E(n \mid n_0^*)} p_{(n_0-1)}(y)\, dy, \tag{1.4.23}$$

as the proportion of time n will not exceed $E(n \mid n_0^*)$. Let $\tilde{n}_0$ be the value of n_0 that minimizes $E(n)$. That is

$$E(n \mid \tilde{n}_0) \leq E(n \mid n_0) \qquad \text{for all } n_0. \tag{1.4.24}$$

Now, consider

$$\psi(n_0) = (1 - p)\{n_p(\tilde{n}_0) - n_p(n_0)\} - \{1 - P(\tilde{n}_0)\}\{E(n \mid n_0) - E(n \mid \tilde{n}_0)\} \tag{1.4.25}$$

and select n_0 as the integer for which $\psi(n_0)$ is a maximum. In effect, (1.4.25) weights the expected changes in $E(n)$ and n_p by the probability of exceeding those values. It is expected that p is chosen independently from nonstatistical considerations. Moshman (1958) has tabulated for $\lambda = .1$, some values of $E(n)$ and $n_{.95}$ (Table 1.4.1). From this table it is clear that $E(n \mid n_0)$ is fairly constant over a considerable range; on the other hand, $n_{.95}(n_0)$ decreases sharply where $E(n \mid n_0)$ is relatively constant.

Table 1.4.1

Dependence of $E(n)$ and $n_{.95}$ on n_0 for $\lambda = .1$

n_0	$E(n \mid n_0)$	$n_{.95}(n_0)$	n_0	$E(n \mid n_0)$	$n_{.95}(n_0)$
2	10.45	38.41	12	12.92	17.89
4	10.45	26.05	14	14.35	17.20
6	10.63	22.14	16	16.15	16.66
8	11.18	20.10	18	18.02	18.00
10	11.84	18.80	20	20.02	20.00

Example 1.4.1 If one takes $\lambda = .1$, where in the ordinary application considered by Stein (1945) $n = n_0 - 1$, then $E(n)$ is a minimum for $\tilde{n}_0 = 3$. It may readily be verified from (1.4.23) that $P(n_0) = P(3) \approx .64$. Rapidly one may evaluate $\psi(n_0)$ from (1.4.25), taking $p = .95$ and find that $\psi(6) = .2686$ is the maximum. Hence, the rule suggested specifies $n_0 = 6$ as the proper choice.

If the initial sample size is chosen poorly in relation to the unknown σ^2, the expected sample size of Stein's procedure will be large in comparison to the sample size which would be used if σ^2 were known. Moreover, the two-stage sampling procedure is asymptotically inefficient as $\sigma \to \infty$, irrespective of the choice of n_0. Robbins and Starr (1965) give a sequential procedure which, while satisfying the specifications on the error probabilities only approximately, is asymptotically efficient and seems to be reasonably efficient (with expected sample size about the same as in Stein's procedure when the optimal value of n_0 is used) for all finite σ^2. It should, however, be noted that the sequential procedure of Robbins and Starr (1965) or its modification is sequential only in that it merely attempts to estimate sequentially the nuisance parameter σ^2 of the test, and not in the sense of Wald's sequential probability ratio test. The reader is referred to the Technical Report for further details on the sequential procedure.

CHAPTER 2

The Sequential Probability Ratio Test

2.1 The Sequential Probability Ratio Test (SPRT)

During World War II, Abraham Wald started working on sequential procedures and he developed what is called the sequential probability ratio test procedure.

Neyman and Pearson (1933) have provided a method of constructing a most powerful test for a simple hypothesis versus simple alternative-testing problem.

Suppose X has pdf $f(x;\,\theta)$. Then, we are interested in testing $H_0\colon \theta = \theta_0$ against $H_1\colon \theta = \theta_1$.

Lemma 2.1.1 (Neyman and Pearson, 1933) Let $X_1, \ldots, X_n$ be a random sample and also let

$$\Lambda_n = \prod_{i=1}^{n} f(X_i;\,\theta_1) \Big/ \prod_{i=1}^{n} f(X_i;\,\theta_0).$$

Then the most powerful test of H_0 against H_1 is given by

$$\begin{array}{lll} \text{rej } H_0 & \text{if} & \Lambda_n \geq K, \\ \text{acc } H_0 & \text{if} & \Lambda_n < K, \end{array}$$

where K is determined by the level of significance. Wald proposed the following sequential probability ratio test: Choose two constants A and B such that $0 < B < A < \infty$ and accept H_0 if $\Lambda_n \leq B$, reject H_0 if $\Lambda_n \geq A$; continue sampling if $B < \Lambda_n < A$, if the experiment has proceeded up to stage n $(n = 1, 2, \ldots)$.

Example 2.1.1 For the exponential family

$$f(x; \theta) = B(\theta)e^{-Q(\theta)R(x)}h(x),$$

the rule becomes: continue sampling if

$$C_1 + Dn < \sum_1^n R(X_i) < C_2 + Dn$$

where C_1, C_2, and D depend on θ_0 and θ_1.

One can then plot two parallel lines corresponding to the linear functions of n in the extremes of the last inequality and simply cumulate the $R(X_i)$ graphically as the sampling proceeds (Fig. 2.1.1a). For the exponential family the graph is shown in Fig. 2.1.1b.

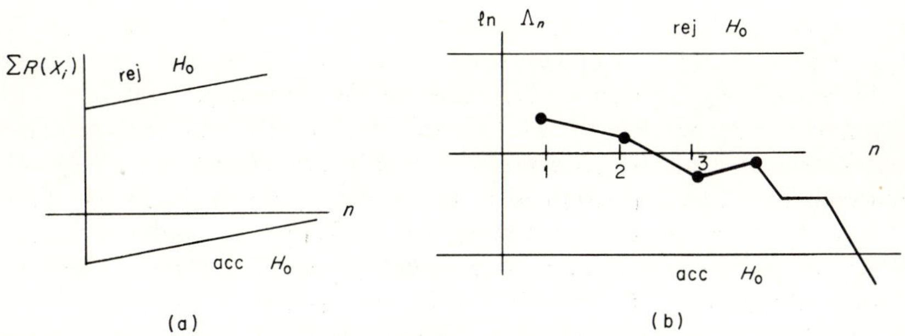

Fig. 2.1.1

Example 2.1.2 $f(x; \theta) = \theta \exp(-x\theta)$, $\theta > 0$,

$$\ln \Lambda_n = n \ln(\theta_1/\theta_0) - (\theta_1 - \theta_0) \sum X_i .$$

The inequality for continuing sampling after the nth observation is given by

$$(\theta_0 - \theta_1)^{-1}\{\ln B + n \ln(\theta_0/\theta_1)\} < \sum X_i < (\theta_0 - \theta_1)^{-1}\{\ln A + n \ln(\theta_0/\theta_1)\} \quad (\theta_0 > \theta_1).$$

Example 2.1.3 For the binomial distribution, a SPRT is defined for a given θ_0 and θ_1 by two constants B and A. At the nth step we continue sampling if $B < \theta_1^m(1 - \theta_1)^{n-m}/\theta_0^m(1 - \theta_0)^{n-m} < A$ where m is the number of defectives among the n observations. The inequalities can be rewritten as

$$c_0 + sn < m < c_1 + sn$$

where

$$s = \ln\{(1 - \theta_0)/(1 - \theta_1)\}/\ln K,$$

$$c_0 = \ln B/\ln K,$$

$$c_1 = \ln A/\ln K,$$

$$K = \theta_1(1 - \theta_0)/\theta_0(1 - \theta_1).$$

In the plane of n and m, the continue-sampling region lies between two lines, each with slope s and with intercepts c_0 and c_1, respectively. Each sample point can be plotted in this plane as a point (n, m) with integer-valued coordinates. Two procedures, defined by a pair of intercepts (c_0, c_1) and (c_0^*, c_1^*) are equivalent if there is no point (n, m), $n \geq m \geq 0$, between the lines $y = c_0 + sx$ and $y = c_0^* + sx$ or between the lines $y = c_1 + sx$ and $y = c_1^* + sx$.

Anderson and Friedman (1960) have shown that if the slope is rational, there is a denumerable number of SPRTs, and if the slope is irrational, there is a nondenumerable number. Suppose $s = M/R$ where M and R are relatively prime integers. Then a point (n, m) is on the line $y = c + sx$ for a value of $c = (Rm - Mn)/R$, that is, a rational value of c. The only lines needed in defining SPRTs in this case are those with intercepts of the form shown. There is a denumerable number of such lines, and hence a denumerable number of pairs of such lines. For the proof when s is irrational, the reader is referred to Anderson and Friedman (1960, p. 65).

EXISTENCE OF A SPRT FOR GIVEN α AND β For the curtailed single-sample plan C studied in Section 1.2, Anderson and Friedman (1960) ask whether there exists any H_0 ($\theta = \theta_0 < \theta_1$) which when combined with the H_1 ($\theta = \theta_1$) of the curtailed single-sampling plan C will yield a sequential probability ratio test identical with the plan C. The answer to this question is affirmative as can readily be proved as follows.

In order that C be a SPRT in the sense just defined, there must exist some $\theta_0 < \theta_1$, and some numbers A and B with $B < 1 < A$,† such that

$$B < (1-\theta_1)^m/(1-\theta_0)^m < A, \qquad m = 1, \ldots, n-1; \tag{2.1.1}$$

$$(1-\theta_1)^n/(1-\theta_0)^n \le B; \tag{2.1.2}$$

$$A \le \theta_1(1-\theta_1)^{m'-1}/\theta_0(1-\theta_0)^{m'-1}, \qquad m' = 1, \ldots, n. \tag{2.1.3}$$

These inequalities correspond to the inequalities of the SPRT for the binomial case where $f_i(x) = \theta_i^x(1-\theta_i)^{1-x}$, $i = 0, 1$, under the tentative assumption that the plan C is a SPRT. Inequality (2.1.1) indicates continuation at the mth step (all 0s), (2.1.2) indicates acceptance of the lot and H_0 (n 0s), and (2.1.3) indicates rejection of the lot and H_0 (one 1). (2.1.1) and (2.1.2) imply that

$$(1-\theta_1)^n/(1-\theta_0)^n \le B < (1-\theta_1)^m/(1-\theta_0)^m, \qquad m = 1, \ldots, n-1. \tag{2.1.4}$$

Since the right-hand side of (2.1.4) is smallest when m has its largest value, the limits (2.1.4) are tightest when $m = n - 1$. Clearly for any value of p_0 one can find a B to satisfy (2.1.4). Also from (2.1.1) and (2.1.3) we have

$$(1-\theta_1)^m/(1-\theta_0)^m < A \le \theta_1(1-\theta_1)^{m'-1}/\theta_0(1-\theta_0)^{m'-1},$$
$$m = 1, \ldots, n-1, \quad m' = 1, \ldots, n. \tag{2.1.5}$$

Since $(1-\theta_1)^m/(1-\theta_0)^m$ is a decreasing function of m, the limits in (2.1.5) are tightest when $m = 1$, and $m' = n$. However, we require $1 < A$. Hence A must satisfy

$$1 < A \le \theta_1(1-\theta_1)^{n-1}/\theta_0(1-\theta_0)^{n-1}. \tag{2.1.6}$$

† The conditions $B < 1 < A$ are not necessary for the definition of the Wald sequential procedure but are necessary for the proof of the optimum property of the SPRT which will be considered in Section 2.9. Usually SPRTs take at least one observation and have $B < 1 < A$. Only in Section 2.9 will we consider SPRTs, which may not take any observations.

The problem then is to determine the set of θ_0 for which the third expression in (2.1.6) is greater than the first expression. The function $(1-\theta_0)^{n-1}\theta_0$ has derivative $(1-\theta_0)^{n-2}(1-n\theta_0)$. Hence, it increases from a value of 0 at $\theta_0 = 0$ to a maximum at $\theta_0 = 1/n$; then it decreases until $\theta_0 = 1$. Thus, if $\theta_1 \le 1/n$, $(1-\theta_1)^{n-1}\theta_1 > (1-\theta_0)^{n-1}\theta_0$ for every $\theta_0 < \theta_1$ and an A can be found that satisfies (2.1.6). If $\theta_0 > 1/n$, the third expression in (2.1.6) is greater than 1 for sufficiently small θ_0 but not for all $\theta_0 < \theta_1$. Thus, there exist values of θ_0 for which an A can be found, but there will be some values of θ_0 for which A cannot be found. The conclusion is that, given θ_1, there always exists a θ_0 such that the plan C is a SPRT for θ_1 and this θ_0.

The next question would be: Is plan C a special case of a SPRT? In other words, whether for every H_0 that is relevant (namely, for which $\theta_0 < \theta_1$), there exists a SPRT that is identical with the plan C? The answer is no and is proved in the following theorem.

Theorem 2.1.1 (Anderson and Friedman, 1960) Given a curtailed single-sampling plan C with error probabilities β at θ_1 and α at θ_0 $(<\theta_1)$, if this procedure is not a sequential probability ratio test defined in terms of θ_0 and θ_1, then there is no sequential probability ratio test that achieves these probabilities of errors.

Proof Suppose a SPRT to be denoted by W is not the same as C for a given n, θ_0, and θ_1, but yields the same value of $\beta = (1-\theta_1)^n$. Then, it must require continuation of sampling for a sample of n items having a value of zero (otherwise, it would lead to acceptance for this and some other cases as well; hence the probability of acceptance when θ_1 is true would be higher than β). However, this means that $B < (1-\theta_1)^n/(1-\theta_0)^n$, since the expression on the right-hand side is the probability ratio for such a sample. Also, a standard inequality for a SPRT [see Wald (1947, p. 41)] is given by

$$\pi_W(\theta_1) \le B\pi_W(\theta_0) \tag{2.1.7}$$

(for a proof refer to Theorem 2.1.2). Hence,

$$\pi_W(\theta_1) \le \{(1-\theta_1)^n/(1-\theta_0)^n\}\pi_W(\theta_0). \tag{2.1.8}$$

Since $\pi_W(\theta_1) = (1-\theta_1)^n = \beta$, we have

$$1-\alpha = (1-\theta_0)^n < \pi_W(\theta_0). \tag{2.1.9}$$

The main point of the theorem is that at some θ_0 and θ_1 plan C can achieve an α and β that cannot be achieved by a SPRT. The SPRT for the given θ_0, θ_1, and β must achieve a smaller α, and this improvement is paid for by greater expected sample sizes at θ_0 and θ_1.

Then, we are led to ask for what values of α and β do the stopping bounds A and B exist? Although Wald's sequential probability ratio test was proposed in the 1940s, the question of existence of stopping bounds A and B given the error probabilities α, β was not answered until 1958. Wijsman (1958) has shown the existence of stopping bounds as follows. In the case of continuous probability ratio, and in the discrete case with suitable randomization, α and β are continuous functions of the stopping bounds. Let C be nonincreasing (and convex) curve of points (α, β) generated by coincident stopping bounds and let E be the set in the α, β plane bounded by C and the coordinate axes. Consider a point (α^*, β^*) on C and separate the stopping bounds in a way that keeps α constant. Since β is a continuous function of the separation d between the bounds, with $\beta(0) = \beta^*$, $\beta(\infty) = 0$, every value β between 0 and β^* is assumed for some d. It follows that for each (α, β) in E there exist stopping bounds. In the continuous case it was shown by Weiss (1956) that β decreases monotonically from β^* to 0 as d increases from 0 to ∞. In that case, for the existence of stopping bounds, it is also necessary that (α, β) be in E.

UNIQUENESS OF THE SPRT We have discussed the existence of a SPRT with specified probabilities of error at the two hypotheses. A related question is the uniqueness of the SPRT. That is, given two probability density functions or discrete probability functions $f_0(x)$ and $f_1(x)$ is there more than one SPRT that achieves given error probabilities α and β? The answer is that there is not. Weiss (1956) has shown this for the case in which $f_1(x)/f_0(x)$ has a continuous distribution with positive probability on every interval in $(0, \infty)$. Anderson and Friedman (1960) have used the optimum property of the SPRT in order to prove its uniqueness. Wijsman (1960) has shown that a certain monotonicity property of the SPRT implies uniqueness. We shall return to the property of uniqueness of the SPRT in Section 2.9 and provide some of the proofs.

We shall explore whether it is possible to express A and B explicitly in terms of α and β. We have

$$\alpha = P_{H_0}(\Lambda_1 \geq A) + P_{H_0}(B < \Lambda_1 < A \text{ and } \Lambda_2 \geq A) + \cdots,$$

$$\beta = P_{H_1}(\Lambda_1 \leq B) + P_{H_1}(B < \Lambda_1 < A \text{ and } \Lambda_2 \leq B) + \cdots.$$

They are not easy to calculate; there is no hope of solving the equations for A and B in terms of α and β.

Theorem 2.1.2 For Wald's SPRT $A \leq (1-\beta)/\alpha$ and $B \geq \beta/(1-\alpha)$.

Proof Let $\mathbf{X} = (X_1, \ldots, X_k)$ and also let E_k be the set of all points in k-dimensional Euclidean space R_k, for which we reject H_0 using the SPRT. Also, let F_k be the set of all points in R_k for which we accept H_0. Notice that $(E_k, k = 1, 2, \ldots)$ are mutually disjoint and $(F_k, k = 1, 2, \ldots)$ are also mutually disjoint. Assume that $P_{H_i}(\{\bigcup E_k\} \cup \{\bigcup F_k\}) = 1$, $i = 0$ and 1. Then

$$\alpha = P_{H_0}(\text{rej } H_0) = P_{H_0}\left(\bigcup E_k\right) = \sum_1^{\infty} P_{H_0}(E_k),$$

$$1 - \beta = P_{H_1}(\text{rej } H_0) = \sum P_{H_1}(E_k).$$

Since $f_1(\mathbf{x}) \geq A f_0(\mathbf{x})$ holds at every point $\mathbf{x} \in E_k$, we obtain

$$\alpha = \sum_{k=1}^{\infty} P_{H_0}(E_k) = \sum \int_{E_k} f_0(\mathbf{x})\, d\mathbf{x} \leq \sum A^{-1} \int_{E_k} f_1(\mathbf{x})\, d\mathbf{x} \leq A^{-1}(1-\beta).$$

Hence

$$1/A \geq \alpha/(1-\beta). \tag{2.1.10}$$

Similarly,

$$\beta = \sum P_{H_1}(F_k) = \sum \int_{F_k} f_1(\mathbf{x})\, d\mathbf{x}$$

$$\leq B \sum \int_{F_k} f_0(\mathbf{x})\, d\mathbf{x} = B(1-\alpha),$$

since

$$1 - \alpha = P_{H_0}(\text{acc } H_0) = \sum P_{H_0}(F_k).$$

We have

$$B \geq \beta/(1-\alpha). \tag{2.1.11}$$

Thus, $1/A \geq \alpha/(1-\beta)$ and $B \geq \beta/(1-\alpha)$. Notice that throughout this theorem, we have assumed that a decision must eventually be made.

The inequalities obtained here are almost equalities since Λ_n does not usually attain the value far above A or the value far below B. So, suppose we take $A = (1-\beta)/\alpha$ and $B = \beta/(1-\alpha)$. Now using the approximate values for A and B, we may not have the same probabilities of type I and type II errors, α and β. Let the true probabilities of error be α' and β'.

Then

$$\alpha/(1-\beta) = 1/A \geq \alpha'/(1-\beta')$$

and

$$\beta/(1-\alpha) = B \geq \beta'/(1-\alpha');$$

i.e.,

$$\alpha(1-\beta') \geq \alpha'(1-\beta)$$

and

$$\beta(1-\alpha') \geq \beta'(1-\alpha).$$

Adding these two we get

$$\alpha + \beta \geq \alpha' + \beta'.$$

That is, at most one of the error probabilities could be larger than the nominal error probabilities. Further,

$$\alpha' \leq \alpha'/(1-\beta') \leq \alpha/(1-\beta) \doteq \alpha(1+\beta)$$

and

$$\beta' \leq \beta'/(1-\alpha') \leq \beta/(1-\alpha) \doteq \beta(1+\alpha).$$

So the increase in error size in α' (β') is not beyond a factor of $1+\beta$ $[1+\alpha]$. These factors are close to unity when α and β are rather small. If $\alpha = \beta = .05$, then $\alpha' = \beta' \leq .0525$. If both $\alpha' < \alpha$ and $\beta' < \beta$, it would be surprising if the statistician could get away with less sampling and still be within the desired α and β. In fact, it would result in an appreciable increase in the number of observations required by the test.

Since $B \geq \beta/(1-\alpha)$ and $A \leq (1-\beta)/\alpha$, we have increased the continue sampling region [see also Wald (1947, pp. 46–48)].

We shall give reasons for the increase in the necessary number of observations caused by the approximations to be only slight.

(i) The sequential process may terminate at the nth stage if $f_1(\mathbf{x})/f_0(\mathbf{x}) \geq A$ or $f_1/f_0 \leq B$. If at the final stage f_1/f_0 were exactly equal to A or B, then the inequalities for A and B would be exact equalities. A possible excess of f_1/f_0 beyond the boundaries A and B at the termination of the test procedure is caused by the discontinuity of the number of observations. If n were continuous, then f_1/f_0 would be continuous in n and the ratio could exactly achieve A and B at the time of the termination. Wald (1947) has shown, in Section (3.9), that the increase in the expected sample number using the inequalities is slight.

(ii) If $f(x; \theta_1)$ is near $f(x; \theta_0)$, the increase in the expected sample size is small the reason being, the nearer the $f(x; \theta_1)$ to $f(x; \theta_0)$, the smaller the expected excess of f_1/f_0 over the boundaries A and B. If $f(x; \theta_1)$ tends to $f(x; \theta_0)$ the exact values of A and B tend to the approximate values.

A nice feature of the SPRT is that A and B are functions of only α and β and can be computed once and for all since they are approximately free of f provided overshoot over the stopping bounds is neglected; whereas the critical values in Neyman and Pearson formulations of fixed-sample procedures depend on f and α. So, in SPRT no distribution problems are involved, except where one is interested in finding the distribution of the number of trials required to terminate the experiment. However this is of secondary importance as long as we know that the sequential test on the average leads to a saving in the number of observations.

When $B = \beta/(1 - \alpha)$ and $A = (1 - \beta)/\alpha$, it is trivial to show that $B < 1 < A$.

Example 2.1.4 Let $\theta_0 = .5$ and $\theta_1 = .8$ in Example 2.1.3. Then

$$\Lambda_n = (.8)^r(.2)^{n-r}/(.5)^r(.5)^{n-r}.$$

Also, letting $\alpha = \beta = .2$ we have

$$\Lambda_{n+1} = \begin{cases} \frac{8}{5}\Lambda_n & \text{if } (n+1)\text{th trial results in a defective,} \\ \frac{2}{5}\Lambda_n & \text{if } (n+1)\text{th trial results in a good one,} \end{cases}$$

$$B = .2/.8 = \tfrac{1}{4}, \qquad A = .8/.2 = 4.$$

Suppose we observe $D\ G\ G\ D\ D\ D\ D\ D\ D$, where D denotes a defective item and G denotes a nondefective or good item. The continuation region is

$$\frac{n \ln \frac{5}{2}}{2 \ln 2} - 1 < r < n \frac{\ln \frac{5}{2}}{2 \ln 2} + 1,$$

or

$$0.65n - 1 < r < .65n + 1.$$

Hence, we reject H_0 on the ninth observation.
Fixed-sample size procedures for the above problem:

$$P(r > k \mid \theta_0 = .5) = .2,$$

$$P(r \le k \mid \theta_1 = .8) = .2;$$

i.e.

$$\Phi\left\{\frac{k - .5n}{(.25n)^{1/2}}\right\} \simeq .8 = \Phi(.84)$$

and

$$\Phi\left\{\frac{k - .8n}{(.16n)^{1/2}}\right\} \simeq 0.2 = \Phi(-.84);$$

i.e.

$$k - .5n = .84n^{1/2}/2, \qquad k - .8n = -.84n^{1/2}(0.4).$$

Hence

$$.3n = .84^{1/2}(.9), \qquad n^{1/2} = .84(3) = 2.52.$$

That is, $n \doteq 7$.

2.2 Finite Termination of the SPRT

We will show that although we might obtain a sequence of observations that called for continuing the sampling at each stage, with no decision reached, the probability of such an occurrence is zero.

Theorem 2.2.1 Let $Z = \ln\{f(X; \theta_1)/f(X; \theta_0)\}$. Then, Wald's SPRT terminates finitely with probability one provided $P(Z = 0) < 1$.

Proof First we shall show that the number of observations has tail probabilities that tend exponentially to zero. Z will be zero if and only if f_1 and f_0 agree at each point where $f_i = f(x; \theta_i)$ $(i = 0, 1)$. If they are to define essentially different distributions, then the set of points at which Z is zero must have probability less than unity for every state of nature. Assume that $P(Z = 0) < 1$. Then it follows that there is a $d > 0$ and a $q > 0$ so that either $P(Z > d) \geq q$ or $P(Z < -d) \geq q$. In the former case choose an integer r such that $rd > \ln(A/B)$ and write, for $n = kr$ (k an integer),

$$Z_1 + \cdots + Z_n = (Z_1 + \cdots + Z_r) + \cdots + (Z_{k-r+1} + \cdots + Z_{kr}).$$

Then

$$P\left(\left|\sum_1^r Z_i\right| > \ln\frac{A}{B}\right) \geq P\left(\sum_1^r Z_i > \ln\frac{A}{B}\right) \geq P\left(\sum_1^r Z_i > rd\right)$$
$$\geq P(Z_i > d, i = 1, \ldots, r) \geq q^r > 0.$$

If, on the other hand, $P(Z < -d) \geq q$, the integer r is chosen such that $-rd < -\ln(A/B)$ with the same conclusion, namely that

$$\begin{aligned} P\left(\left|\sum_1^r Z_i\right| > \ln(A/B)\right) &\geq P\left(\sum_1^r Z_i < -\ln(A/B)\right) \\ &\geq P\left(\sum_1^r Z_i < -rd\right) \\ &\geq P(Z_i < -d, i = 1, \ldots, r) \\ &\geq q^r > 0. \end{aligned}$$

Let N be the random sample size necessary to reach a decision. The events $\{N > n\}$ form a descending sequence whose intersection is the event that N is not finite and then

$$P(N \text{ is not finite}) = \lim_{n\to\infty} P(N > n) = \lim_{k\to\infty} P(N > kr).$$

The last equality follows from the fact that the sequence of probabilities $\{P(N > n)\}$ is monotone, bounded below, and so has a limit; this limit is also the limit of any subsequence, such as the sequence consisting of every rth element.

Next, since the Zs are independent and identically distributed random variables, no group of rZs in $\ln \Lambda_n$ could exceed $\ln(A/B)$ in magnitude if all Λ_n fall within the range for continuing sampling, that is, if the sampling process never terminates. For, if the continue-sampling inequality were always satisfied, and if the jth group of Zs were greater than $\ln(A/B)$, then

$$\begin{aligned}\ln \Lambda_{jr} &= \ln \Lambda_{jr-r} + (Z_{jr-r+1} + \cdots + Z_{jr}) \\ &> \ln B + \ln(A/B) = \ln A.\end{aligned}$$

On the other hand, if the jth group were less than $-\ln(A/B)$,

$$\ln \Lambda_{jr} \leq \ln A - \ln(A/B) = \ln B.$$

So, in either case, the inequality for continuing sampling would be violated, at the (jr)th stage, contradicting the assumption that it is always satisfied. Hence, for any k,

$$\begin{aligned}P(N > kr) &\leq P(|Z_{jr-r+1} + \cdots + Z_{jr}| < \ln(A/B),\ j = 1, \ldots, k) \\ &= \prod_{j=1}^{k} P[|Z_{jr-r+1} + \cdots + Z_{jr}| < \ln(A/B)] \\ &\leq (1 - q^r)^k \to 0 \qquad \text{as} \quad k \to \infty.\end{aligned}$$

$P(N$ not finite$) \leq \lim_{k\to\infty}(1 - q^r)^k = 0$, because $q > 0$ and hence $1 - q^r < 1$. This completes the proof of Theorem 2.2.1.

The above method of proof is due to Stein (1946); Wald proved this proposition under the assumption that var $Z > 0$.

2.3 The Operating Characteristic Function (OC Function)

The operating characteristic (probability of accepting H_0) of a SPRT can be obtained, somewhat indirectly, as follows. Consider a test defined by given constants A and B, with $B < 1 < A$, in order to test $H_0 : f = f_0(x) = f(x; \theta_0)$ against $H_1 : f = f_1(x) = f(x; \theta_1)$. If θ_0 and θ_1 are the only two states of nature, then there is no point in considering the OC function. However, if the above hypothesis-testing is a simplification of $H_0 : \theta \leq \theta^*$ versus $H_1 : \theta > \theta^*$, then one would be interested in the OC as a function of all possible values of θ.

Let θ be fixed and determine as a function of that θ a value of h other than 0 for which

$$E_\theta\{[f(X; \theta_1)/f(X; \theta_0)]^h\} = 1.$$

This expectation is 1 when $h = 0$, but often there is just one other value of h for which it is also 1. Notice that $h = 1$ if $\theta = \theta_0$ and $h = -1$ if $\theta = \theta_1$. The above formula can be written as

$$\int_{-\infty}^{\infty} [f(x; \theta_1)/f(x; \theta_0)]^h f(x; \theta)\, dx = 1.$$

Define

$$f^*(x; \theta) = [f(x; \theta_1)/f(x; \theta_0)]^h f(x; \theta).$$

Consider the auxiliary problem of testing

$$H: \quad f = f(x; \theta) \qquad \text{against} \qquad H^*: \quad f = f^*(x; \theta)$$

which are simple for fixed h and θ. So, one continues sampling in testing H against H^* if

$$B^* = B^h < \prod f^*(x_i; \theta)/\prod f(x_i; \theta) = \prod [f(x_i; \theta_1)/f(x_i; \theta_0)]^h < A^h = A^*.$$

Upon taking the $1/h$ power (assuming $h > 0$) we obtain the same inequality as that used for continuing sampling in testing H_0 against H_1 (if $h < 0$, the lower boundary would have been A^h and the upper boundary would have been B^h, yielding the same net result. Hence $P_\theta(\text{acc } H_0) = P_\theta(\text{acc } H^*) = P_H(\text{acc } H) = 1 - \alpha^*$ where α^* is the size of the type I error for the auxiliary problem. However solving the equations

$$B^h = \beta^*/(1 - \alpha^*) \qquad \text{and} \qquad A^h = (1 - \beta^*)/\alpha^*$$

we find that $\alpha^* = (1 - B^h)/(A^h - B^h)$, hence $\text{OC}(\theta) \doteq (A^h - 1)/(A^h - B^h)$, which is expressed in terms of h. However, θ depends on h and these two relations define the operating characteristic curve, $\text{OC}(\theta)$ parametrically. Each choice of h determines a θ and a value of $P_\theta(\text{acc } H_0)$, namely a point on the OC curve. One gets a useful parametric representation for the OC in terms of h only for exponential models.

The expression relating h and θ does not define θ for $h = 0$ since it is automatically satisfied for all θ. However, one can determine the point on the OC curve corresponding to $h = 0$ by passing to the limit as $h \to 0$. So using L'Hospital's rule, we obtain

$$\lim_{h \to 0} \text{OC}(\theta) = \ln A/(\ln A - \ln B).$$

We know $\mathrm{OC}(\theta_0) = 1 - \alpha$, $\mathrm{OC}(\theta_1) = \beta$,

$$\lim_{h\to\infty} \mathrm{OC}(\theta) = 1, \qquad \text{and} \qquad \lim_{h\to-\infty} \mathrm{OC}(\theta) = 0.$$

Thus we get the following table of values.

h	$-\infty$	-1	0	1	∞
θ		θ_0		θ_1	
OC	0	$1-\alpha$	$\ln A/\{\ln A - \ln B\}$	β	1

Example 2.3.1 Considering the problem of testing $\theta = \theta_0$ against $\theta = \theta_1 > \theta_0$ in a Bernoulli population, we obtain

$$E_\theta\left\{\frac{f(X;\theta_1)}{f(X;\theta_0)}\right\}^h = E_\theta\left\{\frac{\theta_1{}^X(1-\theta_1)^{1-X}}{\theta_0{}^X(1-\theta_0)^{1-X}}\right\}^h$$

$$= \theta\left(\frac{\theta_1}{\theta_0}\right)^h + (1-\theta)\left(\frac{1-\theta_1}{1-\theta_0}\right)^h.$$

Setting this equal to 1 and solving for θ we obtain

$$\theta = \frac{1 - \{(1-\theta_1)/(1-\theta_0)\}^h}{(\theta_1/\theta_0)^h - \{(1-\theta_1)/(1-\theta_0)\}^h};$$

as $h \to 0$ this becomes

$$\lim_{h\to 0} \theta = \frac{-\ln\{(1-\theta_1)/(1-\theta_0)\}}{\ln(\theta_1/\theta_0) - \ln\{(1-\theta_1)/(1-\theta_0)\}}.$$

If $\theta_0 = 0.5$, $\theta_1 = 0.9$, and $\alpha = \beta = .05$, we obtain

$$\theta = (1 - 5^{-h})/\{(1.8)^h - 5^{-h}\} = (5^h - 1)/(9^h - 1)$$

and we obtain the following table:

h	$-\infty$	-1	0	1	∞
θ	1	.9	.733	.5	0
OC	0	.05	.5	.95	1

Problems

1. Obtain the relation between θ and h in the SPRT for $\theta = 0$ against $\theta = 1$ in a normal population with unit variance. Plot the OC function of the test with $\alpha = \beta = .01$.
2. Show that in testing $\theta = \theta_0$ against $\theta = \theta_1$ in a Poisson population the relation between θ and h is

$$\theta = (\theta_0 - \theta_1)^h / \{1 - (\theta_1/\theta_0)^h\}.$$

3. Obtain the graph of the OC(θ) in the SPRT of the density $\theta_0 e^{-\theta_0 x}$ against the density $\theta_1 e^{-\theta_1 x}$ $(x > 0)$ using $\theta_0 = 2$ and $\theta_1 = 1$, $\alpha = .05$ and $\beta = .01$.

2.4 Average Sample Number

The sample size needed to reach a decision in a sequential or a multiple sampling plan is a random variable. The distribution of this random variable depends on the distribution that actually obtains during the sampling process, i.e., on the state of nature.

In Section 2.2 it was shown that N is finite with probability one. So that N can take on values 1, 2, 3, ... with probabilities $p_1, p_2, \ldots$ where $\sum_1^\infty p_i = 1$. The moments of N cannot explicitly be computed. However, one can show that $E(N^i) < \infty$ for all i. Toward this, consider

$$\begin{aligned} E(N^i) &= (1^i p_1 + 2^i p_2 + \cdots + r^i p_r) + \{(r+1)^i p_{r+1} + \cdots + (2r)^i p_{2r}\} + \cdots \\ &\le r^i\left(\sum_1^r p_j\right) + (2r)^i\left(\sum_{j=1}^r p_{r+j}\right) + (3r)^i\left(\sum_{j=1}^r p_{2r+j}\right) + \cdots. \end{aligned}$$

Now,

$$\sum_1^r p_j \le 1, \qquad \sum_1^r p_{r+j} < 1 - q^r, \qquad \sum_1^r p_{2r+j} < (1 - q^r)^2, \qquad \text{etc.}$$

which follow from the inequality obtained in Section 2.2, namely that $P(N > kr) \le (1 - q^r)^k$, $k = 1, 2, \ldots$. Consequently,

$$\begin{aligned} E(N^i) &\le r^i + (2r)^i(1 - q^r) + (3r)^i(1 - q^r)^2 + \cdots \\ &= r^i\{1 + 2^i(1 - q^r) + 3^i(1 - q^r)^2 + \cdots\}. \end{aligned}$$

Now, the series in brackets can be shown to be convergent for $0 < q \le 1$ by using the ratio test. Hence, $E(N^i) < \infty$. In fact one can show that the moment-generating function of N is finite. Toward this, consider

$$
\begin{aligned}
M_N(t) = E(e^{Nt}) &= (e^t p_1 + e^{2t} p_2 + \cdots + e^{rt} p_r) + (e^{(r+1)t} p_{r+1} + \cdots \\
&\quad + e^{2rt} p_{2r}) + \cdots \\
&\le e^{rt}\left(\sum_1^r p_j\right) + e^{2rt}\left(\sum_{j=1}^r p_{r+j}\right) + \cdots \\
&\le e^{rt} + e^{2rt}(1 - q^r) + e^{3rt}(1 - q^r)^2 + \cdots \\
&= e^{rt}/\{1 - e^{rt}(1 - q^r)\}
\end{aligned}
$$

provided $e^{rt}(1 - q^r) < e^{rt} < 1$. That is, $t \in [-\theta, -\infty]$, for some $\theta \ge 0$.

If a decision is reached at the nth stage, Λ_n is approximately distributed as a Bernoulli variable with values B and A and

$$
\begin{aligned}
E(\ln \Lambda_N) &\doteq (\ln B)P(\ln \Lambda_N = \ln B) + (\ln A)P(\ln \Lambda_N = \ln A) \\
&= (\ln B)P(\text{acc } H_0) + (\ln A)P(\text{rej } H_0).
\end{aligned}
$$

So

$$E_{\theta_0}(\ln \Lambda_N) \doteq (\ln B)(1 - \alpha) + (\ln A)\alpha$$

and

$$E_{\theta_1}(\ln \Lambda_N) \doteq (\ln B)\beta + (\ln A)(1 - \beta).$$

However $\ln \Lambda_N = Z_i + \cdots + Z_N$ where $Z_i = \ln\{f(X_i; \theta_1)/f(X_i; \theta_0)\}$. In the sequel, using the simple method of Johnson (1959), we shall show that $E(\ln \Lambda_N) = E(N)E(Z)$. Let $Z, Z_1, Z_2, \ldots$ be a sequence of independent, identically distributed random variables and N a random variable with values $1, 2, \ldots$ such that the event $\{N \ge i\}$ is independent of any events involving $Z_i, Z_{i+1}, \ldots$. Let Y_i be zero if $N < i$ and 1 if $N \ge i$. Then

$$
\begin{aligned}
E(Z_1 + \cdots + Z_N) = E\left(\sum_{i=1}^{\infty} Y_i Z_i\right) &= \sum_{i=1}^{\infty} E(Y_i Z_i) \\
&= E(Z) \sum_{i=1}^{\infty} E(Y_i)
\end{aligned}
$$

provided the interchange of summation and expectation is justified. Then

$$\sum_{i=1}^{\infty} E(Y_i) = \sum_{i=1}^{\infty} P(N \geq i) = \sum_{i=1}^{\infty} \sum_{j=i}^{\infty} P(N = j)$$
$$= \sum_{j=1}^{\infty} \sum_{i=1}^{j} P(N = j) = \sum_{j=1}^{\infty} jP(N = j) = E(N),$$

which completes the proof of the assertion that

$$E(\ln \Lambda_N) = E(Z)E(N). \tag{2.4.1}$$

The interchange of summation and expectation is valid if the series is absolutely convergent. Consider

$$\sum_{i=1}^{\infty} |E(Y_i Z_i)| \leq \sum_{i=1}^{\infty} E(|Y_i|)E(|Z_i|) = E(|Z|)E(N) < \infty$$

provided $E(|Z|)$ and $E(N)$ are finite. Thus it follows as an application of the last result to the sequence

$$Z_i = \ln\{f(X_i; \theta_1)/f(X_i; \theta_0)\}$$

that $E(N) = E(\ln \Lambda_N)/E(Z)$. Hence

$$E_{\theta_0}(N) \doteq \{\alpha \ln A + (1 - \alpha) \ln B\}/E_{\theta_0}(Z) \tag{2.4.2}$$

and

$$E_{\theta_1}(N) \doteq \{(1 - \beta) \ln A + \beta \ln B\}/E_{\theta_1}(Z). \tag{2.4.3}$$

Example 2.4.1 Let X be normal with mean θ and variance 1. Let $\theta_0 = 0$ and $\theta_1 = 1$ and $\alpha = \beta = 0.1$. Then

$$A = 99 = 1/B, \qquad \ln A = 4.595.$$

So $E_0(\ln \Lambda_N) \doteq (1 - 2\alpha) \ln 99 = 4.5031$.

$$Z = \ln\{f(X; \theta_0)/f(X; \theta_0)\} = X - \tfrac{1}{2},$$
$$E(Z) = \theta - \tfrac{1}{2},$$
$$E_{\theta_0}(N) \doteq 4.5031/\tfrac{1}{2} \doteq 9 \qquad \text{and} \qquad E_{\theta_1}(N) \doteq 9.$$

For a fixed-sample size procedure, $n \doteq 22$. The expected sample size can be computed for states of nature other than H_0 and H_1:

$$E_\theta(N) \doteq [(\ln B)\pi(\theta) + (\ln A)(1 - \pi(\theta))]/E_\theta(Z), \tag{2.4.4}$$

where $\pi(\theta) = P_0(\text{acc } H_0) = \text{OC}(\theta)$.

Problems

1. Obtain Wald's sequential probability ratio test for the following hypothesis-testing problem:

$$f(x; \sigma) = \sigma^p e^{-\sigma x} x^{p-1}/\Gamma(p), \qquad x, \sigma > 0, \quad (p \text{ is specified})$$

$$H_0: \quad \sigma = 1, \qquad H_1: \quad \sigma > 1.$$

2. Let $f(x; \sigma) = \sigma^x(1 - \sigma)$, $x = 0, 1, \ldots, 0 < \sigma < 1$. We wish to test $H_0: \sigma = \sigma_0$ against $H_1: \sigma = \sigma_1$. Tabulate some values of the OC function with $\sigma_0 = \frac{1}{2}$, $\sigma_1 = \frac{3}{4}$, $\alpha = \beta = .05$. Obtain an expression for the OC function. Also find $E_\sigma(N)$.

3. Let the common distribution of $X_1, X_2, \ldots$ have a density $f(x; \sigma) = \frac{1}{2}(1 - \sigma^2) \times \exp(-|x| + \sigma x)$, $|\sigma| < 1$. Obtain the sequential probability ratio test of the hypothesis $H_0: \sigma = -\frac{1}{2}$, against $H_1 : \sigma = \frac{1}{2}$.

4. Let X have the following probability density function:

$$f(x; \theta) = (2\pi\theta^2)^{-1/2} \exp(-x^2/2\theta^2),$$

we wish to test $H_0: \theta = 1$ against $\theta = 2$. Obtain Wald's SPRT, ASN, and OC(θ).

5. Let X have density function $\theta \exp(-\theta x)$ $(x > 0)$. We wish to test $H_0: \theta = 2$ against $H_1: \theta = 1$ with $\alpha = .05$ and $\beta = .10$. Set up Wald's SPRT.

6. Let .7, .8, .9, .9, .85 denote the squares of the observations drawn randomly from the density $\theta x^{\theta-1}$ $(0 < x < 1)$. We wish to test $H_0: \theta = 1$ against $H_1: \theta = 3$ with $\alpha = \beta = .10$. Carry out the SPRT with the above data and see whether you accept H_0 or H_1.

7. Let X be distributed uniformly on $[\theta, \theta + 2]$. We wish to test $H_0: \theta = 0$ against $H_1: \theta = 1$. Let 1.2, 1.5, 1.8, 1.4, .9 denote a random sample drawn sequentially from the above density. Carry out the SPRT.

8. Let $\{X_i\}$ be independent and identically distributed according to the Pareto density $\theta a^\theta/x^{\theta+1}$ for $x \geq a$. We wish to test $H_0: \theta = \theta_0$ against $H_1: \theta = \theta_1$ $(0 < \theta_0 < \theta_1)$ for a known $a > 0$. Construct the SPRT, and obtain ASN and OC(θ).

9. The number X of children in a family has the probability function

$$f(x; \theta) = \begin{cases} 1 - \theta a(1 - a)^{-1} & \text{for } x = 0, \\ \theta a^x & \text{for } x = 1, 2, \ldots, \end{cases}$$

where $\frac{1}{2} < a < 1$ is a known constant [according to Lotka (1939), $a = 0.7358$ in American families]. We wish to test $H_0: \theta = \theta_0$ against $H_1: \theta = \theta_1$ $[0 < \theta_0 < \theta_1 < (1 - a)/a]$. Compare Wald's SPRT with the SPRT based on $\{Y_i\}$ where Y_i is the number of boys in the ith family sampled and the binomial proportion $\theta a/(1 - a)$.

Higher moments of randomly stopped sums have been derived by Chow *et al.* (1965). In the following we shall give their results for the second moment;

the reader is kindly referred to their paper for the proofs that are based on the theory of martingales.

Let $(\Omega, \mathscr{F}, P)$ be a probability space and let $Z_1, Z_2, \ldots$ be a sequence of random variables on Ω and let $\mathscr{F}_n$ be the σ-algebra generated by $Z_1, \ldots, Z_n$, with $\mathscr{F}_0 = (\varnothing, \Omega)$. A stopping variable (of the sequence $Z_1, Z_2, \ldots$) is a random variable N with positive integer values such that the event $[N = n]$ belongs to $\mathscr{F}_n$ for every $n \geq 1$. Let $S_n = \sum_1^n Z_i$; then $S_N = \sum_1^N Z_i$ is a randomly stopped sum. We shall assume that

$$E|Z_n| < \infty, \qquad E(Z_{n+1}|\mathscr{F}_n) = 0 \qquad (n \geq 1). \tag{2.4.5}$$

The moments of S_N have been investigated in the literature. Wald's (1947) theorem states that for independent and identically distributed (iid) Z_i with $EZ_i = 0$, $EN < \infty$ implies that $ES_N = 0$. For higher moments of S_N, Chow *et al.* (1965) point out that the known results are not entirely satisfactory. We have the following results of Chow *et al.* (1965).

Theorem 2.4.1 Let $Z_1, Z_2, \ldots$ be independent with $EZ_n = 0$, $E|Z_n| = a_n$, $EZ_n^2 = \sigma_n^2 < \infty$ $(n \geq 1)$ and let $S_n = \sum_1^n Z_i$. Then if N is a stopping variable, either of the two relations

$$E\sum_1^N a_i < \infty, \qquad E\sum_1^N \sigma_i^2 < \infty, \tag{2.4.6}$$

implies that $ES_N = 0$ and

$$ES_N^2 = E\sum_1^N Z_i^2 = E\sum_1^N \sigma_i^2. \tag{2.4.7}$$

If $\sigma_n^2 = \sigma^2 < \infty$, then $EN < \infty$ implies that

$$ES_N^2 = E\sum_1^N Z_i^2 = \sigma^2 EN. \tag{2.4.8}$$

REMARK 2.4.1.1 Some stronger sufficient conditions for (2.4.8) have been given by earlier investigators.

Corollary 2.4.1.1 Let $Z_1, Z_2, \ldots$ be independent with $EZ_n = 0$, $EZ_n^2 = 1$, and define N^* (respectively, N_*) = first $n \geq 1$ such that $|S_n| > n^{1/2}$ (respectively $<$) ($= \infty$ otherwise). Then $EN^* = EN_* = \infty$.

REMARK 2.4.1.2 Note that both N^* and N_* are genuine stopping variables if the Z_n are, in addition, identically distributed. The example $P(Z_n = 1) = P(Z_n = -1) = \frac{1}{2}$ shows that the $>$ ($<$) cannot be replaced by $\geq$ ($\leq$), since

$EZ_n = 0$, $EZ_n^2 = 1$ and $N^* = N_* = 1$. On the other hand, if N^* is redefined as the first $n > 1$ for which $|S_n| \geq n^{1/2}$, EN^* and EN_* are again infinite.

Corollary 2.4.1.2 If $Z_1, Z_2, \ldots$ are iid with $EZ_n = 0$, $EZ_n^2 = \sigma^2$,

$$P(|Z_n| \leq a < \infty) = 1,$$

and if $ES_N^2 < \infty$ for a stopping variable N, then $EN < \infty$ if and only if

$$\liminf nP(N > n) = 0. \tag{2.4.9}$$

Although in Theorem 2.4.1 we have provided the second moment of a sum of a random number of random variables, it is not readily applicable to our problem since in Theorem 2.4.1 expectation of each random component, namely Z is assumed to have mean zero. In the following, we shall derive in an elementary fashion the first four moments of S_N. For the sake of completeness, we shall include $E(S_N)$ in the statement of the results. The subscript θ for E is suppressed for the sake of simplicity.

Theorem 2.4.2 (i) If $E|Z| < \infty$ and $EN < \infty$, then

$$ES_N - (EZ)(EN), \tag{2.4.10}$$

where

$$S_N = \sum_1^N Z_i, \qquad Z_i = \ln\{f(X_i; \theta_1)/f(X_i; \theta_0)\}, \qquad i = 1, 2, \ldots$$

(ii) If $EZ^2 < \infty$ and $EN^2 < \infty$, we have

$$ES_N^2 = (EN)(EZ^2) + 2(EZ) \sum_{j=2}^{\infty} E(S_{j-1} | Y_j = 1)P(Y_j = 1).$$

Proof Result (i) has already been proved. As before, let $Y_i = 1$ if $N \geq i$ and zero otherwise. Then one can write $S_N = \sum_1^\infty Y_i Z_i$ and recall that Y_i is independent of $Z_i, Z_{i+1}, \ldots$. Then

$$E(S_N^2) = E\left(\sum_1^\infty Y_i^2 Z_i^2 + 2\sum\sum_{i<j} Y_i Y_j Z_i Z_j\right)$$

$$= (EZ^2) \sum_1^\infty EY_i^2 + 2(EZ) \sum\sum_{i<j} E(Z_i | Y_j = 1)P(Y_j = 1)$$

termwise integration being permissible because of the hypothesis. Also,

$$\sum_1^\infty EY_i^2 = \sum_1^\infty EY_i = \sum_1^\infty P(N \geq i) = EN. \tag{2.4.11}$$

Corollary 2.4.2.1 If $EZ = 0$, $EN = E(S_N^2)/E(Z^2)$.

Theorem 2.4.3 If $E(|Z|^3) < \infty$ and $E(N^3) < \infty$, we have

$$\begin{aligned} E(S_N^3) = (EZ^3)(EN) &+ 3(EZ^2) \sum_{i=2}^\infty E(S_{i-1} \mid Y_i = 1)P(Y_i = 1) \\ &+ 3(EZ) \sum_{i=2}^\infty E(S_{i-1}^2 \mid Y_i = 1)P(Y_i = 1). \end{aligned} \tag{2.4.12}$$

Proof

$$E(S_N^3) = E\left(\sum_1^\infty Y_i^3 Z_i^3 + 3\sum_{i \neq j}\sum Y_i^2 Y_j Z_i^2 Z_j + \sum_{i \neq j \neq k}\sum\sum Y_i Y_j Y_k Z_i Z_j Z_k\right).$$

Now

$$\begin{aligned} E\left(\sum_{i \neq j}\sum Y_i^2 Y_j Z_i^2 Z_j\right) &= (EZ)\sum_{i<j}\sum E(Z_i^2 \mid Y_j = 1)P(Y_j = 1) \\ &\quad + (EZ^2)\sum_{i>j}\sum E(Z_j \mid Y_i = 1)P(Y_i = 1). \end{aligned}$$

Further,

$$E\left(\sum_{i \neq j \neq k}\sum\sum Y_i Y_j Y_k Z_i Z_j Z_k\right) = 3(EZ)\sum_{k=2}^\infty E\left(\sum_{i \neq j}^{k-1}\sum^{k-1} Z_i Z_j \mid Y_k = 1\right)P(Y_k = 1),$$

where

$$E\left(\sum_{i \neq j}^{k-1}\sum^{k-1} Z_i Z_j \mid Y_k = 1\right) = E(S_{k-1}^2 \mid Y_k = 1) - E\left(\sum_{i=1}^{k-1} Z_i^2 \mid Y_k = 1\right).$$

Using the above computations one can complete the proof of Theorem 2.4.3.

Theorem 2.4.4 Let $U_{r,n} = \sum_{i=1}^{n-1} E(Z_i^r \mid Y_n = 1)$, $r = 1, 2, \ldots$, and $n = 2, 3, \ldots$. Then $U_{1,n} = E(S_{n-1} \mid Y_n = 1)$. If $EZ^4 < \infty$ and $EN^4 < \infty$, we have

$$E(S_N{}^4) = (EZ^4)(EN) + 4(EZ^3)\sum_{i=2}^{\infty} U_{1,i} P(Y_i = 1)$$
$$+ 6(EZ^2)\sum_{i=2}^{\infty} E(S_{i-1}^2 \mid Y_i = 1)P(Y_i = 1)$$
$$+ (EZ)\sum_{i=2}^{\infty} E(S_{i-1}^3 \mid Y_i = 1)P(Y_i = 1). \tag{2.4.13}$$

Proof Consider

$$E(S_N{}^4) = E\Bigl(\sum Y_i{}^4 Z_i{}^4 + 4\sum_{i \neq j}\sum Y_i{}^3 Y_j Z_i{}^3 Z_j + 3\sum_{i \neq j}\sum Y_i{}^2 Y_j{}^2 Z_i{}^2 Z_j{}^2$$
$$+ 6\sum_{i \neq j \neq k}\sum\sum Y_i^2 Y_j Y_k Z_i{}^2 Z_j Z_k + \sum_{i \neq j \neq k \neq l}\sum\sum\sum Y_i Y_j Y_k Y_l Z_i Z_j Z_k Z_l\Bigr). \tag{2.4.14}$$

Then, straightforward computations yield: $E\sum Y_i{}^4 Z_i{}^4 = (EZ^4)(EN)$.

$$E\sum_{i \neq j}\sum Y_i{}^3 Y_j Z_i{}^3 Z_j = E\sum_{i<j}\sum + E\sum_{i>j}\sum$$
$$= (EZ)\sum_{j=2}^{\infty} U_{3,i} P(Y_i = 1) + (EZ^3)\sum_{j=2}^{\infty} U_{1,i} P(Y_i = 1).$$

$$E\sum_{i \neq j}\sum Y_i{}^2 Y_j{}^2 Z_i{}^2 Z_j{}^2 = 2E\sum_{i<j}\sum E(Y_i{}^2 Y_j{}^2 Z_i{}^2 Z_j{}^2) = 2(EZ^2)\sum_{j=2}^{\infty} U_{2,i}\, P(Y_i = 1).$$

$$E\sum_{i \neq j \neq k}\sum\sum Y_i{}^2 Y_j Y_k Z_i{}^2 Z_j Z_k = E\Biggl(\sum_{\substack{j \neq k \\ (j,k)<i}}\sum\sum + \sum_{\substack{i \neq k \\ (i,k)<j}}\sum\sum + \sum_{\substack{i \neq j \\ (i,j)<k}}\sum\sum\Biggr).$$

$$E\sum_{\substack{j \neq k \\ (j,k)<i}}\sum\sum = (EZ^2)\sum^{i-1}_{j \neq k}\sum^{i-1} E(Z_j Z_k \mid Y_i = 1)$$
$$= (EZ^2)\sum_{i=2}^{\infty}\{E(S_{i-1}^2 \mid Y_i = 1) - U_{2,i}\}P(Y_i = 1).$$

$$E\sum_{\substack{i \neq k \\ (i,k)<j}}\sum\sum = (EZ)\sum^{j-1}_{i \neq k}\sum^{j-1} E(Z_i{}^2 Z_k \mid Y_j = 1)$$
$$= (EZ)\sum_{j=2}^{\infty}\{E(S_{j-1} U_{2,j} \mid Y_j = 1) - U_{3,j}\}P(Y_j = 1).$$

Similarly

$$E \sum_{\substack{i \neq j \\ (i,j) < k}} \sum \sum = (EZ) \sum_{k=2}^{\infty} \{E(S_{k-1} U_{2,k} \mid Y_k = 1) - U_{3,k}\} P(Y_k = 1).$$

Finally,

$$\begin{aligned}
E & \sum_{i \neq j \neq k \neq l} \sum \sum \sum Y_i Y_j Y_k Y_l Z_i Z_j Z_k Z_l \\
&= 4E \sum_{\substack{i \neq j \neq k \\ (i,j,k) < l}} \sum \sum \\
&= 4(EZ) \sum^{l-1} \sum_{i \neq j \neq k}^{l-1} \sum^{l-1} E(Z_i Z_j Z_k \mid Y_l = 1) P(Y_l = 1) \\
&= 4(EZ) \sum_{l=2}^{\infty} \left\{ E(S_{l-1}^3 \mid Y_l = 1) - 3E\left[S_{l-1} \left(\sum_{i=1}^{l-1} Z_i^2 \right) \mid Y_l = 1 \right] + 2U_{3,l} \right\} \\
&\quad \times P(Y_l = 1)
\end{aligned}$$

since

$$\sum^{l-1} \sum_{i \neq j \neq k}^{l-1} \sum^{l-1} Z_i Z_j Z_k = S_{l-1}^3 - 3S_{l-1} \left(\sum_{i=1}^{l-1} Z_i^2 \right) + 2 \sum_{i=1}^{l-1} Z_i^3.$$

Now, when we use these computations in (2.4.14) many terms cancel off leaving the desired terms.

REMARK 2.4.1 The first four moments of S_N involve $E(S_{i-1}^k \mid Y_i = 1)$ for $k = 1$, 2, and 3. In SPRTs, especially when the underlying hypotheses are about normal or exponential populations, one can evaluate $E(S_{i-1}^k \mid Y_i = 1)$. Thus, Theorem 2.4.2 will enable one to obtain approximate expressions for EN and var N in the case of Wald's SPRT.

One could ask: is it possible to obtain approximate expressions for the average sample numbers given that one has accepted or rejected the null hypothesis? That is, one desires approximate expressions for $E_\theta(N \mid \text{acc } H_0)$ and $E_\theta(N \mid \text{rej } H_0)$.

In order to obtain approximate expressions for $E(N \mid \text{acc } H_0)$ and $E(N \mid \text{rej } H_0)$, we consider the following:

$$\begin{aligned}
E(\ln \Lambda_N \mid \text{acc } H_0) &= \sum_{n=1}^{\infty} E_\theta(\ln \Lambda_N \mid N = n, \text{acc } H_0) P(N = n \mid \text{acc } H_0) \\
&= \sum_{n=1}^{\infty} \int_{F_n} \frac{(\ln \Lambda_n)}{\text{OC}_n(\theta)} \prod_{i=1}^{n} f(x_i; \theta)\, dx_i, \qquad (2.4.15)
\end{aligned}$$

where

$$\mathrm{OC}_n(\theta) = \int_{F_n} \prod_{i=1}^{n} f(x_i; \theta)\, dx_i,$$

and F_n is that part of R_n that calls for acceptance of H_0. An analogous expression holds for $E_\theta(\ln \Lambda_N | \text{rej } H_0)$. However, one can approximately set

$$E_\theta(S_N{}^k | \text{acc } H_0) \doteq (\ln B)^k \tag{2.4.16}$$

and

$$E_\theta(S_N{}^k | \text{rej } H_0) \doteq (\ln A)^k, \qquad k = 1, 2, \ldots. \tag{2.4.17}$$

In Theorems 2.4.2–2.4.4, interpreting the expectations as conditional expectations given that H_0 is accepted [rejected], one can derive approximate expressions for the mean and variance of N given that H_0 is accepted [rejected].

One disturbing feature of the expected sample size of the Wald SPRT is the following: if one is really interested in testing H_0: $\theta \leq \theta^*$ against the alternative $H_1{}^*: \theta > \theta^*$, then one would set up H_0: $\theta = \theta_0$ ($\theta_0 \in H_0$) and $H_1: \theta = \theta_1$ with $\theta_1 \geq \theta_0$, the zone between θ_0 and θ_1 being the "indifference zone." If the population is indexed by a θ belonging to this indifference zone, i.e. near θ^*, $E(N)$ tends to be largest. Thus the test tends to take longer to reach a decision when θ is near θ^* where there is little concern as to which decision is made. This is clear intuitively also, because it should not take so long to discover that a population is overwhelmingly of one kind or the other as it does to discover which kind it is when it is near the borderline. What is annoying then is that wrong decisions are less likely to be costly in the borderline case, and it seems unreasonable to invest in the cost of the large sample required to reach a decision.

Example 2.4.2 For the Bernoulli problem of θ_0 against θ_1, we have shown that

$$\theta = \left[1 - \left(\frac{1-\theta_1}{1-\theta_0}\right)^h\right] \Big/ \left[\frac{\theta_1}{\theta_0} - \left(\frac{1-\theta_1}{1-\theta_0}\right)^h\right],$$

$$\pi(\theta) = \mathrm{OC}(\theta) = (A^h - 1)/(A^h - B^h).$$

Now using $\theta_0 = .5$, $\theta_1 = .9$, $\alpha = \beta = .05$, one obtains

$$\pi(\theta) = (19^h - 1)/(19^h - 19^{-h})$$

and

$$E_\theta(Z) = \theta \ln\left\{\frac{\theta_1(1-\theta_0)}{\theta_0(1-\theta_1)}\right\} + \ln\left(\frac{1-\theta_1}{1-\theta_0}\right)$$
$$= \theta \ln 9 - \ln 5.$$

As $h \to 0$, both EZ and $E(\ln \Lambda_N)$ tend to zero, but their ratio can be computed by evaluating its limit as $h \to 0$. We have the table of values given in Table 2.4.1. So let us find $\lim_{h\to 0} E(N|\theta)$:

Table 2.4.1

The OC Function and the Expected Sample Size of the SPRT for the Bernoulli Problem

h	$-\infty$	-1	0	1	∞
θ	1	.9	.733	.5	0
$\pi(\theta)$	0	.05	.5	.95	1
$E(Z\|\theta)$	ln 1.8	.9 ln 9-ln 5	0	ln .6	−ln 5
$E(N\|\theta)$	$\dfrac{\ln 19}{\ln 1.8} = 5.01$	$\dfrac{.9 \ln 19}{.9 \ln 9 - \ln 5} = 5.19$	9.17	$\dfrac{-.9 \ln 19}{\ln .6} = 7.2$	$\dfrac{\ln 19}{\ln 5} = 1.83$

$$\lim_{h\to 0} E(N|\theta) = \lim_{h\to 0} \frac{(\ln 19)[(1-19^{-h}) - (19^h - 1)]/(19^h - 19^{-h})}{\{[1-(.2)^h]/[(1.8)^h - (.2)^h]\}\ln 9 - \ln 5}.$$

Using the expansion

$$a^h - b^h = e^{h \ln a} - e^{h \ln b}$$
$$= h(\ln a - \ln b) + \frac{h^2}{2}[(\ln a)^2 - (\ln b)^2] + \cdots,$$

we have

$$\frac{(1-19^{-h}) - (19^h - 1)}{19^h - 19^{-h}}$$
$$= \left[-h^2(\ln 19)^2 - \frac{h^4}{12}(\ln 19)^4 - \cdots\right] \Big/ 2\left[h \ln 19 + \frac{h^3}{6}(\ln 19)^3 + \cdots\right]$$
$$= -(h \ln 19)\left\{1 + \frac{h^2}{12}(\ln 19)^2 + \cdots\right\} \Big/ 2\left\{1 + \frac{h^2}{6}(\ln 19)^2 + \cdots\right\}$$
$$= -\frac{h}{2}(\ln 19)\left\{1 - \frac{h^2}{12}(\ln 19)^2 + \cdots\right\}.$$

Also,

$$\frac{1-(.2)^h}{(1.8)^h-(.2)^h}=\frac{5^h-1}{9^h-1}$$

$$=(h\ln 5)\left(1+\frac{h}{2}\ln 5+\cdots\right)(h\ln 9)^{-1}\left(1+\frac{h}{2}\ln 9+\cdots\right)^{-1}$$

$$=\frac{\ln 5}{\ln 9}\left(1+\frac{h}{2}\ln 5-\frac{h}{2}\ln 9+\cdots\right).$$

Hence

$$\frac{1-(.2)^h}{(1.8)^h-(.2)^h}(\ln 9)-\ln 5=\ln 5\left\{1+\frac{h}{2}\ln\left(\frac{5}{9}\right)+\cdots\right\}-\ln 5$$

$$=\frac{h}{2}(\ln 5)\ln\left(\frac{5}{9}\right)+\cdots.$$

Thus

$$\frac{[(1-19^{-h})-(19^h-1)]/(19^h-19^{-h})}{(5^h-1)(9^h-1)^{-1}(\ln 9)-\ln 5}$$

$$=-\frac{h}{2}(\ln 19)\left\{1-\frac{h^2}{12}(\ln 19)^2+\cdots\right\}\Big/\left\{\frac{h}{2}\ln 5\ln\left(\frac{5}{9}\right)+\cdots\right\}$$

$$=\frac{-\ln 19}{\ln 5\ln\left(\frac{5}{9}\right)}\{1+O(h)\},$$

$$\lim_{h\to 0}\text{L.H.S.}=\frac{-\ln 19}{\ln 5\ln\left(\frac{5}{9}\right)}.$$

So

$$\lim_{h\to 0}E(N|\theta)=(\ln 19)^2/\ln 5\ln\left(\frac{9}{5}\right)=\frac{(2.944)^2}{(.5878)(1.6094)}$$

$$=\frac{8.64}{.95}=9.1.$$

Although the SPRT terminates finitely, in any single experiment N could be very large. Hence, one might establish a bound n_0 such that one terminates the procedure at n_0. If no decision is reached by the n_0th stage, sampling is stopped anyway and H_0 accepted if $\Lambda_{n_0} < 1$ and rejected if $\Lambda_{n_0} > 1$. The truncation of the procedure would certainly affect the error sizes of the test and this aspect has been studied by Wald (1947, §3.8) and the effect is slight if n_0 is fairly large.

Problems

1. Find the ASN curve for testing H_0: $f(x;\,\theta_0) = \theta_0 e^{-x\theta_0}$ against H_1: $f(x;\,\theta) = \theta_1 e^{-x\theta_1}$ $(x > 0)$ using $\theta_0 = 2$, $\theta_1 = 1$, $\alpha = .05$, and $\beta = .10$. Notice that for one of the five points used $E(N) = 0$. In view of the fact that $N \geq 1$, this result must be wrong. Explain.
2. Find the ASN for the SPRT for testing $\theta = 0$ against $\theta = 1$ in a normal population with unit variance (use $\alpha = \beta = .01$).

2.5 Wald's Fundamental Identity

In this section we shall give an identity of Wald (1947) which plays a fundamental role in deriving the moments of the sample size required in order to terminate the SPRT for testing H_0: $\theta = \theta_0$ against H_1: $\theta = \theta_1$, where X has the probability density function given by $f(x;\,\theta)$.

Theorem 2.5.1 (Wald, 1947) Let $Z = \ln\{f(X;\,\theta_1)/f(X;\,\theta_0)\}$ and let $P(Z = 0) < 1$. Then

$$E\{\exp(S_N t)[C(t)]^{-N}\} = 1$$

for every t in D, where

$$S_N = \sum_1^N Z_i, \qquad C(t) = E(e^{Zt}),$$

and D is the set of points t in the complex plane such that $C(t)$ is finite and $|C(t)| \geq 1$.

Proof Consider the identity

$$E(\exp[S_N t + (S_n - S_N)t]) = \{C(t)\}^n, \tag{2.5.1}$$

where n denotes a positive integer. Let p_n be the probability that $N \leq n$ and let $E_n(U)$ denote the conditional expectation of U given that $N \leq n$, and

$E_n^*(U)$ denote the conditional expectation of U given that $N > n$. Then the identity in (2.5.1) can be written as

$$p_n E_n(\exp[S_N t + (S_n - S_N)t]) + (1 - p_n)E_n^*(\exp(S_n t)) = \{C(t)\}^n. \quad (2.5.2)$$

In the subpopulation defined by the condition $N \leq n$ the expression $S_n - S_N$ is independent of S_N. Consequently

$$E_n(\exp[S_N t + (S_n - S_N)t]) = E_n\{(\exp(S_N t))[C(t)]^{n-N}\}. \quad (2.5.3)$$

Hence (2.5.1) can be rewritten [after dividing throughout by $\{C(t)\}^n$] as

$$p_n E_n\{\exp(S_N t)[C(t)]^{-N}\} + (1 - p_n)E_n^*(\exp(S_n t))[C(t)]^{-n} = 1. \quad (2.5.4)$$

$P(Z = 0) < 1$ implies that $\lim_{n\to\infty}(1 - p_n) = 0$ and because $n < N$ implies that $\ln B < S_n < \ln A$, we have that $|E_n^*(\exp(S_n t))|$ is a bounded function of n. Hence we have, for every t in D,

$$\lim_{n\to\infty} (1 - p_n)E_n^*(\exp(S_n t))[C(t)]^{-n} = 0. \quad (2.5.5)$$

Also since $\lim_{n\to\infty} p_n E_n(\exp(S_N t)[C(t)]^{-N}) = E\{\exp(S_N t)[C(t)]^{-N}\}$, the fundamental identity follows from (2.5.4) and (2.5.5). Wald (1946, 1947) has shown that the fundamental identity can be differentiated under the expectation sign any number of times with respect to t at any real value t such that $C(t) \geq 1$. The proof is based on the following propositions:

(i) All derivatives of $\exp(S_N t)C^{-N}(t)$ wrt t exist in the domain of t for which $C(t) \geq 1$.

(ii) For any positive integer r and for any finite interval I in which $C(t) \geq 1$, it is possible to find a function $D(S_N, N)$ such that

$$D(S_N, N) \geq |(d^r/dt^r)[\exp(S_N t)C^{-N}(t)]|$$

for all values of t in I, and $E[D(S_N, N)] < \infty$.

Applications of the Fundamental Identity

Differentiating wrt t and setting $t = 0$, we obtain

$$E[S_N - E(N)C'(0)] = 0.$$

That is

$$E(S_N) = E(N)E(Z). \quad (2.5.6)$$

Differentiating twice and setting $t = 0$ we obtain

$$E\left\{\left[S_N - N\frac{C'(t)}{C(t)}\right]^2 - N\left[\frac{C''(t)C(t) - (C'(t))^2}{C^2(t)}\right]\exp(S_N t)C^{-N}(t)\right\}\Big|_{t=0} = 0,$$

i.e.

$$E\{[S_N - NC'(0)]^2 - NC''(0) + N[C'(0)]^2\} = 0.$$

If $E(Z) = C'(0) = 0$, then

$$E(S_N{}^2) = E(N)E(Z^2).$$

That is

$$E(N) = E(S_N{}^2)/E(Z^2), \tag{2.5.7}$$

where

$$E(S_N{}^2) \sim (\ln A)^2 P(S_N \geq \ln A) + (\ln B)^2 P(S_N \leq \ln B),$$

$$P(S_N \geq \ln A) \doteq (1 - B^h)/(A^h - B^h)$$

and

$$P(S_N \leq \ln B) \doteq (A^h - 1)/(A^h - B^h)$$

where h is the nonzero root of the equation $C(t) = 1$. The above formula was derived under the assumption that $E(Z) \neq 0$. If $E(Z)$ approaches zero, then $h \to 0$ since

$$\frac{d}{dh}\int\left[\frac{f(x;\theta_1)}{f(x;\theta_0)}\right]^h f(x;\theta)\,dx = \frac{d}{dh}(1) = 0;$$

i.e.

$$\int \ln\left[\frac{f(x;\theta_1)}{f(x;\theta_0)}\right]\left[\frac{f(x;\theta_1)}{f(x;\theta_0)}\right]^h f(x;\theta)\,dx = 0.$$

Hence,

$$E(Z) = \int\left[\ln\frac{f(x;\theta_1)}{f(x;\theta_0)}\right] f(x;\theta)\,dx = 0.$$

[Also see the discussion after the proof of Wald's (1947) lemma A.1.1.] Then

$$\lim_{h\to 0} [(1 - B^h)/(A^h - B^h)] = \ln B/(\ln A - \ln B)$$

and

$$\lim_{h\to 0} [(A^h - 1)/(A^h - B^h)] = \ln A/(\ln A - \ln B)$$

therefore

$$\begin{aligned} E(S_N{}^2) &= [-(\ln A)^2 \ln B + (\ln B)^2 \ln A]/(\ln A - \ln B) \\ &= -\ln A \ln B. \end{aligned}$$

Hence

$$E(N) \doteq -\ln A \ln B/E(Z^2), \qquad \text{when} \quad E(Z) = 0. \tag{2.5.8}$$

Problem

Let the common density of $X_1, X_2, \ldots$ be $\frac{1}{2}(1-\theta^2)\exp(-|x|+\theta x)$ for $|\theta|<1$, $-\infty < x < \infty$. Consider the SPRT of H_0: $\theta = -\frac{1}{2}$ against H_1: $\theta = \frac{1}{2}$ with $\alpha = \beta = .1$. Evaluate EN.

Bahadur (1958) points out that the fundamental identity of sequential analysis (see Theorem 2.5.1) can be regarded as a special case of a formula for the probability that the sampling terminates finitely. In the following are presented Bahadur's results.

Let Z be a random variable with distribution function F. It is assumed that the moment-generating function $C(t) = E(e^{tZ})$ exists for every real t in some neighborhood of $t = 0$. Let $Z_{(\infty)} = (Z_1, Z_2, \ldots)$ denote a sequence of independent and identically distributed observations on Z. Consider a fixed sequential sampling procedure, namely, a set of rules for observing the components $Z_1, Z_2, \ldots$ one by one, such that at each stage, the decision whether experimentation is to be continued is a (possibly randomized) function of the observed values in hand at that stage. Let N denote the total number of components Z_n observed in a given instance. Assume that the sampling procedure is closed under F, that is

$$P(N < \infty | F) = 1. \tag{2.5.9}$$

The procedure is otherwise arbitrary. Write $S_N = Z_1 + \cdots + Z_N$ and

$$\psi(t, N, S_N) = \begin{cases} [C(t)]^{-N}e^{tS_N}, & N < \infty, \\ 1, & N = \infty. \end{cases} \tag{2.5.10}$$

Then we have the following theorem.

Theorem 2.5.2 (Bahadur, 1958). For every t,

$$E[\psi(t, N, S_N) | F] = P(N < \infty | G_t), \tag{2.5.11}$$

where

$$dG_t = [C(t)]^{-1}e^{tx}\, dF. \tag{2.5.12}$$

Proof For each t, G_t defined by (2.5.12) is a probability distribution function, so that $\{G_t\}$ is an exponential family of alternative distributions of Z, with F a member of this family. The family $\{G_t\}$ is introduced here purely as a device in the study of the given sampling rule when F obtains. To establish Theorem 2.5.2, for each $n = 1, 2, \ldots$ let R_n denote the nonsequential sample space of exactly n observations, that is, of points $(Z_1, \ldots, Z_n) = Z_{(n)}$. For each n, let $p_n(Z_{(n)})$ be the conditional probability of the event $N = n$ given $Z_{(\infty)}$. The sequence $p_1, p_2, \ldots$ of functions on $R_1, R_2, \ldots$ characterizes the given sampling procedure, and is, of course, independent of the distribution of Z. For each n, let $F^{(n)}$ denote the distribution of $Z_{(n)}$ when F obtains. That is, $F^{(n)}(Z_1, \ldots, Z_n) = \prod_{i=1}^{n} F(Z_i)$.

Let V denote the outcome of the sequential experiment, that is, $V = (Z_1, \ldots, Z_N)$ if $N < \infty$ and $V = Z_{(\infty)}$ if $N = \infty$. Notice that if g is a real-valued function of V such that $E(|g||F) < \infty$ and (2.5.9) holds, then

$$E(g|F) = \sum_{n=1}^{\infty} \int_{R_n} p_n g_n \, dF^{(n)} \tag{2.5.13}$$

where $g_1, g_2, \ldots$ is the (essentially unique) sequence of functions on R_1, $R_2, \ldots$ such that $g_n = g$ when $N = n$. The right-hand side of (2.5.13) is an absolutely convergent series, in fact g is integrable if and only if

$$\sum_n \int p_n |g_n| \, dF^{(n)} < \infty.$$

With the above notation, for each $n = 1, 2, \ldots$, let S_n denote the nonsequential random variable $Z_1 + \cdots + Z_n$. Then,

$$\begin{aligned} P(N < \infty \mid G_t) &= \sum_{n=1}^{\infty} P(N = n \mid G_t) \\ &= \sum_{n=1}^{\infty} \int_{R_n} p_n \, dG_t^{(n)} \\ &= \sum_{n=1}^{\infty} \int_{R_n} p_n C^{-n} e^{tS_n} \, dF^{(n)}, \qquad \text{by (2.5.12)}, \\ &= E(\psi \mid F), \qquad \text{by (2.5.9), (2.5.10), and (2.5.13).} \end{aligned}$$

It follows from Theorem 2.5.2 that Wald's identity namely

$$E[\psi(t, N, S_N)] = 1 \tag{2.5.14}$$

holds for a given value of t if and only if

$$P(N < \infty \mid G_t) = 1 \tag{2.5.15}$$

for the same value of t. It follows easily that a sufficient condition for the validity of (2.5.14) for all t is that there exist a finite k such that $P(N < k \mid F) = 1$.

Next, we shall describe an alternative sufficient condition for the validity of (2.5.15) [and hence of (2.5.14)] in an assigned neighborhood of $t = 0$, given (2.5.9) and (2.5.12). This condition does not depend on the detailed structure of the sampling rule; it requires the existence of the joint moment-generating function of N and S_N under F in a sufficiently large neighborhood of the origin. The condition also assures the validity of differentiation under the expectation sign in (2.5.14), that is, $d^k\psi/dt^k$ is integrable and

$$d^k\psi/dt^k = 0, \qquad k = 1, 2, \ldots, \tag{2.5.16}$$

and each t in the neighborhood. Let I be an open interval including $t = 0$ such that $C(t)$ exists for each t in I.

Theorem 2.5.3 (Bahadur, 1958) Assume that, for each t in I, there exists a $z > -\ln C(t)$ such that

$$E(\exp(tS_N + zN) \mid F) < \infty.$$

Then (2.5.14), (2.5.15), and (2.5.16) hold for t in I.

Proof First, let us show that (2.5.15) holds in a sufficiently small neighborhood of zero. That is

$$\sum_{n=1}^{\infty} q_n(t) = 1, \qquad q_n(t) = P(N = n \mid G_t), \tag{2.5.17}$$

so that

$$q_n(t) = \int_{R_n} p_n [C(t)]^{-n} \exp(tS_n)\, dF^{(n)} \qquad \text{by (2.5.12).}$$

Write

$$\beta_n = 1 - \sum_{i=1}^{n} p_i \qquad \text{and} \qquad \rho(t) = C(2t)/C^2(t).$$

Then, for any t in I and any n

$$\begin{aligned}
P(N > n \mid G_t) &= \int_{R_n} \beta_n\, dG_t^{(n)}, \qquad \text{since} \quad \beta_n = P(N > n \mid Z_{(n)}),\\
&= C^{-n} \int_{R_n} \beta_n \exp(tS_n)\, dF^{(n)}\\
&\le C^{-n} \left\{ \int_{R_n} \beta_n^{\,2}\, dF^{(n)} \right\}^{1/2} \left\{ \int_{R_n} \exp(2tS_n)\, dF^{(n)} \right\}^{1/2}\\
&= \rho^{n/2} \left\{ \int_{R_n} \beta_n^{\,2}\, dF^{(n)} \right\}^{1/2}\\
&\le \rho^{n/2} \left\{ \int_{R_n} \beta_n\, dF^{(n)} \right\}^{1/2}, \qquad \text{since} \quad 0 \le \beta_n \le 1,\\
&= \{\rho^n(t) P(N > n \mid F)\}^{1/2}.
\end{aligned} \tag{2.5.18}$$

It follows from the hypothesis of Theorem 2.5.3 with $t = 0$ that, for some $\lambda > 1$, $\lambda^n P(N > n \mid F) \to 0$, as $n \to \infty$. Hence by (2.5.18) $P(N > n \mid G_t) \to 0$ as $n \to \infty$ for each t such that $\rho(t) \le \lambda$. Thus, (2.5.17) holds whenever $\rho(t) \le \lambda$. This establishes the desired conclusion since $\lambda > 1$, ρ is continuous, and $\rho(0) = 1$. The next step is to extend the validity of (2.5.17) for all t in I by analytic continuation; the details are omitted.

REMARK 2.5.3.1 The hypothesis of Theorem 2.5.3 can be replaced by the following: For each t in I, the conditional moment-generating function of N given $N < \infty$ exists in some neighborhood of zero when G_t obtains, that is, $E(e^{\delta N} \mid G_t, N < \infty) = \sum e^{\delta n} P(N = n \mid G_t, N < \infty) < \infty$ for some $\delta > 0$.

REMARK 2.5.3.2 Suppose that $C(t)$ exist for all t. If

$$E(e^{zN}|F) < \infty \qquad \text{for some} \quad z > \sup_t\{-\ln C(t)\}, \tag{2.5.19}$$

and if

$$E(\exp(tS_N)|F) < \infty \qquad \text{for all} \quad t, \tag{2.5.20}$$

then (2.5.14), (2.5.15), and (2.5.16) hold for all t. A stronger sufficient condition for the same conclusion is that (2.5.19) hold for all z.

REMARK 2.5.3.3 If (2.5.19) holds for some $z > 0$, then N and S_N possess moments of all orders, and there exists a neighborhood of $t = 0$ in which (2.5.14), (2.5.15), and (2.5.16) hold. A stronger sufficient condition for the same conclusion is that $E(Z|F) \neq 0$ and (2.5.20) hold for some t of the same sign as $E(Z|F)$. These conditions are of interest since the validity of (2.5.16) in a neighborhood of $t = 0$ is sufficient for most applications of differentiation.

REMARK 2.5.3.4 A theorem of Albert's (1947, 1948) states that if $C(t)$ exists for all t, if $P(Z > 0|F) > 0$ and $P(Z < 0|F) > 0$ and if the sampling procedure is a random walk based on the cumulative sums S_n (with fixed barriers a and b, $b < 0 < a$), then (2.5.14) and (2.5.16) hold for all t. It can be shown that in this case the hypothesis of Theorem 2.5.3 is satisfied with $I = (-\infty, \infty)$ so that Albert's theorem is a special case of Theorem 2.5.3.

REMARK 2.5.3.5 It is easy to see that $P(N < k|F) = 1$ for some finite k implies the hypothesis of Theorem 2.5.3. Examples seem to exist indicating that (2.5.15) does not imply the hypothesis of Theorem 2.5.3.

Exact Formulas for the OC and ASN Functions for a Special Class of Densities

Exact formulas for the OC and ASN functions have been derived by Wald (1947, Appendix A4), when $Z = \ln\{f(X;\theta_1)/f(X;\theta_0)\}$ takes on only a finite number of integral multiples of a constant d. This result is somewhat general since any continuous Z can be approximated by a discrete distribution as closely as one wishes. Let g_1 and g_2 be two positive integers such that $P_\theta(Z = -g_1 d)$ and $P_\theta(Z = g_2 d)$ are positive and Z can take only integral multiples of d which are $\geq -g_1 d$ and $\leq g_2 d$. Let $h_j = P_\theta(Z = jd)$. Then the moment-generating function of Z is given by

$$C(t) = E_\theta(e^{Zt}) = \sum_{j=-g_1}^{g_2} h_j e^{jtd}.$$

To obtain the roots of the equation $C(t) = 1$, we set $e^{td} = u$ and solve the equation

$$\sum_{j=-g_1}^{g_2} h_j u^j = 1. \tag{2.5.21}$$

Let $g = g_1 + g_2$ and $u_1, u_2, \ldots, u_g$ denote the g roots where it is assumed that $u_i \neq u_j$ for $i \neq j$. Substituting u_j for e^{td} in the fundamental identity (Theorem 2.5.1), we have

$$E_\theta(u_j^{S_N/d}) = 1 \qquad (j = 1, \ldots, g). \tag{2.5.22}$$

Let k be the smallest integer $\geq a/d$; and l the largest integer $\leq b/d$ where $a = \ln A$, and $b = \ln B$. Then S_N/d can take only the values

$$(l - g_1 + 1), (l - g_2 + 2), \ldots, l, k, k + 1, \ldots, (k + g_2 - 1). \tag{2.5.23}$$

Let $c_1, c_2, \ldots, c_g$ denote the g different values in (2.5.23). Then (2.5.22) can be written as

$$\sum_{i=1}^{g} \xi_i u_j^{c_i} = 1 \qquad (j = 1, \ldots, g), \qquad \text{where} \quad \xi_i = P_\theta(S_N = c_i d). \tag{2.5.24}$$

Let Δ be the determinant of the matrix $(u_j^{c_i})$ $(i, j = 1, \ldots, g)$ and let Δ_i be the determinant obtained by substituting 1 for the elements in the ith column. If $\Delta \neq 0$, it follows from (2.5.24) that

$$\xi_i = \Delta_i/\Delta. \tag{2.5.25}$$

Thus the probability $\pi(\theta)$ that the process will terminate with $S_N \leq \ln B$ is given by

$$\pi(\theta) = \sum \Delta_i/\Delta \tag{2.5.26}$$

where the summation is over all values of i such that $dc_i \leq \ln B$. Thus, we get an exact expression for the OC(θ). Further

$$\begin{aligned} E_\theta(S_N) &= \sum_{i=1}^{g} c_i\, \Delta_i\, d/\Delta, \\ E_\theta(N) &= \sum_{i=1}^{g} c_i\, \Delta_i\, d/\Delta E_\theta(Z). \end{aligned} \tag{2.5.27}$$

Calculation of the probabilities $\xi_1, \ldots, \xi_g$ as described above requires the computation of the roots of the polynomial equation (2.5.21). Girshick (1946) has proposed an alternative method of evaluating $\xi_1, \ldots, \xi_g$ which involves solving a system of $g + k - l - 1$ linear equations.

If $g_2 = 1$, the OC function $\pi(\theta)$ is a simple expression of the roots $u_1, \ldots, u_g$. In fact $\pi(\theta) = P_\theta(S_N \le b) = 1 - P_\theta(S_N \ge a) = 1 - \xi_g$. Using the properties of the Vandermonde determinants, one can easily show [see Wald (1947, p. 183–184) for the details], that

$$\xi_g = \frac{\Delta_g}{\Delta} = \sum_{j=1}^{g} \left[\frac{u_j^{g_1 - 1 - l}}{(u_j - 1)\prod_{i \neq j}(u_j - u_i)}\right] \Big/ \sum_{j=1}^{g} \left[\frac{u_j^{g_1 - 1 + k - l}}{(u_j - 1)\prod_{i \neq j}(u_j - u_i)}\right].$$

We shall illustrate the above derivations by a simple example considered by Wald (1947).

Example 2.5.1 Let X be the Bernoulli random variable taking on values 0 and 1 with probabilities $1 - \theta$ and θ respectively. We wish to test $H_0\colon \theta = \theta_0 = (1 - e^{-2})/(e - e^{-2})$ against $H_1\colon \theta = \theta_1 = (e - e^{-1})/(e - e^{-2})$. We wish to evaluate the OC and ASN functions at $\theta = \frac{3}{7}$.

Since Z takes values 1 and -2 with probabilities $\frac{3}{7}$ and $\frac{4}{7}$ respectively,

$$C(t) = (3e^t + 4e^{-2t})/7.$$

Setting $e^t = u$ and solving the equation $3u + 4u^{-2} = 7$ we obtain the roots $u_1 = 1$, $u_2 = 2$, and $u_3 = -\frac{2}{3}$. Also $c_1 = \ln B - 1$, $c_2 = \ln B$, $c_3 = \ln A$. Hence, with $a = \ln A$, $b = \ln B$,

$$\Delta = \begin{vmatrix} 1 & 1 & 1 \\ 2^{b-1} & 2^b & 2^a \\ (-\frac{2}{3})^{b-1} & (-\frac{2}{3})^b & (-\frac{2}{3})^a \end{vmatrix}$$

and

$$\pi(\tfrac{3}{7}) = (\Delta_1 + \Delta_2)/\Delta \qquad \text{and} \qquad E(N) = \{7(-b + 1)\,\Delta_1 - b\Delta_2 - a\Delta_3\}/5\Delta.$$

Other families of distributions of the Z_i for which $P(S_N \ge a)$ and EN are capable of being expressed by exact formulas have been described by Kemperman (1961) and Ferguson (1967). In the following we shall describe these distributions.

Example 2.5.2 [Ferguson (1967, p. 378)] Let the iid X_i have the common density

$$f(x; \theta) = \tfrac{1}{2}(1 - \theta^2)\exp(-|x| + \theta x), \qquad |\theta| < 1. \tag{2.5.28}$$

Consider the SPRT (B, A) of $H_0: \theta = -\frac{1}{2}$ against $H_1: \theta = \frac{1}{2}$. Here $Z_i = X_i$ and one can readily compute

$$C(t) = (1 - \theta^2)/[1 - (\theta + t)^2],$$

which exists for all t in the interval $-1 - \theta < t < 1 - \theta$. Let

$$\phi_j(Z_1, \ldots, Z_j) = \begin{cases} 0 & \text{if } \; b < S_j < a, \\ 1 & \text{otherwise,} \end{cases}$$

and $I_{(c,d)}(x) = 1$ if $x \in (c, d)$ and zero elsewhere. Then, following Kemperman (1961), we compute

$$\begin{aligned}
E_\theta&\left\{\exp(tS_n) \prod_1^{n-1} [1 - \phi_j(Z_1, \ldots, Z_j)] I_{(a,\infty)}(S_n)\right\} \\
&= E_\theta\left\{\exp(tS_{n-1}) \prod_1^{n-1} [1 - \phi_j(Z_1, \ldots, Z_j)]\right. \\
&\qquad \left. \times E_\theta[\exp(tZ_n) I_{(a - S_{n-1}, \infty)}(Z_n) \,|\, Z_1, \ldots, Z_{n-1}]\right\} \\
&= E_\theta\left\{\exp(tS_{n-1}) \prod_1^{n-1} [1 - \phi_j(Z_1, \ldots, Z_j)]\right. \\
&\qquad \left. \times \int_{a - S_{n-1}}^{\infty} f(z; \theta) e^{tz}\, dz\right\} \\
&= E_\theta\left\{\exp(tS_{n-1}) \prod_1^{n-1} [1 - \phi_j(Z_1, \ldots, Z_j)] \frac{(1 - \theta^2)}{2(1 - \theta - t)}\right. \\
&\qquad \left. \times \exp[-(1 - \theta - t)(a - S_{n-1})]\right\} \\
&= \frac{(1 - \theta)e^{at}}{(1 - \theta - t)} A_n(\theta)
\end{aligned} \tag{2.5.29}$$

where

$$A_n(\theta) = \tfrac{1}{2}(1 + \theta) \exp\{-a(1 - \theta)\} E_\theta\left\{\exp[(1 - \theta)S_{n-1}] \prod_1^{n-1} [1 - \phi_j(Z_1, \ldots, Z_j)]\right\}.$$

Setting $t = 0$ in (2.5.29) shows that $A_n(\theta) = P_\theta(N = n, \; S_N \geq a)$. Similarly, one can compute

$$E_\theta\left\{\exp(tS_N) \prod_1^{n-1} [1 - \phi_j(Z_1, \ldots, Z_j)] I_{(-\infty, b)}(S_n)\right\} = \frac{(1 + \theta)e^{tb}}{(1 + \theta + t)} B_n(\theta),$$

where $B_n(\theta) = P_\theta(N = n,\ S_N \leq b)$. Now, from the fundamental identity,

$$\begin{aligned}
1 &= E_\theta\{\exp(tS_N)C(t)^{-N}\} \\
&= \sum_{n=1}^{\infty} E_\theta\left\{\exp(tS_n) \prod_1^{n-1} [1 - \phi_j(Z_1, \ldots, Z_j)][I_{(a,\infty)}(S_n) + I_{(-\infty,b)}(S_n)]\right\} C(t)^{-n} \\
&= \frac{(1-\theta)e^{ta}}{(1-\theta-t)} \sum_{n=1}^{\infty} A_n(\theta)C(t)^{-n} + \frac{(1+\theta)e^{tb}}{(1+\theta+t)} \sum_{n=1}^{\infty} B_n(\theta)C(t)^{-n}. \qquad (2.5.30)
\end{aligned}$$

Since $P_\theta(S_N \geq a) = \sum_{n=1}^{\infty} A_n(\theta) = 1 - \sum_{n=1}^{\infty} B_n(\theta)$, we can solve (2.5.30) for $P_\theta(S_N \geq a)$ after replacing t with the nonzero real value t_θ, for which $C(t_\theta) = 1$; namely $t_\theta = -2\theta$. This yields

$$P_\theta(S_N \geq a) = \begin{cases} \dfrac{(1-\theta)^2 - (1+\theta)^2 e^{-2\theta b}}{(1-\theta)^2 e^{-2\theta a} - (1+\theta)^2 e^{-2\theta b}} & \text{for} \quad \theta \neq 0, \\[2ex] \dfrac{1-b}{2+a-b} & \text{for} \quad \theta = 0. \end{cases} \qquad (2.5.31)$$

By differentiating the identity in (2.5.30) with respect to t and setting $t = 0$, we obtain

$$\begin{aligned}
0 = &\left(\frac{1}{1-\theta} + a\right) \sum_{n=1}^{\infty} A_n(\theta) - C'(0) \sum_{n=1}^{\infty} nA_n(\theta) \\
&- \left(\frac{1}{1+\theta} - b\right) \sum_{n=1}^{\infty} B_n(\theta) - C'(0) \sum_{n=1}^{\infty} nB_n(\theta).
\end{aligned}$$

Since

$$E_\theta(N) = \sum_1^{\infty} n(A_n(\theta) + B_n(\theta)) \qquad \text{and} \qquad C'(0) = 2\theta/(1-\theta^2),$$

we have

$$E_\theta(N) = \begin{cases} \dfrac{1-\theta^2}{2\theta} \left[\left(\dfrac{1}{1-\theta} + a\right) P_\theta(S_N \geq a) - \left(\dfrac{1}{1+\theta} - b\right)(1 - P_\theta(S_N \geq a))\right], & \theta \neq 0, \\[2ex] [(1+a)(1-b)+1]/2, & \theta = 0. \end{cases} \qquad (2.5.32)$$

where $E_\theta(N)$ is determined at $\theta = 0$ by continuity or by taking a second derivative of (2.5.30) at $t = 0$. Wald's approximations given in Section 2.3 with $h = -2\theta$ give

$$P_\theta(S_N \geq a) = \begin{cases} (1 - e^{-2\theta b})/(e^{-2\theta a} - e^{-2\theta b}) & \text{for} \quad \theta \neq 0, \\ -b/(a - b) & \text{for} \quad \theta = 0. \end{cases} \tag{2.5.33}$$

For $\theta = 0$, the Wald approximation is close to the true value (2.5.32) if $-b$ and a are large; if $a = b$, it is exact. However, as $\theta \to -1$, the true value of $P_\theta(S_N \geq a)$ tends to zero whereas its approximation converges to

$$(1 - e^{2b})/(e^{2a} - e^{2b}) > 0;$$

and as $\theta \to 1$, $P_\theta(S_N \geq a) \to 1$ while its approximation tends to

$$(e^{-2b} - 1)/(e^{-2b} - e^{-2a}) < 1.$$

If $\theta_0 = -\frac{1}{2}$ and $\theta_1 = \frac{1}{2}$,

$$\alpha = P_{-1/2}(S_N \geq a) = (3 - B)/(9A - B)$$

and

$$\beta = P_{1/2}(S_N \leq b) = (3 - A^{-1})/(9B^{-1} - A^{-1}).$$

So, the Wald approximations of α and β are about three times too large when A and B^{-1} are large. That is, the effective error probabilities are about one-third as large as the prescribed ones. This being on the conservative side is excusable. However, it results in a corresponding decrease in the expected sample size. On the basis of some computations carried out, Ferguson (1967) suggests that more care must be exercised in using Wald's approximations to the expected sample sizes. If $E(Z)$ is much larger than a, then so is ES_N; consequently when ES_N is replaced by $bP(S_N \leq b) + aP(S_N \geq a)$, the ratio $EN = ES_N/EZ$ is approximated by something much less than one. In general, the Wald approximations underestimate EN.

Example 2.5.3 [Kemperman (1961), pp. 70–71] Let the iid X_i have the common density

$$f(x; \theta) = \begin{cases} \theta e^{-\theta x} & \text{if} \quad x \geq 0, \\ 0 & \text{if} \quad x < 0. \end{cases} \tag{2.5.34}$$

As before, we wish to test $H_0: \theta = \theta_0$ against $H_1: \theta = \theta_1$ with $\theta_0 < \theta_1$. We obtain

$$E_\theta(e^{Zt}) = C(t) = (\theta_1/\theta_0)^t[1 + \theta^{-1}(\theta_1 - \theta_0)t]^{-1}.$$

Consequently, the density function of Z is given by

$$g_\theta(x) = \begin{cases} B(\theta)\exp[\theta x/(\theta_1 - \theta_0)] & \text{if} \quad x < \delta = \ln(\theta_1/\theta_0), \\ 0 & \text{if} \quad x > \delta, \end{cases} \tag{2.5.35}$$

where $B(\theta)$ denotes a constant. Consequently $Z_n < \delta$, thus

$$S_N \le a + \delta. \tag{2.5.36}$$

Proceeding as in Example 2.5.2, Kemperman (1961) obtains the exact inequalities

$$\varepsilon \frac{1 - \varepsilon B}{A - \varepsilon^2 B} \le \alpha \le \frac{1 - \varepsilon B}{A - \varepsilon B} \tag{2.5.37}$$

and

$$\varepsilon \frac{1 - A^{-1}}{B^{-1} - \varepsilon A^{-1}} \le \beta \le \varepsilon \frac{1 - \varepsilon A^{-1}}{B^{-1} - \varepsilon^2 A^{-1}}$$

where α, β are the error probabilities and $\varepsilon = e^{-\delta} = \theta_0/\theta_1 < 1$. From these considerations, a better choice for B than Wald's approximation $\beta/(1 - \alpha)$ would be $\varepsilon^{-1}\beta/(1 - \alpha)$. One can easily derive upper and lower bounds for the OC and ASN functions.

Problems

1. For the density function considered in Example 2.5.3, derive the exact upper and lower bounds for the OC and ASN functions.

2. Of two given coins, it is known that one is fair and the other has probability $\frac{2}{3}$ for heads; however it is not known which is the fair coin. The coins, tossed simultaneously and randomly, produce the data (X_1, Y_1), (X_2, Y_2), ..., where X_i (Y_i) is 1 or 0 depending on whether the ith toss of the first (second) coin comes up with a head or tail. We wish to test H_0: the second coin is fair against H_1: the first coin is fair. For a given SPRT (A, B), find the exact error probabilities and expected sample sizes. *Answer*: If $A = 2^j$ and $B = 2^{-k}$ with j and k integers, then

$$E(N \mid H_0) = 6\{k(A - 1) - j(1 - B)\}/(A - B).$$

3. Let $X_1, X_2, \ldots$ be independent and identically distributed random variables having the common density function

$$f(x;\theta) = \begin{cases} \theta^x, & x = 1, 2, \ldots, \\ (1-2\theta)/(1-\theta), & x = 0, \end{cases}$$

where $0 < \theta < \frac{1}{2}$. Consider the SPRT (A, B) of $H_0\colon \theta = \frac{1}{5}$ against $H_1\colon \theta = \frac{3}{8}$. Evaluate the exact error probabilities and the expected sample sizes.

2.6 Bounds for the Expected Sample Size in a SPRT

We have given Wald's approximate value for $E_\theta(N)$ and Wald gave upper and lower bounds for the error in his approximation [see Wald (1945) or Wald (1947, Eqs. (A.77) and (A.78))]. Wald mentioned that, in general, the error in approximation will be small when the mean and variance of Z are small. For the general case, M. N. Ghosh (1960) has given a bound for the error which is wider than the bound of Wald, especially when the population is normal. He also provides an example showing that the error in Wald's approximate formula for $E(N)$ is not small even though the mean and variance of Z are small so that no general statement about the validity of the approximation can be made. In the following we shall present M. N. Ghosh's (1960) results.

Lemma 2.6.1 Let Z be any random variable with distribution function $G(z)$. Then $E(Z|Z > x)$ is a nondecreasing function of x provided Z is absolutely continuous.

Proof Since

$$E(Z|Z > x) = \int_x^\infty z\, dG(z)[1 - G(x)]^{-1},$$

$$\frac{d}{dx} E(Z|Z > x) = \frac{-xG'(x)}{1 - G(x)} + \int_x^\infty z\, dG(z)[1 - G(x)]^{-2}G'(x) \tag{2.6.1}$$

$$= G'(x)[1 - G(x)]^{-2}\left[\int_x^\infty (z - x)\, dG(x)\right] \geq 0$$

since $G'(x) \geq 0$.

REMARK 2.6.1 Similarly, $E(Z|Z < x)$ is a nonincreasing function of x.

Theorem 2.6.1 Let $p_{0,i} = P(N = i, S_N < \ln B)$; $p_0 = P(S_N < \ln B)$; $p_{1,i} = P(N = i,\ S_N > \ln\ A)$; $p_1 = P(S_N > \ln\ A)$; and let $E_{0,i}$, E_0, $E_{1,i}$, and E_1 denote the conditional expectations in the corresponding situations. We then have the following inequality for the Wald's SPRT:

$$\left| E(N) - \frac{p_0 \ln B + p_1 \ln A}{E(Z)} \right| \le \frac{\max\{-p_0 E(Z|Z < \ln B - \ln A),\, p_1 E(Z|Z > \ln A - \ln B)\}}{|E(Z)|}. \tag{2.6.2}$$

Proof Notice that

$$p_0 E_0(S_N) = \sum_{i=1}^{\infty} p_{0,i} E_{0,i}(S_N) = \sum_{i=1}^{\infty} p_{0,i}[E_{0,i}(S_{i-1}) + E_{0,i}(Z_i)]. \tag{2.6.3}$$

However, $E_{0,i}(Z_i) = E(Z|Z < \ln B - S_{i-1})$ which according to Lemma 2.6.1 has minimum value for $S_{i-1} = \ln A$. Also, since $S_{i-1} \ge \ln B$

$$\ln B \ge E_{0,i}(S_N) \ge E(Z|Z < \ln B - \ln A) + \ln B. \tag{2.6.4}$$

Similarly, $\ln A \le E_{1,i}(S_N) \le E(Z|Z > \ln A - \ln B) + \ln A$. Multiplying by $p_{0,i}$ and $p_{1,i}$ respectively and summing over i we obtain

$$p_0 \ln B \ge p_0 E_0(S_N) \ge p_0 E(Z|Z < \ln B - \ln A) + p_0 \ln B, \tag{2.6.5}$$

and

$$p_1 \ln A \le p_1 E_1(S_N) \le p_1 E(Z|Z > \ln A - \ln B) + p_1 \ln A,$$

so that

$$p_0 E(Z|Z < \ln B - \ln A) \le p_0 E_0(S_N) + p_1 E_1(S_N) - (p_0 \ln B + p_1 \ln A) \le p_1 E(Z|Z > \ln A - \ln B). \tag{2.6.6}$$

Now set $p_0 E_0(S_N) + p_1 E_1(S_N) = E(S_N)$, divide the inequality in (2.6.6) by $E(Z)$, and notice that $E(S_N)/E(Z) = E(N)$. Thus the proof of Theorem 2.6.1 is complete.

REMARK 2.6.2 The bound in (2.6.2) will be useful when $E(Z)$ is not too small, but it is often too wide and the bounds for $E_{0,i}(S_N)$ and $E_{1,i}(S_N)$ can be made closer together in many situations. Thus we may find

$$\sup_{\ln B < S_{i-1} \leq \ln A} \{S_{i-1} + E(Z|Z > \ln A - S_{i-1})\} \tag{2.6.7}$$

and

$$\inf_{\ln B \leq S_{i-1} \leq \ln A} \{S_{i-1} + E(Z|Z < \ln B - S_{i-1})\}, \tag{2.6.8}$$

and these may give closer bounds for the error in Wald's approximation to $E(N)$ as shown by Wald (1945, p. 144) for the case of normal variables.

When both $\ln A$ and $\ln B$ are finite, Stein (1946) has shown that $E(N)$ is bounded. The formula $E(N) = E(S_N)/E(Z)$ is meaningful provided $E(N)$ is finite. When one of the boundaries, say $\ln A$ is infinite, we cannot show that $E(N)$ is bounded, although when it exists the formula $E(N) = E(S_N)/E(Z)$ will hold either from Blackwell's theorem (1946) or by differentiation of Wald's fundamental identity. M. N. Ghosh (1960) has shown that $E(N)$ is convergent if Z satisfies a certain condition which is given in Theorem 2.6.2.

Theorem 2.6.2 Let the random variable Z having distribution function $G(z)$ be such that $E(Z) < 0$. Then $E(N)$ is bounded provided

$$\int_{-\infty}^{-x} z \, dG(z) \Big/ \int_{-\infty}^{-x} dG(z) = E(Z|Z < -x) \geq -x - c \tag{2.6.9}$$

for some c and k such that $x > k > 0$, $c > 0$.

Proof See M. N. Ghosh (1960, pp. 362–365).

REMARK 2.6.3 M. N. Ghosh (1960) has shown that the condition (2.6.9) is satisfied when Z has either the normal distribution or the double exponential distribution.

Example 2.6.1 Consider the density function:

$$g(z) = \begin{cases} b^{-1}|z|^{-r}, & z < -b, r > 2, \\ 0, & 0 > z \geq -b, \\ f(z), & z \geq 0, \end{cases} \tag{2.6.10}$$

where $b > 0$ and $f(z)$ is any function so that $\int_{-\infty}^{\infty} g(z)\, dz = 1$. Let $G(z)$ be the cumulative distribution function of Z. M. N. Ghosh (1960) has shown that for this distribution the mean and variance are small, however, the excess at the boundary is not small; this contradicts the conjecture of Wald (1947).

Hoeffding (1953) has derived a lower bound for the ASN required by an arbitrary sequential test that satisfies the usual conditions regarding the probabilities of type I and type II errors.

Let $X_1, X_2, \ldots$ be a sequence of independent random variables with a common frequency function $f(x; \theta)$ (either the probability density or the probability function) where θ is confined to a set Ω. Let D be an arbitrary (possibly randomized) sequential test for deciding between two alternatives, $H_0: \theta \in \omega_0$ and $H_1: \theta \in \omega_1$ where ω_0 and ω_1 are two disjoint subsets of Ω. Let α, β be two numbers between 0 and 1 such that

$$\begin{aligned} P_\theta(D \text{ acc } H_1) &\leq \alpha \quad \text{if} \quad \theta \in \omega_0, \\ P_\theta(D \text{ acc } H_0) &\leq \beta \quad \text{if} \quad \theta \in \omega_1, \end{aligned} \tag{2.6.11}$$

where $P_\theta(E)$ denotes the probability of E under the frequency function $f(x; \theta)$. Let N be the number of observations required to terminate the test D (by accepting H_0 or H_1). Then we have the following result of Hoeffding (1953).

Theorem 2.6.3 With the above notation if (i) $\alpha + \beta \leq 1$ and (ii) $P_\theta(D$ accepts $H_0) + P_\theta(D$ accepts $H_1) = 1$ for all $\theta \in \Omega$, then

$$E_\theta(N) \geq \left\{\beta \ln \frac{\beta}{1-\alpha} + (1-\beta) \ln \frac{1-\beta}{\alpha}\right\} \Big/ e_0(\theta) \qquad \text{if} \quad \theta \in \omega_1, \tag{2.6.12}$$

and

$$E_\theta(N) \geq \left\{\alpha \ln \frac{\alpha}{1-\beta} + (1-\alpha) \ln \frac{1-\alpha}{\beta}\right\} \Big/ e_1(\theta) \qquad \text{if} \quad \theta \in \omega_0, \tag{2.6.13}$$

where

$$e_i(\theta) = \inf_{\theta' \in \omega_i} E_\theta\{\ln[f(X; \theta)/(f(X; \theta')]\}, \qquad i = 0, 1. \tag{2.6.14}$$

Proof See Hoeffding (1953, pp. 128–130).

REMARK 2.6.4 Inequalities (2.6.12) and (2.6.13) were obtained by Wald (1945, Section 4.7; or 1947, Appendix A.7) under the assumptions that the sets ω_0 and ω_1 consist of one point each and that equality holds in (2.6.11). Equality in (2.6.12) and (2.6.13) can be achieved by a SPRT for certain special distributions $f(x; \theta)$ and suitable values of θ [see Wald (1945)]. The bounds given in Theorem 2.6.3, although in general not the best ones, have the advantage of being simple. Hoeffding (1953) has also obtained a somewhat complicated lower bound for $E_\theta(N)$.

Example 2.6.2 Let $f(x; \theta) = (2\pi)^{-1/2} \exp\{-\frac{1}{2}(x - \theta)^2\}$, $\omega_0 = \{\theta \leq -\delta\}$, $\omega_1 = \{\theta \geq \delta\}$. Then

$$e_0(\theta) = \begin{cases} 0 & \text{if} \quad \theta \leq -\delta, \\ (\theta + \delta)^2/2 & \text{if} \quad \theta > -\delta, \end{cases} \qquad e_1(\theta) = \begin{cases} (\theta - \delta)^2/2 & \text{if} \quad \theta < \delta, \\ 0 & \text{if} \quad \theta \geq \delta. \end{cases}$$

Suppose $\alpha = \beta$, then

$$E_\theta(N) \geq \begin{cases} 2\left[(1-2\beta)\ln\left(\dfrac{1-\beta}{\beta}\right)\right]\Big/(\theta+\delta)^2 & \text{if} \quad \theta \in \omega_1, \\ 2\left[(1-2\beta)\ln\left(\dfrac{1-\beta}{\beta}\right)\right]\Big/(\theta-\delta)^2 & \text{if} \quad \theta \in \omega_0. \end{cases}$$

Both the numerator and the denominator in the inequalities (2.6.12) and (2.6.13) are positive since $\ln x > 1 - (1/x)$ for $x > 0$ unless $x = 1$, the integral in the denominators can be equal to $+\infty$, in which case the lower bounds take the trivial value 0. Equality in (2.6.12) and (2.6.13) can be attained with a SPRT in the case where the sets ω_0 and ω_1 consist of one point each and the ratio f_1/f_0 [where $f_i = f(x; \theta_i)$, $i = 0, 1$] takes on only the two values C and $1/C$ provided that the values α and β can be achieved as error probabilities in the test. The bounds due to Hoeffding (1953, 1960) will be given in the following.

Let $X_1, X_2, \ldots$ be a sequence of independent random variables having common density f with respect to a σ-finite measure μ (in particular f could be indexed by a parameter θ). One of the two decisions δ_0 and δ_1 is to be made. Let f_0 and f_1 be two probability densities such that the decision δ_0 (δ_1) is considered as wrong if $f = f_1$ (f_0). Consider sequential tests (decision rules) for making decision δ_0 or δ_1 such that the probability of a wrong decision does not exceed a positive number α (β) when $f = f_0$ (f_1). Let N denote the (random) number of observations required by such a test. Hoeffding (1953, 1960) obtained lower bounds for $E_2(N)$, the expected sample size when $f = f_2$ where f_2 is, in general, different from f_0 and f_1.

Theorem 2.6.4 (Hoeffding, 1953) With the above notation and hypotheses (i) and (ii) of Theorem 2.6.3 we have

$$E_2(N) \geq \sup_{0<c<1} \frac{-\ln[\alpha^c(1-\beta)^{1-c} + (1-\alpha)^c\beta^{1-c}]}{c\int f_2\{\ln(f_2/f_0)\}\,d\mu + (1-c)\int f_2\{\ln(f_2/f_1)\}\,d\mu}. \tag{2.6.15}$$

Proof See Hoeffding (1953).

For $f_2 = f_0$ and $c \to 0$, (2.6.15) reduces to (2.6.13). This bound is likely to be close when f_2 is close to f_0 or f_1.

Theorem 2.6.5 (Hoeffding, 1960) Let the sequential test terminate with probability one under each of f_0, f_1, and f_2. Also assume that $E_2(N) < \infty$.

In order to exclude trivialities, also assume that $\alpha + \beta < 1$. Then, we have

$$E_2(N) \geq \frac{1 - \alpha - \beta}{1 - \int \min(f_0, f_1, f_2)\, d\mu} \tag{2.6.16}$$

and

$$E_2(N) \geq \{[(\tau/4)^2 - \zeta \ln(\alpha + \beta)]^{1/2} - \tau/4\}^2 \zeta^{-2}, \tag{2.6.17}$$

where

$$\zeta = \max(\zeta_0, \zeta_1), \qquad \zeta_i = \int f_2 \ln(f_2/f_i)\, d\mu, \quad i = 0, 1, \tag{2.6.18}$$

and

$$\tau^2 = \int [\ln(f_1/f_0) - \zeta_0 + \zeta_1]^2 f_2\, d\mu. \tag{2.6.19}$$

It is assumed that the above integrals exist and $f_2(x) = 0$ implies that $\min[f_0, f_1] = 0$.

Proof See Hoeffding (1960).

Discussion of Inequality (2.6.16) Equality holds in (2.6.16) provided α and β are the error probabilities of the sequential test and

$$f_0(X_i) = f_1(X_i) = f_2(X_i) \qquad \text{for } i = 1, \ldots, N - 1 \text{ and for each } N = n\ (n \geq 1), \tag{2.6.20}$$

and

$$f_2(x) \geq \min[f_0(x), f_1(x)]. \tag{2.6.21}$$

Condition (2.6.21) is satisfied by many common one-parameter families of distributions when θ_2 lies between θ_0 and θ_1. Under this condition inequality (2.6.16) reduces to

$$\begin{aligned} E_2(N) &\geq (1 - \alpha - \beta)\Big/\left[1 - \int \min(f_0, f_1)\, d\mu\right] \\ &= (1 - \alpha - \beta)\Big/\left[\tfrac{1}{2} \int |f_0 - f_1|\, d\mu\right]. \end{aligned} \tag{2.6.22}$$

Condition (2.6.20) is satisfied, and equality holds in (2.6.22), in the following two cases. The first case is where the densities are arbitrary, subject to only (2.6.21) but the values α and β are such that they can be attained as error probabilities with a test requiring at most one observation ($N \leq 1$); and if an observation x is taken decision δ_0 (δ_1) is made if $f_0(x) - f_1(x) > 0$ (<0).

Example 2.6.3 The second case in which equality in (2.6.22) is attained is where, in addition to (2.6.21), the set $E = \{x : f_0(x) = f_1(x) = f_2(x)\}$ has a positive probability. Let $E_2 \subset E$ and let the complement of E_2 be subdivided into two disjoint sets E_0 and E_1 such that $f_0(x) - f_1(x) \geq 0$ if $x \in E_0$ and ≤ 0 if $x \in E_1$. Let N be the least n such that $x_n \notin E_2$. Decision δ_i is made if $x_N \in E_i$, $i = 0, 1$. (Instead, suitable randomized decisions can be made when $x_n \in E$.) Then, it can easily be verified that equality holds in (2.6.22) with $E_2(N) = (1 - p_2)^{-1}$, $\alpha = p_{01}(1 - p_2)^{-1}$, $\beta = p_{10}(1 - p_2)^{-1}$, where p_2 is the probability of E_2 (under any f_i) and p_{ij} is the probability of E_j under f_i. In the particular case where μ is linear Lebesgue measure, $f_i(x) = f(x - \theta_i)$, $f(x) = 1/l$, $-l/2 \leq x \leq l/2$, $f(x) = 0$ otherwise, $0 < \theta_1 - \theta_0 < l$, and $\theta_0 \leq \theta_2 \leq \theta_1$, we have $E = [\theta_1 - l/2;\ \theta_0 + l/2]$. Let $\theta_1 - l/2 \leq c \leq d \leq \theta_0 + l/2$, $E_2 = (c, d)$, $E_0 = (-\infty, c]$, $E_1 = [d, \infty)$. Then, with the test just described, perhaps preceded by a randomized decision as to whether to take at least one observation, any error probabilities can be attained ($\alpha \geq 0$, $\beta \geq 0$, $\alpha + \beta \leq 1$). Moreover, the maximum with respect to all real θ of the expected sample size of this test when the density is $f(x - \theta)$ is attained when θ is between θ_0 and θ_1. Hence the test minimizes the expected sample size. Notice that the present test is a modified version of the SPRT defined by Wald (1947, p. 120). It differs from the latter in one respect: If the probability ratio after n observations equals one of the two numbers A and B (in the case of this example $A = B = 1$), the stopping decision and the terminal decision may depend on the position of the sample point in the corresponding sets, instead of being randomized decisions.

It is of interest to note that the bound in (2.6.16) is always positive whereas the bounds in (2.6.15), (2.6.12), and (2.6.13) take on the trivial value 0 if the integrals in their denominators are equal to ∞. However, in most of the common cases, the bounds (2.6.12), (2.6.13), (2.6.15), and (2.6.17) are better than (2.6.16). For instance, if f_2, f_0, f_1 are normal densities having a common variance and respective means 0, $-\xi$, ξ, the bound in (2.6.16) is of the order ξ^{-1}, but those in (2.6.15) and (2.6.17) are of the order ξ^{-2} and hence better than (2.6.16) for small ξ.

Discussion of Inequality (2.6.17) Equality in (2.6.17) can hold in trivial cases. In order to have an idea of how close the bound in (2.6.17) can come to the minimum attainable value of $E_2(N)$, let us consider the following example.

Example 2.6.4 Let f_2, f_0, f_1 be the normal densities with variance 1 and respective means $0, -\xi, \xi$ $(\xi > 0)$. Then $\zeta_0 = \zeta_1 = \xi^2/2$, $\tau = 2\xi$ and inequality (2.6.17) takes the form

$$E_2(N) \geq \xi^{-2}\{[1 - 2\ln(2\alpha)]^{1/2} - 1\}, \tag{2.6.23}$$

where $\alpha = \beta$. This bound will be compared with the values of $E_2(N)$ for a fixed-sample size test, Wald's SPRT, and Anderson's (1960) modified SPRT with error probabilities $\alpha = \beta$ $(<\frac{1}{2})$ in each case. Let $S_m = X_1 + \cdots + X_m$. For a fixed-sample size test such that decision δ_1 or δ_2 is made according as $S_n < 0$ or $S_n > 0$, the error probabilities at $\pm\xi$ are both equal to $\Phi(-\xi n^{1/2})$, where $\Phi(x)$ denotes the standard normal distribution function. Hence $E_2(N)$ is the least n such that $\Phi(-\xi n^{1/2}) \leq \alpha$. If z_α is defined by $\Phi(-z_\alpha) = \alpha$, we have

$$E_2(N) = \xi^{-2} z_\alpha^{\ 2}, \tag{2.6.24}$$

exactly or with a good approximation. If $\alpha \to 0$, the $z_\alpha \to \infty$ and $\alpha = \Phi(-z_\alpha) = (2\pi)^{-1/2} z_\alpha^{-1} \exp(-z_\alpha^{\ 2}/2)\{1 + O(z_\alpha^{-2})\}$. Hence $z_\alpha^{\ 2} = -2\ln\alpha + O[\ln(-2\ln\alpha)]$. The factor of ξ^{-2} in inequality (2.6.23) is

$$\{[1 - 2\ln(2\alpha)]^{1/2} - 1\}^2 = -2\ln\alpha + O\{(-2\ln\alpha)^{1/2}\}.$$

Thus, if α is sufficiently small, the bound in (2.6.23) is nearly attained with a fixed-sample size test, although the asymptotic approach is extremely slow. It follows that the fixed sample size test nearly minimizes the expected sample size at mean 0 when α is very small.

Next, consider the SPRT that stops as soon as $2\xi|S_n| > \ln A$ (>0). Then $(\ln A)^2 \leq 4\xi^2 E_2(S_N^{\ 2}) = 4\xi^2 E_2(N)$ and $A \leq (1-\alpha)/\alpha$. These inequalities are close approximations for a fixed α and sufficiently small δ (Wald, 1947). With this approximation

$$E_2(N) = \xi^{-2}\left(\frac{1}{2}\ln\frac{1-\alpha}{\alpha}\right)^2. \tag{2.6.25}$$

Set $\alpha = (1-\varepsilon)/2$. Then

$$\left(\frac{1}{2}\ln\frac{1-\alpha}{\alpha}\right)^2 = \varepsilon^2 + \frac{2}{3}\varepsilon^4 + \frac{23}{45}\varepsilon^6 + \cdots$$

and

$$\{[1 - 2\ln(2\alpha)]^{1/2} - 1\}^2 = \varepsilon^2 + \frac{2}{3}\varepsilon^4 - \frac{1}{6}\varepsilon^6 + \cdots.$$

Thus, if α is close to its upper bound $\alpha/2$, and ξ is small enough, the lower bound in (2.6.23) is nearly achieved by the SPRT. Hence, the SPRT nearly minimizes $E_2(N)$ in this case. Table 2.6.1 shows that even for $\alpha = 0.2$ the

Table 2.6.1

Values of $E_2(N)$ and of the Lower Bound in (2.6.23) for $\xi = .1$

α	.0001	.001	.01	.05	.1	.2	.3
Fixed sample size	1383	955	541.2	270.6	164.3	70.8	27.5
Wald's SPRT	2121	1193	527.9	216.7	120.7	48.0	17.9
Anderson's test	—[a]	—	402.2	192.2	—	—	—
Lower bound (2.6.23)	1054	710	388.3	187.0	111.1	46.6	17.8

[a] Not available.

expected sample size exceeds the lower bound by only 3%. [The lower bound in (2.6.15) with $c = \frac{1}{2}$ also approaches $E_2(N)$ for the SPRT as $\alpha \to \frac{1}{2}$.] However, inequality (2.6.23) is better than (2.6.15) in the present example, for all values of α.

For α not close to 0 or $\frac{1}{2}$, we compare the bound in (2.6.23) with the expected sample size of a test considered by Anderson (1960). This test stops as soon as $|S_n| \geq c + dn$, where $d < 0 < c$. Anderson approximated the sequence $\{S_n\}$ by a Wiener process so that his values for the expected stopping time $E_2(\tau)$ when the mean of the process is 0 are approximations to $E_2(N)$. He chose the constants c and d so as to minimize $E_2(N)$ subject to prescribed error probabilities $\alpha = \beta$ at $\pm\xi$ for $\xi = .1$ and $\alpha = .01$ and .05. Anderson's (1960) values are given in Table 2.6.1. The expected sample sizes exceed the lower bounds by only 3.6 and 2.8%, respectively. This shows that both Anderson's test (as judged by the expected sample size at mean 0) and inequality (2.6.23) cannot greatly be improved in these cases.

Hoeffding (1960) has shown also that for each of the two sequential tests considered here, the expected sample size attains its maximum when the mean θ of the normal distribution is zero. In conjunction with the preceding results this implies that each of these tests (as well as the fixed-sample size test) comes close to minimizing the maximum expected sample size for certain α values. Both tests are of the form: stop sampling as soon as $|S_n| \geq c_n$ where $c_1, c_2 \ldots$ are nonnegative constants. The expected value of N at θ is the sum of the probabilities $P_\theta\,(N > n)$. We can write

$$P_\theta\,(N > n) = \int_E f(y - \theta\mathbf{1})\,dy$$

where $y = (y_1, \ldots, y_n)$, $\mathbf{1} = (1, \ldots, 1)$, $f(x)$ is the probability density of n independent normal chance variables having mean 0 and variance 1, and $E = \{y : |y_1 + \cdots + y_m| < c_m,\ m = 1, 2, \ldots, n\}$. The set E is convex, and $y \in E$ implies $-y \in E$. It follows from a theorem of Anderson (1955) that $P_\theta(N > n)$ attains its maximum at $\theta = 0$ (and is monotone for $\theta < 0$ and $\theta > 0$). Thus, the same conclusion is true for the expected value of N.

Distribution of the Sample Size

Wald (1947) has derived an approximate expression for the characteristic functions of the sample size, of the conditional sample size given that H_0 is accepted, and of the conditional sample size given that H_0 is rejected. Wald (1947) claimed that his approximations are valid when $|E(Z)|$ and var(Z) are small, which may not be true in the light of M. N. Ghosh's (1960) results.

Theorem 2.6.6 (Wald, 1947) We have, for all imaginary τ

$$E_\theta(e^{\tau N}) \doteq (A^{h_2} - A^{h_1} + B^{h_1} - B^{h_2})/(B^{h_1}A^{h_2} - A^{h_1}B^{h_2}), \qquad (2.6.26)$$

$$E_\theta{}^*(e^{\tau N}) \doteq (A^{h_2} - A^{h_1})/(B^{h_1}A^{h_2} - A^{h_1}B^{h_2})\pi, \qquad (2.6.27)$$

and

$$E_\theta^{**}(e^{\tau N}) \doteq (B^{h_1} - B^{h_2})/(1 - \pi)(B^{h_1}A^{h_2} - A^{h_1}B^{h_2}) \qquad (2.6.28)$$

where $\pi = \pi(\theta) = P(\text{acc } H_0 | \theta)$, and $h_1 = h_1(\tau)$, $h_2 = h_2(\tau)$ are the solutions of the equation

$$C_\theta(h) = E_\theta(e^{Zh}) = e^{-\tau}$$

such that $\lim_{\tau\to 0} h_1(\tau) = 0$ and $\lim_{\tau\to 0} h_2(\tau) = h_\theta$, h_θ being the unique nonzero root of the equation $C_\theta(h) = 1$. Further, $E^*(e^{\tau N})$ $[E^{**}(e^{\tau N})]$ denotes the characteristic function of the conditional distribution of N given that H_0 is accepted [rejected].

For the special case of the normal density, he has derived an expression for the cumulative distribution function of N by inverting the characteristic function of N. By differentiating the characteristic function of N with respect to τ, B. K. Ghosh (1969) has obtained expressions for the first four moments of N and studies them with reference to the normal, binomial, and exponential distributions. In the normal case, B. K. Ghosh (1969), using a Pearson type III density function gives an approximate probability for $P_\theta(N = n)$ in terms of the first four moments. Bhate (1959) has investigated the upper and lower bounds to the distribution of N, and Cox and Roseberry (1966) have conducted some Monte Carlo studies in order to determine empirically the sampling distribution of N; however, the results in both the cases are of limited scope. The first moment of N can be computed with desired accuracy as demonstrated by Bartlett (1946) and Hoeffding (1960). Recently, Chanda (1971) has derived results pertaining to convergence in probability and asymptotic normality of N when suitably standardized and when the sampling distribution of the observations belongs to the exponential family. In the following we shall state his two main results without proof since the proofs are quite long.

Let $X_1, X_2, \ldots$ be a sequence of independent and identically distributed random variables having pdf $f(x; \theta)$ where

$$f(x; \theta) = B(\theta)e^{\theta R(x)}h(x) \tag{2.6.29}$$

and the range of variation of X_1 is free of θ. Also, assume that ln $B(\theta)$ is thrice differentiable with respect to θ and that

$$d^2[\ln B(\theta)]/d\theta^2 < 0 \qquad \text{and} \qquad d^3[\ln B(\theta)]/d\theta^3$$

is bounded for all θ in Ω (Ω denoting an interval). For the sake of simplicity, further assume that the distribution of $Y_1 = R(X_1)$ is absolutely continuous and has support on the entire real line. Let $g(y; \theta)$ and $G(y; \theta)$ respectively denote the density and distribution function of Y_1. Also assume that for every $\theta \in \Omega$,

$$\begin{aligned} &\text{(i)} \quad \lim_{y\to\infty} \{g(y; \theta)/[1 - G(y; \theta)]\} = \tilde{C}(\theta), \\ &\text{(ii)} \quad \lim_{y\to-\infty} [g(y; \theta)/G(y; \theta)] = D(\theta) \end{aligned} \tag{2.6.30}$$

where $\tilde{C}(\theta)$ and $D(\theta)$ are positive constants, which in particular, may assume the value ∞. Notice that (2.6.30) needs some modifications when Y_1 is bounded and/or has a discrete distribution. If Y_1 is bounded below [above] (i) [(ii)] is to be dropped; and if $g(y; \theta)$ is discrete, the limits $y \to \infty$ or $-\infty$ (whichever is admissible) will be assumed to range through those values of y for which $g(y; \theta)$ assumes positive values. One can easily check that the normal, gamma, Poisson, binomial, and negative binomial distributions do satisfy (2.6.29) and (2.6.30). The following results pertaining to the tail-end probabilities of Y_1 are the direct consequence of assumption (2.6.30).

$$\begin{aligned} &\text{(i)} \quad \lim_{y\to\infty} y^s[1 - G(y; \theta)] = 0, \qquad s = 0, 1, 2. \\ &\text{(ii)} \quad \lim_{y\to-\infty} y^s G(y; \theta) = 0, \qquad s = 0, 1, 2. \end{aligned} \tag{2.6.31}$$

For every θ belonging to Ω and real constant β such that $\theta + \beta$ belongs to Ω,

(iii) $\lim_{y \to \infty} e^{\beta y}[1 - G(y; \theta)] = 0,$

and

(iv) $\lim_{y \to -\infty} [e^{\beta y} G(y; \theta)] = 0.$

With the above assumptions and notation, we have the following results. We are interested in testing $H: \theta = \theta_0$ against the alternative: $\theta = \theta_1 = \theta_0 + \Delta$ where Δ is small.

Theorem 2.6.7 (Chanda, 1971) Let θ_2 be the true value of θ where $\theta_2 \neq \theta_0$. Let N be the stopping variable for Wald's SPRT. Define

$$\alpha = [\ln B(\theta_0 + \Delta) - \ln B(\theta_0)]/\Delta,$$
$$\alpha_j = \frac{d}{d\theta} \ln B(\theta)|_{\theta = \theta_j}, \qquad j = 0, 2. \tag{2.6.32}$$

Then

$$p \lim_{\Delta \to 0} N\Delta(\alpha - \alpha_2) = \begin{cases} \ln A, & \theta_2 > \theta_0, \\ \ln B, & \theta_2 < \theta_0. \end{cases} \tag{2.6.33}$$

REMARK 2.6.7.1 Notice that it is assumed that the sequential sampling is always done under θ_2 which being different from θ_0 is outside the range $(\theta_0, \theta_0 + \Delta)$ for sufficiently small Δ. θ_1 and θ_2 can otherwise be any pair of values of θ lying inside Ω.

Proof of Theorem 2.6.7 The proof consists of using Wald's fundamental identity in order to get an asymptotic expression for the characteristic function of $N\Delta(\alpha - \alpha_2)$.

Theorem 2.6.8 (Chanda, 1971) Let θ_0, θ_2, α, α_0, and α_2 be as defined in Theorem 2.6.7. Then:

(i) if $\theta_2 > \theta_0$, the distribution of

$$N^* = c_1^{-1}[N - \ln A/\Delta(\alpha - \alpha_2)]$$

where $c_1 = \Delta^{-1/2}\sigma_2[\ln A/(\alpha_0 - \alpha_2)^3]^{1/2}$ and $\sigma_2{}^2 = -d^2 \ln B(\theta)/d\theta^2|_{\theta=\theta_2} > 0$ tends to the standard normal distribution as $\Delta \to 0$; and

(ii) if $\theta_2 < \theta_0$, the distribution of

$$N^{**} = c_2^{-1}\{N - \ln B/\Delta(\alpha - \alpha_2)\}$$

where $c_2 = \Delta^{-1/2}\sigma_2\{\ln\ B/(\alpha_0 - \alpha_2)^3\}^{1/2}$ tends to the standard normal as $\Delta \to 0$.

Proof The proof consists of using the fundamental identity and obtaining an asymptotic expression for the characteristic functions of N^* and N^{**}. Theorems 2.6.7 and 2.6.8 imply that to the order of $1/\Delta$ one can obtain

$$E_{\theta_2}(N) = \begin{cases} \ln A/\delta\Delta, & \theta_2 > \theta_0, \\ \ln B/\delta\Delta, & \theta_2 < \theta_0; \end{cases} \tag{2.6.34}$$

$$\mathrm{var}_{\theta_2}(N) = \begin{cases} \sigma_2^{\,2} \ln A/\delta^3\Delta = c_1^{\,2}, & \theta_2 > \theta_0, \\ \sigma_2^{\,2} \ln B/\delta^3\Delta = c_2^{\,2}, & \theta_2 < \theta_0; \end{cases} \tag{2.6.35}$$

where θ_2 is the true value and $\delta = \alpha - \alpha_2$. In Table 2.6.2 we tabulate certain numerical calculations given by Chanda (1971). Notice that $N_2 \simeq E_{\theta_2}(N) + 1.645\ \{\mathrm{var}_{\theta_2}(N)\}^{1/2}$.

Table 2.6.2

The First Two Moments of the Sample Size When Sampling Is from the Exponential Family with $\ln A = 1$, $\ln B = -2$

Distribution	Δ	θ_2	θ_0	$E_{\theta 2}(N)$	$\mathrm{Var}_{\theta 2}(N)$	N_2
Normal (unknown mean)	.10($\delta = 2.00$)			5	1.25	7
	.20($\delta = -.50$)			20	80	35
Normal (unknown variance)	.25	.50	1.00	40	80	55
	.055	1.00	0.50	20	160	41
Gamma with $\beta = 1$, (exponential with unknown mean θ)	.026	2.00	0.50	6.67	11.85	13
	.50	.50	2.00	13.33	1.48	16
Poisson	.21	4.00	2.00	5.00	5.00	9
	.88	2.00	4.00	5.00	2.50	8
Binomial	.024	.6	.4	50.00	300.00	79
	.30	.4	.8	25.00	37.50	36

In order to bring the probability density to the form (2.6.29), a certain (monotonic) transformation of θ (to γ) is required for each of the above distributions except the normal with unknown mean.

In concluding Chanda's work, it is worthwhile to point out the open question: how small should Δ be so that the asymptotic normality of N holds reasonably? There exists no analytical method that answers this question satisfactorily.

2.7 Formulas for Calculating the Operating Characteristic and the Average Sample Number of Some Sequential Tests

Wald (1947, A.2.3) has obtained upper and lower limits for the OC functions. Page (1954) has suggested an improvement to Wald's approximations to the OC function and the ASN, and Kemp (1958) has provided even better approximations using the same method as Page (1954) but making different assumptions. In the following we shall present Kemp's results.

2.7.1 Operating Characteristic Function

Let Z_i be the ith observation drawn from a population with probability density function $g(z)$ [in our case $Z_i = \{f(X_i;\, \theta_1)/f(X_i;\, \theta_0)\}$] and consider a sequential scheme with horizontal decision lines a distance ω apart. The lower line is taken as the line of reference. Also, let $P(z)$ denote the probability that a sequential test which starts at a point a distance z from this line will end on or below the lower boundary. Then $P(z)$ satisfies

$$P(z) = \int_{-\infty}^{-z} g(x)\, dx + \int_0^{\omega} P(x) g(x - z)\, dx. \tag{2.7.1}$$

If $P(x) \doteq 1$ when $x \le 0$ and if either $P(x) = 0$ or $\int_{\omega}^{\infty} g(x - z)\, dx = 0$ when $x \ge \omega$, we can rewrite Eq. (2.7.1) as

$$\begin{aligned} P(z) &\doteq \int_{-\infty}^{0} P(x)g(x - z)\, dx + \int_0^{\omega} P(x)g(x - z)\, dx + \int_{\omega}^{\infty} P(x)g(x - z)\, dx \\ &= \int_{-\infty}^{\infty} P(x)g(x - z)\, dx. \end{aligned} \tag{2.7.2}$$

Notice that a term $\int_{-z}^{0} g(x)\, dx = 0$ is added to the right-hand side. If h is the nonzero root of the equation formed by equating the moment-generating function of Z [or $g(z)$] to unity, then a solution of (2.7.2) is

$$P(z) = A + Be^{hz}. \tag{2.7.3}$$

Substituting $z = 0$ and $z = \omega$, we obtain

$$A = [P(\omega) - P(0)e^{\omega h}]/(1 - e^{h\omega}) \qquad \text{and} \qquad B = [P(0) - P(\omega)]/(1 - e^{h\omega}).$$

The quantity $P(z)$ can be calculated for any z from Eq. (2.7.3) when $P(0)$ and $P(\omega)$ are known. If the above expressions for A and B are substituted in (2.7.3) and the resulting expression for $P(z)$ is substituted in Eq. (2.7.1) and the integration carried out, we can obtain two simultaneous equations to solve for $P(0)$ and $P(\omega)$ by putting $z = 0$ and $z = \omega$.

Application to Normal Populations Conditions analogous to those assumed in deriving Eq. (2.7.2) are reasonably satisfied when Z_i is normally distributed with mean θ and unit standard deviation and $\omega \geq 3$. Here $h = -2\theta$ and we have that

$$P(z) = P(0) + [P(\omega) - P(0)](1 - e^{-2\theta z})(1 - e^{-2\omega\theta})^{-1}. \qquad (2.7.4)$$

Substituting for $P(z)$ from (2.7.4), integrating, and putting $z = 0$ and $z = h$ as described, the simultaneous equations in $P(0)$ and $P(\omega)$ are

$$K_1 P(0) + K_2 P(\omega) = \Phi(-\theta), \qquad (2.7.5)$$

$$e^{-2\omega\theta} K_2 P(0) + K_1 P(\omega) = \Phi(-\omega - \theta), \qquad (2.7.6)$$

where

$$K_1 = 1 - \Phi(\omega - \theta) + \Phi(-\theta) - K_2,$$

$$K_2 = [1 - \Phi(\omega + \theta) - \Phi(\omega - \theta) - 2\Phi(\theta)](1 - e^{-2\omega\theta})^{-1}, \qquad (2.7.7)$$

and Φ is the standard normal distribution function. It is not necessary to solve Eqs. (2.7.5) and (2.7.6) for both positive and negative θ since

$$P(0 | \theta) = 1 - P(\omega | -\theta). \qquad (2.7.8)$$

When $\theta = 0$, the limiting form of the operating characteristic is

$$P(z) = P(0) + [1 - 2P(0)]z/\omega$$

where

$$P(0) = \{\tfrac{1}{2}\omega(2\pi)^{1/2} + (1 - e^{-\omega^2/2})\}/\{h(2\pi)^{1/2}[\tfrac{3}{2} - \Phi(\omega)] + 2(1 - e^{-\omega^2/2})\}.$$

Table 2.7.1 presents a comparison of the values of $P(\omega)$ given by Eqs. (2.7.5) and (2.7.6), the true values, and those obtained by Page (1954) when $\omega = 10$. Table 2.7.2 presents a comparison of the values of $P(z)$ given by (2.7.4), the true values, and those given by Wald (1947) when $\omega = 10$. Kemp (1958) has drawn figures in which $P(0)$ and $P(\omega)$ are plotted against θ for several values of ω. From these curves and Eq. (2.7.4) it is possible to calculate the operating characteristics of a variety of sequential schemes.

Table 2.7.1

Values of $P(\omega)$ When $\omega = 10$

θ	Page's approximation	Kemp's approximation	True value
−1.00	.73	.81	.80
−.50	.42	.55	.53
.00	.043	.069	.063
.125	.0096	.0128	.012
.25	.0014	.0015	.0015

Table 2.7.2

Values of $P(z)$ for $z = 2.5$ and $z = 5.0$ When $\omega = 10$

θ	$z = 2.5$			$z = 5.0$		
	Wald	Kemp	True	Wald	Kemp	True
−1.00	1.00	1.00	1.00	1.00	1.00	1.00
−.50	.99	.9998	.9997	.993	.997	.996
.00	.75	.7161	.724	.500	.500	.500
.125	.494	.4058	.4276	.223	.190	.199
.250	.282	.1896	.2113	.076	.052	.058

2.7.2 Average Sample Number

If $n(z)$ is the expected sample number of a scheme which starts at z, then $n(z)$ satisfies the equation

$$n(z) = 1 + \int_0^{\omega} n(x)g(x - z)\, dx. \tag{2.7.9}$$

If $n(x) = 1$ for $x \leq 0$ and $\int_{-\infty}^{-z} g(x)\, dx$ and $\int_{\omega-z}^{\infty} g(x)\, dx$ are small enough to be neglected, we can write Eq. (2.7.9) in the form

$$\begin{aligned} n(z) &= 1 + \int_{-\infty}^{0} n(x)g(x - z)\, dx + \int_0^{\omega} n(x)g(x - z)\, dx + \int_{\omega}^{\infty} n(x)g(x - z)\, dx \\ &= 1 + \int_{-\infty}^{\infty} n(x)g(x - z)\, dx. \end{aligned} \tag{2.7.10}$$

A solution of (2.7.10) is given by

$$n(z) = [(C + De^{hz}) - z]/E(Z) \tag{2.7.11}$$

where h is as defined in Section 2.7.1, C and D are constants, and $E(Z)$ is the expected value of Z.

APPLICATION TO NORMAL POPULATION The above assumptions are again reasonably satisfied when Z_i is normally distributed with mean θ and unit standard deviation and when $\omega \geq 3$. So, in the normal case

$$C = \{\theta[n(\omega) - n(0)e^{-2\omega\theta}] + \omega\}/(1 - e^{-2\omega\theta}),$$
$$D = \{\theta[n(0) - n(\omega)] - \omega\}/(1 - e^{-2\omega\theta}),$$

and

$$n(z) = n(0) - z/\theta + \{[n(\omega) - n(0)]\theta - \omega\}(1 - e^{-2\theta z})/\theta(1 - e^{-2\omega\theta}). \tag{2.7.12}$$

The simultaneous equations in $n(0)$ and $n(\omega)$ obtained after substituting for $n(z)$ from Eq. (2.7.12) into Eq. (2.7.10), integrating, and putting $z = 0$ and $z = \omega$, are given by

$$K_1 n(0) + K_2 n(\omega) = 1 - \Phi(\omega - \theta) + \Phi(-\theta) - K_2\omega/\theta$$
$$+ \{\exp[-(\omega - \theta)^2/2] - \exp(-\theta^2/2)\}/\theta(2\pi)^{1/2}, \tag{2.7.13}$$

$$e^{-2\omega\theta}K_2 n(0) + K_1 n(\omega) = 1 + \frac{h}{\theta} - [\Phi(\omega + \theta) - \Phi(\theta)]\left(1 + \frac{\omega}{\theta}\right)$$
$$- \frac{K_1\omega}{\theta} + \{\exp(-\theta^2/2) - \exp[-(\omega + \theta)^2/2]\}/\theta(2\pi)^{1/2}. \tag{2.7.14}$$

The values of K_1 and K_2 are the same as those given for (2.7.5) and (2.7.6). It is necessary to calculate $n(0)$ and $n(\omega)$ only for positive or negative θ since $n(0|\theta) = n(\omega|-\theta)$. For $\theta = 0$, the limiting forms of the above expressions are

$$n(z) = n(0) + (\omega - z)/z,$$
$$n(0) = 1 + \omega(2\pi)^{-1/2}\{\tfrac{3}{2} - \Phi(\omega)\}^{-1}.$$

Table 2.7.3

Values of $n(0)$ When $\omega = 10$

θ	Page	Kemp	True
-1.00	1.08	1.28	1.25
$-.50$	1.40	2.02	1.89
$+0.00$	4.49	8.98	7.89
.50	7.26	11.62	10.78
1.00	6.74	9.06	8.60

Table 2.7.3 compares the values of $n(0)$ obtained by Page (1954) and Kemp (1958) with the true values. Table 2.7.4 compares the true values of $n(z)$, Kemp's values, and Wald's (1947) values. Kemp (1958) has drawn curves for $n(0)$ and $n(z)$ for various values of θ and ω. Notice that the true values in Tables 2.7.1–2.7.4 are obtained by solving the exact equations for $P(z)$ and $n(z)$.

Table 2.7.4

Values of $n(z)$ for $z = 2.50$ and $z = 5.00$ When $\omega = 10$

θ	$z = 2.50$			$z = 5.00$		
	Wald	Kemp	True	Wald	Kemp	True
-1.00	2.5	3.8	3.4	5.0	6.3	5.9
$-.50$	5.0	7.0	6.4	9.9	12.0	11.4
.00	18.8	27.7	25.2	25.0	34.0	31.4
.50	12.3	16.2	15.4	9.9	12.0	11.4
1.00	9.3	8.8	8.4	5.0	6.3	5.9

2.7.3 Formulas to Improve Wald's Approximation to OC and Average Sample Number Functions

Tallis and Vagholkar (1965) provide an alternative proof of Wald's fundamental identity (Theorem 2.5.1). They also obtain improvements to the OC function and the average sample number comparable to those of Page (1954) and Kemp (1958). Let $X_1, X_2, \ldots$ be a sequence of iid random variables having F and $f(x; \theta)$ for the common distribution function and the density function respectively. Also, let

$$S_n = \sum_{i=1}^{n} Z_i, \qquad Z_i = \ln\{f(X_i; \theta_1)/f(X_i; \theta_0)\}, \tag{2.7.15}$$

where $H_0: \theta = \theta_0$ and $H_1: \theta = \theta_1$. Also, let $a = \ln A$ and $b = \ln B$ so that the continue-sampling inequality is $b < S_n < a$. Also let

$$S = \sum_{i=1}^{\infty} S_n I_{(N=n)}, \tag{2.7.16}$$

where I_E denotes the indicator function of the set E. Then we have the following expression for the characteristic function of S.

Theorem 2.7.1 (Tallis and Vagholkar, 1965) With the above notation, let $S_n = V_n + W_n$ where V_n is a Bernoulli random variable such that

$$P(V_n = b) = L_n(\theta) \qquad \text{and} \qquad P(V_n = a) = 1 - L_n(\theta)$$

and $L_n(\theta) = P(\text{acc } H_0 \text{ for a fixed } n)$. Then

$$\Psi(t) = E(e^{itS}) = \{1 - F(0)\}^{-1}\{1 - \pi(\theta)\}e^{ait} \int_0^{\infty} f(z)\left(\frac{e^{itz} - 1}{itz}\right) dz$$
$$- \{F(0)\}^{-1}\pi(\theta)e^{bit} \int_{-\infty}^{0} f(z)\left(\frac{1 - e^{itz}}{itz}\right) dz \tag{2.7.17}$$

where $\pi(\theta) = \sum_{n=1}^{\infty} L_n(\theta)p_n$, $p_n = P(N = n)$, N denoting the stopping variable.

Proof First consider the boundary value a. Here we have $W_n = S_n - V_n \geq 0$. Since $S_{n-1} < a \leq S_n$, Z_n is a random variable with the restriction that it be greater than or equal to zero. For fixed n let the pdf of S_{n-1} be $g(s)$. Then the conditional distribution of Z_n for given $S_{n-1} = s$ is of the form

$$h(z \mid S_{n-1} = s) = \begin{cases} f(z)/\{1 - F(a - s)\} & \text{if } \quad z > a - s, \\ 0 & \text{otherwise.} \end{cases}$$

Hence, the joint pdf of S_{n-1} and Z_n is of the form

$$p(s, z) = \begin{cases} g(s)f(z)/\{1 - F(a - s)\} & \text{if } \quad z > a - s, \\ 0 & \text{otherwise.} \end{cases}$$

It now follows that the marginal density of Z_n is of the form $h(z) = k(a)f(z)$ where $k(a)$ is a function of a only. However,

$$\int_0^{\infty} h(z)\, dz = k(a) \int_0^{\infty} f(z)\, dz = 1$$

implies that $k(a) = \{1 - F(0)\}^{-1}$. Hence, $h(z) \doteq f(z)/\{1 - F(0)\}$†. Further, the conditional distribution of W_n for fixed Z_n can reasonably be assumed to be approximately uniform, so that we have, for the conditional pdf of W_n,

† The authors seem to ignore the fact that the range of variation of Z_n is dependent on S_{n-1}. Hence the expression for $h(z)$ is only approximately true under the assumption that $a - S_{n-1} \doteq 0$ or $S_{n-1} - b \doteq 0$.

$$q(w \mid Z_n = z) = \begin{cases} 1/z, & w \le z, \\ 0, & w > z. \end{cases}$$

Thus, the joint distribution of V_n, W_n, and Z_n is

$$\begin{aligned} p(V_n = a, w, z) &\doteq \{1 - F(0)\}^{-1}\{1 - L_n(\theta)\}f(z)/z, && \text{for } 0 \le w \le z, \\ p(V_n = b, w, z) &\doteq \{F(0)\}^{-1}L_n(\theta)f(z)/(-z), && \text{for } z \le w \le 0, \\ p(v_n, w, z) &= 0, && \text{for } v_n \neq a \text{ or } b. \end{aligned} \tag{2.7.18}$$

From (2.7.18) we obtain

$$\Psi_n(t) = E\{\exp[it(V_n + W_n)]\} \doteq \{1 - F(0)\}^{-1}\{1 - L_n(\theta)\}e^{ita} \iint_{0 < w \le z} e^{itw}\frac{f(z)}{z} \times dw\, dz + \{F(0)\}^{-1}L_n(\theta)e^{itb} \iint_{z \le w < 0} e^{itw}\frac{f(z)}{-z}\, dw\, dz$$

which reduces to

$$\Psi_n(t) \doteq \{1 - F(0)\}^{-1}\{1 - L_n(\theta)\}e^{ait} \int_0^\infty f(z)\left(\frac{e^{itz} - 1}{itz}\right) dz - \{F(0)\}^{-1}L_n(\theta)e^{bit} \int_{-\infty}^0 f(z)\left(\frac{1 - e^{itz}}{itz}\right) dz. \tag{2.7.19}$$

The characteristic function $\Psi(t)$ of $S = \bigcup_1^\infty S_n$ is obtained from (2.7.19) by replacing $L_n(\theta)$ by $\pi(\theta)$ where $\pi(\theta) = \sum_{n=1}^\infty L_n(\theta)P(N = n)$. Thus $\Psi(t)$ is given by (2.7.17).

Corollary 2.7.1.1 From Eq. (2.7.17) we obtain

$$ES \doteq \Psi'(0) = \{1 - \pi(\theta)\}a + \pi(\theta)b + \tfrac{1}{2}\{1 - F(0)\}^{-1}\{1 - \pi(\theta)\} \times \int_0^\infty zf(z)\, dz + \tfrac{1}{2}\{F(0)\}^{-1}\pi(\theta) \int_{-\infty}^0 zf(z)\, dz \tag{2.7.20}$$

and

$$E(S^2) \doteq \{1 - \pi(\theta)\}a^2 + \pi(\theta)b^2 + \{1 - F(0)\}^{-1}\{1 - \pi(\theta)\} \times \int_0^\infty (az + \tfrac{1}{3}z^2)f(z)\, dz + \{F(0)\}^{-1}\pi(\theta) \int_{-\infty}^0 (bz + \tfrac{1}{3}z^2)f(z)\, dz. \tag{2.7.21}$$

The authors point out that using an argument similar to the one employed in the alternative proof of Wald's identity, it can be shown that

$$\Psi\{h(\theta)\} = \sum_{n=1}^{\infty} \Psi_n\{h(\theta)\}P(N = n) = 1 \tag{2.7.22}$$

where $h(\theta)$ is the nonzero solution of the equation $E(e^{Zh(\theta)}) = 1$. Solving for $\pi(\theta)$ from (2.7.22) we obtain

$$\pi(\theta) \doteq \frac{F(0)\{1 - F(0) - C^*(\theta)\}}{D(\theta)\{1 - F(0)\} - F(0)C^*(\theta)} \tag{2.7.23}$$

where $$C^*(\theta) = e^{ah(\theta)} \int_0^{\infty} f(z) \left(\frac{e^{zh(\theta)} - 1}{h(\theta)z}\right) dz$$

and $$D(\theta) = e^{bh(\theta)} \int_{-\infty}^{0} f(z) \left(\frac{1 - e^{h(\theta)z}}{h(\theta)z}\right) dz.$$

Formula (2.7.23) can be used to obtain better approximations for the boundary values of a and b so that the two types of error probabilities for a SPRT are α and β. The equations for a and b will be

$$\pi(\theta_0) = 1 - \alpha \quad \text{and} \quad \pi(\theta_1) = \beta. \tag{2.7.24}$$

Application to the Normal Case. Let z be normal with mean θ and variance unity. Then

$$\pi(\theta) \doteq \frac{\Phi(-\theta)\{1 - \Phi(-\theta) - C^*(\theta)\}}{D(\theta)\{1 - \Phi(-\theta)\} - \Phi(-\theta)C^*(\theta)} \tag{2.7.25}$$

and

$$\begin{aligned} E(S) \doteq\ & \{1 - \pi(\theta)\}a + \pi(\theta)b \\ & + \tfrac{1}{2}\{1 - \Phi(-\theta)\}^{-1}\{1 - \pi(\theta)\}\{\phi(\theta) + \theta\Phi(\theta)\} \\ & + \tfrac{1}{2}\{\Phi(-\theta)\}^{-1}\pi(\theta)\{\theta\Phi(-\theta) - \phi(\theta)\} \end{aligned} \tag{2.7.26}$$

where Φ and ϕ denote the standard normal distribution and density function respectively, and

$$C^*(\theta) = -(e^{-2a\theta}/2\theta)I(\theta), \qquad D(\theta) = -(e^{-2b\theta}/2\theta)I(\theta),$$

and

$$I(\theta) = \int_0^{\infty} \left(\frac{e^{-2\theta z} - 1}{z}\right) \phi(z - \theta)\, dz.$$

Except for $I(\theta)$ all other quantities involved in formulas (2.7.25) and (2.7.26) can be determined from standard tables. The function $I(\theta)$ can be evaluated as follows.

$$
\begin{aligned}
I(\theta) &= \int_0^\infty \left(\frac{e^{-2\theta z}-1}{z}\right)\frac{1}{(2\pi)^{1/2}}\exp[-(z-\theta)^2/2]\,dz \\
&= -\frac{\exp(-\theta^2/2)}{(2\pi)^{1/2}}\int_0^\infty \exp[-(z^2/2)]\left(\frac{e^{\theta z}-e^{-\theta z}}{z}\right)dz \\
&= -\frac{\exp(-\theta^2/2)}{2(2\pi)^{1/2}}\int_{-\infty}^\infty \exp[-(z^2/2)]\left(\frac{e^{\theta z}-e^{-\theta z}}{z}\right)dz \\
&= -\frac{\exp(-\theta^2/2)}{2(2\pi)^{1/2}}\int_{-\infty}^\infty \exp[-(z^2/2)]\,dz\int_{-\theta}^{\theta} e^{tz}\,dt \\
&= -\frac{\exp(-\theta^2/2)}{2(2\pi)^{1/2}}\int_{-\theta}^{\theta} dt\int_{-\infty}^\infty \exp[-(z^2/2)+tz]\,dz \\
&= -\exp(-\theta^2/2)\int_0^\theta \exp(t^2/2)\,dt \\
&= -\exp(-\theta^2/2)\left(\theta+\frac{\theta^3}{6}+\frac{\theta^5}{40}+\cdots\right).
\end{aligned}
\tag{2.7.27}
$$

When $\theta = 0$, $E(Z) = 0$ and in that case (2.5.25) takes the following limiting form:

$$\pi(0) \doteq \{1 + (2\pi)^{1/2}a\}/\{2 + (2\pi)^{1/2}(a-b)\}$$

and

$$E(N) = E(S^2)/E(Z^2)$$

where

$$
\begin{aligned}
E(S^2) \doteq{}& \{1-\pi(0)\}a^2 + \pi(0)b^2 + 2\{\tfrac{1}{6} + a/(2\pi)^{1/2}\}\{1-\pi(0)\} \\
&+ 2\{\tfrac{1}{6} - b/(2\pi)^{1/2}\}\pi(0).
\end{aligned}
$$

Tallis and Vagholkar (1965) tabulate the values of OC and ASN functions for two different tests (for the normal case) with various values of θ and compare them with the approximations of Wald (1947), Page (1954), and Kemp (1958). In Tables 2.7.5 and 2.7.6 we give a few selected values from their tables. These tables show that the formulas of Tallis and Vagholkar (1965) yield results comparable to those of Page and Kemp with less computational effort. Hence, for all practical purposes the formulas (2.7.20) or (2.7.21) and (2.7.23) will give satisfactory approximations for the OC and ASN functions.

Table 2.7.5

Operating Characteristic Function $\pi(\theta)$

θ	Test 1: $a=5, b=-5$					Test 2: $a=7.5, b=-2.5$				
	Wald	Page	Kemp	Present	True	Wald	Page	Kemp	Present	True
−.50	.993	.996	.997	.996	.996	.999	1.000	1.000	1.000	1.000
−.125	.777	.796	.808	.794	.802	.922	.923	.925	.923	.924
.00	.500	.500	.500	.500	.500	.750	.729	.716	.731	.724
.125	.223	.203	.190	.206	.198	.494	.450	.406	.447	.428
.50	.007	.004	.003	.004	.004	.082	.048	*na*[a]	.055	.046

[a] *na* indicates not available.

Table 2.7.6

Average Sample Number $E(N)$

θ	Test 1: $a=5, b=-5$					Test 2: $a=7.5, b=-2.5$				
	Wald	Page	Kemp	Present	True	Wald	Page	Kemp	Present	True
−.50	9.9	11.3	12.0	10.9	11.4	5.0	6.4	7.0	6.0	4.4
−.125	22.2	25.8	*na*[a]	25.6	27.2	13.8	17.0	*na*[a]	16.8	18.1
.00	25.0	29.5	34.0	29.3	31.4	18.8	23.3	27.7	23.1	25.2
.125	22.2	25.8	*na*[a]	25.6	27.2	20.5	24.9	*na*[a]	24.7	26.7
.50	9.9	11.3	12.0	10.9	11.4	12.3	14.2	16.2	14.8	15.4

[a] *na* indicates not available.

2.8 Truncated SPRT

Although the SPRTs enjoy the property of eventually being terminated with probability one, it is desirable to set a definite upper limit, say n_0, for the number of observations. This may be achieved by truncating the SPRT. Thus, the truncated SPRT is formulated as follows: If the sampling procedure has progressed to the nth stage ($n \le n_0 - 1$),

$$\text{rej } H_0 \quad \text{if} \quad \sum_{i=1}^{n} Z_i \ge \ln A,$$

$$\text{acc } H_0 \quad \text{if} \quad \sum_{i=1}^{n} Z_i \le \ln B,$$

and take one more observation if $\ln B < \sum_{i=1}^{n} Z_i < \ln A$. If the above rule does not lead to a final decision at $n = n_0$, reject H_0 if $0 < \sum_{i=1}^{n_0} Z_i < \ln A$ and accept H_0 if $\ln B < \sum_{i=1}^{n_0} Z_i < 0$.

By truncating the sequential process at the n_0th stage, we have changed the probabilities of type I and type II errors. The following theorem of Wald (1947) would provide upper bounds for the modified probabilities of errors.

Theorem 2.8.1 Let α and β be the nominal probabilities of errors of first and second kind for the SPRT. Let $\alpha(n_0)$ and $\beta(n_0)$ respectively denote the modified α and β for the truncated SPRT at n_0. Then, for sufficiently large n_0,

$$\alpha(n_0) \leq \alpha + \Phi\left(\frac{\ln A - n_0 E_0(Z)}{n_0^{1/2}\sigma_0(Z)}\right) - \Phi\left(\frac{-n_0 E_0(Z)}{n_0^{1/2}\sigma_0(Z)}\right), \tag{2.8.1}$$

$$\beta(n_0) \leq \beta + \Phi\left(\frac{-n_0 E_1(Z)}{n_0^{1/2}\sigma_1(Z)}\right) - \Phi\left(\frac{\ln B - n_0 E_1(Z)}{n_0^{1/2}\sigma_1(Z)}\right) \tag{2.8.2}$$

where

$$Z = \ln\{f(X;\theta_1)/f(X;\theta_0)\},$$

$$E_i(Z) = E_{\theta i}(Z), \qquad \sigma_i^2(Z) = \operatorname{var}(Z \mid \theta_i), \qquad i = 0 \text{ or } 1.$$

Proof Let $\rho_0(n_0)$ denote the probability under H_0 of obtaining a sample such that the truncated process leads to rejection of H_0, while the nontruncated process leads to acceptance of H_0. Then, we have

$$\alpha(n_0) \leq \alpha + \rho_0(n_0). \tag{2.8.3}$$

Notice that inequality instead of equality holds in (2.8.3) because there may be sample points for which the truncated process leads to acceptance of H_0 while the untruncated process leads to rejection of H_0. Next, let us derive an upper bound for $\rho_0(n_0)$ which is the probability under H_0 that the sequence of observations $Z_1, Z_2, \ldots$ satisfies the following three conditions:

(i) $\ln B < \sum_{i=1}^{n} Z_i < \ln A$ for $n = 1, \ldots, n_0 - 1$;
(ii) $0 < \sum_{i=1}^{n_0} Z_i < \ln A$;
(iii) when the sequential process is continued beyond n_0, it terminates with the acceptance of H_0. Hence

$$\rho_0(n_0) \leq \bar{\rho}_0(n_0) = P_{H_0}\left(0 < \sum_{i=1}^{n_0} Z_i < \ln A\right)$$

$$\doteq \Phi\left(\frac{\ln A - n_0 E_0(Z)}{n_0^{1/2}\sigma_0(Z)}\right) - \Phi\left(\frac{-n_0 E_0(Z)}{n_0^{1/2}\sigma_0(Z)}\right). \tag{2.8.4}$$

Then (2.8.3) and (2.8.4) would yield (2.8.1). Analogously one can obtain that $\beta(n_0) \le \beta + \rho_1(n_0)$ where $\rho_1(n_0)$ denotes the probability under H_1 that the truncated process leads to the acceptance of H_0, while the untruncated process leads to the rejection of H_0. Hence

$$\rho_1(n_0) \le \bar{\rho}_1(n_0) = P_{H_1}\left(\ln B < \sum_{i=1}^{n_0} Z_i < 0\right)$$

$$\doteq \Phi\left(\frac{-n_0 E_1(Z)}{n_0^{1/2}\sigma_1(Z)}\right) - \Phi\left(\frac{\ln B - n_0 E_1(Z)}{n_0^{1/2}\sigma_1(Z)}\right). \tag{2.8.5}$$

Consequently, (2.8.5) would lead to (2.8.2). This completes the proof of Theorem 2.8.1.

2.9 Optimal Properties of the SPRT

The sequential probability ratio test (SPRT) for testing a simple hypothesis H_0 against a simple alternative H_1 was first proved to be optimal by Wald and Wolfowitz (1948) in a certain sense. A much simpler proof has been given by LeCam [see Lehmann (1959, pp. 107–110)]. Matthes (1963) has given a proof of this optimality which relies primarily on a simple mapping theorem.

Theorem 2.9.1 (Wald and Wolfowitz, 1948) Among all tests (fixed sample or sequential) for which $P(\text{rej } H_0 | \theta_0) \le \alpha$, $P(\text{acc } H_0 | \theta_1) \le \beta$ and for which $E(N | \theta_i) < \infty$, $i = 0, 1$ the SPRT with error probabilities α and β minimizes both $E(N | \theta_0)$ and $E(N | \theta_1)$.

Proof The main part of the proof of Theorem 2.9.1 consists of finding the solution of the following auxiliary problem: Let w_i denote the losses resulting from wrong decision under H_i ($i = 0, 1$), and let c denote the cost of each observation. Then the risk (expected loss) of a sequential procedure is

$$\alpha w_0 + cE(N | \theta_0) \quad [\beta w_1 + cE(N | \theta_1)]$$

when H_0 $[H_1]$ holds; α and β are the error probabilities. If the state of nature θ is a random variable such that $P(\theta = \theta_0) = \pi$ and $P(\theta = \theta_1) = 1 - \pi$, then the total average risk of a procedure δ is

$$r(\pi, w_0, w_1, \delta) = \pi[\alpha w_0 + cE(N | \theta_0)] + (1 - \pi)[\beta w_1 + cE(N | \theta_1)].$$

Then the proof consists of determining the Bayes procedure for this problem, that is, the procedure that minimizes $r(\pi, w_0, w_1, \delta)$, and of showing that Wald's SPRT is Bayes in the following sense: Given any SPRT, given $c > 0$ (fixed), and any π with $0 < \pi < 1$, there exist positive constants w_0 and w_1 such that the SPRT is Bayes relative to π, c, w_0, w_1. It is important to note that π can be arbitrarily chosen.

Now, starting with the Bayes character of Wald's SPRT, one can show its optimum property as follows: Let δ^* be any other competitive procedure having error probabilities $\alpha^* < \alpha$, $\beta^* \le \beta$, and expectations of sample size $E^*(N|\theta_i) < \infty$ $(i = 0, 1)$. Since Wald's SPRT minimizes the Bayes risk, it satisfies

$$\pi[\alpha w_0 + cE(N|\theta_0)] + (1-\pi)[\beta w_1 + cE(N|\theta_1)] \\ \le \pi[\alpha^* w_0 + cE^*(N|\theta_0)] + (1-\pi)[\beta^* w_1 + cE^*(N|\theta_1)],$$

and hence

$$\pi E(N|\theta_0) + (1-\pi)E(N|\theta_1) \le \pi E^*(N|\theta_0) + (1-\pi)E^*(N|\theta_1).$$

Since this inequality is valid for all $0 < \pi < 1$, it implies

$$E(N|\theta_0) \le E^*(N|\theta_0) \quad \text{and} \quad E(N|\theta_1) \le E^*(N|\theta_1),$$

which establishes the optimum property of Wald's SPRT. For the details of the proof, the reader is referred to Lehmann (1959, pp. 104–110) or B. K. Ghosh (1970, pp. 93–98).

Thus, the optimal property (OP) can be expressed in words as follows: Among all tests whose error probabilities do not exceed those of the SPRT, the latter has the smallest expected sample size under both hypotheses. However, the validity of the OP has been demonstrated only under two conditions. The first is that $B \le 1 \le A$ and the second is that the expected sample sizes are finite. Burkholder and Wijsman (1963) have shown that the second condition is not necessary but the first one is. The superfluousness of the second condition is shown for a general cost function in the following form: If a test has OP among all tests with finite expected sample sizes, then it has OP among all tests. If $B \le 1 \le A$ does not hold, then the SPRT is not admissible in some sense. However, if one considers only tests that take at least one observation, then there are no restrictions on A and B (other than $B \le A$) for a SPRT to have OP.

In Section 2.1, we briefly discussed the existence and uniqueness of a SPRT. We need the following terminology. Let α and β denote the error probabilities corresponding to the two densities f_0, f_1 under consideration: Let v_i $(i = 0, 1)$ denote the expected sample sizes. In passing from one test to another, let $\Delta\alpha$, $\Delta\beta$, and Δv_i denote the changes in α, β, and v_i $(i = 0, 1)$. A test will be called *inadmissible* if there exists another test such that $\Delta\alpha \le 0$, $\Delta\beta \le 0$, $\Delta v_i \le 0$, $i = 0, 1$ with strict inequality in at least one of the four. We further need the following concept of "a monotonicity property" (property M) of a SPRT.

Definition 2.9.1 (Wijsman, 1960) A SPRT is said to have the monotonicity property (property M) if when the upper stopping bound of the SPRT is increased and the lower stopping bound decreased, then at least one of the error probabilities decreases, unless the new test is equivalent to the old one, in which case the error probabilities are unchanged. (Two tests are said to be equivalent if their sample sequences differ on a set of probability zero under both the hypothesized distributions.)

Clearly, the optimum property (OP) implies admissibility, and the restricted optimum property (namely, that every SPRT has the optimum property among all SPRTs) implies restricted admissibility, that is, admissibility within the class of SPRTs.

Let $X_1, X_2, \ldots$ be a sequence of independent and identically distributed random variables (or vectors) having common density f_i with respect to some sigma-finite measure where $i = 0, 1$ corresponding to the two hypotheses under consideration. The trivial case $f_0 = f_1$ almost everywhere will be excluded. Let Λ_n denote the probability ratio at the nth observation, that is, $\Lambda_n = \prod_{j=1}^{n} \{f_1(X_j)/f_0(X_j)\}$. If some stopping rule is defined, let N denote the random number of observations. Of fundamental importance in what follows is the basic double inequality

$$dP_\theta(d < \Lambda_N < c) \le P_1(d < \Lambda_N < c) \le cP_0(d < \Lambda_N < c) \qquad (2.9.1)$$

for any real numbers d and c including ∞. The strict inequality signs within the parentheses in (2.9.1) may be replaced by less-than-or-equal signs, and we will do so whenever this is convenient. The following inequalities are special cases of (2.9.1):

$$P_1(\Lambda_N \ge d) \ge dP_0(\Lambda_N \ge d), \qquad P_1(\Lambda_N \le c) \le cP_0(\Lambda_N \le c), \qquad (2.9.2)$$

which have been used by Wald (1947) (see Section 2.1). An important consequence of (2.9.1) is that either

$$P_0(d < \Lambda_N < c) = P_1(d < \Lambda_N < c) = 0 \qquad (2.9.3)$$

or

$$P_0(d < \Lambda_N < c) > 0 \quad \text{and} \quad P_1(d < \Lambda_N < c) > 0. \qquad (2.9.4)$$

As an application, compare the SPRTs (B, A) and (B, A^*) with $B < A < A^*$. In (2.9.3) and (2.9.4) identify d with A, c with A^*, and N with the random number of observations if test (B, A) is used. If (2.9.3) holds, the two tests are clearly equivalent. If (2.9.4) prevails, we can conclude that $\Delta\alpha < 0$, $\Delta\beta > 0$, $\Delta v_i > 0$ for $i = 0, 1$. Now, suppose that both stopping bounds are changed such that $B^* \leq B < A \leq A^*$. Again, unless the two tests are equivalent, we have $\Delta v_i > 0$ for both i. Then we are led to the following result.

Theorem 2.9.2 (Wijsman, 1960) The restricted optimum property (ROP) implies and is implied by the monotonicity property (M).

Proof Assume ROP, which implies restricted admissibility, which in turn implies that at least one of $\Delta\alpha$ and $\Delta\beta$ is negative. In other words, one of the error probabilities has to decrease, which is property M. Conversely assume property M, and compare tests (B, A) and (B^*, A^*) that are not equivalent and for which $\Delta\alpha \leq 0$ and $\Delta\beta \leq 0$. Then, we cannot have $B^* \leq B$, $A^* \leq A$, for in that case $\Delta\alpha > 0$ and $\Delta\beta < 0$. Similarly, the cases $B^* \geq B$, $A^* \geq A$, and $B < B^* < A^* < A$ are excluded since otherwise, by the monotonicity property, one of $\Delta\alpha$ and $\Delta\beta$ would be positive. Hence, the only remaining possibility is $B^* < B < A < A^*$, which implies $\Delta v_i > 0$ for $i = 0, 1$, that is, the restricted optimum property.

Next, let us consider uniqueness of a SPRT. Toward this we have the following theorem which combines the results of Weiss (1956), Anderson and Friedman (1960), and Wijsman (1960).

Theorem 2.9.3 There is at most one sequential probability ratio test for testing $H_0: f(x) = f_0(x)$ against $H_1: f(x) = f_1(x)$, that achieves a given α and β provided

(i) $f_1(x)/f_0(x)$ has a continuous distribution with positive probability on every interval in $(0, \infty)$; or

(ii) the SPRT has stopping bounds that satisfy $B < 1 < A$; or

(iii) the SPRT has the monotonicity property M.

Proof For (i), see Weiss (1956). For the proof of (ii) see Anderson and Friedman (1960). It suffices to prove (iii) since $B < 1 < A$ implies that the SPRT has the optimum property which implies the monotonicity property (see Theorem 2.9.2). Now, property M implies uniqueness because if the stopping bounds are changed in the same direction, then both error probabilities change (in opposite directions), whereas, if the stopping bounds are changed in opposite directions, then according to property M, at least one of the error probabilities changes; unless, of course, the two tests are equivalent.

Since the monotonicity property is a consequence of the optimum property, a separate proof of the property may not be necessary. However, the proof of the optimum property requires a sizeable machinery, whereas, property M has been proved by Wijsman (1960) in a somewhat elementary way using the basic inequalities (2.9.1). Further, since the uniqueness and the restricted optimum property are immediate consequences of property M, it is worthwhile to present the proof of the latter. Moreover, the methods used by Wijsman (1960) yield a stronger result, which does not follow from the optimum property, and which has, besides property M, some other interesting consequences. For the proof of property M, we need the following further notation.

In a SPRT with stopping bounds A and B $(A < B)$ and random number of observations N, the error probabilities α and β are functions of A and B,

$$\alpha(B, A) = P_0(\Lambda_N \geq A) = 1 - P_0(\Lambda_N \leq B), \tag{2.9.5}$$

$$\beta(B, A) = P_1(\Lambda_N \leq B) = 1 - P_0(\Lambda_N \geq A). \tag{2.9.6}$$

It is convenient to introduce the functions U_i and V_i defined by

$$U_i(B, A) = P_i(\Lambda_N \leq B), \qquad V_i(B, A) = P_i(\Lambda_N \geq A), \tag{2.9.7}$$

if B, A are the stopping bounds. We have

$$U_i(B, A) + V_i(B, A) = 1, \qquad i = 0, 1, \tag{2.9.8}$$

and the relation between the α, β, U_i, and V_i is simply

$$\alpha(B, A) = V_0(B, A), \qquad \beta(B, A) = U_1(B, A).$$

Theorem 2.9.4 [Property M: Wijsman (1960)] Let (u, v) and (u', v') define two nonequivalent SPRTs, with $0 < u \leq u' < v' \leq v < \infty$, and let $\Delta\alpha = \alpha(u, v) - \alpha(u', v')$, $\Delta\beta = \beta(u, v) - \beta(u', v')$. Then at least one of $\Delta\alpha$, $\Delta\beta$ must be negative.

Proof Let N denote the random number of observations in the test (u', v') and define $G_i(y) = P_i(\Lambda_N \leq y)$. Then using (2.9.7) we have

$$\Delta\alpha = V_0(u, v) - V_0(u', v'), \qquad \Delta\beta = U_1(u, v) - U_1(u', v'). \tag{2.9.9}$$

Then we compute

$$V_0(u', v') = \int_v^\infty dG_0(y) + \int_{v'}^{v-} dG_0(y) \tag{2.9.10}$$

and

$$V_1(u, v) = \int_v^{\infty} dG_0(y) + \int_{v'}^{v-} V_0(u/y, v/y)\, dG_0(y) + \int_u^{u'} V_0(u/y, v/y)\, dG_0(y). \tag{2.9.11}$$

Notice that in (2.9.11) we have used the fact that Λ_{n+1}/Λ_n are independent and identically distributed. Substitution in (2.9.9) and using (2.9.8) gives

$$\Delta\alpha = \int_u^{u'} V_0(u/y, v/y)\, dG_0(y) - \int_{v'}^{v-} U_0(u/y, v/y)\, dG_0(y). \tag{2.9.12}$$

Analogously one can obtain

$$\Delta\beta = \int_{v'}^{v} U_1(u/y, v/y)\, dG_1(y) - \int_u^{u'} V_1(u/y, v/y)\, dG_1(y). \tag{2.9.13}$$

If $v' < v$, then for y in the interval $[v', v)$, we have, using (2.9.7) and (2.9.2),

$$U_1(u/y, v/y) \leq (u/y) U_0(u/y, v/y) \leq (u/v') U_0(u/y, v/y), \tag{2.9.14}$$

and using (2.9.1)

$$dG_1(y) \leq v\, dG_0(y) \tag{2.9.15}$$

so that

$$\int_{v'}^{v-} U_1(u/y, v/y)\, dG_1(y) \leq (uv/v') \int_{v'}^{v-} U_0(u/y, v/y)\, dG_0(y). \tag{2.9.16}$$

In (2.9.15) and consequently in (2.9.16) there is strict inequality unless

$$\int_{v'}^{v-} dG_i(y) = 0 \qquad \text{for} \quad i = 0, 1. \tag{2.9.17}$$

If $v' = v$, (2.9.16) remains valid (trivially). Similarly,

$$\int_u^{u'} V_1(u/y, v/y)\, dG_1(y) \geq (uv/u') \int_u^{u'} V_0(u/y, v/y)\, dG_0(y), \tag{2.9.18}$$

and again this inequality is strict unless

$$\int_u^{u'} dG_i(y) = 0, \qquad i = 0, 1. \tag{2.9.19}$$

The tests are equivalent if and only if both (2.9.17) and (2.9.19) hold. Hence, if the tests are not equivalent, then at least one of the inequalities in (2.9.16) and (2.9.18) is strict. From (2.9.12), (2.9.13), (2.9.16), and (2.9.18), it follows that

$$uv\,\Delta\alpha + v'\,\Delta\beta < 0, \tag{2.9.20}$$

$$uv\,\Delta\alpha + u'\,\Delta\beta < 0. \tag{2.9.21}$$

The conclusion of Theorem 2.9.4 is an immediate consequence of either of the inequalities (2.9.20) and (2.9.21).

Wijsman (1960) has also strengthened Theorem 2.9.4 as follows.

Theorem 2.9.5 (Wijsman, 1960) Under the hypothesis of Theorem 2.9.4, we have

$$u\,\Delta\alpha + \Delta\beta < 0, \tag{2.9.22}$$

$$v\,\Delta\alpha + \Delta\beta < 0. \tag{2.9.23}$$

Proof See the proof of Theorem 2 of Wijsman (1960, pp. 683–684).

From Theorem 2.9.5, Wijsman (1960) draws the following consequences:

(i) Consider u', v' fixed and u, v varying subject to $u < u' < v' < v$. Then, (α, β) lies in a cone with vertex (α', β'), containing the point $(0, 0)$ and bounded by two lines with slopes $-u'$, and $-v'$. This cone does not contain any point of the triangle with vertices (α', β'), $(0, 1)$, and $(1, 0)$.

(ii) Let

$$\gamma_0 = \alpha/(1-\beta), \qquad \gamma_1 = \beta/(1-\alpha), \tag{2.9.24}$$

and the quantities γ_0' and γ_1' are defined in a similar manner. Then, since $\Delta\alpha < 0$ implies $\Delta\gamma_0 < 0$ and $\Delta\beta < 0$ implies $\Delta\gamma_1 < 0$, we obtain that under the hypothesis of Theorem 2.9.4, at least one of the $\Delta\gamma_i$ must be negative.

(iii) Now, $\Delta\alpha < 0$ is in particular satisfied when $u = u'$, and $\Delta\beta < 0$ when $v = v'$. Using this and (ii) we infer that if the lower stopping bound u of a SPRT is fixed, γ_0 is a monotonic nonincreasing function of the upper stopping bound v. The function is strictly monotonic except at any point v for which there is a $v^* > v$ such that the tests (u, v) and (u, v^*) are equivalent. A completely analogous statement for γ_1 is obtained by fixing v and decreasing u.

Also, notice that γ_0 is the tangent of the angle the line through (α, β) and $(0, 1)$ makes with the β axis; γ_1 has a similar interpretation.

REMARK 2.9.1 It has been assumed throughout that the continuation region of the SPRT is an open interval. Theorems 2.9.4 and 2.9.5 remain valid if the continuation intervals contain one or both of their end points. If I' and I are the continuation intervals of two SPRTs, all we need is that $I' \subset I$. Then, the inequalities (2.9.22) and (2.9.23) remain valid, except that one of the inequalities may be an equality.

Existence and uniqueness questions concerning SPRTs for testing a simple hypothesis against a simple alternative in the case of independent and identically distributed random variables are also considered by Wijsman (1963) under general conditions. Using the characterization of SPRTs as Bayes tests, it is shown [see Wijsman (1963, Theorem 2)] that there is a SPRT with given error point α if and only if $\alpha + \beta \leq 1$ in the case of tests that need not take an observation ($n \geq 0$) and α belonging to a certain set in the case $n \geq 1$. A SPRT is, in general, not unique. Wijsman (1963) claims (see his Theorem 5) only uniqueness of the stopping bounds up to an equivalence, in the case of a SPRT with given error point.

Next, we shall turn to the monotonicity of the SPRT when the underlying distributions have a monotone likelihood ratio (MLR).† Toward this we need the following lemmas.

Lemma 2.9.1 (Karlin, 1957) Let $f(x;\theta)$ be a family of densities on the real line with monotone likelihood ratio in x.

(i) If $\psi(x)$ is a nondecreasing function of x, then $E_\theta \psi(X)$ is a nondecreasing function of θ;

(ii) for any $\theta < \theta'$, the distribution functions of X under θ and θ' are such that $F(x;\theta') \leq F(x;\theta)$, for all x.

Proof Let $\theta < \theta'$ and let E and $\bar{E}$ be the sets for which $f(x;\theta') < f(x;\theta)$ and $f(x;\theta') > f(x;\theta)$ respectively. If $a = \sup_E \psi(x)$ and $b = \inf_{\bar{E}} \psi(x)$, then $b - a \geq 0$ since $E = \{x : x < k\}$ for some k.

$$\begin{aligned}\int \psi[f(x;\theta') - f(x;\theta)]\,dx &\geq a \int_E [f(x;\theta') - f(x;\theta)]\,dx \\ &\quad + b \int_{\bar{E}} [f(x;\theta') - f(x;\theta)]\,dx \\ &= (b-a) \int_{\bar{E}} [f(x;\theta') - f(x;\theta)]\,dx \geq 0.\end{aligned}$$

Now (ii) follows by setting $\psi(x) = 1$ for $x > x_0$ and $\psi(x) = 0$ otherwise.

† A density or probability function $f(z;\theta)$ where θ is real is said to have MLR in $t(z)$ if the distributions indexed by distinct θs are distinct and if $f(z;\theta)/f(z;\theta')$ is a nondecreasing function of $t(z)$ for $\theta > \theta'$.

Lemma 2.9.2 (Lehmann, 1959) Let F_0 and F_1 be two distribution functions on the real line. Then $F_1(x) \le F_0(x)$ for all x if and only if there exist two nondecreasing functions g_0 and g_1 and a random variable V such that (i) $g_0(v) \le g_1(v)$ for all v, and (ii) the distributions of $g_0(V)$ and $g_1(V)$ are F_0 and F_1 respectively.

Proof Suppose that g_0, g_1, and V exist. Then

$$F_1(x) = P[g_1(V) \le x] \le P[g_0(V) \le x] = F_0(x),$$

for all x. Conversely, suppose that $F_1(x) \le F_0(x)$, then define

$$g_i(y) = \inf\{x : F_i(x - 0) \le y \le F_i(x)\}, \qquad i = 0, 1.$$

These functions are nondecreasing and for $g_i = g$, $F_i = F$ satisfy $g[F(x)] \le x$ and $F[g(y)] \ge y$, for all x and y. Notice that g acts as an inverse function. It follows that $y \le F(x_0)$ implies $g(y) \le g[F(x_0)] \le x_0$ and also $g(y) \le x_0$ implies $F[g(y)] \le F(x_0)$ and hence $y \le F(x_0)$ so that the statements $g(y) \le x_0$ and $y \le F(x_0)$ are equivalent. Let V be uniformly distributed on (0, 1). Then

$$P[g_i(V) \le x] = P[V \le F_i(x)] = F_i(x), \qquad i = 0, 1.$$

Hence $F_1(x) \le F_0(x)$ for all x implies that $g_0(y) \le g_1(y)$ for all y. Then, we can state the following result of Lehmann (1959, p. 101) pertaining to the monotonicity of the power function (or the OC function).

Theorem 2.9.6 (Lehmann, 1959) Let $X_1, X_2, \ldots$ be independent random variables having the probability density $f(x; \theta)$ which has monotone likelihood ratio in $T(x)$. Then any SPRT for testing H_0: $\theta = \theta_0$ against $H_1 : \theta = \theta_1$ $(\theta_0 < \theta_1)$ has a nondecreasing power function.

Proof Let $Z_i = \ln[f(X_i;\ \theta_1)/f(X_i;\ \theta_0)] = h(T_i)$, $T_i = T(X_i)$, where h is nondecreasing, and let $\theta < \theta'$. By Lemma 2.9.1 the distribution function of T, $F_T(t;\ \theta)$, satisfies $F_T(t;\ \theta') \le F_T(t;\ \theta)$ for all t and by Lemma 2.9.2 there exist a random variable V and functions g and g' such that $g(v) \le g'(v)$ for all v and that the distributions of $g(V)$ and $g'(V)$ are $F_T(\cdot\,;\ \theta)$ and $F_T(\cdot\,;\ \theta')$ respectively. The SPRT is such that sampling is continued as long as the sample points fall inside the bands formed by the parallel straight lines;

$$\sum_{i=1}^{n} h(t_i) = \ln B \quad \text{or} \quad \ln A.$$

The hypothesis is rejected if the path formed by the points $(1, h(t_1)), (2, h(t_1) + h(t_2)), \ldots,$ leaves the band through the upper boundary. The probability of this event is therefore the probability of rejection, for θ $[\theta']$ when each T_i is replaced by $g(V_i)$ $[g'(V_i)]$. Since $g(V_i) \leq g'(V_i)$ for all i, the path generated by the $g'(V_i)$ leaves the upper boundary whenever the path generated by the $g(V_i)$ leaves the upper boundary. Hence $p(\theta) \leq p(\theta')$, where $p(\theta)$ denotes the power of the SPRT at θ. This completes the proof of the assertion.

REMARK 2.9.6.1 Notice that if $f(x; \theta)$ has MLR in $T(x)$, then the SPRT is unbiased [that is, $\mathrm{OC}(\theta_0) \geq \mathrm{OC}(\theta_1)$].

J. K. Ghosh (1960b) has generalized Theorem 2.9.6 in two directions: (i) when the boundaries change from stage to stage and (ii) the variables are not independent. This result is given by B. K. Ghosh (1970, Lemma 3.3, pp. 100–102).

2.10 Generalized SPRT (GSPRT)

Weiss (1953) has introduced a generalized SPRT which differs from Wald's SPRT only in that the same limits A, B are not necessarily used at each stage of the sampling, but at the ith stage A_i, B_i are used, where these numbers are predetermined constants. It is shown that for any given test T, there is a generalized SPRT G whose probabilities of incorrectly accepting either hypothesis are no larger than the corresponding probabilities for T, and such that the distribution function of the number of observations required to come to a decision when using G is never below the corresponding distribution function when using T, under either hypothesis. Thus, one may say that G is uniformly better than T.

Let $f_i(x)$ denote the probability (density) function of X under H_i $(i = 0, 1)$. Assume that $f_0(x)$ and $f_1(x)$ are bounded everywhere and have at most a finite number of discontinuities. Assume that $X_1, X_2, \ldots$ represent an iid sequence. Assume that for any n and any finite nonzero c,

$$\int \cdots \int \prod_{j=1}^{n} f_i(x_j)\, dx_j \to 0 \qquad \text{as} \quad \Delta c \to 0 \quad \text{for} \quad i = 0, 1$$

the region of integration being

$$c \leq \prod_{j=1}^{n} \{f_1(x_j)/f_0(x_j)\} \leq c + \Delta c.$$

Only nonrandomized tests are considered and such that in the space of the first n chance variables, the regions where H_0 is accepted and H_1 is accepted are Borel sets, for any n. Then, the generalized SPRT is defined as follows:

Let $(\bar{A}_1, \bar{A}_2, \ldots)$ and $(\bar{B}_1, \bar{B}_2, \ldots)$ be two sequences of predetermined nonnegative constants such that $\bar{A}_i \geq \bar{B}_i$ for all i. The value ∞ is not excluded. The sampling is continued as long as

$$\bar{B}_n < \prod_{j=1}^{n} [f_1(x_j)/f_0(x_j)] < \bar{A}_n. \tag{2.10.1}$$

The first time (2.10.1) is violated, we accept H_1 if the upper bound was violated, and accept H_0 if the lower bound was violated. If $\bar{A}_m = \bar{B}_m$, while for all $i < m$ we have $\bar{A}_i > \bar{B}_i$, the test is truncated at the mth step.

Let us use the following further notation:

$T: D_i(n)$ is the probability that the sample size required to come to a decision is less than or equal to n, when the test T is used and H_i is true.

$T: E_i(n)$ is the region in the space of $(X_1, X_2, \ldots, X_n)$ where we accept H_i when the test T is used. To be in this region, we must have taken an nth observation. $T: E_i$ is the region in the infinite dimensional space where H_i is accepted when using T.

Then, we have the following results of Weiss (1953) which we shall state without proofs.

Theorem 2.10.1 (Weiss, 1953) If T is any test of H_0 against H_1 such that

$$\lim_{n\to\infty} T: D_i(n) = 1 \qquad \text{for} \quad i = 0, 1, \tag{2.10.2}$$

then there is a sequence $(G_1, G_2, \ldots)$ of generalized sequential probability ratio tests such that

$$G_j: D_i(n) \geq T: D_i(n) \qquad \text{for all } n, \text{ all } j \text{ and } i = 0, 1,$$

and (2.10.3)

$$\lim_{j\to\infty} G_j: P_i(E_i) \geq T: P_i(E_i) \qquad \text{for} \quad i = 0, 1.$$

Theorem 2.10.2 (Weiss, 1953) If T is any test of H_0 against H_1 satisfying the above assumptions, then there is a generalized SPRT G such that

$$G: D_i(n) \geq T: D_i(n) \qquad \text{for all } n, \text{ and } i = 0, 1,$$

and (2.10.4)

$$G: P_i(E_i) \geq T: P_i(E_i) \qquad \text{for} \quad i = 0, 1.$$

Kiefer and Weiss (1957) have considered certain properties of generalized SPRTs (GSPRTs). They show that under certain regularity conditions, the distributions of the sample size under the two hypotheses uniquely determine a GSPRT. Many GSPRTs are shown to be inadmissible (admissibility being defined in terms of the probabilities of the two types of error and the distribution of the sample size required to come to a decision). They also provide some fine characterizations of the generalized SPRTs.

2.11 Restricted SPRT

Although the SPRT has the optimum property, in general, the expected sample size is relatively large for values of the parameter between the two specified ones; that is, in cases in which one does not care greatly which decision is taken, a large number of observations is expected. This raises the question whether there are other sequential procedures that will reduce the expected number of observations for parameter values in the middle of the range without increasing it much at the hypothesized values of the parameter. Another difficulty with the SPRT is that for most cases the number of observations (which is random) is unbounded and has a positive probability of being greater than any given constant. Since it is economically unfeasible to take an arbitrarily large number of observations, often the SPRT is truncated. This modified procedure may considerably increase the expected sample size at the hypothesized values of the parameter. Weiss (1953) has proposed certain generalized SPRTs. Armitage (1957) has proposed certain restricted SPRTs, leading to closed boundaries in order to test for the mean of a normal population. He converted the boundaries to the Wiener process but only approximated the probabilities and expected time. Donnelly (1957) has proposed straight-line boundaries that meet and converted to the Wiener process and obtained certain results. Anderson (1960) has also considered the modified SPRT in order to test for the mean of the normal population with known variance and derived approximations to the operating characteristic (or power function) and the average sample number. In the following we shall present Anderson's procedure which is similar to Armitage's and Donnelly's procedures. Anderson's (1960) method consists of replacing the straight-line boundaries by boundaries that are linear in the sample size. The method is easily applicable to the exponential family of distributions because the SPRT can be based on sums of iid random variables each of which has a distribution belonging to the exponential family.

Before we describe the procedure, assume that observations are drawn sequentially from a normal distribution with mean θ and variance σ^2. We wish to test $H_0: \theta = \theta_0$ against the alternative $H_1: \theta = \theta_1$ $(\theta_1 > \theta_0)$ and wish to have a procedure that minimizes $E_\theta(N)$ at $\theta = (\theta_0 + \theta_1)/2$ or alternatively to minimize the supremum of $E_\theta(N)$. It is convenient to replace the observation X by the transformed observation $[X - (\theta_0 + \theta_1)/2]/\sigma$ and define $\theta^* = (\theta_1 - \theta_0)/2\sigma$. Then the hypotheses become H_0: $\theta = -\theta^*$ and H_1: $\theta = \theta^*$ $(\theta^* > 0)$ when sampling from the normal population having mean θ and variance 1.

Procedure Take observations (transformed) $X_1, X_2, \ldots$ sequentially. At the nth stage, reject H_0 if $\sum_1^n X_i \geq c_1 + d_1 n$ and accept H_0 if $\sum_1^n X_i \leq c_2 + d_2 n$ and take one more observation if none of the above happens. If n_0 observations are drawn, stop sampling and reject H_0 if $\sum_1^{n_0} X_i \geq k$ and accept H_0 if $\sum_1^{n_0} X_i < k$. We take $c_1 > 0 > c_2$. To avoid intersection of the lines before the truncation point, we require

$$c_2 + d_2(n_0 - 1) < c_1 + d_1(n_0 - 1).$$

Also, since we wish the lines to converge we require $d_1 < 0 < d_2$. Because of the symmetry of the hypotheses about $\theta = 0$, we consider the case when the error probabilities are equal. Since the problem is symmetric, it is reasonable to consider only symmetric procedures. That is $c_1 = -c_2 = c$ and $d_1 = -d_2 = d$ and $k = 0$.

The calculation of the probabilities and expected values that are of interest is extremely complicated and involved. However, one can calculate such quantities if $\sum_1^n X_i$ $(n = 1, 2, \ldots)$ is replaced by an analogous $X(t)$ $(0 \leq t < \infty)$, the Wiener process with $EX(t) = \theta t$ and variance of $X(t) = t$. Anderson (1960) has derived the probability of rejecting H_0 as a function of θ, and the expected length of time is also found. That is, he proposed obtaining a specified significance level at $-\theta^*$ and a specified power at θ^* with some minimization of the expected time. Then the OC and expected time are approximations to the OC and expected sample number when observations are taken discretely. One might hope that the approximations are as good as for the SPRT, which is the special case of $d_1 = d_2 = 0$ and $T = \infty$ where T was the truncation point of the Wiener process considered here.

Anderson (1960) obtained the probabilities and expected times as infinite series of terms involving Mill's ratios. Subject to the condition that the error probabilities are the same, the constants c and d were varied so as to obtain the smallest expected observation time at $\theta = 0$. The line $x = c + dt$ has intercept c at $t = 0$ and $c + dT$ at $t = T$. When $c + dT = 0$, the two lines converge to a point. For each of several values of the ratio of these two intercepts, $[(c + dT)/c = 0, .1, .2]$ Table 2.11.1 gives the combination of c and T and hence d that approximately minimizes the expected observation time at $\theta = 0$. In Table 2.11.1, the values in parentheses are for error probabilities equal to .01 each. The numerical computations suggest that the convergent line procedures show a considerable improvement over the SPRT

Table 2.11.1

The Pair (c, T) That Approximately Minimizes the Expected Observation Time at $\theta = 0$ Together with the Expected Times for $\alpha = \beta = .05$ (.01), $\theta = -.1$ and .1

Condition	c	T	Expected time at $\theta = 0$	Expected time at $\theta = -.1$ and .1
Fixed size		270.6(541.2)	270.6(541.2)	270.6(541.2)
SPRT	14.7(23.0)	$\infty(\infty)$	216.7(528.0)	132.5(225.0)
$c + dT = 0$	19.9(35.5)	600.3(870.3)	192.2(402.2)	139.2(294.4)
$c + dT = .1c$	20.1(35.5)	529.0(783.2)	192.2(402.2)	139.3(249.4)
$c + dT = .2c$	20.3(35.5)	457.1(700.0)	192.2(402.8)	139.8(249.8)

at $\theta = 0$ with a moderate decrease in efficiency at $\theta = -.1$ and $\theta = +.1$. One can say that when one operates at the 5% levels he is better off with the modified SPRT if intermediate values of θ occur at least one-quarter of the time and when one operates at the 1% level if intermediate values occur at least one-sixth of the time. The difference in the expected times at $\theta = 0$ when the ratios of the intercepts are 0 and .1 is not significant, because in the latter case the probability of reaching a decision at $t = T$ is almost zero.

Approximate expressions derived by Armitage (1957) for the OC function and the expected sample number are given by B. K. Ghosh (1970, p. 237) and the reader is kindly referred to the same.

Bartlett (1946) has obtained the probability of absorption before time n (which is approximately equal to the probability of crossing the upper boundary with not more than n observations) and is given by

$$P_0(\theta, T) = 1 - \Phi\left\{\frac{c - (\theta - d)n}{\sigma n^{1/2}}\right\} + \exp\left\{\frac{2c(\theta - d)}{\sigma^2}\right\}\Phi\left\{\frac{-c - (\theta - d)n}{\sigma n^{1/2}}\right\}, \tag{2.11.1}$$

with $P_0(\theta_0, T) = \alpha$, where Φ denotes the standard normal distribution. Armitage (1957) suggested using (for the original observations)

$$c = \{\sigma^2/(\theta_1 - \theta_0)\} \ln \{(1 - \beta)/\alpha\}, \qquad d = (\theta_1 - \theta_0)/2. \tag{2.11.2}$$

Let $p(\theta, d, c)$ denote the probability of accepting H_0 (for the transformed observations). It helps to think of $p(\theta, d, c)$ for $\theta \geq 0$ as a probability of an incorrect decision. Analogously let $q(\theta, d, c)$ denote the probability that the continuous Brownian motion $X(t)$ on $[0, \infty)$ exits $I(t)$: $(-c + dt,\ c - dt)$ through the lower boundary. The quantity $q(\theta, d, c)$ is an approximation to $p(\theta, d, c)$; Anderson (1960, p. 170) remarks that it is actually an upper bound for $p(\theta, d, c)$ when $\theta \geq 0$. Fabian (1974) has derived a simple and explicit expression for $q(\theta, d, c)$ when θ and d are such that θ/d is an even integer. Interpolation might complete the results for other values of θ/d. There is also a way of choosing d. In general, when testing that the normal mean θ is $-\theta^*$ against $\theta = \theta^*$ $(\theta^* > 0)$ at a prescribed level, the asymptotically optimal value of d (making the expected sample size at $\theta = 0$ minimal) is $d = \theta^*/2$ [this is easy to see by using the strong law of large numbers, or using, for example, Perng's (1969) Theorem 2.2]. In the following, we shall present Fabian's (1974) result.

Theorem 2.11.1 (Fabian, 1974)

$$q(\theta, d, c) = e^{-2c(\theta-d)}q(2d-\theta, d, c) = e^{-2a(\theta-d)}\{1 - q(\theta - 2d, d, c)\} \text{ if } \quad \theta \geq d, \tag{2.11.3}$$

$$q(\theta, d, c) = \sum_{r=1}^{j} (-1)^{r+1}\Psi^{(2j-r)r/(2j-1)}(1 - \tfrac{1}{2}\delta_{rj}) \quad \text{if} \quad \theta = 2jd, \tag{2.11.4}$$

where δ_{ij} denotes Kronecker's delta and

$$\Psi = e^{-2c(\theta-d)}. \tag{2.11.5}$$

Proof See Theorem 2.3 of Fabian (1974).

Often with a preassigned α and given θ we wish to determine d and c so that

$$q(\theta, d, c) = \alpha. \tag{2.11.6}$$

Table 2.11.2

Values of Ψ for Which $q(\theta, d, c) = \alpha$

θ/d \ α	.1	.05	.01	.005	.001
2	.2	.1	.02	.010	.002
4	.13443	.06237	.01126	.00548	.00105
6	.12367	.05742	.01055	.00518	.00101
8	.11957	.05571	.01035	.00511	.00101
10	.11745	.05487	.01027	.00508	.00100
12	.11617	.05439	.01023	.00506	.00100
∞	.11111	.05263	.01010	.00502	.00100

If we also specify d, then the value of c satisfying (2.11.6) is uniquely determined and is given by

$$c = \tfrac{1}{2} \ln \Psi^{-1}/(\theta - d) \tag{2.11.7}$$

where Ψ is the solution of $q(\theta, d, c) = \alpha$ with $q(\theta, d, c)$ given by (2.11.4). Notice that Paulson's (1964a) bound: $q(\theta, d, c) \le e^{-2c(\theta-d)}$ yields c given by (2.11.7) with Ψ replaced by α. Fabian (1974) has computed the values of Ψ for given values of α and θ/d. These are given in Table 2.11.2.

Problem

Let θ denote the probability of obtaining a head in a single toss of a coin. Suppose we wish to test H_0: $\theta = \frac{1}{4}$ against $\theta = \frac{3}{4}$. Can you extend the above considerations to this binomial problem? [*Hint*: The binomial tends to the normal distribution when suitably standardized. See also Armitage (1957).]

2.12 Extended SPRTs

2.12.1 Grouped Data

If observations are made in batches of equal or unequal sizes m_i $(i = 1, 2, \ldots)$ one can formulate Wald's SPRT for the nth batch and one can show that if $m_i \equiv m$ $(i = 1, 2, \ldots)$ the extended SPRT terminates in a finite number of batches with probability one. One can find the expected number of batches required to reach a decision and the OC function. For further details, see B. K. Ghosh (1970, Section 5.2.2).

Phatarfod (1965) has considered Wald's SPRT for a sequence of random variables which form a Markov chain. Wald's SPRT is extended with approximate boundary points based on the error probabilities. He also derives expressions for the fundamental identity, expected sample number, and OC function. The termination of SPRT for the case when the underlying random variables are neither independent nor identically distributed has been considered in the literature by Ifram (1965) and Wijsman (1967a) who provide sufficient conditions for the finite termination of the SPRT with probability one. For the formulation of this problem and certain results, the reader is referred to B. K. Ghosh (1970, Section 3.4.2). In the following we shall present a result of B. K. Ghosh (1970).

Theorem 2.12.1 (B. K. Ghosh, 1970) Let the SPRT be based on a sequence of random variables $X_1, X_2, \ldots$ where the X_i are not necessarily independent or identically distributed. Suppose that for $n >$ some n_0, the continue sampling inequality can be represented as $\underline{T}_n < T_n < \overline{T}_n$ where T_n, $\overline{T}_n$, and, $\underline{T}_n$ $(\overline{T}_n > \underline{T}_n)$ may involve the unknown parameter θ. Then the SPRT terminates finitely with probability one (that is, $P(\underline{T}_n < T_n < \overline{T}_n) \to 0$ as $n \to \infty$) provided the following condition holds.

CONDITION 1 The sequence $\{T_n\}$ converges in probability to ψ which may involve θ and $\lim \underline{T}_n = \lim \overline{T}_n \neq \psi$.

Proof. It suffices to show that, under the hypothesis for every $\varepsilon > 0$ there exists a positive integer $m \geq n_0$ such that $P(\underline{T}_n < T_n < \overline{T}_n | \theta) < \varepsilon$ for all $n > m$. Let $\lim \underline{T}_n = \lim \overline{T}_n = c$ as $n \to \infty$, where c may be $\pm\infty$. By the first part of condition I we can choose $m' \geq n_0$ depending on c, ψ, and ε, such that

$$P(|T_n - \psi| < \tfrac{1}{2}|\psi - c| \,|\theta) > 1 - \varepsilon \qquad \text{whenever} \qquad n > m'.$$

By the second part of condition I there exists $m'' \geq n_0$ depending on (c, x) such that $|\underline{T}_n - c| < |\psi - c|/2$ and $|\overline{T}_n - c| < |\psi - c|/2$ whenever $n \geq m''$. Hence for all $n > m = \max(m', m'')$,

$$\begin{aligned} P(T_n < T_n < \overline{T}_n|\theta) &\leq P(|T_n - \psi| > \tfrac{1}{2}|\psi - c| \,|\theta) \\ &\leq 1 - P(|T_n - \psi| < \tfrac{1}{2}|\psi - c| \,|\theta) < \varepsilon. \end{aligned}$$

This completes the proof of the assertion.

REMARK 2.12.1.1 B. K. Ghosh (1970) has also shown the finite termination of the SPRT when the distribution of T_n converges to some distribution function $H(t)$ which may involve θ and $\lim \overline{T}_n = \lim \underline{T}_n = c$ where c is a point of continuity of $H(t)$. If $c = \pm\infty$, then $H(t)$ need not be continuous. However, the latter condition is more stringent and it may lead to stronger conclusions, like termination of the SPRT with probability one for all θ. For the proof of this and an interesting example involving normal populations, see B. K. Ghosh (1970, pp. 121–123).

REMARK 2.12.1.2 It should be noted that we do not know of any results pertaining to the optimality of the extended SPRTs.

Savage and Savage (1965) have considered the finite termination of the SPRTs based on dependent and nonidentically distributed random sequences of variables. In the following we shall present their results without proofs.

Assume that $Y_i = (X_1, \ldots, X_i)$ has the density f_i and g_i under the null and alternative hypotheses respectively, where the X_i are not necessarily independent and identically distributed. Further, the range of X_i may vary with i. Let $R_i(Y_i)$ denote the ith likelihood ratio $g_i(Y_i)/f_i(Y_i)$. Corresponding to the constants A and B $(0 < B \leq 1 \leq A)$, the generalized SPRT is based on $R_i(Y_i)$ and the continuation region is given by $B \leq R_i(Y_i) \leq A$.

Theorem 2.12.2 (Savage and Savage, 1965) If there exists a sequence $\{F_i\}$ of sets with F_i in the range of Y_i for which $P_f(F_i) \to 1$ and $P_g(F_i) \to 0$ (when the subscripts on f and g are omitted), then

$$\text{(a)} \quad P_f(N = \infty) = 0; \qquad \text{(b)} \quad P_g(N = \infty) = 0$$

where N denotes the stopping variable for the generalized SPRT.

Proof See Savage and Savage (1965, p. 286).

Theorem 2.12.3 (Savage and Savage, 1965) If there exists a sequence $\{F_i\}$ of sets with F_i in the range of Y_i for which $\sum_1^\infty P_f(\bar{F}_i)$ converge where $\bar{F}_i$ is the complement of F_i, then $E_f(N)$ and $E_g(N)$ are both finite.

Proof See Savage and Savage (1965, pp. 286–287).

REMARK 2.12.3.1 For many problems the F_i can be formed as follows. Associated with the random variable Y_i is a real-valued random variable Z_i satisfying

(i) $E_f(Z_i) = m < M = E_g(Z_i)$.
(ii) The fourth central moment of Z_i under f_i and under g_i is of order i^{-2}.

If F_i is the event $\{-\infty < Z_i < (m + M)/2\}$, the Chebyshev inequality yields

$$P_f(\bar{F}_i) = P_f((Z_i - m) > K) \leq E_f(Z_i - m)^4/K^4 \qquad \text{where} \quad K = (M - m)/2.$$

Hence

$$\sum_1^\infty P_f(\bar{F}_i) \leq \sum_1^\infty P_f(|Z_i - m| > (M - m)/2) = \sum_1^\infty O(i^{-2}).$$

An analogous argument applies to $\sum_1^\infty P_g(\bar{F}_i)$.

Theorem 2.12.4 (Savage and Savage, 1965) If there exists a sequence $\{\bar{F}_i\}$ of sets F_i in the range of Y_i, a number ρ in the interval (0, 1), and a positive number K such that (i) $P_f(\bar{F}_i) < K\rho^i$ and (ii) $P_g(F_i) < K\rho^i$, then (a) $E_f(t^N)$ is finite for t in the interval $[0, \rho^{-1}]$; (*b*) $E_g(t^N)$ is finite for t in the interval $[0, \rho^{-1}]$.

Proof See Savage and Savage (1965, p. 288).

REMARK 2.12.4.1 The statements and proofs of Theorems 2.12.2–2.12.4 did not require that Y_i actually be $(X_1, \ldots, X_i)$. Thus, the results are applicable to procedures based on the likelihood ratio of only part of the data. Situations where some of the data is missing (due to limited storage facilities or otherwise) do arise in many practical problems.

REMARK 2.12.4.2 Nonparametric considerations have motivated the above results. Savage and Savage (1965) indicate that their results are applicable to sequential procedures based on the sign statistic and the Mann–Whitney–Wilcoxon statistic.

When the underlying variables are not independent, the OC function of the generalized SPRT based on the ratios of the joint densities is monotone provided the joint density possesses a monotone likelihood ratio [see J. K. Ghosh (1960)]. However, there may be situations where Ghosh's condition is not met. Hoel (1970) has provided an alternative sufficient condition for the monotonicity of the OC function.

Let $Y_i = (X_1, \ldots, X_i)$ be a random vector with probability density function $f_i(y_i; \theta)$ which depends on the real parameter θ. Let

$$T_i = \ln\{f_i(Y_i; \theta)/f_i(Y_i; \theta_0)\}, \qquad i = 1, 2, \ldots \tag{2.12.1}$$

be the sequence of log-likelihood ratios which with the boundaries $\ln A$ and $\ln B$ define the SPRT of the hypothesis $H_0: \theta = \theta_0$ against $H_1: \theta = \theta_1$.

Let $F = \{F_n\}$ and $G = \{G_n\}$ denote two sequences of distribution functions. We are interested in comparing the probability of accepting H_0 when the distribution of $(T_1, \ldots, T_n)$ is F_n with the probability of accepting H_0 when the distribution of $(T_1, \ldots, T_n)$ is G_n. Toward this we have the following result of Hoel (1970).

Theorem 2.12.5 (Hoel, 1970) Let $F = \{F_n\}$ and $G = \{G_n\}$ be two sequences of absolutely continuous distribution functions, F_n and G_n being defined on the n-space, satisfying,

(i) $G_n(t_n | t_1, \ldots, t_{n-1}) \geq F_n(t_n | t_1, \ldots, t_{n-1})$; and
(ii) $G_n(t_n | t_1, \ldots, t_{n-1})$ is nonincreasing in $t_1, \ldots, t_{n-1}$ or $F_n(t_n | t_1, \ldots, t_{n-1})$ is nonincreasing in $t_1, \ldots, t_{n-1}$ for all n.

Then $P_G(\text{acc } H_0) > P_F(\text{acc } H_0)$.

Proof See Hoel (1970, pp. 311–312).

REMARK 2.12.5.1 In applying this theorem to establish that an SPRT has a monotone OC function, we see that the requirement of a stochastically increasing "conditional" likelihood ratio replaces the usual stochastically increasing condition used in the independence case (see Lemma 2.9.1). As another application of the theorem suppose we wish to know $P_F(\text{acc } H_0)$, which we cannot approximate directly. It may be possible, however, to find a distribution G for the likelihood ratios for which $P_G(\text{acc } H_\theta)$ is either known or can be approximated. Further, if the conditions of the theorem are satisfied, we then have $P_G(\text{acc } H_\theta)$ as an upper bound to $P_F(\text{acc } H_0)$. The author provides two nonparametric situations where the conditions of Theorem 2.12.5 are met and not J. K. Ghosh's (1960b) condition.

REMARK 2.12.5.2 It seems that the densities of Y_i need not be indexed by a real parameter. In fact, we may be interested in testing that the density is $f_i(y_i)$ against the alternative that the density is $g_i(y_i)$ at the ith stage. Theorem 2.15.1 seems to be valid in this case also.

Problem

Consider a sequence of random variables $\{X_i\}$ each of which can take values 0 or 1 only. The sequence is said to occur randomly if $\{X_i\}$ are independent and identically distributed Bernoulli variables. The sequence is said to occur according to a homogeneous Markov chain if $P(X_i = 1) = p$ and for $i > 1$

$$P(X_i = 1 \mid X_{i-1} = 0) = p_0, \qquad P(X_i = 1 \mid X_{i-1} = 1) = p_1,$$

$$P(X_i = 1 \mid X_i = x_1, \ldots, X_{i-1} = x_{i-1}) = P(X_i = 1 \mid X_{i-1} = x_{i-1}).$$

We wish to test H_0: the sequence is random with a specified $0 < p < 1$ against H_1: the sequence is Markovian with specified p, p_0, p_1 $(0 < p, p_0, p_1 < 1)$.

2.13 Asymptotic Properties of SPRTs

In this section we shall give the asymptotic behavior for the error rates and ASN of the Wald SPRT which was studied by Berk (1973). We assume that $X, X_1, \ldots$ are iid with common pdf f_i under hypotheses H_i, $i = 0, 1$. The Wald SPRT of H_0 versus H_1 gives the stopping time,

$$N = \inf\{n : S_n \notin (b, a)\}, \tag{2.13.1}$$

where $S_n = \sum_1^n Z_j$, $Z_j = \ln[f_1(X_j)/f_0(X_j)]$, and (b, a) are two positive numbers. [We assume that Z is finite with probability one (wp1).] The error rates are denoted by α and β. That is, $\alpha = P_0(S_N \geq a)$ and $\beta = P_1(S_N \leq b)$. We shall call (α, β) the strength of the test. Wald's (1947) inequalities (see Theorem 2.1.1) for (α, β) may be written as

$$\beta \leq (1 - \alpha)e^b, \qquad \alpha \leq (1 - \beta)e^{-a}. \tag{2.13.2}$$

Notice that these inequalities are quite general and do not depend on the iid structure. Further, it is proved in Theorem 2.2.1 that if $X, X_1, \ldots$ are iid and $P(Z = 0) < 1$, then $E\{e^{tN}\} < \infty$ for some $t > 0$. (Here P and E refer to the true distribution of X, which need not be either f_0 or f_1.) In particular $EN < \infty$. To avoid trivialities, we assume throughout this section that $P(Z = 0) < 1$. Suppose now $c = \min(-b, a) \to \infty$. We write $\lim_c$ for $\lim_{c\to\infty}$. Then $\lim_c \alpha = \lim_c \beta = 0$ and wp1 $\lim_c N = \infty$. (Consequently $\lim_c EN = \infty$.)

The following theorem states more precisely the asymptotic behavior of N and EN.

Theorem 2.13.1 (Berk, 1973) Suppose $X, X_1, \ldots$ are iid and $\mu = EZ$ exists. Then if $\mu > 0$, wpl

(i) $\lim_c I_{(S_n \ge a)} = \lim_c P(S_N \ge a) = 1$,
(ii) $\lim_c N/a = \lim_c EN/a = 1/\mu$.

If $\mu < 0$, wpl

(iii) $\lim_c I_{(S_n \le b)} = \lim_c P(S_N \le b) = 1$,
(iv) $\lim_c\{N/b\} = \lim_c\{EN/b\} = 1/\mu$.

REMARK 2.13.1.1 We can have $|\mu| = \infty$. The case $\mu = 0$ will be covered by a different theorem. Govindarajulu (1968) has conjectured Theorem 2.13.1. We present Berk's (1973) method of proof.

Proof We treat the case $\mu > 0$. Since $\lim_n S_n/n = \mu$ wpl, also $\lim_n S_n = \infty$ wpl. Thus, $S_* = \min_n S_n$ is finite wpl. We then have $I_{(S_N \le b)} \le I_{(S^* \le b)} \to 0$ wpl as $c \to \infty$. Thus, $\lim_c I_{(S_N \ge a)} = 1$ and by dominated convergence, $\lim_c P(S_N \ge a) = 1$. Since wpl $\lim_c N = \infty$, $\lim_c S_N/N = \mu$ wpl. By the definition of N,

$$S_{N-1} I_{(S_N \ge a)} < a I_{(S_N \ge a)} \le S_N I_{(S_N \ge a)}.$$

On dividing throughout by N and letting $c \to \infty$, the extreme terms both approach μ wpl; thus wpl $\lim_c a/N = \mu$ or $\lim_c N/a = 1/\mu$. By Fatou's lemma, $\liminf_c EN/a \ge 1/\mu$. Now, let $t = \inf\{n : S_n \ge a\}$. Clearly $N \le t$. It follows from the results of Siegmund (1967) (for $\mu < \infty$, plus an easy truncation argument if $\mu = \infty$) that under our assumptions $\lim_c EN/a = 1/\mu$. Hence also $\limsup_a EN/a \le 1/\mu$. The proof of the theorem for $\mu < 0$ is analogous.

Theorem 2.13.1 shows that Wald's approximation for the ASN is asymptotically correct. This approximation [see Wald (1947, p. 53)] applies when $0 < |\mu| < \infty$ and may be written as

$$EN \doteq [bP(S_N \le b) + aP(S_N \ge a)]/\mu. \tag{2.13.3}$$

According to Theorem 2.13.1, the ratio of the two sides of (2.13.3) approaches one as $c \to \infty$. From Wald's inequalities (2.13.2) one can get the yet cruder inequalities

$$\alpha \le e^{-a}, \qquad \beta \le e^{b}. \tag{2.13.4}$$

The next theorem shows that asymptotically, the inequalities in (2.13.4), in some sense, are equalities.

Theorem 2.13.2 (Berk, 1973) Let $X, X_1, \ldots$ be iid and $E_i|Z| < \infty$ for $i = 0, 1$. Then

$$\lim_c a^{-1} \ln \alpha^{-1} = 1 = \lim_c b^{-1} \ln \beta^{-1}.$$

Proof Let $\mu_i = E_i Z$. Necessarily $\mu_0 < 0 < \mu_1$. Wald (1947, p. 197) gives the following inequality:

$$E_1 S_N = \mu_1 E_1 N \geq (1 - \beta) \ln[(1 - \beta)/\alpha] + \beta \ln[\beta/(1 - \alpha)]. \quad (2.13.5)$$

By (2.13.4), $\alpha = o(1) = \beta$ as $c \to \infty$. Upon dividing throughout in (2.13.5) by a, we obtain

$$\mu_1 E_1 N/a \geq a^{-1} \ln \alpha^{-1}[1 + o(1)].$$

From Theorem 2.13.1, we have that $\lim_c \mu_1 E_1 N/a = 1$; hence $\limsup_c a^{-1} \times \ln \alpha^{-1} \leq 1$. By (2.13.4) $a^{-1} \ln \alpha^{-1} \geq 1$, and hence $\lim_c a^{-1} \ln \alpha^{-1} = 1$. The proof of the result for β is analogous.

REMARK 2.13.2.1 This result also shows that Wald's approximations for the error rates [obtained by treating the relations in (2.13.2) as equalities and solving for (α, β)] are asymptotically correct in the sense of the theorem.

An approximation for the power curve of the SPRT is obtained (see Section 2.3) under the additional assumption that for some (necessarily unique) real number $h \neq 0$, $E\{e^{hZ}\} = 1$. In the above notation the approximation to the power may be written as

$$P(S_N \geq a) = (1 - e^{hb})/(e^{ha} - e^{hb}).$$

Theorem 2.13.2, in conjunction with Wald's device of considering the SPRT as being generated by $S_n' = hS_n$ with stopping boundaries

$$(b', a') = (hb, ha) \quad \text{if } h > 0 \quad [\text{respectively } (b', a') = (ha, hb) \quad \text{if} \quad h < 0]$$

establishes the following corollary.

Corollary 2.13.2.1 Let $X_1, X_1, \ldots$ be iid, $E|Z| < \infty$ and for some $h \neq 0$, $E(e^{hZ}) = 1$. Then

$$\lim_c (-ha)^{-1} \ln P(S_N \geq a) = 1 \qquad \text{if} \quad h > 0,$$
$$\lim_c (hb)^{-1} \ln P(S_N \leq b) = 1 \qquad \text{if} \quad h < 0.$$

Proof Recall that S_n' is a log-likelihood ratio for $X_1, \ldots, X_n$ under two iid distributions, with the true distribution of X in the denominator. Thus, Theorem 2.13.2 applies directly.

Notice that given the existence of h, we have $EhZ < \ln E(e^{hZ}) = 0$, so that $\mu = EZ \neq 0$ and is opposite in sign to h. When $EZ = 0$, the asymptotic behavior of EN is substantially different from that given in Theorem 2.13.1.

Theorem 2.13.3 (Berk, 1973) Let $X, X_1, \ldots$ be iid with $\sigma^2 = EZ^2 < \infty$ and $EZ = 0$. Let $d = a - b$ and for $j > 0$, let

$$\phi_j(d) = \max[\sup\{E([Z^+ - u]^j \mid Z^+ \geq u) : 0 \leq u \leq d\},\\ \sup\{E([Z^- - u]^j \mid Z^- \geq u) : 0 \leq u \leq d\}].$$

If either $d = O(c)$ or $\phi_1(d) = o(c)$, then, letting $p = P(S_N \geq a)$ and $\pi = -b/d$,

$$\lim_c p/\pi = 1 = \lim_c (1 - p)/(1 - \pi). \tag{2.13.6}$$

If the above conditions are strengthened to $d = O(c)$ or $\phi_2(d) = o(c^2)$, then also

$$\lim_c \sigma^2 EN/(-ab) = 1 \tag{2.13.7}$$

where $Z^+ = Z$ ($Z^- = Z$) if $Z > 0$ ($Z < 0$) and zero otherwise.

Proof The proof is somewhat complicated and hence is not presented here. The reader is referred to Berk (1973).

Theorem 2.13.3 shows that when $EZ = 0$ and $EZ^2 < \infty$, Wald's approximations for the power and ASN (see Sections 2.3 and 2.4) are, under certain regularity conditions, asymptotically exact.

CHAPTER 3

Sequential Tests for Composite Hypotheses

3.1 Wald's (1947) Method of Weight Functions

In Chapter 2 we have considered SPRTs to test a simple hypothesis against a simple alternative. However, in practical situations, the simple null hypothesis is only a representative of a set of simple hypotheses and the same thing can be said about the simple alternative. Thus, we are faced with the problem of testing a composite hypothesis against a composite alternative. The compositeness of the hypotheses can arise from two situations: (i) the composite hypotheses are concerned with the parameters of interest and there are no nuisance parameters, and (ii) the hypotheses may be simple or composite in the presence of nuisance parameters.

Let $f(x; \theta)$ denote the probability function (or probability density function) of X indexed by the unknown parameter θ which could be vector valued.

We wish to test the composite hypothesis $H_0: \theta \in \omega_0$ against the composite alternative $H_1: \theta \in \omega_1$. Let S_1 denote the surface of the boundary of ω_1. Wald (1947, pp. 199–209) proposed the method of weight functions in order to construct an optimal SPRT. Assume that it is possible to construct two weight functions $g_0(\theta)$ and $g_1(\theta)$ of θ such that

$$\int_{\omega_0} g_0(\theta)\, d\theta = 1, \qquad \int_{S_1} g_1(\theta)\, dS_1 = 1. \tag{3.1.1}$$

Then the SPRT is based on the ratio

$$\frac{f_{1,n}}{f_{0,n}} = \int_{S_1} g_1(\theta) \prod_{i=1}^{n} f(x_i\,; \theta)\, dS_1 \Big/ \int_{\omega_0} g_0(\theta) \prod_{i=1}^{n} f(x_i\,; \theta)\, d\theta \tag{3.1.2}$$

satisfying the conditions:

(i) the probability of type I error, $\alpha(\theta)$ is constant on ω_0;
(ii) the probability of type II error, $\beta(\theta)$ is constant over S_1; and
(iii) for any point θ in the interior of ω_1 the value of $\beta(\theta)$ does not exceed its constant value of S_1.

Wald (1947, Section A.9) claims that the weight functions $g_i(\theta)$ $(i = 0, 1)$ are optimum in the sense that for any other weight functions $h_0(\theta)$ and $h_1(\theta)$, the associated error probabilities $\alpha^*(\theta)$, $\beta^*(\theta)$ satisfying (as good approximations)

$$\int_{\omega_0} \alpha^*(\theta) h_0(\theta)\, d\theta = (1 - B)/(A - B),$$

$$\int_{\omega_1} \beta^*(\theta) h_1(\theta)\, d\theta = B(A - 1)/(A - B) \tag{3.1.3}$$

are such that

$$\max_{\theta\in\omega_0} \alpha^*(\theta) \geq (1 - B)/(A - B) = \max_{\theta\in\omega_0} \alpha(\theta), \tag{3.1.4}$$

and

$$\max_{\theta\in\omega_1} \beta^*(\theta) \geq B(A - 1)/(A - B) = \max_{\theta\in\omega_1} \beta(\theta). \tag{3.1.5}$$

3.1.1 Applications of Weight Functions

(a) *Sequential binomial test* Let X take the value 1 or 0 with probability θ and $1 - \theta$ respectively. We wish to test $H_0: \theta = \frac{1}{2}$ against the two-sided alternative $H_1: |\theta - \frac{1}{2}| \geq \delta > 0$. So, let $g_0(\frac{1}{2}) = 1$ and $g_0(\theta) = 0$, for $\theta \neq \frac{1}{2}$, and $g_1(\theta_1) = g_1(1 - \theta_1) = \frac{1}{2}$, and $g_1(\theta) = 0$ otherwise, where $\theta_1 = \frac{1}{2} + \delta$. Then, if m denotes the number of positive observations in a total sample of size n, the continuation region of the SPRT is given by

$$B < 2^{n-1}\{\theta_1{}^m(1 - \theta_1)^{n-m} + \theta_1^{n-m}(1 - \theta_1)^m\} < A. \tag{3.1.6}$$

(*b*) *Sequential chi-square test* Let X_i be independent and normally distributed having unknown mean θ and variance σ^2. We wish to test $H_0: \sigma = \sigma_0$ against $H_1: \sigma = \sigma_1 > \sigma_0$. Here choose $g(\theta) = 1/2c$, $-c \le \theta \le c$, and zero otherwise. The limit of the ratio of the modified likelihoods under H_1 and H_0 tends to

$$\left(\frac{\sigma_0}{\sigma_1}\right)^{n-1} \exp\{(n-1)s^2(\sigma_0^{-2} - \sigma_1^{-2})/2\}. \tag{3.1.7}$$

It should be noted that the ratio in (3.1.7) indicates that the problem of testing H_0 against H_1 with θ as a nuisance parameter is equivalent to the problem of testing a simple hypothesis about σ against a simple alternative with known mean zero from a sample of size $n - 1$. This can be established using Helmert's transformation

$$\begin{aligned} Y_1 &= (X_1 + \cdots + X_n)/n^{1/2} \\ Y_j &= [X_1 + \cdots + X_{j-1} - (j-1)X_j]/[j(j-1)]^{1/2}, \qquad j = 2, \ldots, n. \end{aligned} \tag{3.1.8}$$

Hence, $s^2 = \sum_2^n Y_j{}^2/(n-1)$. Thus the properties of the SPRT based on the ratio (3.1.7) can be studied via the properties of Wald's SPRT as considered in Chapter 2.

3.2 Sequential t- and t^2- Tests and Sequential Analogues of Stein's Two-Stage Test

3.2.1 Sequential t- (t^2-) Test

Let X be normally distributed with mean θ and unknown variance σ^2. We wish to test $H_0 : \theta = \theta_0$ against $H_1 : |\theta - \theta_0| \ge \delta\sigma$ where $\delta > 0$. Then the boundary S_1 consists of all points (θ, σ) for which $|\theta - \theta_0| = \delta\sigma$. That is, it consists of two points. Define $g_0(\theta, \sigma) = 1/c$ if $0 \le \sigma \le c$, $\theta = \theta_0$, and zero elsewhere; and $g_1(\theta, \sigma) = 1/2c$ if $0 \le \sigma \le c$ and $\theta = \theta_0 \pm \delta\sigma$, and zero elsewhere. Then, Wald (1947, pp. 205–207) has shown that $f_{1,n}/f_{0,n}$ is a strictly increasing function of $|\bar{X} - \theta_0|/s_n$ where $\bar{X}$ and s_n denote the sample mean and sample standard deviation based on n observations. Wald (1947) has further shown that $\alpha(\theta, \sigma)$ is constant on ω_0 and $\beta(\theta, \sigma)$ is a function of $|\theta - \theta_0|/\sigma$ alone and Wald (1947) claims that $\beta(\theta, \sigma)$ is monotonically decreasing with $|\theta - \theta_0|/\sigma$. However, he has not proved this point satisfactorily.

Thus the sequential procedure can be based on t_n where $t_n = n^{1/2}(\bar{X} - \theta_0)/s_n$. That is, the sequential t- (or t^2-) test of H_0: $\theta = \theta_0$ against the alternative $H_1: \theta - \theta_0 \geq \delta\sigma$ $[|\theta - \theta_0| \geq \delta\sigma]$ can be described as follows: If the experiment has proceeded until the *nth* stage, the sampling continuation region is given by

$$B_n < t_n < A_n \quad (B_n < |t_n| < A_n), \qquad n = 2, 3, \ldots \tag{3.2.1}$$

where the constants B_n and A_n are chosen so that the procedure terminates finitely; otherwise they are quite arbitrary. However, when the constants A_n and B_n are obtained by equating the ratio of the noncentral and central distributions of t to given constants A and B, it has been shown by David and Kruskal (1956) that this sequential test terminates finitely with probability one.

Rushton (1950) has considered asymptotic approximations to the log of the likelihood ratio arising in the sequential t-test. Let

$$\begin{aligned} u_n &= \sum_1^n (X_i - \theta_0) \Big/ \left[\sum_1^n (X_i - \theta_0)^2\right]^{1/2} \\ &= n^{1/2} t (n - 1 + t^2)^{1/2}. \end{aligned} \tag{3.2.2}$$

If the sampling continuation region for the sequential t-test is (approximately) of the form

$$\ln[\beta/(1 - \alpha)] < S_n < \ln[(1 - \beta)/\alpha], \tag{3.2.3}$$

then Rushton (1950) has obtained, for large n,

$$S_n \approx \tfrac{1}{4}\delta^2 u_n^{\,2} + n^{1/2}\delta u_n[1 - 1/(4n) + \delta^2 u_n^{\,2}/(24n)]. \tag{3.2.4}$$

From numerical study, Rushton (1950) concludes that one can use the approximation (3.2.4) with confidence and that one should add the terms $1/4n$ and $\delta^2 u_n^{\,2}/24n$ only when one is about to reach a decision.

The sequential t has the following properties.

Property I The (one-sided) sequential t-test terminates finitely with probability one.

Let $\beta_T(\theta, \sigma)$ denote the probability of type II error when the true parameter values are θ and σ. If $\beta_T(\theta, \sigma)$ depends only on a single function $\psi(\theta, \sigma)$ of the parameters, we shall write it as $\beta_T(\psi)$. The same convention is adopted for $\alpha_T(\theta, \sigma)$.

Property II For any sequential t-test T of H_0 against a one-sided H_1, $\beta_T(\delta)$ is a decreasing function of $\delta = (\theta - \theta_0)/\sigma > 0$.

Proof Since the joint distribution of $t_1, t_2, \ldots$ depends only on δ, we may without loss of generality assume that the X_i have common mean δ and unit variance. Consider two values δ' and δ'' of δ where $\delta'' = \delta' + \eta$, $\eta > 0$. Let $Y_i = X_i + \eta$, $i = 1, 2, \ldots$, and let the statistics $t_2', t_3', \ldots$ be the same functions of Ys as $t_2, t_3, \ldots$ are of Xs. Then

$$t_n' = t_n + \eta n^{1/2}/s_n > t_n, \qquad \text{all} \quad n \geq 2. \tag{3.2.5}$$

Let W_n be the region in the n-dimensional space such that $(X_1, \ldots X_n)$ lies in W_n if H_0 is accepted in n or fewer stages when the test T is employed. Let W_n' be the analogous region when $t_2', t_3', \ldots, t_n'$, are employed instead of t_2, $t_3, \ldots, t_n$; however the same boundaries on ts are retained. We claim $W_n \supset W_n'$. To establish this, assume that the set $\{t_2', t_3', \ldots, t_m'\}$ leads to the acceptance of H_0 at the mth stage. Since none of $t_2', \ldots, t_m'$ could have crossed the upper boundaries, none of $t_2, \ldots, t_m$ could have crossed them either. Further, $t_m \leq t_m' \leq B_m$. Hence, the set $(t_1, \ldots, t_m)$ leads to acceptance of H_0 in m or fewer stages. So, $(X_1, \ldots, X_n) \in W_n'$ implies that $(X_1, \ldots, X_n) \in W_n$. Consequently,

$$\begin{aligned} P_{\delta'}[(X_1, \ldots, X_n) \in W_n] &> P_{\delta'}[(X_1, \ldots, X_n) \in W_n'] \\ &= P_{\delta'}[(Y_1, \ldots, Y_n) \in W_n] \\ &= P_{\delta''}[(X_1, \ldots, X_n) \in W_n]; \end{aligned} \tag{3.2.6}$$

the last part of the equality in (3.2.6) follows from the fact that the Ys have the same distribution when $\delta = \delta'$ as the Xs have when $\delta = \delta''$. This completes the proof of property II.

Let $C_{\alpha,\beta}$ denote the class of tests with average error of type I equal to α and average error of type II equal to β, the two averages being with respect to any two weight (prior density) functions defined over the regions $\theta = \theta_0$ and $\theta - \theta_0 \geq \delta\sigma$ respectively. Let $C'_{\alpha,\beta}$ be the class of weighted probability ratio tests belonging to $C_{\alpha,\beta}$. Further, let $C_{A,B}$ be the class of weighted probability ratio tests with boundaries A, B. Thus, we have the following theorem pertaining to the optimum property of the sequential t-test.

Theorem 3.2.1 (J. K. Ghosh, 1960a) Let T be any sequential t-test of $H_0: \theta = \theta_0$ against $H_1: (\theta - \theta_0) \geq \delta\sigma > 0$ with error probabilities $\alpha_T(\theta_0) = \alpha$ and β with respect to $\theta - \theta_0 = \delta\sigma$. Then, T has the double minimax property in the class $C_{\alpha,\beta}$, namely: for $T' \in C_{\alpha,\beta}$,

$$\sup_{\theta=\theta_0} \alpha_T(\theta, \sigma) \leq \sup_{\theta=\theta_0} \alpha_{T'}(\theta, \sigma) \tag{3.2.7}$$

$$\sup_{(\theta-\theta_0)/\sigma \geq \delta} \beta_T(\theta, \sigma) \leq \sup_{(\theta-\theta_0)/\sigma \geq \delta} \beta_{T'}(\theta, \sigma). \tag{3.2.8}$$

Proof $\alpha_T(\theta, \sigma) = \alpha$ when $\theta = \theta_0$ and since $T' \in C_{\alpha,\beta}$

$$\begin{aligned}\sup_{\theta=\theta_0} \alpha_{T'}(\theta, \sigma) &\geq \text{average } \alpha_{T'}(\theta, \sigma) \text{ [with respect to some weight function defined over the region } \theta = \theta_0] \\ &= \alpha.\end{aligned}$$

Next, from property II, when $(\theta - \theta_0)/\sigma > \delta$, $\beta_T(\theta, \sigma)$ exceeds $\beta_T(\theta, \sigma)$ evaluated at $(\theta - \theta_0)/\sigma = \delta$. Hence, $\sup_{(\theta-\theta_0)/\sigma \geq \delta} \beta_T(\theta, \sigma) = \beta$. Also

$$\begin{aligned}\sup_{(\theta-\theta_0)/\sigma \geq \delta} \beta_{T'}(\theta, \sigma) &\geq \text{average } \beta_{T'}(\theta, \sigma) \text{ [with respect to some weight function defined over } (\theta - \theta_0)/\sigma \leq \delta] \\ &= \beta.\end{aligned}$$

This completes the proof of Theorem 3.2.1.

Corollary 3.2.1.1 If Wald's approximations to the boundaries of the probability ratio in terms of the error probabilities are allowed, T has the double minimax property in the class $C_{A,B}$ where $A = (1 - \beta)/\alpha$ and $B = \beta/(1 - \alpha)$.

PROPERTIES OF SEQUENTIAL t^2-TEST (OR t-TEST FOR TWO-SIDED ALTERNATIVES)

Let us study the properties of the sequential t^2 procedure as given in (3.2.1).

Lemma 3.2.1 (J. K. Ghosh, 1960). The integrals

$$\int_0^c \sigma^{-m} \exp\left[-\sum_1^m (X_i - \theta_0)^2/2\sigma^2\right] d\sigma$$

and

$$\tfrac{1}{2}\int_0^c \sigma^{-m}\left\{\exp\left[-\sum_1^m (X_i - \theta_0 - \delta\sigma)^2/2\sigma^2\right] + \exp\left[-\sum_1^m (X_i - \theta_0 + \delta\sigma)^2/2\sigma^2\right]\right\} d\sigma$$

are uniformly convergent in Xs as $c \to \infty$ for $m \geq 2$.

Proof Consider

$$\int_{c'}^{c} \sigma^{-m} \exp\left[-\sum_{1}^{m} (X_i - \theta_0)^2 / 2\sigma^2 \right] d\sigma \leq \int_{c'}^{c} \sigma^{-m} \, d\sigma < \varepsilon$$

for sufficiently large c and c' provided $m \geq 2$. A similar argument holds for the second integral.

Property I The sequential t^2-test terminates finitely with probability one.

Proof This follows as a special case of Ray's (1957a) result after using Lemma 3.2.1 in order to fill a gap in Ray's argument.

Property II The sequential t^2-test for $H_0\colon \theta = \theta_0$ against $H_1\colon |\theta - \theta_0| = \delta\sigma$ has constant error probabilities α and β where

$$A \leq (1 - \beta)/\alpha \qquad \text{and} \qquad B \geq \beta/(1 - \alpha).$$

Furthermore, if Wald's approximations are used

$$\alpha \sim (1 - B)/(A - B) \qquad \text{and} \qquad \beta \sim B(A - 1)/(A - B).$$

Proof See J. K. Ghosh, (1960a, p. 83).

Theorem 3.2.2 (J. K. Ghosh, 1960a) For $H_0\colon \theta = \theta_0$ against $H_1\colon |\theta - \theta| = \delta\sigma$, the sequential t^2-test has the double minimax property [which is obtained by replacing T by T^2, T' by T'^2 and $(\theta - \theta_0) \geq \delta\sigma$ by $|\theta - \theta_0| = \delta\sigma$ in (3.2.7) and (3.2.8)] in $C_{\alpha,\beta}$, if A and B are so chosen that T^2 has error probabilities α and β. Moreover, if $A \sim (1 - \beta)/\alpha$ and $B \sim \beta/(1 - \alpha)$, then T^2 is double minimax in $C_{A,B}$.

Proof The proof is analogous to the proof of Theorem 3.2.1 and one should use property II.

Theorem 3.2.3 (J. K. Ghosh, 1960) For $H_0\colon \theta = \theta_0$ against $H_1\colon (\theta - \theta_0) \geq \delta\sigma$, the sequential t^2-test has the double minimax property provided $\beta_{T^2}(\delta)$ is a decreasing function of $\delta = |\theta - \theta_0|/\sigma$.

Proof The proof is analogous to that of Theorem 3.2.1.

REMARK 3.2.3.1 Wald (1947) claimed but did not complete the proof of the monotonicity of $\beta_{T^2}(\delta)$. Hoel (1954) criticized Wald's assertion (namely Theorems 3.2.2 and 3.2.3) by comparing the sequential t^2-test with Arnold's (1951) sequential t-test by employing the same constants A and B for both the tests. However, the use of the same A and B in both cases does not fix the error probabilities at α and β nor does it ever lead to the same average values for α and β with the appropriate weight functions. Thus, Hoel's (1954) criticism is not applicable to Theorem 3.2.2 and it is relevant to Theorem 3.2.3 for a wrong reason.

Recently, Sacks (1965) has proposed a sequential t-test for a one-sided alternative with possible extensions to the two-sided situations. His procedure is as follows. Let $H_0: \theta = 0$, $\sigma > 0$, $H_1: \theta = \sigma$, $\sigma > 0$.

Let

$$f_\sigma(X_1, \ldots, X_n) = \sigma^{-n} \exp\left[-\sum_1^n (X_i - \sigma)^2/2\sigma^2\right],$$

and

$$g_\sigma(X_1, \ldots, X_n) = \sigma^{-n} \exp\left[-\sum_1^n X_i^2/2\sigma^2\right].$$

Then take an $(n + 1)$th observation if

$$\int_0^\infty f_\sigma(X_1, \ldots, X_n)\,(d\sigma/\sigma) < A \int_0^\infty g_\sigma(X_1, \ldots, X_n)\,(d\sigma/\sigma) \tag{3.2.9}$$

and if

$$\int_0^\infty f_\sigma(X_1, \ldots, X_n)\,(d\sigma/\sigma) > A^{-1} \int_0^\infty g_\sigma(X_1, \ldots, X_n)\,(d\sigma/\sigma), \tag{3.2.10}$$

stop and accept H_0 if (3.2.10) is violated, and stop and accept H_1 if (3.2.9) is violated.

Notice that the procedure is symmetric with symmetric bounds A and A^{-1} (that is, $\alpha = \beta$). Sacks (1965) has also obtained the following result.

Theorem 3.2.4 (Sacks, 1965) For the above procedure, let N denote the stopping time. Then

$$E(N \mid \mu = \sigma = 1) \sim \ln A/\ln 2^{1/2} \quad \text{as} \quad A \to \infty. \tag{3.2.11}$$

Proof See Sacks (1965, pp. 1868–1869).

REMARK 3.2.4.1 Similar ideas would work if $H_1 = \theta = \delta\sigma$ or $|\theta| = \delta\sigma$. Since the distribution of N depends only on μ/σ, (3.2.11) is valid for any θ and σ with $\theta = \sigma$ (that is, for any point in H_1). Sacks (1965) points out that the moment-generating function of N is finite for t in some neighborhood of zero. The sequential t-test has bad asymptotic properties when $\mu = \sigma/2$. When $\mu/\sigma = 0$ or 1 the author claims that the error probability is $o(\ln A/A)$.

REMARK 3.2.4.2 Obviously Sacks (1965) has employed the following class of weight functions:

$$g_0(\theta, \sigma) = \begin{cases} 1/(2\sigma \ln c), & \theta = \delta_0 \sigma \quad \text{and} \quad c^{-1} < \sigma < c, \\ 0, & \text{elsewhere in } \omega_0, \end{cases} \tag{3.2.12}$$

and

$$g_1(\theta, \sigma) = \begin{cases} 1/(2\sigma \ln c), & \theta = \delta_1 \sigma \quad \text{and} \quad c^{-1} < \sigma < c, \\ 0, & \text{elsewhere in } \omega_1. \end{cases} \tag{3.2.13}$$

Problem

Let X be normally distributed with mean θ and unknown variance σ^2. We wish to test H_0: $\theta = 0$ against H_1: $\theta > \delta\sigma$ where $\delta > 0$. The following is a sequence of independent observations from the above population: $-.13$, .34, 41, $-.06$, .94, 1.44, $-.22$, $-.34$. Carry out the sequential t-test [see Eq. (3.2.4)] for $\delta = \frac{1}{4}$ and $\frac{1}{2}$, starting with $n = 2, 3$, etc., and using $\alpha = \beta = .1$.

3.2.2 Certain Sequential Analogues of Stein's Two-Stage Test

Hall (1962) has presented several analogues of Stein's two-stage test procedures for testing hypotheses about the mean of a normal population with unknown variance and with specified bounds on the error probabilities. These procedures are modifications of a procedure due to Baker (1950). When sequential experimentation is feasible, they provide alternatives to the sequential normal test (variance known) or the sequential t-test. The performance of the tests does not depend on the validity of any assumption regarding the variance. Moreover, unlike the t-test, these procedures do not require that the alternative hypothesis be specified in standard deviation units.

When sampling from a normal population with unknown mean μ and unknown variance σ^2, one may wish to test the composite hypotheses $H_0: \mu \le 0$, $\sigma > 0$ against $H_1: \mu \ge \delta\ (>0)$, $\sigma > 0$ with preassigned bounds α and β on the error probabilities. A common solution is to restate H_1 in (unknown) standard deviation units namely $\mu \ge \delta\sigma$, and use the t-test (sequential or nonsequential), or equivalently allow β to be a function of σ. There are two known solutions to the problem as stated. One is Stein's two-stage procedure described in Section 1.4. Notice that in Stein's two-stage procedure, the size of the second sample depends only on the sample variance of the first sample, so that its distribution depends only on σ^2, and not on μ. The second known solution is due to Baker (1950). Hall's (1962) methods presented here are modifications of those of Baker (1950). The first stage is identical to that of Stein which is used to estimate σ^2, but sampling is then continued, if at all, one observation at a time rather than in a nonsequential fashion. It should be emphasized that the distribution of the sample size in this case depends on μ and σ^2.

THE SEQUENTIAL TEST T

Let $X_1, X_2, \ldots$ be independent normal (μ, σ^2) random variables, $-\infty < \mu < \infty$, $0 < \sigma < \infty$. Let m be a specified integer exceeding 1. Consider a SPRT $\mu = 0$ against $\mu = \delta$ (σ unknown) based on $\overline{X}_m, X_{m+1}, \ldots$ with termination boundaries A_m and B_m and with σ replaced by s_m where $\overline{X}_m = \sum_1^m X_i/m$ $(v = m - 1)$, $s_m{}^2 = \sum_1^m (X_i - \overline{X}_m)^2/(m-1)$, and

$$\begin{aligned} a_m &= \ln A_m = \tfrac{1}{2}v(\alpha^{-2/v} - 1) = (-\ln \alpha)[1 + (-\ln \alpha)/v + o(1/v)], \\ b_m &= \ln B_m = -\tfrac{1}{2}v(\beta^{-2/v} - 1) = -(-\ln \beta)[1 + (-\ln \beta)/v + o(1/v)]. \end{aligned} \tag{3.2.14}$$

The above test will be referred to as test T. Notice that $A_m > A \equiv 1/\alpha$ and $B_m < B \equiv \beta$ with approximate equalities instead of inequalities if m is large, and A, B are the conservative termination boundaries of Wald (1947, p. 42) that are appropriate if σ were known. Let

$$r_n(s_m) = \delta \sum_{i=1}^{n} (X_i - \delta/2)/s_m{}^2 \qquad (n \geq m). \tag{3.2.15}$$

Then the procedure T is:

Observe $(X_1, \ldots, X_m)$ and then $X_{m+1}, X_{m+2}, \ldots$ successively and, for each $n \geq m$, after observing X_n,
stop sampling and make decision d_0 (acc H_0) if $r_n(s_m) \leq b_m$;
stop sampling and make decision d_1 (acc H_1) if $r_n(s_m) \geq a_m$;
continue sampling if $b_m < r_n(s_m) < a_m$.

THE STRENGTH OF TEST T

For given (s_m, σ), consider the conditional SPRT $T(s_m, \sigma)$ of $\mu = 0$ against $\mu = \delta$ based on $\overline{X}_m, X_{m+1}, \ldots$ with termination boundaries $\bar{A}_m$ and $\bar{B}_m$ where

$$\bar{a}_m = \ln \bar{A}_m = a_m s_m{}^2/\sigma^2, \qquad \bar{b}_m = \ln \bar{B}_m = b_m s_m{}^2/\sigma^2. \tag{3.2.16}$$

Computing the relevant probability ratios, noting that $\overline{X}_m, X_{m+1}, \ldots$ are statistically independent of s_m, one finds that decisions are made according as $r_n(\sigma) \leq \bar{b}_m$, $r_n(\sigma) \geq \bar{a}_m$, or $\bar{b}_m < r_n(\sigma) < \bar{a}_m$. Using Wald's conservative bounds on the error probabilities of a SPRT, we have

$$\begin{aligned} P[d_1 \text{ using } T(s_m, \sigma) \,|\, s_m, \sigma, \mu = 0] &< 1/\bar{A}_m = \exp(-a_m s_m{}^2/\sigma^2), \\ P[d_0 \text{ using } T(s_m, \sigma) \,|\, s_m, \sigma, \mu = \delta] &< \bar{B}_m = \exp(b_m s_m{}^2/\sigma^2). \end{aligned} \tag{3.2.17}$$

However, $r_n(\sigma) = r_n(s_n)s_m^2/\sigma^2$ so that $T(s_m, \sigma)$ has precisely the same decision rule at each stage as does the test T, with s_m computed from $X_1, \ldots, X_m$. Hence

$$E[P(d_i \text{ using } T(s_m, \sigma) \,|\, s_m, \sigma, \mu)] = E[P(d_i \text{ using } T \,|\, s_m, \sigma, \mu)]$$
$$= P(d_i \text{ using } T \,|\, \sigma, \mu),$$

and therefore, using (3.2.17) we have

$$P(d_1 \text{ using } T \,|\, \sigma, 0) < E[\exp(-a_m s_m^2/\sigma^2)] = (1 + 2a_m/v)^{-v/2} \quad (3.2.18)$$

for all σ since $vs_m^2/\sigma^2 = \chi_v^2$ and $E[\exp(t\chi_v^2)] = (1 - 2t)^{-v/2}$. Similarly,

$$P(d_0 \text{ using } T \,|\, \sigma, \delta) < E[\exp(b_m s_m^2/\sigma^2)] = (1 - 2b_m/v)^{-v/2}, \quad (3.2.19)$$

for all σ. Since the conditional test $T(s_m, \sigma)$ is a normal-mean SPRT, its OC function is monotone in μ (see property II, p. 105) that is, $P[d_i$ using $T(s_m, \sigma) \,|\, s_m, \sigma, \mu]$ is monotone in μ for every fixed s_m and σ. This, together with (3.2.18), (3.2.19), and (3.2.14) implies

$$\begin{aligned} &P(d_1 \text{ using } T \,|\, \mu, \sigma) < \alpha \qquad \text{for all} \quad \mu \le 0, \sigma > 0, \\ &P(d_0 \text{ using } T \,|\, \mu, \sigma) < \beta \qquad \text{for all} \quad \mu \ge \delta, \sigma > 0 \end{aligned} \quad (3.2.20)$$

which implies that T has strength (α, β). Note also that since the SPRT $T(s_m, \sigma)$ terminates finitely with certainty for every fixed value of s_m, the test T also terminates with certainty.

The OC Function of T

In order to obtain the OC function of the test T, Hall (1962) uses (coincidentally) the same method as Baker (1950). For the conditional test $T(s_m, \sigma)$, Wald's approximation to the OC function may be used, namely

$$P(d_0 \,|\, s_m, \sigma, \mu) \doteq (\bar{A}_m^h - 1)/(\bar{A}_m^h - \bar{B}_m^h) \qquad (h \ne 0) \qquad (3.2.21)$$

where $h(\mu) = 1 - (2\mu/\delta)$ and the excess over the boundary (which should be small if the test is likely to have a sample number substantially larger than m) is neglected. Taking expectations with respect to (3.2.21), using (3.2.16), and dropping the subscripts on a_m, b_m, and s_m, we have, for $h > 0$,

$$P\{d_0 | \sigma, \mu\} \doteq E\left\{[1 - \exp(-ahs^2/\sigma^2)]/\{1 - \exp[-(a-b)hs^2/\sigma^2]\}\right\}$$

$$= E\left\{[1 - \exp(-ahs^2/\sigma^2)] \sum_{i=0}^{\infty} \exp[-i(a-b)hs^2/\sigma^2]\right\}$$

which does not depend on σ. Taking expectations term by term, the right-hand side becomes

$$1 - [1 + 2ah/v]^{-v/2} + [1 + 2(a-b)h/v]^{-v/2}$$
$$- [1 + 2(2a-b)h/v]^{-v/2} + [1 + 4(a-b)h/v]^{-v/2}$$

or using (3.2.14) as

$$1 - (1 - h + h\alpha^{-2/v})^{-v/2} + (1 - 2h + h\alpha^{-2/v} + h\beta^{-2/v})^{-v/2}$$
$$- (1 - 3h + 2h\alpha^{-2/v} + h\beta^{-2/v})^{-v/2}$$
$$+ (1 - 4h + 2h\alpha^{-2/v} + 2h\beta^{-2/v})^{-v/2} + \cdots. \tag{3.2.22}$$

The above expressions are valid for $h > 0$, that is for $\mu < \delta/2$. For $\mu > \delta/2$ ($h < 0$) we obtain analogously

$$P(d_0 | \sigma, \mu) \doteq 1 - E\left\{[1 - \exp(-bhs^2/\sigma^2)]/\{1 - \exp[-(b-a)hs^2/\sigma^2]\}\right\}$$
$$= [1 + 2bh/v]^{-v/2} - [1 + 2(b-a)h/v]^{-v/2}$$
$$+ [1 + 2(2b-a)h/v]^{-v/2} - [1 + 4(b-a)h/v]^{-v/2} + \cdots$$
$$= (1 + h - h\beta^{-2/v})^{-v/2} - (1 + 2h - h\beta^{-2/v} - h\alpha^{-2/v})^{-v/2}$$
$$+ (1 + 3h - 2h\beta^{-2/v} - h\alpha^{-2/v})^{-v/2}$$
$$- (1 + 4h - 2h\beta^{-2/v} - 2h\alpha^{-2/v})^{-v/2} + \cdots. \tag{3.2.23}$$

For $\beta = \alpha$ and setting $j = 0$ if $\mu < \delta/2$ and $j = 1$ if $\mu > \delta/2$, we have,

$$P(d_j | \sigma, \mu) \doteq \sum_{i=0}^{\infty} (-1)^i [1 + i|h|(\alpha^{-2/v} - 1)]^{-v/2} \qquad (\mu \neq \delta/2)$$

where again, as in (3.2.22) and (3.2.23) the partial sums provide successive upper and lower bounds for the OC function. At $\mu = \delta/2$ ($h = 0$), we obtain by taking limits in (3.2.21) as $h \to 0$ and using l'Hospital's rule that $P(d_0 | s, \sigma, \delta/2) \doteq a/(a-b)$ irrespective of s so that

$$P(d_0 | \sigma, \delta/2) \doteq a/(a-b) = (\alpha^{-2/v} - 1)/(\alpha^{-2/v} + \beta^{-2/v} - 2) \quad (3.2.24)$$

which equals $\frac{1}{2}$ if $\beta = \alpha$. The series (3.2.22)–(3.2.23) converge reasonably fast except near $h = 0$ ($\mu = \delta/2$). Numerical computations suggest that the OC function for T would be steeper than that of the analogous SPRT, assuming σ is known correctly.

The ASN Function of T

Analogous methods may be used to study the ASN function. For $\mu \neq \delta/2$ ($h \neq 0$) and for all σ, we have, using (3.2.15)

$$\begin{aligned} E[r_N(\sigma)] &= \delta E \sum_{i=1}^{N} (X_i - \delta/2)/\sigma^2 \\ &= \delta(\mu - \delta/2)(EN)/\sigma^2 = -h\,\delta^2(EN)/2\sigma^2. \end{aligned} \quad (3.2.25)$$

Also, dropping the subscript m,

$$Er_N(\sigma) = E\{E[r_N(\sigma) | s]\} = E\left[\sum_{i=0}^{1} E[r_N(\sigma) | s, d_i] P(d_i | s)\right]. \quad (3.2.26)$$

Now, still dropping the ms and ignoring excess,

$$E[r_N(\sigma) | s, d_1] \doteq \bar{a} = as^2/\sigma^2, \qquad E[r_N(\sigma) | s, d_0] \doteq \bar{b} = bs^2/\sigma^2.$$

The excess should not be significant if N tends to be large relative to m. Equation (3.2.26) thus leads to

$$\begin{aligned} Er_N(\sigma) &\doteq E[bs^2 P(d_0 | s) + as^2 P(d_1 | s)]/\sigma^2 \\ &= a - (a-b)E[s^2 P(d_0 | s)]/\sigma^2. \end{aligned} \quad (3.2.27)$$

Using (3.2.21) and proceeding as in the case of the OC function, we obtain, for $h > 0$

$$Er_N(\sigma) \doteq a - (a-b)E\left\{(s^2/\sigma^2)[1 - \exp(-ahs^2/\sigma^2)] \sum_{i=0}^{\infty} \exp[-i(a-b)hs^2/\sigma^2]\right\}.$$

Noting that $E[\chi^2 \exp(-t\chi^2/v)] = (1 + 2t)^{-1-v/2}$, we find after taking expectations term by term and equating with (3.2.25), that for $h > 0$

$$\begin{aligned}(\delta/\sigma)^2 E(N) &\doteq \frac{-2b}{h} - \frac{2(a-b)}{h}\{[1 + 2ah/v]^{-1-v/2} - [1 + 2(a-b)h/v]^{-1-v/2} \\ &\quad + [1 + 2(2a-b)h/v]^{-1-v/2} - [1 + 4(a-b)h/v]^{-1-v/2} + \cdots\} \\ &= (v/h)\{\beta^{-2/v} - 1 - (\alpha^{-2/v} + \beta^{-2/v} - 2)[(1 - h + h\alpha^{-v/2})^{-1-v/2} \\ &\quad - (1 - 2h + h\alpha^{-v/2} + h\beta^{-2/v})^{-1-v/2} \\ &\quad + (1 - 3h + 2h\alpha^{-2/v} + h\beta^{-2/v})^{-1-v/2} \\ &\quad - (1 - 4h + 2h\alpha^{-2/v} + 2h\beta^{-2/v})^{-1-v/2} + \cdots]\}. \end{aligned} \tag{3.2.28}$$

For $h < 0$, the same formula holds with h replaced by $-h$ and with α and β interchanged. For $\beta = \alpha$ and $h \neq 0$

$$(\delta/\sigma)^2 E(N) \doteq \frac{v}{|h|}(\alpha^{-2/v} - 1)\left\{1 + 2\sum_{i=1}^{\infty}(-1)^i[1 + i|h|(\alpha^{-2/v} - 1)]^{-1-v/2}\right\}. \tag{3.2.29}$$

At $\mu = 0$ and $\mu = \delta$ Wald's conservative bounds (3.2.17) on the error probabilities of the conditional test may be used instead of (3.2.21). Thus (3 2.27) together with (3.2.25) leads to

$$\begin{aligned}(\delta/\sigma)^2 E(N|\mu = 0) &\gtrsim -2b - 2(a-b)E[s^2 \exp(-as^2/\sigma^2)/\sigma^2] \\ &= v[\beta^{-2/v} - 1 - \alpha(1 + \beta^{-2v}\alpha^{2/v} - 2\alpha^{2/v})]\end{aligned}$$

and similarly for $(\delta/\sigma)^2 E(N|\mu = \delta)$ with α and β interchanged, where $\gtrsim$ denotes approximately greater than. If $\beta = \alpha$, these two relations reduce to

$$(\delta/\sigma)^2 E(N|\mu = 0 \text{ or } \delta) \gtrsim v[\alpha^{-2/v} - 1 - 2\alpha(1 - \alpha^{2/v})]. \tag{3.2.30}$$

For $h = 0$ ($\mu = \delta/2$) we have analogous to (3.2.25)–(3.2.27)

$$E[r_N(\sigma)]^2 = (\delta/\sigma)^2 E(N|\mu = \delta/2) \tag{3.2.31}$$

and

$$E[r_N(\sigma)]^2 \doteq a^2 E(s^4/\sigma^4) - (a^2 - b^2)E[s^4 P(d_0|s)]/\sigma^2. \tag{3.2.32}$$

Equating (3.2.31) and (3.2.32) and using (3.2.24) we have

$$(\delta/\sigma)^2 E(N \mid \mu = \delta/2) \doteq [a^2 - a(a+b)]E(\chi_v{}^4/v^2) = -ab[1 + (2/v)]$$
$$= \tfrac{1}{4}v^2[1 + (2/v)](\alpha^{-2/v} - 1)(\beta^{-2/v} - 1). \qquad (3.2.33)$$

Baker has carried out some sampling studies of the approximations for the OC and ASN functions.

Comparison with the OC and ASN Functions of Stein's Two-Stage Test

Let t' (and t'' respectively) be that number which is exceeded by a t-statistic with probability α (and β respectively) with v degrees of freedom. The total sample size N in Stein's (one-sided) procedure with initial sample size $m = v + 1$ and error bounds (α, β) is given by $N = \max\{[s_m{}^2(t' + t'')^2/\delta^2] + 1, m\}$ where here $[\cdot]$ means the largest integer contained in $(\cdot)$, and

$$(\delta/\sigma)^2 EN \doteq (t' + t'')^2 \qquad (3.2.34)$$

the approximation (actually lower bound) being valid if it implies EN is somewhat larger than m (Stein, 1945). The terminal decision rule for Stein's test may be taken to be: decide d_1 (d_0) if $\bar{x}_n > (<)\ \delta t'\ (t' + t'')$ or equivalently

$$\left.\begin{array}{l}\text{decide } d_1\\ \text{decide } d_0\end{array}\right\} \quad \text{if} \quad r_n(s_m) \gtrless \tfrac{1}{2}\,\delta^2(n/s_m{}^2)(t' - t'')/(t' + t'').$$

The OC function is found to be

$$P(d_0 \mid \mu, \sigma) = P\left\{\frac{\sum_1^N (X_i - \mu)}{N^{1/2}s_m} < \frac{N^{1/2}}{s_m(t' + t'')}\,[\delta t' - \mu(t' + t'')] \,\Big|\, \mu, \sigma\right\}.$$

Approximating $N^{1/2}$ by $s_m(t' + t'')/\delta$ in the right-hand side, we have

$$P(d_0 \mid \mu, \sigma) \doteq F_v(t' - (\mu/\delta)(t' + t'')), \qquad (3.2.35)$$

which is free of σ, where F_v is the distribution function of a (central) t-statistic with v degrees of freedom. The true OC function is actually slightly steeper than the approximation in (3.2.35) since the true sample size is larger than the approximating value.

Hall (1962) has made certain comparisons of the approximate power and ASN functions of the sequential test T, and Stein's two-stage test; and also of Wald's SPRT and the fixed-sample size test (FSST) if σ were known for

Table 3.2.1

Approximate Power Functions $P(d_1 \mid \mu)$ of Procedures for $\mu \leq 0$ against $\mu \geq \delta$ $(\alpha = \beta)$

			Test T		SPRT	Stein's test		FSST
α	μ/δ	h	$m = 16$	$m = 31$	(σ known)	$m = 16$	$m = 31$	(σ known)
.05	.5	0	.5	.5	.5	.5	.5	.5
	.1	.8	.073	.078	.087	.091	.092	.094
	0	1.0	.045	.046	.05	.05	.05	.05
	−.1	1.2	.029	.028	.028	.026	.025	.024
	−.5	2.0	.006	.004	.003	.002	.001	.001
.01	.5	0	.5	.5	.5	.5	.5	.5
	.1	.8	.019	.022	.025	.028	.029	.031
	0	1.0	.01	.01	.01	.01	.01	.01
	−.1	1.2	.005	.004	.004	.003	.003	.003
	−.25	1.5	.002	.002	.001	.001	.000	.000

Table 3.2.2

Approximate ASN Functions $(\delta/\sigma)^2 E\{N \mid \mu\}$ of Procedures for Testing $\mu \leq 0$ against $\mu \geq \delta$ with $\alpha = \beta$

			Test T		SPRT	Stein's test		FSST
α	μ/δ	h	$m = 16$	$m = 31$	(σ known)	$m = 16$	$m = 31$	(σ known)
.05	.5	0	15.4	11.7	8.7			
	.1	.8	8.2	7.2	6.1			
	0	1.0	6.9	6.1	5.3	12.3	11.5	10.8
	−.1	1.2	5.9	5.3	4.6			
	−.5	2.0	3.7	3.3	2.9			
.01	.5	0	45.8	31.0	21.1			
	.1	0.8	15.5	13.0	10.9			
	0	1.0	12.6	10.6	9.0	27.1	24.1	21.6
	−.1	1.2	10.5	8.9	7.6			
	−.25	1.5	8.5	7.2	6.1			

Lower bound on $(\delta/\sigma)^2 E(N \mid \mu)$

α	μ/δ	h	Test T'		MPRT	Stein's test		FSST
.05	.5	0	9.3	8.4	7.5	12.3	11.5	10.8
.01	.5	0	22.2	18.5	15.5	27.1	24.1	21.6

$\alpha = \beta = .05$ and .01. If the assumptions about σ were incorrect, the power functions of the SPRT and FSST may be drastically altered. These are presented in Tables 3.2.1 and 3.2.2. Table 3.2.2 suggests that substantial savings may be possible using the sequential test T—at least if one of the hypotheses is correct. The comparison between T and Stein's test is analogous to the comparison between the SPRT and the FSST of the same strength. Actually the comparison would be in closer analogy if the SPRT with conservative boundaries ($A = 1/\alpha$, $B = \beta = \alpha$) were considered, since the test T uses conservative boundaries. If the boundaries of T were modified by increasing α and β in (3.2.14) to achieve error probabilities equal to α [as attempted by Baker (1950)], the ASN of T would be further reduced. The lack of knowledge of σ requires about a 15–50% increase in the number of observations on the average (with these values of m, α, β), and thus the test T or Stein's test may possibly be recommended even if σ is thought to be known (if a one-stage test is not essential).

AN ALTERNATIVE SEQUENTIAL TEST T'

For the symmetric ($\alpha = \beta$) one-sided case, a *minimum probability ratio test* (MPRT) which has converging straight-line boundaries can be adopted in the same manner as the SPRT above. The MPRT is equivalent to one of Anderson's (1960) tests. The test T' is given as follows:

Stop sampling as soon as $\delta\left|\sum_{i=1}^{n}(X_i - \delta/2)\right|/s_m^2 \geq c_m - n\,\delta^2/4s_m^2$ $(n \geq m)$ and choose d_1 or d_0 according as $\sum(X_i - \delta/2)$ is $>$ or <0, where

$$c_m = v[(2\alpha)^{-2/v} - 1] = -2\ln(2\alpha)[1 + (-2\ln(2\alpha))/v + o(1/v)]. \quad (3.2.36)$$

After m observations have been taken and s_m computed, an upper bound on the total sample size is $4c_m s_m^2/\delta^2$. If $\alpha \neq \beta$, the test can still be used with 2α in (3.2.36) replaced by $\alpha + \beta$, but then one can only assert that the sum of the two error probabilities is less than $\alpha + \beta$. No approximations to the OC or ASN functions, other than the bound given below have been obtained. Presumably, T' would have a smaller ASN than T in the neighborhood of $\mu = \delta/2$ at the cost of a slightly larger ASN at (and beyond) $\mu = 0$ and δ. The comparison would be analogous to the comparisons of the SPRT and Anderson's (1960) test (σ known) given by Hoeffding (1960) (see Table 2.6.1, p. 62).

A lower bound on the ASN of T' at $\mu = \delta/2$ may be obtained as follows. Evaluating Hoeffding's [1960, Eq. (1.4)] lower bound [see (2.6.17)] for the conditional MPRT, we find

$$(\delta/\sigma)^2 E(N \mid s_m, \mu = \delta/2, \sigma) > 4\{[1 + (s_m/\sigma)^2 c_m]^{1/2} - 1\}^2.$$

Averaging over s_m and recalling that the expected value of a square root is less than the square root of the expected value, we obtain

$$(\delta/\sigma)^2 E(N \mid \mu = \delta/2, \sigma) > 4\,[(1 + c_m)^{1/2} - 1]^2. \tag{3.2.37}$$

For the MPRT (Anderson's test) when σ is known. Hoeffding's lower bound is the same as in (3.2.37) with c_m replaced by $-2 \ln(\alpha + \beta)$. According to Hoeffding's (1960) calculations, the lower bound in (3.2.37) might be quite close to an equality if $\delta/2\sigma$ is 0.1 and $\alpha = 0.01$ or 0.05. Some evaluations of (3.2.37) are shown in Table 3.2.2.

Two-Sided Case

A two-sided normal test (variance known) based on the weight function method (see Section 3.1) or the invariance method (see Section 3.6) cannot be adapted to the case of unknown variance, as was the one-sided test, since σ^2 does not factor out of the relevant probability ratios. However, the Sobel–Wald (1949) test procedure (see Section 3.8) is easily adapted. To test $\mu = 0$ against $|\mu| \geq \delta$, one can run two T- (T-′) tests (of $\mu = 0$ against $\mu = \delta$ and $\mu = 0$ against $\mu = -\delta$) simultaneously and continue sampling until both tests have terminated.

Heuristic Tests T_n and T_n'

If the procedure is not terminated early, σ would essentially be known. Thus, if one uses a procedure requiring knowledge of σ but replaces it by an estimate, the properties of the procedure should not be greatly affected. The tests T and T' are in this situation. Hall (1962) proposes the following procedure which is based on purely intuitive grounds: reestimate σ^2 by $s_n^{\,2}$ at each stage n and base the test on $r_n(s_n)$ $(n > 1)$ with boundaries (a_n, b_n) given by (3.2.14) with v replaced by $n - 1$. This test, designated T_n, is like a SPRT (σ known) with σ replaced by a new estimate at each stage and with the boundaries widened in an attempt to account for the lack of knowledge of σ. The boundaries converge to Wald's conservative boundaries $(-\ln \alpha, \ln \beta)$. The alternative test T' can be analogously modified, obtaining T_n' with the roles of c_m and s_m replaced by c_n and s_n. Its boundaries depend on $s_n^{\,2}$ and thus cannot be graphed in advance. According to Hall, no theoretical evaluation of these procedures has been possible.

Problem

Let .33, .84, .91, .44, 1.44, 1.94, .28, .16 denote a sequence of independent observations from a normal population with unknown mean μ and variance σ^2. We wish to test $H_0: \mu = 0$ against $\mu = \frac{1}{2}$. Carry out the two-stage test [of Hall (1962)] with $\alpha = \beta = .1$ and evaluate the approximate OC and ASN functions at $\mu = \frac{1}{2}$ assuming that $\sigma = 1$.

3.3 Sequential F-Test

Hoel (1955) has applied Wald's method of weight functions to the general linear hypothesis-testing problem. Let $X_1, \ldots, X_s$ be independent normal variables with means $\theta_1, \ldots, \theta_s$ ($\theta_i = 0$, $i = k+1, \ldots, s$) and common variance σ^2. We wish to test H_0: $\theta_1 = \cdots = \theta_p = 0$ for $p \le k$, $\sigma > 0$ against the alternative H_1: $\sum_1^p \theta_i^2/\sigma^2 \ge \delta_0$ where δ_0 is a given constant. Notice that $\theta_{p+1}, \ldots, \theta_k$ and σ are nuisance parameters and hence make H_0 a composite hypothesis. Let S_1 denote the boundary of the region $\omega_1 = \{(\theta_1, \ldots, \theta_s)$: $\sum_1^p \theta_i^2/\sigma^2 \ge \delta_0\}$. As normalizing weight functions choose,

$$g_0(\theta, \sigma) = \begin{cases} a_1 \sigma^{b_1}(2c)^{p-k}, & 0 \le \sigma \le c,\ |\theta_i| \le c,\ i = p+1, \ldots, k, \\ 0, & \text{elsewhere.} \end{cases}$$

$$g_1(\theta, \sigma) = \begin{cases} a_2 \sigma^{b_2}(2c)^{p-k}, & 0 \le \sigma \le c,\ |\theta_i| \le c,\ i = p+1, \ldots, k, \\ 0, & \text{elsewhere.} \end{cases}$$

where a_1, a_2, b_1, and b_2 are certain related constants to be specified later. Choose $b_2 = b + 1 - p$. Let ω_0 (ω_1) be that part of the parameter space where H_0 (H_1) holds. Let $S_{1,c}$ and $\omega_{0,c}$ be those parts of S_1 and ω_0 where $0 \le \sigma \le c$ and $|\theta_i| \le c$, for $i = p+1, \ldots, k$. Let $f_{1,n,c}$ ($f_{0,n,c}$) denote the density obtained by integrating the joint density of the Xs with respect to the prior density $g_1(\theta, \sigma)$ $[g_0(\theta, \sigma)]$ on the region $S_{1,c}(\omega_{0,c})$. Now let $c \to \infty$. Hoel (1955) obtains, after a considerable amount of integration and simplification,

$$f_{1,n}/f_{0,n} = \exp(-n\,\delta_0/2) M[\tfrac{1}{2}(ns + p - k - b_1 - 1), \tfrac{1}{2}p, \lambda^2/2\,\delta_0] \quad (3.3.1)$$

where $\lambda^2 = n\,\delta_0^2 \sum_1^p \bar{X}_i^2/(\sum_{j=1}^n \sum_{i=1}^s X_{ij}^2 - n \sum_{i=p+1}^k \bar{X}_i^2)$ and

$$M(a, b, u) = \sum_{r=0}^{\infty} [\Gamma(b)\Gamma(a+r)u^r/\Gamma(a)\Gamma(b+r)], \quad (3.3.2)$$

the confluent hypergeometric function. The derivation requires that $b_1 \le 0$. If $b_1 = -1$ and $s = k$, the ratio in (3.3.2) reduces to the one obtained by Johnson (1953) using Cox's (1952a) method. Certain invariance considerations also would lead one to the criterion $\lambda^2/2\,\delta_0$.

Ray (1957a) has shown that Johnson's (1953) and Hoel's (1955) procedures are of the form: Continue sampling at nth stage if

$$\beta/(1-\alpha) < e^{-\lambda(n)/2} M(a(n), b, \tfrac{1}{2}\lambda(n) G^{(n)}/(1 + G^{(n)})) < (1-\beta)/\alpha, \quad (3.3.3)$$

otherwise accept or reject the null hypothesis depending upon whether the left-hand or right-hand inequality is violated, where $\lambda(n)$ characterizes the alternative hypothesis, $a(n)$ is half the sum of the degrees of freedom of the numerator and denominator of the test criterion $G^{(n)}$, and b is half the degrees of freedom of the denominator of $G^{(n)}$. Notice that α, β are the error probabilities and $\lambda(n)$ and $a(n)$ are linear functions in n, and b is a fixed positive constant $[a(n) > b > 0]$. Sampling is terminated whenever $G^{(n)} \leq \underline{G}^{(n)}$ or $\geq \bar{G}^{(n)}$ where $\underline{G}^{(n)}$ and $\bar{G}^{(n)}$ are solutions of the equations

$$h_n(G) = e^{-\lambda(n)/2} M\left(a(n), b, \frac{\frac{1}{2}\lambda(n)G}{(1+G)}\right) = A, B \quad \text{respectively,} \tag{3.3.4}$$

where $A = (1-\beta)/\alpha$ and $B = \beta/(1-\alpha)$.

Theorem 3.3.1 (Ray, 1957a) The generalized SPRT to test H_0 against H_1 terminates finitely with probability one.

Proof. It will be sufficient to show that

$$\lim \bar{G}^{(n)} = \lim \underline{G}^{n} = G_0 \text{ (say)} \qquad \text{as} \quad n \to \infty.$$

Note that

$$h_n'(G) = e^{-\lambda/2} \frac{a}{b} M\left(a+1, b+1, \frac{\frac{1}{2}\lambda G}{1+G}\right) \frac{\frac{1}{2}\lambda}{(1+G)^2}$$

where the argument n in λ and a is suppressed. Also, from a recurrence relation of the confluent hypergeometric function, it can be shown that

$$b/a < M(a+1, b+1, u)/M(a, b, u) < 1 \qquad (u > 0) \tag{3.3.5}$$

from which it follows that

$$\frac{\frac{1}{2}\lambda}{(1+G)^2} < \frac{h_n'(G)}{h_n(G)} < \frac{\frac{1}{2}\lambda a/b}{(1+G)^2} \qquad \text{for all } G > 0. \tag{3.3.6}$$

Since $\lambda \to \infty$ as $n \to \infty$, (3.3.6) indicates that $h_n'(G)/h_n(G) \to \infty$ as $n \to \infty$. Since $\ln h_n(G)$ is a positive strictly increasing continuous function of G, it follows that there can be at most one value of G, say G_0 where $\ln h_n(G)$ does not become infinite as $n \to \infty$. Consequently, $\ln h_n(G) \to -\infty$ for $G < G_0$ and $\ln h_n(G) \to \infty$ for $G > G_0$. Hence $\bar{G}^{(n)} \to \underline{G}^{(n)} \to G_0$ and the sampling must terminate. If there does not exist a finite G_0 for which $\ln h_n(G)$ does not become infinite, then $\ln h_n(G)$ becomes infinite for all $G > 0$. Thus $h_n(G)$ either becomes infinite for all $G > 0$ or approaches zero for all $G > 0$. In the first case, sampling will terminate because $h_n(G) > A$ for sufficiently large n for all $G > 0$; and in the latter case also, since $h_n(G) < B$ for sufficiently large n for all $G > 0$. This completes the proof of Theorem 3.3.1.

It should be pointed out that it is not always possible to find the appropriate weight functions proposed by Wald (1947) because no general method is available for choosing the weight funcions. A second way of getting rid of the nuisance parameters is to transform the original observations such that the transformed variables do not depend on the nuisance parameters. Cox (1952a) proved a general theorem (which has implications in fixed-sample theory) which can be used to test composite hypotheses sequentially. For the statement of the theorem and some applications of the theorem, the reader is referred to Wetherill (1966, pp. 49–53). Since the transformed variables will usually be dependent, very little can be said about the other properties of the resultant sequential procedure, like the OC function and ASN (average sample number). Wald's (1947) method is more appropriate when composite hypotheses involve ranges of the parameters of interest and Cox's (1952a) method is appropriate when nuisance parameters exist. In the latter case, it is appropriate to construct test statistics that are invariant under a suitable group of transformations.

3.4 Certain Two-Sample Sequential Procedures

Suppose we are observing two processes sequentially and we also happen to know (approximately) their labels (that is, the two underlying parameters characterizing the two processes). However, the labels are mixed up and we wish to test a hypothesis about the correct labeling. To be more specific, let Π_0 and Π_1 be two populations belonging to the exponential family having density functions of the form

$$f(x;\theta_i) = B(\theta_i)\exp[Q(\theta_i)R_i(x)]h(x), \qquad i = 0, 1,$$

where $Q(\theta)$ is assumed to be increasing in $\theta \in \Theta$. At each stage, a pair of observations is obtained from the two populations and the independent pairs are denoted by (X_i, Y_i), $i = 1, 2, \ldots$. We wish to test

$$H_0\colon\ \theta_0 = \theta',\ \theta_1 = \theta'' \qquad \text{against} \qquad H_1\colon\ \theta_0 = \theta'',\ \theta_1 = \theta', \tag{3.4.1}$$

subject to α and β being the error probabilities. The SPRT based on the iid

$\{Z_i\}$ can be carried out to test the hypothesis in (3.4.1) where

$$Z_i = \ln \frac{f(X_i\,;\theta'')f(Y_i\,;\theta')}{f(X_i\,;\theta'(f(Y_i\,;\theta'')} = [R_0(X_i) - R_1(Y_i)]\psi_0 \tag{3.4.2}$$

where

$$\psi_0 = \psi(\theta'', \theta'), \qquad \psi(\theta_0, \theta_1) = Q(\theta_0) - Q(\theta_1). \tag{3.4.3}$$

Since Q is increasing, $\psi(\theta_0, \theta_1)$ is a single-valued function and satisfies $\psi(\theta, \theta) = 0$ for every θ in Θ and $\psi(\theta_0, \theta_1) = -\psi(\theta_1, \theta_0) < 0$ for all $\theta_0 < \theta_1$. Since $\psi_0 > 0$, the continuation region of the SPRT for (3.4.1) can be written as

$$\ln B/\psi_0 < \sum_{i=1}^{n} [R_0(X_i) - R_0(Y_i)] < \ln A/\psi_0 . \tag{3.4.4}$$

Then, one can obtain approximate expressions for $\mathrm{OC}(\theta)$ and $E_\theta(N)$. Also, finite termination of the procedure is ensured with probability unity and the optimum property of the SPRT is also valid. Notice that the nonzero solution to Wald's equation [see (2.3.1)] is obtained by direct substitution as

$$h = h(\theta_0, \theta_1) = \psi(\theta_1, \theta_0)/\psi_0 . \tag{3.4.5}$$

Further, the monotonicity of the OC function ensures us that (3.4.4) with $A = (1 - \beta)/\alpha$ and $B = \beta/(1 - \alpha)$ tests

$$H_0\colon\ \psi(\theta_0, \theta_1) \le -\psi_0 \quad \text{against} \quad H_1\colon\ \psi(\theta_0, \theta_1) \ge \psi_0, \qquad \psi_0 > 0, \tag{3.4.6}$$

with maximum error probabilities (α, β). The hypothesis in (3.4.6) is more realistic in the sense that often one is interested in discriminating between $\theta_0 < \theta_1$ and $\theta_0 > \theta_1$. [An equivalent hypothesis-testing problem would be to choose H_0 when $\theta_0 \le \theta'$ and H_1 when $\theta_1 \ge \theta''$ $(\theta' < \theta'')$.] It is worthwhile to note that there is no standard method of obtaining an optimal fixed-sample size procedure for (3.4.6). The above results also apply to continuous time stochastic processes.

Example 3.4.1 (Binomial proportions) Consider two independent Bernoulli populations with parameters θ_0 and θ_1. Here take

$$B(\theta) = 1 - \theta, \qquad Q(\theta) = \ln[\theta/(1 - \theta)], \qquad R_i(X) = X, \qquad h(x) \equiv 1.$$

For further elaboration on this topic and other details, the reader is referred to B. K. Ghosh (1970, pp. 239–246).

3.5 Fraser Sufficiency and Conditional Tests

3.5.1 Generalized or Fraser Sufficiency

In testing composite hypotheses involving nuisance parameters, one can exploit Fraser's sufficiency and reduce the problem to testing a simple hypothesis against a simple alternative. Fraser's (1956) definition of sufficiency is given below. Let $f(x; \gamma, \delta)$ be a given multiparameter family of probability densities in which γ and δ are functionally independent. That is $\delta \in \Delta$, $\gamma \in \Gamma$, and $\theta = (\gamma, \delta) \in \Theta = \Gamma \times \Delta$ (the Cartesian product of Γ and Δ). Notice that γ and δ could be vector valued. Let $\mathbf{X}_n = (X_1, \ldots, X_n)'$ be a sample of size n from $f(x; \gamma, \delta)$.

Definition 3.5.1 (Fraser, 1956) A statistic $T_n = T_n(\mathbf{X}_n)$ is said to have generalized sufficiency (or Fraser's sufficiency) in γ if for every $n \geq 1$,

$$f_n(\mathbf{X}_n; \gamma, \delta) = h_n(T; \gamma)g_n(\mathbf{X}_n \mid T, \delta), \tag{3.5.1}$$

[$h_n(t; \gamma)$ denotes the probability density of T] almost everywhere in R_n where $T = T_n(X_1, \ldots, X_n)$. Notice that the exponential family does not necessarily satisfy (3.5.1). Then consider

$$H_0\colon \ \gamma = \gamma_0 \qquad \text{against} \qquad H_1\colon \ \gamma = \gamma_1 \qquad \gamma_0 \neq \gamma_1, \tag{3.5.2}$$

where δ is a nuisance parameter. If T_n is a generalized sufficient statistic for γ, it follows that

$$Z_n = \ln f_n(\mathbf{X}_n; \gamma_1, \delta)/f_n(\mathbf{X}_n; \gamma_0, \delta) = \ln h_n(T_n; \gamma_1)/h_n(T_n; \gamma_0)$$

is independent of δ for every $n \geq 1$. Hence, the distribution of Z_n and z_n depends only on γ for every n where

$$z_n = Z_n - Z_{n-1} = \ln \frac{h_n(T_n; \gamma_1)h_{n-1}(T_{n-1}; \gamma_0)}{h_n(T_n; \gamma_0)h_{n-1}(T_{n-1}; \gamma_1)}, \tag{3.5.3}$$

with $Z_0 = 0$. Consequently, all the results pertaining to Wald's SPRT (derived in Chapter 2) are valid for testing (3.5.2). In the earlier proofs one replaces $(X_1, \ldots, X_n; \theta)$ by $(T_n; \gamma)$. The OC function and the ASN are free of

δ. If the Z_i are iid, then Wald's approximation for the OC function and the ASN (see Sections 2.3 and 2.4) are valid. In particular, if the X_i are iid then the z_i are iid. Because, if

$$f(x_i; \gamma, \delta) = h(u_i; \gamma)g(x_i | u_i; \delta), \; u_i = u(x_i) \tag{3.5.4}$$

then

$$T_n = T_n(u_1, \ldots, u_n) \qquad \text{and} \qquad z_i = \ln\{h(u_i; \gamma_1)/h(u_i; \gamma_0)\}.$$

If the z_i are not iid, termination of the procedure can be established by resorting to Theorem 2.12.1. Further, if the sequence of distribution functions of T_n has MLR for every n then the sequential procedure with $\gamma_0 < \gamma$ can be extended to test $H_0: \gamma \le \gamma_0$ against $H_1: \gamma \ge \gamma_1$ subject to the restriction that the maximum error probabilites are α and β.

Example 3.5.1 Sequential sign test: Consider a univariate population with unknown distribution function $F(x; \theta)$; $\theta \in \Theta$. Using a sequence of independent observations from this population, we wish to test

$$H_0: \; p(\theta) = F(x_0; \theta) \le p_0 \qquad \text{against} \qquad H_1: \; p(\theta) \ge p_1, \qquad 0 < p_0 < p_1 < 1$$

with error probabilities α and β, where x_0 is a specified real number. If $p_0 = \frac{1}{2} < p_1$, the problem reduces to that of testing whether the median of a population is less than x_0 or greater than x_0. Let $f^-(x)$ and $f^+(x)$ denote the probability density functions associated with the conditional distributions

$$F^-(x) = P(X \le x | X \le x_0), \qquad F^+(x) = P(X \le x | X > x_0).$$

Also, from among $X_1, X_2, \ldots, X_n$, let $X_{i_1}, \ldots, X_{i_m}$ denote those Xs that are less than or equal to x_0 and $X_{j_1}, \ldots, X_{j_{n-m}}$ denote those Xs that exceed x_0. Then the joint density of $X_1, \ldots, X_n$ namely f_n can be written as

$$f_n(\mathbf{x}_n; \theta) = [p(\theta)]^m [1 - p(\theta)]^{n-m} \prod_{k=1}^{m} f^-(x_{i_k}) \prod_{r=1}^{n-m} f^-(x_{j_r}) \tag{3.5.5}$$

where $p(\theta)$ is obviously independent of F^- or F^+ and the latter are also independent of $p(\theta)$. So take $\gamma = p(\theta)$ and $\delta = F^-$ or F^+, $\Gamma = [0, 1]$ and Δ is the collection of all distribution functions on $(-\infty, x_0)$ or (x_0, ∞). Thus, for every n, $m = \sum_1^n Y_i$ where $Y_i = 1$ if $X_i \le x_0$ and zero elsewhere, is Fraser sufficient for $p(\theta)$. It follows that the sequential sign test based on m is optimal not only among tests based on $\{Y_i\}$ but also among all tests based on $\{X_i\}$.

3.5.2 Conditional Tests

Suppose that we are interested in testing composite hypotheses about δ:

$$H_0:\ \delta = \delta_0 \qquad \text{against} \qquad H_1:\ \delta = \delta_1, \qquad \delta_1 \neq \delta_0. \tag{3.5.6}$$

If a Fraser sufficient statistic exists for γ which is a nuisance parameter here, it follows that

$$Z_n = \ln[f_n(\mathbf{X}_n;\gamma,\delta_1)/f_n(\mathbf{X}_n;\gamma,\delta_0)] = \ln[g_n(\mathbf{X}_n \mid T_n,\delta_1)/g_n(\mathbf{x}_n \mid T_n,\delta_0)] \tag{3.5.7}$$

which is independent of γ for every $n > 1$. Hence a SPRT can be carried out in order to test the hypotheses in (3.5.6) on the basis of $\{z_i\}$ where

$$z_i = Z_i - Z_{i-1}, \qquad Z_0 = 0. \tag{3.5.8}$$

Now the distributions of the z_i (which are not necessarily iid) do involve γ and δ. Hence the OC function and the true error probabilities also involve γ. If the SPRT based on Z_n is closed (that is, it terminates finitely with probability unity) and Z_n does not involve any excess over the boundaries, the true error probabilities are independent of γ [see B. K. Ghosh (1970, p. 249, p. 92)]. However, if the excess over the boundaries of Z_n is small, then the true error probabilities are approximately equal to the preassigned α and β. Also, if the conditional distributions of $\mathbf{X}_n$ given T_n have a monotone likelihood ratio in $V_n(\mathbf{X}_n)$ for every $n > 1$, then the SPRT is also valid for testing H_0: $\delta \le \delta_0$ against H_1: $\delta \ge \delta_1$. If $\{X_i\}$ are iid, then $z_i = \ln\{g(X_i \mid u_i, \delta_1)/g(X_i \mid u_i, \delta_0)\}$ are also iid; hence the finite termination and the optimum property of the SPRT are valid.

Example 3.5.2 (Trinomial distribution) Consider an iid vector sequence (X_i, Y_i) $(i = 1, 2, \ldots)$ with a common probability function

$$f(x, y; p_1, p_2) = \frac{m!}{x!\,y!\,(m-x-y)!} p_1{}^x p_2{}^y (1 - p_1 - p_2)^{m-x-y}; \tag{3.5.9}$$

$x, y = 0, 1, \ldots, m$; $x + y \le m$; $0 < p_1 + p_2 \le 1$; and m is a fixed positive integer. A Fraser sufficient statistic does not exist either for p_1 or p_2. However, by setting $\gamma = p_1 + p_2$ and $\delta = p_1/(p_1 + p_2)$, we can write $f(x, y, p_1, p_2)$ as

$$f^*(x, y; \gamma, \delta) = \binom{m}{u}\gamma^u(1-\gamma)^{m-u}\binom{u}{x}\delta^x(1-\delta)^{u-x}, \qquad u = x + y. \tag{3.5.10}$$

So, here $\Gamma = [0, 1]$, $\Delta = [0, 1]$, and $T_n(\mathbf{X}_n, \mathbf{Y}_n) = \sum_1^n (X_i + Y_i)$ is Fraser sufficient for γ and the conditional distribution of $(\mathbf{X}_n, \mathbf{Y}_n)$ given T_n has monotone likelihood ratio in $V_n(\mathbf{X}_n, \mathbf{Y}_n) = \sum_1^n X_i$. Hence the continuation region for the SPRT for testing $H_0: \delta \le \delta_0$ against $\delta \ge \delta_1$ $(0 < \delta_0 < \delta_1 < 1)$ is given by

$$\ln B < \ln\left(\frac{\delta_1}{\delta_0}\right) \sum_1^n X_i - \ln\left(\frac{1-\delta_0}{1-\delta_1}\right) \sum_1^n Y_i < \ln A.$$

One can easily compute the OC function and the ASN since the test is analogous to the binomial testing problem. For other examples, see B. K. Ghosh (1970, pp. 253–255).

3.6 Invariant Sequential Procedures

3.6.1 Principle of Invariance

In this section we shall consider a family of distributions indexed by a vector-valued parameter θ. We may be interested in testing hypotheses about some function of θ, namely $\lambda(\theta)$. If a Fraser sufficient statistic for $\lambda(\theta)$ exists, then one can test a hypothesis about $\lambda(\theta)$ sequentially using the methods proposed in Section 3.5. However, there are many multiparameter families in which such methods are not feasible; for instance, consider the normal population having mean μ and variance σ^2. Then take $\theta = (|\mu|, \text{sign } \mu)$ and $\lambda(\theta) = |\mu|$ or take $\theta = (\mu, \sigma)$ and $\lambda(\theta) = \mu/\sigma$. In such cases, one may be able to find a sequence of functions $\{T_n\}$ of the sample observations $X_1, \ldots, X_n$ such that the distribution of T_n depends only on λ. However, there might be many such sequences of T_n. Hence, we wish to shrink the domain of possible choices of such functions by imposing an additional criterion like the invariance principle which essentially restricts us to the choice of a special sequence of functions of the sample observations. In the following, we shall elaborate some of the ideas underlying the principle of invariance. Let G denote a group of one-to-one transformations of R_n for every $n \ge 1$.

Definition 3.6.1 The family of distributions $\{F_{(n)}(\mathbf{x}_n;\ \theta)\}$ [where $\mathbf{x}_n = (x_1, \ldots, x_n)'$] is said to be invariant under G if (i) for any $\theta \in \Theta$ and any $g \in G$ there exists $\theta' \in \Theta$ such that

$$P(gX_i \le x_i, i = 1, \ldots, n \mid \theta) = F_n(x_1, \ldots, x_n;\ \theta'), \tag{3.6.1}$$

for all $(x_1, \ldots, x_n) \in R_n$, and (ii) for any θ' there exists a $\theta \in \Theta$ such that

(3.6.1) holds. The point $\theta' \in \Theta$ satisfying (i) and (ii) is denoted by $\bar{g}\theta$ and the collection of such $\bar{g}$s, will be denoted by $\bar{G}$.

We shall also say that G induces a (one-to-one) group of transformations $\bar{G}$ from Θ onto itself. One can express (3.6.1) also as $F^{(n)}(x_1, \ldots, x_n; \theta) = F^{(n)}(gx_1, \ldots, gx_n; \bar{g}\theta)$. A hypothesis-testing problem with $H_0: \theta \in \omega$ against $H_1: \theta \in \omega'$, $\omega' \subset \Theta - \omega$ is said to be invariant under G if (i) and (ii) hold for all $\theta, \theta' \in \omega$ or ω'. Any point $\mathbf{x}_n'$ in R_n is called equivalent to x_n under G if $\mathbf{x}_n' = g\mathbf{x}_n$ for at least one $g \in G$. Thus, G partitions R_n into equivalence classes or *orbits*, the orbit $O(\mathbf{x}_n)$ of a point $\mathbf{x}_n \in R_n$ being the set of all points in R_n that are equivalent to $\mathbf{x}_n$ under G. Any function $T_n(\mathbf{x}_n)$ on R_n is said to be invariant under G if it is constant on each orbit, that is, if

$$T_n(g\mathbf{x}_n) = T_n(\mathbf{x}_n) \qquad \text{for all } \mathbf{x}_n \in R_n \quad \text{and} \quad g \in G. \tag{3.6.2}$$

Notice that T_n could be vector valued. A subset E of R_n is said to be invariant under G if $\mathbf{x}_n \in E$ implies $g\mathbf{x}_n \in E$ for every $g \in G$ (that is, $gE \equiv E$). Thus, a test criterion is said to be invariant under G if its critical region is invariant under G. That is, if we restrict ourselves to invariant tests, then the orbits would become the basic elements in the construction of a test of H_0. Now all invariant tests of H_0 can be characterized as functions of a *maximal invariant.*

Definition 3.6.2 A function $T_n(\mathbf{x}_n)$ on R_n is said to be maximal invariant under G if it assumes a different value on each orbit, that is, if it is invariant under G and if $T_n(\mathbf{x}_n) = T_n(\mathbf{x}_n')$ implies that $\mathbf{x}_n' = g\mathbf{x}_n$ for some $g \in G$.

Maximal invariants under a group G always exist and are equivalent in the sense that they assume constant values on each orbit, but their functional representation may not be unique. Further, $\bar{G}$ induces a partition of Θ into equivalence classes or orbits. We denote the maximal invariant function on Θ under $\bar{G}$ by $\lambda(\theta)$. Then, we have the following theorem.

Theorem 3.6.1 Let $U_n(\mathbf{x}_n)$ be a maximal invariant on R_n under G, and let $\lambda(\theta)$ be a maximal invariant on Θ under $\bar{G}$.

(i) If $T(\mathbf{x}_n)$ is any invariant function on R_n under G, the distribution of $T(\mathbf{X}_n)$ depends on θ only through $\lambda(\theta)$ [where $\mathbf{X}_n = (X_1, \ldots, X_n)'$].

(ii) Any function $T(\mathbf{x}_n)$ on R_n is invariant under G if and only if $T(\mathbf{x}_n)$ depends on $\mathbf{x}_n$ only through $U_n(\mathbf{x}_n)$.

Proof Let θ' and θ'' be any two points in Θ such that $\lambda(\theta') = \lambda(\theta'')$. Since $\lambda(\theta)$ is a maximal invariant, $\theta'' = \bar{g}\theta'$ for some $\bar{g} \in \bar{G}$. Hence

$$\begin{aligned} P(T_n(\mathbf{X}_n) \le t \mid \theta') &= P(T_n(g\mathbf{X}_n) \le t \mid \theta') = P(T_n(\mathbf{X}_n) \le t \mid \bar{g}\theta') \\ &= P(T_n(\mathbf{X}_n) \le t \mid \theta''). \end{aligned}$$

That is, for any t, the distribution of $T_n(\mathbf{X}_n)$ under θ remains unchanged for all $\theta' \in \Theta$ which belong to the same orbit as θ. To prove (ii) let ψ be a function such that $\psi(U_n(\mathbf{x}_n)) = T_n(\mathbf{x}_n)$ for all $\mathbf{x}_n \in R_n$. Then for each $g \in G$, $T(g\mathbf{x}_n) = \psi(U_n(g(\mathbf{x}_n)) = \psi(U_n(\mathbf{x}_n)) = T_n(\mathbf{x}_n)$, for all $\mathbf{x}_n \in R_n$. That is T_n is invariant under G. Conversely, if T_n is invariant under G, let $\mathbf{x}_n'$ and $\mathbf{x}_n''$ be any two points in R_n such that $U_n(\mathbf{x}_n') = U_n(\mathbf{x}_n'')$. Since $U_n(\mathbf{x}_n)$ is a maximal invariant on R_n we have $\mathbf{x}_n'' = g\mathbf{x}_n'$ for some $g \in G$ and hence $T_n(\mathbf{x}_n') = T_n(g\mathbf{x}_n') = T_n(\mathbf{x}_n'')$. That is, T_n remains constant for all $\mathbf{x}_n$ for which $U_n(\mathbf{x}_n)$ is constant. Hence T_n is a function of U_n only. This completes the proof of Theorem 3.6.1.

Example 3.6.1 Let $G = \{g^+, g^-\}$, the group of sign changes, defined by $g^+\mathbf{x}_n = \mathbf{x}_n$, $g^-\mathbf{x}_n = -\mathbf{x}_n = (-x_1, \ldots, -x_n)'$. $U_1(x_1) = |x_1|$ is a maximal invariant on R_1, $U_2(\mathbf{x}_2) = (|x_1|, |x_2|, \Delta_2)$ is a maximal invariant on R_2, where $\Delta_2 = 1$ if $x_1x_2 > 0$ and $\Delta_2 = 0$ if $x_1x_2 \leq 0$, since $(|g^+x_1|, |g^+x_2|, g^+\Delta_2) = (|x_1|, |x_2|, \Delta_2) = (|g^-x_1|, |g^-x_2|, g^-\Delta_2)$, for all $-\infty < x_1, x_2 < \infty$. Further, $|x_1| = |x_1'|$, $|x_2| = |x_2'|$ and $\Delta_2 = \Delta_2'$ imply that either $(x_2' = g^+x_1, x_2' = g^+x_2)$ or $(x_1' = g^-x_1, x_2' = g^-x_2)$. Without the Δ_2 component $U_2(\mathbf{x}_2)$ is not a maximal invariant. In general, $U_n(\mathbf{x}_n) = (|x_1|, \ldots, |x_n|, \Delta_n)$ is a maximal invariant on R_n, where Δ_n assumes one of 2^{n-1} distinct values, any value corresponding to a pair of "opposite quadrants" in the n-dimensional coordinate system (that is, where

$$x_{i_1} \geq 0, \ldots, x_{i_k} \geq 0, \qquad x_{j_1} \leq 0, \ldots, x_{j_{n-k}} < 0,$$

or

$$x_{i_1} < 0, \ldots, x_{i_k} < 0, \qquad x_{j_1} \geq 0, \ldots, x_{j_{n-k}} \geq 0$$

for $1 = i_1 < i_2 < \cdots < i_k$, $j_1 < j_2 < \cdots < j_{n-k}$ and any $i \neq$ any $j \leq n$).

Example 3.6.2 If G is the group of translations (location shifts), that is, $g\mathbf{x}_n = (x_1 + c, \ldots, x_n + c)'$, $-\infty < c < \infty$, for every g in G, then

$$U_n(\mathbf{x}_n) = (x_1 - x_2, x_2 - x_3, \ldots, x_{n-1} - x_n)$$

or

$$= (x_1 - x_n, \ldots, x_{n-1} - x_n)$$

or

$$= (x_1 - \bar{x}_n, \ldots, x_n - \bar{x}_n),$$

where $\bar{x}_n$ denotes the arithmetic mean.

Example 3.6.3 Let G be the group of positive scale changes, that is, for every g in G, $g\mathbf{x}_n = (cx_1, \ldots, cx_n)'$, $0 < c < \infty$. For R_1, $U_1(x_1) =$ sign of x_1 is a maximal invariant; for R_n

$$U_n(\mathbf{x}_n) = (x_1/|x_n|, \ldots, x_n/|x_n|)$$

or

$$= (x_1/x_n, \ldots, x_{n-1}/x_n, \text{ sign of } x_n).$$

If G is the group of all scale changes (that is, $c \neq 0$), then $U_1(x_1) =$ constant, and

$$U_n(\mathbf{x}_n) = (x_1/x_n, \ldots, x_{n-1}/x_n).$$

Example 3.6.4 If G is the set of $n!$ permutations of the n coordinates of $\mathbf{x}_n$, then the set of order statistics (ordered coordinates) $x^{(1)} \leq x^{(2)} \leq \cdots \leq x^{(n)}$ is a maximal invariant.

Example 3.6.5 If G is the set of all continuous and strictly increasing transformations (that is, $x_i' = g(x_i)$ where g is continuous and strictly increasing), then the set of ranks $(r_1, \ldots, r_n)$ of $(x_1, \ldots, x_n)$ is maximal invariant. For some details on the existence of such a function g, see Lehmann (1959, p. 217).

3.6.2 Invariant Sufficiency

Let $\Lambda = \Lambda(\theta)$ denote the space of λ, the maximal invariant on Θ. Further, U_m considered as a function of $\mathbf{x}_n = (\mathbf{x}_m, x_{m+1}, \ldots, x_n)'$ for any $m < n$ is invariant on R_n under G and therefore $U_m(\mathbf{x}_m)$ is itself a function of $U_n(\mathbf{x}_n)$. That is, for any $n > 1$, if the observed value U_n is given, the observed values of $U_1, \ldots, U_{n-1}$ are available. The principle of invariance suggests that any decision procedure about λ should also be invariant under G. That is, the decision rule at the nth stage should be based only on U_n. Such rules are called *invariant procedures* for λ, which replace the original data in terms of U_n and replace the original family of distributions with $\{F_{U_n}(u; \lambda), \lambda \in \Lambda\}$. Thus the nuisance parameters are eliminated and the properties of an invariant SPRT (like OC and ASN functions) can be evaluated in terms of λ.

Sometimes we can further reduce the data by confining ourselves to an invariant sufficient statistic which is defined below.

Definition 3.6.3 A function $V_n = V_n(\mathbf{X}_n)$ is said to be invariantly sufficient for $\{F^{(n)}(\mathbf{x}_n; \theta), \theta \in \Theta\}$ under G if

(i) V_n is an invariant function on R_n under G, and
(ii) $P(\mathbf{X}_n \in E \mid V_n = v)$ does not involve θ for any invariant set E in R_n under G and for all v.

Part (i) of Definition 3.6.3 and Theorem 3.6.1 imply that if U_n is any maximal invariant on R_n under G, V_n can always be expressed as a function T_n on the sample space of U_n. Part (ii) of the definition is equivalent to

$$f_{U_n}(u; \lambda) = f_{V_n}(v; \lambda) h_n(u) \tag{3.6.3}$$

where for each n, $f_{U_n}(u; \lambda)$ $[f_{V_n}(v; \lambda)]$ denotes the probability (density) function of U_n $[V_n]$. The reduction of the data $\mathbf{x}_n$ for any n, to the "invariant data" u_n and consequently to the "invariantly sufficient data" v_n can be symbolically expressed as

$$[F^{(n)}(x_n; \theta), \theta \in \Theta] \xrightarrow{G} [F_{U_n}(u; \lambda), \lambda \in \Lambda] \xrightarrow{\text{suff}} [F_{V_n}(v; \lambda), \lambda \in \Lambda]. \tag{3.6.4}$$

It is of interest to note that an invariantly sufficient sequence always exists for any family under a given G. Conversely, however, given some parametric function $\psi(\theta)$ there may not exist any G on R_n such that $\psi(\theta)$ acts as a maximal invariant on Θ under $\bar{G}$. Further, every invariant procedure may be represented by some equivalent decision rule that depends, for each n, only on v_n provided $\{V_n\}$ is also a *transitive* sequence for $\{F_{U_n}(u; \lambda), \lambda \in \Lambda\}$. That is, any procedure that uses only v_n at stage n preserves the principle of invariance; however, for the converse to hold one further requires that, for each n and $\lambda \in \Lambda$,

$$P(V_{n+1} \le v \mid U_n = u, \lambda) = P(V_{n+1} \le v \mid V_n = T_n(u), \lambda), \tag{3.6.5}$$

for all (u, v). It should be noted again that every invariant procedure is necessarily based only on u_n at stage n.

There may be situations where the parameter of interest remains fixed, but the nuisance parameters may vary from stage to stage. The principle of invariance can be extended to embrace such situations. Hall *et al.* (1965) have considered more complicated structures of invariance in their definition of an invariantly sufficient sequence.

3.6.3 The Invariant SPRT

Consider the problem of testing H_0: $\lambda = \lambda_0$ against H_1: $\lambda = \lambda_1$, $\lambda_0 \neq \lambda_1$ where $\lambda = \lambda(\theta)$ is a maximal invariant on Θ under $\bar{G}$. Since V_n is sufficient for the distribution of U_n and the joint distribution of $(V_1, \ldots, V_n)$, we have from the factorization criterion that

$$\frac{f_{U_n}(u_n; \lambda_1)}{f_{U_n}(u_n; \lambda_0)} = \frac{f_{V_n}(v_n; \lambda_1)}{f_{V_n}(v_n; \lambda_0)} = \frac{g_n(v_1, \ldots, v_n; \lambda_1)}{g_n(v_1, \ldots, v_n; \lambda_0)} = \exp(Z_n) \quad \text{say,}$$

where Z_n could be treated as a function of u_n or v_n. Also, any test based on this probability ratio for $n = 1, 2, \ldots$ uses only v_n at stage n and hence is an invariant test. The transitivity of $\{V_n\}$ for $\{F_{U_n}(u; \lambda)\}$ is not necessary for this assertion; however, it is required to assert that all invariant tests about λ are essentially based on v_n at each stage. Thus, one can carry out Wald's SPRT based on Z_n with approximate boundary values obtained from the specified error probabilities. If λ is real valued and $\{F_{V_n}(v; \lambda), \lambda \in \Lambda\}$ or $\{F_{U_n}(u; \lambda), \lambda \in \Lambda\}$ possesses a monotone likelihood ratio for each n, then the SPRT is effectively testing $H_0: \lambda \le \lambda_0$ against $H_1: \lambda \ge \lambda_1$ $(\lambda_0 < \lambda_1)$ at the same error probabilities. Suppose that the z_i are iid where $z_i = Z_i - Z_{i-1}$ for $i \ge 1$ with $Z_0 = 0$. Notice that both z_i and Z_i are functions of U_i only. Then it follows from Theorem 2.2.1 that the SPRT terminates finitely with probability one provided $P(z_i = 0 | \lambda) < 1$. Furthermore, the approximations to the OC function and the ASN are valid and the optimum property of the SPRT holds. The independence of $\{U_i\}$ or $\{V_i\}$, which rarely holds in applications, is necessary for this optimality. However, it should be added that there might be a rival invariant test having smaller ASN than the invariant SPRT at some $\lambda \ne \lambda_0, \lambda_1$, or there may exist a noninvariant test procedure having smaller ASN than the invariant SPRT even at $\lambda = \lambda_0$ or λ_1. One can rule out the latter possibility when $\{V_i\}$ is also Fraser sufficient for λ.

If the $\{z_i\}$ are not iid, which is true in many important applications, then none of the assertions of the preceding paragraph holds. Then, the termination property has to be verified by some other method such as the one given in Theorem 2.12.1. Ifram (1965) discussed this aspect when $\{F_{V_n}(v; \lambda), \lambda \in \Lambda\}$ possesses a monotone likelihood ratio for each n and Wijsman (1967b) considered the case when the $\{X_i\}$ are iid multivariate normal vectors. Both authors provide sufficient conditions under which $P(\ln B < Z_n < \ln A | \lambda) = 0$ for all $\lambda \in \Lambda$ as $n \to \infty$.

Berk (1970) has shown that the maximal invariant on R_n may be identified with an exchangeable sequence of random variables derived from $X_1, X_2, \ldots$. Toward this we have the following theorem of Berk (1970).

Theorem 3.6.2 (Berk, 1970) Suppose there exists a positive integer k and a $\omega_0 \cup \omega_1$ null subset $\mathcal{N}$ (where $\omega_i = \{\theta: \lambda(\theta) = \lambda_i\}$, $i = 0, 1$) of R_k so that every g in G is determined by its values at the k coordinates of any point of $R_k - \mathcal{N}$. Then letting $U(x_1, \ldots, x_{k+1})$ denote a maximally G-invariant function of $x_1, \ldots, x_{k+1}$, a maximally G-invariant function of $x_1, \ldots, x_n$, $n > k$, is $U_n(x_1, \ldots, x_n) = (U(x_1, \ldots, x_{k+1}), \ldots, U(x_1, \ldots, x_k, x_n))$.

Proof It is clear that U_n is invariant; we will show that it is maximal. Let $(x_1, \ldots, x_n)$ and $(x_1', \ldots, x_n')$ represent two possible outcomes of $(X_1, \ldots, X_n)$ for which $U_n(x_1, \ldots, x_n) = U_n(x_1', \ldots, x_n')$. Then $U(x_1, \ldots, x_k, x_i) = U(x_1', \ldots, x_k', x_i')$, $i = k + 1, \ldots, n$, and since U is maximal, there exists $g_i \in G$ such that $(g_i x_1, \ldots, g_i x_k, g_i x_i) = (x_1', \ldots, x_k', x_i')$ $i = k + 1, \ldots, n$. We may assume that $(x_1, \ldots, x_k)$ and $(x_1', \ldots, x_k') \in R_k - \mathcal{N}$; then since $(g_i x_1, \ldots, g_i x_k) = (x_1', \ldots, x_k')$ for all i, $g_{k+1} = \cdots = g_n = g$ (say). Thus $(x_1', \ldots, x_n') = (gx_1, \ldots, gx_n)$, showing that U_n is maximal.

REMARK 3.6.2.1 In typical applications of Theorem 3.6.2, R_k is a vector space and $\mathcal{N}$ is the set of points with linearly dependent coordinates. Letting $Y_i = U(X_1, \ldots, X_k, X_{k+i})$ and $U_{k+n} = (Y_1, \ldots, Y_n)$, the SPRT based on the U_n is seen to depend on the likelihood ratios for the successive initial segments of the exchangeable sequence $Y_1, Y_2, \ldots$. The following sequential t-test would illustrate the invariance reduction.

Example 3.6.6 Suppose under H_0 (H_1), X_i are distributed normally with mean $\mu = \delta\sigma$ and variance σ^2 (mean $\mu = \delta'\sigma$ and variance σ^2) where $0 \leq \delta \leq \delta'$ and $\Theta = (0, \infty)$. [Thus the hypotheses are $(\mu/\sigma) = \delta$ or δ'.] Both hypotheses are invariant under changes of scale ($X \to cX$, $c > 0$) and are generated by these transformations. Theorem 3.6.2 with $k = 1$ is easily seen to hold. Since x_2/x_1 is a maximally invariant function of (x_1, x_2), we see that $(Y_1, Y_2, \ldots) = (X_2/X_1, X_3/X_1, \ldots)$ is a maximally invariant exchangeable sequence guaranteed by the theorem. Thus, the hypotheses may be tested by considering at the $(n + 1)$th stage the likelihood ratio of $U_{n+1} = (Y_1, \ldots, Y_{n+1-k})$ under the hypotheses. Since $\bar{X}_n = \sum_1^n X_i/n$ and $(n - 1)s_n^2 = \sum_1^n (X_i - \bar{X}_n)^2$ are jointly sufficient for $X_1, \ldots, X_n$ (under $H_0 \cup H_1$) the theory of Hall *et al.* (1965) shows that a maximally invariant function of $(\bar{X}_n, s_n)$ is sufficient for U_n. The sufficient invariant may be taken to be $T_n = \bar{X}_n/s_n$. Hence, the procedure reduces to considering a sequence of likelihood ratios for successive t statistics. It is easily checked that the other normal theory SPRTs based on invariance reduction (like sequential χ^2 and F tests) also have this exchangeable structure.

Berk (1970) points out the connection between data reduction using the method of weight functions and the principle of invariance. In either case, the SPRT is based on ratios of probability densities of exchangeable sequence of random variables. This is an instance of de Finetti's representation theorem for exchangeable random variables. The theorem of Hewitt and Savage (1955) shows that in most cases of interest an exchangeable distribution for a data sequence is a mixture of iid distributions. It is well known that the likelihood ratios for certain invariant SPRTs can be obtained by mixing the corresponding joint densities for $X_1, X_2, \ldots$ with appropriate measures. For the sequential t-test, this has implicitly been shown by Wald (1947) and Barnard (1952). The reason for this has been expounded by Wijsman (1967a) who shows under general conditions that the probability density for a maximally invariant function of $X_1, \ldots, X_n$ is proportional to

$$\int_G \prod_{i=1}^{n} f(x_i \mid g\theta)\, d\nu(g)$$

where θ is a reference point in Θ and ν is a right Haar measure on G. (The omitted factors involve constants and Jacobians.) This equation shows that the likelihood ratio would be symmetric in $X_1, \ldots, X_n$.

Thus, Berk (1970) considers the sure termination of SPRTs based on exchangeable likelihood ratios and gives unified results for many of the SPRTs that have been proposed. Instead of discussing his results, we shall discuss Wijsman's (1971) results pertaining to exponentially bounded stopping time of invariant SPRTs.

Let $X_1, \ldots$ be a sequence of iid random variables having common distribution F. The joint distribution of the Xs will also be denoted by F. Stein (1946) showed (see Theorem 2.2.1) that the stopping time N (random sample size) of Wald's SPRT for testing one simple hypothesis against another is exponentially bounded; that is, for some $c < \infty$, $0 < \rho < 1$ we have

$$P(N > n) < c\rho^n, \qquad n = 1, 2, \ldots, \tag{3.6.6}$$

for every F except those F under which the log probability ratio is degenerate at zero. The reason Wald's SPRT can be treated with such relative ease is that $\{S_n, n = 1, 2, \ldots\}$ is a random walk, where S_n is the log probability ratio at the nth stage. If the hypotheses to be tested are composite which, however can be reduced to simple hypotheses after an invariance reduction, an invariant SPRT can be defined on the reduced data. However, the corresponding S_n is no longer a random walk. Consequently it is much harder to establish (3.6.6) for invariant SPRTs. The general results obtained in this regard are somewhat incomplete and do not demonstrate for which Fs (3.6.6) fails. Wijsman (1972) gives a more thorough discussion on this and a list of references. He has provided two examples in which (3.6.6) is demonstrated to hold for all F except possibly for a class of Fs exhibiting a certain degeneracy, thereby showing that at least in those examples the existence of a moment-generating function is irrelevant.

Definition 3.6.4 F is called *obstructive* if (3.6.6) is not satisfied for all stopping bounds.

Let Λ_n denote the probability ratio at the nth stage of the invariant SPRT. Let $S_n = \ln \Lambda_n$. The stopping time of an invariant SPRT is then defined in the usual way:

$$N = \text{smallest } n \geq 1 \text{ such that } b < S_n < a \text{ is violated}, \tag{3.6.7}$$

where $-\infty < b < a < \infty$ are the chosen stopping bounds for $\{S_n\}$. We shall assume that there exists a function s from the range of X_1 into Euclidean k-space R_k for some $k \geq 1$, such that with the notation

$$Y_i = s(X_i), \qquad i = 1, 2, \ldots, \tag{3.6.8}$$

$$\overline{Y}_n = n^{-1} \sum_1^n Y_i, \tag{3.6.9}$$

Λ_n (and consequently S_n) is a function of $X_1, \ldots, X_n$ only through n and $\overline{Y}_n$.

Then, we have the following theorem of Wijsman (1971) which consists of two parts, the second part proving (3.6.6) and the first part proving sure termination,

$$P(N < \infty) = 1, \tag{3.6.10}$$

under weaker conditions on F. Most of the labor that goes into proving (3.6.6) can also be used for (3.6.10) which is obtained at no extra cost. The method of proof utilizes an idea of Stein (1946). Moreover, existence of a certain real-valued function Ψ on R_k possessing certain properties will be assumed. From Wijsman (1972) it is obvious that there is no natural candidate for Ψ, however for the proof of the theorem, it is irrelevant from where Ψ comes. We shall make the following two sets of assumptions; the first used to prove (3.6.10) and both to prove (3.6.6).

Assumption A (i) $E(Y_1 | F) = \xi$ exists and is finite; (ii) there exists a neighborhood V of ξ and a real-valued function Ψ on V and a finite constant B_1 such that

$$|S_n - n\Psi(\overline{Y}_n)| < B_1 \quad \text{if} \quad \overline{Y}_n \in V, \qquad n = 1, 2, \ldots; \tag{3.6.11}$$

(iii) Ψ has continuous first partial derivatives on V; let grad Ψ be the vector of first partials and $\Delta =$ grad Ψ evaluated at ξ, then

$$P\{\Delta'(Y_1 - \xi) = 0\} < 1. \tag{3.6.12}$$

Assumption B For all components Y_{1j} $(j = 1, \ldots, k)$ of Y_1 we have $E[\exp(tY_{1j}) | F] < \infty$ for t in some neighborhood of 0.

Clearly Assumption B implies Assumption A(i).

Theorem 3.6.3 (Wijsman, 1972) Let $-\infty < b < a < \infty$ be arbitrary and N defined by (3.6.7). Then (3.6.10) is true if Assumption A is satisfied and (3.6.6) is true if Assumptions A and B are satisfied. Assumption A(iii) is not needed if $\Psi(\xi) \neq 0$.

Proof Put $D = B_1 + \max(|b|, |a|)$ and

$$\Psi_n = n\Psi(\overline{Y}_n). \tag{3.6.13}$$

In view of (3.6.7) and using (3.6.11) one infers that, if for some n, $\overline{Y}_n \in V$ and $|\Psi_n| \geq D$, then $N \leq n$. We distinguish two cases: $\Psi(\xi) \neq 0$ and $\Psi(\xi) = 0$. In case 1, assume that $\Psi(\xi) > 0$ [the case $\Psi(\xi) < 0$ is subject to an analogous treatment]. Choose $0 < \varepsilon < \Psi(\xi)$ and reduce V, if necessary, to ensure that $\Psi > \varepsilon$ on V (this is possible since Ψ is continuous at ξ). For any $n > D/\varepsilon$, if $\overline{Y}_n \in V$ then $\Psi_n > D$ so that $N \leq n$. Let N' be the first $n > D/\varepsilon$ such that $\overline{Y}_n \in V$. By the above we have $N \leq N'$. By Assumption A(i), $\overline{Y}_n \to \xi$ (almost everywhere F) so that $\overline{Y}_n \in V$ eventually (almost everywhere F). That is, $P(N' < \infty) = 1$ and so a fortiori $P(N < \infty) = 1$. If Assumption B also holds, then it follows essentially from Chernoff (1952, Theorem 1) [see also Wijsman (1968, Section 3)] that

$$P(\overline{Y}_n \notin V) < c_1 \rho_1{}^n \tag{3.6.14}$$

for some $c_1 < \infty$, $0 < \rho_1 < 1$. Since $n > D/\varepsilon$ and $\overline{Y}_n \in V$ imply $N \leq n$, it follows that $P(N > n) < c_1\rho_1{}^n$ for all $n > D/\varepsilon$ and (3.6.6) follows.

Now, for case 2: $\Psi(\xi) = 0$, the proof is somewhat long and involved although not difficult. Hence, it is not presented here. It is based on an idea of Stein (1946).

Example 3.6.7 Let $X_1, X_2, \ldots$ be iid normal with mean μ and variance σ^2, both unknown. We wish to test $H_0: \sigma = \sigma_0$ against $H_1: \sigma = \sigma_1$ where the σ_i are given and distinct (thus, μ is a nuisance parameter). Under the transformations $X_i \to X_i + c$ $(i = 1, 2, \ldots)$, $\mu \to \mu + c$, $\sigma \to \sigma$, $-\infty < c < \infty$, the problem is invariant and one can easily obtain [see, for instance, Wijsman (1972, Eq. (4.1))] that

$$S_n = [(2\sigma_0{}^2)^{-1} - (2\sigma_1{}^2)^{-1}] \sum_{i=1}^{n} (X_i - \overline{X}_n)^2 + (n-1)\ln(\sigma_0/\sigma_1) \tag{3.6.15}$$

in which $\overline{X}_n = (1/n)\sum_1^n X_i$. For the purpose of exponential boundedness, we may multiply S_n by any nonzero constant. We may then take that

$$S_n = \sum_{i=1}^{n} (X_i - \overline{X}_n)^2 - (n-1)\,d^2 \tag{3.6.16}$$

where

$$d^2 = (\ln \sigma_1 - \ln \sigma_0)/[(2\sigma_0{}^2)^{-1} - (2\sigma_1{}^2)^{-1}]. \tag{3.6.17}$$

Now, let the common distribution of the X_i be F and try to establish (3.6.6) with N defined by (3.6.7).

Case 1 (*F unbounded*) We shall establish (3.6.6) for every $-\infty < b < a < \infty$. This will be accomplished by showing that there exists $\varepsilon > 0$ such that for all n

$$P(S_{n+1} > a | X_1, \ldots, X_n, b < S_n < a) \geq \varepsilon. \tag{3.6.18}$$

An elementary computation based on (3.6.16) yields

$$S_{n+1} - S_n = [n/(n+1)](X_{n+1} - \overline{X}_n)^2 - d^2. \tag{3.6.19}$$

Now, set $s = a - b$. Then, given $b < S_n < a$ the event $S_{n+1} > a$ is implied by $S_{n+1} - S_n > s$, that is, using (3.6.19), $(X_{n+1} - \overline{X}_n)^2 > [(n+1)/n](d^2 + s)$. This in turn is implied by $|X_{n+1} - \overline{X}_n| > c$ where $c^2 = 2(d^2 + s)$. Hence, the left-hand side of (3.6.18) is greater than or equal to

$$P(|X_{n+1} - \overline{X}_n| > c | X_1, \ldots, X_n, b < S_n < a). \tag{3.6.20}$$

Thus, it suffices to find a positive lower bound for (3.6.20). Since X_{n+1} is independent of $(X_1, \ldots, X_n)$, the value of (3.6.20) depends on the conditioning only through $\overline{X}_n$. Furthermore,

$$P(|X_{n+1} - \overline{X}_n| > c | \overline{X}_n = x) = P(|X_{n+1} - x| > c) = P(|X_1 - x| > c).$$

We shall show that there exists $\varepsilon > 0$ such that

$$P(|X_1 - x| > c) \geq \varepsilon \qquad \text{for all} \quad -\infty < x < \infty. \tag{3.6.21}$$

Since F is unbounded, there is a number x_0 such that

$$P(X_1 < x_0 - c) = \varepsilon_1 > 0 \qquad \text{and} \qquad P(X_1 > x_0 + c) = \varepsilon_2 > 0.$$

Since $P(X_1 < x - c)$ is a nondecreasing function of x, $P(|X_1 - x| > c) \geq \varepsilon_1$ if $x > x_0$. Similarly, $P(|X_1 - x| > c) \geq \varepsilon_2$ if $x \leq x_0$. Then taking $\varepsilon = \min(\varepsilon_1, \varepsilon_2)$ gives (3.6.21).

Case 2 (*F bounded*) In this case, Theorem 3.6.3 will be used. The function $s(\cdot)$ in (3.6.8) is chosen: $s(x) = (x^2, x)'$, $-\infty < x < \infty$ (thus $k = 2$) so that $Y_i = (X_i^2, X_i)'$. Now choose the function Ψ on R_2 as follows:

$$\Psi(x_1, x_2) = x_1 - x_1^2 - d^2 \tag{3.6.22}$$

where d^2 is given by (3.6.17). Then,

$$n\Psi(\bar{Y}_n) = \sum_1^n (X_i - \bar{X}_n)^2 - nd^2, \tag{3.6.23}$$

and comparing this with (3.6.16) we note that Assumption A(ii) is satisfied, with $B_1 = d^2$, no matter what neighborhood V is chosen. The boundedness of F guarantees the validity of Assumption B. For convenience, we shall write $E(X_1 | F) = \mu$ and $E(X_1^2 | F) = \sigma^2 + \mu^2$. Then $\xi = E(Y_1 | F) = (\sigma^2 + \mu^2, \mu)'$. From the form of S_n given by (3.6.17), it is obvious that all distributions F obtained from a single one by translation produce the same stochastic behavior of $\{S_n\}$. Hence, it suffices to assume that $\mu = 0$ so that $\xi = (\sigma^2, 0)'$. Substituting ξ for x in (3.6.22) we obtain $\Psi(\xi) = \sigma^2 - d^2$ so that $\Psi(\xi) \neq 0$ provided $\sigma \neq d$. For any such F, we can conclude by Theorem 3.6.3 that (3.6.6) holds.

Now, suppose that F is such that $\sigma = d$ so that $\Psi(\xi) = 0$. In order to conclude (3.6.6) we need Assumption A(iii). From (3.6.22) we compute grad Ψ = $(1, -2x_2)'$ so that $\Delta = (1, 0)'$ and $\Delta'(Y_1 - \xi) = X_1^2 - \sigma^2$. Hence, if $P(X_1^2 - d^2 = 0) < 1$, Assumption A(iii) is satisfied and we can conclude (3.6.6). On the other hand, if $P(X_1^2 - d^2 = 0) = 1$, then in order that μ be equal to zero we must have $P(X_1 = \pm d) = \frac{1}{2}$. In this case it was shown by Wijsman (1972, Section 4) (for $d = 1$, but the extension to arbitrary d is trivial) that (3.6.6) fails for sufficiently wide stopping bounds. On the other hand, (3.6.10) is still valid. Thus, in summary, N is exponentially bounded for every F except when

$$P(X_1 = \mu \pm d) = \tfrac{1}{2} \qquad \text{for some} \quad -\infty < \mu < \infty \tag{3.6.24}$$

with d given by (3.6.17). The distributions defined by (3.6.24) are obstructive, but for every such F we still have $P(N < \infty) = 1$. For Example 3.6.7, B. K. Ghosh (1970) points out that the z_i are iid and hence all the properties of an SPRT are valid.

Example 3.6.8 Let $X_1, X_2, \ldots,$ be iid normal with mean μ and variance σ^2, both unknown. Set $\gamma = \mu/\sigma$ and test $H_0\colon \gamma = \gamma_0$ against $H_1\colon \gamma = \gamma_1$ where γ_0 and γ_1 are any two distinct finite numbers. The problem is invariant under the transformations $X_i \to cX_i$ $(i = 1, 2, \ldots)$, $\mu \to \mu c$, $\sigma \to \sigma c$, $c > 0$. Thus, the group G of invariant transformations consists of the positive reals c under multiplication. As the invariant measure on G, we shall take $\nu_G(dg) =$

dc/c. Set $\theta = (\mu, \sigma)$. Then the orbit of θ under G is $G\theta = \{g\theta;\ g \in G\} = \{(c\mu, c\sigma) : c > 0\}$. Taking in particular, $\theta_j = (\gamma_{j,1}), j = 0, 1$ the two orbits $G\theta_j$ are the two composite hypotheses that are to be tested. From Wijsman (1972, Eq. (2.1)) we can represent the probability ratio at the nth stage as

$$S_n = J_n(\theta_1)/J_n(\theta_0) \tag{3.6.25}$$

where

$$J_n(\theta) = \int \prod_{i=1}^{n} f(X_i, g\theta)\, v_G(dg) \tag{3.6.26}$$

and $f(x; \theta)$ is the density of X_1 with respect to Lebesgue measure and is given by

$$f(x; \theta) = (2\pi\sigma^2)^{-1/2} \exp[-(x - \mu)^2/2\sigma^2]. \tag{3.6.27}$$

It is seen that (3.6.26) depends on the X_i only through $\sum_1^n X_i^2$ and $\sum_1^n X_i$. Thus, in (3.6.8) we shall take s as in Example 3.6.7: $s(x) = (x^2, x)'$ so that $Y_1 = (X_1{}^2,\ X_1)'$. After setting $g\theta = (\gamma c,\ c)$, $v_G(dg) = dc/c$ and making a change of variable of integration: $c = 1/t$, we can write (3.6.26) as

$$J_n(\theta) = \int_0^\infty \exp[nr(t, \overline{Y}_n; \gamma)]t^{-1}\, dt \tag{3.6.28}$$

where

$$r(t, x; \gamma) = \tfrac{1}{2}x_1 t^2 + \gamma x_2 t + \ln t - \tfrac{1}{2}\gamma^2 - \tfrac{1}{2}\ln 2\pi \tag{3.6.29}$$

and $x = (x_1, x_2)$. For fixed γ, $x_1 > 0$, and x_2, r as a function of t has a unique maximum at t_{m} where

$$t_{\mathrm{m}} = x_1^{-1/2}\alpha(\gamma x_2 x_1^{-1/2}) \tag{3.6.30}$$

where

$$\alpha(u) = \tfrac{1}{2}[u + (u^2 + 4)^{1/2}], \qquad -\infty < u < \infty. \tag{3.6.31}$$

Set

$$\begin{aligned} \xi(x; \gamma) &= \max_{t>0} r(t, x; \gamma) = r(t_{\mathrm{m}} x; \gamma) \\ &= \beta(\gamma x_2 x_1^{-1/2}) - \tfrac{1}{2}\{\ln x_1 + \gamma^2 + \ln(2\pi) + 1\} \end{aligned} \tag{3.6.32}$$

where

$$\beta(u) = \tfrac{1}{2}u\alpha(u) + \ln \alpha(u). \tag{3.6.33}$$

It can be shown [which is nontrivial, see Wijsman (1971) for the details] that

$$(2\pi)^{-1/2}n^{1/2}(x_1 + t_m^{-2})^{1/2}\, t_m\, e^{-n\zeta(x;\gamma)} \int_0^\infty e^{nr(t,x;\gamma)}t^{-1}\, dt \to 1 \qquad \text{as} \quad n \to \infty. \tag{3.6.34}$$

Thus, S_n is equivalent to

$$\zeta(x;\gamma_1) - \zeta(x;\gamma_0) = \beta(\gamma_1 x_2 x_1^{-1/2}) - \beta(\gamma_0 x_2 x_1^{-1/2}) - \tfrac{1}{2}(\gamma_1^2 - \gamma_0^2). \tag{3.6.35}$$

Now one can show [see Wijsman (1971) for the details] that Assumptions A and B of Theorem 3.6.3 are met.

Thus, one can conclude that in the sequential t-test, when X_1 has distribution F, the stopping time N is exponentially bounded if X_1^2 has a finite moment generating function (mgf) and if F is not one of the two point distributions defined by

$$\begin{aligned} P\{X_1 &= (\sigma^2 + \mu^2)^{1/2}\mu^{-1}((\sigma^2 + \mu^2)^{1/2} \pm \sigma)\} \\ &= \tfrac{1}{2}[1 \mp \sigma(\sigma^2 + \mu^2)^{-1/2}], \qquad \sigma > 0, \quad \mu \neq 0. \end{aligned} \tag{3.6.36}$$

If we do not require X_1^2 to have a finite mgf but only a finite expectation, then $P(N < \infty) = 1$ if F does not satisfy (3.6.36). Berk (1970), by a different method obtained the same conclusions except that his assumptions necessitate the exclusion of all two-point distributions rather than only the proper subclass defined by (3.6.36), which defines a one-parameter family of distributions obtained from a single one by scale transformations. Wijsman (1971) conjectures that the distributions (one-parameter family) given by (3.6.36) are indeed obstructive for the sequential t-test and that they are the only ones (the latter would be proved if it can be shown that any unbounded F is not obstructive, as in Example 3.6.7).

Wijsman (1971) remarks that other classical invariant SPRTs such as the sequential F-test etc. are amenable to the same treatment as afforded the sequential t-test.

As pointed out in Eq. (3.6.4) an invariantly sufficient sequence $\{V_n\}$ is obtained by first making a maximal invariance reduction on R_n for each n, and then making a sufficiency reduction for each n on the sample space of the maximal invariant statistic. However, in practical problems the determination of $f_{V_n}(u; \lambda)$ may be quite cumbersome. In many cases, it may be much easier to first make a sufficiency reduction on R_n and then make a maximal invariance reduction on the sample space of the sufficient statistic. The question is under what conditions the alternative method is equivalent to the first method.

Definition 3.6.5 Let $T_n = T_n(\mathbf{X}_n)$ be a sufficient sequence for the given family of distributions $\{F_n(\mathbf{x}_n; \theta); \theta \in \Theta\}$ and let $F_{T_n}(t; \theta)$ denote the distribution function of T_n (which could be vector valued). The group G under which the given family remains invariant, is said to induce a group G_t of (one-to-one) transformations on the sample space of T_n if for each $g \in G$, $T_n(\mathbf{x}_n) = T_n(\mathbf{x}_n')$ implies that $T_n(g\mathbf{x}_n) = T_n(g\mathbf{x}_n')$. In other words, for each $g \in G$, there exists a transformation $g_t \in G_t$ on the sample space of T_n such that $g_t t = T_n(g\mathbf{x}_n)$ for all $\mathbf{x}_n \in R_n$ satisfying $T_n(\mathbf{x}_n) = t$.

Notice that g_t may depend on n. Henceforth assume that there exists an induced group G_t. Let $U_n = U_n(T_n)$ be a maximal invariant under G_t on the sample space of T_n. As a function of $\mathbf{x}_n$ on R_n we shall write

$$U_n(T_n(\mathbf{x}_n)) = V_n(x_n) \qquad \text{for each } n \text{ and all } \mathbf{x}_n \in R_n.$$

From the definition of G_t it is clear that V_n is invariant on R_n under G for each n and that U_n and V_n are identically distributed as $F_{V_n}(v; \lambda)$ where $\lambda = \lambda(\theta)$ is a maximal invariant on Θ under $\bar{G}$. The reduction of the data $\mathbf{x}_n$ to the sufficient data T_n and thence to the invariant data V_n for the sufficient data can be symbolically expressed as

$$\{F_n(\mathbf{x}_n; \theta), \theta \in \Theta\} \xrightarrow{\text{suff}} \{F_{T_n}(t; \theta), \theta \in \Theta\} \xrightarrow{G_t} \{F_{V_n}(v; \lambda), \lambda \in \Lambda\}. \tag{3.6.37}$$

The following theorems show that under certain conditions V_n is indeed an invariantly sufficient statistic under G.

Theorem 3.6.4 (Stein) Let $T_n = T_n(\mathbf{X}_n)$ be a k-dimensional sufficient statistic for the family $\{F_n(\mathbf{x}_n; \theta), \theta \in \Theta\}$, and let $U_n(T_n)$ be a maximal invariant under G_t, the group induced on the sample space of T_n by G which leaves the given family invariant. If T_n is a discrete random variable, then $V_n = V_n(\mathbf{X}_n) = U_n(T_n(\mathbf{X}_n))$ is invariantly sufficient for the given family under G.

Proof Since T_n is sufficient $P(\mathbf{X}_n \in A \mid T_n = t)$ does not involve θ for any $A \in R_n$ and for any t that satisfies $P(T_n = t \mid \theta) > 0$. Hence, if A is any invariant set under G, we have, for each $g \in G$, $\theta \in \Theta$ and t satisfying $P(T_n = t \mid \theta) > 0$,

$$\begin{aligned} P(\mathbf{X}_n \in A \mid T_n = t) &= \frac{P(\mathbf{X}_n \in A, T_n = t \mid \theta)}{P(T_n = t \mid \theta)} \\ &= \frac{P(\mathbf{X}_n \in gA, T_n = g_t t \mid \bar{g}\theta)}{P(T_n = g_t t \mid \bar{g}\theta)} \\ &= P(\mathbf{X}_n \in gA \mid T_n = g_t t, \bar{g}\theta) \\ &= P(\mathbf{X}_n \in A \mid T_n = g_t t). \end{aligned} \tag{3.6.38}$$

Now,

$$P(\mathbf{X}_n \in A \mid V_n = v, \theta) = \sum_t P(T_n = t \mid V_n = v, \theta) P(\mathbf{X}_n \in A \mid T_n = t, V_n = v),$$

the summation extending over those values of t that satisfy $V_n(t) = v$. Since $U_n(T_n)$ is a maximal invariant, the values of t that go into the sum must be of the form $g_t t_0$ where t_0 is fixed but g_t varies in G_t. Hence, using (3.6.38) in the second step,

$$\begin{aligned} P(\mathbf{X}_n \in A \mid V_n = v, \theta) &= \sum_{g_t} P(T_n = g_t t_0 \mid V_n = v, \theta) P(\mathbf{X}_n \in A \mid T_n = g_t t_0) \\ &= P(\mathbf{X}_n \in A \mid T_n = t_0) \sum_{g_t} P(T_n = g_t t_0 \mid V_n = v, \theta) \\ &= P(\mathbf{X}_n \in A \mid T_n = t_0) \end{aligned}$$

which does not involve θ. This completes the proof.

If T_n is continuous or of mixed type and (3.6.38) holds, then the assertion of Theorem 3.6.4 follows because

$$\begin{aligned} P(\mathbf{X}_n \in A \mid V_n = v, \theta) &= E\{P(\mathbf{X}_n \in A \mid T_n = t_0) \mid V_n = v; \theta\} \\ &= P(\mathbf{X}_n \in A \mid T_n = t_0) \end{aligned}$$

where the expectation is with respect to T_n conditional upon $U_n(T_n) = v$.

Hall *et al.* (1965) have extended Theorem 3.6.4 to the case when T_n is not discrete. For the statement of their result see B. K. Ghosh (1970, Theorem 6.2, p. 297). In the following we shall state one of the assumptions which is easy to verify in nonparametric problems.

Theorem 3.6.5 (Hall *et al.*, 1965) Theorem 3.6.4 holds if T_n is not discrete provided, for every $g \in G$ there exists an invariant conditional probability,

$P(\mathbf{X}_n \in gA \mid T_n = t) = P(\mathbf{X}_n \in gA \mid T_n = g_t t)$, for every invariant set $A \in R_n$ and for all t.

It is worthwhile to note that if Theorem 3.6.4 holds for each n, the transitivity of $\{T_n\}$ for $\{F_n(\mathbf{x}_n; \theta), \theta \in \Theta\}$ implies the transitivity of $\{V_n\}$ for $\{F_{V_n}(v; \lambda), \lambda \in \Lambda\}$. Consequently any invariant procedure for λ is essentially based only on V_n at stage n provided

$$P(T_{n+1} \leq t \mid \mathbf{X}_n = \mathbf{x}_n, \theta) = P(T_{n+1} \leq t \mid T_n = t_n(\mathbf{x}_n), \theta) \qquad (3.6.39)$$

for each n, θ, $\mathbf{x}_n$, and t. It is much easier to verify (3.6.39) than (3.6.5). One will not use the Stein's theorem if $f_{V_n}(v; \lambda)$ can be obtained. Wijsman (1967a) gave a general method of determining $f_{V_n}(v; \lambda)$ from $f_n(\mathbf{x}_n; \theta)$ or $f_{V_n}(v; \lambda)$ from $f_{T_n}(t; \theta)$. The following examples illustrate the alternative route.

Example 3.6.9 Suppose that $\{X_i\}$ are iid normal with mean μ and variance σ^2. Let $\theta = (\mu, \sigma)$ be unknown. Let G be the group of translations on R_n which induces $\bar{G}$ on Θ such that $\bar{g}\theta = (\mu + c, \sigma)$. It follows that $T_n(\mathbf{X}_n) = (\bar{X}_n, s_n^2)$, where $\bar{X}_n = \sum_1^n X_i/n$ and $(n-1)s_n^2 = \sum_1^n (X_i - \bar{X}_n)^2$, is sufficient for θ for each $n \geq 1$. The group G induces G_t on $(-\infty < \bar{x}_n < \infty,\ s_n^2 \geq 0)$ for each $n \geq 1$ such that $g_t t = (\bar{x}_n + c, s_n^2)$, $-\infty < c < \infty$, and a maximal invariant under G_t is $U_n(T_n) = (n-1)s_n^2$ for $n > 1$. Then

$$V_n(\mathbf{X}_n) = U_n(T_n(\mathbf{X}_n)) = \sum_1^n (X_i - \bar{X}_n)^2$$

is invariant on R_n under G for each $n > 1$. If $n = 2$, one directly verifies that $V_2(\mathbf{X}_2) = (X_1 - X_2)^2$ is sufficient for the pdf of the maximal invariant (on R_2 under G) $U_2(\mathbf{X}_2) = X_1 - X_2$. For each $n > 2$, Assumption III of Theorem 6.2 in B. K. Ghosh (1970) [which is due to Hall *et al.* (1965)] holds. Hence $\{(n-1)s_n^2\}$ is an invariantly sufficient sequence for the given family. Since $(n-1)s_n^2/\sigma^2$ is distributed as chi-square with $n-1$ degrees of freedom for each $n > 1$, the pdf $f_{V_n}(v; \sigma)$ can be readily obtained, and the invariant SPRT of $\sigma = \sigma_0$ against $\sigma = \sigma_1$ is given by: at the nth stage $(n = 1, 2, \ldots)$ take one more observation if

$$\ln B < S_n < \ln A, \qquad S_n = (n-1)\ln(\sigma_1/\sigma_0) + \tfrac{1}{2}(\sigma_0^{-2} - \sigma_1^{-2})V_n.$$

Also, since $\bar{x}_{n+1} = (n\bar{x}_n + x_{n+1})/(n+1)$ and

$$s_{n+1}^2 = \frac{n-1}{n}\left[s_n^2 + \frac{n(\bar{x}_n - x_{n+1})^2}{n^2 - 1}\right],$$

it can be shown that $\{(n-1)s_n^2\}$ satisfies (3.6.39). Thus, $\{(n-1)s_n^2\}$ is also a transitive sequence for $[F_{V_n}(v; \sigma), \sigma > 0]$.

Example 3.6.10 Let the joint pdf of $(X_1, \ldots, X_n)$ be for each $n > 1$

$$f_n(x_1, \ldots, x_n; \mu, \delta) = \mu^{n\delta}(x_1 \cdots x_n)^{\delta-1} \exp[-(\mu x_n)^\delta/\delta]$$

for $0 < x_1 < \cdots < x_n$, where $\theta = (\mu, \delta)$ is unknown and $\Theta = \{\mu > 0, \delta > 0\}$. The model arises when events occur randomly in time x at a "rate" $r(x) = \mu^\delta x^{\delta-1}$, and X_i denotes the random instant of the ith occurrence. The given family remains invariant under the group G of positive scale changes, and $g\theta = (\mu/c, \delta)$, $c > 0$. It follows that $T_n(\mathbf{X}_n) = (X_1, \ldots X_n, X_n{}^n)$ is sufficient for θ for each $n \geq 1$. The group G induces G_t on $(0 < x_1 \cdots < x_n \leq x_n{}^n)$ such that $g_t t = (c^n x_1 \cdots x_n, c^n x^n)$, $c > 0$ and a maximal invariant under G_t is $U_n(T_n) = X_1 \cdots X_n/X_n{}^n$ for each $n > 1$. Then $V_n(\mathbf{X}_n) = X_1 \cdots X_n/X_n{}^n$ is invariant on R_n under G. For $n = 2$, $V_2(\mathbf{X}_2) = X_1/X_2$ is obviously sufficient for the pdf of the maximal invariant (on R_2 under G) statistic $V_2(\mathbf{X}_2) = X_1/X_2$. For each $n > 2$, Assumption III of Theorem 6.2 of B. K. Ghosh (1970) holds. Hence $\{X_1 \cdots X_n/X_n{}^n\}$ is an invariantly sufficient sequence for $f_n(x_1, \cdots, x_n; \mu, \delta)$ and it can be shown that $-2\delta \ln V_n(\mathbf{X}_n)$ is distributed as chi-square with $2(n-1)$ degrees of freedom for each $n > 1$. So, the invariant SPRT for $\delta = \delta_0$ against $\delta = \delta_1$ is given by: at the nth stage take one more observation if

$$\ln B < (n-1) \ln (\delta_1/\delta_0) + (\delta_1 - \delta_0) \ln V_n < \ln A, \qquad n = 1, 2, \ldots.$$

It should be pointed out that the invariant SPRT, and more generally, the principle of invariance do not apply to many interesting decision problems [especially when $\{X_i\}$ are discrete] about a real parameter in the presence of one or more nuisance parameters. For instance, let

$$f(x, y; p_x, p_y) = p_x{}^x(1-p_x)^{1-x} p_y{}^y(1-p_y)^{1-y}, \qquad x, y = 0, 1$$

and let $\psi(p_x, p_y) = \ln[p_x(1-p_y)/p_y(1-p_x)]$. Then there do not exist any transformations that leave f and ψ invariant, so that invariance considerations do not apply in testing hypotheses about ψ.

B. K. Ghosh (1970) has given several applications of the invariant SPRT namely, the sequential t-test, the sequential F-test, the sequential chi-square test, the sequential (Hotelling's) T^2-test, etc. He employs Stein's theorem (Theorem 3.6.4) and also Theorem 2.12.1 for establishing finite termination of the SPRTs with probability one. The reader is referred to B. K. Ghosh (1970, pp. 300–324).

3.6.4 Asymptotic Properties of Invariant SPRTs

Some of the asymptotic properties of the SPRTs have been discussed in Section 2.13. Here we shall give analogous results for the invariant SPRTs.

Let H_0 and H_1 be the two invariant composite hypotheses about an iid data sequence $X, X_1, X_2, \ldots$ both generated by the same group G. That is G is a group of one-to-one bimeasurable transformations acting on range X and under H_i, the distribution of X belongs to $\mathscr{P}_i = GP_i$, $i = 0, 1$. (P_i can be any distribution in $\mathscr{P}_i$.) For further elaboration, $\mathscr{I}_n$ denotes the G-invariant subsets of $\mathscr{F}_n = \mathscr{B}(X_1, \ldots, X_n)$, namely the Borel field induced by $X_1, \ldots, X_n$. (G acts coordinatewise on each X_i.) For $P_i \in \mathscr{P}_i$, let P_{in} denote the induced product measure on $\mathscr{F}_n$ and Q_{in} denote the restriction of P_{in} to $\mathscr{I}_n$. (Note that because G generates $\mathscr{P}_i$, every distribution in $\mathscr{P}_i$ gives the same Q_{in}.) We assume that there exists P_i in $\mathscr{P}_i$, $i = 0, 1$ so that $P_{1n} = P_{0n}$ for all n. Then the likelihood ratio under H_0 and H_1 of the G-maximal invariant for $(X_1, \ldots, X_n)$ is

$$\Lambda_n = dQ_{1n}/dQ_{0n} = E_0(dP_{1n}/dP_{0n} \mid \mathscr{I}_n).$$

Since $\{\Lambda_n\}$ is a sequence of likelihood ratios on increasing σ-fields ($\mathscr{I}_n \subset \mathscr{I}_{n+1}$) an SPRT of H_0 against H_1 can be based on $\{\Lambda_n\}$. Letting $S_n = \ln \Lambda_n$, as pointed out earlier, one can use Wald's SPRT. Wald's inequalities (2.13.1) for the error rates remain valid. Let P denote the actual distribution of X which need not be in either hypothesis. The following result does not use the invariance structure; thus it applies to any SPRT satisfying the hypotheses.

Theorem 3.6.6 (Berk, 1973) Suppose that wp1, $S_n/n \to \mu \in (0, \infty]$. Then

(i) $\lim_c I_{S_N \geq a} = \lim_c P(S_N \geq a) = 1$,
$\lim_c N/a = 1/\mu \qquad (\lim_c = \lim_{c \to \infty})$

where (b, a) are the boundaries of the SPRT and $c = \min(-b, a)$. If also, for some $\delta \in (0, \mu)$, the "large-deviation" probabilities $p_n = P(S_n/n < \delta)$ satisfy $\lim_n np_n = 0$ and $\sum_n p_n < \infty$, then also

(ii) $\lim_c EN/a = 1/\mu$.

Analogous statements hold if $S_n/n \to \mu < 0$.

Proof The argument in the proof of Theorem 2.13.1 carries over verbatim to establish (i). To prove (ii), it then suffices to show that N/a is uniformly integrable. For this, we need only show that

$$\sup_{c>1}\{nP(N > na) + \sum_{k>n} P(N > ka)\} = o(1) \quad \text{as} \quad n \to \infty. \tag{3.6.40}$$

It is sufficient to establish (3.6.40) when a ranges in the positive integers [since N does not exceed the stopping time obtained by replacing a with $\{a\} = \inf\{n : n \geq a\}$]. For any integer $n > 1/\delta$, letting $r = na$, we have

$$P(N > na) \leq P(S_r < a) = P(S_r/r < 1/n) \leq p_r. \tag{3.6.41}$$

It follows from the hypothesis that $d_n = \sup_{k \geq n} kp_k \downarrow 0$ as $n \to \infty$. Thus (3.6.41) entails

$$nP(N > na) \leq np_r \leq d_n \tag{3.6.42}$$

and also that

$$\sum_{k>n} P(N > ka) \leq \sum_{k>r} p_k \leq \sum_{k>n} p_k. \tag{3.6.43}$$

Now (3.6.42) and (3.6.43) entail (3.6.40). Berk (1973) has also obtained the following results which we state without proofs.

Theorem 3.6.7 (Berk, 1973) Suppose that for $i = 0, 1$, $P_i(S_n/n \to \mu_i) = 1$ where $-\infty < \mu_0 < 0 < \mu_1 < \infty$ and that for some $\delta_i \in (0, |\mu_i|)$, $p_n = P_0(S_n/n > -\delta_0)$, $+ P_1(S_n/n < \delta_1)$ satisfies $\lim_n np_n = 0$ and $\sum_n p_n < \infty$. Suppose also that (P_0, P_1) can be chosen so that $E_i|Z| < \infty, i = 0, 1$ where $Z = dP_1/dP_0(X)$ and that invariance and almost invariance are equivalent to $\mathscr{P}_i, i = 1, 2$. Then

$$\lim_c (-b)^{-1} \ln \beta^{-1} = 1 = \lim_c a^{-1} \ln \alpha^{-1}.$$

[Conditions for the equivalence of invariance are given by Berk and Bickel (1968).]

Theorem 3.6.8 (Berk, 1973) If $P(S_n/n \to 0) = 1$, then wp1

$$\lim_c N/c = \lim_c EN/c = \infty.$$

Proof We have $|S_N| \geq c$. The result follows upon noting that $P(\lim_c S_N/N = 0) = 1$. Hence $P(\lim_c N/c = \infty) = 1$ and then, by Fatou's lemma, $\lim_c EN/c = \infty$.

The preceding results provide an asymptotic justification of Bhate's conjecture for invariant sequential tests, which is as follows. By neglecting the overshoot, we have the approximate equality

$$E_0 S_N = (1 - \alpha)b + \alpha a. \tag{3.6.44}$$

Letting $\lambda_0(n) = E_0 S_n$, Bhate (1955) [see also Wetherill (1966, p. 69)] conjectured that a reasonable approximation is given by

$$E_0 S_N \doteq \lambda_0(E_0 N), \tag{3.6.45}$$

where $\lambda_0(\cdot)$ is supposed extended to the real line in a convenient manner. This yields an approximation n_0 to $E_0 N$, obtained by equating the right-hand expressions in (3.6.44) and (3.6.45).

Theorem 3.6.9 (Berk, 1973) Under the regularity conditions of Theorem 3.6.7, Bhate's (1955) approximation is asymptotically correct, that is, $\lim_c n_i/E_i N = 1$, $i = 0, 1$.

Proof We provide the proof for n_0. First we show that $\lim_n \lambda_0(n)/n = \lim_n E_0 S_n/n = \mu_0 < 0$. Since $P_0(S_n/n \to \mu_0) = 1$, we see that $P_0(S_n^+/n \to 0) = 1$ and $P_0(S_n^-/n \to -\mu_0) = 1$. We note that for $x > 0$, $P_0(S_n^+ > x) = P_0(\Lambda_n > e^x) \le e^{-x}$ since $E_0 \Lambda_n = 1$. Thus $\{S_n^+\}$ is uniformly integrable, and, a fortiori, so is $\{S_n^+/n\}$; thus $E_0 S_n^+/n \to 0$. It follows from the proof of Theorem 3.6.6 that $\{S_n^-/n\}$ is also uniformly integrable. Thus $E_0(S_n^-/n) \to -\mu_0$ and hence $\lim_n \lambda_0(n)/n = \mu_0$.

The equation for n_0 is

$$\lambda_0(n_0) = b[1 + o(1)]. \tag{3.6.46}$$

Thus, $\lim_c n_0 = \infty$ and consequently

$$\lim_c \lambda_0(n_0)/n_0 = \mu_0. \tag{3.6.47}$$

From (3.6.46) we see that $\lim_c \lambda_0(n_0)/b = 1$ and hence by Theorem 3.6.5 and (3.6.47) we have

$$\lim_c n_0/b = 1/\mu_0 = \lim_c E_0 N/b. \tag{3.6.48}$$

It follows that $\lim_c n_0/E_0 N = 1$.

REMARK 3.6.9.1 Since $\lambda_0(n) \sim n\mu_0$, a modified Bhate's (1955) approximation would replace $\lambda(E_0 N)$ by $\mu_0 E_0 N$ in (3.6.45). Similarly the right-hand side of (3.6.44) can be replaced by the no-overshoot approximation

$$(1 - \alpha) \ln[\beta/(1 - \alpha)] + \alpha \ln[(1 - \beta)/\alpha]$$

with (α, β) then being replaced by appropriate asymptotic expressions; for example (e^{-a}, e^b).

Since the approximations obtained for $E_i N$ have only an asymptotic justification, it seems simplest to use the asymptotic expression given by Theorem 3.6.5: $E_0 N = b/\mu_0$ and $E_1 N = a/\mu_1$. However, one has to determine μ_i for this, although the necessity of inverting $\lambda_i(\cdot)$ is avoided.

REMARK 3.6.9.2 The proof of Theorem 3.6.7 shows that $\lim_c E_0 S_N/b = 1$. Since $\lim_c \lambda_0(E_0 N)/E_0(N) = \mu_0$, it follows from (3.6.48) that

$$\lim_c E_0 S_N/\lambda_0(E_0 N) = 1.$$

That is, Bhate's (1955) conjecture (3.6.45) is asymptotically correct.

Problems

1. Let $\{X_i\}$ be independent and identically distributed normal random variables having unknown μ and variance σ^2. $\lambda = \mu/\sigma$ is a maximal invariant on $\Theta = \{(\mu, \sigma): -\infty < \mu < \infty, \sigma > 0\}$. Set up the sequential invariant test of $H_0: \lambda = 0$ against $H_1: \lambda = 2$. Let the data be .33, .85, .92, .46, 1.44, .31, .20. Carry out the test using $\alpha = \beta = .1$.

2. Let $X_1, \ldots, X_n$ denote a random sample of size n from a normal population with unknown mean μ and variance σ^2. Let $\bar{X}_n$ and s_n^2 denote the sample mean and sample variance based on n observations. Also let $t_{n-1} = (n(n-1))^{1/2}\bar{X}_n/s_n$. Let $\delta = n^{1/2}\mu/\sigma$. If $p_\delta(t_1, \ldots, t_{n-1})$ denotes the joint density of $t_1, \ldots, t_{n-1}$ show that $p_{\delta_1}(t_1, \ldots, t_{n-1})/p_{\delta_0}(t_1, \ldots, t_{n-1}) = p_{\delta_1}(t_{n-1})/p_{\delta_0}(t_{n-1})$ where $p_\delta(t_{n-1})$ is the density of the noncentral t-distribution. (That is, the sequence $\{t_n\}$ is transitive.)

3. Let (X_i, Y_i) $i = 1, 2, \ldots$ denote a sequence of independent and identically distributed random variables where X_i and Y_i are mutually independent and X_i has density $\sigma^{-1} \exp[-(x-\xi)/\sigma]$ for $x \geq \xi$ and Y_i has density $\tau^{-1} \exp[-(y-\eta)/\tau]$ for $y \geq \eta$. Set up an invariant SPRT for $H_0: \tau/\sigma = \Delta_0$ against $H_1: \tau/\sigma = \Delta_1$ $(\Delta_1 > \Delta_0 > 0)$.

4. Let (X_i, Y_i) $i = 1, 2, \ldots$ denote a sequence of independent and identically distributed random variables having a bivariate normal distribution with unknown means, variances and correlation coefficient. Develop an invariant SPRT for testing H_0: $\rho = \rho_0$ against H_1: $\rho = \rho_1$ with $\rho_1 > \rho_0$ where ρ denotes the correlation coefficient between X_i and Y_i. (*Hint*: The sample correlation coefficient is invariantly sufficient for the bivariate normal family of distributions.)

3.7 Sequential Likelihood Ratio Test Procedures

3.7.1 Maximum Likelihood SPRT's

If the method of weight functions of Wald (1947) is not feasible and if Fraser sufficiency and invariance do not apply, then one should look for a procedure in which sample estimates replace the true values of the nuisance parameters. Any estimates like the BAN (best asymptotically normal) would suffice. In particular, we confine ourselves to maximum likelihood estimators which have desirable large sample properties. In this section we implicitly assume that we are testing hypotheses that are close to each other and hence the need for large sample sizes for terminating the experiments. Large sample properties of maximum likelihood estimators (mle) in fixed-sample size case have been studied by Cramér (1946) and the regularity conditions have been relaxed to a large extent by LeCam (1970) whose two basic results are given in Appendix 5.

Let $f(x; \gamma, \delta)$ be the common probability density (probability function) of a sequence of iid random variables $\{X_i\}$, where $\gamma \in \Gamma$ and $\delta \in \Delta$ and let $\Theta = \Gamma \times \Delta$.

Lemma 3.7.1 For each $(\gamma, \delta) \in \Theta$, assume that

(i) $\partial \ln f/\partial\gamma$ and $\partial \ln f/\partial\delta$ exist and are absolutely bounded by some functions that are integrable with respect to the density, and

(ii) $\partial^2 \ln f/\partial\gamma^2$, $\partial^2 \ln f/\partial\gamma\, \partial\delta$, and $\partial^2 \ln f/\partial\delta^2$ exist, are absolutely bounded by some integrable functions of x, and their expectations exist.

Then, we have

$$\begin{aligned} n^{-1}\, \partial^2 \ln f(\mathbf{X}_n; \gamma, \delta)/\partial\gamma^2 &\to -I_{\gamma\gamma} && \text{in probability as } n \to \infty; \\ n^{-1}\, \partial^2 \ln f(\mathbf{X}_n; \gamma, \delta)/\partial\gamma\, \partial\delta &\to -I_{\gamma\delta} && \text{in probability as } n \to \infty; \\ n^{-1}\, \partial^2 \ln f(\mathbf{X}_n; \gamma, \delta)/\partial\delta^2 &\to -I_{\delta\delta} && \text{in probability as } n \to \infty. \end{aligned} \tag{3.7.1}$$

where $f(\mathbf{X}_n; \gamma, \delta) = \prod_{i=1}^n f(X_i; \gamma, \delta)$,

$$\begin{aligned} E(\partial \ln f/\partial\gamma) &= 0 = E(\partial \ln f/\partial\delta), \\ \operatorname{var}(\partial \ln f/\partial\gamma) &= -E(\partial^2 \ln f/\partial\gamma^2) = I_{\gamma\gamma}, \\ \operatorname{cov}(\partial \ln f/\partial\gamma, \partial \ln f/\partial\delta) &= -E(\partial^2 \ln f/\partial\gamma\, \partial\delta) = I_{\gamma\delta}, \\ \operatorname{var}(\partial \ln f/\partial\delta) &= -E(\partial^2 \ln f/\partial\delta^2) = I_{\delta\delta}. \end{aligned} \tag{3.7.2}$$

Proof Assumptions (i) and (ii) enable one to perform differentiation underneath the integral sign and assume the existence of $I_{\gamma\gamma}$, $I_{\gamma\delta}$ and $I_{\delta\delta}$. Also, $\ln f(\mathbf{X}_n; \gamma, \delta) = \sum_1^n \ln f(X_i; \gamma, \delta)$ is the sum of iid random variables since the Xs are. Hence, application of Khitchine's theorem, pertaining to the convergence of an arithmetic mean of an iid sequence of random variables to their mean, yields (3.7.1).

Lemma 3.7.2 (Cramér, 1946; LeCam, 1970) If the assumptions of Lemma

3.8.1 are satisfied and if

$$\lim_{\varepsilon \to 0} \sup_{t \le \varepsilon} E \left| \frac{\partial^2}{\partial \gamma^2} \ln f(\mathbf{X}_n; \gamma + t, \delta) - \frac{\partial^2}{\partial \gamma^2} \ln f(\mathbf{X}_n; \gamma, \delta) \right| = 0 \tag{3.7.3}$$

and analogous results hold for the mixed second derivative and the second partial derivative of $\ln f$ with respect to δ, then the maximum likelihood estimators $\hat{\gamma}_n$, $\hat{\delta}_n$ based on a random sample of size n, have an asymptotic bivariate normal distribution with means γ and δ and variances $I_{\delta\delta}/n(I_{\gamma\gamma} I_{\delta\delta} - I_{\gamma\delta}^2)$, $I_{\gamma\gamma}/n(I_{\gamma\gamma} I_{\delta\delta} - I_{\gamma\delta}^2)$ and covariance $I_{\gamma\delta}/n(I_{\gamma\gamma} I_{\delta\delta} - I_{\gamma\delta}^2)$.

Proof Use Theorem A.5.1.

REMARK 3.7.1.1 Uniform continuity of the second partial derivatives of $\log f(\mathbf{X}_n; \gamma, \delta)$ with respect to γ and δ or the existence and absolute boundedness by integrable functions in x of the third-order partial derivatives could replace the assumptions in (3.7.3).

Maximum likelihood SPRTs have been proposed by Bartlett (1946) and by Cox (1963). First we shall give Cox's procedure which is a slight modification of Bartlett's (1946) procedure. Suppose we are interested in testing $H_0: \gamma = \gamma_0$ against $H_1: \gamma = \gamma_1$. where δ is the nuisance parameter, no prior information about which is available. We further assume that $|\gamma_1 - \gamma|$ and $|\gamma_0 - \gamma|$ are of the order $n^{-1/2}$ where γ denotes the true value. Let $\hat{\gamma}_n$ and $\hat{\delta}_n$ denote the mles of γ and δ based on $(X_1, \ldots, X_n)$. Then consider the sequential procedure based on

$$Z_n' = \ln\{f(\mathbf{X}_n; \gamma_1, \hat{\delta}_n)/f(\mathbf{X}_n; \gamma_0, \hat{\delta}_n) \tag{3.7.4}$$

which is called the sequential likelihood ratio. The method of Bartlett (1946) involved the use of two mles of δ, one assuming $\gamma = \gamma_1$ and the other with $\gamma = \gamma_0$. This method will be described later. Notice that $\exp Z_n'$ is no longer a ratio of probability (density) functions, and hence the approximate boundaries in terms of error probabilities cannot be inserted without proper justification.

The Taylor's expansions for $\ln f(\mathbf{X}_n, \gamma_i; \hat{\delta}_n)$ for $i = 0, 1$ about the true (γ, δ) yields

$$\begin{aligned} Z_n' = (\gamma_1 - \gamma_0) \frac{\partial \ln f(\mathbf{X}_n; \gamma, \delta)}{\partial \gamma} + \frac{1}{2}(\gamma_1 - \gamma_0)(\gamma_1 + \gamma_0 - 2\gamma) \frac{\partial^2 \ln f(\mathbf{X}_n; \gamma, \delta)}{\partial \gamma^2} \\ + (\gamma_1 - \gamma_0)(\hat{\delta}_n - \delta) \frac{\partial^2 \ln f(X_n; \gamma, \delta)}{\partial \gamma \, \partial \delta} + R_n(\mathbf{X}_n) \end{aligned} \tag{3.7.5}$$

where R_n involves the differences of the second-order derivatives, and it converges to zero in probability when $|\gamma_i - \gamma|$, $i = 0, 1$, are sufficiently small and the second derivatives are smooth (see Remark 3.7.1.1). Next, expanding the following functions about $\hat{\gamma}_n$ and $\hat{\delta}_n$, which are the solutions of

$$\partial \ln f(\mathbf{X}_n; \hat{\gamma}_n, \hat{\delta}_n)/\partial\hat{\gamma}_n = 0 = \partial \ln f(\mathbf{X}_n; \hat{\gamma}_n, \hat{\delta}_n)/\partial\hat{\delta}_n, \tag{3.7.6}$$

we obtain

$$\begin{aligned}\partial \ln f(\mathbf{X}_n; \gamma, \delta)/\partial\gamma = &-(\hat{\gamma}_n - \gamma)\, \partial^2 \ln f(\mathbf{X}_n; \gamma, \delta)/\partial\gamma^2 \\ &-(\hat{\delta}_n - \delta)\, \partial^2 \ln f(\mathbf{X}_n; \gamma, \delta)/\partial\gamma\, \partial\delta + R_n'(\mathbf{X}_n)\end{aligned} \tag{3.7.7}$$

and

$$\begin{aligned}\partial \ln f(\mathbf{X}_n; \gamma, \delta)/\partial\delta = &-(\hat{\gamma}_n - \gamma)\, \partial^2 \ln f(\mathbf{X}_n; \gamma, \delta)/\partial\gamma\, \partial\delta \\ &-(\hat{\delta}_n - \delta)\, \partial^2 \ln f(\mathbf{X}_n; \gamma, \delta)/\partial\delta^2 + R_n''(\mathbf{X}_n)\end{aligned} \tag{3.7.8}$$

where R_n' and R_n'' converge to zero in probability and hence can be neglected for sufficiently large n. Substituting (3.7.7) in (3.7.5) and ignoring terms of order $O(1)$ we obtain

Theorem 3.7.1 (Cox, 1963) Under suitable regularity conditions, we have

$$Z_n' \approx (\gamma_1 - \gamma_0) I_{\gamma\gamma}\, n(\hat{\gamma}_n - \tfrac{1}{2}(\gamma_0 + \gamma_1)), \tag{3.7.9}$$

where $n\hat{\gamma}_n$ is asymptotically normal with mean $n\gamma$ and variance

$$nI_{\delta\delta}/(I_{\gamma\gamma} I_{\delta\delta} - I_{\gamma\delta}^2).$$

Also, using (3.7.1) in (3.7.7) and (3.7.9) we see that for large n, $n\hat{\gamma}_n$ is the sum of iid $\{Y_i\}$, $i = 1, \ldots, n$, where

$$Y_i - \gamma = [I_{\gamma\gamma} I_{\delta\delta} - I_{\gamma\delta}^2]^{-1} [I_{\delta\delta}\, \partial \ln f/\partial\gamma - I_{\gamma\delta}\, \partial \ln f(\mathbf{X}_n; \gamma, \delta)/\partial\delta]. \tag{3.7.10}$$

Now, we can use Wald's (1947) approximations for the boundary values in terms of error probabilities. In other words, the expressions for the OC function and the ASN are also valid for sufficiently large n. Thus

$$\begin{aligned}\ln B &= (1 - I_{\gamma\delta}^2/I_{\gamma\gamma} I_{\delta\delta})^{-1} \ln[\beta/(1-\alpha)] \doteq c_n \ln[\beta/(1-\alpha)], \\ \ln A &= (1 - I_{\gamma\delta}^2/I_{\gamma\gamma} I_{\delta\delta})^{-1} \ln[(1-\beta)/\alpha] \doteq c_n \ln[(1-\beta)/\alpha]\end{aligned} \tag{3.7.11}$$

where $c_n = (1 - I_{\hat{\gamma}_n, \hat{\delta}_n}^2 / I_{\hat{\gamma}_n, \hat{\gamma}_n} I_{\hat{\delta}_n, \hat{\delta}_n})^{-1}$.

It should be pointed out that the above procedure can be carried out even if $\delta = (\delta_1, \ldots, \delta_s)'$, $s > 1$; appropriate modifications should be made in the justification.

Example 3.7.1 Suppose that X_i are iid normal with mean θ and variance σ^2. We wish to test H_0: $\gamma = \theta/\sigma = \gamma_0$ against H_1: $\gamma = \gamma_1$ $(\gamma_0 < \gamma_1)$. Here $\hat{\gamma}_n = \bar{X}_n/s_n$, $\hat{\delta}_n = \hat{\sigma}_n = s_n$. Thus

$$Z_n' = n(\gamma_1 - \gamma_0)\{(\bar{X}_n/s_n) - \tfrac{1}{2}(\gamma_0 + \gamma_1)\}, \qquad n = 2, 3, \ldots,$$

and $I_{\gamma\gamma} = 1$, $I_{\sigma\sigma} = (2 + \gamma^2)/\sigma^2$, and $I_{\gamma\sigma} = \gamma/\sigma$. So we take $c_n = (1 + \bar{X}_n{}^2/2s_n{}^2)$ and carry out the procedure. It is of interest to note that the above procedure is asymptotically equivalent to the sequential t-test [to the asymptotic form of Rushton (1950), see Eq. (3.2.4)].

One can obtain sequential likelihood ratio tests in order to test H_0: $\theta = \theta_0$ against H_1: $\theta = \theta_1$ $(\theta_1 > \theta_0)$ when σ is unknown. Here, Fraser sufficiency and invariance do not apply. Further one can also derive a sequential likelihood ratio procedure for the Behrens–Fisher problem (in which we wish to test the equality of the means of two normal populations having unequal and unknown variances). For details on these two problems, the reader is referred to B. K. Ghosh (1970, pp. 334–335).

There might be some situations where it is easier to calculate the mles of δ when $\gamma = \gamma_1$ and $\gamma = \gamma_0$ than to compute the joint mles $\hat{\gamma}_n$ and $\hat{\delta}_n$. Bartlett (1946) proposed that the sequential likelihood ratio procedure be based on

$$Z_n'' = \ln[f(\mathbf{X}_n; \gamma_1, \hat{\delta}_1)/f(\mathbf{X}_n; \gamma_0, \hat{\delta}_0)] \tag{3.7.12}$$

when $\hat{\delta}_1$ $(\hat{\delta}_0)$ denotes the mle of δ when $\gamma = \gamma_1$ (γ_0). Notice that the subscript n on δ is omitted for the sake of simplicity.

Theorem 3.7.2 (Bartlett, 1946) Under suitable regularity conditions we have

$$\begin{aligned} Z_n'' \approx (\gamma_1 - \gamma_0)&\left[\frac{\partial \ln f(\mathbf{X}_n; \gamma, \delta)}{\partial\gamma} - I_{\gamma\delta}\frac{\partial \ln f(\mathbf{X}_n; \gamma_1, \delta)/I_{\delta\delta}}{\partial\delta}\right] \\ &+ \tfrac{1}{2}n(\gamma_1 - \gamma_0)^2[I_{\gamma\gamma} - I_{\gamma\delta}^2/I_{\delta\delta}]. \end{aligned} \tag{3.7.13}$$

Proof Write

$$\begin{aligned} Z_n'' = {} & [\ln f(\mathbf{X}_n; \gamma_1, \hat{\delta}_1) - \ln f(\mathbf{X}_n; \gamma_0, \delta)] \\ & - [\ln f(\mathbf{X}_n; \gamma_0, \hat{\delta}_0) - \ln f(\mathbf{X}_n; \gamma_0, \delta)]. \end{aligned}$$

Expanding the two differences in Taylor series, we obtain

$$\begin{aligned} Z_n'' \approx (\gamma_1 - \gamma_0)\, \partial \ln f(\mathbf{X}_n; \gamma_1, \delta)/\partial \gamma_1 + (\hat{\delta}_1 - \hat{\delta}_0)\, \partial \ln f(\mathbf{X}_n; \gamma, \delta)/\partial \delta \\ + \tfrac{1}{2} n\{(\gamma_1 - \gamma_0)^2 I_{\gamma\gamma} - 2(\gamma_1 - \gamma_0)(\hat{\delta}_0 - \delta) I_{\gamma\delta} \\ + [(\hat{\delta}_0 - \delta)^2 - (\hat{\delta}_1 - \delta)^2] I_{\delta\delta}\}. \end{aligned} \tag{3.7.14}$$

Also since $\delta \ln f(\mathbf{X}_n; \gamma_0, \hat{\delta}_0)/\partial \delta \simeq \partial \ln f(\mathbf{X}_n, \gamma_1, \hat{\delta}_1)/\partial \delta + (\gamma_1 - \gamma_0) I_{\gamma\delta}$, we have

$$(\hat{\delta}_0 - \hat{\delta}_1) \approx (\gamma_1 - \gamma_0) I_{\gamma\delta}/I_{\delta\delta}.$$

Using this in (3.7.14) we obtain (3.7.13) which is a sum of iid random variables, and hence Wald's approximations to the boundary values, OC function, and ASN are applicable.

REMARK 3.7.1 Cox (1963) has applied a sequential likelihood ratio procedure in order to test certain hypotheses about two binomial proportions. If θ_1 and θ_2 are the two proportions, let $\gamma = \theta_2 - \theta_1$ and $\delta = \theta_1$ or $\gamma = \ln[\theta_2/(1 - \theta_2)] - \ln[\theta_1/(1 - \theta_1)]$ and $\delta = \theta_1/(1 - \theta_1)$. We will be interested in testing a simple hypothesis about γ against a simple alternative.

Problems

For each of the following problems carry out a sequential likelihood ratio test procedure.

1. Let 6, 15, 3, 12, 6, 21, 15, 18, 12, ... denote a sequence of independent observations from the normal population having unknown mean μ and variance σ^2. We wish to test sequentially $H_0: \mu = 8$ against $\mu = 14$.

2.† As part of a learning experiment twelve subjects recited a series of digits. The number of correct responses out of 180 is given below for two groups of subjects, one group having had 20 practice trials and the oher group 30 practice trials.

Group 1	169	97	16	113	61	77
Group 2	100	141	151	169	100	166

Assuming that the proportion of correct responses is approximately normal and that the two groups have the same variability, test sequentially $H_0: \mu_1 = \mu_2 = \frac{1}{2}$ against $H_1: \mu_2 = \frac{3}{4}$, $\mu_1 = \frac{1}{2}$.

3.† The following data give the hemoglobin content in milligrams in six patients having pernicious anemia before and after three months treatment with vitamin B_{12}.

† The data in Problems 2 and 3 are taken from Kurtz, (1963). The data in 2 were provided to Dr. Kurtz by Professor Donald C. Butler and the data in 3 were taken from *Southern Med. J.* **43**, 678 (1950).

Patient	Before	After
1	12.2	13.0
2	11.3	13.4
3	14.7	16.0
4	11.4	13.6
5	11.5	14.0
6	12.7	13.8

Assume that the data are normal and let μ_d denote the expectation of the difference between "before" and "after" hemoglobin content. We wish to test sequentially $H_0: \mu_d = 0$ against $H_1: \mu_d = -1$.

4. Let 40, 47, 35, 60, 54, 42, 66, 51 denote independent observations from the exponential-density $\sigma^{-1} \exp\{-(x-\theta)/\sigma\}$, $x \geq \theta$. Test sequentially H_0: $\sigma = 5$ against H_1: $\sigma = 8$.

3.7.2 Open-Ended Likelihood Ratio Tests for Koopman–Darmois Families

Before we present the open-ended tests for Koopman–Darmois families we need the following result of Lorden (1970) pertaining to the excess over the boundary. In SPRTs cumulative sums play an important role. Let $S_n = X_1 + \cdots + X_n$ denote the sum of n independent random variables with common mean $\mu > 0$. The stopping time is given by $N(t) = \inf\{n : S_n > t\}$. Wald's (1947) equation, i.e. $\mu EN(t) = ES_{N(t)}$ [see Eq. (2.4.16)] whenever $\sup_n E|X_n|$ and $EN(t)$ are finite, can be rewritten as $\mu EN(t) = t + ER_t$, where $R_t = S_{N(t)} - t$.

Definition 3.7.1 Let a random walk $\{S_n\}_{n=0}^{\infty}$ having positive drift and starting at the origin be stopped the first time $S_n > t \geq 0$. Let $N(t) = \inf\{n : S_n > t\}$. Then $R_t = S_{N(t)} - t$ is called the *excess over the boundary*.

The excess over the boundary is often presumed to be negligible. Wald (1947) gave an upper bound for $\sup_{t \geq 0} ER_t$ in the case of independent and identically distributed Xs, namely the sufficient quantity $\sup_{r \geq 0} E(X - r | x > r)$ [see Wald (1947, p. 172, Eq. (A.73))]. This bound is exact for the exponential distribution and is quite good in many cases (for example, normal distributions) where the common distribution function F has monotone increasing hazard rate, $(d/dx)\{-\ln[1 - F(x)]\}$, for $x \geq 0$. In fact for this class of distributions the supremum of $E(X - r | X > r)$ is attained at $r = 0$ and since $\sup_{t \geq 0} ER_t \geq ER_0 \geq [1 - F(0)]E(X | X > 0)$, evidently Wald's bound is too large by at most a factor of $[1 - F(0)]^{-1}$, which is typically less than two. Also, Wald's bound may be difficult to calculate and it is frequently much too large (for instance, when the distribution of X has large "gaps"). In fact, it may be infinite even if $E(X^+)^2$ is finite (where $X^+ = X$ if $X > 0$ and $X^+ = 0$ if $X < 0$) which is well known to be necessary and sufficient for the finiteness of $\sup_{t \geq 0} ER_t$. Finding other useful bounds for $\sup_{t \geq 0} ER_t$ has proved troublesome because the excess over the boundary seems to behave in a pleasantly regular fashion only in the limit as t becomes large. The approach of Farrell (1964a) is of much interest in the case where $E(X^+)^2 = \infty$ since Farrell's bounds on ER_t necessarily tend to infinity as t becomes large. The powerful methods of renewal theory yield impressive asymptotic results in the case of nonnegative Xs, where $EN(t)$, usually denoted by $U(t)$, is the expected number of renewal epochs in $[0, t]$, counting the zero epoch. For an excellent exposition of these results, the reader is referred to the work of Feller (1966, pp. 355–357). For nonnegative X with nonarithmetic distribution, a refinement of the renewal theorem yields $ER_t \to EX^2/2\mu$ and $P(R_t > \mu) \to E(X - u)^+/\mu$ as t gets large. The asymptotic expansion of $EN(t)$ has been further refined under suitable regularity conditions [see, for instance, Stone (1966)].

For arithmetical distributions with span h, lim sup $ER_t = EX^2/(2\mu) + h/2$ and h may be as large as EX^2/μ. Considering $t = 0$ and still assuming X is nonnegative, we find $ER_0 = EX = (EX^2 - \text{var } X)/\mu$ and if var X is small compared to μ, then $\sup_{t \geq 0} ER_t$ is at best nearly EX^2/μ. Thus, in the nonnegative case, for t at both ends of $[0, \infty]$ we may find ER_t as large as EX^2/μ. The proof that this quantity is an upper bound on ER_t for all t is based upon the fact that, just as the asymptotic behavior of ER_t is fairly regular, so is its average behavior over intervals. Toward this, we have the following theorem of Lorden (1970).

Theorem 3.7.3 (Lorden, 1970) Let $X_1, X_2, \ldots$ be independent and identically distributed random variables with $EX = \mu > 0$ and $E(X^+)^2 < \infty$ where $X^+ = X$ if $X > 0$ and $X^+ = 0$ if $X < 0$. Let $S_n = X_1 + \cdots + X_n$, $N(t) = \inf\{n : S_n > t\}$, and $R_t = S_{N(t)} - t$. Then

$$\sup_{t \geq 0} ER_t \leq E(X^+)^2/\mu. \tag{3.7.15}$$

Proof Whatever the values of $X_1, X_2, \ldots$, the random function $\{R_t;\ t \geq 0\}$ is piecewise linear, all pieces having slope -1. First let us assume that the Xs are nonnegative. It is easy to see that for $c \geq 0$

$$\begin{aligned}\int_0^c R_t\,dt &= \int_0^{S_{N(c)}} R_t\,dt - \int_c^{S_{N(c)}} R_t\,dt \\ &= \sum_{n=1}^{N(c)} X_n^2/2 - R_c^2/2,\end{aligned} \tag{3.7.16}$$

because

$$\int_0^{S_{N(c)}} R_t\,dt = \sum_{i=1}^{N(c)} \int_{S_{i-1}}^{S_i} S_i\,dt - \tfrac{1}{2} S_{N(c)}^2 = \sum_{i=1}^{N(c)} X_i^2/2$$

and

$$\int_c^{S_{N(c)}} R_t\,dt = \int_c^{S_{N(c)}} S_{N(c)}\,dt - \int_c^{S_{N(c)}} t\,dt = R_c S_{N(c)} - \tfrac{1}{2}(S_{N(c)}^2 - c^2)$$
$$= R_c{}^2/2.$$

Since $R_t \geq 0$ for all t, we have by Fubini's theorem and Wald's equation [(2.4.16)]

$$\int_0^c ER_t\,dt = (EX^2)EN(c)/2 - ER_c{}^2/2. \tag{3.7.17}$$

[Since $EN(c)$ is finite under our hypothesis, the summation term in (3.7.16) has finite expectation by Wald's equation; and since the other terms are nonnegative, they also have finite expectations.] By Schwarz's inequality and Wald's equation we have

$$\int_0^c ER_t\,dt \leq EX^2(c + ER_c)/2\mu - (ER_c)^2/2. \tag{3.7.18}$$

It is easy to see that for all t, $u \geq 0$, $EN(t+u) \leq EN(t) + EN(u)$, since the conditional expectation of $N(t+u) - N(t)$ given $N(t) = n$, $X_1 = x_1, \ldots, X_n = x_n$, equals $EN(u-r)$ where $r = x_1 + \cdots + x_n - t > 0$ and $N(u-r)$ is zero if $r > u$, so that $EN(u-r) \leq EN(u)$ in any case. It follows from Wald's equation that ER_t is likewise a subadditive function of t and consequently

$$cER_c/2 \leq (c/2) \inf_{0 \leq t \leq c/2} (ER_t + ER_{c-t}) \leq \int_0^{c/2} (ER_t + ER_{c-t})\,dt$$
$$= \int_0^c ER_t\,dt. \tag{3.7.19}$$

Combining (3.7.18) and (3.7.19) and rewriting we have

$$(ER_c)^2 + (c - EX^2/\mu)ER_c - c(EX^2/\mu) \leq 0. \tag{3.7.20}$$

The left-hand side of (3.7.20) is quadratic in ER_c which is nonpositive only between its roots, $-c$ and EX^2/μ. Hence $ER_c \leq EX^2/\mu$ and since c is arbitrary the proof is complete for the nonnegative case. The case where X_1, $X_2, \ldots$ may take negative values reduces to the nonnegative case through consideration of the associated sequence of positive ladder variables which are independent and identically distributed as $S_{N(0)}$ [see Blackwell (1953)]. Let

N_1 be the smallest n for which $S_n > 0$ and write $Y_1 = S_{N_1}$; let N_2 be the smallest $n > 0$, for which $S_{N_1+n} - S_{N_1} > 0$, and write $Y_2 = S_{N_1+N_2} - S_{N_1}$, and so on. Continuing in this way, we obtain sequences $N_1, N_2, \ldots; Y_1, Y_2, \ldots$ of independent, positive, identically distributed random variables such that $S_{N_1+\cdots+N_k} = Y_1 + \cdots + Y_k$. Also notice that each Y_i is distributed as $S_{N(0)}$ where $N(0) = \inf\{n : S_n > 0\}$. Since R_t is pointwise the same for $X_1, X_2, \ldots$ and the sequence of ladder variables $Y_1, Y_2, \ldots$, and clearly $0 < S_{N(0)} \le X_{N(0)}^+$ (because $S_{N(0)-1} \le 0$), the result for the nonnegative case implies

$$\sup_{t\ge 0} ER_t \le \frac{ES_{N(0)}^2}{ES_{N(0)}} \le \frac{E(X_{N(0)}^+)^2}{ES_{N(0)}}$$

$$\le \frac{E[(X_1{}^+)^2 + \cdots + (X_{N(0)}^+)^2]}{E(X_1 + \cdots + X_{N(0)})} = E(X^+)^2/\mu$$

by Wald's equation and the proof is complete.

Corollary 3.7.3.1 Under the hypothesis of Theorem 3.7.3, if $b \le 0 \le a$ and $N^* = \inf\{n : S_n \notin [b, a]\}$, then

$$EN^* \le [(1-\beta)a + \beta b]/\mu + E(X^+)^2/\mu^2 \tag{3.7.21}$$

where $\beta = P(S_{N^*} < b)$.

Proof It is easy to verify that

$$S_{N^*} \le \min(S_{N^*}, a) + S_{N^*} - a.$$

Hence

$$ES_{N^*} \le a(1-\beta) + b\beta + E(S_{N^*} - a) \le (1-\beta)a + b\beta + E(X^+)^2/\mu$$

and Wald's equation gives the desired result.

In applying Corollary 3.7.3.1, it is desirable to have a lower bound on β. For SPRTs, Wald (1947) established only upper bounds for α and β (see Theorem 2.1.1). However, the same argument demonstrates that

$$\alpha/(1-\beta) = E[\exp(-S_{N^*}) \mid S_{N^*} > a]$$

where α is the probability of type I error. [The upper bound for β is obtained from the observation that the conditional expectation is at most e^{-a}.] By applying the conditional version of Jensen's equality [which states that if $g(x)$ is convex then $Eg(X) \ge gE(X)$] and an argument similar to the one used in Corollary 3.7.3.1, we obtain

$$\alpha/(1-\beta) \geq \exp\{-E[S_N^* \mid S_N^* > a]\} \geq \exp\left\{-\left[a + \frac{E(X^+)^2}{(1-\beta)\mu}\right]\right\}.$$

Using the standard upper bound $\beta \leq e^b$

$$\alpha \geq (1-e^b)\exp\left\{-\left[a + \frac{E(X^+)^2}{(1-e^b)\mu}\right]\right\} \tag{3.7.22}$$

and a similar calculation using the alternative distribution yields a lower bound for β. Bounds for higher moments and the distribution of R_t have been obtained by Lorden. We shall state these without proofs.

Theorem 3.7.4 (Lorden, 1970) Under the assumptions of Theorem 3.7.3

$$\sup_{t\geq 0} E(R_t)^p \leq (p+2)E(X^+)^{p+1}/\mu(p+1) \tag{3.7.23}$$

for all $p > 0$.

Proof See Lorden (1970, Theorem 3, p. 524).

Theorem 3.7.5 (Lorden, 1970) If $X_1, X_2, \ldots$ are independent and identically distributed random variables with $EX = \mu > 0$ and $E(X^+)^2 \leq r\mu < \infty$, then for all $t, z \geq 0$ we have

$$P(R_t \geq z) \leq \mu^{-1}E[(2X - z)I\{X \geq z\}]\left(\frac{t+r}{t+z}\right) \tag{3.7.24}$$

where $I\{X \geq z\}$ is the indicator function of the set where $X \geq z$.

Proof See Lorden (1970, Theorem 4).

Now we shall consider the open-ended likelihood ratio test procedures. Recently, H. Robbins and his collaborators have been advocating "open-ended tests" which like the one-sided SPRT, continue sampling indefinitely (with prescribed probability) when the null hypothesis is true and stop only if the alternative is to be accepted. Their methods are most effective in the case of testing a normal mean, although less-sharp estimates are obtained for other cases. Lorden (1973) has investigated the generalized likelihood ratio approach to the problem of open-ended tests for Koopman–Darmois families. As shown by Schwarz (1962), the approach leads to easily computed procedures. Since the tests are equivalent to simultaneous one-sided SPRTs, it is easy to obtain an upper bound on expected sample sizes. The heart of the investigation concerns the problem of bounding error probabilities, the simple Wald approach not being applicable. For the one-parameter Koopman–Darmois family, Wong (1968) has shown that the error probabilities go to zero faster than $c \ln c^{-1}$ (as $c \to 0$) for simultaneous one-sided SPRTs with c ($c < 1$) being the cost of each observation; however explicit bounds are not obtained. Lorden (1973a) obtains an explicit bound which is of the order of $c \ln c^{-1}$.

Let $X_1, X_2, \ldots$ denote independent and identically distributed random variables having the density

$$f(x; \theta) = \exp[\theta x - b(\theta)] \qquad \text{for some } \theta \text{ in } (\underline{\theta}, \bar{\theta}) \tag{3.7.25}$$

with respect to a nondegenerate σ-finite measure. Stopping times N (possibly randomized) will be required to satisfy

$$P(N < \infty \mid H_0) \leq \gamma \qquad \text{for some } 0 < \gamma < \tfrac{1}{3}, \tag{3.7.26}$$

where $H_0: \theta = \theta_0$ and $H_1: \theta > \theta_0$. (Reparametrize the density if necessary to shift the boundary point between null and alternative hypotheses to $\theta = 0$.) Also, without loss of generality assume that $b(0) = 0$. Let $S_n = X_1 + \cdots + X_n$, $n = 1, 2, \ldots$, and note that

$$\ln\left[\prod_1^n \{f(X_i; \theta)/f(X_i; 0)\}\right] = \theta S_n - nb(\theta) \tag{3.7.27}$$

so that one-sided SPRTs of $f(x; 0)$ against $f(x; \theta)$, $\theta > 0$ are given by: stop as soon as

$$S_n > [\ln \alpha^{-1} + nb(\theta)]/\theta \tag{3.7.28}$$

for prescribed type I error probability α ($0 < \alpha < 1$). [Notice that here we set $\beta = 0$ in $A = (1 - \beta)/\alpha$.] The function $b(\cdot)$ is necessarily convex and infinitely differentiable in $(\underline{\theta}, \bar{\theta})$ which need not be the entire natural space of the family of densities considered here. One can easily establish that $E_\theta(X) = b'(\theta)$ and $\operatorname{var}_\theta X = b''(\theta)$ where primes give the order of the derivatives. An easy calculation shows that the information number $E_\theta\{\ln[f(X;\theta)/f(X;0)]\}$ equals

$$\theta b'(\theta) - b(\theta) = I(\theta), \tag{3.7.29}$$

while the variance of $\ln[f(X; \theta)/f(X; 0)]$ under θ equals $\theta^2 b''(\theta)$. Define a likelihood ratio open-ended test of H_0: $\theta = 0$ against H_1: $0 < \theta_1 \leq \theta < \bar{\theta}$

as a stopping time, $N(\theta_1, \alpha)$: the smallest $n \geq 1$ (or ∞ if there is no n) such that

$$S_n > \inf_{\theta_1 \leq \theta < \bar{\theta}} \left[\frac{\ln \alpha^{-1}}{\theta} + n\frac{b(\theta)}{\theta}\right]. \tag{3.7.30}$$

For the alternative $\theta \leq \theta_2 < 0$, $N(\theta_2, \alpha)$ can similarly be defined. Although the infimum in (3.7.29) is readily computed in some cases, for example, the case of the normal mean, it is simpler in many cases such as the normal scale parameter and negative exponential distribution to formulate the critical inequality in terms of $\bar{X}_n = S_n/n$ and n, as in Schwarz (1962). First note that (3.7.30) is equivalent to

$$\sup_{\theta_1 \leq \theta \leq \bar{\theta}} [\theta S_n - nb(\theta)] > \ln \alpha^{-1}. \tag{3.7.31}$$

In case $b'(\theta_1) \leq \bar{X}_n < b'(\bar{\theta})$, the supremum is attained at $q(\bar{X}_n)$ where q is the inverse of the increasing function b'. In this case (3.7.31) is equivalent to

$$\bar{X}_n\, q(\bar{X}_n) - b(q(\bar{X}_n)) > (\ln \alpha^{-1})/n. \tag{3.7.32}$$

In case $\bar{X}_n < b'(\theta_1)$, the supremum is obtained at θ_1; and if $\bar{X}_n > b'(\bar{\theta})$, the supremum is approached as $\theta \to \bar{\theta}$ (and attained at $\bar{\theta}$ if the latter belongs to the natural parameter space). Then we have the following theorem of Lorden (1973).

Theorem 3.7.6 (Lorden, 1973) If $\hat{N} = N(\theta_1, \alpha)$ with $\alpha = \gamma/3[I(\theta_1)^{-1} + 1]^2 \times \ln \gamma^{-1}$, then $\hat{N}$ satisfies (3.7.26) and

$$E_\theta(\hat{N}) \leq \frac{\ln \gamma^{-1} + \ln \ln \gamma^{-1}}{I(\theta)} + 2\frac{\ln\{3^{1/2}[I(\theta_1)^{-1} + 1]\}}{I(\theta)} + \frac{\theta^2 b''(\theta)}{[I(\theta)]^2} + 1 \tag{3.7.33}$$

for all θ in $[\theta_1, \bar{\theta}]$. If N satisfies (3.7.26), then

$$E_\theta(\hat{N}) \geq (\ln \gamma^{-1})/I(\theta) \qquad \text{for} \quad \theta \neq 0. \tag{3.7.34}$$

Proof The lower bound (3.7.34) comes from Wald's inequality for competitors of the (two-sided) SPRT [see Eq. (2.4.16)] by letting $\beta \to 0$ in that inequality. Inequality (3.7.33) follows at once from the fact that

$$E_\theta N(\theta_1, \alpha) \le \frac{\ln \alpha^{-1}}{I(\theta)} + \frac{\theta^2 b''(\theta)}{[I(\theta)]^2} + 1 \qquad \text{for } \theta \text{ in } [\theta_1, \bar{\theta}] \tag{3.7.35}$$

by virtue of the choice of α in the theorem. Relation (3.7.35) is valid because $E_\theta N(\theta_1, \alpha)$ is not larger than the expected time until (3.7.28) holds, which is bounded by the right-hand side of (3.7.35), because of Wald's equation, (2.4.16), and Theorem 3.7.3. [Notice that Theorem 3.7.3 states that the expected "excess over the boundary $= S_n^* - \ln \alpha^{-1} = \sum_1^n Y_i - \ln \alpha^{-1}$" is at most the second moment of $Y = \ln[f(X; \theta)/f(X; 0)]$ divided by the first moment of Y.]

The remainder of the proof consists in showing that the choice of α is sufficient for $\hat{N}$ to satisfy (3.7.26). To determine the infimum in (3.7.30) note that the derivative with respect to θ of the bracketed quantity is

$$\theta^{-2}[-\ln \alpha^{-1} + \theta b'(\theta) - b(\theta)] = \theta^{-2}[nI(\theta) - \ln \alpha^{-1}], \tag{3.7.36}$$

which is nonnegative on $[\theta_1, \bar{\theta}]$ if $nI(\theta_1) \ge \ln \alpha^{-1}$. Thus for $n \ge \ln \alpha^{-1}/I(\theta_1)$ the infimum is attained at θ_1 and hence

$$\begin{aligned} &P[\ln \alpha^{-1}/I(\theta) \le N < \infty \mid H_0] \\ &\qquad \le P\left[S_n > \frac{\ln \alpha^{-1}}{\theta_1} + n\frac{b(\theta_1)}{\theta_1}, \text{ for some } n \ge 1 \,\middle|\, H_0\right] \le \alpha, \end{aligned} \tag{3.7.37}$$

the last inequality coming from Wald's upper bound on SPRT error probabilities. For fixed $n < \ln \alpha^{-1}/I(\theta_1)$, a sequence $\{\theta_k\}$ can be chosen along which the infimum in (3.7.30) is approached, and the probability that strict inequality holds in (3.7.30) is the limit of the probabilities that (3.7.28) holds for $\theta = \theta_k$. The latter probabilities are at most α, again by Wald's upper bound, and thus

$$P\left(S_n > \inf_{\theta_1 \le \theta < \bar{\theta}} \left[\frac{\ln \alpha^{-1}}{\theta} + n\frac{b(\theta)}{\theta}\right] \middle| H_0\right) \le \alpha \tag{3.7.38}$$

or equivalently

$$P(\hat{N} = n \mid H_0) \le \alpha. \tag{3.7.39}$$

From (3.7.37) and (3.7.39) we have

$$P(\hat{N} < \infty \mid H_0) < \alpha\left[\frac{\ln \alpha^{-1}}{I(\theta_1)} + 1\right] < \alpha[\ln \alpha^{-1}][I(\theta_1)^{-1} + 1] \tag{3.7.40}$$

since $\ln \alpha^{-1} > \ln \gamma^{-1} > 1$.

The choice $\alpha = \gamma/3[I(\theta_1)^{-1} + 1]^2 \ln \gamma^{-1}$ suffices to make the extreme right-hand member of (3.7.40) less than γ since

$$\frac{\gamma \ln\{3[I(\theta_1)^{-1} + 1]^2\gamma^{-1} \ln \gamma^{-1}\}}{3[I(\theta_1)^{-1} + 1] \ln \gamma^{-1}} \le \frac{2\gamma \ln\{3^{1/2}[I(\theta_1)^{-1} + 1]\}}{3[I(\theta_1)^{-1} + 1]} + \frac{\frac{3}{2}\gamma \ln \gamma^{-1}}{3[I(\theta_1)^{-1} + 1] \ln \gamma^{-1}}$$

$$\le \frac{2\alpha}{3^{1/2}e} + \frac{\alpha}{2} \le \alpha,$$

using the estimates $\ln x < x^{1/2}$ and $\ln x/x \le 1/e$ for $x \ge 1$.

REMARK 3.7.6.1. By letting $\theta_1 \to 0$ as $\gamma \to 0$ [for example, set $I(\theta_1)^{-1} = \ln \gamma^{-1}$] one can obtain a class of procedures $\hat{N}(\gamma) = N(\theta_1(\gamma), \alpha(\gamma))$ such that for all $\theta > 0$ there is an $M(\theta)$ such that

$$E_\theta \hat{N}(\gamma) - \frac{\ln \gamma^{-1}}{I(\theta)} \le \frac{3 \ln \ln \gamma^{-1}}{I(\theta)} + M(\theta) \tag{3.7.41}$$

for all $\gamma < \exp[-1/I(\theta)]$. Since $0 = b(0) = b(\theta) - \theta b'(\theta) + \frac{1}{2}\theta^2 b''(\xi)$ for some $0 < \xi < \theta$, and b'' is continuous, $I(\theta) = \theta b'(\theta) - b(\theta) \sim \frac{1}{2}b''(0)\theta^2$ as $\theta \to 0$, so that the choice of θ_1 just made results in $\theta_1 \sim$ (const) $(\ln \gamma^{-1})^{-1/2}$ as $\gamma \to 0$. Other negative powers of $\ln \gamma^{-1}$ yield similar results, and lead to bounds of the form (3.7.41) with the factor 3 replaced by any number larger than unity.

REMARK 3.7.6.2 The open-ended test $N(\theta_1, \alpha)$ has $E_\theta N = \infty$ for $\theta > 0$ closer (in terms of information numbers) to θ than to θ_1. This may be acceptable in certain applications, and the interval of positive θs where this happens can be regarded as an "indifference zone." However, if one desires to consider procedures that satisfy $E_\theta N < \infty$ for all $\theta > 0$, then one can simply use combinations of the form $N^* = \min_k N(\theta_k, \alpha_k)$ where $\theta_k \to 0$ and (say) $\alpha_k = \gamma 2^{-k}/3[I(\theta_k)^{-1} + 1]^2 \ln(\gamma^{-1}2^k)$ so that $P(N^* < \infty \mid H_0) \le \gamma(2^{-1} + 2^{-2} + 2^{-3} + \cdots) = \gamma$. Then (3.7.33) holds for all N^* if γ is replaced by $\gamma/2$ and similar inequalities hold for θ in (θ_{k+1}, θ_k), $k = 1, 2, \ldots$. Notice that bounds of the form (3.7.41) hold for N^* also. The stopping boundaries for N^* can be calculated "piecemeal" as n increases, passing from $N(\theta_k, \alpha_k)$ to $N(\theta_{k+1}, \alpha_{k+1})$ as soon as

$$\inf_{\theta_{k+1} \le \theta < \theta_k} \left[\frac{\ln \alpha_{k+1}^{-1}}{\theta} + n\frac{b(\theta)}{\theta}\right] < \frac{\ln \alpha_k^{-1}}{\theta_k} + n\frac{b(\theta_k)}{\theta_k}.$$

Suppose it is desired to test $\theta < 0$ against $\theta > 0$ with error probabilities less than γ and β. An obvious procedure is to stop at $\tilde{N} = \min(N_1^*, N_2^*)$ where $N_1^* = \min_k N(\theta_k, \alpha_k)$ with $\theta_k \downarrow 0$ and $N_2^* = \min_k N(-\theta_k, \alpha_k')$ where α_k' is defined using β in place of γ. If $\theta < 0$ or $\theta > 0$ is chosen according as N_2^* (respectively) N_1^* stops first, then clearly

$$P(\text{stop and decide } \theta > 0 \,|\, \theta) \leq \gamma \qquad \text{for all } \theta \leq 0 \tag{3.7.42}$$

and

$$P(\text{stop and decide } \theta < 0 \,|\, \theta) \leq \beta \qquad \text{for all } \theta \geq 0. \tag{3.7.43}$$

Upper bounds of the form (3.7.40) also hold for $E_\theta \tilde{N}$ for $\theta \neq 0$ and are to be compared with Wald's (1947, p. 197) lower bound

$$\begin{aligned} I(\theta)E_\theta N &\geq (1-\beta)\ln[(1-\beta)/\gamma] + \beta \ln[\beta/(1-\gamma)] \\ &\geq (1-\beta)\ln \gamma^{-1} + (1-\beta)\ln(1-\beta) + \beta \ln \beta \\ &\geq (1-\beta)\ln \gamma^{-1} - \ln 2 \end{aligned}$$

[since $(1-x)\ln(1-x) + x \ln x$ has minimum value, $-\ln 2$ at $x = \frac{1}{2}$] for $\theta > 0$ (and an analogous bound holds for $\theta < 0$) which is satisfied by the stopping time N of any test satisfying (3.7.41) and (3.7.42). Omitting $\theta = 0$ in (3.7.42) and (3.7.43) is only an apparent weakening of these requirements because it can be shown (for instance by the dominated convergence theorem) that $P(N \leq n \,|\, \theta)$ is continuous in θ for each n, and letting $n \to \infty$, that $P(N < \infty \,|\, \theta)$ is lower semicontinuous.

Lorden (1973) has also obtained the following result which we shall state without proof.

Theorem 3.7.7 (Lorden, 1973) We have

$$P(N < \infty \,|\, H_0) \lesssim \alpha + \alpha(\ln \alpha^{-1})^{1/2} \int_{\theta_1}^{\bar{\theta}} [b''(\theta)/2\pi I(\theta)]^{1/2} \, d\theta. \tag{3.7.44}$$

Corollary 3.7.7.1 If θ is the mean of a normal distribution with variance unity, then (3.7.44) becomes

$$P(N < \infty \,|\, H_0) \lesssim \alpha[1 + (\ln \alpha^{-1}/\pi)^{1/2} \ln(\bar{\theta}/\theta_1)]. \tag{3.7.45}$$

Further, Lorden (1973) verifies that the asymptotic inequality (3.7.45) is an actual inequality in this case.

3.8 Sequential Tests Between Three Hypotheses

In the preceding sections we have discussed a theory of sequential tests that is appropriate for distinguishing between two simple (or composite) hypotheses. However, there are many practical situations wherein a choice among three or more courses of action is required and the theory described earlier is not appropriate. For example, the tolerance limits set to a machine require a choice among the possibilities that the setting is too high, too low, or acceptable. As a further example let X, Y denote the breaking strength of two yarns. We may be interested in testing whether X and Y are approximately equal or that one or the other was stochastically greater. Both the above examples could be formulated in terms of a sequential test among three hypotheses. If the three hypotheses could be ordered in terms of the expected results from samples, sequential tests may be obtained by performing two SPRTs simultaneously, one between each pair of neighboring hypotheses in the ordering. This was proposed by Armitage (1947, 1950), and Sobel and Wald (1949) and they show how to obtain sequential tests which approximately satisfy certain conditions on the probabilities of misclassification.

In the following let us formulate the k-hypotheses problem. Consider a given family. $[F_n(\mathbf{X};\, \theta),\ \theta \in \Theta]$ of distributions and $k > 2$ hypotheses:

$$H_j\colon\ \theta \in \omega_j, \quad j = 0, 1, \ldots, k-1 \tag{3.8.1}$$

where $\{\omega_j\}$ are disjoint nonempty subsets of Θ. A statistical test δ_k of (3.8.1) is a set of rules based on $\{\mathbf{X}_n\}$ $(n = 1, 2, \ldots)$ to accept one and only one of the k hypotheses. For every n, the space R_n is partitioned into k acceptance regions and one continuation region. Letting

$$L_j(\theta) = P(\delta_k \text{ accepts } H_j \mid \theta) \qquad \text{for any } \theta \in \Theta,$$

$\{L_j(\theta)\}$ will be called the OC functions of δ_k and $1 - L_j(\theta)$ is a true risk if $\theta \in \omega_j$. The stopping variable N of δ_k is the random variable corresponding to the number of observations required by the test procedure in order to reach a final decision. One can also define the ASN as $E_\theta(N)$. The test is said to be *closed* if and only if

$$\sum_{j=0}^{k-1} L_j(\theta) = 1 \qquad \text{for every } \theta \in \Theta. \tag{3.8.2}$$

Notice that the test δ_k can generally lead to k distinct types of decision errors, namely to reject H_j even though $\theta \in \omega_j$ for $j = 0, 1, \ldots, k-1$. In certain situations one may wish to consider the acceptance of a certain H_j a serious error when θ belongs to a certain subset of $\Theta - \omega_0 - \cdots - \omega_{k-1}$; however, we shall ignore such possibilities for the moment. The class $\mathscr{C}_k$ of all closed tests satisfying

$$1 - L_j(\theta) \leq \alpha_j \qquad \text{for every } \theta \in \omega_j \text{ and for } j = 0, 1, \ldots, k-1 \tag{3.8.3}$$

is said to be a *valid* test where $(0 \leq \alpha_j \leq 1)$ are preassigned *error probabilities*. Our objective should be to choose the uniformly most economical (UME) test $\delta_k^* \in \mathscr{C}_k$ in the sense that

$$\inf_{\delta \in \mathscr{C}_k} E_\theta(N \mid \delta_k) = E_\theta(N \mid \delta_k^*) \qquad \text{for all } \theta \in \Theta. \tag{3.8.4}$$

However, it is not known whether such a test exists, even when the $\{H_j\}$ are all simple hypotheses. In the following, we shall describe a reasonably good test in $\mathscr{C}_k$ proposed by Sobel and Wald (1949) for $k = 3$ and for a general family of probability models. The test often turns out to be more efficient for all θ than the corresponding optimum fixed-sample test. For certain values of θ, the ASN of this test also compares favorably with some general lower bounds given by Simons (1967a).

THE SOBEL–WALD TEST Let $\{X_i\}$ be iid chance variables whose distribution belongs to the Darmois–Koopman family (or exponential family) having the density

$$f(x; \theta) = B(\theta) e^{Q(\theta)R(x)} h(x), \qquad -\infty < x < \infty$$

where it is assumed that $Q(\theta)$ is strictly increasing in θ. Sobel and Wald (1949) have considered the test for the normal mean with known variance. B. K. Ghosh (1970) has given their results in an extended form so as to be applicable to the exponential family. So, in the following we shall present the results in the extended form.

Consider the three hypotheses:

$$H_0\colon \theta = \theta_0, \quad H_1\colon \theta = \theta_1, \quad H_2\colon \theta = \theta_2, \quad \underline{\theta} < \theta_0 < \theta_1 < \theta_2 < \bar{\theta} \tag{3.8.5}$$

and the restriction (3.8.3) with $k = 3$. The optimum fixed-sample (nonrandomized) test procedure is given by

$$\begin{aligned} &\text{acc } H_0 \quad \text{if } T_n \leq \underline{t} \\ &\text{acc } H_1 \quad \text{if } \underline{t} < T_n < \bar{t} \\ &\text{acc } H_2 \quad \text{if } T_n > \bar{t} \end{aligned} \tag{3.8.6}$$

where $T_n = \sum_1^n R(X_i)$, $\underline{t} = \underline{t}(\alpha_0, \alpha_1, \alpha_2) < \bar{t} = \bar{t}(\alpha_0, \alpha_1, \alpha_2)$ and the smallest positive integer $n = n(\alpha_0, \alpha_1, \alpha_2)$ is determined by

$$F_n(\underline{t}; \theta_0) \geq 1 - \alpha_0, \quad F_n(\bar{t}; \theta_1) - F_n(\underline{t}; \theta_1) \geq 1 - \alpha_1, \quad F_n(\bar{t}; \theta_2) \leq \alpha_2, \tag{3.8.7}$$

where $F_n(t; \theta) = P_\theta(T_n \leq t)$. Let, for $j = 1, 2$, $\delta(b_j, a_j)$ be a SPRT of H_{j-1} against H_j with $b_j < a_j$, and let

$$\begin{aligned} Z_i(j) &= \ln[f(X_i; \theta_j)/f(X_i; \theta_{j-1})] = \psi_j + \Delta_j R(X_i), \\ S_n(j) &= \sum_{i=1}^{n} Z_i(j), \end{aligned} \tag{3.8.8}$$

where $\psi_j = \ln[B(\theta_j)/B(\theta_{j-1})]$, $\Delta_j = Q(\theta_j) - Q(\theta_{j-1}) > 0$. Then the critical inequality of $\delta(b_j, a_j)$ at stage $n \geq 1$ is

$$b_j < S_n(j) < a_j, \qquad \text{or equivalently,} \qquad c_n(j) < T_n < d_n(j)$$

where $c_n(j) = (b_j - n\psi_j)/\Delta_j < d_n(j) = (a_j - n\psi_j)/\Delta_j$. It is easy to see that for every $n \geq 1$, $c_n(1) \leq c_n(2)$ and $d_n(1) \leq d_n(2)$ implies that it is impossible for $\delta(b_1, a_1)$ to accept H_0 when $\delta(b_2, a_2)$ leads to the acceptance of H_2. However, the inequalities $c_n(1) \leq c_n(2)$ and $d_n(1) \leq d_n(2)$ may not be necessary in order to ensure that not both H_0 and H_2 are accepted.

Lemma 3.8.1 A necessary and sufficient condition for both inequalities $c_n(1) \leq c_n(2)$ and $d_n(1) \leq d_n(2)$ to hold for every $n \geq 1$ is that

$$\psi_1 \Delta_1 - \psi_2 \Delta_2 \geq \max[b_1 \Delta_2 - b_2 \Delta_1, a_1 \Delta_2 - a_2 \Delta_1]. \tag{3.8.9}$$

A necessary and sufficient condition for $d_n(1) \leq c_n(2)$ to hold for every $n \geq 1$ is that $\psi_1 \Delta_2 - \psi_2 \Delta_1 > a_1 \Delta_2 - b_2 \Delta_1$. Also, $d_n(1) < c_n(2)$ for all n.

Proof Since $Q(\theta)$ is increasing in θ, we can express $B(\theta)$ as $B^*(Q)$ and one can write

$$dF(x; \theta) = B^*(Q) e^{QR(x)}\, dG(x), \qquad Q(\underline{\theta}) < Q < Q(\bar{\theta}) \tag{3.8.10}$$

where $F(x; \theta)$ denotes the distribution of X. Also $Q_1 = Q(\theta_1)$ can be expressed as $\lambda Q_0 + (1 - \lambda) Q_2$ for some $0 < \lambda < 1$ where $Q_i = Q(\theta_i)$, $i = 0, 1, 2$. Now, by Hölder's inequality [see for instance, Hardy *et al.* (1959, pp. 119–

121)] we obtain

$$\begin{aligned}[B^*(Q_1)]^{-1} &= \int \exp[\{\lambda Q_0 + (1-\lambda)Q_2\}R(x)]\, dG(x) \\ &< \left\{\int \exp[Q_0 R(x)]\, dG(x)\right\}^{\lambda}\left\{\int \exp[Q_2 R(x)]\, dG(x)\right\}^{1-\lambda} \\ &= [B^*(Q_0)]^{-\lambda}[B^*(Q_2)]^{\lambda-1},\end{aligned}$$

which implies that $\psi_1\Delta_2 > \psi_2\Delta_1$ whenever $\underline{\theta} < \theta_0 < \theta_1 < \theta_2 < \bar{\theta}$. It follows from the definition of $c_n(j)$ and $d_n(j)$ that, if (3.8.9) holds, then $c_1(1) \le c_1(2)$ and $d_1(1) \le d_1(2)$ while $c_n(1) \le c_n(2)$ and $d_n(1) < d_n(2)$ for every $n > 1$. Conversely if $c_n(1) \le c_n(2)$ and $d_n(1) \le d_n(2)$, then (3.8.9) holds. The second assertion of the lemma is immediate and it implies (3.8.9) if $b_1 \le a_1$ and $b_2 \le a_2$. Finally, letting

$$n^* = (a_1\Delta_2 - b_2\Delta_1)/(\psi_1\Delta_2 - \psi_2\Delta_1), \tag{3.8.11}$$

it is clear that $d_n(1) \ge c_n(2)$ for every $n \le n^*$ and $d_n(1) < c_n(2)$ for every $n > n^*$. This completes the proof of the lemma.

The Sobel–Wald test for testing the hypotheses in (3.8.5) is given as follows:

Choose $b_1 < a_1$ and $b_2 < a_2$ such that (3.8.9) is satisfied and carry out $\delta(b_1, a_1)$ and $\delta(b_2, a_2)$ simultaneously by observing $\{X_i\}$ successively. At stage n $(n \ge 1)$ make one of the following decisions:

(i) accept H_0 if $\delta(b_1, a_1)$ accepts H_0 [that is, if $T_n < c_n(1)$];
(ii) accept H_1 if both $\delta(b_1, a_1)$ and $\delta(b_2, a_2)$ accept H_1 [that is if $d_n(1) < T_n < c_n(2)$]; (3.8.12)
(iii) accept H_2 if $\delta(b_2, a_2)$ accepts H_2 [that is, if $T_n > d_n(2)$];
(iv) otherwise take one more observation [that is, if $c_n(1) < T_n < \min(d_n(1), c_n(2))$].

Notice that the restriction in (3.8.3) excludes the possibility of $\delta(b_1, a_1)$ accepting H_0 and $\delta(b_2, a_2)$ accepting H_2 and this feature of the test facilitates the evaluation of the properties of the test. The reader is advised to draw a figure in a typical case with n on the x axis and T_n on the y axis in order to locate the regions of acceptance of H_0, H_1, and H_2. Here (b_1, b_2, a_1, a_2) will be called the constants of the Sobel–Wald procedure $\delta(b_1, b_2, a_1, a_2)$. Before we explain how one goes about choosing the constants that satisfy (3.8.3), let us explore some of the basic properties of $\delta(b_1, b_2, a_1, a_2)$. Let $L^{(j)}(\theta)$ and $E(N_j; \theta)$ denote the OC and ASN functions of $\delta(b_j, a_j)$. Then the stopping variable of $\delta(b_1, b_2, a_1, a_2)$ is $N = \max(N_1, N_2)$ and, consequently, for every $n \geq 1$ and $\underline{\theta} < \theta < \bar{\theta}$.

$$P_\theta(N > n) = P_\theta(N_1 > n) + P_\theta(N_2 > n) - P_\theta(N_1 > n, N_2 > n). \quad (3.8.13)$$

It follows from Theorem 2.2.1 that each of the quantities on right-hand side tends to zero as n tends to infinity. Hence $\delta(b_1, b_2, a_1, a_2)$ is closed and all moments of N are finite.

Theorem 3.8.1 For every $\underline{\theta} < \theta < \bar{\theta}$, the OC functions of $\delta(b_1, b_2, a_1, a_2)$ satisfy the identities:

$$L_0(\theta) = L^{(1)}(\theta), \quad L_1(\theta) = L^{(2)}(\theta) - L^{(1)}(\theta), \quad L_2(\theta) = 1 - L^{(2)}(\theta) \quad (3.8.14)$$

and $L_0(\theta_0) + L_1(\theta_1) + L_2(\theta_2) > 1$. For every $\underline{\theta} < \theta < \bar{\theta}$, the ASN of $\delta(b_1, b_2, a_1, a_2)$ satisfies

$$\max[E_\theta(N_1), E_\theta(N_2)] \leq E_\theta(N) \leq \min[E_\theta(N'), E_\theta(N'')] \quad (3.8.15)$$

where N' is the stopping variable of $\delta(b_1, \infty)$ for testing H_0 against H_1 and N'' is the stopping variable of $\delta(-\infty, a_2)$ for testing H_1 against H_2.

Proof $\delta(b_1, b_2, a_1, a_2)$ accepts H_0 if and only if $\delta(b_1, a_1)$ accepts H_0 and consequently $L_0(\theta) \equiv L^{(1)}(\theta)$. Analogously, $\delta(b_1, b_2, a_1, a_2)$ accepts H_2 if and only if $\delta(b_2, a_2)$ accepts H_2, and thus $L_2(\theta) = 1 - L^{(2)}(\theta)$ for all θ. Since $\delta(b_1, b_2, a_1, a_2)$ is closed, $L_1(\theta) = 1 - L_0(\theta) - L_2(\theta)$ which leads to the middle of the identities in (3.8.14). From the inequalities in Theorem 2.1.1, it easily follows that

$$L^{(1)}(\theta_1) - L^{(1)}(\theta_0) < 0, \qquad L^{(2)}(\theta_2) - L^{(2)}(\theta_1) < 0$$

which together with (3.8.14) lead to $L_0(\theta_0) + L_1(\theta_1) + L_2(\theta_2) > 1$. Next, since for every $n \geq 1$ and $\underline{\theta} < \theta < \bar{\theta}$

$$\begin{aligned} P_\theta(N \geq n) &= P_\theta(N_1 \geq n) + P_\theta(N_1 < n \text{ and } N_2 \geq n) \\ &= P_\theta(N_2 \geq n) + P_\theta(N_2 < n \text{ and } N_1 \geq n) \end{aligned}$$

the left-hand inequality in (3.8.15) readily follows. When $\psi_1 \Delta_2 - \psi_2 \Delta_1 \geq a_1 \Delta_2 - b_2 \Delta_1$ we must have by Lemma 3.8.1 [inequality (3.8.9)] that either $P_\theta(N_1 = 1) = 1$ or $P_\theta(N_2 = 1) = 1$ for all θ, and

$$\begin{aligned} P_\theta(N = 1) = 1 &- P_\theta[(b_1 - \psi_1)/\Delta_1 < R(X) < (a_1 - \psi_1)/\Delta_1] \\ &- P_\theta[(b_2 - \psi_2)/\Delta_2 < R(X) < (a_2 - \psi_2)/\Delta_2] \end{aligned}$$

$P_\theta(N = n) = \max[P_\theta(N_1 = n), P_\theta(N_2 = n)]$ if $n > 1$, so that the strict inequality in (3.8.15) holds. Finally, since termination of $\delta(b_1, \infty)$ or $\delta(-\infty, a_2)$ at any stage $n \geq 1$ implies acceptance under $\delta(b_1, b_2, a_1, a_2)$ at or before stage n, the right-hand side inequality in (3.8.15) follows. This completes the proof of Theorem 3.8.1.

Let $\zeta_\theta = E_\theta\{T(X)\}$. Then, since, for $j = 1$ or 2, $S_n(j)$ is the sum of iid random variables, Wald's approximations for $L_j(\theta)$ yield

$$L_0(\theta) \simeq \begin{cases} \{\exp[a_1 h_1(\theta)] - 1\}\{\exp[a_1 h_1(\theta)] - \exp[b_1 h_1(\theta)]\}^{-1} & \text{if } \zeta_\theta \neq -\psi_1/\Delta_1, \\ a_1/(a_1 - b_1) & \text{if } \zeta_\theta = -\psi_1/\Delta_1, \end{cases} \tag{3.8.16}$$

$$L_2(\theta) \simeq \begin{cases} \{1 - \exp[b_2 h_2(\theta)]\}\{\exp[a_2 h_2(\theta)] - \exp[b_2 h_2(\theta)]\}^{-1} & \text{if } \zeta_\theta \neq -\psi_2/\Delta_2, \\ -b_2/(a_2 - b_2) & \text{if } \zeta_\theta = -\psi_2/\Delta_2, \end{cases}$$

$$L_1(\theta) = 1 - L_0(\theta) - L_2(\theta)$$

where $h_j(\theta) \neq 0$ is determined by

$$E(\exp[Z_i(j) h_j(\theta)]) = 1, \quad \text{that is,} \quad E(\exp[\Delta_j h_j(\theta) R(X)]) = \exp[-\psi_j h_j(\theta)]. \tag{3.8.17}$$

Furthermore,

$$E_\theta(N_1) \simeq \begin{cases} [a_1 - (a_1 - b_1)L_0(\theta)](\psi_1 + \Delta_1 \zeta_\theta)^{-1} & \text{if } \zeta_\theta \neq -\psi_1/\Delta_1, \\ -a_1 b_1/\Delta_1{}^2 \operatorname{var}_\theta(R(X)) & \text{if } \zeta_\theta = -\psi_1/\Delta_1; \end{cases}$$

$$E_\theta(N_2) \simeq \begin{cases} [(a_2 - b_2)L_2(\theta) + b_2](\psi_2 + \Delta_2 \zeta_\theta)^{-1} & \text{if } \zeta_\theta \neq -\psi_2/\Delta_2, \\ -a_2 b_2/\Delta_2{}^2 \operatorname{var}_\theta(R(X)) & \text{if } \zeta_\theta = -\psi_2/\Delta_2; \end{cases} \tag{3.8.18}$$

$$E_\theta(N') \simeq b_1(\psi_1 + \Delta_1 \zeta_\theta)^{-1} \qquad \text{if } \zeta_\theta < -\psi_1/\Delta_1,$$

$$E_\theta(N'') \simeq a_2(\psi_2 + \Delta_2 \zeta_\theta)^{-1} \qquad \text{if } \zeta_\theta > -\psi_2(\Delta_2).$$

It should be noted that the approximations in (3.8.16) and (3.8.18) are reasonable provided the excess over the boundaries of b_j and a_j ($j = 1, 2$) are small. Also, since $f(x; \theta)$ has monotone likelihood ratio in $R(x)$, $L_j(\theta)$ ($j = 0, 1, 2$) will be monotone in θ (see Theorem 2.9.2) and hence the test is valid for $H_0' : \theta \leq \theta_0$, $H_1' : \theta_0 < \theta < \theta_2$, $H_2' : \theta \geq \theta_2$.

Next, let us consider the problem of determining the constants of the Sobel–Wald procedure such that (3.8.3) is satisfied. Since by Theorem 3.8.1, the sum of the three true error probabilities of $\delta(b_1, b_2, a_1, a_2)$ must be less than two, it seems logical to assume that $\alpha_0 + \alpha_1 + \alpha_2 < 2$. Given $(\alpha_0, \alpha_1, \alpha_2)$ we can equate the right-hand expressions of (3.8.16) at $\theta = \theta_0, \theta_2, \theta_1$ respectively to $1 - \alpha_0$, $1 - \alpha_2$, $1 - \alpha_1$ and then solve for $b_1 < 0$, $b_2 < 0$, $a_1 > 0$, and $a_2 > 0$. Observing that $h_1(\theta_0) = h_2(\theta_1)$ and $h_1(\theta_1) = h_2(\theta_2) = -1$ and setting $(1 - e^{b_2})/(e^{a_2} - e^{b_2}) = \xi$ the solutions are

$$\begin{aligned} b_1(\xi) &= \ln[(\gamma_1 - \xi)/(1 - \alpha_0)], & b_2(\xi) &= \ln[\alpha_2/(1 - \xi)], \\ a_1(\xi) &= \ln[(1 - \xi + \gamma_1)/\alpha_0], & a_2(\xi) &= \ln[(1 - \alpha_2)/\xi], \end{aligned} \tag{3.8.19}$$

where ξ is any real number satisfying

$$\max[0, \alpha_0 + \alpha_1 - 1] < \xi < \min[\alpha_1, 1 - \alpha_2].$$

If the excess of $S_n(j)$ over $b_j(\xi)$ and $a_j(\xi)$ is small for $j = 1, 2$, the test $\delta(b_1(\xi), b_2(\xi), a_1(\xi), a_2(\xi))$ will satisfy the inequalities (3.8.3) approximately, provided the four constants also fulfill (3.8.9). Also, there exists a considerable flexibility in the choice of ξ which is an advantage rather than a drawback of the Sobel–Wald test procedure. For instance, if one wants to control $L_0(\theta_1)$ which is not restricted by (3.8.3), at a preassigned small $\alpha_3 > 0$, this additional requirement yields $\xi = \alpha_1 - \alpha_3$. Such a control is not possible by the fixed-sample size procedure (3.8.6). Alternatively, one may decide to choose the value of ξ that minimizes $\sup_\theta E_\theta(N)$. Such a choice of ξ is not feasible to evaluate analytically, although numerical calculations of (3.8.18) for various ξ may provide a rough guideline. On the other hand, if no additional restrictions are imposed on the OC and ASN functions of $\delta(b_1, b_2, a_1, a_2)$, then using the fundamental inequalities pertaining to the error probabilities of a SPRT (see Theorem 2.1.1) and (3.8.19) we obtain

$$L_0(\theta_0) = 1 - \alpha(\theta_0) \geq 1 - \frac{\alpha(\theta_0)}{1 - \beta(\theta_1)} \geq 1 - e^{-a_1} = 1 - \frac{\alpha_0}{1 + \xi - \alpha_1}$$

$$\begin{aligned} L_1(\theta_1) = L^{(2)}(\theta_1) - L^{(1)}(\theta_1) &= L^{(2)}(\theta_1) - L_0(\theta_1) \\ &= 1 - P_{\theta_1}(\text{rej } H_1) - P_{\theta_1}(\text{acc } H_0) \\ &= 1 - \alpha(\theta_1) - \beta(\theta_1) \\ &\geq 1 - \frac{\alpha(\theta_1)}{1 - \beta(\theta_2)} - \frac{\beta(\theta_1)}{1 - \alpha(\theta_0)} \\ &\geq 1 - e^{-a_2} - e^{b_1} \\ &= 1 - \frac{\xi}{1 - \alpha_2} - \frac{\alpha_1 - \xi}{1 - \alpha_0} \end{aligned} \tag{3.8.20}$$

$$L_2(\theta_2) = 1 - L^{(2)}(\theta_2) = 1 - \beta(\theta_2) \geq 1 - \frac{\beta(\theta_2)}{1 - \alpha(\theta_1)} \geq 1 - e^{b_2} = 1 - \frac{\alpha_2}{1 - \xi}.$$

The inequalities in (3.8.20) suggest that a reasonable choice of ξ is midway between $\max[0, \alpha_0 + \alpha_1 - 1]$ and $\min[\alpha_1, 1 - \alpha_2]$. For instance, one would choose $\xi = \alpha_0/2$ if $0 < \alpha_0 = \alpha_1 = \alpha_2 < \frac{2}{3}$, and $\xi = \alpha_0$ if $0 < \alpha_0 = \alpha_1/2 = \alpha_2 < \frac{1}{2}$.

Except for the trivial case when $\psi_1\Delta_2 - \psi_2\Delta_1 \geq a_1\Delta_2 - b_2\Delta_1$, it is quite difficult to derive any approximate formulas for the ASN function of the Sobel–Wald procedure. The bounds in (3.8.15) along with (3.8.18) are often helpful for this purpose. If $L_0(\theta_0)$ is large, both $E_\theta(N_1)$ and $E_\theta(N')$ will be close to $E_\theta(N)$ for $\theta \leq \theta_0$. For then by Theorem 2.9.2, $L_0(\theta)$ is large for $\theta < \theta_0$ and therefore the decision under $\delta(b_1, b_2, a_1, a_2)$ is largely dominated by $\delta(b_1, a_1)$ or $\delta(b_1, \infty)$. Similarly, if $L_2(\theta_2)$ is large, both $E_\theta(N_2)$ and $E_\theta(N'')$ will be close to $E_\theta(N)$ for $\theta \geq \theta_2$. However, the upper bound in (3.8.15) is useless for values of θ satisfying $-\psi_1/\Delta_1 \leq \zeta_\theta \leq -\psi_2/\Delta_2$. In order to obtain alternative upper bounds in such cases let m denote the smallest integer that is equal to or exceeds $(a_1\Delta_2 - b_2\Delta_1)/(\psi_1\Delta_2 - \psi_2\Delta_1)$. Consider two procedures W_1 and W_2 each of which starts by drawing m observations and let $T_m = R(X_1) + \cdots + R(X_m)$.

Procedure W_1

(i) accept H_0 if $T_m \leq c_m(1)$;

(ii) carry out (by observing $X_{m+1}, \ldots$) the conditional SPRT $\delta(b_1 - m\psi_1 - \Delta_1 T_m, a_1 - m\psi_1 - \Delta_1 T_m)$ of H_0 against H_1 if $c_m(1) < T_m < d_m(1)$;

(iii) accept H_1 if $d_m(1) \leq T_m \leq c_m(2)$; and

(iv) carry out the conditional SPRT $\delta(b_2 - m\psi_2 - \Delta_2 T_m, \infty)$ of H_1 against H_2 if $T_m > c_m(2)$.

Procedure W_2

(i) accept H_2 if $T_m \geq d_m(2)$;

(ii) carry out the conditional SPRT $\delta(b_2 - m\psi_2 - \Delta_2 T_m, a_2 - m\psi_2 - \Delta_2 T_m)$ of H_1 against H_2 if $c_m(2) < T_m < d_m(2)$;

(iii) accept H_1 if $d_m(1) \leq T_m \leq c_m(2)$; and

(iv) carry out the conditional SPRT $\delta(-\infty, a_1 - m\psi_1 - \Delta_1 T_m)$ of H_0 against H_1 if $T_m < d_m(1)$.

It can easily be verified that

$$E_\theta(N) \leq \min[E_\theta(N_{W_1}), E_\theta(N_{W_2})] \qquad \text{for every } \theta. \tag{3.8.21}$$

Wald's approximations to the ASN functions of the various conditional SPRTs yield [see B. K. Ghosh (1970, p. 163)]

$$\begin{aligned} E_\theta(N_{W_1}) &\simeq m - \frac{\Delta_2}{\psi_2 + \Delta_2\zeta_\theta}\int_{y > c_m(2)} [y - c_m(2)]\, dH_m(y;\theta) + K_1(\theta) \\ &\qquad\qquad \text{if} \quad \zeta_\theta < -\psi_2/\Delta_2 \\ E_\theta(N_{W_2}) &\simeq m + \frac{\Delta_1}{\psi_1 + \Delta_1\zeta_\theta}\int_{y < d_m(1)} [d_m(1) - y]\, dH_m(y;\theta) + K_2(\theta) \\ &\qquad\qquad \text{if} \quad \zeta_\theta > -\psi_1/\Delta_1 \end{aligned} \tag{3.8.22}$$

where $H_m(y;\theta) = P_\theta[T_m \le y]$ and for $j = 1, 2$

$$K_j(\theta) = \frac{\Delta_j}{\psi_j + \Delta_j \zeta_\theta} \int_{c_m(j)<y<d_m(j)} \left[d_m(j) - y - \frac{a_j - b_j}{\Delta_j} L_j(y, \theta) \right] dH_m(y;\theta)$$

$$\text{if} \quad \zeta_\theta \ne -\psi_j/\Delta_j$$

$$= \frac{1}{\operatorname{var}_\theta(R(X))} \int_{c_m(j)<y<d_m(j)} [d_m(j) - y][y - c_m(j)]\, dH_m(y;\theta)$$

$$\text{if} \quad \zeta_\theta = -\psi_j/\Delta_j$$

$$L_j(y, \theta) = \{\exp[d_m(j)\Delta_j h_j(\theta)] - \exp[y\Delta_j h_j(\theta)]\}\{\exp[d_m(j)\Delta_j h_j(\theta)] - \exp[c_m(j)\Delta_j h_j(\theta)]\}^{-1},$$

where $h_j(\theta)$ is defined by (3.8.16). In practical applications $H_m(y;\theta)$ assumes a simple form. The upper bound in (3.8.15) will, in general, be sharper than the one in (3.8.21) when $\zeta_\theta < -\psi_1/\Delta_1$ or $> -\psi_2/\Delta_2$. For θ around θ_1 (which implies $-\psi_1/\Delta_1 < \zeta_\theta < -\psi_2/\Delta_2$), $E_\theta(N_{W_1})$ and $E_\theta(N_{W_2})$ will be close to each other.

The ASN of the Sobel–Wald procedure attains a local minimum when θ is around θ_1, and a local maximum on each of the two intervals (θ_0, θ_1) and (θ_1, θ_2). The overall behavior can be studied from the bounds in (3.8.15) and (3.8.21) along with (3.8.18) and (3.8.22). Since the stopping variable of the test procedure depends on the stopping variables of both $\delta(b_1, a_1)$ and $\delta(b_2, a_2)$, the decision rules in (3.8.6) do not depend *completely* on the sequence of sufficient statistics $\{T_n\}$. Thus, intuitively, it would seem that $\delta(b_1(\xi), b_2(\xi), a_1(\xi), a_2(\xi))$ may not be the most efficient test of (3.8.5) even at $\theta = \theta_0, \theta_1, \theta_2$. However, if $L_0(\theta_0)$ [or $L_2(\theta_2)$] is very large, the dominant role of $\delta(b_1(\xi), a_1(\xi))$ [or $\delta(b_2(\xi), a_2(\xi))$] would imply that $E_{\theta_0}(N)$ [or $E_{\theta_2}(N)$] and $E_{\theta_1}(N)$ will be nearly minimized by the Sobel–Wald test procedure, with an appropriate choice of ξ, among all valid tests. In particular, if the preassigned error probabilities are not too small ($\ge .01$ say) the ASN of the Sobel–Wald test often turns out to be smaller than the optimum fixed-sample size for all θ.

Example 3.8.1 (The normal mean with a known variance) Suppose $\{X_i\}$ are iid as normal with unknown mean θ and known variance, which without loss of generality can be taken to be unity. Consider the hypotheses

$$H_0: \theta = \theta_0, \qquad H_1: \theta = \theta_1, \qquad H_2: \theta = \theta_2, \qquad -\infty < \theta_0 < \theta_1 < \theta_2 < \infty.$$

We are dealing with the special case where

$$B(\theta) = \exp(-\theta^2/2), \qquad Q(\theta) = \theta, \qquad R(x) = x, \qquad h(x) = (2\pi)^{-1/2} \exp(-x^2/2).$$

In the Sobel–Wald test we have for $j = 1, 2$,

$$\psi_j = -\Delta_j(\theta_j + \theta_{j-1})/2, \qquad \Delta_j = \theta_j - \theta_{j-1}.$$

Condition (3.8.9) can be restated as

$$\Delta_1\Delta_2(\Delta_1 + \Delta_2) \geq 2\max[b_1\Delta_2 - b_2\Delta_2, a_1\Delta_1 - a_2\Delta_2]$$

and one easily verifies that this condition is also necessary in order that not both H_0 and H_2 be accepted when the two SPRTs are carried out simultaneously. In (3.8.16) we have

$$h_j(\theta) = (\theta_j + \theta_{j-1} - 2\theta)/\Delta_j, \qquad \zeta_\theta = \theta, \quad \mathrm{var}_\theta(R(X)) = 1.$$

B. K. Ghosh (1970, p. 265) has given explicit (approximate) expressions for $E_\theta(N_{W_1})$ and $E_\theta(N_{W_2})$. For the optimum fixed-sample test in (3.8.7) we have $\underline{t} = n\theta_0 + n^{1/2}z_{1-\alpha_0}$, $\bar{t} = n\theta_2 + n^{1/2}z_{\alpha_2}$ and $n = n(\alpha_0, \alpha_1, \alpha_2)$ is determined by the single equation

$$\Phi(\Delta_1 n^{1/2} - z_{1-\alpha_0}) + \Phi(\Delta_2 n^{1/2} + z_{\alpha_2}) = 2 - \alpha_1$$

where z_α is defined by $\alpha = \Phi(z_\alpha)$. Consider the numerical example:

$$\theta_0 = -.1, \qquad \theta_1 = 0, \qquad \theta_2 = .1, \qquad \alpha_0 = \alpha_1 = \alpha_2 = .05.$$

Let $\xi = .025$, then $b_1 = -3.64$, $a_1 = 2.97$, $b_2 = -2.97$, $a_2 = 3.64$. If $L_0(\theta_1) = \alpha_3 = \frac{1}{60}$, then $\xi = \alpha_1 - \alpha_3 = \frac{1}{30}$, and then $b_1 = -4.04$, $a_1 = 2.80$, $b_2 = -2.96$, and $a_2 = 3.35$. Simons (1967a) has computed certain general lower bounds (not necessarily the greatest) for the ASN function of any test that satisfies:

$$L_0(\theta_0) = 0.95 = L_1(\theta_1) = L_2(\theta_2), \qquad L_0(\theta_1) = \tfrac{1}{60}, \qquad \text{and} \qquad L_2(\theta_1) = \tfrac{1}{30}.$$

B. K. Ghosh (1970) has tabulated the determinations of the ASN for various values of θ and these are given in Table 3.8.1.

From Table 3.8.1, it appears that the Sobel–Wald procedures are more efficient than the optimum fixed-sample test for most values of θ. Also, Simons' (1967a) lower bounds indicate some possible optimality of the Sobel–Wald test at $\theta = \theta_0, \theta_1, \theta_2$ among all tests whose $L_0(\theta_0)$, $L_0(\theta_1)$, $L_1(\theta_1)$, $L_2(\theta_1)$, and $L_2(\theta_2)$ are held fixed at prescribed levels. Finally, a comparison between the two Sobel–Wald procedures shows that, if the restriction $L_0(\theta_1) =$

Table 3.8.1

ASN Functions of Two Sobel–Wald Procedures ($\xi = .025$ and $\frac{1}{30}$) and the Fixed-Sample Test in the Normal Case ($\sigma = 1$, $\theta_0 = -.1$, $\theta_1 = 0$, $\theta_2 = .1$, $\alpha_0 = \alpha_1 = \alpha_2 = .05$)[a]

θ		$-.20$	$-.10$	$-.05$	0	.05	.10	.20
Sobel–Wald ASN	$\geq$	242.4	661.4	1080.5	561.0	1080.5	661.4	242.4
($\xi = .025$)	$\leq$	242.5	727.5	1269.6	769.9	1269.6	727.5	242.5
Sobel–Wald ASN	$\geq$	269.5	738.4	1204.4	572.4	992.2	606.9	223.3
($\xi = \frac{1}{30}$)	$\leq$	269.5	806.6	1379.6	731.4	1194.8	670.0	223.3
ASN of any test	$\geq$	109.3	738.4	940.0	572.4	867.5	606.9	99.3
Fixed-sample size					1299.5			

[a] This is Table 5.6 of B. K. Ghosh (1970, p. 267), which is produced here with kind permission of the author and Addison Wesley.

α_3 is not imposed, then $\xi = \alpha_0/2$ should be a good choice [because $\sup_\theta E_\theta(N)$ is minimized].

B. K. Ghosh (1970) also illustrates the Sobel–Wald procedure by applying it to the binomial case. The reader is referred to B. K. Ghosh (1970, pp. 267–269).

REMARK 3.8.1 Because of the monotonicity of the OC functions $L_0(\theta)$, $L_2(\theta)$ [$L_0(\theta)$ is decreasing in θ and $L_2(\theta)$ is increasing in θ], the Sobel–Wald procedure is applicable to test two situations:

$$H_0\colon\ \theta \leq \theta_0, \quad H_1\colon\ \theta = \theta_1, \quad H_2\colon\ \theta \geq \theta_2, \quad \underline{\theta} < \theta_0 < \theta_1 < \theta_2 < \bar{\theta} \tag{3.8.23}$$

$$H_0\colon\ \theta \leq \theta_0, \quad H_1\colon\ \theta' \leq \theta \leq \theta'', \quad H_2\colon\ \theta \geq \theta_2, \quad \underline{\theta} < \theta_0 < \theta' < \theta'' < \theta_2 < \bar{\theta}. \tag{3.8.24}$$

For details on these, see B. K. Ghosh (1970, pp. 269–272).

Simons (1967b) developed a sequential procedure for testing (3.8.5) in the particular case when $\{X_i\}$ are iid as normal with unknown mean θ and known variance σ^2. Numerical computations indicate that the overall efficiency of his test is about the same as the Sobel–Wald procedure for a suitable choice of ξ. Although his procedure is a little more flexible, it would be somewhat difficult to extend it to nonnormal populations. Billard and Vagholkar (1969) have proposed a sequential test for two-sided alternatives.

The Sobel–Wald procedure can be applied to populations other than Darmois–Koopman family. B. K. Ghosh (1970) applies it to the exponential translation parameter case. One would usually require some condition of the type (3.8.9) to ensure (3.8.'4) under (3.8.12).

Example 3.8.2 (Uniform distribution with unknown mean) Let $f(x;\theta) = 1$ if $\theta - \frac{1}{2} \le x \le \frac{1}{2} + \theta$ and zero elsewhere. Then, consider the hypotheses

$$H_0\colon\ \theta = \theta_0, \qquad H_1\colon\ \theta = \theta_1, \qquad H_2\colon\ \theta = \theta_2$$

with $0 < \theta_1 - \theta_0 < 1$, $0 < \theta_2 - \theta_1 < 1$.

If $b_j < 0 < a_j$ for $j = 1, 2$, the critical inequality of $\delta(b_j, a_j)$ for testing H_{j-1} against H_j is $\theta_j - \frac{1}{2} < X_{(n)} < \theta_{j-1} + \frac{1}{2}$ for $n \ge 1$ where $X_{(n)} = \max(X_1, \ldots, X_n)$. Evidently, it is impossible for $\delta(b_1, a_1)$ to accept H_0 if $\delta(b_2, a_2)$ accepts H_2, and conversely. Hence, if $\delta(b_1, b_2, a_1, a_2)$ is carried out with any $b_1 < 0 < a_1$, $b_2 < 0 < a_2$, the identities in (3.8.14) hold. Thus, one can obtain

$$L_0(\theta) = \begin{cases} 1 & \text{if } \theta \le \theta_0, \\ (\theta_1 - \theta)/(\theta_1 - \theta_0) & \text{if } \theta_0 < \theta < \theta_1, \\ 0 & \text{if } \theta \ge \theta_1; \end{cases}$$

$$L_1(\theta) = \begin{cases} 0 & \text{if } \theta \le \theta_0, \\ (\theta - \theta_0)/(\theta_1 - \theta_0) & \text{if } \theta_0 < \theta \le \theta_1, \\ (\theta_2 - \theta)/(\theta_2 - \theta_1) & \text{if } \theta_1 < \theta < \theta_2, \\ 0 & \text{if } \theta \ge \theta_2; \end{cases}$$

$$L_2(\theta) = \begin{cases} 0 & \text{if } \theta \le \theta_1, \\ (\theta - \theta_1)/(\theta_2 - \theta_1) & \text{if } \theta_1 < \theta < \theta_2, \\ 1 & \text{if } \theta \ge \theta_2. \end{cases}$$

In particular, $L_0(\theta_0) + L_1(\theta_1) + L_2(\theta_2) = 3$ so that the test in the present case always satisfies (3.8.3) for testing (3.8.5) or (3.8.23). The ASN function of $\delta(b_1, b_2, a_1, a_2)$ can be evaluated exactly without much difficulty.

REMARK 3.8.2 A decision procedure similar to that of Sobel and Wald was proposed by Armitage (1947) as an alternative to the sequential t-test (with two-sided alternatives). Essential inequalities on the OC functions are pointed out, but no attempt is made to determine the complete OC and ASN functions.

The problem of testing k ($k > 3$) hypotheses will be taken up in Section 3.9.

Problems

Using $\alpha = \beta = .1$ carry out the Sobel–Wald sequential procedure for:

1. The binomial distribution with $H_0\colon \theta = \frac{1}{4}$, $H_1\colon \theta = \frac{1}{2}$, and $H_2\colon \theta = \frac{3}{4}$.

2. The Poisson distribution with $H_0\colon \theta = 1$, $H_1\colon \theta = 2$, and $H_2\colon \theta = 3$.

3. The negative exponential distribution having density $f(x;\theta) = \theta^{-1}\exp(-x/\theta)$ for $x > 0$ for $H_0\colon \theta = 1$, $H_1\colon \theta = 2$, and $H_2\colon \theta = 3$.

4. The Weibull distribution having density $f(x;\theta) = \theta x^{\theta-1}\exp(-x^\theta)$, $x > 0$ for $H_0\colon \theta = 1$, $H_1\colon \theta = 2$, and $H_2\colon \theta = 3$.

3.9 Sequential k-Decision Problems

In this section we shall consider the problem of testing k ($k > 3$) hypotheses. Suppose that the hypotheses in (3.8.1) are all simple, $\omega_j = \{\theta_j\}$, and that $k > 3$. If the given family of densities is of the form

$$f_n(\mathbf{x}_n; \theta) = B_n(\theta) \exp[Q_n(\theta) T_n(\mathbf{x}_n)] h_n(\mathbf{x}_n)$$

where $\mathbf{x}_n = (x_1, \ldots, x_n)$ and $Q_n(\theta)$ is strictly monotone in θ on $[\underline{\theta}, \bar{\theta}]$, one can obtain without any difficulty an optimum fixed-sample procedure that satisfies (3.8.2) [see B. K. Ghosh (1970, Problem 5.35)]. On the other hand, if $\{X_i\}$ are iid according to the Darmois–Koopman form, any extension (by operating simultaneously $k-1$ SPRTs) of the Sobel–Wald procedure does not seem to possess simple properties such as those of Theorem 3.8.1. Armitage (1950) and, for special problems, Paulson (1963) and Simons (1967c) have developed sequential procedures to solve such k-decision problems. In the following we shall briefly describe Armitage's (1950) solution which is intuitively appealing and quite general in that no assumption regarding the form of the underlying distribution or independence of $\{X_i\}$ will be made.

3.9.1 Armitage's Procedure

Let $I_{jl}(n)$ denote the inequalities

$$-a_{jl} < S_n(j, l) = \ln[f_n(\mathbf{X}_n; \theta_l)/f_n(\mathbf{X}_n; \theta_j)] < a_{lj}$$

for $j \neq l = 0, 1, \ldots, k-1$ and $n \geq 1$, where $\{a_{jl}\}$ are $k(k-1)$ positive constants. Notice that $I_{jl}(n) = I_{lj}(n)$ for every $n \geq 1$ and that $I_{jl}(n)$ is also the critical inequality of the SPRT $\delta(-a_{jl}, a_{lj})$ for testing H_j: $\theta = \theta_j$ against H_l: $\theta = \theta_l$. Armitage's procedure $\delta(\{a_{jl}\})$ for $k > 2$ is then defined as follows: Observe $\{X_i\}$ successively and at stage $n \geq 1$

(i) acc H_j, for any $j \neq l$ if all $k-1$ lower inequalities in $I_{j0}, \ldots, I_{j,j-1}, I_{j,j+1}, \ldots, I_{j,k-1}$ are violated;
(ii) acc H_l if all $k-1$ upper inequalities in $I_{0,l}, \ldots, I_{l-1,l}, I_{l+1,l}, \ldots, I_{k-1,l}$ are violated; (3.9.1)
(iii) continue sampling otherwise by observing X_{n+1}.

It is easily verified that when $k = 2$, $\delta(\{a_{jl}\})$ is identical to the SPRT $\delta(-a_{01}, a_{10})$ of H_0 against H_1, but for $k = 3$, $\delta(\{a_{jl}\})$ is generally not the same as the Sobel–Wald procedure. The rule in (3.9.1) is equivalent to stating that the $k(k-1)/2$ SPRTs, $\{\delta(-a_{jl}, a_{lj})\}$ $(j < l = 1, \ldots, k-1)$, are carried out simultaneously and H_j is accepted at stage n if and only if the $k-1$ SPRTs pertaining to H_j accept H_j at stage n. It may, of course, happen that $\delta(-a_{jl}, a_{lj})$ has already accepted H_j or H_l at stage m but is carried out beyond that stage because $\delta(\{a_{jl}\})$ did not reach a decision by stage m. A sample path can be constructed so as to illustrate the different regions.

Next, if $\delta(\{a_{jl}\})$ does not reach a terminal decision at stage $m \le n$, at least one of the $k(k-1)/2$ inequalities $\{I_{jl}(m)\}$ $(j < l = 0, 1, \ldots, k-1)$ must hold for every $1 \le m \le n$. Hence, if N denotes the stopping variable,

$$P_\theta(N > n) \le P_\theta(\text{at least one of } I_{01}(n), \ldots, I_{k-2,k-1}(n) \text{ holds})$$
$$\le \sum_{j<l}\sum P_\theta(-a_{jl} < S_n(j, l) < a_{lj}).$$

Hence, $\delta(\{a_{jl}\})$ will certainly terminate if the individual SPRTs terminate with probability one. In particular, if the $\{X_i\}$ are iid, then by Theorem 2.2.1 $\delta(\{a_{jl}\})$ is closed. Next, using the same reasoning as in Theorem 2.1.1 (considering all values of $\mathbf{X}_n$, $n \ge 1$, which lead to the acceptance of H_j), we obtain

$$L_j(\theta_l) \le L_j(\theta_j)\exp(-a_{jl}) \le \exp(-a_{jl}) \qquad \text{for any} \quad j \ne l,$$

whether $\delta(\{a_{jl}\})$ is closed or open. Further, if $\delta(\{a_{jl}\})$ is closed, then $L_0(\theta) + \cdots + L_{k-1}(\theta) = 1$ for any θ, and consequently,

$$L_j(\theta_j) \ge 1 - \sum_{l \ne j} \exp(-a_{lj}) \qquad \text{for} \quad j = 0, 1, \ldots, k-1. \tag{3.9.2}$$

The inequalities in (3.9.2) would enable one to choose the constants of $\delta(\{a_{jl}\})$ such that, for all j, the requirements $L_j(\theta_j) \ge 1 - \alpha_j$ are guaranteed for specified $\{\alpha_j\}$. For instance, if $\alpha_j = \alpha > 0$ for all j, then the choice $a_{jl} = \ln[(k-1)/\alpha]$ for all $j \ne l$ will ensure that $L_j(\theta_l) \le \alpha/(k-1)$ for every $j \ne l$. Unfortunately, (3.9.2) may not be close to equality and hence such a choice of $\{a_{jl}\}$ will lead to an unnecessarily large ASN. Further, we have no idea of the ASN function of the Armitage procedure for $k > 3$. Simons (1967b) has explored the particular case when $k = 3$ and $\{X_i\}$ are iid as normal with unknown mean θ and known variance σ^2. His results seem to indicate that for $k > 3$, and for nonnormal $\{X_i\}$, $\delta(\{a_{jl}\})$ may possess some optimum properties. Since the procedure is characterized by $k(k-1)$ constants, $\delta(\{a_{jl}\})$ can be designed to fulfill $k(k-2)$ additional requirements on the OC and (or) ASN functions, in addition to the ones in (3.8.2). It is worthwhile to explore Armitage's procedure.

3.9.2 Paulson's Procedure Concerning the Unknown Mean of a Normal Population

Let θ denote the unknown mean and σ^2 the given variance of the normal population. Let $\{I_j\}$ ($j = 1, 2, \ldots, k$) denote k nonoverlapping intervals whose union $\bigcup_{j=1}^{k} I_j$ is the real line. It is of considerable practical interest to decide on the basis of a sample $\{X_i\}$ of independent observations on X to which of the k intervals the unknown mean θ belongs. For example, a manufacturer might wish to make k different decisions concerning the selling price of his output, according to which of the k intervals contains the unknown θ. A sequential solution to the above problem was given by Sobel and Wald (1949) for the special case when $k = 3$ and σ was assumed known. Paulson (1963) has given a solution for the general case (both when σ is known and unknown) using a different method.

Case 1 (*σ is known*) Let $\lambda_1 < \lambda_2 < \cdots < \lambda_{k-1}$, and let I_1 denote the interval $(-\infty, \lambda_1]$; for $s = 2, 3, \ldots, k-1$, let I_s denote the half-open interval $(\lambda_{s-1}, \lambda_s]$ and let I_k denote the interval (λ_{k-1}, ∞). Let δ_j ($j = 1, \ldots, k$) denote the decision that θ is an element of I_j. We can also assume [see Sobel and Wald (1949)] that one can specify on the basis of practical considerations an indifference zone (β_{1j}, β_{2j}) about each point λ_j, in which it is a matter of indifference whether decision δ_j or δ_{j+1} is made. Also, let $r(\delta_j; \theta)$ be a function taking values 0 or 1 which represents the error made in selecting δ_j when θ is the true parameter. Further, we assume that $r(\delta_j; \theta)$ satisfies

$$r(\delta_1; \theta) = 0 \qquad \text{if} \quad -\infty < \theta < \beta_{21}$$

and otherwise is equal to unity. For $s = 2, 3, \ldots, k-1$,

$$r(\delta_s; \theta) = 0 \qquad \text{if} \quad \beta_{1,s-1} < \theta < \beta_{2,s}$$

and otherwise is equal to 1.

$$r(\delta_k; \theta) = 0 \qquad \text{if} \quad \beta_{1,k-l} < \theta < \infty$$

and otherwise is equal to 1. We now proceed to find a sequential procedure for choosing one of the k decisions $(\delta_1, \ldots, \delta_k)$ so that the probability of making an error does not exceed a preassigned value α for all values of θ. We shall repeatedly make use of the following lemma.

Lemma 3.9.1 Let $Y_1, Y_2, \ldots$ be iid as Y with $E(Y) < 0$. Then, if $a > 0$ and

$j = 1, 2, \ldots$

$$P\left(\sum_{i=1}^{j} Y_i > a \text{ for at least one value of } j\right) \leq \exp(-t_0 a) \qquad (3.9.3)$$

where $t_0 \neq 0$ satisfies the relation, $E[\exp(t_0 Y)] = 1$.

Proof From Remark A-1.1 (see Appendix A-1) we have $t_0 > 0$ since t_0 and EY are of opposite signs.

Let $f(x)$ be the true probability (density) function of Y_i and $g(x)$ be any other probability (density) function. Since $A \leq (1 - \beta)/\alpha \leq 1/\alpha$ (see Theorem 2.1.1 giving the fundamental inequalities of Wald's SPRT), we have

$$P_f\left(\prod_{i=1}^{r} \{g(Y_i)/f(Y_i)\} \geq 1/\alpha \text{ for at least one value of } r,\ r = 1, 2, \ldots\right)$$
$$\leq P_f\left(\prod_{i=1}^{r} \{g(Y_i)/f(Y_i)\} \geq A \text{ for at least one value of } r,\ r = 1, 2, \ldots\right)$$
$$\leq \alpha.$$

Now, set $g(x) = \exp(t_0 x) f(x)$ and let $\alpha = \exp(-t_0 a)$. Then, the preceding inequality becomes

$$P\left(\exp\left(t_0 \sum_{1}^{r} Y_i\right) \geq \exp(t_0 a), \text{ for at least one value of } r,\ r = 1, 2, \ldots\right)$$
$$\leq \exp(-t_0 a).$$

That is, since $t_0 \neq 0$

$$P\left(\sum_{i=1}^{r} Y_i \geq a \text{ for at least one value of } r,\ r = 1, 2, \ldots\right) \leq \exp(-t_0 a).$$

This completes the proof Lemma 3.9.1.

Applying Lemma 3.9.1 to $Y_i = (X_i - \theta - d)/\sigma^2$ where $d > 0$ we find that, since $t_0 = 2d$

$$P\left(\sum_{i=1}^{j} [(X_i - \theta - d)/\sigma^2] > a \text{ for at least one value of } j\right) \leq e^{-2da}, \qquad (3.9.4)$$

which is equivalent to

$$P(\overline{X}_j - d - a\sigma^2/j > \theta \text{ for at least one value of } j) \leq e^{-2da} \qquad (3.9.5)$$

where $\overline{X}_j = \sum_{i=1}^{j} X_i/j$. Applying (3.9.3) again with $Y_i = (-X_i + \theta - d)/\sigma^2$ we find, in an analogous manner

$$P(\overline{X}_j + d + a\sigma^2/j < \theta \text{ for at least one value of } j) \leq e^{-2da}. \quad (3.9.6)$$

Now setting $e^{-2da} = \alpha/2$ and applying the inequality

$$P(E_1 \cap E_2) \geq 1 - P(\overline{E}_1) - P(\overline{E}_2)$$

to (3.9.5) and (3.9.6) we find that

$$P(\overline{X}_j - d - \sigma^2 \ln(2/\alpha)/2jd \leq \theta \leq \overline{X}_j + d + \sigma^2 \ln(2/\alpha)/2jd,$$
$$\text{for every } j, j = 1, 2, \ldots) \geq 1 - \alpha \quad (3.9.7)$$

where all the logarithms are to the base e.

For the application of (3.9.7) it is desirable to choose d so as to minimize the value of j under the condition

$$2[d + \sigma^2 \ln(2/\alpha)/2jd] = \Delta.$$

Rewriting this as

$$2jd(\Delta/2 - d) = \sigma^2 \ln(2/\alpha)$$

and setting the derivative with respect to d equal to zero we find the solution to be $d = \Delta/4$.

First let $\Delta_j = \beta_{2j} - \beta_{1j}$, the length of the jth indifference zone, let $\Delta = \min(\Delta_1, \ldots, \Delta_{k-1})$ and let

$$u_r = \max_{1 \leq i \leq r} [\overline{X}_i - \Delta/4 - 2\sigma^2 \ln(2/\alpha)/i\Delta],$$

$$v_r = \min_{1 \leq i \leq r} [\overline{X}_i + \Delta/4 + 2\sigma^2 \ln(2/\alpha)/i\Delta].$$

We now describe the sequential procedure as follows:

At the rth stage of the experiment ($r = 1, 2, \ldots$) we obtain X_r and then calculate u_r and v_r on the basis of the observed values of $X_1, X_2, \ldots, X_r$. The experiment is terminated at the rth stage if either $u_r > v_r$ or if $u_r \leq v_r$ and the interval (u_r, v_r) falls entirely within one of the k intervals

$$I_1' = (-\infty, \beta_{21}), \qquad I_2' = (\beta_{11}, \beta_{22}), \qquad I_3' = (\beta_{12}, \beta_{23}), \ldots,$$
$$I_{k-1}' = (\beta_{1,k-2}, \beta_{2,k-1}), \qquad I_k' = (\beta_{1,k-1}, \infty).$$

If the experiment terminates at the rth stage with $u_r > v_r$ (the probability of this will usually be quite small), we make the decision corresponding to the interval among $\{I_j\}$ to which $\bar{X}_r$ belongs. If the experiment terminates at the rth stage with $u_r \le v_r$, we make a decision corresponding to the interval among $\{I_j'\}$ in which (u_r, v_r) lies so that if (u_r, v_r) falls in, say, I_t', we select δ_t, while if (u_r, v_r) falls inside two intervals, say I_t' and I_{t+1}' we can choose between δ_t and δ_{t+1} by tossing a coin. If no decision is reached at the rth stage, we go on to the $(r + 1)$th stage by observing X_{r+1} and repeat the above procedure.

Let N denote the stopping variable for the sequential procedure. It is clear that the experiment must have terminated on or before the jth stage if

$$2[\Delta/4 + 2\sigma^2 \ln(2/\alpha)/j\Delta] < \Delta,$$

so that we must have $N < 1 + 8\sigma^2 \ln(2/\alpha)/\Delta^2$. The sequential procedure is consequently closed. From the definition of $r(\delta_j;\, \theta)$ and the specification of the sequential procedure, it is clear that no error can be committed when θ is really contained in (u_N, v_N). Since it follows from (3.9.7) that $P(u_N \le \theta \le v_N) \ge 1 - \alpha$, we can conclude that the probability the sequential procedure will lead to an error does not exceed α for all θ.

Paulson (1963) points out that when $k = 3$ his procedure may be less efficient than the Sobel–Wald procedure. However, there is one compensation, namely, when the sequential procedure terminates, in addition to making a decision regarding the interval in which θ lies, we can also assert that $u_n \le \theta \le v_n$ with probability $\ge 1 - \alpha$. This remark also suggests a modification of the original problem which might be of interest in itself and which can be formulated as follows:

We want a sequential procedure that will simultaneously select one of the decisions $(\delta_1, \delta_2, \ldots, \delta_k)$ with the probability of an error $\le \alpha$, and also provide a confidence interval for θ of length $\le l$ with confidence coefficient $\ge 1 - \alpha$ (l is a constant which is assumed to be $> \Delta$).

To solve this modified problem, we use the sequential procedure described earlier with the modification that we continue sampling until not only (u_r, v_r) falls inside one of the intervals $\{I_j'\}$ but in addition $v_r - u_r \le l$. (The minor complication caused by the fact that it is possible to have $u_r > v_r$ can be handled by taking $\bar{X}_r \pm l/2$ as the confidence limits for θ when this unlikely event occurs.) It is clear that with this modification the probability is still $\le \alpha$ of making an incorrect decision and the probability is $\ge 1 - \alpha$ that $u_n \le \theta \le v_n$ and that $0 \le v_n - u_n \le l$.

Case 2 (*σ is unknown*) Let us start the experiment by first taking a preliminary sample of n_0 measurements $X_1, \ldots, X_{n_0}$. Let $m = n_0 - 1$ and $s^2 = \sum_{i=1}^{n_0} (X_i - \bar{X}_{n_0})^2/m$. Also, let $\{Z_i\}$ be a sequence of independent and normally distributed random variables with mean $-d/\sigma^2$ and variance $1/\sigma^2$ which is independent of the sequence $\{X_i\}$. Making use of the fact that $\overline{X}_{n_0}$ is independent of s^2, we obtain

$$\begin{aligned} P&\left[\sum_{i=1}^{r} \{(X_i - \theta - d)/s^2\} > a \text{ for at least one } r,\ r = n_0, n_0 + 1, \ldots \mid s^2\right] \\ &= P\left[\sum_{i=1}^{r} Z_i \geq a(s^2/\sigma^2) \text{ for at least one } r,\ r = n_0, n_0 + 1, \ldots \mid s^2\right] \\ &< P\left[\sum_{i=1}^{r} Z_i \geq a(s^2/\sigma^2) \text{ for at least one } r,\ r = 1, 2, \ldots \mid s^2\right] \\ &\leq \exp(-2ads^2/\sigma^2) \qquad (\text{since } t_0 = 2d \text{ in Lemma 3.9.1}). \end{aligned} \tag{3.9.8}$$

Consequently

$$\begin{aligned} P&\left(\sum_{i=1}^{r} [(X_i - \theta - d)/s^2] > a \text{ for at least one } r,\ r = n_0, n_0 + 1, \ldots\right) \\ &< E[\exp(-2ads^2/\sigma^2] = (1 + 4ad/m)^{-m/2}. \end{aligned} \tag{3.9.9}$$

If we set $d = \Delta/4$ and let $\bar{a} = (m/\Delta)[(2/\alpha)^{2/m} - 1]$ denote the value of a for which $[1 + 4ad/m]^{-m/2} = \alpha/2$ we obtain from (3.9.9)

$$P(\overline{X}_r - \Delta/4 - \bar{a}s^2/r > \theta \text{ for at least one } r,\ r = n_0, n_0 + 1, \ldots) \leq \alpha/2. \tag{3.9.10}$$

Similarly, starting with $(-X_i + \theta - d)/s^2$, we find that

$$P(\overline{X}_r + \Delta/4 + \bar{a}s^2/r < \theta \text{ for at least one } r,\ r = n_0, n_0 + 1, \ldots) \leq \alpha/2. \tag{3.9.11}$$

Therefore, combining (3.9.10) and (3.9.11) we have

$$\begin{aligned} P(\overline{X}_r - \Delta/4 - \bar{a}s^2/r \leq \theta \leq \overline{X}_r + \Delta/4 + \bar{a}s^2/r, \\ \text{for all values of } r,\ r = n_0, n_0 + 1, \ldots) > 1 - \alpha. \end{aligned} \tag{3.9.12}$$

Let

$$u_r' = \max_{n_0 \leq j \leq r} (\overline{X}_j - \Delta/4 - \bar{a}s^2/j)$$

and

$$v_r' = \min_{n_0 \leq j \leq r} (\overline{X}_j + \Delta/4 + \bar{a}s^2/j).$$

Now, we can specify the sequential procedure when σ^2 is unknown. We start by taking n_0 observations, compute s^2, and for $r = n_0, n_0 + 1, \ldots,$ we then use the sequential procedure described earlier with u_r and v_r replaced by u_r' and v_r' respectively. It follows from (3.9.12) that the probability of making an error is still $<\alpha$ for all θ. The practical value of the procedure described in case 2 is somewhat limited by the lack of an efficient rule for selecting n_0. This seems to be still an unsolved problem.

Comparing the Means of k Experimental Categories with an Unknown Standard Let Π_0 denote the standard (or control) category, let $\Pi_1, \Pi_2, \ldots, \Pi_k$ denote the k experimental categories, and let X_{jr} denote the rth observation from Π_j. Assume that for each r, X_{jr} is normally distributed with mean θ_j and variance σ_j^2, and that for all r and j ($j = 0, 1, \ldots, k$ and $r = 1, 2, \ldots$) all observations are independent. Also, Π_v is said to be superior to Π_0 if $\theta_v > \theta_0$ ($v = 1, 2, \ldots, k$). First we shall consider the problem of finding a closed sequential procedure for classifying each of the k experimental categories as superior or inferior which will satisfy the requirement that the probability is at least $1 - \alpha$ that all experimental categories with means $\leq \theta_0$ or $\geq \theta_0 + \Delta$ are classified correctly. First, let $Z_{jr} = X_{jr} - X_{0r}$, let $\bar{Z}_{jn} = \sum_{r=1}^{n} Z_{jr}/n$, let $\tilde{\sigma}_j^2 = \sigma_j^2 + \sigma_0^2$. Let

$$u_j(d, n, \bar{Z}_{jn}, \alpha) = \bar{Z}_{jn} + d + [\tilde{\sigma}_j^2 \ln(1/\alpha)]/2dn,$$

$$v_j(d, n, \bar{Z}_{jn}, \alpha) = \bar{Z}_{jn} - d - [\tilde{\sigma}_j^2 \ln(1/\alpha)]/2dn,$$

$$u_{jn}(d, \alpha) = \min_{1 \leq r \leq n} [u_j(d, r, \bar{Z}_{jn}, \alpha)],$$

$$v_{jn}(d, \alpha) = \max_{1 \leq r \leq n} [v_j(d, r, \bar{Z}_{jn}, \alpha)].$$

When the variances are known the desired closed sequential procedure can be obtained as follows. At the first stage of the experiment start with one observation on all $k + 1$ categories. At the rth stage ($r = 2, 3, \ldots$) take one observation on the standard category and one observation on each of the experimental categories not yet classified. Classify category Π_j as inferior as soon as $u_j(d, n, \bar{Z}_{jn}, \alpha/k) < \Delta$ and classify Π_j as superior as soon as $v_j(d, n, \bar{Z}_{jn}, \alpha/k) > 0$. The experiment is terminated as soon as all categories are classified. Paulson (1964b) recommends using $d = 3\Delta/8$ in order to minimize the number of observations required to classify all the categories.

Let us next consider a mixed decision and estimation problem. In addition to classifying all k experimental categories subject to the requirement that all categories Π_j with θ_j either $\leq \theta_0$ or $> \theta_0 + \Delta$ are classified correctly with probability $\geq 1 - \alpha$, we also want to have a simultaneous confidence interval for $\{\theta_j - \theta_0\}$ ($j = 1, 2, \ldots, k$) when the experiment is terminated with joint confidence $1 - \gamma$ so that the width of the confidence interval for each category classified as superior shall not exceed L $(L > \Delta)$. This can be accomplished as follows. Stop as before, taking observations on any category when it is classified as inferior, but continue taking observations on any category Π_j after it has been classified as superior until $u_{jn}(d, \gamma/2k) - v_{jn}(d, \gamma/2k) \leq L$. If n_j is the number of observations taken with category Π_j when the experiment is terminated, one can easily see that $P[v_{j,n_j}(d, \gamma/2k) \leq \theta_j - \theta_0 \leq u_{j,n_j}(d, \gamma/2k)$ for every j, $j = 1, 2, \ldots, k] \geq 1 - \gamma$, and obviously the width of the confidence interval for each category classified as superior cannot exceed L.

When the $k + 1$ categories have a common unknown variance, start by taking n_0 measurements from each of the $k + 1$ categories, let

$$s^2 = \sum_{j=0}^{k} \sum_{r=1}^{n_0} (X_{jr} - \overline{X}_{j,n_0})^2/(k+1)(n_0 - 1),$$

take $v = (k + 1)(n_0 - 1)$, and take observations one at a time, use the procedure just described with $\tilde{\sigma}_j^{\,2}$ replaced by $2s^2$, $\ln(1/\alpha)$ replaced by $a(\alpha) = [(1/\alpha)^{2/v} - 1](v/2)$, and $u_{jn}(d, \alpha)$ and $v_{jn}(d, \alpha)$ defined as the minimum and maximum on $n_0 \leq r \leq n$ instead of $1 \leq r \leq n$. Although an optimal choice of n_0 is not available, Paulson (1964b) recommends selecting n_0 so that $a(\alpha) \leq 1.1 \ln(1/\alpha)$.

Deciding Which of k Nonoverlapping Intervals Contains σ^2 Let $\{X_r\}$ $(r = 1, 2, \ldots)$ denote a sequence of independent and normally distributed random variables having mean θ and variance σ^2. Let

$$s_n^{\,2} = \sum_{r=1}^{n} (X_r - \overline{X}_n)^2/(n-1).$$

Also, let $I_1 = [0, c_1)$, $I_2 = [c_1, c_2)$, $\ldots I_k = [c_{k+1}, \infty)$ denote the k nonoverlapping intervals whose union is $(0, \infty)$. Let D_j denote the decision that σ^2 falls in I_j $(j = 1, 2, \ldots, k)$. Assume that on the basis of practical considerations we can find an indifference zone (a_j, b_j) about each end point c_j in which it does not matter whether decision D_j or D_{j+1} is made. Let $W(D_j, \sigma^2)$ denote the loss incurred in taking decision D_j when σ^2 is the true value of the unknown variance. Also let $a_0 = 0$ and $b_n = \infty$ and assume that for each j $(j = 1, 2, \ldots, k)$

$$W(D_j, \sigma^2) = \begin{cases} 0 & \text{if } \; a_{j-1} < \sigma^2 < b_j, \\ 1 & \text{otherwise.} \end{cases}$$

We desire a closed sequential procedure for choosing one of the k decisions $D_1, D_2, \ldots, D_k$ so that the probability of making an error shall not exceed α.

Let $w^2 = \min(b_1/a_1, \ldots, b_{k-1}/a_{k-1})$, let I_j' denote the interval (a_{j-1}, b_j) for $j = 1, 2, \ldots, k$ and take $\lambda = 1 + .7(w - 1)$. Then, the solution is as follows: Take measurements one at a time, and stop sampling and make a decision as soon as the interval $[v_{1n}(\alpha/2), u_{1n}(\alpha/2)]$ falls inside one of the intervals I_j', $j = 1, 2, \ldots, k$. If $[v_{1n}(\alpha/2), u_{1n}(\alpha/2)]$ falls in interval I_j', make decision D_j, and if it falls in $I_j' \cap I_{j+1}'$, then randomize between D_j and D_{j+1}, where

$$u_{1n}(\alpha) = \begin{cases} \infty & \text{for } n \le n^*(\alpha) = [1 + \ln(1/\alpha)/\ln \lambda] \\ \min\limits_{n^*(\alpha) < r \le n} \{s_r^2(\lambda^2 - 1)/2[\ln \lambda - (\ln(1/\alpha))/(r-1)]\}, & \text{for } n > n^*(\alpha), \end{cases}$$

$$v_{1n}(\alpha) = \max_{1 \le r \le n} \{s_r^2(\lambda^2 - 1)/2\lambda^2[\ln \lambda + (\ln(1/\alpha))/(r-1)]\}.$$

It can easily be shown [see, Eq. (4.10.12)] that the probability of making an error is at most α.

Comparing the Variances of k Experimental Categories with an Unknown Standard Let Π_0 (standard), $\Pi_1, \ldots, \Pi_k$ denote the $k + 1$ categories having variances $\sigma_0^2, \sigma_1^2, \ldots, \sigma_k^2$ respectively. An experimental category Π_j is said to be superior to the standard if $\sigma_j^2 < \sigma_0^2$. We shall consider the problem of finding a closed sequential procedure for classifying the k experimental categories as superior or inferior with respect to the standard so that the probability should be $\ge 1 - \alpha$ of correctly classifying all experimental categories Π_j such that $\sigma_j^2 \ge \sigma_0^2$ or $\sigma_j^2 \le \sigma_0^2/R^2$, where $R > 1$. First, let $\rho_j^2 = \sigma_j^2/\sigma_0^2$ and $T_{jn} = \sum_{r=1}^{n}(X_{jr} - \bar{X}_{jn})^2/\sum_{r=1}^{n}(X_{0r} - \bar{X}_{0n})^2$. The closed sequential procedure for this problem is given as follows: Start by taking $n^*(\alpha/k) + 1$ observations on each of the $k + 1$ categories. At each subsequent stage we take one observation on the standard, and one on each of the categories that are not yet classified. A category Π_j $(j = 1, \ldots, k)$ is classified as superior as soon as $u(n, T_{jn}, \alpha/k)$ is < 1, and is classified as inferior as soon as $v(n, T_{jn}, \alpha/k) > 1/R^2$. The sampling terminates as soon as all categories are classified.

3.9.3 Likelihood Ratio Tests for k-Decision Problems

Sequential tests based on the generalized likelihood ratio appeared in a striking way in a paper of Schwarz (1962). He showed that Bayes procedures for testing two hypotheses separated by an indifference zone, in the case of Darmois–Koopman families, have regions in the plane of n, S_n (the cumulative sum sufficient statistic) which when scaled down by the $\ln(1/c)$ (c denoting the cost per observation) approach a limiting region as $c \to 0$. This limiting region, scaled up by the factor $\ln(1/c)$, is given by the rule: stop when the maximum likelihood in either of the hypotheses is less than c times the unrestricted maximum likelihood. Schwarz also showed how the boundaries of the likelihood ratio test regions can be computed explicitly. These results strongly suggest the use of sequential likelihood ratio tests in applications and raise the question of how close to optimal is their performance. Wong (1968) has demonstrated two forms of asymptotic optimality. The first is the asymptotic Bayes property: the average risk of the likelihood ratio procedure is asymptotic to the Bayes risk as $c \to 0$. The second is reminiscent of the optimality property of the SPRT. The expected sample size of the likelihood ratio test at every point in the indifference zone was shown to be asymptotically minimum as $c \to 0$ compared to tests whose error probabilities at the end points of the indifference interval are not larger and which also satisfy a restriction on the growth of the expected sample sizes at the end points as $c \to 0$. In Section 3.7 we have presented Lorden's (1973) results pertaining to open-ended likelihood ratio tests for the Darmois–Koopman family of densities. Here, we shall present his extensions of these results to k-decision problems.

Let $X_1, X_2, \ldots$ be iid random variables having the probability (density) function

$$f(x; \theta) = \exp[\theta x - b(\theta)] \qquad \underline{\theta} < \theta < \bar{\theta}$$

with respect to a nondegenerate σ-finite measure. As noted in Section 3.7, $b(\cdot)$ is necessarily convex and infinitely differentiable on $(\underline{\theta}, \bar{\theta})$, which need not be the entire natural parameter space of the family. Also $E_\theta(X) = b'(\theta)$ and $\text{var}_\theta(X) = b''(\theta)$. Let $S_n = X_1 + \cdots + X_n$, $n = 1, 2, \ldots$, and define the log-likelihood function $L_n(\theta) = \theta S_n - nb(\theta)$. For any n and S_n, the log-likelihood is a strictly concave function of θ and hence is either monotonic over $(\underline{\theta}, \bar{\theta})$ or else has a unique maximum at $\hat{\theta}$, the maximum likelihood estimate of θ. In the latter case, $b'(\hat{\theta}) = S_n/n$. If the log-likelihood function is monotonic on $(\underline{\theta}, \hat{\theta})$, set $\hat{\theta} = \bar{\theta}$ or $\underline{\theta}$ according as $L_n(\theta)$ is increasing or decreasing. In any case, denote the supremum of $L_n(\theta)$ for $\underline{\theta} < \theta < \bar{\theta}$ by $L_n(\hat{\theta})$.

The statistical problem is specified by $s + 1$ $(s \geq 2)$ intervals:

$$H_1\colon (\theta_0, \theta_1],\ H_2\colon [\theta_1, \theta_2], \ldots, H_{s+1}\colon [\theta_s, \theta_{s+1})$$

where $\underline{\theta} = \theta_0 < \theta_1 < \cdots < \theta_{s+1} = \bar{\theta}$, together with $k \geq 2$ decisions and the requirement

$$P_\theta(j\text{th decision}) \leq \alpha_{ij} \qquad \text{for all } \theta \in H_i (i = 1, \ldots, s+1; j = 1, \ldots, k), \tag{3.9.13}$$

where the α_{ij} belong to (0, 1] and satisfy the assumptions:

(I) $\underset{\sim}{\alpha} = \min \alpha_{ij} \leq \frac{1}{3}$
(II) $\max_j \min(\alpha_{ij}, \alpha_{i+1,j}) = 1$ for $i = 1, \ldots, s$.

Assumption (I) is merely a convenience which makes $(\ln)^2 \underset{\sim}{\alpha}^{-1} > 1$. Assumption (II) is a "separation of hypotheses" type of condition ensuring that adjacent H_i have at least one "correct" decision in common($(\ln)^2 = \ln \ln$). Define

$$\bar{\alpha} = \max\{\alpha_{ij} | \alpha_{ij} < 1\} \qquad \text{and} \qquad r = \ln \bar{\alpha}/\ln \underset{\sim}{\alpha} \geq 1. \tag{3.9.14}$$

The bounds obtained in the proof of the theorem can be chosen to depend on the values of the $\alpha_{ij} < 1$ only through $\underset{\sim}{\alpha}$ and r. To obtain bounds in the theorem which are uniform on $(\underline{\theta}, \bar{\theta})$ it is necessary that, for example, in a one-sided SPRT of θ_2 versus a θ near $\underline{\theta}$ the extra sampling due to "excess over the boundary" be bounded uniformly in θ. A sufficient condition in order to meet requirements of this type is the following assumption.

(III) $(\theta - \theta_1)^2 b''(\theta)/[(\theta - \theta_1)b'(\theta) - (b(\theta) - b(\theta_1))]^2$ is bounded above as $\theta \to \underline{\theta}, \bar{\theta}$.

Notice that in (III) one could replace θ_1 by any point in $(\underline{\theta}, \bar{\theta})$. For $\theta, \theta' \in (\underline{\theta}, \bar{\theta})$, define the information number

$$I(\theta, \theta') = E_\theta\{\ln[f(X; \theta)/f(X; \theta')]\} = (\theta - \theta')b'(\theta) - (b(\theta) - b(\theta')) \tag{3.9.15}$$

and note also that

$$\operatorname{var}_\theta\{\ln[f(X; \theta)/f(X; \theta')]\} = (\theta - \theta')^2 b''(\theta),$$

so that assumption (III) concerns the ratio of the variance of $\ln\{f(X; \theta)/f(X; \theta')\}$ to the square of its mean when θ is true. Since the ratio is continuous, (III) is implied by

(III′) $\underline{\theta}$ and $\bar{\theta}$ belong to the interior of the natural parameter space of the Darmois–Koopman family.

If θ is the mean of a normally distributed variable with known variance, then (III) is satisfied with $\underline{\theta} = -\infty$, $\bar{\theta} = \infty$. Also, assumption (III) is valid with $(\underline{\theta}, \bar{\theta})$ taken as the full natural parameter space in the case of normal scale parameter, Bernoulli, Poisson, and negative exponential distributions.

Before stating the theorem, let us define the likelihood ratio tests and examine their structure. For specified $\gamma_{ij} \in (0, 1)$ satisfying assumptions (I) and (II), the likelihood ratio test $(\hat{N}, \hat{\delta})$ consists of a terminal decision rule

$\hat{\delta}$ and a stopping time (or variable) $\hat{N} = \min_j \max_i \hat{N}_{ij}$, where $\hat{N}_{ij}$ is the smallest n (or ∞ if there is no n) such that

$$L_n(\hat{\theta}) \geq \sup_{H_i} L_n(\theta) + \ln \gamma_{ij}^{-1}, \tag{3.9.16}$$

and δ is the smallest j such that $N = \max_i N_{ij}$. Note that for fixed j, values of i such that $\gamma_{ij} = 1$ play no role since (3.9.16) always holds. Also, by the concavity of $L_n(\cdot)$, either $\hat{\theta} \in H_i$ and hence $\sup_{H_i} L_n(\theta) = L_n(\hat{\theta})$, or else the supremum of $L_n(\cdot)$ on H_i is attained at the point closest to $\hat{\theta}$, evidently an end point.

The structure and performance of likelihood ratio tests is illuminated by investigating the asymptotic behavior of $E_\theta(\hat{N})$ as $\underline{\gamma} = \min \gamma_{ij} \to 0$ in the case where the γ_{ij} less than one are equal (applicable to the testing problem in the case where $\bar{\alpha} = \underline{\alpha}$). Fix θ belonging to the interior of H_m. Clearly it follows from (3.9.16) and the remark thereof that $\hat{N}$ stops no later than the first time that for some j with $\gamma_{mj} = 1$

$$L_n(\theta) \geq \max_{i:\gamma_{ij} < 1} [L_n(\psi_i) + \ln \gamma_{ij}^{-1}], \tag{3.9.17}$$

where ψ_i is the end point of H_i closer to θ. Now, if the γ_{ij} appearing in (3.9.17) are all equal to $\underline{\gamma}$, Theorem 3.3 of Lorden (1967) implies that

$$I(\theta) E_\theta(\hat{N}) \leq \ln \underline{\gamma}^{-1} + O(\{\ln \underline{\gamma}^{-1}\}^{1/2}) \qquad \text{as} \qquad \underline{\gamma} \downarrow 0, \tag{3.9.18}$$

where

$$I(\theta) = \max_{j:\gamma_{mj} < 1} \min_{i:\gamma_{ij} < 1} I(\theta, \psi_i). \tag{3.9.19}$$

If θ is the common end point of H_m and H_{m+1}, the above remarks apply with (3.9.17) and (3.9.19) restricted to j such that $\gamma_{mj} = \gamma_{m+1,j} = 1$. Note that by assumption (II) there is for each θ a j such that the closest to θ of the θ_i does not occur as a ψ_i in (3.9.17) or (3.9.19). Hence $\underline{I} = \min_{\underline{\theta} < \theta < \bar{\theta}} I(\theta) > 0$. The minimum is attained by virtue of the continuity of $I(\cdot, \theta_i)$ for each $i = 1, \ldots, s$ and the obvious fact that $I(\theta)$ is decreasing on H_1 and increasing on H_{s+1}. The determination of $\underline{I}$ in a specific problem is straightforward since $I(\theta)$ is alternately increasing and decreasing, with local extrema at values of θ such that $I(\theta, \psi_{i_1}) = I(\theta, \psi_{i_2})$ where i_1 and i_2 are minimizing choices of i in (3.9.18) for different js. The preceding relation between the I functions is equivalent to $b'(\theta) = [b(\psi_{i_1}) - b(\psi_{i_2})]/(\psi_{i_1} - \psi_{i_2})$. As a consequence of Theorem 3.9.1 obtained by choosing (N, δ) to be $(\hat{N}, \hat{\delta})$, the inequality (3.9.18) is actually an equality, in the case of equal error probabilities. Note that the square root term in (3.9.18) is of larger order of magnitude than the difference between $E_0(N)$ and the minimum attainable. It is also well known in the case $s = 2$ [for instance, see Schwarz (1962) and Wong (1968)] that sequential likelihood ratio tests are truncated and this result is true in the present context by virtue of the semiindifference assumption (II). In fact

$$\hat{N} \leq [\underline{I}^{-1} \ln \gamma^{-1}] + 1 \text{ with probability one} \tag{3.9.20}$$

where $[x]$ denotes here the greatest integer $\leq x$. In the case where all $\gamma_{ij} (\gamma_{ij} < 1)$ are equal, (3.9.20) follows from the fact that if $n > \underline{I}^{-1} \ln \gamma^{-1}$ and $\hat{\theta} \in (\underline{\theta}, \bar{\theta})$, then $S_n = nb'(\hat{\theta})$ and hence, for all θ

$$L_n(\hat{\theta}) - L_n(\theta) = n[(\hat{\theta} - \theta)b'(\hat{\theta}) - (b(\hat{\theta}) - b(\theta))] = nI(\hat{\theta}, \theta). \tag{3.9.21}$$

Thus, $L_n(\hat{\theta})$ exceeds each of the $L_n(\psi_i)$ in (3.9.17) by $nI(\hat{\theta}, \psi_i)$ which is at least $n\underline{I}$, hence greater than $\ln \gamma^{-1}$ and sufficient for (3.9.17). If $\hat{\theta} = \underline{\theta}$, then $S_n \leq nb'(\theta_1)$ when $L_n(\theta_1) - L_n(\theta) \geq nI(\theta_1, \theta)$ for $\theta > \theta_1$ and (3.9.17) is satisfied for $\theta = \theta_1$, just as if $\hat{\theta} = \theta_1$. The case $\hat{\theta} = \bar{\theta}$ is analogous. If the $\gamma_{ij} < 1$ are not all equal, decreasing them all to γ evidently cannot decrease $\hat{N}$, so (3.9.20) is valid in this case also.

The proof of Theorem 3.9.1 relies on two simple inequalities which will be given below. The first is Wald's (1947, p. 197) lower bound on the expected sample sizes of competitors of the SPRT. Recall that this was also used in the proof of Theorem 3.7.6. For applications in the sequel, this result can be stated as follows.

Lemma 3.9.2 (Wald, 1947) Suppose N_1 and N_2 are (possibly infinite) stopping times for $X_1, X_2, \ldots$ such that

$$P_{\theta'}(N_1 < \infty) \leq \alpha < 1 \qquad \text{and} \qquad P_\theta(N_2 < \infty) \leq \beta < 1.$$

If $I(\theta, \theta') = E_\theta\{\ln[f(X; \theta)/f(X; \theta')]\} > 0$, then

$$\begin{aligned} I(\theta, \theta')E_\theta\{\min(N_1, N_2)\} &\geq (1 - \beta) \ln [(1 - \beta)/\alpha] + \beta \ln [\beta/(1 - \alpha)] \\ &\geq (1 - \beta) \ln \alpha^{-1} - \ln 2 \end{aligned} \tag{3.9.22}$$

and

$$I(\theta, \theta')E_\theta(N_1) \geq \ln \alpha^{-1}. \tag{3.9.23}$$

The last inequality in (3.9.22) is obvious from the fact that $(1 - \alpha) \ln(1 - x) + x \ln x$ has minimum value $-\ln 2$ at $x = \frac{1}{2}$. Relation (3.9.23) can be derived from the first inequality in (3.9.22) by setting $N_2 = +\infty$ and letting $\beta \downarrow 0$.

The second key inequality (given in the following lemma) is needed in cases where $I(\theta, \theta')$ is small.

Lemma 3.9.3 (Lorden, 1972) If $P_{\theta'}(N < \infty) \le \alpha$ and $I(\theta, \theta') > 0$, then for all $\varepsilon > 0$

$$E_\theta\{\ln \alpha^{-1} - (I(\theta, \theta') + \varepsilon)N\}^+ \le e^{-1} + (2\varepsilon)^{-1} \operatorname{var}_\theta\{\ln(f(X; \theta)/f(X; \theta'))\} \tag{3.9.24}$$

where $e = 2.718\ldots$ and $x^+ = x$ if $x \ge 0$ and $x^+ = 0$ if $x < 0$.

Proof Let $m = \ln \alpha^{-1}/[I(\theta, \theta') + \varepsilon]$. Clearly

$$\begin{aligned}\int_{\{N<m\}} \ln[f(x_N; \theta)/f(x_N; \theta')]\, dP_\theta - \int_{\{N<m\}} (I(\theta, \theta') + \varepsilon)N\, dP_\theta \\ \le E_\theta \sup_{n\ge 0} \{\ln[f(X_n; \theta)/f(X_n; \theta')] - n(I(\theta, \theta') + \varepsilon)\} \\ \le (2\varepsilon)^{-1} \operatorname{var}_\theta\{\ln[f(X; \theta)/f(X; \theta')]\},\end{aligned} \tag{3.9.25}$$

where the last inequality is due to Kingman (1962) and states that a random walk with negative mean has expected supremum at most one-half the ratio of its variance to the absolute value of its mean. (For the result one may require the existence of the third absolute moment, however, all moments exist for the Darmois–Koopman family of distributions.) Let $f(\mathbf{x}_n; \theta)$ denote the joint density of $X_1, \ldots, X_m$. Then by Jensen's inequality

$$E_\theta\{\ln[f(\mathbf{X}_N; \theta')/f(\mathbf{X}_N; \theta) \mid N < m] \le \ln E_\theta[f(\mathbf{X}_N; \theta')/f(\mathbf{X}_N; \theta) \mid N < m] \tag{3.9.26}$$

and using relations of the form $dP_{\theta'} = [f(\mathbf{x}_N; \theta')/f(\mathbf{x}_N; \theta)]\, dP_\theta$,

$$\begin{aligned}P_\theta(N < m)E_\theta[\{f(\mathbf{X}_N; \theta')/f(\mathbf{X}_N; \theta)\} \mid N < m] \\ = \sum_{n=1}^{m-1} \int_{\{N=n\}} (f(\mathbf{x}_N; \theta')/f(\mathbf{x}_N; \theta))\, dP_\theta \\ = \sum_{n=1}^{m-1} \int_{\{N=n\}} dP_{\theta'} = P_{\theta'}(N < m) \le \alpha.\end{aligned} \tag{3.9.27}$$

Multiplying both sides of (3.9.26) by $P_\theta(N < m)$ and applying (3.9.27),

$$\begin{aligned}\int_{\{N<m\}} \ln[f(\mathbf{x}_N; \theta')/f(\mathbf{x}_N; \theta)]\, dP_\theta \le P_\theta(N < m) \ln(\alpha/P_\theta(N < m)) \\ \le \int_{\{N<m\}} \ln \alpha\, dP_\theta + e^{-1}\end{aligned} \tag{3.9.28}$$

since $x \ln x^{-1} \le e^{-1}$ for $x \ge 0$. The desired result follows upon adding (3.9.25) and (3.9.28).

Now we shall state the basic result of Lorden (1972) without proof since the proof is quite long, although not difficult to follow.

Theorem 3.9.1 (Lorden, 1972) There is a $D \ge 1$ such that the sequential likelihood ratio test $(\hat{N}, \hat{\delta})$ with

$$\gamma_{ij} = \begin{cases} \alpha_{ij}/D \ln \underline{\alpha}^{-1} & \text{if} \quad \alpha_{ij} < 1, \\ 1 & \text{if} \quad \alpha_{ij} = 1 \end{cases} \tag{3.9.29}$$

satisfies (3.9.13) whenever the α_{ij} satisfy assumptions (I) and (II). Denote by $n(\theta)$ the infimum of $E_\theta N$ over all tests (N, δ) for which (3.9.13) holds. When assumptions (I)–(III) are satisfied, there is an increasing function $M(\cdot)$ on $(1, \infty)$ such that

$$E_\theta \hat{N} - n(\theta) \le M(r)(\ln)^2 \underline{\alpha}^{-1} \qquad \text{for all } \theta \in (\underline{\theta}, \bar{\theta}) \tag{3.9.30}$$

where $r = \ln \bar{\alpha}/\ln \underline{\alpha}$.

Proof See Lorden (1972, Theorem 1).

The following remarks deal with modifications of the above formulations and other refinements.

REMARK 3.9.1 The proof of Theorem 3.9.1 is valid (with slightly larger bounds) if the restriction on (N, δ) is weakened by multiplying all α_{ij} that are less than one by k, with the choice of $(\hat{N}, \hat{\delta})$ unchanged. This observation suffices to show that a similar theorem holds if the testing problem is reformulated with (3.9.13) replaced by

$$\sum_{j \in J_i} P_\theta(j\text{th decision}) \le \alpha_i \qquad \text{for all } \theta \in H_i, \quad i = 1, \ldots, s+1, \tag{3.9.13$'$}$$

where J_i is the subset of decisions "incorrect" for H_i (corresponding to the set of js with $\alpha_{ij} < 1$ in the original formulation), because (3.9.13′) is stronger than (3.9.13) with $\alpha_{ij} = \alpha_i$ for $j \in J_i$, $=1$ for $j \notin J_i$, but is weaker than (3.9.13) with $\alpha_{ij} = \alpha_i/k$ for $j \in J_i$, $=1$ for $j \notin J_i$. Of course, when $s = 2$, the two formulations are identical.

REMARK 3.9.2 A natural definition of $\hat{N}$ is the smallest n such that, for some j, (3.9.16) holds for all i. This "memoryless" version of $\hat{N}$ requires (possibly) more sampling, since (3.9.16) may not hold for all i simultaneously at time $n = \max_i \hat{N}_{ij}$. However, unlike the original $\hat{N}$, it makes the decision to stop on the basis of the sufficient statistic (n, S_n) only. Lorden (1972) points out that Theorem 3.9.1 holds for this modification of $\hat{N}$, but an additional estimate is needed to bound the expected time required, after N_θ stops, for the "memoryless" version of N_θ [defined by (3.9.17)] to stop. Evidently this is no larger than the time required for $L_n(\theta)$ to increase by at least Y relative to every $L_n(\psi_i)$ such that $\alpha_{ij} < 1$ for some $j \in J_i$ where

$$Y = \sum_i [L_{\hat{N}_{ij}}(\theta) - L_{\hat{N}_{ij}}(\psi_i) - (L_{N_\theta}(\theta) - L_{N_\theta}(\psi_i))]^+.$$

Given Y, the (conditional) expected time required for this increase is bounded by $2Y \min_i(I(\theta, \psi_i))^{-1} + B$ [where B can be calculated by applying Theorem 3.7.3 for $\theta < \theta_1$, $\theta > \theta_s$ and Theorem 3.3 of Lorden (1967) for $\theta_1 < \theta \leq \theta_s$]. Clearly each term in Y is stochastically smaller than $\inf_{n \geq 0}[L_n(\theta) - L_n(\psi_i)]$ whose expectation is bounded above by $(\theta - \psi_i)^2/2I(\theta, \psi_i)$ by applying Kingman's (1962) bound, used for (3.9.25). Thus, the expectation of the extra time is bounded above by s times $b''(\theta) \max_i(\theta - \psi_i)^2/\{\min_i I(\theta, \psi_i)\}^2 + B$, which is easily seen to be bounded uniformly in θ by virtue of assumption (III). Adding this bound in the derivations of Eqs. (38) and (47) of Lorden (1972) is the only change required in the proof of Theorem 3.9.1 in order to handle the "memoryless" $\hat{N}$. Once again, the discussion does not apply to the case $s = 2$ where the two versions of $\hat{N}$ are identical.

REMARK 3.9.3 Another alternative to the error probability (3.9.13) is

$$P_{\theta_i}(j\text{th decision}) \leq \alpha_{ij} \qquad (i = 1, \ldots, s, j = 1, \ldots, k), \qquad (3.9.13'')$$

which does not restrict the probabilities of the decisions when $\theta \neq \theta_1, \ldots, \theta_s$. Theorem 3.9.1 holds provided $\hat{N}$ is changed by defining $\hat{N}_{ij}$ as the smallest n such that $L_n(\hat{\theta}) \geq L_n(\theta_i) + \ln \gamma_{ij}^{-1}$. The proof carries over with certain modifications indicated by Lorden (1972, Remark 3).

Another refinement of the proof of Theorem 3.9.1 leads to the following result.

Theorem 3.9.2 (Lorden, 1972) Under assumptions (I) and (II), for every $\theta \in (\underline{\theta}, \bar{\theta})$ there is an M_θ such that

$$I(\theta)(E_\theta \hat{N} - n(\theta)) \leq (\ln)^2 \underline{\alpha}^{-1} + ((\ln)^2 \underline{\alpha}^{-1})^{1/2} M_\theta, \qquad (3.9.31)$$

provided $\bar{\alpha} = \underline{\alpha}$.

Proof See Lorden (1972, Theorem 2).

The following result gives an optimality of the sequential likelihood ratio test procedures.

Theorem 3.9.3 (Lorden, 1972) Under assumptions (I), (II), and (III′), there is an $M^*(r)$ such that any procedure satisfying (3.9.13) has expected sample sizes for all $\theta \in (\underline{\theta}, \bar{\theta})$ which are larger than the likelihood ratio test with $\gamma_{ij} = M^*(r)\alpha_{ij}$ when $\alpha_{ij} < 1$, and equal to one when $\alpha_{ij} = 1$.

Proof The proof of Theorem 3.9.3 leans on some of the computations made in the proof of Theorem 3.9.1. Hence, the proof is suppressed although it is not long nor complicated. The reader is referred to Lorden (1972, Theorem 3).

3.9.4 Open-Ended Tests and Error Probability Approximations

The sequential likelihood ratio test for the case $s = 2$ can be described as follows:

Stop and reject $\theta \le \theta_1$ the first time

$$L_n(\hat{\theta}) \ge L_n(\theta_1) + \ln \gamma_1^{-1} \qquad \text{and} \qquad \hat{\theta} > \theta_1, \tag{3.9.32}$$

and stop and reject $\theta \ge \theta_2$ the first time

$$L_n(\hat{\theta}) \ge L_n(\theta_2) + \ln \gamma_2^{-1} \qquad \text{and} \qquad \hat{\theta} < \theta \tag{3.9.33}$$

where γ_1 and γ_2 are chosen to yield error probabilities less than prescribed α, β. The criterion (3.9.32) for rejecting $\theta \le \theta_1$ defines what is called an "open-ended test" of the hypothesis $\theta \le \theta_1$ against the alternative $\theta > \theta_1$. The prototype for open-ended tests is the one-sided SPRT "reject f_0 when $f_0/f \le A'$." Open-ended sequential likelihood ratio tests are studied by Lorden (1973) and are presented in Section 3.7. Just as an ordinary SPRT is obtained by performing two one-sided SPRTs simultaneously, so also the sequential likelihood ratio test for $s = 2$ is obtained by performing two open-ended tests, namely, the ones defined by (3.9.32) and (3.9.33). In fact, the general k-decision likelihood ratio tests of Section 3.9.3 are equivalent to performing simultaneously k open-ended tests, one for each decision, and stopping the first time any one of them stops. This reduction to open-ended tests is important for the following reasons. First it is a useful way to set up the application of a sequential likelihood ratio test and to compute stopping boundaries. Second, it facilitates the approximation of error probabilities. In the case of an SPRT, Wald's approximations of the error probabilities differ only by factors of $1 - \alpha$ and $1 - \beta$ from the error probability approximations for each of the one-sided SPRTs involved. This suggests bounding the error probabilities of sequential likelihood ratio tests in terms of open-ended tests. From (3.9.20) there is an upper bound M on the number of observations.

Thus, letting N be the extended stopping variable defined by (3.9.32), the error probability under θ_1, of the test defined by (3.9.32) and (3.9.33) is at most $P_{\theta_1}(N \leq M)$. An approach to the problem of approximating probabilities of this type is the following.

Assume that $\bar{\theta}$ belongs to the natural parameter space and $I(\bar{\theta}, \theta_1) < \ln \gamma_1^{-1}$. Let $m \geq 1$ be the largest integer such that $mI(\bar{\theta}, \theta_1) \leq \ln \gamma_1^{-1}$. For $n > m$, define $\theta_n{}^*$ as the solution in $\theta \in (\theta_1, \bar{\theta})$ of $nI(\theta, \theta_1) = \ln \gamma_1^{-1}$. Also, let $N(\theta)$ be the smallest n (or ∞ if there is no such n) such that

$$L_n(\theta) \geq L_n(\theta_1) + \ln \gamma_1^{-1}. \tag{3.9.34}$$

Note that (3.9.32) is equivalent to

$$S_n \geq \inf_{\theta_1 < \theta < \bar{\theta}} \left[\frac{\ln \gamma_1^{-1}}{\theta - \theta_1} - n\frac{b(\theta) - b(\theta_1)}{\theta - \theta_1}\right] \tag{3.9.35}$$

and for $n > m$ the infimum is attained at $\theta_n{}^*$. From this and the fact that $N \leq N(\theta_n{}^*)$ by definition, it follows that $N = n > m$ only if $N(\theta_n{}^*) = n$. Therefore

$$P_{\theta_1}(N = n) \leq P_{\theta_1}(N(\theta_n{}^*) = n) \qquad \text{for} \quad n > m. \tag{3.9.36}$$

If $n \leq m$, the infimum in (3.9.35) is attained at $\bar{\theta}$, so that

$$P_{\theta_1}(N \leq m) = P_{\theta_1}(N(\bar{\theta}) \leq m). \tag{3.9.37}$$

By the cancellation of densities argument used for Wald's (1947) upper bound on SPRT error probabilities

$$P_{\theta_1}(N(\theta) = n) \leq \gamma_1 P_\theta(N(\theta) = n) \qquad \text{for all} \quad \theta > \theta_1, n. \tag{3.9.38}$$

From (3.9.36)–(3.9.38) we have

$$P_{\theta_1}(N \leq M) \leq \gamma_1\left\{P_{\bar{\theta}}(N(\bar{\theta}) \leq m) + \sum_{n=m+1}^{M} P_{\theta_n{}^*}(N(\theta_n{}^*) = n)\right\}. \tag{3.9.39}$$

As shown by Wald (1947) when $\theta_n{}^*$ is true and γ_1 is small, $N(\theta_n{}^*)$ is approximately normally distributed with mean $\ln \gamma_1^{-1}/I(\theta_n{}^*, \theta_1) = n$ and variance

$$\frac{(\ln \gamma_1^{-1}) \operatorname{var}_{\theta_n{}^*}(\ln[f(X; \theta_n{}^*)/f(X; \theta_1)])}{\{E_{\theta_n{}^*} \ln[f(X; \theta_n{}^*)/f(X; \theta_1)]\}^3} = \frac{(\ln \gamma_1^{-1})(\theta_n{}^* - \theta_1)^3 b''(\theta_n{}^*)}{[I(\theta_n{}^*, \theta_1)]^3}.$$

This suggests the approximations

$$P_{\theta_n^*}(N(\theta_n^*) = n) \approx \frac{[I(\theta_n^*, \theta_1)]^{3/2}}{(\theta_n^* - \theta_1)[2\pi b''(\theta_n^*) \ln \gamma_1^{-1}]^{1/2}} \tag{3.9.40}$$

and

$$P_{\bar{\theta}}(N(\bar{\theta}) \leq m) \approx \gamma_1/2. \tag{3.9.41}$$

Regarding n as a continuous variable on the interval from $(\ln \gamma_1^{-1})/I(\bar{\theta}, \theta_1)$ to M, with $nI(\theta_n^*, \theta_1) \equiv \ln \gamma_1^{-1}$, one obtains from (3.9.40) and (3.9.41) the following approximation to the upper bound (3.9.39)

$$P_{\theta_1}(N \leq M) \lesssim \frac{\gamma_1}{2} + \gamma_1 \int_{(\ln \gamma_1^{-1})/I(\bar{\theta}, \theta_1)}^{M} \frac{[I(\theta_n^*, \theta_1)]^{3/2}}{(\theta_n^* - \theta_1)[2\pi b''(\theta_n^*) \ln \gamma_1^{-1}]^{1/2}}\, dn. \tag{3.9.42}$$

Making the change of variable $\theta = \theta_n^*$ and differentiating the relation defining θ_n^* one obtains

$$(\theta_n^* - \theta_1) b''(\theta_n^*) \frac{d\theta_n^*}{dn} = -\frac{\ln \gamma_1^{-1}}{n^2} = -\frac{[I(\theta_n^*, \theta)]^2}{\ln \gamma_1^{-1}}$$

and hence

$$P_{\theta_1}(N \leq M) \lesssim \frac{\gamma_1}{2} + \gamma_1 (\ln \gamma_1^{-1})^{1/2} \int_{\theta_M^*}^{\bar{\theta}} \left[\frac{b''(\theta)}{2\pi I(\theta, \theta_1)}\right]^{1/2} d\theta. \tag{3.9.43}$$

In the case of a normal mean θ (variance unity), (3.9.43) becomes

$$P_{\theta_1}(\text{rej } \theta \leq \theta_1) \lesssim \gamma_1 \left[\frac{1}{2} + \pi^{-1} (\ln \gamma_1^{-1})^{1/2} \ln \frac{\bar{\theta} - \theta_1}{\theta_M^* - \theta_1}\right]. \tag{3.9.44}$$

By using Anderson's (1960) results pertaining to the (defective) distribution of the time for a Wiener process to reach a straight line boundary, it is shown by Lorden (1973) that inequality does hold in (3.9.44). The numerical examples of Lorden (1972) indicate that choosing $\bar{\theta}$ equal to or not much larger than θ_2 significantly reduces the error probability. In case $s > 2$, approximations similar to (3.9.43) can be derived by considering, for example, the probability under θ_1 that (3.9.32) holds for some n in a restricted range $[M_1, M_2]$ and transforming sums like the one in (3.9.39) into the one in (3 9.43).

3.10 Sequential Tests for Two-Sided Alternatives

3.10.1 Introduction

Let us consider the problem of testing H_0: $|\psi| \le \psi_0$ against H_1: $|\psi| \ge \psi_1$ $(0 \le \psi_0 < \psi_1)$ in a given family $[F_n(\mathbf{x}_n; \theta), \theta \in \Theta]$, where $\psi = \psi(\theta)$ is some real-valued function of θ which can take both positive and negative values. The problem can be stated as testing for H_0: $\psi^2 \le \psi_0{}^2$ against H_1: $\psi^2 \ge \psi_1{}^2$. If there exists a group G under which $|\psi|$ or ψ^2 acts as a maximal invariant on Θ, tests of $|\psi| = \psi_0$ against $|\psi| = \psi_1$ can be constructed using the corresponding invariantly sufficient sequence $\{V_n\}$ on $\{R_n\}$. In particular, the invariant SPRT of Section 3.6 where $\lambda = |\psi|$ or ψ^2 is applicable in order to test H_0 against H_1 if $[F_{V_n}(v; |\psi|), |\psi| > 0]$ also possesses a monotone likelihood ratio for each n. In one-parameter families to which the method applies, G typically involves changes in $\{x_i\}$ or in functions thereof. In multiparameter families, G involves some additional subgroup of transformations so as to eliminate the nuisance parameters. Some of these cases will be discussed in Section 3.10.2. The sequential F, chi-square, and T^2 tests for some special values of k (where k denotes the dimensionality of the underlying random variable) did solve such problems on two-sided alternatives [see B.K. Ghosh (1970, Sections 6.2.3–6.2.6)]. Some of these problems can also be solved via the method of weight functions. Wald (1947) gave a sequential procedure for testing a null hypothesis against a two-sided alternative hypothesis, although he did not obtain the OC and ASN functions.

Often, one wishes to know, after H_0 has been rejected, whether $\psi \le -\psi_1$ or $\psi \ge \psi_1$ subject to some preassigned risks. The present procedures do not provide any answer to this inquiry. In fact such a problem can be formulated in terms of a test among three hypotheses: H_0: $-\psi_0 \le \psi \le \psi_0$, H_1: $\psi \le -\psi_1$ and H_2: $\psi \ge \psi_1$. Recall that for some one-parameter families the solution to such problems is given by the Sobel–Wald procedure described in Section 3.9. On the other hand, if one is indifferent as to whether $\psi \le -\psi_1$ or $\psi \ge \psi_1$, the present procedures should be preferred to Sobel–Wald type of procedures.

3.10.2 The Darmois–Koopman Family

Suppose that $\{X_i\}$ are iid having the density

$$f(x; \theta) = B(\theta) e^{Q(\theta) R(x)} h(x) \qquad \text{for} \quad x \in R_1$$

under the restriction $Q(\theta) = \theta$, $B(\theta) = B(-\theta)$, $R(x) = -R(-x)$ and $h(x) = h(-x)$. We wish to test H_0: $|\theta| \le \lambda_0$ against H_1: $|\theta| \ge \lambda_1$ subject to α and β being the upper bounds on the error probabilities. Although the above family involves a single parameter, we can treat the underlying family as having two parameters, namely $|\theta|$ and sign of θ. Thus, in order to obtain invariant procedures about $|\theta|$ we may resort to the group $G = \{g^+, g^-\}$ such that $g^+\mathbf{x}_n = (x_1, \ldots, x_n)'$, $g^{-1}\mathbf{x}_n = (-x_1, \ldots, -x_n)'$. Then the given family remains invariant under G with $\bar{g}^+\theta = \theta$, $\bar{g}^-\theta = -\theta$, and $\lambda = |\theta|$ is a maximal invariant on $\Theta \equiv (|\theta| \ge 0$, signs of $\theta)$ under $\bar{G}$. Now, $T_n = R(X_1) + \cdots + R(X_n)$ is sufficient for the family for each n with $g_t{}^+t = t$ and $g_t{}^-t = -t$. Then by assumption (I) of a theorem of Hall *et al.* (1965) [see B. K. Ghosh (1970 Theorem 6.2)]

$$V_n = \left| n^{-1} \sum_{i=1}^{n} R(X_i) \right| \quad \text{for} \quad n \ge 1 \tag{3.10.1}$$

is invariantly sufficient for the given family under G. Since the probability density of $R(X)$ is of the form $B(\theta)\, e^{\theta t} J(t)$ where $J(t)$ is a symmetric function of t, we obtain

$$f_{V_n}(v; \lambda) = B^n(\lambda)[e^{n\lambda v} + e^{-n\lambda v}]h_n(v) \qquad \text{for} \quad v \ge 0 \tag{3.10.2}$$

where $h_n(v)$ is some positive function of v. Given $0 \le \lambda_0 < \lambda_1$

$$\ln \frac{f_{V_n}(v; \lambda_1)}{f_{V_n}(v; \lambda_0)} = S_n = n \ln \frac{B(\lambda_1)}{B(\lambda_0)} + \ln \frac{\cosh(n\lambda_1 v)}{\cosh(n\lambda_0 v)} \tag{3.10.3}$$

which can easily be verified to be strictly increasing in v. Hence, the invariant SPRT of H_0 against H_1 uses the critical inequality $b^* < S_n < a^*$, where S_n is calculated from (3.10.1) and (3.10.3) or equivalently $\underline{v}_n < V_n < \bar{v}_n$ where $\underline{v}_n$ and $\bar{v}_n$ are the predetermined solutions of $S_n(\underline{v}_n) = b^*$ and $S_n(\bar{v}_n) = a^*$. Although the finite termination of the invariant SPRTs proposed in this section can be established using Theorem 3.6.3 which is due to Wijsman (1972), in the following we shall provide more elementary methods of establishing the same. Write

$$S_n(v) = n[\ln B(\lambda_1) - \ln B(\lambda_0) + (\lambda - \lambda_0)v + o(n^{-1})].$$

We obtain

$$\lim_{n\to\infty} \underline{v}_n = \lim_{n\to\infty} \bar{v}_n = \frac{\ln B(\lambda_0) - \ln B(\lambda_1)}{\lambda_1 - \lambda_0} > 0$$

because of the following considerations:

$$E_\theta(R(X)) = \frac{-d}{d\theta} \ln B(\theta) = -E_{-\theta}(R(X)),$$

$$\text{var}_\theta\{R(X)\} = -d^2 \ln B(\theta)/d\theta^2$$

and, given $\underline{\theta} < \theta_0 < \theta_1 < \bar{\theta}$

$$Z(x) = \ln[f(x; \theta_1)/f(x; \theta_0)] = \ln[B(\theta_1)/B(\theta_0)] + (\theta_1 - \theta_0)R(x)$$

increases with $R(x)$ and $E_\theta(Z(X))$ strictly increases with θ since, for any $\underline{\theta} < \theta < \bar{\theta}$

$$\frac{dE_\theta(T)}{d\theta} = \frac{B'(\theta)}{B(\theta)} E_\theta(R(X)) + E_\theta(R^2(X)) = \text{var}_\theta(R(X)) > 0.$$

By the central limit theorem $\sum_1^n R(X_i)/n$ is asymptotically normal with mean μ_θ and variance $\sigma_\theta{}^2/n$ where

$$\mu_\theta = -d \ln B(\theta)/d\theta, \qquad \sigma_\theta{}^2 = -d^2 \ln B(\theta)/d\theta^2.$$

If $\theta = 0$, then $B(\theta)$ is a constant (that is, $\mu_\theta = 0$) and hence $\sum_1^n R(X_i)/n$ and consequently V_n converges to zero in probability, and by Theorem 2.12.1 the procedure terminates finitely with probability one. If $\theta > 0$, then $P[\sum_1^n R(X_i)/n > 0]$ approaches unity so that $n^{1/2}(V_n - \mu_\lambda)/\sigma_\lambda$ is asymptotically normal. By Theorem 2.12.1 again, the procedure terminates finitely, except perhaps when

$$d \ln B(\lambda)/d\lambda = [\ln B(\lambda_1) - \ln B(\lambda_0)]/(\lambda_1 - \lambda_0).$$

For this special value of λ, the proof for finite termination follows after using the mean value theorem technique employed by B. K. Ghosh (1970, p. 305) in establishing the finite sure termination of the sequential t-test. The case when $\theta < 0$ (that is, $\lambda = -\theta$) can be handled analogously. From the exponential property of the underlying densities and the symmetry of the distribution of T_n, it follows that the uniformly most powerful unbiased (UMPU) test of H_0: $-\lambda_0 < \theta < \lambda_0$ at level α accepts H_0 if $-c \leq T_n \leq c$, where $c > 0$ is solved from $P(-c \leq T_n \leq c | \lambda_0) = 1 - \alpha$. From the invariant sufficiency of $V_n = |T_n|$ and the monotone likelihood ratio property of V_n it follows that the same procedure is also uniformly most powerful invariant (UMPI). Hence, the optimum fixed-sample test of H_0 against H_1 accepts H_0 if $V_{n(\alpha, \beta)} \leq c(\alpha, \beta)$ and rejects H_0 otherwise, where $c(\alpha, \beta) > 0$ and the smallest integer $n(\alpha, \beta)$

are determined by $P_{\lambda_0}(V_{n(\alpha,\beta)} \le c(\alpha, \beta)) \ge 1 - \alpha$, $P_{\lambda_1}(V_{n(\alpha,\beta)} \le c(\alpha, \beta)) \le \beta$. One needs the knowledge of $h_n(v)$ in (3.10.2), in order to calculate the above probabilities. In certain specific cases $h_n(v)$ can be determined. A comparison of $n(\alpha, \beta)$ with the ASN of the invariant SPRT discussed above would give some idea about the relative efficiency of the invariant SPRT.

Example 3.10.1 If $\{X_i\}$ are iid as normal with mean μ and variance σ^2 and one wishes to test $H_0\colon |\mu| \le \lambda_0$ against $H_1\colon |\mu| \ge \lambda_1$, one identifies θ above with μ and

$$B(\mu) = \exp(-\mu^2/2\sigma^2), \qquad R(x) = x/\sigma^2, \qquad h(x) = (\sigma^2 2\pi)^{-1/2} \exp(-x^2/2\sigma^2).$$

Thus, the restrictions imposed on $f(x; \theta)$ are satisfied and the invariant SPRT applies with $V_n = |\bar{X}_n|/\sigma^2$ for $n \ge 1$ where $\bar{X}_n$ denotes the sample mean of n observations. The procedure is identical to the sequential chi-square test with $k = 1$. Notice that $h_n(v)$ for this case is given by $\sigma(n/2\pi)^{1/2} \exp(-nv^2\sigma^2/2)$.

Example 3.10.2 If $\{X_i\}$ are iid as Bernoulli with parameter p, the given family can be equivalently expressed in terms of the joint probability function of the iid $\{X_i^* = X_i - \frac{1}{2}\}$ where $f(x^*; \theta)$ is of the desired form for $x^* = \pm\frac{1}{2}$ with $\theta = Q(\theta) = \ln\{p/(1-p)\}$, $B(\theta) = (e^{\theta/2} + e^{-\theta/2})^{-1}$, $R(x^*) = x^*$, $h(x^*) = 1$. Hence, the invariant SPRT applies in order to test $|\theta| \le \lambda_0$ against $|\theta| \ge \lambda_1$, and one uses $V_n = |\sum_1^n X_i^*/n| = |\bar{X}_n - \frac{1}{2}|$ in the procedure for $n \ge 1$. In terms of p the two hypotheses are $|p - \frac{1}{2}| \le \gamma_0$ against $|p - \frac{1}{2}| \ge \gamma_1$ where $\gamma_j = (\exp \lambda_j - 1)/2(\exp \lambda_j + 1)$. The particular case $\gamma_0 = 0$ (that is $\lambda_0 = 0$) often arises in practical situations.

Example 3.10.3 (Two-sided t-test) Let $\{X_i\}$ be iid as normal with unknown mean and variance σ^2. We wish to test $H_0\colon |\mu/\sigma| \le \lambda_0$ against $H_1\colon |\mu/\sigma| \ge \lambda_1$. Then, the family is invariant under G, where $g \in G$ implies that $g\mathbf{x}_n = (cx_1, \ldots, cx_n)'$, $c \ne 0$ so that $\bar{g}\theta = (c\mu, |c/\sigma|)'$. A maximal invariant on Θ under the subgroup $c > 0$ is μ/σ and under the subgroup $c < 0$ is $-\mu/\sigma$, so that $\lambda = |\mu/\sigma|$ is a maximal invariant under $\bar{G}$. As seen in Section 3.6 a maximal invariant on R_n is $U_n = (X_1/X_n, \ldots, X_{n-1}/X_n)$ for $n > 1$ and $f_{U_n}(u; \lambda)$ is found easily. From this joint probability density one concludes that $|\sum_1^n X_i|/(\sum_1^n X_i^2)^{1/2}$ or equivalently $V_n = \bar{X}_n^2/s_n^2$ is invariantly sufficient for the given family. Consequently, the invariant SPRT is identical to the sequential t^2 [see B. K. Ghosh (1970, Section 6.2.3)] where the earlier λ is replaced by the present λ^2. Notice that the present method does not make use of the Stein theorem.

If $\{X_{1i}\}$ are iid as normal with mean μ_1 and variance σ^2, and $\{X_{2i}\}$ are iid as normal with mean μ_2 and variance σ^2, then an invariant SPRT for $|\mu_1 - \mu_2|/\sigma \le \lambda_0$ against $|\mu_1 - \mu_2|/\sigma \ge \lambda_1$ can be derived under the transformation $x_{1i} \to cx_{1i} + d$ and $x_{2i} \to cx_{2i} + d$ for $i = 1, \ldots, n$, $c \ne 0$,

$-\infty < d < \infty$. The test is, in fact, the sequential F-test discussed by B. K. Ghosh (1970, Section 6.2.4) for $k = 2$, where the earlier λ^* is replaced by the present $\lambda^2/2$. The procedure can also be called a two-sided sequential t-test because

$$2(n-1)V_n^* \sim F_{1,\,2(n-1),\,n\lambda^2/2} \equiv t^2_{2(n-1),\,\lambda\sqrt{n/2}}$$

where $F_{1,k,\eta}[t_{k,\eta}]$ denotes the noncentral $F[t]$ distribution with $1, k[k]$ as degrees of freedom and η as the noncentrality parameter.

For other examples pertaining to testing for the correlation coefficient, ratio of two normal variances and ratio of Poisson means, the reader is referred to B. K. Ghosh (1970, pp. 343–347).

3.10.3 Two-Sided Normal Alternatives for the Mean with Known Variance

Let $\{X_i\}$ be a sequence of iid normal random variables having unknown mean μ and known variance σ^2. We wish to test $H_0\colon \mu = \mu_0$ against the alternative hypothesis $\bar{H}\colon \mu \neq \mu_0$. The composite alternative hypothesis is usually replaced in practice by two simple hypotheses $H_i\colon \mu = \mu_i$ $(i = \pm 1)$ where $\mu_{-1} < \mu_0 < \mu_1$. A test procedure is then sought which has a high probability, say $1-\alpha$, of accepting H_0 when $\mu = \mu_0$ and a high probability, say $1-\beta$, of rejecting H_0 (accepting $\bar{H}$) when $\mu = \mu_i$ $(i = \pm 1)$. There are a few sequential procedures in the literature defined for this purpose [for instance, Armitage's (1950) procedure]. However, none of these procedures have either the OC or the ASN function and further, it is not known whether any one of these is optimum in any sense. Recently, Billard and Vagholkar (1969) have proposed a sequential procedure which overcomes both of these short-comings for the above hypothesis-testing problem.

The test procedure in general can be defined as follows (see Fig. 3.10.1). Let $S_n = \sum_1^n X_i$ denote the sum of n observations at the nth stage. Let the sample path be defined as the path traced out by the point (n, S_n) as successive observations are taken. Then continue taking observations until the sample path crosses either of the boundary lines LA, AP when H_1 is accepted or either of the lines BC, RQ, and CR when H_0 is accepted or it crosses either of the lines MD and DS when H_{-1} is accepted. The lines AP, BQ and CR, DS are pairs of parallel lines. In the existing procedures, it is theoretically possible to make a terminal decision after only as few as one or two observations have been taken. However, in practice most experimenters would not be content to terminate the experiment so soon and would take at least a few more observations even if only to confirm the terminating decision, especially when the difference between the value of the mean for the null and alternative hypotheses is small. The procedure due to Billard and Vagholkar (1969)

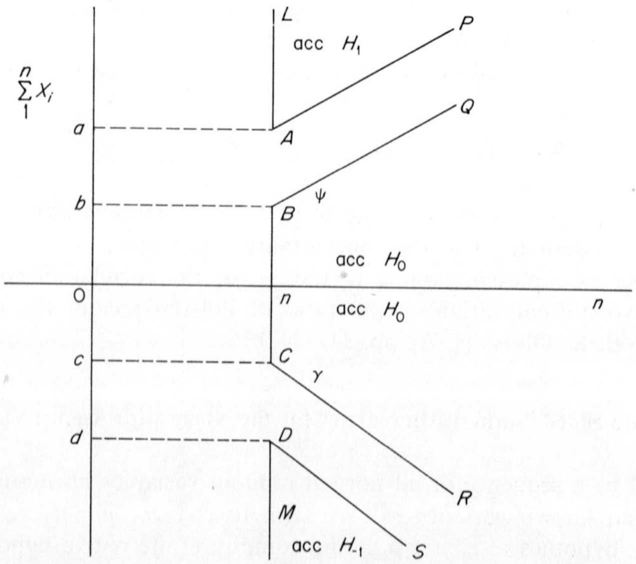

Fig. 3.10.1. The test procedure of Billard and Vagholkar (1969).

takes care of such a situation by stipulating that a few observations be taken before any serious comparison of the hypotheses is undertaken. Such comparison is then carried out by employing the Wald SPRT. The size of the initial sample n_0 is itself a random variable which would decrease (increase) as $|\mu_i - \mu_0|$ increases (decreases) for $i = \pm 1$. The present procedure terminates finitely with probability one since $N = \min(N_1, N_{-1})$ where N_i is the stopping variable of the Wald SPRT for testing H_0 against H_i ($i = \pm 1$) and we know that N_i are finite with probability one (see Section 2.2). It is easily seen that the procedure is completely specified in terms of the geometric parameters (n_0, a, b, c, d, ψ, γ). The values of these parameters are determined so as to minimize the ASN function subject to certain constraints on the OC functions.

THE OPERATING CHARACTERISTIC FUNCTION The OC function $L(\mu)$ is defined as the probability that the null hypothesis is accepted when the true mean is μ. Under the present procedure the null hypothesis may be accepted when either

(i) $c \le S_n \le b$ at $n = n_0$; or

(ii) $b < S_n < a$ at $n = n_0$ with the boundary BQ being the first boundary to be crossed at some $n > n_0$; or

(iii) $d < S_n < c$ at $n = n_0$ with the boundary CR being the first boundary to be crossed at some $n > n_0$.

Let $L_0(\mu)$, $L_1(\mu)$, and $L_{-1}(\mu)$ be the respective probabilities that each of these disjoint events happen. Then,

$$L(\mu) = L_0(\mu) + L_1(\mu) + L_{-1}(\mu). \tag{3.10.4}$$

Clearly

$$\begin{aligned} L_0(\mu) &= \int_c^b \phi[(x - n_0\mu)/\sigma n_0^{1/2}]\, dx \\ &= \Phi[(b - n_0\mu)/\sigma n_0^{1/2}] - \Phi[(c - n_0\mu)/\sigma n_0^{1/2}] \end{aligned} \tag{3.10.5}$$

where $\phi(x)$ and $\Phi(x)$ respectively denote the standard normal density and distribution functions. $L_1(\mu)$ is given by

$$L_1(\mu) = \int_b^a \phi[(x - n_0\mu)/\sigma n_0^{1/2}]\Pi(x)\, dx \tag{3.10.6}$$

where $\Pi(x)$ is the probability that a random walk starting from a point x on AB when $n = n_0$ and operating between the absorbing boundary lines AP and BQ is ultimately absorbed by BQ when H_0 is accepted.

Lemma 3.10.1 [(Cox and Miller) 1965, p. 56]. For a random walk that operates between the parallel absorbing boundaries $S_n = a^*$ and $S_n = -b^*$ and which starts at the point $S_0 = 0$ the probability p_{-b^*} that the walk ceases with absorption by the lower boundary $S_n = -b^*$ $(S_n = X_1 + \cdots + X_n)$ is given by

$$\begin{aligned} p_{-b^*} &= [\exp(-t_0 a^*) - 1]/[\exp(-t_0 a^*) - \exp(t_0 b^*)], \qquad t_0 \neq 0 \\ &= a^*/(a^* + b^*), \qquad t_0 = 0 \end{aligned} \tag{3.10.7}$$

where t_0 is the nonzero solution of

$$E_\mu(e^{-tX}) = 1. \tag{3.10.8}$$

Proof Let p_{a^*} (p_{-b^*}) denote the probability that the random walk is absorbed at the upper barrier a^* (lower barrier $-b^*$). We shall use Wald's identity (see Theorem 2.5.1) to obtain an approximation to these quantities. Wald's identity can be written as

$$\begin{aligned} &p_{a^*} E[\exp(-tS_N)\{C(-t)\}^{-N} \mid S_N \geq a^*] \\ &\quad + p_{-b^*} E[\exp(-ts_N)\{C(-t)\}^{-N} \mid S_N \leq -b^*] = 1 \end{aligned} \tag{3.10.9}$$

where $C(t) = E_\mu\{e^{tX}\}$, since $S_N \geq a^*$ and $S_N \leq -b^*$ are mutually exclusive events. If we neglect the excess over the boundaries and write

$$S_N \simeq a^* \quad \text{when } S_N \geq a^* \qquad \text{and} \qquad S_N \simeq -b^* \quad \text{when } S_N \leq -b^*, \tag{3.10.10}$$

then we have

$$p_a{}^* \exp(-t_0 a^*) + p_{-b^*} \exp(t_0 b^*) \simeq 1. \tag{3.10.11}$$

Finally, using the fact that $p_{a^*} + p_{-b^*} = 1$ (because of sure termination) we obtain the approximate probabilities for absorption when $\mu \neq 0$

$$p_{a^*} \simeq [1 - \exp(t_0 b^*)]/[\exp(-t_0 a^*) - \exp(t_0 b^*)] \tag{3.10.12}$$

and p_{-b^*} given by (3.10.7), which ought to be good for large a^* and b^* because with reasonable distributions any excess over the barrier will then be small compared with a^* and b^*. When $\theta_0 = 0$, then $t_0 = 0$ and the expressions in (3.10.7) and (3.10.12) become indeterminate. However, when $\mu = 0$, we let $\theta_0 \to 0$ and obtain (by using L'Hospital's rule)

$$p_{a^*} \simeq b^*/(a^* + b^*), \qquad p_{-b^*} \simeq a^*/(a^* + b^*). \tag{3.10.13}$$

Notice that the approximations in (3.10.13) are independent of the particular distribution involved. This completes the proof of the lemma.

In order to apply these formulas to our case, it is necessary to make a transformation that will rotate the boundary lines through an angle ψ and which will shift the origin of the S_n axis from 0 to x. Thus, the required formulas are found to be

$$\Pi(x) = \begin{cases} [\exp(-h_0 a) - \exp(-h_0 x)]/[\exp(-h_0 a) - \exp(-h_0 b)], & h_0 \neq 0, \\ (a - x)/(a - b), \quad h_0 = 0, \end{cases} \tag{3.10.14}$$

where h_0 is the nonzero solution of the equation

$$E_\mu \exp[-h(x - \tan\psi)] = 1. \tag{3.10.15}$$

When X is a normal variable with mean μ and variance σ^2, then (3.10.15) gives

$$h_0 = 2(\mu - \tan\psi)/\sigma^2. \tag{3.10.16}$$

Substituting the value of $\Pi(x)$ from (3.10.14) into (3.10.16) and after some algebraic manipulation, we have for $h_0 \neq 0$

$$L_1(\mu) = \left[\frac{\exp(-h_0 a)}{\exp(-h_0 a) - \exp(-h_0 b)}\right]\left[\Phi\left(\frac{a - n_0 \mu}{\sigma n_0^{1/2}}\right) - \Phi\left(\frac{b - n_0 \mu}{\sigma n_0^{1/2}}\right)\right]$$
$$- \left[\frac{\exp(-h_0 n_0 \mu + \frac{1}{2}h_0{}^2 n_0 \sigma^2}{\exp(-h_0 a) - \exp(-h_0 b)}\right]\left[\Phi\left(\frac{a - n_0 \mu}{\sigma n_0^{1/2}} + h_0 \sigma n_0^{1/2}\right)\right.$$
$$\left. - \Phi\left(\frac{b - n_0 \mu}{\sigma n_0^{1/2}} + h_0 \sigma n_0^{1/2}\right)\right] \tag{3.10.17}$$

and for $h_0 = 0$

$$L_1(\mu) = \left(\frac{a - n_0 \mu}{a - b}\right)\left[\Phi\left(\frac{a - n_0 \mu}{\sigma n_0^{1/2}}\right) - \Phi\left(\frac{b - n_0 \mu}{\sigma n_0^{1/2}}\right)\right]$$
$$+ \left(\frac{n_0 \sigma^2}{a - b}\right)\left[\phi\left(\frac{a - n_0 \mu}{\sigma n_0^{1/2}}\right) - \phi\left(\frac{b - n_0 \mu}{\sigma n_0^{1/2}}\right)\right]. \tag{3.10.18}$$

In an analogous manner, one can obtain an expression for $L_{-1}(\mu)$ by replacing h_0 by h_0', a by d, b, by c, and multiplying each of the differences in Φ and ϕ functions by -1, where $h_0' = 2(\mu - \tan\gamma)/\sigma^2$ is the nonzero solution of

$$E_\mu\{\exp[-h'(X - \tan\gamma)]\} = 1. \tag{3.10.19}$$

Hence by adding together the relevant terms of $L_0(\mu)$, $L_1(\mu)$, and $L_{-1}(\mu)$ the OC function as defined by (3.10.4) is obtained.

THE AVERAGE SAMPLE NUMBER FUNCTION The ASN function, $E_\mu(N)$ (where N is the random stopping variable) is defined as the expected number of observations required to reach a decision when the true mean is μ. For the present procedure this function comprises of three parts, namely,

(i) a decision is reached at $n = n_0$;

(ii) $b < S_n < a$ at $n = n_0$; with a further additional $n_1(\mu)$ expected observations before either H_0 or H_1 is accepted; and

(iii) $d < S_n < c$ at $n = n_0$ with a further additional $n_{-1}(\mu)$ expected observations before either H_0 or H_{-1} is accepted.

Hence, we have

$$E_\mu(N) = n_0 + n_1(\mu) + n_{-1}(\mu). \tag{3.10.20}$$

Consider first the term $n_1(\mu)$. It is clear that

$$n_1(\mu) = \int_b^a \phi((x - n_0\,\mu)/\sigma n_0^{1/2}) n(x)\, dx \tag{3.10.21}$$

where $n(x)$ is defined as the expected number of observations for absorption of a random walk starting from the point $S_n = x$ at $n = n_0$ and operating between the parallel absorbing boundaries AP and BQ. The expected number of steps $E_\mu(N)$ required for absorption by parallel boundaries when the starting point is zero is given by Cox and Miller (1965, p. 58), and we shall give this in the following lemma.

Lemma 3.10.2 (Cox and Miller, 1965) Let N denote the stopping variable for absorption by parallel boundaries of the form $S_n = a^*$ and $S_n = -b^*$ when the starting point of the random walk is zero. Then $E_\mu(N)$ is approximately given by

$$E_\mu(N) \simeq \begin{cases} \dfrac{(a^* + b^*) - a^* \exp(t_0 b^*) - b^* \exp(-t_0 a^*)}{\mu[\exp(-t_0 a^*) - \exp(t_0 b^*)]}, & t_0 \neq 0, \\ a^* b^*/\sigma^2, & t_0 = 0, \end{cases} \tag{3.10.22}$$

where t_0 is defined by (3.10.8).

Proof Let $1/s = C(t)$ where $C(t)$ is defined in (3.10.9). By the convexity property of $C(t)$ [since $\{d^2/dt^2\}C(t) > 0$ for all t] the above equation will have two real roots in t, provided $1/s > C(t_1)$ where t_1 is the unique value of t at which $C(t)$ attains its minimum value [for instance, see Cox and Miller (1965, p. 49) for an elaboration of t_1]. Let us denote the roots by $\lambda_1(s)$ and $\lambda_2(s)$, where $\lambda_1(1) = 0$ and $\lambda_2(1) = t_0$. We can express Wald's identity in another form, by the following two equations:

$$E[\exp\{-\lambda_i(s) S_N\} s^N] = 1 \qquad (i = 1, 2). \tag{3.10.23}$$

Again using the approximation (3.10.10), we have

$$p_{a^*} \exp[-\lambda_i(s) a^*]\, E^*(s^N) + p_{-b^*} \exp[\lambda_i(s)]\, E^{**}(s^N) \tag{3.10.24}$$

where E^* (E^{**}) denotes expectation conditional on absorption at the upper boundary a^* (lower boundary $-b^*$). Using the expressions for p_{a^*} and p_{-b^*} given by (3.10.7) and (3.10.12), the two equations (3.10.24) may be solved in order to obtain approximately $E^*(s^N)$ and $E^{**}(s^N)$ and consequently the probability generating function of N, namely

$$E(s^N) = p_{a^*} E^*(s^N) + p_{-b^*} E^{**}(s^N).$$

From Theorem 2.4.9 or by differentiating underneath the expectation sign in Wald's identity, we have

$$E_\mu(N) = \begin{cases} E_\mu(S_N)/\mu & \text{if} \quad \mu \neq 0, \\ E_\mu(S_N{}^2)/\sigma^2 & \text{if} \quad \mu = 0, \end{cases} \tag{3.10.25}$$

where μ and σ^2 are the mean and variance of X respectively. Now, $E_\mu(S_N) \simeq a^* p_{a^*} - b^* p_{-b^*}$ and $E_\mu(S_N{}^2) \simeq a^{*2} p_{a^*} + b^{*2} p_{-b^*}$ so that from (3.10.25) we obtain (3.10.22) as an approximation for the expected number of steps to absorption. This completes the proof of Lemma 3.10.2.

Formulas (3.10.22), when adapted to the present case in the same way as described in obtaining the OC function, become for $h_0 \neq 0$

$$n(x) = \left(\frac{(a-b)\exp(-h_0 x) - [a \exp(-h_0 b) - b \exp(-h_0 a)]}{\exp(-h_0 a) - \exp(-h_0 b)} \right) (\mu - \tan \psi)^{-1} \tag{3.10.26}$$

and for $h_0 = 0$

$$n(x) = (a - x)(x - b)/\sigma^2 \tag{3.10.27}$$

where $h_0 = 2(\mu - \tan \psi)/\sigma^2$ as before. This expression for $n(x)$ may now be substituted into Eq. (3.10.21). Then after considerable algebraic manipulations, we have for $h_0 \neq 0$

$$\begin{aligned} n_1(\mu) = \Bigg\{ & \frac{(a-b)\exp(-h_0 n_0 \mu + \frac{1}{2} h_0{}^2 n_0 \sigma^2)}{\exp(-h_0 a) - \exp(-h_0 b)} \\ & \times \left[\Phi\left(\frac{a - n_0 \mu}{\sigma n_0^{1/2}} + h_0 \sigma n_0^{1/2} \right) - \Phi\left(\frac{b - n_0 \mu}{\sigma n_0^{1/2}} + h_0 \sigma n_0^{1/2} \right) \right] \\ & - \left[\frac{a \exp(-h_0 b) - b \exp(-h_0 a)}{\exp(-h_0 a) - \exp(-h_0 b)} + n_0 \mu \right] \\ & \times \left[\Phi\left(\frac{a - n_0 \mu}{\sigma n_0^{1/2}} \right) - \Phi\left(\frac{b - n_0 \mu}{\sigma n_0^{1/2}} \right) \right] \\ & - n_0 \sigma^2 \left[\phi\left(\frac{b - n_0 \mu}{\sigma n_0^{1/2}} \right) - \phi\left(\frac{a - n_0 \mu}{\sigma n_0^{1/2}} \right) \right] \Bigg\} (\mu - \tan \psi)^{-1} \end{aligned} \tag{3.10.28}$$

and for $h_0 = 0$

$$n_1(\mu) = n_0\left[\left(\frac{a - n_0\,\mu}{n_0\,\sigma^2}\right)\phi\left(\frac{b - n_0\,\mu}{\sigma n_0^{1/2}}\right) - \left(\frac{b - n_0\mu}{n_0\,\sigma^2}\right)\phi\left(\frac{a - n_0\,\mu}{\sigma n_0^{1/2}}\right)\right]$$
$$- [(a - n_0\,\mu)(b - n_0\,\mu)\sigma^{-2} + n_0]\left[\Phi\left(\frac{a - n_0\,\mu}{\sigma n_0^{1/2}}\right) - \Phi\left(\frac{b - n_0\,\mu}{\sigma n_0^{1/2}}\right)\right]. \tag{3.10.29}$$

In a similar manner $n_{-1}(\mu)$ can be derived. This is obtained from (3.10.28) and (3.10.29) by replacing h_0 by h_0', a by d, b by c, and by changing the algebraic sign of each of the differences in Φ and ϕ functions and also the weighted difference of two ϕ functions.

Notice that $h_0' = 2(\mu - \tan\gamma)/\sigma^2$. Thus, by adding the relevant terms of Eqs. (3.10.28) and (3.10.29) to n_0, the ASN for $E_\mu(N)$ can be obtained.

Numerical Results and Comparison In the general case there are seven geometrical parameters (n_0, a, b, c, d, ψ, γ) that completely define the test procedure. However, for numerical results and comparison purposes attention will be centered only on the symmetrical case, that is, when $\mu_0 = 0$ and $-\mu_{-1} = \mu = \mu_1$ and σ^2 is taken to be equal to 1. It is easy to see that in such a case $c = -b$, $d = -a$, and $\gamma = -\psi$. Therefore there remain only four parameters, namely (n_0, a, b, ψ) which values need to be specified in order to carry out the procedure. The optimal set of values was obtained by Billard and Vagholkar (1969) for these parameters so that this set minimized the ASN function $E_\mu(N)$ for some specified μ, subject to the following constraints on the OC function $L(\mu)$, namely

$$L(\mu_0) \geq 1 - \alpha, \qquad L(\mu_i) \leq \beta, \qquad i = \pm 1, \tag{3.10.30}$$

where α and β are the preassigned error probabilities. This minimization process was carried out by using a simplex method presented by Nelder and Mead (1965). Using this method, an optimum solution set of the geometrical parameters (n_0, a, b, ψ) was obtained by minimizing $E_\mu(N)$ under the null hypothesis, that is, when $\mu = \mu_0 = 0$. This was done separately for two different values of μ_1, namely $\mu_1 = 1.0$ and $\mu_1 = .8$ with $\alpha = \beta = .05$ in each case. These optimal solution sets together with the OC function for several values of μ obtained by Billard and Vagholkar are given in columns (i) and (v) of Table 3.10.1. In Table 3.10.2 the ASN function is tabulated for the same values of μ and the corresponding results are given in columns (i) and (v) of Table 3.10.2. In Tables 3.10.1 and 3.10.2 columns (ii), (vi), and (iii), (vii) give the corresponding results which were obtained when the ASN function was minimized at $\mu = \mu_1$ and $\mu = (\mu_0 + \mu_1)/2$ respectively.

Table 3.10.1

Operating Characteristic Function $L(\mu)$

	$\mu_0 = 0,\ \mu_1 = 1.0,\ \alpha = \beta = 0.05$				$\mu_0 = 0,\ \mu_1 = 0.8,\ \alpha = \beta = 0.05$			
Procedure	Present		Sobel and Wald		Present		Sobel and Wald	
	(i)	(ii)	(iii)	(iv)	(v)	(vi)	(vii)	(viii)
$E_\mu(N)$ minimized at $\mu =$	0.0	1.0	0.5	—	0.0	0.8	0.4	—
n_0	5.1465	5.1261	5.2358	—	8.1352	8.1692	8.5752	—
a	6.0242	6.0424	5.9750	—	7.4903	7.5863	7.3377	—
b	0.1377	0.1415	0.1362	—	0.1733	0.2566	0.1314	—
ψ	25°15′	24°56′	26°18′	—	21°15′	20°30′	24°5′	—
μ	$L(\mu)$				$L(\mu)$			
0.0	0.950	0.950	0.950	0.950	0.950	0.950	0.950	0.950
0.2	0.893	0.892	0.895	0.898	0.860	0.857	0.896	0.866
0.4	0.693	0.691	0.701	0.703	0.551	0.543	0.580	0.550
0.6	0.391	0.389	0.400	0.389	0.207	0.204	0.220	0.197
0.8	0.158	0.157	0.162	0.152	0.050	0.050	0.050	0.050
1.0	0.050	0.050	0.050	0.050	0.009	0.009	0.008	0.012
1.2	0.013	0.013	0.013	0.016	0.001	0.001	0.001	0.003

Table 3.10.2[a]

Average Sample Number, $E_\mu(N)$

	$\mu_0 = 0,\ \mu_1 = 1.0,\ \alpha = \beta = 0.05$				$\mu_0 = 0,\ \mu_1 = 0.8,\ \alpha = \beta = 0.05$			
Procedure	Present		Sobel and Wald		Present		Sobel and Wald	
	(i)	(ii)	(iii)	(iv)	(v)	(vi)	(vii)	(viii)
$E_\mu(N)$ minimized at $\mu =$								
μ	$E_\mu(N)$				$E_\mu(N)$			
0.0	8.06	8.07	8.06	9.83	12.59	12.61	12.68	15.36
0.2	8.73	8.74	8.69	10.30	14.10	14.15	14.02	16.82
0.4	10.02	10.04	9.96	12.22	16.08	16.10	16.05	19.84
0.6	10.15	10.16	10.15	12.43	14.41	14.38	14.67	17.08
0.8	8.80	8.79	8.83	10.32	11.35	11.37	11.56	12.80
1.0	7.24	7.24	7.27	8.19	9.41	9.45	9.65	7.42
1.2	6.18	6.17	6.22	5.05	8.55	8.59	8.89	5.66

[a] The portion of Table 3.10.1 above $L(\mu)$ also applies to Table 3.10.2.

The OC and ASN values for the Sobel–Wald (1949) procedure are given in columns (iv) and (viii) of Tables 3.10.1 and 3.10.2. The OC function is calculated from the formula of their paper [see Eq. (3.8.20)] while the ASN function values given are upper bounds when $\mu_0 \le |\mu| \le \mu_1$ and lower bounds for $|\mu| > \mu_1$. This is in accordance with their suggestion that these particular bounds are close to the exact ones in the respective regions of μ [see the discussion after Eq. (3.8.22)]. A comparison between the ASN values for the Sobel–Wald procedure and those for the present procedure clearly shows the present procedure to be the better for $|\mu| < \mu_1$ without any loss of efficiency. In fact values of the OC function are almost equal. For values of $|\mu| > \mu_1$, the ASN values are somewhat greater for the Billard–Vagholkar procedure. This is to be expected since the present procedure always takes at least n_0 observations before any hypothesis can be accepted. Billard and Vagholkar (1969) carried out Monte Carlo studies on both the Sobel–Wald procedure and their procedure specified by the optimal solution set obtained when $E_\mu(N)$ was minimized at $\mu = \mu_0$. In all, 1000 trials were carried out in each case and every trial was continued until a hypothesis was accepted. The theoretical results of Tables 3.10.1 and 3.10.2 are also substantiated by the Monte Carlo results presented in their Tables 3 and 4. Further, the Monte Carlo results show that the standard deviations of the ASN values of Billard–Vagholkar procedure are much smaller than for the Sobel–Wald procedure. This is so because, the procedure depends on ΣX_i only. Furthermore, after n_0 observations have been taken and if no hypothesis is immediately accepted, then the experimenter thereafter will be mainly testing either H_0 against H_1 or H_0 against H_{-1} and not all three hypotheses simultaneously. This will be so because the information in the first n_0 observations will be sufficient to discard either H_1 or H_{-1} in favor of the other. Hence, in that case the optimal test procedure after n_0 observations should be SPRTs (proved optimal by Wald and Wolfowitz (1948, Theorem 2.9.1) which in fact they are in the present procedures. A close examination of the optimal solution sets of the geometrical parameters (n_0, a, b, ψ) obtained by minimizing $E_\mu(N)$ at $\mu = \mu_0$, μ_1 and $(\mu_0 + \mu_1)/2$ shows that there is only a slight difference between them, and also between the values of their OC and ASN functions. This fact makes Billard and Vagholkar conjecture that their procedure may be close to being optimal for $|\mu| \le \mu_1$.

Problems

1.† The following data consists of a sequence of independent observations measuring the quantity of vitamin B_2 in millograms per gram for turnip greens:

110.4, 102.8, 101.0, 108.4, 100.7, 100.3, 102.0, 93.7,
98.9, 96.6, 99.4, 96.2, 99.0, 88.4, ···

† The data in Problem 1 are from Anderson and Bancroft (1952).

As in Example 3.10.1, assuming normality, carry out a two-sided sequential test of H_0: $|\mu| \leq 95$ against H_1: $|\mu| \geq 100$. Use $\alpha = \beta = 1$.

2. Let HTTTHHTHHT $\cdots$ denote a sequence of independent trials of a Bernoulli experiment with a coin. As in Example 3.10.2, carry out a sequential test of H_0: $|p - \frac{1}{2}| \leq \frac{1}{8}$ against $|p - \frac{1}{2}| \geq \frac{1}{4}$ where $P(X = 1) = p$ and X denotes the Bernoulli variable and head (H) is coded as unity and tail (T) is coded as zero.

3.11 Efficiency of the SPRT

3.11.1 Relative Efficiency at θ_0 and θ_1

Let $\{X_i\}$ be iid having $f(x; \theta)$ for the common probability (density) function and $\theta \in \Theta$. For the sake of simplicity, assume that θ is real and Θ is a part of the real line. We are interested in testing H_0: $\theta = \theta_0$ against the alternative $H_1: \theta = \theta_1$ when $\theta_0 \neq \theta_1$, subject to the prescribed error probabilities α and β. By Theorem 2.9.1 SPRT is the uniformly most economical (UME) test. Given any rival test δ, one can study the efficiency of the SPRT relative to δ in minimizing the ASN at $\theta = \theta_0, \theta_1$. We shall, in particular, investigate the precise amount of saving achieved by the SPRT over the corresponding optimum fixed-sample test for the same hypotheses. It is assumed that the optimum fixed-sample size $n(\alpha, \beta)$ can be determined for various combinations of (α, β), $\alpha + \beta < 1$. We define the relative efficiency of the SPRT at $\theta \in \Theta$ by

$$Re(\theta) = Re(\alpha, \beta, \theta) = n(\alpha, \beta)/E_\theta(N) \tag{3.11.1}$$

where $E_\theta(N)$ is the ASN of the SPRT at θ. Obviously $100[Re(\theta) - 1]/Re(\theta)$ is the average percentage saving resulting from the use of the SPRT instead of the optimum fixed-sample test when θ is the true value of the parameter. For example, let us assume that X_i is normally distributed with mean μ and known variance σ^2 and try to compute $Re(\mu_0)$ and $Re(\mu_1)$. We shall also assume that $|\mu_1 - \mu_0|$ is small so that the approximations for $E_{\mu_i}(N)$ $(i = 0, 1)$ in (2.4.2) and (2.4.3) are reasonable. For the fixed-sample size case, one can easily establish that the sample size and the critical point are given by

$$\begin{aligned} n(\alpha, \beta) &= \left[\left(\frac{z_{1-\alpha} - z_\beta}{\mu_1 - \mu_0}\sigma\right)^2\right], \\ c(\alpha, \beta) &= n(\alpha, \beta)\mu_0 + \sigma z_{1-\alpha}[n(\alpha, \beta)]^{1/2} \end{aligned} \tag{3.11.2}$$

provided $\alpha + \beta < 1$ where z_γ is defined by $\Phi(z_\gamma) = \gamma$ and $[\cdot]$ denotes largest integer contained in $(\cdot)$. Thus, it follows from (2.4.2), (2.4.3), and (2.12.2) that

$$Re(\mu_0) \simeq \frac{n(\alpha, \beta)}{E_{\mu_0}(N)} \simeq \frac{(z_{1-\alpha} - z_\beta)^2}{2\{(1-\alpha)\ln[(1-\alpha)/\beta] - \alpha \ln[(1-\beta)/\alpha]\}} \tag{3.11.3}$$

and

$$Re(\mu_1) = \frac{n(\alpha, \beta)}{E_{\mu_1}(N)} \simeq \frac{(z_{1-\alpha} - z_\beta)^2}{2\{(1-\beta)\ln[(1-\beta)/\alpha] - \beta \ln[(1-\alpha)/\beta]\}}, \tag{3.11.4}$$

which are independent of the actual values of (μ_0, μ_1, σ). The formulas in (3.11.3) and (3.11.4) become exact as $\mu_0 \to \mu_1$. B. K. Ghosh (1970) has tabulated some typical values of $Re(\mu_0)$ and $Re(\mu_1)$. It can easily be verified that $Re(\alpha, \beta, \mu_0) = Re(\beta, \alpha, \mu_1)$ for all (α, β), and $Re(\alpha, \alpha, \mu_0) = Re(\alpha, \alpha, \mu_1)$. The values in Table 3.11.1 indicate that the actual saving accomplished by

Table 3.11.1

Approximate Values of the Relative Efficiency of the SPRT for the Normal Mean (σ Known)[a]

β \ α		.01	.05	.10
.01	$Re(\mu_0)$	2.40	1.89	1.70
	$Re(\mu_1)$	2.40	2.71	2.93
.05	$Re(\mu_0)$	2.71	2.04	1.80
	$Re(\mu_1)$	1.89	2.04	2.15
.01	$Re(\mu_0)$	2.93	2.15	1.87
		1.70	1.80	1.87

[a] This constitutes a part of Table 3.4 of B. K. Ghosh (1970, p. 138), which is produced here with the kind permission of the author and Addison-Wesley.

the SPRT ranges from 40 to 70 percent. In fact, the true saving by using the SPRT will be slightly more than what is indicated in Table 3.11.1, because the approximate boundary values used in the SPRT would overestimate the true values of $E_{\mu_0}(N)$ and $E_{\mu_1}(N)$.

B. K. Ghosh (1970) has shown that, for real or vector parameter families, $Re(\theta_0)$ and $Re(\theta_1)$ are essentially the same as the right-hand expressions of (3.11.3) and (3.11.4) when $\theta_0 \to \theta_1$ under certain general regularity assumptions about the common probability (density) function $f(x; \theta)$.

Theorem 3.11.1 [B. K. Ghosh (1970, p. 139–140)]. Let $\{X_i\}$ be iid random variables having $f(x; \theta)$ for the common probability (density) function, where without loss of generality it is assumed that θ is real. For θ in the neighborhood of θ_0 and θ_1, assume that $\int f(x; \theta)\, dx$ can be differentiated twice with respect to θ under the integral sign, and that $\partial^2 \ln f(x; \theta)/\partial\theta^2$ is uniformly continuous in θ. Then the formulas for $Re(\theta_0)$ and $Re(\theta_1)$ of the SPRT of $\theta = \theta_0$ against $\theta = \theta_1$ relative to the optimum fixed-sample size procedure are approximately given by (3.11.3) and (3.11.4).

Proof Expanding $\ln f(x; \theta_j)$ for $j = 0, 1$ in Taylor's series about θ_0 or θ_1, we can write $Z = \ln[f(X; \theta_1)/f(X; \theta_0)]$ in two alternative forms as follows)

$$\begin{aligned} z &= \Delta \frac{\partial \ln f(x; \theta_0)}{\partial \theta_0} + \frac{1}{2}\Delta^2 \frac{\partial^2 \ln f(x; \theta_0)}{\partial {\theta_0}^2} + o(\Delta^2) \\ &= \Delta \frac{\partial \ln f(x; \theta_1)}{\partial \theta_1} - \frac{1}{2}\Delta^2 \frac{\partial^2 \ln f(x, \theta_1)}{\partial {\theta_1}^2} + o(\Delta^2), \end{aligned}$$

where $\Delta = (\theta_1 - \theta_0)$. However, under the differentiability assumption, we have

$$E_{\theta_j}\left\{\frac{\partial \ln f(X; \theta_j)}{\partial \theta_j}\right\} = 0,$$

and

$$E_{\theta_j}\left\{-\frac{\partial^2 \ln f(X; \theta_j)}{\partial {\theta_j}^2}\right\} = E_{\theta_j}\left\{\left[\frac{\partial \ln f(X; \theta_j)}{\partial \theta_j}\right]^2\right\} = I(\theta_j), \qquad j = 0, 1.$$

Hence, the iid Z_i have

$$\begin{aligned} E_{\theta_0}(Z) &= -\tfrac{1}{2}\Delta^2 I(\theta_0) + o(\Delta^2), & E_{\theta_1}(Z) &= \tfrac{1}{2}\Delta^2 I(\theta_1) + o(\Delta^2), \\ \operatorname{var}_{\theta_0}(Z) &= \Delta^2 I(\theta_0) + o(\Delta^2), & \operatorname{var}_{\theta_1}(Z) &= \Delta^2 I(\theta_1) + o(\Delta^2). \end{aligned}$$

If $|\Delta|$ is small, then $|E_\theta(Z)|$ and $\operatorname{var}_\theta Z$ are small at $\theta = \theta_0, \theta_1$, and (2.4.2) and (2.4.3) yield

$$\begin{aligned} E_{\theta_0}(N) &\simeq \frac{(1-\alpha)\ln[(1-\alpha)/\beta] - \alpha \ln[(1-\beta)/\alpha]}{\tfrac{1}{2}\Delta^2 I(\theta_0) + o(\Delta^2)}, \\ E_{\theta_1}(N) &\simeq \frac{(1-\beta)\ln[(1-\beta)/\alpha] - \beta \ln[(1-\alpha)/\beta]}{\tfrac{1}{2}\Delta^2 I(\theta_1) + o(\Delta^2)}. \end{aligned} \tag{3.11.5}$$

Formulas (3.11.5) become exact as $\theta_0 \to \theta_1$. Next, by the Neyman–Pearson lemma, the optimum fixed-sample size test of $\theta = \theta_0$ against $\theta = \theta_1$ is defined by the critical region $S_{n(\alpha,\beta)} > c(\alpha, \beta)$ where $c(\alpha, \beta)$ and the smallest integer $n(\alpha, \beta)$ satisfy

$$P_{\theta_0}(S_{n(\alpha,\beta)} \le c(\alpha, \beta)) \ge 1 - \alpha \qquad \text{and} \qquad P_{\theta_1}(S_{n(\alpha,\beta)} \le c(\alpha, \beta)) \le \beta,$$

where, as before, $S_n = \sum_{i=1}^{n} Z_i$. If $|\Delta|$ is small, then $n(\alpha, \beta)$ is large and by the central limit theorem, $S_{n(\alpha,\beta)}$ is approximately normal with mean $n(\alpha, \beta)E_{\theta_j}(Z)$ and variance $n(\alpha, \beta) \operatorname{var}_{\theta_j}(Z)$. Consequently,

$$\begin{aligned} n(\alpha, \beta) &\simeq \frac{[z_{1-\alpha}[\operatorname{var}_{\theta_0}(Z)]^{1/2} - z_\beta[\operatorname{var}_{\theta_1}(Z)]^{1/2}]^2}{[E_{\theta_1}(Z) - E_{\theta_0}(Z)]^2} \\ &\simeq \frac{[z_{1-\alpha}\{\Delta^2 I(\theta_0) + o(\Delta^2)\}^{1/2} - z_\beta\{\Delta^2 I(\theta_1) + o(\Delta^2)\}^{1/2}]^2}{[\frac{1}{2}\Delta^2 I(\theta_0) + \frac{1}{2}\Delta^2 I(\theta_1) + o(\Delta^2)]^2}. \end{aligned} \tag{3.11.6}$$

The preceding formula becomes exact as $\theta_0 \to \theta_1$. The three formulas in (3.11.5) and (3.11.6) lead to (3.11.3) and (3.11.4) for $Re(\theta_0)$ and $Re(\theta_1)$ as $\theta_0 \to \theta_1$. The uniform continuity assumption ensures that, as $\theta_0 \to \theta_1$, $\lim o(\Delta^2)/\Delta^2 = 0$ and $\lim I(\theta_0)/I(\theta_1) = 1$. Thus, the percentage savings indicated by Table 3.11.1 will also apply to the SPRT of $\theta = \theta_0$ against $\theta = \theta_1$ for arbitrary $f(x; \theta)$ provided $|\theta_0 - \theta_1|$ is fairly small. It should, of course, be understood that the relative efficiency can be estimated, if necessary, more precisely by obtaining $n(\alpha, \beta)$ without the regularity and normality assumptions.

3.11.2 Relative Efficiency at $\theta \neq \theta_0, \theta_1$

Although the null and alternative hypotheses are simple, the unknown value of θ may conceivably be different from θ_0 and θ_1. The performance of OC function of the SPRT at values of θ lying between θ_0 and θ_1 may not be of much practical interest because one is indifferent as to whether $\theta = \theta_0$ is accepted or rejected. On the other hand, the performance of the ASN function at such values is of considerable importance since the optimum property of the SPRT hold only at $\theta = \theta_0$ or θ_1. If $E_\theta(N)$ is a continuous function of θ, one would expect that the discussion of the preceding subsection should also apply to $Re(\theta)$ for θ in the neighborhood of θ_0 and θ_1. In general, if α and β are not too small, then $\sup_\theta E_\theta(N) < n(\alpha, \beta)$. Whether the maximum ASN of the SPRT is really less than $n(\alpha, \beta)$ can be verified in any given situation using

Wald's approximations. On the other hand, it can happen that $E_\theta(N) > n(\alpha, \beta)$ for all $\theta \in \Theta$ where Θ is a subset of Θ. When this happens in one-parameter families, Θ is typically an interval of values located between θ_0 and θ_1. This feature can be illustrated by the UME SPRT and the optimum fixed-sample test of the normal mean μ and with known variance σ^2, namely H_0: $\mu \leq \mu_0$, against $H_1: \mu \geq \mu_1$, $-\infty < \mu_0 < \mu_1 < \infty$. Let $\alpha = \beta < \frac{1}{2}$. Observe that $E_\mu(N) < E_{\mu_0}(N) = E_{\mu_1}(N) < n(\alpha, \alpha)$ for $\mu < \mu_0$ or $\mu > \mu_1$, and $\sup_\mu E_\mu(N) = E_{\tilde{\mu}}(N)$, $\tilde{\mu} = (\mu_0 + \mu_1)/2$.

From (3.11.2) and the relation

$$E_\mu(N) \simeq \frac{L(\mu) \ln [\beta/(1-\alpha)] + [1 - L(\mu)] \ln[(1-\beta)/\alpha]}{E_\mu(z)}$$

where Z is normally distributed with mean $(\mu_1 - \mu_0)(2\mu - \mu_1 - \mu_0)/2\sigma^2$ and variance $(\mu_1 - \mu_0)^2/\sigma^2$, we obtain

$$n(\alpha, \alpha) = 4z_{1-\alpha}^2[(\mu_1 - \mu_0)/\sigma]^2, \qquad E_{\tilde{\mu}}(N) \simeq \frac{\{\ln[(1-\alpha)/\alpha]\}^2}{[(\mu_1 - \mu_0)/\sigma]^2}.$$

Consequently

$$\inf_\mu Re(\mu) = \frac{n(\alpha, \alpha)}{\sup_\mu E_\mu(N)} \simeq \left[\frac{2z_{1-\alpha}}{\ln[(1-\alpha)/\alpha]}\right]^2 = \psi(\alpha) \qquad \text{for} \quad 0 < \alpha < \tfrac{1}{2}.$$

We present in Table 3.11.2 the values of $\psi(\alpha)$ for some values of α. It is easy to verify that

$$\lim_{\alpha \to 0} \inf_\mu Re(\mu) = 0 \qquad \text{and} \qquad \lim_{\alpha \to 1/2} \inf_\mu Re(\mu) = \pi/2.$$

Table 3.11.2

Approximate Values of $\inf_\mu Re(\mu)$ for Testing the Normal Mean (σ Known) when $\alpha = \beta$[a]

α	.005	.01	.05	.10
$\inf_\mu Re(\mu)$	.947	1.03	1.25	1.36

[a] This constitutes a part of Table 3.5 of B. K. Ghosh (1970, p. 141), which is produced here with the kind permission of the author and Addison Wesley.

In order to study the behavior of $\inf_\mu Re(\mu)$, it is sufficient to examine the derivative of $[\psi(\alpha)]^{1/2}$. Hence, consider

$$h(\alpha) = z\phi(z) - \alpha(1-\alpha)\ln[(1-\alpha)/\alpha] \qquad \text{where} \quad z = z_{1-\alpha}$$
$$= \tfrac{1}{2} - \alpha - \int_0^z x^2\phi(x)\,dx - \alpha(1-\alpha)\ln[(1-\alpha)/\alpha],$$

after performing integration by parts. Notice the error in Ghosh's (1970, p. 141) computation and his consequent argument. Hence

$$h'(\alpha) = (1-2\alpha)\ln[(1-\alpha)/\alpha] - 2 + z^2,$$
$$h''(\alpha) = -2\ln[(1-\alpha)/\alpha] - (1-2\alpha)/\alpha(1-\alpha) - 2z/\phi(z) < 0.$$

That is, $h'(\alpha)$ is decreasing in α. Since $h(0) = h(\frac{1}{2}) = 0$, h cannot change sign for α in $(0, \frac{1}{2})$. Since $h(\alpha)$ is not identically zero, it is strictly positive for α in $(0, \frac{1}{2})$.

The solution of the equation $\psi(\alpha) = 1$ lies between $\alpha = .0080$ and $\alpha = .0081$. Consequently, for $\alpha \leq .008$, the UME SPRT is less efficient than the optimum fixed-sample test when the true value of μ is in the neighborhood of $(\mu_0 + \mu_1)/2$. A similar comparative study can be made for any given pair (α, β), $\alpha \neq \beta$.

The UME SPRT may suffer, for $\theta \neq \theta_0, \theta_1$, from a similar lack of efficiency relative to some other valid test.

3.11.3 Limitations of the Optimum Property of the SPRT

The UME SPRT may not always be the best test for a given hypothesis-testing problem subject to certain prescribed bounds on the error probabilities. B. K. Ghosh (1970, pp. 142–144) has pointed out some reasonable grounds on which some suitable rival (non-SPRT) test δ satisfying the restrictions on the error probabilities may be preferable. Suppose we are interested in testing $H_0: \theta = \theta_0$ against $H_1: \theta = \theta_1$ on the basis of iid sequence $\{X_i\}$ or $\{Z_i\} = \{f(X_i; \theta_1)/f(X_i; \theta_0)\}$. Although the SPRT is optimal and preferable to the fixed-sample size $[n(\alpha, \beta)]$ test δ (say) it is quite possible that the SPRT may have a larger value for

$$P_{\theta_0}(N > n(\alpha, \beta)) \qquad \text{and/or} \qquad P_{\theta_1}(N > n(\alpha, \beta)).$$

This would offset the optimum property of the SPRT relative to δ because for many observed sequences $\{x_i\}$ one may actually need more observations to carry out the SPRT than δ. In order to make a more careful study of this aspect, let us derive a lower bound for $P_\theta(N \le n)$ for the SPRT where n is a given positive integer and $\theta \in \Theta$ is specified. For any arbitrary SPRT with stopping bounds (b, a), we have

$$\begin{aligned} P_\theta(N \le n) &= 1 - P_\theta(b < S_i < a, i = 1, 2, \ldots, n) \\ &\ge P_\theta(S_n \le b) + P_\theta(S_n \ge a) \\ &\simeq \Phi\left[\frac{b - nE_\theta(Z)}{[n \operatorname{var}_\theta(Z)]^{1/2}}\right] + 1 - \Phi\left[\frac{a - nE_\theta(Z)}{[n \operatorname{var}_\theta(Z)]^{1/2}}\right] \\ &= p_\theta(n) \qquad \text{(say).} \end{aligned} \tag{3.11.7}$$

since, S_n is asymptotically normal with mean $nE_\theta(Z)$ and variance $n \operatorname{var}_\theta(Z)$ provided the latter is finite. If $|E_\theta Z|$ and $\operatorname{var}_\theta(Z)$ are small then with $b = \ln[\beta/(1-\alpha)]$ and $a = \ln[(1-\beta)/\alpha]$ (3.11.7) gives a lower bound to the probability that the SPRT requires at most n observations in order to reach a decision. In particular if $\{X_i\}$ are iid as normal with mean μ and known variance σ^2 and the hypotheses are $H_0\colon \mu = \mu_0$, $H_1\colon \mu = \mu_1$ we have

$$E_{\mu_0}(Z) = -\Delta^2/2 = -E_{\mu_1}(Z), \quad \operatorname{var}_\mu(Z) = \Delta^2, \qquad \Delta = (\mu_1 - \mu_0)/\sigma.$$

The lower bound in (3.11.7) reduces to, for every $n \ge 1$,

$$\begin{aligned} p_{\mu_0}(n) &= \Phi\left(\frac{b + n\Delta^2/2}{\Delta n^{1/2}}\right) + \Phi\left(\frac{-a - n\Delta^2/2}{\Delta n^{1/2}}\right), \\ p_{\mu_1}(n) &= \Phi\left(\frac{b - n\Delta^2/2}{\Delta n^{1/2}}\right) + \Phi\left(\frac{-a + n\Delta^2/2}{\Delta n^{1/2}}\right). \end{aligned} \tag{3.11.8}$$

Given (α, β), $\alpha + \beta < 1$ one can choose Δ for given $n(\alpha, \beta)$ from

$$\delta = |z_{1-\alpha} - z_\beta| / [n(\alpha, \beta)]^{1/2},$$

B. K. Ghosh (1970, p. 143) has tabulated the lower bounds given in (3.11.8) with $n(\alpha, \beta) = 100$ and various combinations of α and β. Some of these are presented in Table 3.11.3. It appears from Table 3.11.3 that both $P_{\mu_0}(N \le n(\alpha, \beta))$ and $P_{\mu_1}(N \le n(\alpha, \beta))$ of the SPRT are fairly large for commonly used values of (α, β). In Table 3.11.3, $p_{\mu_0}(n)$ and $p_{\mu_1}(n)$ are not the greatest lower bounds, which may favor the SPRT appreciably. The pattern indicated by Table 3.11.3 seems to hold in other problems also. However, the position of $P_\theta(N \le n(\alpha, \beta))$ for $\theta \ne \theta_0, \theta_1$ is not quite clear-cut when θ is

Table 3.11.3

Lower Bounds to $P_{\mu_0}(N \le n)$ and $P_{\mu_1}(N \le n)$ of the UME SPRT of H_0: $\mu = \mu_0$ against H_1: $\mu = \mu_1$ for a Normal Population with Known σ When the Optimum Fixed-Sample Test Requires 100 Observations[a]

n	$\alpha = \beta = .01$		$\alpha = .01, \beta = .05$		$\alpha = \beta = .05$		$\alpha = .005, \beta = .1$	
	$p_{\mu_0}(n)$	$p_{\mu_1}(n)$	$p_{\mu_0}(n)$	$p_{\mu_1}(n)$	$p_{\mu_0}(n)$	$p_{\mu_1}(n)$	$p_{\mu_0}(n)$	$p_{\mu_1}(n)$
$E_{\mu_j}(N)$	.50	.50	.49	.49	.46	.46	.49	.41
100	.91	.91	.89	.80	.78	.78	.91	.72
200	.995	.995	.99	.98	.96	.96	.99	.96
300	1.00	.100	1.00	1.00	.99	.99	1.00	1.00

[a] This constitutes a part of Table 3.6 of B. K. Ghosh (1970, p. 143), which is produced here with the kind permission of the author and Addison-Wesley.

real or vector valued. For one-parameter families, $P_\theta(N \le n(\alpha, \beta))$ may not be so large for θ lying in the midway between θ_0 and θ_1.

3.11.4 The Efficiency of an Invariant SPRT

The optimum property of the SPRT does not generally hold for invariant SPRTs with maximum errors (α, β). However, one can compare the ASN of the invariant SPRT with the ASN of the rival invariant test procedure and, in particular, with $n(\alpha, \beta)$ of the optimum fixed-sample invariant test. Such comparisons have been made by research workers for the sequential F-test, sequential t^2-test and sequential chi-square and T^2-tests. For these and other examples, the reader is referred to B. K. Ghosh (1970, pp. 350–353).

3.11.5 Limiting Relative Efficiency of the SPRT

The limiting relative efficiency of the SPRT has been studied by Paulson (1947), Chernoff (1956), Aivazian (1959), and Bechhofer (1960). The familiar statement that the SPRT often results in an average saving of 50 percent in sample size needs to be qualified. Bechhofer (1960) has studied the limiting relative efficiency of the SPRT for the normal case when the error probabilities are related to each other and they tend to zero. He brings out some of the surprises which were never anticipated. In the following we shall present his results.

Let $X_1, X_2, \ldots$ be an iid sequence of normal variables having unknown mean μ and known variance σ^2. We wish to test the hypothesis H_0: $\mu = \mu_0$

against the alternative H_1: $\mu = \mu_1$ ($\mu_1 > \mu_0$). Let α and β be the bounds on the error probabilities such that $0 < \alpha, \beta < 1$ and $\alpha + \beta < 1$. We shall study the efficiency (measured in terms of ratio of average sample size to fixed sample size) of the SPRT to the best competing fixed-sample procedure as (α, β) approach zero in a prescribed manner. In Section 3.11.2, we studied the same problem for fixed $\alpha = \beta$. However, for simplicity, let

$$\theta = [2\mu - (\mu_0 + \mu_1)]/2\sigma, \qquad \theta^* = (\mu_1 - \mu_0)/2\sigma. \tag{3.11.9}$$

Then $\mu = \mu_0$ and $\mu = \mu_1$ are equivalent to $\theta = -\theta^*$ and $\theta = \theta^*$ respectively. One can easily obtain Wald's approximations to the OC and ASN functions [see Section 2.3 and Eq. (2.4.1)] as

$$L(\theta) \sim (A^h - 1)/(A^h - B^h), \qquad h = -\theta/\theta^* \tag{3.11.10}$$

and

$$E_\theta(N) \sim \{L(\theta) \ln B + [1 - L(\theta)] \ln A\}/2\theta\theta^* \tag{3.11.11}$$

where $A \doteq (1 - \beta)/\alpha$ and $B \doteq \beta/(1 - \alpha)$. The ratios of these approximations (which neglect the "excess over the boundary") to the corresponding true values are known to approach unity very rapidly as the ASN becomes large. [It has been shown by Dvoretsky *et al.* (1953) that the right-hand members of (3.11.10) and (3.11.11) are actually lower bounds to the true OC and ASN respectively.] Also from (3.11.2), the fixed-sample size which guarantees the same bounds on the error probabilities, is the smallest integer equal to or greater than

$$(z_\alpha - z_{1-\beta})^2/4(\theta^*)^2 \qquad \text{where} \quad \Phi(z_\gamma) = \gamma, \qquad 0 \le \gamma \le 1. \tag{3.11.12}$$

Then, the relative efficiency (neglecting excess) is given by

$$Re(\alpha, \beta, \theta, \theta^*) \sim \frac{2\theta^*}{\theta} \frac{L(\theta) \ln B + [1 - L(\theta)] \ln A}{(z_\alpha - z_{1-\beta})^2}. \tag{3.11.13}$$

We shall obtain the limit of (3.11.13) when $\beta = c\alpha^d$ ($c, d > 0$) and $\alpha \to 0$. Now the second element in the numerator of the right-hand side expression of (3.11.13) becomes

$$L(\theta)[\ln(c\alpha^d) - \ln(1 - \alpha)] + [1 - L(\theta)][\ln(1 - c\alpha^d) - \ln \alpha]$$

which approaches $-\ln \alpha + L(\theta)[\ln \alpha + \ln(c\alpha^d)]$ as $\alpha \to 0$. It is straightforward but tedious to show, using l'Hospital's rule, that $L(\theta)[\ln \alpha + \ln(c\alpha^d)]$

approaches $\ln \alpha + \ln(c\alpha^d)$ for $\theta < 0$ and zero for $\theta > 0$ as $\alpha \to 0$. Hence, we have

$$\lim_{\alpha\to 0} Re(\alpha, c\alpha^d, \theta, \theta^*) = \begin{cases} (2\theta^*/\theta) \lim\limits_{\alpha\to 0}[\ln(c\alpha^d)/(z_\alpha + z_{c\alpha^d})^2], & \theta < 0, \\ (2\theta^*/\theta) \lim\limits_{\alpha\to 0}[-\ln \alpha/(z_\alpha + z_{c\alpha^d})^2], & \theta > 0. \end{cases} \tag{3.11.14}$$

Now, for small α, integrating by parts we have

$$\alpha = \int_{-\infty}^{z_\alpha} \phi(y)\,dy = -[\phi(z_\alpha)/z_\alpha] - \int_{-\infty}^{z_\alpha} y^{-2}\phi(y)\,dy.$$

Hence, $\ln \alpha \sim -\frac{1}{2}z_\alpha^{\ 2} - \ln(-z_\alpha) - \ln 2(\pi)^{1/2} \sim -(\frac{1}{2})z_\alpha^{\ 2}$. Thus, for α close to zero we have $z_\alpha \sim \{-2\ln\alpha\}^{1/2}$ and $z_{c\alpha^d} \sim \{-2\ln(c\alpha^d)\}^{1/2}$. Substituting these in (3.11.14) we obtain

$$\lim_{\alpha\to 0} Re(\alpha, c\alpha^d, \theta, \theta^*) = \begin{cases} (-\theta^*/\theta)(1 + d^{-1/2})^{-2} & \text{for } \theta < 0, \\ (\theta^*/\theta)(1 + d^{1/2})^{-2} & \text{for } \theta > 0. \end{cases} \tag{3.11.15}$$

When $d = 1$, that is, $c\alpha = \beta$, (3.11.15) becomes

$$\lim_{\alpha\to 0} Re(\alpha, c\alpha, \theta, \theta^*) = \theta^*/4|\theta| = (\mu_1\mu_0)/4|\mu_0 + \mu_1 - 2\mu|. \tag{3.11.16}$$

Thus, in the limit ($c\alpha = \beta \to 0$) the efficiency of the fixed-sample procedure relative to the SPRT $\to 0$ as $\mu \to \pm\infty$, it is equal to $\frac{1}{4}$ if $\mu = \mu_0$ or $\mu = \mu_1$, it is greater than unity if $(5\mu_0 + 3\mu_1)/8 < \mu < (3\mu_0 + 5\mu_1)/8$, and it is infinite if $\mu = (\mu_0 + \mu_1)/2$. The limiting relative efficiency of $\frac{1}{4}$ obtained when $\mu = \mu_0$ or $\mu = \mu_1$ has previously been noted by Chernoff (1956, p. 19); the result (3.11.15) for $\theta = \theta^*$ has been obtained by Aivazian (1959). Both Chernoff (1956) and Aivazian considered the general problem of testing a simple hypothesis versus a simple alternative, and studied the limiting relative efficiency as the two hypotheses approach each other. For earlier work on this see Paulson (1947).

DISCUSSION If $H_0: \mu \le \mu_0$ and $H_1: \mu \ge \mu_1$ and Wald's SPRT is applied to test the above hypothesis, then for a range of values of μ between μ_0 and μ_1 the advantage of the SPRT over the fixed-sample size procedure may be reversed. That is, the fixed-sample size procedure may yield a many-to-one saving in ASN over the SPRT. In fact, if we let $\theta \to 0$, we have $E_\theta(N) \sim -\ln A \ln B/4(\theta^*)^2$ so that

$$Re(\alpha, \beta, \theta, \theta^*) \simeq [\ln((1-\alpha)/\beta)][\ln((1-\beta)/\alpha)]. \tag{3.11.17}$$

when $\alpha = \beta$, the relative efficiency curve (3.11.13) is symmetric about $\theta = 0$

and assumes its maximum at this point. For this special case, (3.11.17) is greater than unity if $\alpha = \beta < .008$ (approximately). This may momentarily provide comfort to the reader since in practice one seldom uses values of α that are less than .008. However, it should be noted that (3.11.17) is only an approximation which neglects "excess" in both the ASN and the fixed-sample size. ("Excess" in the latter comes about because the smallest integer equal to or greater than (3.11.2) is used.) Monte Carlo studies conducted by Bechhofer and Sobel (1958) indicate that the ratio of the true ASN associated with the SPRT [which sets $A = (1 - \alpha)/\alpha = 1/B$] to the fixed-sample size is greater than unity when $\theta^* = .2$ and $\theta = 0$ for values of α less than .015 (approximately). (This critical value of α probably increases with θ^* since the ASN decreases and the "excess" increases.) It should be pointed out that similar comments can be made of the SPRT for testing the mean of any distribution (where μ denotes the mean of the population) since the central limit theorem comes into play, especially when $\mu_1 - \mu_0$ and $|\mu - \mu_i|$ are small $(i = 0, 1)$.

These results point up the great practical need for sequential procedures that guarantee α and β at μ_0 and μ_1, respectively, and which minimize the maximum value of the ASN. (Such a procedure is called a *minimax* procedure.) A characterization of procedures that provide this guarantee and which minimize the ASN at a third point (other than μ_0 or μ_1) has been obtained by Kiefer and Weiss (1957, Section 4). For the problem considered here, if $\alpha = \beta$ the ASN curve is symmetric about $\mu = (\mu_0 + \mu_1)/2$; hence if one minimizes the ASN at this point, then the characterization of Kiefer and Weiss (1957) applies to the minimax procedure. The restricted procedure proposed by Anderson (1960) is an important contribution in the direction of minimizing the ASN.

3.11.6 Relative Efficiency of the SPRT for the Exponential Family

Sakaguchi (1967) has extended Bechhofer's (1960) results to the exponential family. He also gives the relation between the relative efficiency and Chernoff's (1952) information number discriminating between the two densities. In the following, we shall describe his results pertaining to the relative efficiency of the SPRT.

First, let us consider the classical problem of testing $H_0: f(x) = f_0(x)$, against $H_1: f(x) = f_1(x)$ when X is distributed with the generalized probability density function $f(x)$. The test has α and β for the error probabilities. The most powerful fixed-sample critical region is of the form: $\sum_1^n Z_i > k$ where $Z_i = \ln[f_1(X_i)/f_0(X_i)]$. For sufficiently large n, $\sum_1^n Z_i$ is approximately normally distributed with mean $nE(Z)$ and variance $n\sigma^2(Z)$, under any distribution with density $f(x)$. Hence, k and n are determined from†

$$\Phi\{(k + nI(0{:}\ 1))/n^{1/2}\sigma_0(Z)\} \doteq \alpha, \qquad \Phi\{(k - nI(1{:}\ 0))/n^{1/2}\sigma_1(Z)\} \doteq 1 - \beta$$

where $\sigma_i^2(Z) = \text{var}(Z \mid H_i)$ and

$$I(i{:}\ 1 - i) = \int f_i(x) \ln[f_i(x)/f_{1-i}(x)]\, dv(x) \qquad (i = 0, 1) \tag{3.11.18}$$

is the Kullback–Leibler information number (Kullback, 1959), where $v(x)$ denotes some nondegenerate, σ-finite measure. If z_α is such that $\Phi(z_\alpha) = \alpha$, then eliminating k we obtain

$$n \doteq [z_\alpha \sigma_0(Z) + z_\beta \sigma_1(Z)]^2/J^2 = (z_\alpha + z_\beta)^2 \bar{\sigma}^2/J^2 \tag{3.11.19}$$

where

$$J \equiv I(0{:}\ 1) + I(1{:}\ 0) \qquad \text{and} \qquad \bar{\sigma} = [z_\alpha \sigma_0(Z) + z_\beta \sigma_1(Z)]/(z_\alpha + z_\beta). \tag{3.11.20}$$

Now, let us assume that $f(z) = f(x; \theta) = e^{\theta x - b(\theta)}$, the density with respect to a nondegenerate σ-finite measure $v(x)$ where $\int e^{\theta x} dv(x) = e^{b(\theta)}$. Then, recall that $E_\theta(X) = b'(\theta)$, $\text{var}_\theta(X) = b''(\theta)$. For any parameter values θ_0, θ_1 with $\theta_0 < \theta_1$ and θ in the domain of the natural parameter space, we find that

$$\begin{aligned}
I(\theta_i{:}\ \theta_{1-i}) &= E_{\theta_i}(\ln[f(X; \theta_i)/f(X; \theta_{1-i})]) \\
&= (\theta_i - \theta_{1-i})b'(\theta_i) - \{b(\theta_i) - b(\theta_{1-i})\},
\end{aligned}$$

$$J(\theta_0, \theta_1) = I(\theta_0{:}\ \theta_1) + I(\theta_1{:}\ \theta_0) = (\theta_1 - \theta_0)[b'(\theta_1) - b'(\theta_0)],$$

$$\sigma_i^2(Z) = \text{var}_{\theta_i}(\ln[f(X; \theta_1)/f(X; \theta_0)]) = (\theta_1 - \theta_0)^2 b''(\theta_i),$$

$$\begin{aligned}
E_\theta(Z) &= (\theta_1 - \theta_0)b'(\theta) - [b(\theta_1) - b(\theta_0)], \\
E_\theta(N) &= \{L(\theta) \ln B + [1 - L(\theta)] \ln A\}/E_\theta(Z), \\
L(\theta) &= (A^h - 1)/(A^h - B^h),
\end{aligned} \tag{3.11.21}$$

where $h = h(\theta)$ is, for each θ, the unique nonzero value satisfying

$$b(\theta + (\theta_1 - \theta_0)h) - b(\theta) = [b(\theta_1) - b(\theta_0)]h, \tag{3.11.22}$$

$A = (1 - \beta)/\alpha$, and $B = \beta/(1 - \alpha)$.

† Berk (1973, p. 1137) points out that the central limit theorem is not adequate and that a large deviation result is required in order to obtain the correct error rates, asymptotically.

Substituting these expressions into (3.11.19) and (3.11.21), the efficiency of the SPRT relative to the best competing fixed-sample procedure with fixed error probabilities α, β is given by

$$[Re(\alpha, \beta, \theta)]^{-1}$$
$$= E_\theta(N)/n$$
$$= \frac{L(\theta)\ln B + (1 - L(\theta))\ln A}{(\theta_1 - \theta_0)b'(\theta) - [b(\theta_1) - b(\theta_0)]} \frac{[b'(\theta_1) - b'(\theta_0)]^2}{\{z_\alpha[b''(\theta_0)]^{1/2} + z_\beta[b''(\theta_1)]^{1/2}\}^2}. \tag{3.11.23}$$

Next, we shall obtain the limit of (3.11.23) when $\beta = c\alpha^d$ $(c, d > 0)$ and $\alpha \to 0$. Then

$$L(\theta) \doteq (1 - \alpha^h)/(1 - c^h\alpha^{(1+d)h}) \underset{\alpha\to 0}{\to} \begin{cases} 1 & \text{if } h > 0, \\ 0 & \text{if } h < 0. \end{cases}$$

Also, since $\alpha \doteq -\phi(z_\alpha)/z_\alpha$ gives $\ln \alpha \doteq -z_\alpha{}^2/2$, we find that

$$\frac{(\theta_1 - \theta_0)b'(\theta) - [b(\theta_1) - b(\theta_0)]}{[b'(\theta_1) - b'(\theta_0)]^2} [\lim Re(\alpha, c\alpha^d, \theta)]^{-1}$$
$$= \begin{cases} -(b/2)\{[b''(\theta_0)]^{1/2} + [db''(\theta_1)]^{1/2}\}^{-2} & \text{if } h > 0, \\ -\frac{1}{2}\{[b''(\theta_0)]^{1/2} + [db''(\theta_1)]^{1/2}\}^{-2} & \text{if } h < 0. \end{cases} \tag{3.11.24}$$

This shows that the limiting relative efficiency depends on the particular choice of θ and the relative rate at which α and β approach zero in a particular manner. Define as a function of y, for each fixed θ,

$$K(y; \theta) = \begin{cases} [b(\theta + y) - b(\theta)]/y & \text{if } y \neq 0, \\ b'(\theta) & \text{if } y = 0. \end{cases} \tag{3.11.25}$$

By the strict convexity of $b(\theta)$ this is strictly increasing both in θ and y and is continuous at $y = 0$. From (3.11.22), $h(\theta)$ is explicitly given by

$$h(\theta) = (\theta_1 - \theta_0)^{-1}K^{-1}(b(\theta_1) - b(\theta_0)/(\theta_1 - \theta_0); \theta) \tag{3.11.26}$$

where $K^{-1}(y, \theta)$ is the inverse function of $K(y; \theta)$ for each fixed θ. Since $K^{-1}(b'(\theta); \theta) = 0$ we have $h(\theta) = 0$ if and only if

$$b'(\theta) = [b(\theta_1) - b(\theta_0)]/(\theta_1 - \theta_0). \tag{3.11.27}$$

Let us denote this value of θ by θ_{01}. This value exists uniquely between θ_0 and θ_1, and by the increasing property of $K(y; \theta)$ both in θ and y, we have

$$h(\theta) \gtreqless 0 \qquad \text{according as} \qquad \theta \gtreqless \theta_{01}. \tag{3.11.28}$$

Thus, if $d = 1$ (that is, if $\beta = c\alpha$ and $\alpha \to 0$) we have from (3.11.24) and (3.11.28)

$$[\lim_{\alpha \to 0} Re(\alpha, c\alpha, \theta)]^{-1}$$

$$= \frac{1}{2|(\theta_1 - \theta_0)b'(\theta) - (b(\theta_1) - (b(\theta_0))|} \left\{ \frac{b'(\theta_1) - b'(\theta_0)}{[b''(\theta_0)]^{1/2} + [b''(\theta_1)]^{1/2}} \right\}^2 \tag{3.11.29}$$

which is infinite if $\theta = \theta_{01}$ and is greater than unity in the intervals of θ in which

$$\left| b'(\theta) - \frac{b(\theta_1) - b(\theta_0)}{\theta_1 - \theta_0} \right| < \frac{1}{2(\theta_1 - \theta_0)} \left\{ \frac{b'(\theta_1) - b'(\theta_0)}{[b''(\theta_0)]^{1/2} + [b''(\theta_1)]^{1/2}} \right\}^2 .$$

Another limiting value of the relative efficiency is obtained by fixing α and β and letting $\theta \to \theta_{01}$. By applying l'Hospital's rule we find that

$$\frac{\partial L}{\partial h} = \frac{\partial}{\partial h} \left(\frac{A^h - 1}{A^h - B^h} \right) \underset{h \to 0}{\to} \frac{(-\ln B)(\ln A)}{2(\ln A - \ln B)},$$

and from (3.11.25) and (3.11.26) we have $h'(\theta_{01}) = -\alpha/(\theta_1 - \theta_0)$, After differentiating implicitly [by setting $y = (\theta_1 - \theta_0)h(\theta)$] the equation

$$k_\theta((\theta_1 - \theta_0)h(\theta)) = [b(\theta_1) - b(\theta_0)]/(\theta_1 - \theta_0)$$

one obtains

$$(\theta_1 - \theta_0)h'(\theta_{01}) = \lim_{\theta \to 0} \frac{dy}{d\theta} = \lim_{\theta \to 0} \frac{b'(\theta_{01} + y) - b'(\theta_{01})}{y^{-1}[b(\theta_{01} + y) - b(\theta_{01})] - b'(\theta_{01} + y)}.$$

Since the first factor in the right-hand side of (3.11.23) approaches

$$\frac{h'(\theta_{01})(\ln B - \ln A)}{(\theta_1 - \theta_0)b''(\theta_{01})} \left[\frac{\partial}{\partial h} L \right]_{h=0}$$

as $\theta \to \theta_{01}$, we thus have

$$Re(\alpha, \beta, \theta_{01}) = \left\{\frac{b'(\theta_1) - b'(\theta_0)}{(\theta_1 - \theta_0)[b''(\theta_{01})]^{1/2}}\right\}^2 \frac{(-\ln B)(\ln A)}{\{z_\alpha[b''(\theta_0)]^{1/2} + z_\beta[b''(\theta_1)]^{1/2}\}^2}. \tag{3.11.30}$$

If $\theta_1 = \theta_0 + \Delta$ and $\Delta \to 0$ then, from (3.11.27), θ_{01} also tends to θ_0. Therefore, from (3.11.30)

$$\left[\lim_{\theta_1 \to \theta_0} Re(\alpha, \beta, \theta_{01})\right]^{-1} = (-\ln B)(\ln A)/(z_\alpha + z_\beta)^2, \tag{3.11.31}$$

independently of both $b(\theta)$ and θ_0.

Notice that (3.11.24) specializes to (3.11.15) for the normal case. Paulson (1947) has shown that, for arbitrary densities $f_i(x)$ $(i = 0, 1)$ and θ_0,

$$\left[\lim_{\theta_1 \to \theta_0} Re(\alpha, \beta, H_0)\right]^{-1} = \frac{\alpha \ln[\alpha/(1-\beta)] + (1-\alpha)\ln[(1-\alpha)/\beta]}{\frac{1}{2}(z_\alpha + z_\beta)^2} \tag{3.11.32}$$

and

$$\left[\lim_{\theta_1 \to \theta_0} Re(\alpha, \beta, H_1)\right]^{-1} = \frac{\beta \ln[\beta/(1-\alpha)] + (1-\beta)\ln[(1-\beta)/\alpha]}{\frac{1}{2}(z_\alpha + z_\beta)^2} \tag{3.11.33}$$

independently of the $f_i(x)$ and θ_0, provided that conditions ensuring the central limit property of $\sum_1^n \ln[f_1(X_i)/f_0(X_i)]$ are satisfied. The limits in (3.11.32) and (3.11.33) which coincide with (3.11.3) and (3.11.4) do not exceed unity, because of the optimality of the SPRT (see Theorem 2.9.1).

Examples In the following examples, let (i) $=\theta_{01}$, (ii) $=$the reciprocal of the limiting relative efficiency (3.11.29), (iii) $= [Re(\alpha, \beta, \theta_{01})]^{-1}/(-\ln B)(\ln A)$.

Normal Distribution with Known Variance σ^2

$$f(x; \theta) = (2\pi\sigma^2)^{-1/2} \exp[-(x-\mu)^2/2\sigma^2], \qquad \theta = \mu/\sigma^2, \qquad b(\theta) = \sigma^2\theta^2/2.$$

(i) $=(\theta_0 + \theta_1)/2$,
(ii) $=(\theta_1 - \theta_0)/2|2\theta - (\theta_0 + \theta_1)|$,
(iii) $=(z_\alpha + z_\beta)^{-2}$.

Poisson Distribution

$$f(x;\theta) = e^{-\lambda}\lambda^x/x!, \qquad \theta = \ln\lambda, \qquad b(\theta) = e^\theta.$$

(i) $= \ln(e^{\theta_1} - e^{\theta_0})/(\theta_1 - \theta_0)$,
(ii) $= \frac{1}{2}(\lambda_1^{1/2} - \lambda_0^{1/2})^2/|\lambda\ln(\lambda_1/\lambda_0) - (\lambda_1 - \lambda_0)|$,
(iii) $= [(\lambda_1 - \lambda_0)/\ln(\lambda_1/\lambda_0)][\lambda_0^{1/2} z_\alpha + \lambda_1^{1/2} z_\beta]^{-2}$.

Exponential Distribution

$$f(x;\theta) = \mu^{-1}e^{-x/\mu}, \quad \theta = -1/\mu, \quad b(\theta) = -\ln(-\theta).$$

(i) $= (\theta_1 - \theta_0)\{\ln(\theta_1/\theta_0)\}^{-1}$
(ii) $= \frac{1}{2}[(\mu_1 - \mu_0)/(\mu_0 + \mu_1)]^2/|(\mu_0^{-1} - \mu_1^{-1})\mu - \ln(\mu_1/\mu_0)|$,
(iii) $= [(\mu_1 - \mu_0)/\ln(\mu_1/\mu_0)]^2(\mu_0 z_\alpha + \mu_1 z_\beta)^{-2}$.

Problem

Compute θ_{01}, the relative efficiency as given in (3.11.29) and (3.11.30) for the following cases:

(i) Bernoulli distribution: $f(x;\theta) = p^x(1-p)^{1-x}$, $\theta = \ln[p/(1-p)]$, $b(\theta) = \ln(1+e^\theta)$; and

(ii) Geometric distribution: $f(x;\theta) = (1+\mu)^{-(x+1)}\mu^x$, $\theta = \ln[\mu/(1+\mu)]$, $b(\theta) = -\ln(1-e^\theta)$.

3.11.7 Asymptotic Efficiency of Wald's SPRT in the General Case

Recently, Berk (1973) has obtained elegant results for the asymptotic efficiency of Wald's SPRT relative to the best nonsequential test having the same strength (α, β). Theorem 2.13.2 gives the asymptotic behavior of (α, β) as $c = \min(a, -b) \to \infty$, while Theorem 2.13.1 and 2.13.3 give the corresponding behavior of EN. In order to study the asymptotic efficiency of SPRT, we need an asymptotic expression for the sample size required by the best nonsequential test of strength (α, β). In making the computation, we assume that

$$\lim_c\,[(\ln\beta)/(\ln\alpha)] = \lambda, \qquad 0 < \lambda < \infty, \tag{3.11.34}$$

or in view of Theorem 2.13.2 that $\lim_c[(-b)/a] = \lambda$.

For a sample of size n, the most powerful test of H_0 against H_1 rejects H_0 when $S_n > d_n$. We must choose (n, d_n) so that asymptotically, the nonsequential test has strength (α, β). That is, we must have

$$\text{(i)}\quad P_0(S_n > d_n) \doteq \alpha, \qquad \text{(ii)}\quad P_1(S_n \le d_n) \doteq \beta \tag{3.11.35}$$

in the sense that the ratios of the corresponding logarithms tend to unity. [Thus (n, d_n) depends on (b, a).] We argue that $d_n = n\zeta$, where ζ is a real number depending on H_0, H_1, and λ. The reason for this is to be found in Chernoff's (1952) large deviation result for a series of iid summands. Let $c_0(t) = \ln E_0(\exp tZ)$. Since Z is a log-likelihood ratio, $c_0(t) < \infty$ at least for $0 \leq t \leq 1$. Chernoff's theorem then says that for $z > E_0 Z$,

$$\lim_n n^{-1} \ln P_0(S_n > nz) = -k_0(z) \tag{3.11.36}$$

where

$$k_0(z) = \sup_{-\infty < t < \infty} [tz - c_0(t)]. \tag{3.11.37}$$

Similarly if $z < E_1 Z$

$$\lim_n n^{-1} \ln P_1(S_n \leq nz) = -k_1(z). \tag{3.11.38}$$

In view of (3.11.34), (3.11.36), and (3.11.38) we see that a nonsequential test asymptotically of strength (α, β) is obtained by choosing $d_n = n\zeta(\lambda)$ where $\mu_0 < \zeta(\lambda) < \mu_1$ is the unique solution of

$$k_1(z)/k_0(z) = \lambda \tag{3.11.39}$$

and n is chosen so that (3.11.35), (ii) holds. Thus (3.11.35) and (3.11.36) give for n the relation $nk_0(\zeta) = \ln \alpha^{-1}$, or in view of Theorem 2.13.2

$$nk_1(\zeta) = a. \tag{3.11.40}$$

Hence the sample size required to obtain strength (α, β) asymptotically is

$$v(c, \lambda) = a/k_0(\zeta), \tag{3.11.41}$$

and the corresponding critical value for S_v is $\zeta(\lambda)a/k_0(\zeta)$. Toward a solution of (3.11.39), we note that

$$c_1(t) = \ln \int [f_1(x)/f_0(x)]^t f_1(x)\, dx = c_0(t+1)$$

from which it follows that

$$k_1(z) = k_0(z) - z. \tag{3.11.42}$$

Moreover, when $z = \mu_0 < 0$, $t\mu_0 - c_0(t)$ is maximum at $t = 0$ (since $c_0'(0) = \mu_0$ where $c_0'(t) = dc_0(t)/dt$]; hence $k_0(\mu_0) = 0$. Similarly $k_1(\mu_1) = 0$, so that $k_0(\mu_1) = \mu_1 > 0$. In view of (3.11.42), (3.11.39) becomes $(1 - \lambda)k_0(z) = z$ which has the solution $\zeta = 0$ if $\lambda = 1$. Otherwise (3.11.39) gives the equation

$$k_0(z) = z/(1 - \lambda). \tag{3.11.43}$$

Since, as shown by Chernoff (1952), if $z > \mu_0$, $k_0(z) = \sup_{t>0}[tz - c_0(t)]$, it follows that k_0 is convex in t and increasing for $z > \mu_0$. Since also $k_0(\mu_0) = 0$, $k_0(\mu_1) = \mu_1$ and $0 < \lambda < \infty$ entails $(1 - \lambda)^{-1} \notin (0, 1)$, it follows that the curve defined by the two sides of (3.11.43) intersect in a single point, whose abscissa $\zeta(\lambda)$ (say) is in (μ_0, μ_1) and is thus the unique solution of (3.11.39). In view of (3.11.43), (3.11.41) becomes

$$v(c, \lambda) = \begin{cases} a(1 - \lambda)/\zeta(\lambda), & \lambda \neq 1, \\ a/\kappa, & \lambda = 1, \end{cases} \tag{3.11.44}$$

where, as shown by Chernoff (1952),

$$\kappa = k_0(0) = k_1(0) = -\ln \inf_{0<t<1} \int f_0{}^t(x) f_1^{1-t}(x)\, dx.$$

The asymptotic efficiency of the SPRT relative to the corresponding best nonsequential test is now obtained via (3.11.44) and the results of Section 2.13. If $EZ = \mu \neq 0$, then

$$EN \sim \begin{cases} a/\mu, & \mu > 0, \\ b/\mu, & \mu < 0; \end{cases}$$

while the corresponding nonsequential test requires $v(c, \lambda)$ observations. Hence the ARE (asymptotic relative efficiency) of the SPRT is given by

$$\begin{aligned} \text{ARE} &= \lim_c v(c, \lambda)/EN \\ &= \begin{cases} (1 - \lambda)\mu/\zeta(\lambda), & \mu > 0,\ \lambda \neq 1, \\ -\mu(1 - \lambda)/\lambda\zeta(\lambda), & \mu < 0,\ \lambda \neq 1, \\ |\mu|/\kappa, & \mu \neq 0,\ \lambda = 1. \end{cases} \end{aligned} \tag{3.11.45}$$

If $EZ = 0$, then we see from Theorem 2.13.3 that $EN \sim -ab$, so that $\lim_c v(c, \lambda)/EN = 0$. That is, the SPRT has ARE zero under distributions for which $EZ = 0$. This phenomena has been pointed out by Bechhofer (1960) (see Section 3.11.5) for the normal case.

3.12 Bayes Sequential Procedures

3.12.1 Bayes Sequential Binomial SPRT

Suppose that batches of items are presented for inspection, in which the items must be classified as either effective or defective. Let p denote the fraction defective contained in a batch. Assume that there exists a known critical fraction defective p_0 such that a batch with a fraction defective p_0 may without loss be accepted or rejected. Batches with $p > p_0$ are considered bad and should be rejected. Let $P(p)$ denote the prior distribution (also called the process curve) for the fraction defective contained in batches submitted for inspection and assume that it is of the following type:

$$P(p = p_1) = a_1, \qquad P(p = p_2) = a_2, \quad a_1 + a_2 = 1, \qquad p_1 < p_0 < p_2. \tag{3.12.1}$$

Let W_{21} (W_{12}) denote the decision cost incurred if a batch with fraction defective p_1 (p_2) is rejected (accepted). Let c denote the cost of inspecting a single item. The problem is to determine the most economical sampling inspection plan.

Vagholkar and Wetherill (1960) give a method based on the basic theory developed by Barnard (1954). Because of the special form of the prior distribution, the problem of acceptance sampling is reduced to that of testing two simple hypotheses H_i ($i = 1, 2$) where H_i means that a batch comes from a population with a fraction defective p_i ($i = 1, 2$). We wish to accept a batch if H_1 is accepted and we wish to reject the batch if H_2 is accepted. In the sampling plan one item is inspected at a time, and inspection is stopped as soon as sufficient evidence is obtained in favor of either of the hypotheses. If the cost of the inspection depends merely on the total number of items inspected and no extra cost is involved on account of sequential sampling, a sequential plan will be the most economical one. Also because of the optimum property of the SPRT, the latter will be the optimum test procedure as long as the cost of inspection is linearly related to the number of items inspected. In the following we describe the optimum procedure:

(i) continue inspection as long as

$$\lambda_2 < \lambda = (a_1/a_2)\, l(X, Y) < \lambda_1, \tag{3.12.2}$$

(ii) stop inspection and accept the batch as soon as $\lambda > \lambda_1$,
(iii) stop inspection and reject the batch as soon as $\lambda > \lambda_2$,

where X and Y denote respectively the number of effectives and defectives obtained at any stage, and $l(X, Y)$ is the likelihood ratio and is equal to $(p_1/p_2)^Y(q_1/q_2)^X, q_i = 1 - p_i\ (i = 1, 2)$. It is more convenient to write inequality (3.12.2) as

$$\underline{v} < l(X, Y) < \bar{v} \tag{3.12.3}$$

where $\bar{v} = a_2\lambda_1/a_1$ and $\underline{v} = a_2\lambda_2/a$. Let $L(\underline{v}, \bar{v}, p)$ denote the probability that a batch with fraction defective p will be rejected, and $n(\underline{v}, \bar{v}, p)$ denote the average number of items required to be inspected. The decision boundary (as indicated by Barnard (1954) is the locus of points such that the expected cost of taking an immediate decision is equal to the expected cost of taking at least one more observation and continuing the test. For example, when $\lambda = \lambda_1$, we write down the expected cost of accepting as a function of the prior probabilities and decision costs: If we take another observation, then an effective item leads to acceptance of the batch, incurring a decision cost if H_2 is true, and a defective item will take us to a point where we either continue the test or reject immediately.

We determine λ_1 to be that value for which these two costs—of accepting immediately and of taking one more observation and continuing the test if necessary—are equal. It should also be noted that, in general, the difference between the expected cost of an immediate decision and the expected cost of continuing the test should decrease as the sample size increases. For any λ, it can easily be verified that the posterior probabilities associated with H_1 and H_2 are $\lambda/(1 + \lambda)$ and $1/(1 + \lambda)$ respectively. Now, we are ready to derive the equations for the optimum values of λ_1 and λ_2. Consider $\lambda = \lambda_1$ so that the probability that H_1 is true is $\lambda_1/(1 + \lambda_1)$ and the probability that H_2 is true is $1/(1 + \lambda_2)$. Now the immediate decision to be taken is to accept H_1 and thereby incur a loss W_{12} if H_2 is true. Thus, the loss due to an immediate decision is $W_{12}/(1 + \lambda_1)$. On the other hand, if we inspect one more item, incurring a cost c, and continue the procedure thereafter, we shall accept the batch if the next item is an effective one since $\lambda_1 q_1/q_2 > \lambda_1$. The cost W_{12} will be incurred if H_2 is true, so that the loss is

$$W_{12}\, P(H_2 \text{ is true})\ P(\text{next item is effective} \mid H_2 \text{ is true}) = W_{12}q_2/(1 + \lambda_1).$$

If the next item is defective, the likelihood ratio will be $\lambda_1 p_1/p_2$, and we shall continue with a (new) SPRT starting at this point. If the number of effective and defective items obtained with this new test is (X', Y'), the test is

$$\lambda_2 < \lambda_1 p_1\, l(X', Y')/p_2 < \lambda_1 \qquad \text{or} \qquad \lambda_2 p_2/\lambda_1 p_1 < l(X', Y') < p_2/p_1.$$

Clearly, if $\lambda_2/\lambda_1 \geq p_1/p_2$ we shall reject immediately, so that sampling continues only if $\lambda_2/\lambda_1 < p_2/p_1$. The cost due to this continuation is

$$\{cn(\delta, \rho\delta; p_1) + W_{21}[1 - L(\delta, \delta\rho, p_1)]\}p_1\lambda_1/(1 + \lambda_1)$$

if H_1 is true and $\rho < \delta$ where $\delta = p_2/p_1$ and $\rho = \lambda_2/\lambda_1$. If H_2 is true and $\rho < \delta$, the continuation cost is

$$[cn(\delta, \rho\delta; p_2) + W_{12}L(\delta, \rho\delta; p_2)]p_2/(1 + \lambda_1).$$

If $\rho > \delta$, we reject the batch, and the cost is $W_{21}p_1\lambda_1/(1 + \lambda_1)$. Now, equating the expected cost of an immediate decision to the expected cost if at least one more item is inspected, we have for $\rho < \delta$

$$\begin{aligned} W_{12}/(1 + \lambda_1) = c + \{&cn(\delta, \rho\delta; p_1) \\ &+ W_{21}[1 - L(\delta, \rho\delta; p_1)]\}p_1\lambda_1/(1 - \lambda_1) + W_{12}q_2/(1 + \lambda_1) \\ &+ [cn(\delta, \rho\delta; p_2) + W_{12}L(\delta, \rho\delta; p_2)]p_2/(1 + \lambda_1) \end{aligned} \tag{3.12.4}$$

and, for $\rho > \delta$

$$W_{12}/(1 + \lambda_1) = c + W_{21}p_1\lambda_1/(1 + \lambda_1) + W_{12}q_2/(1 + \lambda_1). \tag{3.12.5}$$

Solving λ_1 from (3.12.4) and (3.12.5) we get, for $\rho < \delta$

$$\lambda_1 = \frac{W_{12}p_2[1 - L(\delta, \rho\delta; p_2)] - c - p_2\, cn(\delta, \rho\delta; p_2)}{W_{21}p_1[1 - L(\delta, \rho\delta; p_1)] + c + p_1\, cn(\delta, \rho\delta, p_1)} \tag{3.12.6}$$

and, for $\rho > \delta$,

$$\lambda_1 = (W_{12}p_2 - c)/(W_{21}p_1 + c). \tag{3.12.7}$$

A similar argument leads to the following equation for λ_2:

$$\lambda_2 = \frac{W_{12}q_2\, L(\gamma/\rho, \gamma; p_2) + c + q_2\, cL(\gamma/\rho, \gamma; p_2)}{W_{21}q_1\, L(\gamma/\rho, \gamma; p_1) - c - q_1 cL(\gamma/\rho, \gamma; p_1)} \tag{3.12.8}$$

where $\gamma = q_2/q_1$ and $\rho < q_2/q_1$. Dividing (3.12.8) by (3.12.6) we have, for $\rho < p_1/p_2$

$$\rho = h(\rho) \qquad \text{(say)} \tag{3.12.9}$$

and dividing (3.12.8) by (3.12.7) we have, for $\rho > p_1/p_2$

$$\rho = h_1(\rho) \qquad \text{(say)}. \tag{3.12.10}$$

Equations (3.12.9) and (3.12.10) can be solved for ρ by iteration. The usual way of determining the boundaries of a SPRT is to use Wald's approximate formulas which necessarily assume the error probabilities to be small, which is not always true in the case of sampling inspection plans designed on a minimum cost basis. Hence, we would like to use some kind of exact formulas. Such formulas that are practically manageable are those given by Burman (1946) and these will be discussed next.

The SPRT for a binomial population as defined by (3.12.3) can be reduced (without much loss in accuracy) to a scoring procedure given by

$$-M_2 = \frac{\ln \underline{v}}{\ln(q_1/q_2)} < X - Yb < \frac{\ln \bar{v}}{\ln(q_1/q_2)} = M_1 \qquad \text{(say)} \tag{3.12.11}$$

where $b = \ln(p_2/p_1)/\ln(q_1/q_2)$. If b, M_1, and M_2 are rounded off to the nearest integers, the SPRT reduces to the following scoring scheme. Start with a score M_2, and add one to the score for each effective item found, subtracting b for each defective item found. Reject the batch if the score falls to zero or less, and accept the batch if the score reaches $2M = M_1 + M_2$. Formulas for the ASN and OC functions of such a scheme have been given by Burman (1946) and are exact if b, M_1, and M_2 are integers. The error involved in rounding them off is small if b is greater than ten, which is often the case in practice. One can express $h(\rho)$ and $h_1(\rho)$ given by Eqs. (3.12.9) and (3.12.10) in terms of the score notation by replacing $L(\gamma/\rho, \gamma; p_j)$ by $L(2M - 1, 1; p_j)$, $n(\gamma/\rho, \gamma; p_j)$ by $n(2M - 1, 1; p_j)$, $L(\delta, \rho\delta; p_j)$ by $L(b, 2M - b; p_j)$, and $n(\delta, \rho\delta; p_j)$ by $n(b, 2M - b, p_j)$ for $j = 1$, 2, where b and $2M = -\ln \rho/\ln(q_1/q_2)$ are rounded off to the nearest integers.

The Procedure for Calculation In practice, we have to fix values for $a_1, a_2, p_1, p_2, W_{12}, W_{21}$, and c. The quantity b can then be calculated. In order to solve Eqs. (3.12.9) and (3.12.10) (which are expressed in terms of the score notation) we start with some guessed value for ρ and proceed with iteration until we get the same value of ρ. In the right-hand side expressions of the modified versions of (3.12.9) and (3.12.10) ρ enters through $2M$, which is an integer, so that when we get a ρ which gives rise to the same value of $2M$ as was used in the previous iteration, we stop and take that as our final solution. Once ρ is obtained, λ_1 and λ_2 can be calculated from (3.12.6) and (3.12.8) or (3.12.7) and (3.12.8) as the case may be. The value of ρ is always less than one, and a lower limit for ρ has been derived by Vagholkar (1955) from which we have

$$c^2/[W_{12}(q_1 - q_2) - c][W_{21}(q_1 - q_2) - c] < \rho < 1. \qquad (3.12.12)$$

The lower limit or any number a little higher than the lower limit, can serve as a good first guess for ρ, in order to start the iteration.

Example 3.12.1 (Vagholkar and Wetherill, 1960) Let $a_1 = \frac{5}{9}$, $a_2 = \frac{4}{9}$, $p_1 = .01$, $p_2 = .10$, $W_{21} = 400$, $W_{12} = 500$, and $c = 1$. We get $b = 24$ and by (3.12.12) we have $.00065 < \rho < 1$. Starting with a guess of .001 for ρ the successive values of ρ obtained are

$$\begin{array}{cccccc} \rho \rightarrow & .001 & \rightarrow & .001607 & \rightarrow & .001600 \\ & \downarrow & & \downarrow & & \downarrow \\ & 2M = 72 & & 2M = 68 & & 2M = 68. \end{array}$$

This gives $\lambda_1 = 30.5$, $\lambda_2 = .0489$, and the optimum test procedure (3.12.11) is given by $M_2 = 34$, $2M = M_1 + M_2 = 68$, $b = 24$.

REMARK 3.12.1 It has been assumed that the cost of inspection is proportional to the number of items inspected. Also, the difficulty in carrying out a SPRT arises in that a decision is required after every item inspected as to whether to continue sampling or sentence the batch. This can be taken account of in the cost function by an extra term, namely Dd where D denotes the number of times the decision to continue sampling or sentence the batch has to be made and d denotes the cost associated with stopping sampling in order to make this decision. Further, one can add on the cost of processing the items sampled. This may be equal to Tt where T denotes the number of times the process is carried out, and t denotes the cost associated with each processing. Thus, the total cost becomes

$$\text{cost} = nc + Dd + Tt. \qquad (3.12.13)$$

In this case the optimum test will be a SPRT with items sampled in groups (not necessarily of a constant size). If the constant group size is k, then (3.12.13) becomes

$$\text{cost} = nc + \frac{n}{k}d + \frac{n}{k}t = nc', \qquad c' = c + \frac{d}{k} + \frac{t}{k} \qquad (3.12.14)$$

which is linear in n. Two particular results apply to group sequential sampling which follow from the results of Wald and Wolfowitz (1948). These are stated as lemmas.

Lemma 3.12.1 The risk of a group sequential sampling scheme is greater than or equal to the risk for a unit-step sequential, for given prior probability and loss functions.

Lemma 3.12.2 The optimum boundaries for group sequential sampling are within the optimum boundaries for unit-step sequential sampling with the same probability and loss functions.

REMARK 3.12.2 The practical utility of the above lemmas is that for group sequential sampling we can derive the optimum boundaries from unit-step sequential sampling using c', the artificial cost per item given in (3.12.14). These boundaries will contain the true optimum boundaries. If the group size is large, and we can replace the two-point binomial distribution by an equivalent two-point normal distribution, a more exact solution is possible.

REMARK 3.12.3 The three-point binomial distribution is difficult to handle mathematically. However, Vagholkar and Wetherill (1960) and Wetherill (1957) provide a result pertaining to the three-point binomial distribution.

3.12.2 Dynamical Programming Method for the Binomial Case

In the previous work of Vagholkar and Wetherill (1960) the prior distribution over a spectrum of values of p is replaced by a distribution concentrated on two special values of p, namely p_1 and p_2 such that one decision is clearly appropriate for p_1 and the second decision is appropriate for p_2. This approach is somewhat unsatisfactory and is an oversimplification of the problem.

For the binomial problem, Lindley and Barnett (1965) have given an optimal Bayes sequential procedure that can be solved numerically by the backward induction techniques of dynamic programming. This will be described in the following (see also Appendix A-8). Without loss of generality, assume that the losses are given by

$$L_1(p) = k_r, \qquad L_2(p) = p, \tag{3.12.15}$$

where, without affecting the problems, the scales of losses have been changed and $L_i(p)$ is the loss associated with the terminal decision d_i ($i = 1, 2$). We shall also assume that $0 < k_r < 1$, otherwise the problem is trivial (if k_r lies outside this range, then one decision is optimum for all p and no sampling is necessary). k_r is called the critical value or the break-even value of p. If $p < k_r$, d_2 is the optimum decision, and if $p > k_r$, d_1 is optimum where for convenience, let d_1 be the decision to reject and d_2 the decision to accept. This agrees with the industrial application to sampling inspection where $X_i = 1$ if the ith individual item in the batch is defective. Let c be the constant cost of sampling an item. Notice that we are not free to scale the cost of sampling since we have already scaled the loss functions. The prior distribution that would be used will be the conjugate prior namely the beta family given by

$$(a + b - 1)!\, p^{a-1}(1 - p)^{b-1}/(a - 1)!\,(b - 1)!, \qquad a, b > 0. \quad (3.12.16)$$

Essentially, if the optimum scheme is tabulated with $a = b = 1$ (that is, a uniform prior density of p), then the optimum scheme is known for any beta prior distribution with positive integer values of a and b. This follows since the tabulated scheme for $a = b = 1$ will tell one what to do if n samples have been taken and r of them are found to be defective. The likelihood of this is proportional to $p^r(1 - p)^{n-r}$ so the distribution of p after the samples have been taken is proportional to $p^r(1 - p)^{n-r}$. Consequently, the situation is exactly as if one were starting with $a = r$, $b = n - r$, and no further tabulation for this prior distribution is necessary. Therefore the tables depend only on two parameters k_r and c.

THE DYNAMIC PROGRAMMING EQUATION If the current probability distribution of p has the beta distribution (3.12.16), then the expected loss is clearly k_r if d_1 is taken and is equal to $E(p) = a/(a + b)$ if d_2 is taken. Hence, if only two terminal decisions were considered one would reject (take d_1) if $k_r > a/(a + b)$ and otherwise accept (take d_2). If the prior distribution has $a = b = 1$, then this current distribution is obtained by observing $r = a - 1$ defectives ($X_i = 1$) and $n - r = b - 1$ effectives ($X_i = 0$). In this case one would reject if $k_r < (r + 1)/(n + 2)$. It is usual to plot the schemes on a diagram with horizontal axis $a + b$ and vertical axis a. The line $k_r = a/(a + b)$ is called the critical line. Above it the optimum among all the terminal decisions is to reject; below it is better to accept. Then, the loss incurred due to this terminal decision is given by

$$D(a, b) = \min[k_r, a/(a + b)]. \quad (3.12.17)$$

One of the other possibilities (beside making a terminal decision) is to take one observation and then choose among the terminal decisions. Among all the possibilities, there is one procedure that has the smallest expected loss and which will be called best. Let $B(a, b)$ be the expected loss of the best possible procedure when the prior distribution has values (a, b). If $B(a, b) = \min(k_r, a/(a + b))$, then the best procedure is to stop and take that terminal decision with the smaller loss. If $B(a, b) < \min(k_r, a/(a+b))$, then take at least one more observation and then proceed in a way that would be described below. Let $B^*(a, b)$ denote the expected loss if one further observation is made followed by the optimum decision. If the observation is of a defective then a and $a + b$ will increase by 1. Hence, the expected loss obtained by adopting the optimum procedure after the observation will be $B(a + 1, b)$. If the observation is an effective one, the expected loss will be $B(a, b + 1)$. Consequently, if one observation is taken, when the state of knowledge is (a, b), we have

$$B^*(a, b) = c + \frac{a}{a+b} B(a+1, b) + \frac{b}{a+b} B(a, b+1). \tag{3.12.18}$$

Knowing $D(a, b)$ and $B^*(a, b)$, the equation for $B(a, b)$ is, since stopping or taking one more observation are the only possibilities,

$$B(a, b) = \min[D(a, b), B^*(a, b)]. \tag{3.12.19}$$

If $B(a, b)$ is known for all a, b, with $a + b = x_0$ (say), then (3.12.19) enables one to find $B(a, b)$ for all a, b with $a + b = x_0 - 1$. Consequently, $B(a, b)$ is known for all a, b with $x_0 - a - b$ equal to an integer. Once $B(a, b)$ is known, the optimum procedure at (a, b) is easily found by accepting if it is equal to $a/(a + b)$, rejecting if equal to k_r, and otherwise taking a further observation. Thus, in the diagram, each point is labeled as acceptance, rejection, or continuation. Lindley and Barnett (1965) claim that it can be shown that for fixed $a + b$ the continuation region is an interval (possibly empty) between the two boundaries with the acceptance region below and the rejection region above, and that for all sufficiently large $a + b$ the interval of the continuation region is empty. Therefore, there is a least upper bound to the values of $a + b$ in the continuation region, at which the rejection and continuation boundaries meet on the critical line. This meeting point [call it $(\bar{a} + \bar{b}, \bar{b})$] will satisfy the relation

$$k_r = \frac{\bar{a}}{\bar{a}+\bar{b}} = c + \frac{\bar{a}}{\bar{a}+\bar{b}} k_r + \frac{\bar{b}}{\bar{a}+\bar{b}} \cdot \frac{\bar{a}}{\bar{a}+\bar{b}+1} \cdot \tag{3.12.20}$$

Letting $\bar{x} = \bar{a} + \bar{b}$, and solving we obtain

$$\bar{a} = k_r \bar{x} \qquad \text{and} \qquad \bar{x} = [k_r(1 - k_r)/c] - 1. \tag{3.12.21}$$

Equation (3.12.21) gives the upper bound beyond which it is never worth taking further observations. From a practical point of view, it is sufficient to start from the *highest reachable* point. The authors discuss how to find the highest reachable point. They also provide a computation method. They include a discussion about the OC and ASN functions of their procedure.

THE EFFICIENCY OF THEIR PROCEDURE If c were equal to zero, we could sample indefinitely until p were known. Then the value of $B(1, 1)$, would be

$$\int_0^1 \min[L_1(p), L_2(p)]\, dp = \int_0^{k_r} p\, dp + \int_{k_r}^1 k_r\, dp = k_r(1 - k_r/2).$$

The expected loss initially when $a = b = 1$ is k_r (if $k_r \le 0.5$) obtained by rejecting. The difference between these two is $(k_r^2/2)$, the expected value of perfect information (EVPI) in the sense of Raiffa and Schlaifer (1961). Similarly, for any other value of c, $k_r - B(1, 1)$ is the expected value of sample information (EVSI). A reasonable measure of the efficiency of the optimum scheme is the ratio of the EVSI to EVPI given by

$$E = 2[k_r - B(1, 1)]/k_r^2. \tag{3.12.22}$$

This measure is a better criterion for the performance of the scheme than $B(1, 1)$ which depends for its interpretation on k_r and c. Table 3.12.1 lists the

Table 3.12.1

Showing the Values of $E = 2[k_r - B(1, 1)]/k_r^2$.

k_r \ c	.05	.01	$.0^21$	$.0^31$	$.0^41$	$.0^51$	$.0^61$
.01					.00	.59	.90
.1			.00	.66	.91	.98	1.00
.2		.00	.52	.86	.97	.99	1.00
.3	.00	.15	.72	.93	.98	1.00	
.4	.00	.45	.82	.95	.99	1.00	
.5	.27	.61	.88	.97	.99	1.00	

efficiencies for various values of k_r and c. For fixed k_r, the efficiency naturally increases as the sampling cost c decreases. The limiting value of 1 is approached much more slowly for small k_r than for values of k_r near $\frac{1}{2}$.

Lindley and Barnett (1965) provide a normal approximation to the boundaries. These agree with the results of Moriguti and Robbins (1962) and Chernoff (1960). The authors also consider the normal case with $H_0: \mu > 0$ against $H_1: \mu \le 0$ and known variance where μ has a normal prior distribution. The reader is referred to Lindley and Barnett (1965, Section 16) for the normal case. The problem of sequential sampling has been considered in great generality for the exponential family by Mikhalevich (1956); in particular, he investigates the circumstances under which the optimum schemes terminate.

3.12.3 Bayes Sequential Procedures for the Normal Mean

This problem has been considered by Chernoff in a series of papers. In his review paper, Chernoff (1968) has summarized the results pertaining to the Bayes sequential testing procedure for the mean of a normal population. We shall present these results and for the details the reader is referred to the references given at the end of the review paper by Chernoff (1968). The problem can be formulated as follows. Let X be a normal random variable having unknown mean μ and known variance σ^2. We wish to test $H_0\colon \mu \geq 0$ against $H_1\colon \mu < 0$ with the cost of an incorrect decision being $k|\mu|$, $k > 0$. Let c denote the cost per observation. The total cost is cn if the decision is correct and $cn + k|\mu|$ if it is wrong where n is the number of observations taken. Thus, the total cost is a random variable whose distribution depends on the unknown μ and the sequential procedure used. The problem is to select an optimal sequential procedure. After much data have been accumulated one is either reasonably certain of the sign of μ or that $|\mu|$ is so small that the loss of deciding wrong is less than the cost of another observation. Here, one expects the proper procedure to be such that one stops and makes a decision when the current estimate of μ is sufficiently large and continues sampling otherwise. What constitutes sufficiently large depends on, and should decrease with, the number of observations or equivalently the precision of the estimate. It can be shown that after a certain sample size, it pays to stop no matter what the current estimate of $|\mu|$. For specific values of the constants, this problem can be solved numerically by the backward induction techniques of dynamic programming employed by Lindley and Barnett (1965). Care must be taken to initiate the backward induction at a sample size n sufficiently large so that no matter what the estimate of μ is, the optimal procedure will lead to a decision rather than to additional sampling. The technique of the backward induction can be summarized by the equation

$$R_n(\xi_n) = \inf_{\delta_n} E\{R_{n+1}[\xi_{n+1}(\delta_n, \xi_n)]\} \tag{3.12.23}$$

where $R_n(\xi_n)$ is the expected cost of an optimal procedure given the history ξ_n up to stage n, $\xi_{n+1}(\delta_n, \xi_n)$ describes the history up to stage $n+1$ which may be random, with distribution depending on ξ_n and the action δ_n taken at stage n. It is possible to show that ξ_n is adequately summarized by the mean and variance of the posterior distribution of μ. The calculus of posterior distributions when dealing with normal random variables and normal priors permits the problem to be treated with no great difficulty. If it is desired to derive some overall view of how the solutions depend on the various parameters, the simple though extensive numerical calculations of the backward induction are not adequate.

An approach which seems to have particular relevance to large sample theory is that of replacing the discrete time random variables by analogous continuous time stochastic processes. The use of the Wiener process converts the problem to one in which the analytic methods of partial differential equations can serve fruitfully. (See Appendix 4 for a discussion of the Wiener process.) So, let us assume that the data consist of a Wiener process $X(t)$ with unknown drift μ and known variance σ^2 per unit time. Also assume that the unknown value of μ has prior normal distribution with mean μ_0 and variance $\sigma_0{}^2$. One can easily verify [see Chernoff (1968, Lemma 4.1)] that the posterior distribution of μ is again normal with mean $Y(s)$ and variance s where

$$Y(s) = [\mu_0 \sigma_0^{-2} + X(t)\sigma^{-2}]/[\sigma_0^{-2} + t\sigma^{-2}], \tag{3.12.24}$$

$$s = \{\sigma_0^{-2} + t\sigma^{-2}\}^{-1}, \tag{3.12.25}$$

and $Y(s)$ is a Wiener process in the $-s$ scale originating at $(y_0, s_0) = (\mu_0, \sigma_0{}^2)$ that is

$$E(dY(s)) = 0, \qquad \operatorname{var}(dY(s)) = -ds. \tag{3.12.26}$$

Notice that s decreases from $s_0 = \sigma_0{}^2$ as information accumulates. Since the X process can be recovered from the Y process, it suffices to deal with the latter, which measures the current estimate of μ and which is easier to analyze.

The posterior expected cost associated with deciding in favor of H_1 at time t [when $Y(s) = y$] is

$$\int_{-\infty}^{0} k|\mu|\phi((\mu - y)/s^{1/2})\, d\mu = ks^{1/2}\psi^+(y/s^{1/2}) \tag{3.12.27}$$

where $\psi^+(\mu) = \phi(\mu) - \mu[1 - \Phi(\mu)]$. Similarly the posterior expected cost associated with deciding $\mu < 0$ is $ks^{1/2}\psi^-(y/s^{1/2})$ where $\psi^-(\mu) = \phi(\mu) + \mu\Phi(\mu)$. It is easy to see that if sampling is stopped at $Y(s) = y$, the decision should be made on the basis of the sign of y and the expected cost of deciding plus the cost of sampling is given by

$$d(y, s) = c\sigma_0{}^2 s^{-1} + ks^{1/2}\psi(y/s^{1/2}) - c\sigma^2/\sigma_0{}^2 \tag{3.12.28}$$

where

$$\psi(u) = \begin{cases} \phi(u) - u[1 - \Phi(u)], & u \geq 0, \\ \phi(u) + u\Phi(u), & u < 0. \end{cases} \tag{3.12.29}$$

Thus the continuous time version of the problem can be viewed as the following *stopping problem*: The Wiener process $Y(s)$ is observed. The statistician may stop at any value of $s > 0$ and pay $d(Y(s), s)$. Find the stopping procedure which minimizes the expected cost. In this version using $Y(s)$, the posterior Bayes estimate, the statistical aspects involving the unknown parameter μ have been abstracted.

The original discrete time problem can also be described in terms of this stopping problem by adding the proviso that allowable stopping values of s are restricted to $s_0, s_1, s_2, \ldots$ where $s_n = (\sigma_0^{-2} + n\sigma^{-2})^{-1}$. At this point it should be straightforward for the reader to see that the discrete version can be treated numerically by backward induction in terms of the $Y(s)$ process starting from any $s_n = c^2/k^2\psi^2(0) = 2\pi c^2/k^2$, where k is the constant occurring in the cost function.

3.12.4 Asymptotically Bayes Sequential Test Procedures

Several significant contributions have been made to the problem of asymptotic optimality of certain sequential test procedures for composite hypotheses by Chernoff (1959), Albert (1961), Bessler (1960), Schwarz (1962), and Kiefer and Sacks (1963), to mention a prominent few. Kiefer and Sacks (1963) have generalized the results of Chernoff, Albert, Bessler, and Schwarz. Bickel and Yahav (1965) have provided some results concerning pointwise optimal procedures. We shall present the main results of Schwarz (1962) for the exponential family with the indifference zone and the main results of Kiefer and Sacks (1963) for the case of no indifference zone.

Schwarz (1962) has provided a very elegant characterization of the asymptotic (as c, the cost per observation tends to zero) shape of the Bayes continuation region when the density function $f(x; \theta)$, $\theta \in \Theta$ belongs to the one-parameter exponential family and the parametric sets Θ_1 and Θ_2 are separated by an "indifference zone" Θ_0. Let H_1 and H_2 be the hypotheses associated with the disjoint sets Θ_1 and Θ_2 in some Euclidean parameter space Θ. Let

$$f(x; \theta) = \exp[\theta x - b(\theta)], \qquad -\infty < x < \infty$$

be the density function of X with respect to a sigma-finite measure ν. The function $L_1(\theta)$ $[L_2(\theta)]$ is the loss function associated with accepting H_1 $[H_2]$.

Let $G(\theta)$ denote a prior distribution on Θ. Let $X_1, X_2, \ldots$ be a sequence of iid variables having common density function $f(x; \theta)$. Let $\mathbf{X}_n = (X_1, \ldots, X_n)'$, $\mathbf{X}_0 \equiv 0$, and let $\rho_0(G, \mathbf{X}_n)$ denote the posterior Bayes risk and $\rho_0(\mathbf{X}_0) \equiv \rho_0$ be the prior Bayes risk.

In the case of the one-parameter family, the natural parameter space Θ is an interval on the real line over which $\int_{-\infty}^{\infty} e^{\theta x}\, dv(x) < \infty$, $\theta \in \Theta$. The function $b(\theta)$ is defined over Θ and satisfies

$$E_\theta\{\exp(tX)\} = \exp[b(t+\theta) - b(\theta)]$$

and hence

$$E_\theta(X) = b'(\theta), \qquad \mathrm{var}_\theta(X) = b''(\theta).$$

Hence, if the distribution of X is nondegenerate, $b(\theta)$ is a convex function on Θ. Since $b'(\theta)$ is strictly increasing, there is a unique inverse $\theta(\mu)$ for $E_\theta(X) = \mu = b'(\theta)$. In particular, consider the composite hypothesis $H_1\colon\ \theta \le \theta_1$ versus the composite hypothesis $H_2\colon\ \theta \ge \theta_2$, where $-\infty < \theta_1 < \theta_2 < \infty$. The interval $\Theta_0 = (\theta_1, \theta_2)$ is an indifference region. We assume that $L_i(\theta)$ is bounded and that $L_i(\theta) = 0$ if $\theta \in (\Theta_0 \cup \Theta_i)$ $(i = 1, 2)$. The (posterior) Bayes risk $\rho_0(G, \mathbf{X}_n)$ depends on $\mathbf{X}_n$ through $T_n = \sum_{i=1}^n X_i$ which is a minimal sufficient statistic for each $n \ge 1$; we denote it by $\hat{R}(n, T_n)$. The continuation region for a stopping rule $s^*(c)$ defines for each n an interval $C_n^*(c)$ such that if $T_n \in C_n^*(c)$, an additional observation is taken. The continuation region is denoted by $C^*(c)$ which is the union of the corresponding intervals $C_n^*(c)$. Let

$$C^*(\gamma) = \{(n, T_n) : \hat{R}(n, T_n) \ge \gamma\}, \qquad 0 < \gamma < \infty, \tag{3.12.30}$$

be a family of continuation regions for which sampling is continued as long as $\hat{R}(n, T_n) \ge \gamma$. The following theorem of Schwarz (1962) provides an upper and lower bound for the Bayes continuation region $\mathscr{B}(c)$ where

$$\mathscr{B}(c) = \{(n, T_n) : \hat{R}(n, T_n) > c + E[\rho(n+1, T_n + X) \mid \mathscr{F}_n]\} \tag{3.12.31}$$

where $\rho(n, T_n)$ is the minimal expected Bayes risk, and $\mathscr{F}_n$ is the Borel sigma-subfield generated by (θ, T_n) [see Zacks (1971, Section 9.4)].

Theorem 3.12.1 (Schwarz, 1962) For all c sufficiently small, $0 < c < c_0$,

$$C^*(c) \supset \mathscr{B}(c) \supset C^*(3c|\ln c|/\Delta(\theta_1, \theta_2)) \tag{3.12.32}$$

where $\Delta(\theta_1, \theta_2) = b(\theta_1) + b(\theta_2) - 2b\{(\theta_1 + \theta_2)/2\}$, which is nonnegative for all θ_1, θ_2, $\theta_1 < \theta_2$ since $b(\theta)$ is convex.

Proof See Schwarz (1962) or Zacks (1971, pp. 487–488).

The region $C^*(3c|\ln c|/\Delta(\theta_1, \theta_2))$ grows to infinity in each direction as $c \to 0$. Thus the Bayes continuation region $\mathscr{B}(c)$ grows to infinity in each direction and is represented asymptotically by $C^*(c)$. It is somewhat difficult to draw the boundary $C^*(c)$. Schwarz (1962) suggested considering the asymptotic shape of a continuation region based on a generalized probability likelihood ratio test, which is also asymptotically optimal. Thus, define the likelihood ratio statistics

$$\lambda_i(n, T_n) = (\xi) \operatorname*{ess\,sup}_{\theta \in \Theta_i} \{\exp[\theta T_n - nb(\theta)]\}/(\xi) \operatorname*{ess\,sup}_{\theta \in \Theta} \{\exp[\theta T_n - nb(\theta)]\}, \quad i = 1, 2, \tag{3.12.33}$$

where ξ denotes the prior probability measure on $(\Theta, \mathscr{C})$ corresponding to the prior distribution G on Θ. The quantity (ξ) ess $\sup_{\theta \in \Theta_i}(\cdot)$ denotes the almost sure supremum of $(\cdot)$ over the set Θ_i, with respect to the prior probability measure ξ. Also, let

$$\mathscr{B}_0 = \{(n, T_n) : \ln \lambda_i(n, T_n) \geq -1, i = 1, 2\}. \tag{3.12.34}$$

Hence

$$\mathscr{B}_0|\ln \gamma| = \{(n, T_n) : \ln \lambda_i(n, T_n) \geq -|\ln \gamma|, i = 1, 2\}. \tag{3.12.35}$$

Theorem 3.12.2 (Schwarz, 1962) Let Γ denote the set of all (n, T_n) for which n is a positive integer. Let $C^*(\gamma)$ be the region defined by (3.12.30). Then

$$C^*(\gamma) = \{\mathscr{B}_0|\ln \gamma| + O(|\ln \gamma|)\} \cap \Gamma \quad \text{as} \quad \gamma \to 0.$$

Proof See Schwarz (1962) or Zacks (1971, pp. 491–492).

We now give the main theorem.

Theorem 3.12.3 (Schwarz, 1962) The continuation region of the Bayes stopping rule, $\mathscr{B}(c)$ satisfies

$$\mathscr{B}(c) = \{\mathscr{B}_0|\ln c| + o(|\ln c|)\} \cap \Gamma \quad \text{as} \quad c \to 0.$$

Proof See Schwarz (1962) or Zacks (1971, p. 493).

The asymptotic shape of $\mathscr{B}(c)$ is thus given by that of $\mathscr{B}_0|\ln c|$ and is represented in the $(n/|\ln c|, T_n/|\ln c|)$ plane by $\mathscr{B}_0$. Since we are mostly concerned with geometric properties of the region $\mathscr{B}_0$, we rename the coordinates in the (n, T_n) plane and denote them by x and y instead of n and T_n respectively.

Theorem 3.12.4 (Schwarz, 1962) When G is a measure that dominates the Lebesgue measure on the θ line (that is, G gives positive weight to every set of positive Lebesgue measure) the boundary of $\mathscr{B}_0$ consists of two curves, whose equations are given by

$$\begin{aligned} xb(\theta(y/x)) - xb(\theta_1) - y\theta(y/x) + y\theta_1 + 1 &= 0, \\ xb(\theta(y/x)) - xb(\theta_2) - y\theta(y/x) + y\theta_2 + 1 &= 0. \end{aligned} \qquad (3.12.36)$$

Proof The almost sure supremum of a bounded continuous function with respect to the Lebesgue measure is simply the maximum of the functions. The function $y\theta - xb(\theta)$ obtains its maximum at the solution of

$$y - xb'(\theta) = 0, \qquad (3.12.37)$$

and that solution is $\theta = \theta(y/x)$ where $\theta(k)$ is the inverse function of $b'(\theta)$. Now, when (ξ)ess $\sup(y\theta - xb(\theta))$ is replaced by $y\theta(y/x) - xb(\theta(y/x))$, Eq. (3.12.36) is readily obtained.

Schwarz (1962) has tabulated the explicit formulas for the boundary of $\mathscr{B}_0$ when the support G is the whole real line, for several cases of interest. This is presented in Table 3.12.2.

Table 3.12.2

Example	Parameter	$b(\theta)$	Boundary of $\mathscr{B}_0$
Normal $(\mu,1)$	$\theta = \mu$	$\theta^2/2$	$y = \theta_i x + (2x)^{1/2}(-1)^i$
Bernoulli (p)	$\theta = \ln[p/(1-p)]$	$\ln(1+e^\theta)$	$(x-y)\ln(x-y)$ $= 1 + x\ln[x/(1+e^{\theta_i})]$ $-y\ln(y/e^{\theta_i})$
Poisson (λ)	$\theta = \ln\lambda$	e^θ	$y\ln(y/x) = 1 - xe^{\theta_i} + y(1+\theta_i)$
Geometric (μ)	$\theta = \ln[\mu/(1+\mu)]$	$-\ln(1-e^\theta)$	$(x+y)\ln(x+y) =$ $-1 + x\ln[x/(1-e^{\theta_i})]$ $+y\ln(y/\theta_i)$
Exponential	$\theta = -1/\mu$	$-\ln(-\theta)$	$x\ln(-x/e^{\theta_i}) = 1 + x\ln y + \theta_i y$

Kiefer and Sacks (1963) have generalized and extended the above results of Schwarz (1962) for testing two composite hypotheses that are separated by an indifference zone. Since they do not confine to the one-parameter exponential family, we cannot obtain a geometric representation similar to that of Schwarz.

Next, consider the case of no indifference zone. That is, $\Theta_1 \cup \Theta_2 = \Theta$ and let $f(x;\,\theta)$ be the common probability density of X with respect to a sigma-finite measure ν. The cost per observation c is independent of $X_1, X_2, \ldots$. Whenever sampling terminates, a Bayes decision is made for the acceptance of H_1 or H_2. Toward a stopping rule, consider the simplified rule: Stop at the smallest integer n, $n \geq 0$ for which $\rho_0(G, \mathbf{X}_n) \leq c$, where $\rho_0(G, \mathbf{X}_n)$ is the posterior Bayes risk with respect to G, the prior distribution of θ, $\mathbf{X}_0 = 0$ and $\rho_0(\mathbf{X}_0) \equiv \rho_0$, the prior Bayes risk. Let $s^*(c)$ denote this stopping rule which, when combined with the Bayes acceptance decision function, can be reformulated as follows:

accept H_1 if

$$\int L_1(\theta) f(\mathbf{x}_n;\,\theta)\,dG(\theta) \leq c \int L_2(\theta) f(\mathbf{x}_n;\,\theta)\,dG(\theta); \tag{3.12.38}$$

accept H_2 if

$$\int L_1(\theta) f(\mathbf{x}_n;\,\theta)\,dG(\theta) > c^{-1} \int L_2(\theta) f(\mathbf{x}_n;\,\theta)\,dG(\theta); \tag{3.12.39}$$

take an additional observation otherwise. The function $f(\mathbf{x}_n;\,\theta)$ denotes the joint density of $\mathbf{X}_n$ for given θ. In the case of k composite hypotheses ($k > 2$) the stopping rule $s^*(c)$ remains the same. We will establish that $s^*(c)$ is asymptotically Bayes as $c \to 0$. That is, if $R_c(G, s^*(c))$ is the expected (or prior) risk associated with $s^*(c)$ and G, and if $\rho_c(G)$ is the Bayes risk, $\rho_c(G) = \inf_s R_c(G, s)$, then

$$\lim_{c \to 0} \{R_c(G, s^*(c))/\rho_c(G)\} = 1. \tag{3.12.40}$$

In order to attain this, we need the following notation and assumptions. Let

$$\underline{L}_i = \inf_{\theta} L_i(\theta), \qquad \bar{L}_i = \sup_{\theta} L_i(\theta).$$

Let $Z(\theta, \psi) = \ln\{f(X;\,\theta)/f(X;\,\psi)\}$ and let $I(\theta, \psi)$ denote the Kullback–Liebler information number. That is $I(\theta, \psi) = E_\theta\{Z(\theta, \psi)\}$. Note that $I(\theta, \psi) > 0$ for all $\theta \neq \psi$. We will be interested in knowing how $I(\theta, \psi)$ behaves when $\theta \in \Theta_1$ and $\psi \in \Theta_2$ where Θ_1 and Θ_2 have a part of their boundaries in common. There is a danger that if $\theta \in \Theta_1$, and $\psi \in \Theta_2$, but both are very close to the common boundary, the information available in the sample in order to distinguish between Θ_1 and Θ_2 will not be sufficient for obtaining an efficient test. We therefore assume the following regularity conditions:

(A.1)

$$0 \leq \underline{L}_i \leq \bar{L}_i < \infty, \qquad i = 1, 2, \tag{3.12.41}$$

(A.2)

$$I^{(1)} = \inf_{\theta \in \Theta_1} \inf_{\psi \in \Theta_2} I(\theta, \psi) > 0, \tag{3.12.42}$$

$$I^{(2)} = \inf_{\theta \in \Theta_2} \inf_{\psi \in \Theta_1} I(\theta, \psi) > 0. \tag{3.12.43}$$

$I(\theta, \psi)$ and $I_1(\theta) = \inf_{\psi \in \Theta_2} I(\theta, \psi)$ are both continuous on Θ_1. Similarly, $I(\theta, \psi)$ and $I_2(\theta) = \inf_{\psi \in \Theta_1} I(\theta, \psi)$ are both continuous on Θ_2.

(A.3) For each $\theta \in \Theta_1$ and $\psi \in \Theta_2$, the following hold

(i) $E_\theta\{Z^2(\theta, \psi)\} < \infty$
(ii) Same as (i) holds with the roles of θ and ψ interchanged,
(iii) $\lim_{\varepsilon \to 0} E_\theta\{[\ln \sup_{|\psi' - \psi| \leq \varepsilon} f(X; \psi') - \ln f(X; \psi)]^2\} = 0$,
(iv) Same as (iii) with the roles of θ and ψ and θ' and ψ' interchanged;
(v) $\lim_{\theta' \to \theta} E_\theta\{Z^2(\theta, \theta')\} = 0$
(vi) Same as (v) with the roles of θ and θ' replaced by ψ and ψ' respectively.

Recall that Θ_i is said to be compact if every open covering of Θ_i has a finite subcovering. Let $N^*(c)$ denote the stopping variable associated with the stopping rule $s^*(c)$. Then the following result of Kiefer and Sacks (1963) is an extension of the results of Chernoff (1959).

Theorem 3.12.5 (Kiefer and Sacks, 1963) Let (A.1)–(A.3) hold. (i) If Θ_1 is compact, then for each $\theta_0 \in \Theta_1$

$$E_{\theta_0}\{N^*(c)\} \leq \{1 + o(1)\}|\ln c|/I_1(\theta_0) \quad \text{as} \quad c \to 0. \tag{3.12.44}$$

(ii) If Θ_2 is compact, then for each $\psi_0 \in \Theta_2$

$$E_{\psi_0}\{N^*(c)\} \leq \{1 + o(1)\}|\ln c|/I_2(\psi_0) \quad \text{as} \quad c \to 0. \tag{3.12.45}$$

Proof See Kiefer and Sacks (1963, Lemma 2).

REMARK 3.12.4 The term $o(1)$ occurring on the right-hand side of (3.12.44) depends on θ_0 and is not uniform in $\theta \in \Theta_1$. A similar comment can be made about the $o(1)$ term in (3.12.45) since it depends on ψ_0. In order to obtain uniformity of $o(1)$ we need an additional condition, namely:

(A.4) The functions $E_\theta\{Z^2(\theta, \psi)\}$ and $E_\theta\{[\ln \sup_{|\psi'-\psi|\le\varepsilon} f(X; \psi') - \ln f(X, \psi)]^2\}$ are continuous on Θ_1 for each ψ, and ε $(0 < \varepsilon)$ sufficiently small. Also, an analogous condition obtained by replacing ψ by θ and ψ' by θ' holds. Moreover, $E_\theta\{Z^2(\theta, \psi)\}$ is jointly continuous in θ and ψ.

We remark that the expected sample size required to test H_1 against H_2 according to the stopping rule $s^*(c)$ is of the order of $|\ln c|$ as $c \to 0$, and is inversely proportional to the minimal information number.

Theorem 3.12.6 (Kiefer and Sacks, 1963) If Θ_1 and Θ_2 are compact and assumptions (A.1)–(A.4) are satisfied, then $s^*(c)$ is asymptotically optimal.

Proof See Kiefer and Sacks (1963).

Corollary 3.12.6.1 If $s'(c)$ is any stopping rule with an associated stopping variable $N'(c)$, and if

$$\sup_{\theta\in\Theta_1} R_c(\theta, s'(c)) + \sup_{\theta\in\Theta_2} R_c(\psi, s'(c)) \le Kc|\ln c| \tag{3.12.46}$$

for some constant K, $0 < K < \infty$, then

$$E_\theta\{N'(c)\} \ge \{1 + o(1)\}|\ln c|/I_1(\theta) \qquad \text{for all } \theta \in \Theta_1, \tag{3.12.47}$$

$$E_\psi\{N'(c)\} \ge \{1 + o(1)\}|\ln c|/I_2(\psi) \qquad \text{for all } \psi \in \Theta_2. \tag{3.12.48}$$

Proof The result follows from Theorem 3.12.6.

REMARK 3.12.5 The maximal error probabilities are shown by Kiefer and Sacks (1963) to be approximately of the order of $O(c|\ln c|)$ as $c \to 0$, which is also the order of magnitude of the (prior) Bayes risk $\rho_c(G)$.

Problem

Carry out the scoring scheme described after Eq. (3.12.11) for the sequence of observations GGDGGGDDGGGG... using $M_1 = 34 = M_2$ where G denotes a good item and D denotes a defective item.

3.13 Iterated Logarithm Inequalities and Their Application to Sequential Testing Hypotheses

Recently Darling and Robbins (1967b, 1967c) have written a series of papers dealing with the iterated logarithm inequalities since they are useful in asserting certain properties of the sequential procedures proposed by them. First we shall present the iterated logarithm inequalities and then their applications to certain stopping rules. We need the following basic definitions and inequalities.

Let $\{X_n\}$ be a stochastic process such that the joint distribution of $(X_0, \ldots, X_n)$ has a strictly positive continuous density f_n $(n = 0, 1, \ldots)$. Conditional densities and expectations are then defined everywhere in an elementary way. Assume that the variables X_n and Y_n have finite expectations.

Definition 3.13.1 The sequence $\{X_n\}$ will be called *absolutely fair* if for $n = 1, 2, \ldots$

$$E(X_1) = 0, \qquad E(X_{n+1} | X_1, \ldots, X_n) = 0. \tag{3.13.1}$$

Definition 3.13.2 A sequence $\{Y_n\}$ is a *martingale* if

$$E(Y_{n+1} | Y_0, \ldots, Y_n) = Y_n. \tag{3.13.2}$$

There is a connection between the two types. Given an absolutely fair sequence $\{X_n\}$, let $Y_n = X_1 + \cdots + X_n + c$ where c is a constant, $n = 1, 2, \ldots$. Then

$$E(Y_{n+1} | X_1, \ldots, X_n) = Y_n. \tag{3.13.3}$$

The conditional variables X_j may be replaced by Y_k so that (3.13.3) is equivalent to (3.13.2). On the other hand, given a martingale $\{Y_n\}$, put $X_1 = Y_1 - EY_1$ and $X_{n+1} = Y_{n+1} - Y_n$. Then, $\{X_n\}$ is absolutely fair and the representation for Y_n as $X_1 + \cdots + X_n + c$ holds with $c = EY_1$. Thus we are led to the following result.

Theorem 3.13.1 $\{Y_n\}$ is a martingale if and only if $Y_n = X_1 + \cdots + X_n + c$ with $\{X_n\}$ absolutely fair.

Example 3.13.1 Let $\{X_n\}$ be a sequence of random variables. Let Y_n denote the likelihood ratio based on $(X_1, \ldots, X_n)$. That is

$$Y_n = f_n(X_1, \ldots, X_n)/g_n(X_1, \ldots, X_n).$$

The asymptotic behavior of $\{Y_n\}$ is of much interest in statistical decision theory. For simplicity assume that f_n are strictly positive and continuous. If the f_n represent the true densities, then the conditional density of X_{n+1} for given $X_1, \ldots, X_n$ equals the ratio f_{n+1}/f_n and hence

$$\begin{aligned} E(Y_{n+1} \mid X_j = x_j, j = 1, \ldots, n) &= \int_{-\infty}^{\infty} \frac{g_{n+1}(x_1, \ldots, x_n, y)}{f_{n+1}(x_1, \ldots, x_n, y)} \frac{f_{n+1}(x_1, \ldots, x_n, y)}{f_n(x_1, \ldots, x_n)} \, dy \\ &= g_n(x_1, \ldots, x_n)/f_n(x_1, \ldots, x_n) = Y_n. \end{aligned} \tag{3.13.4}$$

Definition 3.13.3 The sequence $\{Y_n\}$ is said to be a submartingale (supermartingale) if the equality sign in (3.13.2) is replaced by $\geq$ ($\leq$).

Let $X_1, X_2, \ldots$ be a sequence of iid random variables having mean 0 and variance 1 and moment-generating function $C(t) = E(\exp(tX))$ finite in some neighborhood of $t = 0$. Put $S_n = X_1 + \cdots + X_n$, $\bar{X}_n = S_n/n$. For any sequence of constants a_n, $n \geq 1$, let $p_m = P(|\bar{X}_n| \geq a_n$ for some $n \geq m)$. The law of the iterated logarithm gives conditions on the sequence a_n which guarantee that $p_m \to 0$ as $m \to \infty$. For example, it suffices that $a_n \geq \{2c(\ln)^2 n/n\}^{1/2}$ for some $c > 1$ (we write $\ln \ln n = (\ln)^2 n$, etc.) but does not tell about the rate at which p_m tends to zero. Darling and Robbins (1967a) give explicit bounds for p_m as a function of m and for which $p_m = O(1/(\ln)^2 m)$. Such bounds have interesting statistical applications.

Theorem 3.13.2 (Darling and Robbins, 1967b). With the above notation, let $h(t) = \frac{1}{2} + \ln C(t)/t$ and $b_n = t_i h(t_i)$ for all n such that $m_i \leq n < m_i + 1$ $(i \geq 1)$. Then, we have

$$P(\bar{X}_n \geq b_n \text{ for some } n \geq m_j) \leq \sum_{i=j}^{\infty} \exp(-m_i t_i^2/2) = Q_j \qquad \text{(say)}. \tag{3.13.5}$$

Proof Let $Z_n = \exp(tS_n)/C^n(t)$ for any fixed t for which $C(t) < \infty$. Then, Z_n is a nonnegative martingale with expected value 1. A simple martingale inequality asserts that for any positive constant b

$$P(Z_n \geq b \text{ for some } n \geq 1) \leq 1/b. \tag{3.13.6}$$

Set $b = \exp(mt^2/2)$ for any fixed m and $t > 0$. Then (3.13.6) becomes

$$P\left(S_n \geq \frac{mt}{2} + \frac{n \ln C(t)}{t} \quad \text{for some } n \geq 1\right) \leq \exp(-mt^2/2). \tag{3.13.7}$$

Since $h(t) = \frac{1}{2} + \ln C(t)/t^2$ $(\to 1$ as $t \to 0)$,

$$P(\bar{X}_n \geq th(t) \text{ for some } n \geq m) \leq \exp(-mt^2/2). \tag{3.13.8}$$

Let $\{m_i\}$ $(m_i \to \infty)$ be any increasing sequence of positive constants and t_i be any sequence of positive constants, $i \geq 1$. From (3.13.8) we have for any integer $j \geq 1$,

$$P(\bar{X}_n \geq t_i h(t_i) \text{ for some } m_i < n < m_{i+1}, i \geq j) \leq \sum_{i=j}^{\infty} \exp(-m_i t_i^2/2). \quad (3.13.9)$$

If the sequence of constants b_n for $n \geq m$ are as defined in the hypothesis of the theorem, we can write (3.13.9) as (3.13.5).

Special cases of (3.13.5):

(i) Let $m_i = \exp(i/\ln i)$ and, for any positive constant A,

$$t_i = (2 \ln i + 4 (\ln)^2 i + 2 \ln A)^{1/2} m_i^{-1/2}.$$

Then from (3.13.9)

$$Q_j = \frac{1}{A} \sum_{i=j}^{\infty} \frac{1}{i(\ln i)^2} < \frac{1}{A} \frac{1}{\ln(j-1/2)} \to 0 \quad \text{as} \quad j \to \infty.$$

We shall now find an upper bound for the b_n. A little algebra shows that for $m_i \leq n < m_{i+1}$

$$\ln i \leq (\ln)^2 n + (\ln)^3 n + \ln 2$$
$$(\ln)^2 i \leq (\ln)^3 n + \ln 2$$
$$m_i > n \exp\{-1/(\ln)^2 n\} \qquad (i \geq 3).$$

Hence, from the choice of t_i,

$$t_i < \{2(\ln)^2 n + 6 (\ln)^3 n + 6 \ln 2 + 2 \ln A\}^{1/2} n^{-1/2} \exp[\{2 (\ln)^2 n\}^{-1}] = g(n) \text{ (say)} \quad (3.13.10)$$

and hence

$$b_n \leq g(n) h(v_n) \quad (3.13.11)$$

where v_n (the t_i in the definition of b_n) is some constant such that

$$0 < v_n \leq g(n) \sim \{2(\ln)^2 n/n\}^{1/2} \quad \text{as} \quad n \to \infty. \quad (3.13.12)$$

For the standard normal distribution $h(t) \equiv 1$. For coin tossing with $p = \frac{1}{2}$ and more generally whenever $C(t) \leq \exp(t^2/2)$ $h(t) \leq 1$ and we can omit the term $h(v_n)$ in (3.13.11). In any case since $h(t) \to 1$ as $t \to 0$ and $g(n) \to 0$ as $n \to \infty$, $h(v_n) \to 1$ as $n \to \infty$. From (3.13.9), (3.13.11), and the definition of b_n, we have for $j \geq 3$

$$P(\overline{X}_n \geq g(n)h(v_n) \text{ for some } n \geq \exp(j/\ln j)) \leq Q_j \tag{3.13.13}$$

where v_n satisfies (3.13.12). By continuing (3.13.13) with the analogous inequality for $-X_n$ we obtain

$$P(|\overline{X}_n| \geq g(n)h(v_n) \text{ for some } n \geq \exp(j/\ln j)) \leq 2Q_j \tag{3.13.14}$$

where now (3.13.12) is replaced by

$$0 < |v_n| \leq g(n). \tag{3.13.15}$$

Putting $a_n = g(n)h(v_n)$ we have an inequality for p_m with $a_n \sim [2(\ln)^2 n/n]^{1/2}$ and for which $p_m = O\{1/(\ln)^2 m\}$.

REMARK 3.13.2.1 Darling and Robbins (1967b) have extended the inequality in (3.13.14) to additive processes.

APPLICATIONS

Example 3.13.2 (Optional stopping) Let $Y_1, Y_2, \ldots$ be iid with distributions having an unknown mean μ, variance 1, and a finite moment-generating function. We wish to test $H_0: \mu = \mu_0$ with a given type I error without fixing the sample size in advance. Let $X_n = (Y_n - \mu)$ and define the interval $I_n = (\overline{Y}_n - a_n, \overline{Y}_n + a_n)$. Then reject H_0 if for any desired $N \geq m$, $\mu_0 \notin I_N$. Then $P_{H_0}(\text{rej } H_0) \leq p_m$ which can be made arbitrarily small by taking m sufficiently large. The power function $P_\mu(\text{rej } H_0)$ is bounded below by a calculable function of μ which tends rapidly to one as $|\mu - \mu_0| \to \infty$.

Example 3.13.3 (Tests with uniformly small error probability) In order to test $H_0: \mu < \mu_0$ against $H_1: \mu > \mu_0$, stop sampling with $N = \text{first } n \geq m$ such that $\mu_0 \notin I_n$, and accept H_0 or H_1 according as I_N is to the left or to the right of μ_0. The error probability is then $\leq p_m$ and $E_\mu(N) > \infty$ for all $\mu \neq \mu_0$.

Example 3.13.4 (Tests with zero type II error) In order to test H_0: $\mu = \mu_0$ against $H_1: \mu > \mu_0$, reject H_0 with $N = \text{first } n \geq m$ such that I_n is to the right of μ_0. Then, $P_{H_0}(\text{rej } H_0) < p_m$ and for $\mu > \mu_0$, $P_\mu(\text{rej } H_0) \equiv 1$. [$E_\mu(N) < \infty$ for $\mu > \mu_0$ but $P_{H_0}(N = \infty) > 0$, which may be an advantage.] The case in which σ is unknown can also be treated by the preceding methods.

Theorem 3.13.3 (Darling and Robbins, 1967a) Let $X_1, X_2, \ldots$ be iid random variables and let $\bar{X}_n = \sum_1^n X_i/n$. Set $C(t) = E[\exp(tX)]$. Assume that $C(t) \le \exp(t^2/2)$ for all $t \ge 0$. Let $\gamma > 1$, $c > 1$ be arbitrary constants and set (with $(\ln)^2 n = \ln \ln n$)

$$g_n(A) = \left(\frac{1+\gamma}{2\gamma^{1/2}}\right) [(2c(\ln)^2 n - 2c(\ln)^2 \gamma + 2 \ln A)/n]^{1/2},$$

$$Q_m = \frac{1}{c-1} \left(\frac{\ln m}{\ln \gamma} - \frac{1}{2}\right)^{c-1}.$$

Then for any $m \ge \gamma$, $A \ge 1$, we have

$$P(\bar{X}_n \ge g_n(A) \text{ for some } n \ge m) > Q_m/A. \tag{3.13.16}$$

Proof See Darling and Robbins (1967a).

Theorem 3.13.4 (Darling and Robbins, 1967c) Let $X_1, X_2, \ldots$ be iid random variables such that $EX = \mu$ so that $\bar{X}_n = \sum_1^n X_i/n \to \mu$ with probability one. For any positive function $a(t)$, $t > 0$, with $a(t) \to 0$ as $t \to \infty$, put

$$p_m = P(|\bar{X}_n - \mu| \ge a(n) \text{ for some } n \ge m).$$

Then

(i) For any $A > 0$, $\delta > 0$ and $1 \le p \le 2$ if

$$a(n) = A^{1/p}(\ln n)^{(1+\delta)/p}/n^{1-1/p},$$

then

$$p_m \le \frac{2\mu_p}{A\delta(\ln m)^\delta}\left(1 + \frac{\delta}{\ln m}\right)$$

where $\mu_p = E|X - \mu|^p < \infty$ for some $1 < p < \infty$. (When $p = 2$, the factor 2 on the right-hand side of the inequality for p_m can be deleted.)

Example 3.13.5 As one example of the use of such upper bounds of Theorem 3.13.4 in statistics, suppose that μ is unknown and there is a certain action d such that when $X_1, X_2, \ldots$ are observed sequentially,

(i) if $\mu = 0$, it is desirable never to take action d, while
(ii) if $\mu > 0$, it is desirable to take action d as soon as possible.

Define the stopping time $N =$ first $n \geq m$ such that $\overline{X}_n \geq a(n)$ ($N = \infty$ if no such n occurs). Then, given $\alpha > 0$, there exists m such that:

(i′) if $\mu = 0$, $P(\text{take action } d) = P(N < \infty) = p_m^+ \leq \alpha$

[where $p_m^+ = P(\overline{X}_n - \mu \geq a(n)$ for some $n \geq m)$]. That is, the type I error is $\leq\alpha$; in many cases, for example, if μ is a location parameter, this will automatically hold for all $\mu < 0$, and

(ii′) if $\mu > 0$, $P(\text{take action } d) \equiv 1$ (that is, the type II error is $\equiv 0$).

That $N = \infty$ with probability >0 when $\mu = 0$ is no hardship if the observations $X_1, X_2, \ldots$ are occurring without cost to the experimenter.

Corollary 3.13.4.1 Let N be as defined in Example 3.13.4. Then, for $\mu \to 0^+$ we have

$$EN \leq \{2\mu^{-2}(\ln)^2\mu^{-2}\}\{1 + o(1)\}.$$

Proof See Darling and Robbins (1967c, p. 1579).

In a similar context, Farrell (1964a) has established the existence of certain sequential procedures having asymptotic expected sample size of this form.

By specializing the constants in Theorem 3.13.3 Darling and Robbins (1967c) obtain that if

$$a(n) = [3/(2n)^{1/2}]\{(\ln)^2 n + 1.236\}^{1/2} \tag{3.13.17}$$

then

$$p_m^+ \leq \frac{7}{40}\left(\frac{\ln m}{\ln 2} - \frac{1}{2}\right)^{-1}, \qquad m \geq 2, \tag{3.13.18}$$

where $p_m^+ = P(\overline{X}_n - \mu > a(n)$ for some $n \geq m)$.

Theorem 3.13.5 (Darling and Robbins, 1968a) Suppose the iid sequence $\{X_i\}$ has a probability density $f(x; \theta)$ where $f(x; 0) \neq f(x; \theta)$ for $\theta \neq 0$ and $\theta = 0$ is a possible value for θ. Let N, $1 \leq N \leq \infty$, be any stopping time of the sequence $X_1, X_2, \ldots$. Then for any $\theta \neq 0$

$$E_\theta(N) \geq -\ln P_0(N < \infty)/E_\theta\left[\ln\left(\frac{f(X;\theta)}{f(X;0)}\right)\right]. \tag{3.13.19}$$

Proof Use Jensen's inequality and Wald's (1947) theorem on cumulative sums.

Corollary 3.13.5.1 If $f(x; \theta) = \text{normal}(\theta, 1)$, then (3.3.39) becomes

$$E_\theta(N) \geq -2 \ln P_0(N < \infty)/\theta^2, \qquad \theta \neq 0. \tag{3.13.20}$$

Darling and Robbins (1968a, Eq. (4.2)) also give an inequality which when inverted will provide an upper bound for $E_\theta(N)$.

Example 3.13.6 Let $X_1, X_2, \ldots$ be independent normal variables having unknown mean θ and variance 1. We wish to test $H_0: \theta \leq 0$ against $H_1: \theta > 0$. In the preceding theorems it has been shown how to find in various ways a sequence $a(n)$, $n \geq 1$ such that $a(n)/n \to 0$ as $n \to \infty$ and such that for any $0 < \alpha < 1$ there exists a computable m such that for $S_n = X_1 + \cdots + X_n$ and $\theta = 0$

$$p_0(m) = P_0(S_n \geq a(n) \text{ for some } n \geq m) \leq \alpha. \tag{3.13.21}$$

Setting

$$N = \text{first } n \geq m \text{ such that } S_n \geq a(n) \qquad (N = \infty \text{ if no such } n \text{ occurs}), \tag{3.13.22}$$

we agree to reject H_0 in favor of H_1 if and only if $N < \infty$. Then $P_\theta(\text{rej } H_0) = P_\theta(N < \infty) = P_\theta(S_n \geq a(n), \text{ some } n \geq m) \leq P_0(S_n \geq a(n) - n\theta, \text{ some } n \geq m)$ is clearly a nondecreasing function of θ, so that

$$P_\theta(\text{rej } H_0) \leq P_0(\text{rej } H_0) = P_0(N < \infty) = p_0(m) \leq \alpha \qquad \text{for every } \theta \leq 0; \tag{3.13.23}$$

while by the strong law of large numbers

$$P_\theta(\text{rej } H_0) = P_\theta(S_n/n \geq a(n)/n, \quad \text{some } n \geq m) = 1 \qquad \text{for every } \theta > 0. \tag{3.13.24}$$

Thus, the proposed test has type I error probability $\leq \alpha$ (uniformly) and type II error probability $\equiv 0$. By the same token, while the sample size N is finite with probability 1 when H_1 is true, N is equal to ∞ with probability $\geq 1 - \alpha$ when H_0 is true. The latter property is usually regarded in statistical decision theory as intolerable. In the view of Darling and Robbins (1968a) this is a prejudice, arising perhaps from the fact that Wald's SPRT was originally applied to acceptance sampling, where nonterminating tests would be useless. Consider, however, the following hypothetical situation.

Doctor Smith is using a certain treatment T_1 on his patients, and from experience knows that the benefit obtained from it is a random variable which is normal (0, 1) under proper choice of origin and unit of measurement. Unless some better treatment becomes available, he will continue to use T_1 on all future patients. Doctor John claims to have found a better treatment T_2, and to substantiate his claim, publishes regularly the sequence $X_1, X_2, \ldots$ of benefits obtained by his patients under T_2. Assume that $X_1, X_2, \ldots$ are independent normal $(\theta, 1)$ and let H_0 and H_1 be as defined above.

(i) If H_0 is true (T_1 is better than T_2), Dr. Smith would be content to observe the sequence $X_1, X_2, \ldots$ as long as it continues, forever if need be, without changing from T_1 to T_2 for his own patients. In fact if H_0 is true and N denotes the number of X_i that he observes before changing to T_2, he wants to have $P_\theta(N < \infty) \leq \alpha$ for every $\theta \leq 0$.

(ii) If H_1 is true (T_2 is better than T_1), Dr. Smith wants to be sure of eventually changing to T_2 for his patients, and the sooner the better, no matter how small $\theta > 0$ may be. Hence, he wishes to minimize $E_\theta(N)$ for all $\theta > 0$ subject to the condition imposed in (i) that $P_\theta(N < \infty) \leq \alpha$ for every $\theta \leq 0$. Dr. Smith may set up a typical Bayes–Neyman–Pearson–Wald minimization problem. Among all stopping times $1 \leq N \leq \infty$ of the sequence $X_1, X_2, \ldots$ find N that minimizes $\int_1^\infty E_\theta(N)\, dG(\theta)$ subject to $P_\theta(N < \infty) \leq \alpha$, $\theta \leq 0$, where $G(\theta)$ denotes his prior idea of the distribution function of θ given that it is positive. Now, from (3.13.20) we have $E_\theta(N) \geq -2 \ln \alpha/\theta^2$ so that if G is such that $\int_1^\infty \theta^{-2}\, dG(\theta) = \infty$, then the average value of $E_\theta(N)$ is always infinite. Then we could minimize $\int_1^\infty [E_\theta(N) - \psi(\theta)]\, dG(\theta)$ or $\int_1^\infty [E_\theta(N)/\psi(\theta)]\, dG(\theta)$ where $\psi(\theta)$ is chosen such that the minimum value is finite.

An explicit solution to this problem is known only when G concentrates all its mass at some $\theta > 0$. The authors claim that in this case, it can be shown that the minimum value of $E_\theta(N)$ is given by $N =$ first $n \geq 1$ such that $S_n \geq (n\theta/2) + c$ where c is so chosen as to make $<P_0(N \infty) = \alpha$. This N, however, does not have the property that $P_\theta(N < \infty) = 1$ for every $\theta > 0$.

Although an "optimal" N may be lacking, Dr. Smith can use the $N =$ first $n \geq m$ such that $S_n \geq a(n)$ ($N = \infty$ if no such n occurs), for example, for $m = 1$, $a(n) = \{(n+1)[\ln(n+1) - 2 \ln \alpha]\}^{1/2}$. Perhaps the lower and upper bounds for $E_\theta(N)$, $\theta > 0$, given in (3.13.19) and Section 4 of Darling and Robbins (1968a) may appear to him to be reasonable. He may take comfort in the fact that the type I error of such a test is unaffected by any form of optional stopping, that is, no matter when the sequence $X_1, X_2, \ldots$ may be terminated, the criterion "reject H_0 in favor of H_1 if and only if for some $n \geq m$ $S_n \geq a(n)$" will never lead to rejection of H_0 with probability greater than α for any $\theta \leq 0$.

Example 3.13.7 As a second example, let $Y_1, Y_2, \ldots$ be iid normal $(\mu_1, 1)$ and $Z_1, Z_2, \ldots$ be iid normal $(\mu_2, 1)$, independent of the Ys, with μ_1, μ_2 unknown. We wish to test H_0: $\mu_1 < \mu_2$ against H_1: $\mu_1 > \mu_2$ such that the probability of error is uniformly $\leq \alpha$ for all $\mu_1 \neq \mu_2$. To achieve this, let $X_i = Y_i - Z_i$ $(i = 1, 2, \ldots)$, $S_n = X_n + \cdots + X_n$, $n \geq 1$, and observe (Y_1, Z_1), $(Y_2, Z_2), \ldots$ until $N =$ first $n \geq m$ such that $|S_n| \geq 2^{1/2}a(n)$, accepting H_0 $[H_1]$ if $S_N \leq -2^{1/2}a(N)$ $[S_N \geq 2^{1/2}a(N)]$. Here, $a(n)$ is again any sequence of constants such that $a(n)/n \to 0$ and $P_0(S_n \geq a(n)$, some $n \geq m) \leq \alpha$ where P_0 is the distribution under which $X_1, X_2, \ldots$ are iid normal (0, 1). Suppose that H_0 is true. Then $S_n/n \to \mu_1 - \mu_2 \neq 0$, so $P(N < \infty) = 1$. Further, $P(\text{error}) \geq P(S_n \leq 2^{1/2}a(n)$, some $n \geq m) = P_0(S_n \geq a(n) + n2^{-1/2}(\mu_2 - \mu_1)$, some $n \geq m) \leq P_0(S_n \geq a(n)$, some $n \geq m) \leq \alpha$. Thus, $P(N < \infty) = 1$ and $P(\text{error}) \leq \alpha$ for all $\mu_1 < \mu_2$, and the same holds for all $\mu_1 > \mu_2$, no matter how small $\mu_1 - \mu_2 \neq 0$ [when $\mu_1 = \mu_2$, $P(N < \infty) = P(\text{any decision}) \leq 2\alpha$]. The expected sample size $E_\theta(N)$ is a function of the parameter $\theta = (\mu_1 - \mu_2)/2^{1/2}$ and is bounded by the inequality (with $m = 1$)

$$\theta E_\theta(N) \leq a(E_\theta(N)) + \theta + [\phi(\theta)/\Phi(\theta)] \qquad (3.13.25)$$

where $a(n) = \{(n+1)[\ln(n+1) - 2\ln\alpha]\}^{1/2}$.

3.14 Certain Nonparametric Sequential Test Procedures

In this section we shall consider certain nonparametric test procedures having power one [which were proposed by Darling and Robbins (1968b)] and rank order test procedures for Lehmann alternatives. First we shall consider the test procedures proposed by Darling and Robbins (1968b).

3.14.1 Nonparametric Sequential Tests with Power One

For any two distribution functions (dfs) F, G denote

$$D^+(F, G) = \sup_{-\infty < t < \infty} (F(t) - G(t)), \quad D(F, G) = \sup_t |F(t) - G(t)|.$$

Let $0 < \alpha < 1$ and a positive integer m be given, and let g denote any continuous, positive, nondecreasing function defined for $x \geq m$ and such that (i) $g(x) \leq x$; (ii) $g(x)/x$ is strictly decreasing to zero as $x \to \infty$; (iii) $g(x)$ is concave; and (iv) $\sum_{n=m}^{\infty} \exp\{-g^2(n)/(n+1)\} \geq \alpha$. [Not all these restrictions are needed in some of the results that follow.] Possible choices of g satisfying conditions (i)–(iv) are given by the family

$$g(x) = [(x+1) + (a \ln x + \ln b)]^{1/2} \qquad (a > 1, b \geq 1) \qquad (3.14.1)$$

for suitably large values of a and b, and m; for example, if $x \geq 1$ satisfies $\alpha \exp(x-1) \geq x+1$, we can take $a=2$, $b=1/\alpha(x^*-1)$, $m=x^*$ where x^* denotes the smallest integer $\geq x$. In particular, for $\alpha = .05$ we can take $a=2$, $b=4$, $m=4$ in (3.14.1).

For $0 < x \leq g(m)/m$ define $h(x)$ to be the inverse function of $g(x)/x$. Let $X_1, X_2, \ldots$ be iid as $F(x)$ and let $Y_1, Y_2, \ldots$ be iid as $G(y)$, the Xs and Ys being mutually independent. Denote by $F_n(t)$ $[G_n(t)]$ the empirical df of $X_1, \ldots, X_n$ $[Y_1, \ldots, Y_n]$ $(n \geq 1)$. [That is, $nF_n(t) =$ (number of Xs that are $\leq t$)].

(a) Consider the hypothesis $H_1 : F(t) \leq G(t)$, for all t $(-\infty < t < \infty)$.

Theorem 3.14.1 (Darling and Robbins, 1968b) Define N as the smallest $n \geq m$ such that $D^+(F_n, G_n) \geq g(n)/n$ and reject H_1 if and only if $N < \infty$. Then

$$P(\text{rej } H_1 \mid H_1 \text{ is true}) \leq \alpha, \; P(\text{rej } H_1 \mid H_1 \text{ is false}) = 1. \tag{3.14.2}$$

If H_1 is false and $D^+(F, G) = d > 0$ with $d \leq g(m)/m$, then

$$EN \leq h\{d - [m/h(d)]\}. \tag{3.14.3}$$

Proof For integers $0 \leq r \leq n$, define

$$A(r, n) = (n!)^2/(n-r)!\,(n+r)!. \tag{3.14.4}$$

If F, G are arbitrary dfs such that H_1 holds, then it can be shown that

$$P(D^+(F_n, G_n) \geq r/n) \leq A(r, n). \tag{3.14.5}$$

If $F \equiv G$ is continuous, it was shown by Gnedenko and Korolyuk (1951) that the equality sign holds between the two terms of (3.14.5). The validity of the inequality of (3.14.5) for arbitrary F and G satisfying H_1 is easily seen by representing the Xs and Ys as functions of independent random variables uniformly distributed on (0, 1). We shall now show that

$$A(r, n) \leq \exp[-r^2/(n+1)]. \tag{3.14.6}$$

From Whittaker and Watson (1947, Chapter 12, Problem 31) we have

$$-\ln A(r, n) = \int_0^1 [x^n(1-x^r)(1-x^{-r})/(1-x) \ln x] \, dx,$$

and by setting $x = e^{-t}$ and using the inequalities $\cosh y - 1 \geq y^2/2$, and $\sinh y/y \leq e^y$, $y \geq 0$, we obtain (3.14.6) from

$$-\ln A(r, n) = \int_0^\infty \exp[-(n + \tfrac{1}{2})t] \frac{\cosh rt - 1}{\sinh(t/2)} \frac{dt}{t}$$

$$\geq r^2 \int_0^\infty \exp[-(n + \tfrac{1}{2})t - t/2]\, dt = r^2/(n + 1).$$

Hence, denoting x^* as the smallest integer $\geq x$, we have

$$P(N < \infty \mid H_1) \leq \sum_{n=m}^{\infty} P(D^+(F_n, G_n) \geq g(n)/n) \leq \sum_{n=m}^{\infty} A(g(n)^*, n)$$

$$\leq \sum_{n=m}^{\infty} \exp[-g^2(n)/(n + 1)] \leq \alpha, \tag{3.14.7}$$

which proves the first part of (3.14.2). The second part of (3.14.2) follows from the fact that if H_1 is false, then $D^+(F_n, G_n) \to D^+(F, G) = d > 0$ with probability one as $n \to \infty$, while by hypothesis $g(n)/n \to 0$, so that $P(N < \infty \mid H_1$ is false) $= 1$. To prove (3.14.3) assume for simplicity that t is a number such that $F(t) - G(t) = d$. The distribution of $n(F_n(t) - G_n(t))$ is that of $S_n = U_1 + \cdots + U_n$ where the U_i are iid with $EU_i = d$ and $|U_i| \leq 1$. Let N' denote the smallest $n \geq m$ such that $S_n \geq g(n)$; then $N \leq N'$ and

$$S_{N'} \leq \max(m, g(N') + 1) \leq m + g(N').$$

It can easily be shown that $EN' < \infty$ and by Wald's lemma on cumulative sums and Jensen's inequality, we obtain

$$dEN' \leq m + g(EN'), \tag{3.14.8}$$

which gives an implicit upper bound for EN' and hence for EN. To obtain an explicit bound for EN', we observe that the equation

$$dx = m + g(x) \qquad (x \geq m) \tag{3.14.9}$$

has a unique root x_0 and $EN \leq EN' \leq x_0$. If we write $x_0 = h(y_0)$, Eq. (3.14.9) becomes

$$y_0 = d - m/h(y_0) \tag{3.14.10}$$

and iterating (3.14.10) we obtain $y_0 = d - [m/h(d - m/h(y_0))]$ so that

$$y_0 \geq d - m/h(d), \tag{3.14.11}$$

and hence, $EN \leq x_0 \leq h\{d - m/h(d)\}$ which completes the proof of Theorem 3.14.1.

REMARK 3.14.1 Some of the inequalities used here are quite crude, particularly the first inequality of (3.14.7), and the error probabilities are surely much smaller than indicated. In addition to the bound (3.14.3), it can be shown that $E[\exp(tN)] < \infty$ for some $t > 0$.

(b) Consider the hypothesis H_2: $F(t) = G(t)$ for every t $(-\infty < t < \infty)$.

Theorem 3.14.2 (Darling and Robbins, 1968b) Define N as the smallest $n \geq m$ such that $D(F_n, G_n) \geq g(n)/n$ and reject H_2 if and only if $N < \infty$. Then

$$P(\text{rej } H_2 \mid H_2 \text{ is true}) \leq 2\alpha, \qquad P(\text{rej } H_2 \mid H_2 \text{ is false}) = 1. \tag{3.14.12}$$

If H_2 is false, and $D(F, G) = d > 0$ with $d \leq g(m)/m$, then (3.14.3) holds.

Proof Theorem 3.14.2 follows by arguments similar to those used in the proof of Theorem 3.14.1.

REMARK 3.14.2 It would be easy to extend Theorem 3.14.2 to test H_2': $D(F, G) \leq c$ for any given $0 \leq c < 1$.

(c) Let F_0 be any specified df and consider the hypothesis H_3: $F(t) \leq F_0(t)$ for every $-\infty < t < \infty$.

Theorem 3.14.3 Define N as the smallest $n \geq m$ such that $D^+(F_n, F_0) \geq g(n)/n$, and reject H_3 if and only if $N < \infty$. Then

$$P(\text{rej } H_3 \mid H_3 \text{ is true}) \leq 2^{3/2}\alpha, \qquad P(\text{rej } H_3 \mid H_3 \text{ is false}) = 1. \tag{3.14.13}$$

If H_3 is false and $D^+(F, F_0) = d > 0$ with $d \leq g(m)/m$, then (3.14.3) holds.

Proof We have

$$P_{H_3}(D^+(F_n, F_0) \geq r/n) \leq B(r, n) \equiv \sum_{j=r}^{n} \binom{n}{j} \frac{r}{n-j+r} \frac{(j-r)^j(n-j+r)^{n-j}}{n^n}. \tag{3.14.14}$$

[If $F \equiv F_0$, a continuous df, then equality holds in (3.14.4) as was shown independently by Smirnov (1944) and, Birnbaum and Tingey (1951).] Because of the method of proof of Theorem 3.14.1, it suffices to show that $B(r, n) \leq$

$2^{3/2}A(r, n)$. Using the solution of the classical ballot problem, we establish the curious identity

$$A(r, n) = \sum_{j=r}^{n} \frac{r}{2(n-j)+r}\left[\binom{2j-r}{j}\binom{2(n-j)+r}{n-j}\Big/\binom{2n}{n}\right]. \tag{3.14.15}$$

[In an election in which candidates C_1 and C_2 each receive n votes, let E_j be the event that when C_1 has j votes he is, for the last time, r votes ahead of C_2. The jth term in the sum of (3.14.15) is $P(E_j)$.] Denote by R_j the ratio of the jth term of (3.14.14) to that of (3.14.15). The proof will be complete by showing that $R_j \le 2^{3/2}$. After some algebra, we have

$$R_j = ac/b(j-r, j)b(n-j+r, n-j)$$

with

$$a = [2(n-j)+r]/(n-j+r) = 2 - r/(n-j+r) \le 2,$$
$$b(x; i) = (1 + x^{-1})(1 + 2x^{-1}) \cdots (1 + ix^{-1}), \qquad c = (2n)!/n!\,n^n.$$

Now,

$$\ln b(x; i) \ge \int_0^i \ln(1 + y/x)\, dy = (x+i)\ln(1 + ix^{-1}) - i,$$
$$b(x; i) \ge (1 + ix^{-1})^{x+i}e^{-i},$$
$$b(j-r; j)b(n-j+r; n-j)$$
$$\ge [1 + j(j-r)^{-1}]^{2j-r}[1 + (n-j)/(n-j+r)]^{2(n-j)+r}e^{-n}.$$

If we set $\gamma = 2j - r$, $\delta = 2(n-j) + r$, further algebra and the inequality $1 + x \le e^x$ yield

$$b(j-r, j)b(n-j+r, n-j) \ge (1 - r/\gamma)^{-\gamma}(1 + r/\delta)^{-\delta}2^{2n}e^{-n} \ge 2^{2n}e^{-n}.$$

Hence

$$R_j \le ac2^{-2n}e^n \le 2(2n)!\,e^n/n!\,n^n2^{2n} < 2^{3/2},$$

after using Stirling's inequalities for the factorials.

REMARK 3.14.3 We could test H_3 by generating an artificial iid sequence $Y_1, Y_2, \ldots$ with df F_0 and by applying Theorem 3.14.1. However, the procedure of Theorem 3.14.3 seems more natural.

(d) Let $\mathcal{G}$ be any class of dfs closed under the D metric; for example, $\mathcal{G}$ may consist of a single df F_0 or of the set of normal distributions with mean μ and variance σ^2 $(-\infty < \mu < \infty,\ 0 < \sigma^2 < \infty)$. Consider the hypothesis $H_4\colon F \in \mathcal{G}$. Define

$$W_n = \inf_{G \in \mathcal{G}} D(F_n, G).$$

Theorem 3.14.4 (Darling and Robbins, 1968b) Define N as the smallest $n \geq m$ such that $W_n \geq g(n)/n$ and reject H_4 if and only if $N < \infty$. Then

$$P(\text{rej } H_4 \mid H_4 \text{ is true}) \leq 2^{5/2}\alpha \qquad \text{and} \qquad P(\text{rej } H_4 \mid H_4 \text{ is false}) = 1. \quad (3.14.16)$$

If H_4 is false and $\inf_{G \in \mathcal{G}} D(F, G) = d > 0$ with $d \leq g(m)/m$, then (3.14.3) is valid.

Proof Assume that H_4 is true. Then

$$P_{H_4}(W_n \geq r/n) \leq P_{H_4}(D(F_n, F) \geq r/n) \leq 2B(r, n) \leq 2^{5/2} A(r, n)$$

and the proof of the first part of (3.14.16) proceeds as in Theorem 3.14.1. If H_4 is false and $\inf_{G \in \mathcal{G}} D(F, G) = d > 0$, then for any G in $\mathcal{G}$, $D(F_n, G) \geq D(F, G) - D(F_n, F) \geq d - D(F_n, F)$ so that $W_n \geq d - D(F_n, F) \to d$ with probability 1 as $n \to \infty$ and hence $P(N < \infty \mid H_4 \text{ is false}) = 1$. The assertion concerning EN follows as before.

REMARK 3.14.4 The values of $g(x)$ for large x govern the behavior of EN for small d. More precisely, from (3.14.10) and (3.14.11) we have $1 \geq y_0/d \geq 1 - m/g(h(d)) \to 1$ as $d \to 0$, so that $y_0 \sim d$ and $EN \leq x_0 = h(y_0) \sim h(d)$ as $d \to 0$. If as in (3.14.1) $g(x) = c[x \ln x]^{1/2}$ as $x \to \infty$, it is simple to deduce that $h(x) \sim (c/x^2) \ln(1/x)$ as $x \to 0$ and hence $EN = O\,(d^{-2} \ln(1/d))$ as $d \to 0$. It is of interest to compare this with the EN of a parametric test. Let the X_i be normal $(\mu, 1)$, with $\mu > 0$. Then $D(\Phi(x - \mu), \Phi(x)) \sim \mu/(2\pi)^{1/2}$ as $\mu \to 0^+$ and hence, if H is false, the test of Theorem 3.14.3 applied to the hypothesis $H\colon \mu \leq 0$ would have $EN = O\,[\mu^{-2} \ln(1/\mu)]$ as $\mu \to 0^+$. However from (3.13.19), the parametric test has $EN = O\,[\mu^{-2}\,(\ln)^2(1/\mu)]$ as $\mu \to 0^+$, and this is the minimum possible order of magnitude.

REMARK 3.14.5 One cannot have $g(x) = O\,[(x \ln x)^{1/2}]$ as $x \to \infty$ if inequality (iv) is to hold. The existence of tests for which $g(x) = O\,[(x \ln x)^{1/2}]$ follows from the results of Chung (1949), however, owing to a lack of inequalities similar to those of Darling and Robbins, the latter are unable to exhibit tests of that type.

Robbins (1970) uses an extension of Wald's (1947) inequality instead of Theorem 3.13.2 in order to construct tests of power 1. It seems that one does not have to use the martingale inequality. Robbins (1970) replaces the random walk by the approximating Wiener process in order to get certain limiting results. For further details, the reader is referred to Robbins' (1970) paper.

3.14.2 Sequential Sign Test

Assume that X has df $F(x)$ and Y has df $G(y)$. We wish to test H_0: $F = G$, against the alternative H_1: $F(x) \leq G(x)$, all x. We observe pairs of observations (X_1, Y_1), (X_2, Y_2), $\ldots$ sequentially. Assume that $p = P(X < Y)$ is constant from stage to stage. Then the hypotheses can be replaced by H_0': $p = \frac{1}{2}$ versus H_1': $p > \frac{1}{2}$. The sequential sign test procedure has been proposed by Hall (1965, p. 40). Reduce the data to the signs of the differences $X_i - Y_i$. This may be justified by invariance under monotone transformations g_n applied to each of the observations in stage n [see Hall *et al.* (1965,) Section 1.8). The reduced data constitute a Bernoulli sequence with success (positive sign) probability p. The simple SPRT for Bernoulli data [see Example 2.1.2 or Wald (1947, Chapter 5)] is applicable and the likelihood ratio at the nth stage is that of a binomial random variable with parameters n and p. In the context of normal data, this test was originally described by Armitage (1947).

3.14.3 Rank Order SPRTs Based on Lehmann Alternatives: Two-Sample Case

Nonparametric alternatives known as Lehmann (1953) alternatives are given in terms of powers of the df under the null case. For example, in the two-sample case if H_0: $F = G$, then the Lehmann alternative is given by H_1: (i) $G = F^\Delta$ or (ii) $G = 1 - (1 - F)^\Delta$ $(\Delta > 0)$ or (iii) $G = h(F)$ such that $h(0) = 0$, $h(1) = 1$. Let $(X_1, \ldots, X_{n_1})$ denote a random sample from F and $(Y_1, \ldots, Y_{n_2})$ denote a random sample from G where Xs and Ys are mutually independent, and F and G are assumed to be continuous. Let $s_1 < s_2 < \cdots < s_{n_2}$ be the ranks of $Y_1, Y_2, \ldots, Y_{n_2}$ in the combined sample of size $n = n_1 + n_2$. Then $S = (s_1, s_2, \ldots, s_{n_2})$ is called the rank order. One can derive an explicit expression for the probability of a rank order under Lehmann alternatives [for instance, see Lehmann (1953)]. Using this result, Wilcoxon *et al.* (1963) [see also Bradley (1967)] have developed two sequential two-sample grouped rank tests called sequential *configural rank test* (SCR test) and the *rank sum test*. For both tests, observations are taken in groups of m Xs and m Ys with the observations ranked within the groups. The sequential procedures relate to the usual decisions at the end of each group, namely to accept H_0, to accept H_1 in (i) or to take another group of observations. The sequential methods are direct applications of Wald's (1947) SPRT. The configural rank test depends on a probability ratio statistic based on probabilities of the particular configuration of ordered observations obtained; the rank sum test uses the within-group rank sum for the Y sample and probability ratio based on the probabilities for the rank sums. These procedures (as one might expect) are somewhat inefficient since overall ranking is not used. However, the general theory of Wald's SPRT holds and the approximate boundaries, approximations to OC and ASN functions can be computed. Bradley *et al.* (1965) have carried out a Monte Carlo study of the sequential, two-sample, rank-sum test. They conclude that it is reasonable to use the sequential rank-sum test procedure derived on the basis of Lehmann alternatives, for data from normal populations differing only in locations. Bradley *et al.* (1966) provide a modified version of the configural group rank test proposed earlier, which is based on rerankings of observations as new groups of observations are obtained sequentially. Monte Carlo studies carried on the modified, sequential, configural rank test (MSCR) indicates a formal superiority of the MSCR test over the SCR test. Let us briefly describe the MSCR-test procedure. Suppose that X and Y observations are taken in groups of m and n and that no group or block effects are present. Then, at the tth stage of such a process, mt X observations and nt of Y observations are ranked in joint array. Then the likelihood ratio is the ratio of the probabilities of the ranks $s_j^{(t)}$, $j = 1, 2, \ldots, nt$ which constitute the rank of nt Ys in the joint reranking at stage t. The likelihood ratio at tth stage is given by

$$\frac{\Delta^{nt}[(m+n)t]!}{(s_1^{(t)}-1)!} \prod_{j=1}^{nt} \frac{\Gamma(s_j^{(t)} + j\Delta - j)}{\Gamma(s_{j+1}^{(t)} + j\Delta - j)}. \tag{3.14.17}$$

By invariant sufficiency the likelihood ratio only for the last reranking is equivalent to the joint likelihood ratio for all rerankings up to stage t. Hall *et al.* (1965) discuss the MSCR test and note that its finite termination with probability one under H_0 and H_1 follows from the work of Wirjosudirdjo (1961).

Next we shall turn to the work of Savage and Sethuraman (1966). Let $(X_1, Y_1), (X_2, Y_2), \ldots$ be independent and identically distributed bivariate random variables with a joint distribution $H(\cdot, \cdot)$ which has continuous marginal distributions F and G. We wish to test H_0: X and Y are independent and $G = F$ against the alternative hypothesis H_1: X, Y are independent and $G = F^{\Delta}$ where $\Delta > 0$, $\Delta \neq 1$ is a known constant. At the nth stage of experimentation the available information is the ranks of $(Y_1, \ldots, Y_n)$ among $(X_1, \ldots, X_n, Y_1, \ldots, Y_n)$. Let the combined sample be denoted by $W_1, W_2, \ldots, W_{2n}$ and the ordered combined sample by $W_{n,1}, \ldots, W_{n,2n}$. Let F_n (G_n) denote the empirical distribution function of $X_1, \ldots, X_n$ $(Y_1, \ldots, Y_n)$. Let $s_1 < s_2 \cdots < s_n$ be the ranks of the ordered $Y_1, Y_2, \ldots, Y_n$ in the combined sample. Notice that the statistic $(s_1, s_2, \ldots, s_n)$ is equivalent to $\{G_n(W_{ni}),$ $i = 1, \ldots, 2n\}$ which in turn is equivalent to $\{F_n(W_{ni}), G_n(W_{ni}), i = 1, \ldots, 2n\}$.

Lemma 3.14.1 (Savage and Sethuraman, 1966) With the above notation we have

$$P_{H_1}(s_1, \ldots, s_n) = \Delta^n (n!)^2 / n^{2n} \prod_{i=1}^{2n} \{F_n(W_{ni}) + \Delta G_n(W_{ni})\}. \quad (3.14.18)$$

Proof Savage and Sethuraman (1966) point out that it can easily be deduced from Savage (1956, Corollary 7.a.1). However, in the following, we shall give an independent proof of the same. Let $\mathbf{Z} = (Z_{1n}, \ldots, Z_{2n})$ where $Z_i = 1$ if $W_{n,i}$ is a Y observation and $Z_i = 0$ if W_{ni} is an X $(i = 1, \ldots, 2n)$. Savage (1956) called $\mathbf{Z}$ the two-sample rank order. Notice that $\mathbf{Z}$ uniquely determines $(s_1, \ldots, s_n)$ and conversely. Hence $P_{H_1}(\mathbf{Z} = \mathbf{z}) = P_{H_1}(s_1, \ldots, s_n)$. One can easily establish that

$$\begin{aligned} P_{H_1}(\mathbf{Z} = \mathbf{z}) &= (n!)^2 \int_{-\infty < w_1 < \cdots} \cdots \int_{< w_{2n} < \infty} \prod_{i=1}^{2n} [dF(w_i)]^{1-z_i} [dG(w_i)]^{z_i} \\ &= (n!)^2 \int_{0 < u_1 < \cdots} \cdots \int_{< u_{2n} < 1} \prod_{i=1}^{2n} (du_i)^{1-z_i} (\Delta u_i^{\Delta - 1}\, du_i)^{z_i} \\ &= \Delta^n (n!)^2 \int_{0 < u_1 < \cdots} \cdots \int_{< u_{2n} < 1} \prod_{i=1}^{2n} u_i^{(\Delta - 1) z_i} \prod_{i=1}^{2n} du_i . \end{aligned}$$

Integrating on u_1 first, then on u_2, etc., we obtain

$$P_{H_1}(\mathbf{Z} = \mathbf{z}) = \Delta^n (n!)^2 / \prod_{i=1}^{2n} [(\Delta - 1)(z_1 + \cdots + z_i) + i]. \quad (3.14.19)$$

Note that $z_1 + \cdots + z_i = nG_n(W_{ni})$ and $i = n[F_n(W_{ni}) + G_n(W_{ni})]$. Substituting these in (3.14.19) the lemma follows. Clearly, $P_{H_0}(\mathbf{Z} = \mathbf{z}) = P_{H_0}(s_1, \ldots, s_n) = 1/\binom{2n}{n}$, which follows by setting $\Delta = 1$ in (3.14.18) or from the observation that each rank order $\mathbf{Z}$ is equally likely under H_0 and $\binom{2n}{n}$ denotes the number of possible distinct rank orders. Let

$$\begin{aligned} \Lambda_n(\Delta, F_n, G_n) &= P_{H_1}(s_1, \ldots, s_n) / P_{H_0}(s_1, \ldots, s_n) \\ &= \Delta^n (2n)! / n^{2n} \prod_{i=1}^{2n} [F_n(W_{ni}) + \Delta G_n(W_{ni})]. \end{aligned} \quad (3.14.20)$$

It should be remarked that, since the product on the right-hand side of (3.14.20) is symmetric in the W_{ni}, one can replace the W_{ni} by the W_i. Then the

SPRT based on ranks for testing H_0 against H_1 is given by: Take one more observation if $B < \Lambda_n(\Delta, F_n, G_n) < A$ accept H_0 if $\Lambda_n \le B$, reject H_0 if $\Lambda_n \ge A$, $n = 1, 2, \ldots$.

Let N be the stopping variable for the above SPRT and let

$$nS_n = \ln \Lambda_n(\Delta, F_n, G_n). \tag{3.14.21}$$

Then, from (3.14.18) we have

$$S_n = \ln 4\Delta - 2 - T_n + O((\ln n)/n), \tag{3.14.22}$$

where $T_n = n^{-1} \sum_1^{2n} \ln[F_n(W_i) + \Delta G_n(W_i)]$. Also, let

$$\Omega_1(n) = \sup_x |F_n(x) - F(x)|, \qquad \Omega_2(n) = \sup_x |G_n(x) - G(x)|$$

and

$$\Omega(n) = \Omega_1(n) + \Delta\Omega_2(n). \tag{3.14.23}$$

From Sethuraman (1964, Theorem 1), for each $\delta > 0$ there is a $\rho_1(\delta) < 1$ such that

$$P(\Omega_i(n) \ge \delta) \le \rho_1{}^n(\delta), \qquad i = 1, 2, \ldots, \tag{3.14.24}$$

for sufficiently large n. Hence, for large n

$$\begin{aligned} P(\Omega(n) \ge \delta) &\le P(\Omega_1(n) \ge \delta/(1 + \Delta)) + P(\Delta\Omega_2(n) \ge \Delta\, \delta/(1 + \Delta)) \\ &\le 2\rho_1{}^n(\delta/(1 + \Delta) \le \rho_2{}^n \qquad (\rho_2 < 1). \end{aligned} \tag{3.14.25}$$

Now, let $V_1, V_2, \ldots, V_n$ be iid random variables having mean $E(V_1)$ and finite moment-generating function in a neighborhood of 0. Then, from Chernoff (1952, Theorem 1) we have, for every $\varepsilon > 0$ and sufficiently large n,

$$P\left(\left|n^{-1} \sum_1^n V_i - EV_1\right| \ge \varepsilon\right) \le \rho_3{}^n(\varepsilon) \qquad (\rho_3 < 1). \tag{3.14.26}$$

In (3.14.26) let $V_i = \ln(1 + (\delta/U_i))$ where U_i has the uniform distribution on $[0, 1]$. The moment-generating function of V_1 is finite for $\delta > 0$ and $t < 1$ and $EV_1 \to 0$ as $\delta \to 0$. Thus, from (3.14.26), given $\varepsilon > 0$, we can find δ_0 such that for sufficiently large n

$$P\left[n^{-1} \sum_1^n \ln(1 + (\delta_0/U_i)) \ge \varepsilon\right] < \rho_4{}^n, \qquad (\rho_4 < 1). \tag{3.14.27}$$

Analogous to the random variables S_n and T_n, define the parameters

$$T(\Delta, F, G) = \int \ln[F(x) + \Delta G(x)]\,[dF(x) + dG(x)] \qquad (3.14.28)$$

and

$$S(\Delta, F, G) = \ln 4\Delta - 2 - T(\Delta, F, G). \qquad (3.14.29)$$

Lemma 3.14.2 [Savage and Sethuraman, 1966, 1967)]. Given $\varepsilon > 0$, there exists $\rho < 1$ such that, for sufficiently large n

$$P(|T_n - T(\Delta, F, G)| \geq 3\varepsilon) < \rho^n. \qquad (3.14.30)$$

Proof.

$$\begin{aligned}
T_n &= n^{-1}\sum_1^{2n} \ln[F_n(W_i) + \Delta G_n(W_i)] \\
&= n^{-1}\sum_1^{2n} \ln[F(W_i) + \Delta G(W_i)] \\
&\quad + n^{-1}\sum_1^{2n} \ln\{1 + [(F_n(W_i) - F(W_i)) + \Delta(G_n(W_i) - G(W_i))]/[F(W_i) \\
&\quad + \Delta G(W_i)]\} \\
&= T_{n,1} + T_{n,2}, \qquad \text{say.}
\end{aligned}$$

Let $V_i = \ln[F(X_i) + \Delta G(X_i)] + \ln[F(Y_i) + \Delta G(Y_i)]$. Then V_1 has a finite moment-generating function and $E(V_1) = T(\Delta, F, G)$. Thus, from (3.14.26) for sufficiently large n

$$P(|T_{n,1} - T(\Delta, F, G)| \geq \varepsilon) \leq \rho_3{}^n. \qquad (3.14.31)$$

Now, consider $T_{n,2}$ for $\Delta \geq 1$ (when $\Delta < 1$ the computation is similar). Then

$$\begin{aligned}
-T_{n,2} &\leq n^{-1}\sum_1^{2n} \ln\{1 + \Omega(n)/[F_n(W_i + \Delta G_n(W_i)]\} \\
&\leq n^{-1}\sum_1^{n} [\ln\{1 + \Omega(n)/F_n(X_i)\} + \ln\{1 + \Omega(n)/\Delta G_n(T_i)\}] \\
&\leq 2n^{-1}\sum_1^{n} \ln(1 + n\,\Omega(n)/i),
\end{aligned}$$

where $\Omega(n)$ is given by (3.14.23). Some elementary computation shows that

$$\sum_1^n \ln(1 + K/i) \leq 4 \int_1^n \ln(1 + K/x)\, dx$$
$$= 4[(n + K)\ln(n + K) - n \ln n - (K + 1)\ln(K + 1)].$$

Hence, $-T_{n,2} \leq 8\varepsilon(\Omega(n), n)$, where

$$\varepsilon(\delta, n) = (1 + \delta)\ln(1 + \delta) - \delta \ln(\delta + 1/n) - n^{-1}\ln(1 + n\delta).$$

$\varepsilon(\delta, n) \to 0$ as $\delta \to 0$; in fact, there is a δ_0 such that $\delta < \delta_0$ implies $\varepsilon(\delta, n) < \varepsilon/8$ uniformly in n. Thus

$$P(T_{n,2} < -\varepsilon) < \rho_5{}^n \qquad \text{for some } \rho_5 < 1. \tag{3.14.32}$$

Next consider

$$T_{n,2} \leq n^{-1} \sum_1^{2n} \ln\{1 + \Omega(n)/[F(W_i) + \Delta G(W_i)]\}$$
$$\leq n^{-1} \sum_1^n \ln[1 + \Omega(n)/F(X_i)] + n^{-1} \sum_1^n \ln[1 + \Omega(n)/\Delta G(Y_i)]. \tag{3.14.33}$$

Choose δ_0 such that (3.14.27) holds. From (3.14.27) and (3.14.25)

$$P\left(n^{-1} \sum_1^n \ln\{1 + (\Omega(n)/U_i\} \geq \varepsilon\right) \geq P(\Omega(n) \geq \delta_0)$$
$$+ P\left(n^{-1} \sum_1^n \ln\{1 + (\Omega(n)/U_i\} \leq \varepsilon, \quad \Omega(n) < \delta_0\right)$$
$$\leq \rho_2{}^n + P\left\{n^{-1} \sum_1^n \ln[1 + (\delta_0/U_i)] \geq \varepsilon\right\}$$
$$\leq \rho_2{}^n + \rho_4{}^n \leq \rho_6{}^n,$$

for sufficiently large n and $\rho_6 < 1$. Arguing in a similar fashion for the second term on the right-hand side of (3.14.33) we have, for sufficiently large n

$$P(T_{n,2} > 2\varepsilon) \leq \rho_7{}^n, \qquad 0 < \rho_7 < 1. \tag{3.14.34}$$

Combining (3.14.31), (3.14.32), and (3.14.34) we can find a $\rho < 1$ such that (3.14.30) holds.

Lemma 3.14.3 (Savage and Sethuraman, 1966) Given $\varepsilon > 0$, we can find a $\rho < 1$ such that for sufficiently large n

$$P\{|S_n - S(\Delta, F, G)| \geq 3\varepsilon\} \leq \rho^n. \qquad (3.14.35)$$

Proof Follows readily from (3.14.22) and Lemma 3.14.2.

Now, we are ready to state and prove the main result of Savage and Sethuraman (1966).

Theorem 3.14.5 (Savage and Sethuraman, 1966) With the above notation, let $S(\Delta, F, G) \neq 0$. Then for the SPRT based on $\Lambda_n(\Delta, F_n, G_n)$,

(i) $P(N > n) < \rho^n$ for sufficiently large n, and some $\rho < 1$.
(ii) $P(N < \infty) = 1$.
(iii) $E[\exp(tN)] < \infty$ for t in some interval $(-\infty, \theta)$, $\theta > 0$.

Proof Note that $P(N > n) \leq P\{B < \Lambda_n < A\}$. Next, if $S(\Delta, F, G) \neq 0$, we have

$$\begin{aligned} P(B < \Lambda_n < A) &= P\{(\ln B)/n < S_n < (\ln A)/n\} \\ &\leq P\{|S_n - S(\Delta, F, G)| > \varepsilon\} < \rho^n \qquad (0 < \rho < 1) \end{aligned}$$

for sufficiently large n by Lemma 3.14.3. This completes the proof for (i). Now, (ii) and (iii) are immediate from (i) (for instance, see Theorem 2.2.1 and Section 2.4).

Lemma 3.14.4 lists some of the properties of the parameter $S(\Delta, F, G)$. We will be interested in conditions under which it vanishes.

Lemma 3.14.4

(i) $S(1, F, G) = 0$.
(ii) $S(\Delta, F, h(F))$ is free of F.
(iii) $S(\Delta, F, G) = S(1/\Delta, G, F)$.
(iv) $S(\Delta, F, F) < 0$ for $\Delta \neq 1$.
(v) $S(\Delta, F, F^{\Delta}) > 0$ for $\Delta \neq 1$.
(vi) $S(\Delta, F, F^{\eta})$ is strictly increasing in η for $\Delta > 1$.
(vii) For each Δ, there is a unique $\eta(\Delta)$, lying between 1 and Δ, satisfying the equation

$$S(\Delta, F, F^{\eta}) = 0. \qquad (3.14.36)$$

Further, $\eta(\Delta) = 1/\eta(1/\Delta)$.

Proof (i)–(iv) follow directly from (3.14.30), the definition of $S(\Delta, F, G)$. Now,

$$\begin{aligned} S(\Delta, F, F^\eta) &= \ln 4\Delta - 2 - (1/\Delta)\int_0^1 \ln(t + \Delta t^\eta)\, d(t + \Delta t^\eta) \\ &\quad - (1 - (1/\Delta))\int_0^1 \ln(t + \Delta t^\eta)\, dt \\ &= \ln 4\Delta - 2 - (1/\Delta)[(1+\Delta)\ln(1+\Delta) - (1+\Delta)] \\ &\quad - (1 - (1/\Delta))\int_0^1 \ln(t + \Delta t^\eta)\, dt. \end{aligned} \tag{3.14.37}$$

For $\Delta > 1$, (3.14.37) shows that $S(\Delta, F, F^\eta)$ is a strictly increasing function of η, establishing (vi). Equation (3.14.37) can be further simplified by writing $\ln(t + \Delta t^\eta) = \ln t + \ln(1 + \Delta t^{\eta-1})$ and performing integration by parts once:

$$S(\Delta, F, F^\eta) = -2\ln[(\Delta^{1/2} + \Delta^{-1/2})/2] + (\Delta - 1)(\eta - 1)\int_0^1 (\Delta^{-1} + t^{1-\eta})\, dt \tag{3.14.38}$$

and

$$S(\Delta, F, F^\eta) = \int_0^1 \{[(\Delta - 1)(\eta - 1)/\Delta + t^{1-\eta}] - (\Delta - 1)^2/(4\Delta + (\Delta - 1)^2 t)\}\, dt. \tag{3.14.39}$$

When $\Delta = \eta < 1$, the integrand in the right-hand side of (3.14.39) is >0 for t in $(0, 1)$ and thus $S(\Delta, F, F^\Delta) > 0$ for $\Delta < 1$. From (i) and (iii) $S(\Delta, F, F^\Delta) = S(\Delta, F^{1/\Delta}, F) = S(1/\Delta, F, F^{1/\Delta})$. Hence $S(\Delta, F, F^\Delta) > 0$ for $\Delta > 1$ also. This established (v). Now (vii) follows from all the above.

Savage and Sethuraman (1966) have tabulated pairs of values of Δ and η that satisfy (3.14.36). If the true alternative is of the Lehmann type with exponent η, then finite sure termination of the rank order sequential probability ratio test (ROSPRT) does not hold.

REMARK 3.14.6 Although Theorem 3.14.5 establishes the sure termination of the ROSPRT under very general conditions including H_0 and H_1, the restriction $S(\Delta, F, G) \neq 0$ does not appear to be necessary for finite termination. Theorem 3.14.5 includes the possibility that the pair of random variables X_i and Y_i may be dependent. Although H_0 often occurs in nonparametric work, H_1 is somewhat special. The general case H_1': $F \neq G$ cannot be handled by the methods employed by Savage and Sethuraman (1966). The methods of this paper would certainly apply to situations with other sampling patterns, for example, not taking the same number of observations from each population at each stage or taking more than one observation from each population at some stages [see Bradley *et al.* (1966)]. The continuity of F and G was required in order to define the rank order uniquely and to compute an explicit expression for the rank order probability. Continuity of F and G is not required for the rest of the results, especially for Theorem 3.14.5.

Recently, Sethuraman (1970) was motivated by Stein's proof (1946) for the sure termination of Wald's SPRT. Extending his ideas to take certain "dependencies" into account, he examines the rank order SPRT based on Λ_n [see (3.14.20)] and proves that this test terminates finitely with probability one and that the stopping time has a finite moment-generating function under a mild condition on the bivariate random variables that resembles the Stein condition, namely that a certain random variable is not identically equal to zero. In the following we shall present Sethuraman's basic result without proof which is somewhat long.

Theorem 3.14.6 (Sethuraman, 1970) Let $I(x; z) = 1$ if $x \leq z$ and equal to zero if $x > z$; let $W(z) = F(z) + \Delta G(z)$, $W(x, y; z) = I(x; z) + \Delta I(y; z)$ and let

$$V(X, Y) = -\ln[W(X)W(Y)] + \ln 4\,\Delta - 2 + \int \frac{W(z) - W(X, Y; z)}{W(z)}\, d(F(z) + G(z)). \tag{3.14.40}$$

Assume that $P(V(X, Y) < 0) > 0$. Then the rank order SPRT based on Lehmann alternatives terminates finitely and has a finite moment-generating function.

Proof See Sethuraman (1970, Theorem 4, p. 1331).

The basic condition of Savage and Sethuraman (1966) namely $S(\Delta, F, G) \neq 0$ can be stated as

$$E[-\ln W(X)W(Y) + \ln 4\,\Delta - 2] \neq 0. \tag{3.14.41}$$

Now, $E[\int \{(W(z) - W(X, Y; z))/W(z)\}\, (dF(z) + dG(z))] = 0$. Hence condition (3.14.41) is equivalent to $E[V(X, Y)] \neq 0$. Thus, the case not covered by Theorem 3.14.5 is when $E[V(X, Y)] = 0$. Comparing this uncovered case with the condition of Theorem 3.14.6, we have

Theorem 3.14.7 Let $P(V(X, Y) = 0) \neq 1$. Then the rank order SPRT based on Lehmann alternatives terminates finitely and has a finite moment-generating function.

REMARK 3.14.7 The condition of Theorem 3.14.7, namely that $P(V(X, Y) = 0) \neq 1$ is analogous to the condition in Theorem 2.2.1. Sethuraman (1970) conjectures that for all joint distributions $H(x, y)$, $P(V(X, Y) = 0) < 1$ so that the condition of Theorem 3.14.7 is unnecessary. The following particular case is immediate.

Lemma 3.14.5 (Sethuraman, 1970) Let X, Y be independently distributed. Then $P(V(X, V) = 0) < 1$.

So, had we started with independent bivariate random variables, then the rank order SPRT would terminate finitely and would have a finite moment-generating function.

Govindarajulu (1968a) has considered the asymptotic normality of the rank order SPRT which would enable its finite termination under more general conditions. We shall present these results in the following. Before we consider the asymptotic normality of $S_n = n^{-1} \ln \Lambda_n$, it is worthwhile to make this remark.

REMARK 3.14.8 If F $[G]$ has a finite or denumerable number of jumps, then F $[G]$ can be made continuous by the continuization process described by Govindarajulu *et al.* (1967). Also, the rank orders are, with probability one, well defined provided F and G do not have common discontinuities. However, $\Lambda_n(\Delta, F_n, G_n)$ is well defined if the Xs and Ys have some ties. Consequently, the basic result of Savage and Sethuraman (Theorem 3.14.5) and Sethuraman (Theorem 3.14.6) and hence Theorem 3.14.7 are valid for arbitrary F and G.

ASYMPTOTIC NORMALITY OF $S_n = n^{-1} \ln \Lambda_n(\Delta, F_n, G_n)$

Let us consider a class of statistics of which S_n is a special case and study its asymptotic distribution, when the class of statistics is suitably standardized. Let

$$H(x) = (1 + \Delta)^{-1}F(x) + \Delta(1 + \Delta)^{-1}G(x), \tag{3.14.42}$$

and

$$H_n(x) = (1 + \Delta)^{-1}F_n(x) + \Delta(1 + \Delta)^{-1}G_n(x). \tag{3.14.43}$$

Define the class of statistics T_n by

$$T_n = n^{1/2}\left[\int J_n(H_n(x))\, dF_n(x) - \int J(H)\, dF(x)\right] \tag{3.14.44}$$

where $J_n(u)$ converges to $J(u)$ in Lebesgue measure and $J(u)$ is absolutely continuous in u. Notice that T_n is slightly different from the Chernoff–Savage class of statistics because H_n is a mixture of F_n and G_n and is not the combined empirical df, that is, $H_n \neq (n/2n)F_n + (n/2n)G_n$. Let $f\{g\} \geq 1$ be U-shaped and Lebesgue integrable {square integrable}. We say that $J \in \mathscr{S}$ if J is absolutely continuous with $|J'| \leq fg$ wherever it exists. Then, we have the following theorem.

Theorem 3.14.8 (Govindarajulu, 1968a) If X and Y are independent,

$$\lim_{n\to\infty} \sup_z |P(T_n/\sigma_n \leq z) - \Phi(z)| = 0 \tag{3.14.45}$$

for every $J \in \mathscr{S}$ and for all F and G provided $\sigma_n \neq 0$ where

$$\sigma_n{}^2 = 2\,\Delta^2(1+\Delta)^{-2}\Big\{\iint_{x<y} G(x)[1-G(y)]J'(H(x))J'(H(y))\,dF(x)\,dF(y)$$
$$+ \iint_{x<y} F(x)[1-F(y)]J'(H(x))J'(H(y))\,dG(x)\,dG(y)\Big\}. \tag{3.14.46}$$

Proof By reviewing the proof of the various lemmas and propositions pertaining to the asymptotic normality of the Chernoff–Savage class studied by Govindarajulu *et al.* (1967), one finds that they are valid for the class of statistics involving arbitrary but nontrivial mixtures of the empirical distributions.

Corollary 3.14.8.1 F and G may depend on n; hence the subscript n on σ. Assume that J is not a constant. Let $\{J_k\}$ be a sequence such that $J_k \in \mathscr{S}$ and $J_k' \to J'$ in Lebesgue measure. Let $\{F_k, G_k\}$ be a sequence of pairs converging to a pair (F, G) at all points of continuity of the latter pair. Then if $F = G$, $\sigma_n{}^2[F_k, G_k, J_k]$ converges uniformly in n to

$$2\,\Delta^2(1+\Delta)^{-2}\Big\{\int_0^1 J^2(u)\,du - \Big[\int_0^1 J(u)\,du\Big]^2\Big\}.$$

As a special case of Theorem 3.14.8 let us deduce the asymptotic normality of $S_n = \ln 4\Delta - 2 - n^{-1/2}\sum_1^{2n} \ln\{F_n(W_i) + \Delta G_n(W_i)\} + o(n^{-1/2})$. Let $W = F + \Delta G$, $W_n = F_n + \Delta G_n$. By writing $G_n = (W_n - F_n)/\Delta$ and simplifying one obtains

$$S_n - S(\Delta, F, G) = (\Delta^{-1} - 1)\Big[\int (\ln W_n)\,dF_n(x) - \int (\ln W)\,dF(x)\Big]$$
$$-\Delta^{-1}\Big[\int (\ln W_n)\,dW_n - \int (\ln W)\,dW\Big] + o(n^{-1/2})$$
$$= (\Delta^{-1} - 1)\Big[\int (\ln H_n)\,dF_n(x) - \int (\ln H)\,dF(x)\Big] + o(n^{-1/2}).$$

Then we have

$$\lim_{n\to\infty} \sup_z |P\{n^{1/2}(S_n - S(\Delta, F, G))/\sigma_n^* \le z\} - \Phi(z)| = 0 \qquad (3.14.47)$$

provided $\sigma_n^* \neq 0$ where

$$\sigma_n^{*2} = 2(1-A)^2 \Big[\iint_{x<y} \{G(x)(1-G(y))/W(x)W(y)\}\, dF(x)\, dF(y)$$

$$+ \iint_{x<y} \{F(x)(1-F(y))/W(x)W(y)\}\, dG(x)\, dG(y) \Big]. \qquad (3.14.48)$$

REMARK 3.14.9 Since the asymptotic normality of S_n holds even if $S(\Delta, F, G) = 0$, it follows from Eq. (3.14.47) that the sequential test based on S_n terminates with probability one even if $S(\Delta, F, G) = 0$. Thus, it takes care of the exceptional case not covered by Theorem 3.14.5. However, it should be noted that the asymptotic normality does not ensure the finiteness of the moment-generating function of the stopping variable N. Further notice that

$$\sigma_n^{*2}(F, F) = 2(1-\Delta)^2(1+\Delta)^{-2} \left\{ \int_0^1 (\ln u)^2\, du - \left[\int_0^1 (\ln u)\, du \right] \right\}$$

$$= 2(1-\Delta)^2(1+\Delta)^{-2}.$$

REMARK 3.14.10 Sethuraman (1970) independently has also considered the asymptotic normality of S_n and claims its validity even if X and Y are dependent provided σ_n is approximately modified by considering the co-variance between X and Y.

We also have the following result which implies finite sure termination.

Theorem 3.14.9 (Govindarajulu, 1968a) Let $\{k\}$ denote a sequence of stopping rules and $\{N_k\}$ with $N_1 \le N_2 \le \cdots$ denote the sequence of stopping variables associated with $\{k\}$. Also, let there exist a nonstochastic sequence of positive numbers $\{n_k\}$, $n_1 \le n_2 \le \cdots$, $n_k \to \infty$ as $k \to \infty$, and $N_k/n_k \to 1$ in probability as $k \to \infty$. Define

$$T_k = n_k^{1/2} \Big[\int J_{n_k}(H_{N_k})\, dF_{N_k} - \int J(H)\, dF \Big]. \qquad (3.14.49)$$

Then,

$$\lim_{k\to\infty} \sup_z |P\{T_k/\sigma_k \le z\} - \Phi(z)| = 0,$$

for all F and G and all $J \in \mathscr{S}$ where $J_{n_k}(u) \to J(u)$ for $0 < u < 1$ and σ_k^2 is given by (3.14.46).

Proof Follows from Theorem 3.14.8 and a result of Anscombe (1952).

Corollary 3.14.9.1 If X and Y are independent and the sequences $\{k\}$, $\{n_k\}$, and $\{N_k\}$ are as hypothesized in Theorem 3.14.9, then

$$\lim_{k\to\infty} \sup_z |P\{n_k^{1/2}(S_k - S(\Delta, F, G))/\sigma_k \leq z\} - \Phi(z)| = 0 \qquad (3.14.50)$$

where σ_k^2 is given by (3.14.46).

Example 3.14.1 In the usual SPRT let $\mu = EZ = E[\ln\{f(X; \theta_1)/f(X; \theta_0)]$ > 0 and $\{(B_k, A_k)\}$ be the sequence of boundaries used, then, using the strong law of large numbers it can be shown that N_k converges with probability 1 to $(\ln A_k)/\mu$ if $\mu > 0$ and to $(\ln B_k)/\mu$ if $\mu < 0$.

Berk and Savage (1968) have considered the two-sample problems of sequentially testing the null hypothesis H_0: $G = F$ against the alternative $H_1 = \psi$: $F = \psi_1(K)$, $G = \psi_2(K)$ where ψ_1 and ψ_2 are given dfs on [0, 1] that specify ψ and K ranges through all dfs. [In cases of interest as, in the null hypothesis, ψ is usually of the form $G = \psi(F)$, where ψ is a df on [0, 1].] It is well known that under either hypothesis, the distribution of the rank-order depends only on the corresponding ψ; hence one may compute Λ_n, the likelihood ratio for the rank order based on the two hypotheses. One continues sampling until Λ_n leaves an interval (B, A), $0 < B < 1 < A < \infty$. Since a monotone increasing transformation of the data leaves the rank order $\mathbf{Z}$ invariant, we can assume that all observations are on the unit interval. In particular, let $H = (F + G)/2$ be the uniform df. Justification of this assumption has been shown by Govindarajulu *et al.* (1967) as follows: If the function remains constant over certain intervals, no observations will occur in these intervals. Thus, these intervals can be deleted from the line without affecting the order of the observations. This will leave us with a continuous strictly increasing cumulative distribution which can now be transformed to the uniform cumulative $H(x) = x$ by a strictly increasing continuous transformation. Further, when $H(x)$ is the uniform cumulative distribution the measure dF induced by F possesses a density f with respect to the Lebesgue measure dH on [0, 1]. Then, let f and g be the densities of F and G. Let us define the class Ψ to which ψ belongs.

Definition 3.14.1 (Berk and Savage, 1968) The alternative ψ is in Ψ if representations f and g (on [0, 1]) may both be chosen to satisfy:

(i) f is of bounded variation.

(ii) For δ sufficiently small, $\{t : 0 < f(t) < \delta\}$ is a (possibly empty) interval and $\{t : f(t) = 0\}$ is a (possibly empty) finite union of intervals.

(iii) There exist γ_j, $0 < \gamma_j < 1$ so that

$$\limsup_{\delta \to 0} [\lambda[t : 0 < f < \gamma_j \delta]/\lambda\{t : 0 < f < \delta\}] = 1 - 2\alpha_j < 1$$

where λ denotes Lebesgue measure. In (iii) we take the expression to be zero if the numerator vanishes. Condition (iii) requires that the representative densities not decrease too rapidly to zero. A useful condition that implies (iii) is (iii′): there exist $\eta > 0$ so that in a neighborhood of zero $\lambda\{t : 0 < f < \delta\}/\delta^\eta$ is nondecreasing in δ [under (iii′), (iii) holds for every $0 < \gamma < 1$]. Condition (iii′) is essentially a condition on the likelihood ratio g/f and may be so stated for arbitrary representatives.

Having assumed that $H = (F + G)/2$ is uniform on [0, 1], let $U = FH^{-1}$ and $V = GH^{-1}$ which have bounded derivatives f and g respectively. Similarly if F^* and G^* are the distributions of X and Y, then one can assume that $H^* = (F^* + G^*)/2$ to be uniform on [0, 1] and define $U^* = F^*H^{*-1}$ and $V^* = G^*H^{*-1}$. Suppose that f and g are histograms constant on the M intervals defined by $0 = a_0 < a_1 < \cdots < a_M = 1$. Then the expression for Λ_n is

$$\begin{aligned}\Lambda_n(z) &= (2n)! \int_{0 < x_1 < \cdots < x_{2n} < 1} \prod_1^{2n} [f(x_i)]^{1 - z_i}[g(x_i)]^{z_i}\, dx_i \\ &= \sum_{(k_1, \ldots, k_M)} (2n)! \prod_1^M f_j^{m_j} g_j^{n_j}/k_j! \qquad (3.14.51)\end{aligned}$$

where f_j (respectively, g_j) is the area of f (respectively, g) over (a_j, a_{j+1}) and $n_j = k_j - m_j$ is the sum of the k_j zs in the jth piece of z when partitioned into M pieces of lengths $k_1, k_2, \ldots, k_M$ respectively (from left to right). Since $\sum k_j = 2n$,

$$\Lambda_n(z) = \sum_{(k_1, \ldots, k_M)}^{2n} \exp\left(\sum_1^M m_j \ln f_j + n_j \ln g_j\right). \qquad (3.14.52)$$

Let $V_n(t) = n^{-1} \sum_1^{[2nt]} Z_i$; $0 \le t \le 1$ and $[x]$ is the largest integer equal to or less than x. That is, $V_n(t)$ is the proportion of the Y sample that falls below the t-fractile of the combined sample. $U_n(t)$ is analogously defined. Thus, $U_n(t) + V_n(t) = [2nt]/n$. Let $P_0 = 0$ and $P_j = (k_1 + \cdots + k_j)/2n$; and let

$$n_j/n = V_n(P_j) - V_n(P_{j+1}) = \Delta V_n(P_j).$$

Then

$$n^{-1} \ln \Lambda_n(z) \sim \max_{(k_1,\ldots,k_M)} \left\{ n^{-1} \ln \binom{2n}{k_1, \ldots, k_M} + \sum_1^M [(\ln f_j)\Delta U_n(P_j) + (\ln g_j)\Delta V_n(P_j)] \right\} = I_n'. \quad (3.14.53)$$

Let $p_j = P_j - P_{j-1}$ and

$$i(p, U^*) = \sum [(\ln f_j)\Delta U^*(P_j) + (\ln g_j)\Delta V^*(P_j) - 2p_j \ln p_j], \quad (3.14.54)$$

and

$$i(U^*) = I^* = \sup\{i(P, U^*) : 0 \le P_1 \le \cdots \le P_M = 1\} \quad (3.14.55)$$

where U^* is the true transformed distribution of the Xs. Now, we are ready to state the basic result of Berk and Savage (1968).

Theorem 3.14.10 (Berk and Savage, 1968) Suppose $(X_1, Y_1), (X_2, Y_2)$, are independent and identically distributed bivariate random variables with $P(X = Y) = 0$ and having F^* and G^* for the marginal distributions of X and Y respectively. Then if ψ belongs to a certain class Ψ defined in Definition 3.14.1, there exists a real number I^* defined by (3.14.54) and for all $\delta > 0$, there is a $\rho < 1$ so that for n sufficiently large, $P^*(|n^{-1} \ln \Lambda_n(Z) - I^*| > \delta) < \rho^n$. P^* denotes probability under (F^*, G^*).

Proof [See Berk and Savage (1968, Theorem 2.1)].

REMARK 3.14.11 The convergence of $n^{-1}\Lambda_n(Z)$ to I^* is rapid enough to ensure the finiteness of the moment-generating function of the stopping time defined by the SPRT based on L_n, in a neighborhood of the origin under any (F^*, G^*) for which $I^*(f) \neq 0$ [see Berk and Savage (1968, Sections 7 and 8.1)]. If ψ: $G = F^\Delta$, I^* coincides with $S(\Delta, F^*, G^*)$ obtained by Savage and Sethuraman (1966) [see Berk and Savage (1968, Section 9)]. If f and g are normal densities differing by a shift, it is immediate that $G\{t : \ln(g/f) > x\}$ decreases exponentially. Hence, this alternative belongs to Ψ and Theorem 3.14.10 applies to that particular case of interest also.

When $I^* \neq 0$, the large deviation result given by Theorem 3.14.10 obviously implies the conditions of Theorem 3.6.5, so that the limiting behavior of EN given there applies to this class of sequential rank tests. Letting I_i^* denote the limiting value of $n^{-1} \ln \Lambda_n(Z)$ under H_i $(i = 0, 1)$ it can be verified that $-\infty < I_0^* < 0 < I_1^*$ and the conditions of Theorem 3.6.6 are satisfied by these SPRTs.

3.14.4 One-Sample Rank Order SPRT's for Symmetry

Let $V_1, V_2, \ldots$ be independent and identically distributed random variables observed sequentially and having a continuous df F. We wish to test the hypothesis,

$$H_0\colon \quad F(v) + F(-v) = 1 \qquad \text{for all } v, \tag{3.14.56}$$

that is, F is symmetric about zero. Let $H(v) = P(V \le v \mid V > 0) = [F(v) - F(0)]/[1 - F(0)]$ and $G(v) = P(|V| \le v \mid V < 0) = [F(0) - F(-v)]/F(0)$ for $v \ge 0$, and $H(v)$, $G(v) = 0$, for $v < 0$. Thus $F(v) = F(0)[1 - G(-v)]$ for $v < 0$ and $F(v) = H(v) + F(0)[1 - H(v)]$ for $v \ge 0$. Then, we can rewrite (3.14.56) as

$$H_0\colon \quad H(v) = G(v) \qquad \text{for all } v \text{ and } F(0) = \tfrac{1}{2}$$

and take

$$H_a\colon \quad H(v) \ne G(v) \qquad \text{for some } v.$$

However, probabilities of desired rank orders cannot be derived explicitly under H_a. Assuming that H_a: $H(v) = 1 - [1 - G(v)]^{\Delta}$, $v \ge 0$ with $F(0) = \Delta/(1 + \Delta)$, sampling with groups of fixed size at each stage and ranking the absolute values of the observations within each group at each stage, Weed and Bradley (1971) have proposed two sequential procedures, one based on within-group configuration of signed ranks and the other based on the within-group sums of positive signed ranks. They carried out some Monte Carlo studies in order to test whether the model is appropriate or not when the data are generated from the normal populations. The choice of $F(0)$ is dictated by their desire that $f(v)$ should not have a jump discontinuity at the origin.

Parent (1965) defines sequential rank and signed sequential rank.

Definition 3.14.2 (Parent, 1965) The sequential rank of V_n relative to $V_1, V_2, \ldots, V_n$ is k if V_n is the kth smallest in the set $(V_1, V_2, \ldots, V_n)$. There are $n!$ sequential rank vectors.

For example, the sequential rank of V_1 is 1; the sequential rank of V_2 is 1 or 2, etc. He has also shown that when the Vs are iid, the sequential ranks are independent with distribution

$$P(Z_i = k) = 1/i, \qquad k = 1, 2, \ldots, i, \quad i = 1, 2, \ldots, n, \tag{3.14.57}$$

where Z_i denotes the sequential rank of V_i $(i = 1, 2, \ldots, n)$. Parent (1965) uses the probability ratio of the sequential rank sequence in order to test a hypothesis with Lehmann alternatives in the two-sample case. He also establishes its sure termination.

Definition 3.14.3 (Parent, 1965) The signed sequential rank of V_n relative to $V_1, V_2, \ldots, V_n$ is the product of the sequential rank of $|V_n|$ relative to $|V_1|, \ldots, |V_n|$ and sign of V_n. There are $2^n n!$ possible signed sequential rank vectors corresponding to a sample of size n.

Parent (1965, Lemma 2.2) shows that $|V|$ and sign of V are independent if and only if

$$F(-v) = F(0)[1 - F(v) + F(-v)] \qquad \text{for all } v \geq 0. \tag{3.14.58}$$

That is, $P(|V| \leq v | V > 0) = P(|V| \leq v | V < 0)$. Thus, if $V_1, V_2, \ldots, V_n$ are iid as F where (3.14.58) holds, then the signed sequential ranks of the Vs are mutually independent. Condition (3.14.58) is satisfied by distributions of positive, negative, and symmetric (about 0) random variables. A larger class of distributions satisfies the condition. If we consider all measurable sets $E \subset [0, \infty]$ and define $-E = \{x : -x \in E\}$ then the condition

$$P(V \in E) = kP(V \in -E), \qquad k \geq 0, \quad \text{all } E, \tag{3.14.59}$$

is enough to ensure (3.14.58). Parent (1965) applies signed sequential ranking to a process control problem, the criterion being $S_n = \sum_{i=1}^{n} Y_i/i$ where Y_i are the signed sequential ranks. He also computes $E(N)$ and tabulates some Monte Carlo values of $E(N)$, σ_N when V_i are normal $(\mu, 1)$.

Govindarajulu has considered the alternative

Model II:

$$\begin{aligned} H_{a\text{II}}:\quad & H(v) = G^{\Delta}(v) \qquad \text{for all } v,\ \Delta > 0,\ \Delta \neq 1 \\ & \Delta \text{ is specified}, \qquad F(0) = \lambda_0, \quad \lambda_0 \text{ specified}. \end{aligned} \tag{3.14.60}$$

Weed (1968) has considered the alternative

Model I:

$$\begin{aligned} H_{a\text{I}}:\quad & H(v) = 1 - [1 - G(v)]^{\Delta}, \qquad \text{for all } v,\ \Delta > 0,\ \Delta \neq 1 \\ & \Delta \text{ specified}, \qquad F(0) = \Delta/(1 + \Delta). \end{aligned} \tag{3.14.61}$$

Notice that if $X = -V, V < 0$ and if $Y = V, V \geq 0$, and if X and Y have conditional dfs (G and H respectively) satisfying (3.14.60), then the conditional dfs of $-X$ and $-Y$ would satisfy (3.14.61) provided $\lambda_0 = \Delta/(1+\Delta)$ and the converse statement is also true. Thus, proof would essentially be the same. Hence, the authors have combined their work in Weed *et al.* (1974). In the following we shall present the main results for Model II. Let $X_1, \ldots, X_m$ denote the absolute values of those Vs that are negative and let $Y_1, \ldots, Y_n$ denote the nonnegative Vs, $m + n = t$. Notice that m is binomially distributed with parameters t and $\lambda = F(0)$, $0 < \lambda < 1$. Let the combined sample of Xs and Ys be denoted by $W_1, \ldots, W_t$ and the ordered combined sample by $W_{t1}, < \cdots < W_{tt}$. Let G_m and H_n respectively denote the empirical df of $X_1, \ldots, X_m$ and $Y_1, \ldots, Y_n$ respectively. Further, following Savage (1959), let $\mathbf{Z} = (Z_1, \ldots, Z_t)$ where $Z_i = 1$ or 0 according as W_{ti} corresponds to a negative or nonnegative V respectively. Also, let

$$\Lambda_t(\Delta, \mathbf{Z}) = P_t(\mathbf{Z} = \mathbf{z} \mid \Delta)/P_t(\mathbf{Z} = \mathbf{z} \mid \Delta = 1) = 2^t P_t(\mathbf{Z} = \mathbf{z} \mid \Delta) \quad (3.14.62)$$

where $\mathbf{z}$ denotes a specified value for the rank order. The SPRT for testing H_0 against $H_{a\mathrm{I}}$ or $H_{a\mathrm{II}}$ is given by:

Take one more observation if $B < \Lambda_t < A$; accept H_0 if $L_t \leq B$ and reject H_0 if $L_t \geq A$, $t = 1, 2, \ldots,$

where $0 < B < 1 < A$ are suitable constants (independent of t). Let T denote the stopping variable for the above SPRT.

Lemma 3.14.6 With the above notation and assumptions we have

$$L_t(\Delta, Z) = \begin{cases} 2^t t! \lambda_0{}^m (1-\lambda_0)^n \Delta^n \prod_{i=1}^{t} \{mG_m(W_i) + \Delta n H_n(W_i)\}^{-1} & \text{for Model II,} \\ 2^t t! \{\Delta/(1+\Delta)\}^t \prod_{i=1}^{t} [m\{1 - G_m(W_i)\} + \Delta n\{1 - H_n(W_i)\}]^{-1} & \\ & \text{for Model I.} \end{cases}$$

$$(3.14.63)$$

Proof See Weed *et al.* (1974, Lemma 1).

Proceeding as in the work of Savage and Sethuraman (1966) and extending some of their lemmas to the random sample size situation and also establishing new lemmas, the following basic result can be obtained.

Theorem 3.14.11 Define

$$S_\lambda(\Delta, \lambda_0, G, H) = \ln 2 - 1 + \ln[\Delta(1-\lambda_0)] - \lambda \ln[\Delta(1-\lambda_0)/\lambda_0] - B_\lambda(\Delta, G, H) \quad (3.14.64)$$

where

$$B_\lambda(\Delta, G, H) = \int_0^\infty \ln[\lambda G(v) + \Delta(1-\lambda)H(v)]\, d[\lambda G(v) + (1-\lambda)H(v)]$$
$$= \lambda E_G\{\ln[\lambda G(X) + \Delta(1-\lambda)H(X)]\}$$
$$+ (1-\lambda)E_H\{\ln[\lambda G(Y) + (1-\lambda)H(Y)]\}, \qquad \lambda = F(0). \tag{3.14.65}$$

Let S_λ be not equal to zero, let $0 < \lambda < 1$ and T denote the number of stages before termination of the SPRT under Model II. Then,

(i) $P(T > t) < \rho^t$ for sufficiently large t and some $0 \le \rho < 1$;
(ii) $P(T < \infty) = 1$;
(iii) $E(e^{\theta T}) < \infty$ for θ in some interval $(-\infty, \delta)$, $\delta > 0$.

Proof See Weed *et al.* (1974, Theorem 1).

REMARK 3.14.12 When H_0 or H_{aII} is true, $S_\lambda(\Delta, \lambda_0, G, H) \neq 0$ provided that $\Delta \neq 1$ and/or $\lambda_0 \neq \frac{1}{2}$ and the SPRT terminates with probability one. For example, under H_0, we have $H = G$ and $\lambda = \frac{1}{2}$ and

$$S_{1/2}(\Delta, \lambda_0, G, G) = \tfrac{1}{2}\ln[16\Delta\lambda_0(1-\lambda_0)/(1+\Delta)^2]. \tag{3.14.66}$$

The case $\lambda = 0$ or $\lambda = 1$ can easily be covered. Also, properties of $S_\lambda(\Delta, \lambda_0, G, H)$ are studied.

When λ_0 is not specified, a reasonable procedure would be to replace λ_0 by $\lambda_t = m/t$ in (3.14.63) and base the sequential procedure on the resultant likelihood ratio. This modified rank order SPRT will have the same termination property.

REMARK 3.14.13 In Theorem 3.14.11, the Xs could be dependent on the Ys provided λ_t has the large deviation property given in part (i) of (15) of Weed *et al.* (1974). For instance, m could be a hypergeometric variable. Although continuity of F is assumed in the derivation of $P(\mathbf{Z} = \mathbf{z} \mid \Delta)$ the expressions for $\Lambda_t(\Delta, \mathbf{Z})$ given by (3.14.62) and the conclusion of Theorem 3.14.11 are valid even if F denotes an arbitrary distribution because F can be made continuous by the known continuization process.

It is still an open question whether it is possible to extend and improvise the results of Berk and Savage (1968) to the one-sample situation.

Recently Miller (1970) has proposed a sequential signed rank test for symmetry. Let $X_1, X_2, \ldots$ be a sequence of iid random variables having density $f(x; \theta)$ which is symmetric about θ. We wish to test H_0: $\theta = 0$ against the alternative H_1: $\theta \neq 0$. Notice that the hypothesis regards the location of symmetry and the symmetry is a part of the assumption under both H_0 and H_1.

Thus tests of this hypothesis would be different from the test procedures developed in Section 13.14.4. Let $R_{n,i}$ denote the rank of $|X_i|$ among $|X_1|, \ldots, |X_n|$, and let S denote the sum of the ranks of the positive Xs. That is $S = \sum_1^n R_{n,i} \operatorname{Sign}(X_i)$ where $\operatorname{Sign}(x) = 1$ if $x > 0$ and zero otherwise. Miller's (1970) procedure is as follows: continue sampling as long as

(i) $|S - n(n+1)/4| < z(\alpha, n_0)[n(n+1)(2n+1)/24]^{1/2}$, and
(ii) $n < n_0$.

Stop sampling as soon as (i) or (ii) is violated. If (i) is violated, decide in favor of H_1. If (ii) is violated and not (i), decide in favor of H_0. n_0 and α are selected by the investigator. These determine $z(\alpha, n_0)$ as follows. Let

$$Y_n = [S_n - n(n+1)/4][n(n+1)(2n+1)/24]^{-1/2}, \qquad Y_{n_0} = \max_{1 \le n \le n_0} (|Y_n|).$$

For the rejection boundary on $S_n - n(n+1)/4$ defined by $\pm y(\alpha, n_0)[n(n+1) \times (2n+1)/24]^{-1/2}$, the test will decide in favor of H_1 if and only if $Y_{n_0} \geq y(\alpha, n_0)$. Thus, $y(\alpha, n_0)$ should be the upper α-percentile of the distribution of Y_{n_0}, that is, $P(Y_{n_0} \geq y(\alpha, n_0)) = \alpha$. The behavior of the sequence $\{S_n\}$ for $n = 1, \ldots, n_0$ can be approximated by the Wiener process. In particular, $P\{Y_{n_0} \geq y\}$ can be approximated by the probability that a Wiener process crosses a square root barrier. Miller (1970) has carried out Monte Carlo studies on the distribution of Y_{n_0} for various values of n_0 and has obtained the values of $y(\alpha, n_0)$. These are presented in Table 3.14.1 where K denotes the size of the sample for each n_0.

Table 3.14.1

Values of $y(\alpha, n_0)$

α \ n_0	10	15	20	25	30	40	50	60
.10	2.02	2.16	2.20	2.22	2.28	2.33	2.37	2.39
.05	2.20	2.39	2.40	2.46	2.55	2.55	2.62	2.62
.01	2.55	2.75	2.83	2.91	2.93	3.03	3.03	3.07
$K/10^3$	2	1	3	1	3	2	3	6

The truncation n_0, especially in medical trials, is dictated by limitations on time, money, energy, patients, etc. If S_n denotes the sum of the ranks of the positive Xs among the absolute Xs, then $S_n = S_{n-1} + \sum_1^n \operatorname{Sign}(X_i + X_n)$.

Miller (1970), by means of Monte Carlo studies, tabulates the power and average sample number of his procedure for the translation alternatives of the double exponential distribution and concludes that its power is reasonable and the ASN is less than or equal to n_0 (equals n_0 for large n_0 and small values of the shifts in the location parameter). Miller (1970) also discusses the "inner acceptance boundary."

Recently Groenveld (1971) has proposed a sequential test for the test of symmetry considered by Miller (1970). Groenveld assumes that the density $f(x; \theta) = f(x - \theta)$ where $f(x) > 0$ for all x, $f(x) = f(-x)$ and $f'(x) < 0$ for $x > 0$. In order to test H_0: $\theta = \theta_0$ against H_1: $\theta = \theta_1 > \theta_0$, one can carry out an SPRT by computing $\Lambda_n = \prod_1^n \{f(X_i;\ \theta_1)/f(X_i;\ \theta_0)\}$. Groenveld's (1971) sequential test of the same hypotheses is based on the number of times the ratio $Z = \ln[f(X; \theta_1)/f(X; \theta_0)]$ is positive. Let this statistic be denoted by T_n. It is clear that this ratio is positive if and only if $X > (\theta_0 + \theta_1)/2 = \bar{\theta}$. In this sequential test we are testing two alternative values of a binomial parameter p: H_0: $p = p_0$ against H_1: $p = p_1$ where

$$p_0 = P(X > \bar{\theta} \mid \theta_0) = P(X - \theta_0 > (\theta_1 - \theta_0)/2) = \int_\Delta^\infty f(x)\,dx,$$

and

$$p_1 = P(X > \bar{\theta} \mid \theta_1) = P(X - \theta_1 > (\theta_0 - \theta_1)/2) = \int_{-\Delta}^\infty f(x)\,dx$$

for $\Delta = (\theta_1 - \theta_0)/2$. It is easy to see that $p_0 + p_1 = 1$. It is of interest to compare the efficiency of the sequential sign test (SST) relative to the SPRT. When both the tests have the same error probabilities α and β, then the relative efficiency is $e = E(N)_{\text{SPRT}}/E(N)_{\text{SST}}$. Using Wald's (1947) approximate formula for $E(N)$ given by Eqs. (2.4.2) and (2.4.3) and noting that the random variable Z for the SST is given by $Z = \ln(p_1/p_0)^X[(1-p_1)/(1-p_0)]^{1-X}$ where X takes on the value 1 with probability p_0 and 0 with probability $1 - p_0$. Then, one obtains,

$$e = \{p_0 \ln(p_1/p_0) + (1 - p_0)\ln[(1 - p_0)/(1 - p_1)]\}/E_{\theta_0}(Z)$$

where $Z = \ln[f(X; \theta_1)/f(X; \theta_0)]$. The same value results if θ_0 is replaced by θ_1. Table 3.14.2 gives values of these efficiencies when X is normal or Laplace (double exponential). The parameter Δ is in terms of standard deviations. Both the expressions in the numerator and denominator of e are functions of Δ and the density function $f(x)$. Assuming certain other regularity conditions on $f(x)$, both can be expanded in a Taylor series. The resulting expression for e is

$$e = [-8f^2(0)\,\Delta^2 + O(\Delta^3)]/[-2E\{f'(X)/f(X)\}^2\,\Delta^2 + O(\Delta^3)].$$

Table 3.14.2

Efficiency of SST in Relation to SPRT

2Δ	Normal	Laplace
.1	.634	.978
.2	.635	.959
.4	.634	.927
.5	.632	.914
.6	.630	.902
.8	.624	.880
1.0	.618	.862

Hence

$$\lim_{\Delta\to 0} e = 4f^2(0)/E[(f'/f)^2]$$

which is also the efficiency of the median as an estimator of θ in relation to the best unbiased estimator of θ [see Fisz (1963, Chapter 13)], and is also the asymptotic efficiency of the sign test in relation to the most powerful test for the hypothesis of symmetry if we consider competing tests based on a fixed-sample size [see Hájek and Sidak (1967, Chapter 7)]. Table 3.14.3 gives

Table 3.14.3

Limiting Efficiencies of SST

Normal	Laplace	Logistic	Cauchy
$2/\pi = .637$	1	.75	$8/\pi^2 = .810$

the limiting value for several distributions. One would consider this SST of Groenveld (1971), because the form of $f(x)$ need not be known in order to calculate the statistic T_n. The value p_0 does depend upon the distribution of X. However, the test can be carried out by choosing a value for p_0 (say .4) and the values of A and B (Wald's approximate boundary values) so that the sequential binomial test has errors α and β. If Δ is measured in standard deviation units, then by the improved Chebyshev's inequality for continuous unimodal distributions [see Cramér (1946, p. 183)], $P(X - \theta_0 > \Delta) \le 2/9\,\Delta^2$. Hence, if we use $p_0 = .4$, $\Delta^2 \le 5/9$ and $\Delta \le .745$. That is, the SST corresponds to a SPRT with error sizes (α, β) in which θ_1 (θ_0) differs in absolute value from $\bar{\theta}$ by at most .745 standard deviations, whatever the distribution of X, if σ^2 is finite.

3.14.5 *c*-Sample Rank Order SPRTs

Govindarajulu (1969) has extended Savage and Sethuraman's (1966) work to the c-sample case and studied its asymptotic normality. In the following we shall present the basic result. We need the following notation. Let $(X_{1i}, \ldots, X_{ci})$, $i = 1, 2, \ldots, n$ be iid c-variate random variables with a joint distribution H having marginals $F_1, F_2, \ldots, F_c$ which do not have common discontinuities. We wish to test H_0: $X_1, X_2, \ldots, X_c$ are independent and $F_1 = F_2 = \cdots = F_c(x)$, for all x, against the alternative hypothesis H_1: $X_1, \ldots, X_c$ are independent and $F_i(x) = (F^*)^{\theta i}$, F^* is continuous, $\theta_i > 0$ is known $(i = 1, \ldots, c)$ and at least two θs are different. Let $W_1, \ldots, W_{cn}$ denote the combined sample and $W_{n1}, \ldots, W_{n,cn}$ denote the combined ordered sample. Let $F_{i,n}(\cdot)$ denote the empirical distribution of $(X_{i1}, \ldots, X_{i,n})$ $(i = 1, \ldots, c)$. In the following we shall define the c-sample rank order.

Definition 3.14.4 The c-sample rank order is given by $\mathbf{Z} = (Z_1, \ldots, Z_{cn})$ where $Z_i = j$ when $W_i = X_{jk}$ for some $1 \leq k \leq n$. Then one can easily obtain the probability of $\mathbf{Z}$ under H_1.

Lemma 3.14.7 We have

$$P(\mathbf{Z} = \mathbf{z} \mid H_1) = (n!)^c \prod_{i=1}^{c} \theta_i^{\,n} / n^{cn} \prod_{i=1}^{cn} \left\{ \sum_{k=1}^{n} \theta_k F_{k,n}(W_i) \right\}.$$

Proof See Govindarajulu (1969, Lemma 2.1). Hence,

$$\Lambda_n(\boldsymbol{\theta}, \mathbf{F}_n) = P(\mathbf{Z} = \mathbf{z} \mid H_1) / P(\mathbf{Z} = \mathbf{z} \mid H_0)$$

$$= (cn)! \prod_{1}^{c} \theta_i^{\,n} / n^{cn} \prod_{i=1}^{cn} \left[\sum_{k=1}^{c} \theta_k F_{k,n}(W_i) \right]. \tag{3.14.67}$$

So, we define the c-sample rank order SPRT procedure as follows. If the experiment has proceeded up to the nth stage, that is $(X_{1i}, \ldots, X_{ci})$ have been observed, for $i = 1, \ldots, n$, take one more observation on each of the c variables if $B < \Lambda_n(\boldsymbol{\theta}, \mathbf{F}_n) < A$; accept H_0 if $\Lambda_n \leq B$, and reject H_0 if $\Lambda_n \geq A$, where $0 < B < 1 < A < \infty$ are suitable constants. Let N denote the stopping variable of the above SPRT, let

$$S(\boldsymbol{\theta}, \mathbf{F}) = c \ln c - c + \sum_{i=1}^{c} \ln \theta_i - \int \ln \left[\sum_{k=1}^{c} \theta_k F_k(x) \right] d\left(\sum_{1}^{c} F_k(x) \right). \tag{3.14.68}$$

Then we have the following basic result.

Theorem 3.14.12 If $S(\boldsymbol{\theta}, \mathbf{F}) \neq 0$, where $S(\boldsymbol{\theta}, \mathbf{F})$ is given by (3.14.67) then

(i) $P(N > n) < \rho^n$, for sufficiently large n and some $0 < \rho < 1$,
(ii) $P(N < \infty) = 1$,
(iii) $E[\exp(tN)] < \infty$ for t in some interval $(-\infty, \tau)$, $\tau > 0$.

Proof See Govindarajulu (1969, Theorem 3.1).

Certain properties of $S(\boldsymbol{\theta}, \mathbf{F})$ are also studied and the asymptotic normality of $S_n = n^{-1} \ln \Lambda_n$ has been studied. The latter is summarized in the following result.

Theorem 3.14.13 Let $\lambda_j = \theta_j / \sum_1^c \theta_i$. If $X_1, \ldots, X_c$ are mutually independent and $0 < \lambda_j$ for some j, then

$$\lim_{n \to \infty} \sup_z |P[n^{1/2}(S_n - S(\boldsymbol{\theta}, \mathbf{F}))/\sigma_n \leq z] - \Phi(z)| = 0 \tag{3.14.69}$$

for all $F_1, \ldots, F_c$ where

$$\begin{aligned}
\sigma_n^{\,2} &= 2 \iint_{x<y} [(1 - H(y))/H(y)]\, d\left(\sum_1^c F_k(x)\right) d\left(\sum_1^c F_k(y)\right) \\
&\quad + 2 \iint_{x<y} \left\{\left[\sum_{k=1}^c F_k(x)(1 - F_k(y))\right] \Big/ H(x)H(y)\right\} dH(x)\, dH(y) \\
&\quad - 2 \sum_{k=1}^c \lambda_k \iint_{x<y} \{F_k(x)[1 - F_k(y)]/H(x)H(y)\}\, d\left(\sum_1^c F_k(x)\right) dH(y) \\
&\quad - 2 \sum_{k=1}^c \lambda_k \iint_{x<y} \{F_k(x)[1 - F_k(y)]/H(x)H(y)\}\, dH(y)\, d\left(\sum_1^c F_k(y)\right),
\end{aligned}$$

where

$$H(x) = \sum_{k=1}^c \lambda_k F_k(x).$$

Proof The result follows as a special case of a general theorem of Govindarajulu (1969, Theorem 4.1).

REMARK 3.14.14 Analogously one can extend Theorems 3.14.6 and 3.14.7 (Sethuraman, 1970), Theorem 3.14.10 (Berk and Savage, 1968) to the c-sample case in a reasonably straightforward fashion. The details are omitted.

3.14.6 Rank Order SPRTs for Randomness against Trend

Let $X_1, X_2, \ldots$ be an independent sequence of random variables and X_i has df $F_i(x)$, $i = 1, 2, \ldots$. We wish to test the null hypothesis H_0: $F_1(x) = F_2 = \cdots = F_c$ for all x against the alternative hypothesis H_1: $F_i(x) = (F^*)^{\theta_i}$, $i = 1, 2, \ldots, n$, where $0 < \theta_1 \le \theta_2 \cdots \le \theta_n$ (that is, $F_1 \le F_2 \le \cdots \le F_n$). One can analogously consider the other alternative $\theta_1 > \theta_2 \ge \cdots \ge \theta_n \ge 0$. Savage (1957) has defined the rank order $R = (R_1, \ldots, R_n)$ where R_i = number of X_j that are less than or equal to X_i in the sample ($i = 1, \ldots, n$). We shall define a different rank order which is consistent with the c-sample rank order defined in Section 3.14.5. Let $(W_{n1}, \ldots, W_{nn})$ denote the ordered Xs, let $\mathbf{Z} = (Z_1, \ldots, Z_n)$ be such that $Z_i = j$ when $W_i = X_j$. Then

$$P(\mathbf{Z} = \mathbf{z} | H_1) = \prod_1^n \theta_i \Big/ \prod_{i=1}^n \left[\sum_{k=1}^n \theta_k F_{k,1}(W_i)\right] \tag{3.14.70}$$

and

$$\Lambda_n(\boldsymbol{\theta}, \mathbf{Z}) = n! \prod_{i=1}^n \theta_i \Big/ \left[\sum_1^n \theta_k F_{k,1}(W_i)\right] \tag{3.14.71}$$

where $F_{k,1}(x)$ is the empirical df based on a single observation from $F_k(x)$.

Notice that (3.14.70) follows by interchanging the roles of c and n and setting $n = 1$ in (3.14.66). Now, one can define the SPRT and the stopping time on the basis of $\Lambda_n(\boldsymbol{\theta}, \mathbf{Z})$. In order to show that the resultant SPRT terminates finitely with probability one, let

$$S_n = n^{-1} \ln \Lambda_n = \ln n - 1 + n^{-1} \sum_1^n \ln \theta_i - T_n + O(\ln n/n) \tag{3.14.72}$$

where

$$T_n = n^{-1} \sum_{i=1}^n \ln \left[\sum_{k=1}^n \theta_k F_{k,1}(W_i)\right]. \tag{3.14.73}$$

Let $T_n = T_{n,1} + T_{n,2}$ where

$$T_{n,1} = n^{-1} \sum_{i=1}^n \ln \left[\sum_{k=1}^n \theta_k F_k(X_i)\right] \tag{3.14.74}$$

and

$$T_{n,2} = n^{-1} \sum_{i=1}^n \ln \left\{\left[\sum_{k=1}^n \theta_k F_{k,1}(X_i)\right] \Big/ \left[\sum_{k=1}^n \theta_k F_k(X_i)\right]\right\}. \tag{3.14.75}$$

So we wish to assert that, for every $\varepsilon > 0$, there exists a $0 < \rho_1 < 1$ such that

$$P(|T_{n,1} - T(\boldsymbol{\theta}, \mathbf{F})| \geq \varepsilon) < \rho_1^n \qquad \text{for } n \text{ sufficiently large} \quad (3.14.76)$$

where

$$T(\underline{\theta}, \underline{F}) = n^{-1} \int \ln\left[\sum_1^n \theta_k F_k(x)\right] d\left[\sum_1^n dF_i(x)\right] \qquad (3.14.77)$$

and

$$P[|T_{n,2}| \geq \varepsilon] \leq \rho_2{}^n, \qquad 0 < \rho_2 < 1.$$

3.15 Locally Most Powerful (LMP) Sequential Tests

3.15.1 Introduction

Let $f(x; \theta)$ denote a family of probability densities [with respect to a sigma finite measure on the sample space $(\mathscr{X}, \mathscr{A})$.] Let $\theta \in \Theta$ the parameter space, a subinterval of the real line. Also, let $X, X_1, \ldots$ denote an iid sequence of random variables drawn from $f(x; \theta)$. Based on the data we wish to test the hypothesis $H_0: \theta = \theta^*$ against the one-sided alternative $H_1: \theta > \theta^*$. It is well known that there does not exist a "best" sequential test of the hypotheses, in general. Even among nonsequential tests, uniformly most powerful tests do not necessarily exist. However, under certain regularity conditions, there are LMP nonsequential tests which (for a sample of size n) are of the form: rej H_0 if

$$S_n = \sum_1^n r(X_i) \geq a \qquad (3.15.1)$$

where $r(x) = \partial \ln f(x; \theta)/\partial\theta |_{\theta=\theta^*}$.

Berk (1972) has shown that a similar phenomenon persists in the sequential case by delineating a certain class of sequential tests that are LMP for testing H_0 versus H_1. That is, given a test in the class, let α denote the exact level of significance and $v < \infty$ its expected stopping time under θ^*. Then among all level α sequential tests whose expected stopping times under θ^* do not exceed v, the given test is LMP. These LMP tests constitute a sequential analog of (3.15.1); their stopping times are of the form

$$N = \inf\{n: S_n \notin (b, a)\}. \qquad (3.15.2)$$

Typically, $-b$ and a are both positive. The terminal decision is, as one would expect: rej H_0 when $S_N \geq a$. For short we shall refer to the sequential test as N and denote by $\mathscr{S}$ the class of such tests. The boundary points a and b are determined (in principle) by specifying α and v. That is,

$$\text{(i)} \quad P(S_N \geq a) = \alpha \qquad \text{(ii)} \quad EN = v \tag{3.15.3}$$

where the probabilities and expectations, unless otherwise specified are computed under θ^*. Since such sequential tests are of the form of Wald's SPRT, the various inequalities and approximations that are available for the SPRT apply for the LMP sequential test. Further, Berk (1972) shows that an LMP sequential test is a Wald SPRT of H_0': $\theta = \theta_0$ versus H_1': $\theta = \theta_1$ where $\theta_0 < \theta^* < \theta_1$ when $f(x;\,\theta) = \exp\{\theta x - b(\theta)\}$ [that is, $f(x;\,\theta)$ is the one-parameter exponential model density]. Berk's (1972) work solves a problem posed by Abraham (1969) who showed that sequential tests of the form (3.15.2) are LMP for the symmetric binomial case. However, he left open the question of whether such tests are LMP more generally, to which Berk's answer is affirmative. Berk confines himself to the class of tests for which the expected stopping times are finite. Of course this class excludes from consideration the sequential tests of H_0: $\mu \leq 0$ (where μ denotes the normal mean) proposed by Darling and Robbins (1968a) [see also Robbins (1970)] that have power 1 and have infinite expected stopping time under $\mu \leq 0$. Berk (1972) assumes certain regularity conditions on f which enable the sequential tests to have power curves differentiable at θ^*. Thus, a level α LMP test has a power curve with maximum slope at θ^*.

3.15.2 LMP Sequential Tests

A sequential test of H_0 versus H_1 is given by a stopping time t defined on $X_1, X_2, \ldots$ together with terminal decision rules $\{\phi_n\}$. Here, for each n, ϕ_n is a possibly randomized (measurable) test function depending on $X_1, \ldots, X_n$. As mentioned earlier, we restrict attention to stopping times satisfying $Et < \infty$. For a given sequential test, its level and power at θ are respectively given by

$$\alpha = E\phi_t, \tag{3.15.4}$$

$$1 - \pi(\theta) = E_\theta \phi_t I_{\{t<\infty\}} = \sum_1^\infty \int_{\{t=n\}} \phi_n f_{n,\theta} \tag{3.15.5}$$

where $f_{n,\theta}$ denotes the joint density of $X_1, \ldots, X_n$ under θ. The expression in (3.15.5) allows the possibility of t being infinite when $\theta \neq \theta^*$. In so doing, we interpret sampling for every $t = \infty$ as neither accepting or rejecting. (We get the same result if $t = \infty$ is included in the acceptance region for H_0.) Assuming differentiability, the local slope (the slope of the power curve at θ^*) is

$$m = \partial E_\theta \phi_t I_{\{t<\infty\}}/\partial_\theta |_{\theta=\theta^*}. \tag{3.15.6}$$

We refer to (α, m, Et) as the lower characteristic (LC) of the sequential test. We impose the following regularity conditions on f which, as shown by Abraham (1969) are sufficient to ensure that the derivative in (3.15.6) exists. Let

$$I(\theta; \theta^*) = E \ln\{f(X; \theta^*)/f(X; \theta)\}. \tag{3.15.7}$$

Assumption 1 $\lim \sup_{\theta\to\theta^*}\{I(\theta; \theta^*)/(\theta - \theta^*)^2\} < \infty$.

Assumption 2 $f(X; \theta)$ is wpl differentiable at $\theta = \theta^*$.

Assumption 3 The power function of every nonsequential test of H_0 versus H_1 is differentiable at θ^* under the integral sign.

Assumption 4 Let $0 < Er^2(X) < \infty$, where

$$r(X) = \partial \ln f(X; \theta)/\partial\theta |_{\theta=\theta^*}. \tag{3.15.8}$$

Various conditions that ensure assumption 3 are available in the literature. Two that suffice are: (i) For each θ in a bounded neighborhood U of θ^*, $f(X; \cdot)$ has, wpl, a continuous derivative at θ. (ii) $\sup\{E_\theta | \partial f(X; \theta)/\partial\theta |, \theta \in U\} < \infty$. Notice that condition (ii) is satisfied if $E_U\{\sup_{\theta\in U} | \partial f(X; \theta)/\partial\theta |\} < \infty$, or, more generally, if $\{\partial f(X; \theta)/\partial\theta, \theta \in U\}$ is uniformly integrable. We note that assumptions 3 and 4 imply

$$Er(X) = 0. \tag{3.15.9}$$

Then, we have the following result of Abraham (1969).

Lemma 3.15.1 (Abraham, 1969) Let assumptions 1–3 hold. Then any sequential test $(t, \{\phi_n\})$ with $Et < \infty$ has a power curve differentiable at θ^*. Furthermore, the local slope is given by

$$m = E\phi_t S_t \tag{3.15.10}$$

where

$$S_n = \sum_1^n r(X_i) \tag{3.15.11}$$

Notice that (3.15.10) is obtained formally from (3.15.6) by differentiating underneath the expectation sign. Since we will be concerned with tests for which $Et < \infty$, we assume in the sequel, that (3.15.10) holds. Now, we shall state the main result of Berk (1972).

Theorem 3.15.1 (Berk, 1972) Let assumptions 1–4 hold. Consider a sequential test N in $\mathscr{S}$ and let (α, m, EN) be its local characteristic (LC). If another sequential test of H_0 versus H_1 has LC $(\alpha, \hat{m}, Et)$, then $Et \leq EN$ entails $\hat{m} \leq m$. Moreover, if the first inequality is strict, so is the second. Thus N is LMP among all sequential tests of level $\hat{\alpha} \leq \alpha$ and for which $Et \leq EN$.

Proof The proof is quite long and one has to resort to techniques employed to establish the optimality of the Wald SPRT. Berk (1972) reduces the problem to that of solving a certain auxiliary optimal stopping problem. The reader is referred to Berk (1972) for the details. As a by-product, we note that an LMP sequential test [given by (3.15.2)] is like a SPRT (in that N is the first time a random walk leaves an interval).

Notice that assumptions 1–4 imply $P(r(X) = 0) < 1$ which in turn implies (see Theorem 2.2.1) that the moment-generating function of N is finite in a neighborhood of the origin. In particular, $EN < \infty$.

3.15.3 Exponential Models

We have noted that an LMP sequential test is like a SPRT. For one-parameter exponential models, the LMP sequential tests are SPRTs. To show this, let us parametrize the model so that $f(x;\, \theta) = \exp\{\theta x - b(\theta)\}$ and let Θ denote the natural parameter space:

$$\Theta = \{\theta\colon \text{the integral of } e^{\theta x} \text{ with respect to the sigma finite measure is finite}\}$$

Θ is necessarily an interval (which, to avoid trivialities, we assume to be nondegenerate). For a given θ^* in the interior of Θ, $r(x) = x - b'(\theta^*)$, where $b'(\theta) = db(\theta)/d\theta$. [It is well known (for example, see Section 3.7) that $b(\theta)$ is infinitely differentiable (in the interior of Θ) and strictly convex (unless the sigma finite measure is degenerate, a case which we exclude from consideration). Thus $b'(\theta)$ is strictly increasing. Also note that $b'(\theta) = E_\theta X$ which implies that $E_\theta r(X) = 0$.] As these exponential models have a monotone likelihood ratio, it follows in the usual way [see Lehmann (1959, p. 101)] that a level α LMP sequential test of $H_0\colon \theta = \theta^*$ versus $H_1\colon \theta > \theta^*$ is also level α LMP for testing $H_0\colon \theta \leq \theta^*$ against $H_1\colon \theta > \theta^*$. Given an LMP sequential test of $H_1\colon \theta \leq \theta^*$ versus $H_1\colon \theta > \theta^*$, Berk (1972) shows that there is a continuum of parameter pairs (θ_0, θ_1) in Θ for which a SPRT of $H_0'\colon \theta = \theta_0$ versus $H_1\colon \theta = \theta_1$ coincides with the given LMP test.

It is easy to verify that for $\theta_0 < \theta_1$ in Θ a SPRT of H_0' versus H_1' stops the first time $n \sum_1^n X_i - n[b(\theta_1) - b(\theta_0)]/(\theta_1 - \theta_0)$ leaves a certain interval. The LMP sequential test of H_0 versus H_1 stops the first time $\sum_1^n X_i - nb'(\theta^*)$ leaves a certain interval. The desired result can be established by showing that there exists a continuum of pairs (θ_0, θ_1) of points in Θ for which

$$[b(\theta_1) - b(\theta_0)]/(\theta_1 - \theta_0) = b'(\theta^*), \tag{3.15.12}$$

or equivalently

$$b'(\theta^*) = (\theta_1 - \theta_0)^{-1} \int_{\theta_0}^{\theta_1} b'(\theta)\, d\theta. \tag{3.15.13}$$

Since $b'(\theta)$ is strictly increasing, for any pair (θ_0, θ_1) satisfying (3.15.13) we must also have $\theta_0 < \theta^* < \theta_1$. Let $\Delta = \min(\theta^* - \theta_0, \theta_1 - \theta^*)$; $0 < \Delta \leq \infty$. We show that for any $\delta \in (0, \Delta)$, there is a pair (θ_0, θ_1) in Θ satisfying (3.15.13), with $\theta_1 - \theta_0 = \delta$. For given $\delta \in (0, \Delta)$, let

$$c_\delta(\theta) = \delta^{-1} \int_\theta^{\theta+\delta} b'(u)\, du \tag{3.15.14}$$

(whenever both θ and $\theta + \delta$ are in Θ). Clearly $c_\delta(\theta)$ is continuous and, as $b'(\theta)$ is strictly increasing, so is $c_\delta(\theta)$. Moreover, since $\theta^* - \delta \in \Theta$ it can be seen from (3.15.14) that $c_\delta(\theta^* - \delta) < b'(\theta^*) < c_\delta(\theta^*)$. By the continuity and strict monotonicity of $c_\delta(\theta)$, it follows that there is a unique $\theta_0 = \theta(\delta) \in (\theta^* - \delta, \theta^*)$ for which

$$c_\delta(\theta_0) = b'(\theta^*). \tag{3.15.15}$$

This last relation is precisely (3.15.13) with $\theta_1 = \theta_1(\delta) = \theta_0(\delta) = \delta$. Thus the existence of the desired continuum of pairs is established.

Further, it is easily seen from (3.15.14) that for fixed θ, $c_\delta(\theta)$ is a continuous strictly increasing function of δ. Thus, for fixed v, the inverse function $c_\delta^{-1}(v)$, is continuous and strictly decreasing in δ. As $\theta_0(\delta) = c_\delta^{-1}(b'(\theta^*))$, it follows that $\theta_0(\delta)$ is continuous and strictly increasing in δ. Analogously, $\theta_1(\delta)$ is continuous and strictly increasing in δ. Thus to each $\delta \in (0, \Delta)$ corresponds a unique pair (θ_0, θ_1) satisfying (4.15.12) with $\theta_1 = \theta_0 = \delta$. Moreover, the intervals $(\theta_0(\delta), \theta_1(\delta))$ are nested: If $0 < \delta < \delta' < \Delta$,

$$\theta_0(\delta') < \theta_0(\delta) < \theta^* < \theta_1(\delta) < \theta_1(\delta'). \tag{3.15.16}$$

Further, it follows that $\lim_{\delta \to \infty} |\theta_i(\delta)| = \infty$, $i = 0, 1$.

Lechner (1964) observed the above phenomena in a somewhat different context and with a different goal. His aim was to show that different (θ_0, θ_1) pairs, called by him "conjugate pairs," all lead to the same (given) SPRT (provided the error rates associated with each pair are chosen properly) and to determine these associated error rates in terms of the original specifications defining the SPRT. (Thus, Lechner used the existence of "conjugate pairs" as a device for calculating the power curve of the SPRT.) In Lechner's terminology, θ^* is the exceptional point of the SPRT, the parameter value under which the random walk defining the SPRT has zero drift. The observation that a LMP sequential test is a SPRT thus has a dual formulation: For one-parameter exponential models, a SPRT of two simple (or one-sided) hypotheses is an LMP sequential test for its exceptional point. Also (3.15.13) can be interpreted as the defining relation for conjugate pairs, given the exceptional point of the SPRT.

3.15.4 Asymptotic Results

It is of interest to compare the behavior of a given LMP sequential test N with the corresponding LMP (or UMP) nonsequential test, which has the same level as N and sample size $n = EN$. Berk (1972) makes an asymptotic comparison by letting a and $|b|$ become infinite in such a way that

$$\lim_c \{b/(b-a)\} = \alpha \in (0, 1).$$

Here "$\lim_c$" denotes $-b$ and a tending to infinity in the indicated manner.

For a LMP sequential test N, Wald's no-overshoot approximations give

$$1 - \pi(\theta^*) \doteq b/(b-a) \tag{3.15.17}$$

and

$$EN \doteq -ab/E\{r^2(X)\}. \tag{3.15.18}$$

The following are asymptotically exact [see Berk (1973) or Section 3.11]:

$$\lim_c \{1 - \pi(\theta^*)\} = \gamma \tag{3.15.19}$$

and

$$\lim_c Er^2(X)EN/(-ab) = 1. \tag{3.15.20}$$

Typically, both the sequential and nonsequential LMP tests are consistent; the limiting values in either case are typically 0, α, or 1 according as $\theta < \theta^*$, $\theta = \theta^*$, or $\theta > \theta^*$. The asymptotic relative efficiency (ARE) of the LMP sequential test at θ is $\lim_c n/E_\theta N = \lim_c \{EN/E_\theta N\}$. If $\mu(\theta) = E_\theta r(X)$, then Berk (1973) has shown (see Section 3.11) that

$$E_\theta N \sim \begin{cases} a/\mu(\theta) & \text{if} \quad \mu(\theta) > 0, \\ b/\mu(\theta) & \text{if} \quad \mu(\theta) < 0. \end{cases} \tag{3.15.21}$$

It follows from (3.15.20) and (3.15.21) that

$$E_\theta N \sim o(EN) \tag{3.15.22}$$

provided $\mu(\theta) \neq 0$. [To simplify the exposition, let us assume that $\mu(\theta) \neq 0 \Leftrightarrow \theta \neq \theta^*$, a condition that is met in many models, including the one-parameter exponential models.] Hence

$$\lim_c \{E_\theta N/EN\} = \begin{cases} 1, & \theta = \theta^*, \\ 0, & \theta \neq \theta^*. \end{cases} \tag{3.15.23}$$

In other words, the ARE of a nonsequential LMP test in comparison to the corresponding LMP sequential test is zero for all $\theta \neq \theta^*$. This result differs markedly from the asymptotic comparisons of SPRTs and corresponding nonsequential tests (see Section 3.11). Further, one can interpret (3.15.22) as follows: the ASN a tthe exceptional point is of a larger order (asymptotically) than n, so the ARE of the SPRT is zero at θ^*. The difference in the asymptotic AREs can be attributed to the way one selects a "corresponding" nonsequential test in the two situations. For the SPRT, the conjugate parameter pair is usually held fixed, so the associated error rates tend to zero. This is in contrast to (3.15.19).

The rosy picture presented by (3.15.23) is somewhat offset by another aspect of the asymptotic behavior of the ASN function. For LMP sequential tests, in general, $\sup_\theta \{E_\theta N/EN\} > 1$. (An exceptional case is when $\gamma = \frac{1}{2}$ and $r(X)$ is symmetric under θ^*.) Further, for small γ, the maximum ASN can be appreciably larger than EN, (even asymptotically) despite (3.15.23). These considerations are discussed in the following example.

3.15.5 Example (Symmetric Binomial Case)

Let us test $H_0 : \theta \leq \frac{1}{2}$ against $H_1 : \theta > \frac{1}{2}$. Here $r(X) = 2X - 1$, which assumes values ± 1 and is symmetric when $\theta = \theta^* = \frac{1}{2}$. Thus, in this case, N is the stopping time associated with the classical gambler's ruin problem. Also, without loss of generality, we can take b and a to be integers. We then have the following exact formulas [see, for example, Feller (1966)]:

$$1 - \pi(\theta) = [1 - \{(1-\theta)/\theta\}^{-b}]/[1 - \{(1-\theta)/\theta\}^{a-b}] \tag{3.15.24}$$

and

$$E_\theta N = (a - b)[1 - \pi(\theta) - \alpha]/(2\theta - 1). \tag{3.15.25}$$

Here α is the level of the test and is given by

$$\alpha = b/(b - a) \tag{3.15.26}$$

and

$$EN = -ab. \tag{3.15.27}$$

[Equations (3.15.26) and (3.15.27) can be formally obtained from (3.15.24) and (3.15.25) by continuity at $\theta = \frac{1}{2}$.] Letting $b - a = k$ and for $\theta \neq \frac{1}{2}$, $\psi = \frac{1}{2}\ln(\theta/(1-\theta))$, we have

$$1 - \pi(\theta) = \begin{cases} (1 - e^{-2\psi k\alpha})/(1 - e^{-2\psi k}), & \theta \neq \frac{1}{2}, \\ \alpha, & \theta = \frac{1}{2}, \end{cases} \tag{3.15.28}$$

and

$$E_\theta N = \begin{cases} k[1 - \pi(\theta) - \alpha]/(2\theta - 1), & \theta \neq \frac{1}{2}, \\ k^2\alpha(1-\alpha), & \theta = \frac{1}{2}. \end{cases} \tag{3.15.29}$$

Berk (1972) also shows that

$$\rho(\alpha) = \lim_c \sup_\theta E_\theta N/EN$$

exists and is equal to $\sup_\mu E_\mu \tau/E_0 \tau$ where τ denotes a standard Wiener process with drift μ per unit time and

$$E_\mu \tau = \begin{cases} [1 - \pi_w(\mu) - \alpha]/\mu, & \mu \neq 0, \\ \alpha(1-\alpha), & \mu = 0. \end{cases} \tag{3.15.30}$$

$1 - \pi_w(\mu)$ denotes the probability of the Wiener process hitting $1 - \alpha$ before hitting $-\alpha$ and

$$1 - \pi_w(\mu) = \begin{cases} (1 - e^{-2\alpha\mu})/(1 - e^{-2\mu}), & \mu \neq 0, \\ \alpha, & \mu = 0. \end{cases} \tag{3.15.31}$$

Table 3.15.1 shows some values of $\rho(\alpha)$, together with μ_α, the value of μ for which $E_\mu \tau$ attains its maximum. As suggested by the table, $\rho(\alpha)$ increases

Table 3.15.1

α	.50	.25	.10	.05	.025	.01	.005	.001
μ_α	0	1.12	2.39	3.55	4.8	7.4	10.3	22.7
$\rho(\alpha)$	1	1.09	1.32	1.48	1.61	1.74	1.81	1.91

from 1 as α decreases from $\frac{1}{2}$. Berk (1972) proves that $\lim_{\alpha \to 0} \rho(\alpha) = 2$ and obtains when k is large,

$$\theta_\alpha = \tfrac{1}{2} + \mu_\alpha/2k + o(1/k)$$

where θ_α is the value that maximizes $E_\theta N$.

CHAPTER 4

Sequential Estimation

4.1 Introductory Remarks

In some problems hypothesis-testing seems to be somewhat artificial. For some examples of this, the reader is referred to Wetherill (1966). In fixed-sample size situation there is a close connection between acceptance regions and confidence regions, whereas, such an analogy does not hold in the sequential situation—hence, the need for a theory of sequential estimation. The stopping rules employed in sequential testing may not be meaningful in sequential estimation.

4.2 Certain Two-Stage Procedures

4.2.1 Stein's Two-Stage Procedure for Estimating the Mean of a Normal Distribution with Unknown Variance

Dantzig (1940) has shown that, for a sample of fixed size, there does not exist a test for Student's hypothesis whose power is independent of the variance. Stein (1945) has presented a two-sample test with this property, the size of the second sample depending upon the result of the first sample. The problem of determining confidence intervals of preassigned length and confidence coefficient for the mean of a normal population with unknown variance is solved by the same procedure. In order to make either the power of the test or the length of the confidence interval free of the variance, it seems necessary to waste a small portion of the information contained in the sample. Thus, in practical applications one would not use a procedure with this property, but rather a test that is uniformly most powerful or an interval of the same length, whose confidence coefficient is a function of σ, but always greater than the desired value, the difference being slight, at the same time reducing the expected number of observations by a small amount. The two-sample procedures will be a special case of Wald's sequential analysis. It is further shown by Stein (1945) that if the variance and initial sample size are sufficiently large, the expected number of observations differ only slightly from the number of observations required by a single-sample test (or interval estimation) when the variance is known.

Let $X_i (i = 1, 2, \ldots)$ be independent normal variables having mean θ and variance σ^2 (unknown). We wish to test $H_0: \theta = \theta_0$ so that the power of the test depends only on $\theta - \theta_0$ and not upon σ^2, or to estimate θ by a confidence interval of specified length $2d$ and specified confidence coefficient $1 - \alpha$. Take a sample of size n_0 observations $X_1, \ldots, X_{n_0}$ and compute the sample variance given by

$$s^2 = (n_0 - 1)^{-1}\left(\sum_{i=1}^{n_0} X_i^2 - n_0^{-1}\left(\sum_1^{n_0} X_i\right)^2\right). \tag{4.2.1}$$

Then, define n by

$$n = \max\left\{\left[\frac{s^2}{z}\right] + 1, n_0 + 1\right\} \tag{4.2.2}$$

where z is a previously specified positive constant and $[\cdot]$ here denotes the largest integer less than $(\cdot)$. Now, take additional observations $X_{n_0+1}, \ldots, X_n$. Also, choose real numbers $a_i (i = 1, \ldots, n)$ such that

$$\text{(i)} \quad \sum_1^n a_i = 1, \qquad a_1 = a_2 = \cdots = a_{n_0}$$

and

$$\text{(ii)} \quad s^2 \sum_1^n a_i^2 = z.$$

This is clearly possible since

$$\min \sum_1^n a_i^2 = 1/n \leq z/s^2 \tag{4.2.3}$$

by (4.2.2), the minimum being taken subject to condition (i). Then define the statistic T' by

$$T' = \frac{\sum_1^n a_i X_i - \theta_0}{z^{1/2}} = \frac{\sum_1^n a_i (X_i - \theta)}{z^{1/2}} + \frac{(\theta - \theta_0)}{z^{1/2}}$$

$$= U + (\theta - \theta_0)/z^{1/2} \qquad \text{(say)}. \tag{4.2.4}$$

Now $\sum_1^{no} a_i X_i$ $(a_1 = a_2 = \cdots a_{n_0})$ is independent of s^2 because of normal samples. Writing

$$U = \frac{\sum a_i (X_i - \theta)/(\sum a_i^2)^{1/2} \sigma}{z^{1/2}/(\sum a_i^2)^{1/2} \sigma},$$

we notice that U is distributed as Students t with $n_0 - 1$ degrees of freedom (see Section 1.4). So the test based on T' is unbiased (because of symmetry considerations) and its power is free of σ. A confidence interval for θ of specified length $2d$ and confidence coefficient $1 - \alpha$ is then given by

$$\left(\sum a_i X_i - d < \theta < \sum_1^n a_i X_i + d\right) \tag{4.2.5}$$

where

$$d = z^{1/2} t_{no-1,1-\alpha/2} \tag{4.2.6}$$

$t_{no-1,1-\alpha/2} = 100(1 - \alpha/2)$ percentage point of the t distribution with $n_0 - 1$ degrees of freedom. Stein (1945) also gave the exact distribution of the sample size and derived the following bounds for the same:

$$(n_0 + 1)P(\chi^2_{no-1} < y) + \frac{\sigma^2}{z} P(\chi^2_{no+1} > y)$$

$$< E(n)$$

$$< (n_0 + 1)P(\chi^2_{no-1} < y) + \frac{\sigma^2}{z} P(\chi^2_{no+1} > y) + P(\chi^2_{no-1} > y) \tag{4.2.7}$$

where

$$y = (n_0{}^2 - 1)z/\sigma^2 \tag{4.2.8}$$

and $\chi^2_{n_0-1}$ denotes a chi-squared variable with $n_0 - 1$ degrees of freedom. Consequently, $E(n)$ is a function of σ^2 and can be evaluated from tables of the incomplete gamma function.

As mentioned earlier, these tests and confidence intervals will not be used exactly in this form since they waste information in order to make the power of the test or the length of the confidence interval strictly free of the variance. Instead of (4.2.2) we take a total of

$$n = \max\left\{\left[\frac{s^2}{z}\right] + 1, n_0\right\} \tag{4.2.9}$$

observations and define

$$T'' = \left(n^{-1}\sum_1^n X_i - \theta_0\right)n^{1/2}/s = U' + (\theta - \theta_0)n^{1/2}/s \tag{4.2.10}$$

where U' has the t distribution with $n_0 - 1$ degrees of freedom. By (4.2.2), $n > s^2/z$, so that although $|(\theta - \theta_0)n^{1/2}/s|$ is random,

$$|(\theta - \theta_0)n^{1/2}/s| \geq |\theta - \theta_0|/z^{1/2}. \tag{4.2.11}$$

Hence the power of the test based on T'' is larger than the power of the test based on T'. Also $E(n)$ will be reduced from that in (4.2.7) by $P(X^2_{n_0-1} < y)$. Similarly, if z is defined by (4.2.6) the interval

$$\left(n^{-1}\sum_1^n X_i - d, n^{-1}\sum_1^n X_i + d\right) \tag{4.2.12}$$

has length $2d$ and the probability that it covers the true mean θ is a function of σ but is always greater than $1 - \alpha$ and differs only slightly from $1 - \alpha$ provided $\sigma^2 > n_0 z$. Thus (4.2.12) can by used instead of the confidence interval (4.2.5). From (4.2.7) it follows that

$$\limsup_{\sigma\to\infty} \{E(n) - \sigma^2/z\} \leq 1$$

and (4.2.13)

$$\liminf_{\sigma\to\infty} \{E(n) - \sigma^2/z\} \geq 0.$$

Thus, the approximation $E(n) \doteq \sigma^2/z$ is fair provided $\sigma^2 > zn_0$. The length of the confidence interval is given by

$$2d = 2t_{n_0-1,1-\alpha/2}\, z^{1/2} \simeq \frac{2\sigma t_{n_0-1,1-\alpha/2}}{[E(n)]^{1/2}}.$$

When σ^2 is known, the length of the single-sample confidence interval of confidence coefficient $1-\alpha$ obtained on the basis of n observations is given by

$$2d = 2t_{\infty,1-\alpha/2}\,\sigma/n^{1/2}.$$

Hence, if n_0 is moderately large (say > 30) the expected number of observations of a confidence interval of given length and confidence coefficient is only slightly larger than the fixed number of observations required in the single-sample case when the variance is known provided the variance is moderately large.

4.2.2 Two-Stage Procedures for Estimating the Difference between Means

Putter (1951) has considered the problem of estimating the mean of a population composed of a known number of normally distributed strata whose relative proportions are known [see also Robbins (1952, p. 528)]. Ghurye and Robbins (1954) have considered the problem of estimating the difference between means by a two-stage procedure and in the following we shall present their results.

Let Π_i be a population with unknown mean θ_i and variance $\sigma_i^2 (i = 1, 2)$; we wish to estimate the difference $\theta_1 - \theta_2$. Let $T_i(n)$ be the mean $(X_{i1} + \cdots + X_{in})/n$ of a sample of size n from Π_i. Then $T_1(n_1) - T_2(n_2)$ is an unbiased estimate of $\theta_1 - \theta_2$, with variance $\sum_1^2 (\sigma_i^2/n_i)$. Let the cost of sampling be an unknown linear function of the number of observations. The cost of taking a sample of size n_1 from Π_1 and a sample of n_2 from Π_2 is $a_1 n_1 + a_2 n_2 + a_3$. If there is a prescribed upper bound A_0 to the cost of sampling, n_1 and n_2 are subject to the restriction

$$a_1 n_1 + a_2 n_2 \leq A = A_0 - a_3. \tag{4.2.14}$$

The quantity $\sum(\sigma_i^2/n_i)$, equal to the variance of $T_1(n_1) - T_2(n_2)$ for integral values of the n_i, is minimized for continuous $n_i > 0$ subject to (4.2.14) by taking $n_i = n_i^0$ where

$$n_i^0 = (A/a_i)a_i^{1/2}\sigma_i / \sum_j a_j^{1/2}\sigma_j; \tag{4.2.15}$$

the minimum value being equal to

$$V^0(A) = \sum_i (\sigma_i^2/n_i^0) = A^{-1}\left(\sum_i a_i^{1/2}\sigma_i\right)^2. \tag{4.2.16}$$

When the ratio $\rho = \sigma_2/\sigma_1$ on which the optimum value (4.2.16) depends is unknown, one can use a two-stage procedure for estimating $\theta_1 - \theta_2$, first by taking a sample of $m_1 + m_2$ observations, m_i from Π_i, and then using estimates of σ_i obtained from this preliminary sample to distribute the remaining observations between the Π_i. We shall investigate the performance of this estimation procedure when Π_i are normal as well as nonnormal.

Normal Populations When the Π_i are known to be normal, choose positive integers m_i such that $a_1 m_1 + a_2 m_2 < A$ and take m_i observations from Π_i. Let

$$s_i^2(m_i) = \left\{\sum_{j=1}^{m_1} X_{ij}^2 - m_i T_i^2(m_i)\right\}/(m_i - 1), \tag{4.2.17}$$

$$U(m_1, m_2) = a_1^{1/2} s_1(m_1)/\sum a_i^{1/2} s_i(m_i); \tag{4.2.18}$$

$$n_1^* = \begin{cases} m_1 & \text{if } \; AU(m_1, m_2)/a_1 < m_1, \\ (A - a_2 m_2)/a_1 & \text{if } \; (A - a_2 m_2)/a_1 < AU(m_1, m_2)/a_1, \\ AU(m_1, m_2)/a_1 & \text{if } \; m_1 \le AU(m_1, m_2)/a_1 \le (A - a_2 m_2)/a_1, \end{cases} \tag{4.2.19}$$

$$n_2^* = (A - a_1 n_1^*)/a_2, \tag{4.2.20}$$

$$\tilde{n}_i = [n_i^*] \qquad (i = 1, 2). \tag{4.2.21}$$

where $[x]$ here denotes the largest integer contained in x. Having computed $\tilde{n}_i$ we take $(\tilde{n}_i - m_i)$ additional observations $(X_{ij}, j = m_i + 1, \ldots, \tilde{n}_i)$ from Π_i, and estimate $\theta_1 - \theta_2$ by

$$T_1(\tilde{n}_1) - T_2(\tilde{n}_2) = \sum_{j=1}^{\tilde{n}_1} X_{1j}/\tilde{n}_1 - \sum_{j=1}^{\tilde{n}_2} X_{2j}/\tilde{n}_2. \tag{4.2.22}$$

Let $\tilde{V}(A) = \text{var}\{T_1(\tilde{n}_1) - T_2(\tilde{n}_2)\}$. Now, one can write

$$T_i(\tilde{n}_i) = m_i T_i(m_i)/\tilde{n}_i + (\tilde{n}_i - m_i) T_i(\tilde{n}_i - m_i)/\tilde{n}_i,$$

where $T_i(\tilde{n}_i - m_i) = \sum_{j=m_i+1}^{\tilde{n}_i} X_{ij}/(\tilde{n}_i - m_i)$. Since the $\tilde{n}_i$ depend only on the $s_i(m_i)$ for fixed s_i, the random variables $T_1(m_1)$, $T_2(m_2)$, $T_1(\tilde{n}_1 - m_1)$, $T_2(\tilde{n}_2 - m_2)$ are mutually independent, the conditional distributions of

$T_i(m_i)$ and $T_i(\tilde{n}_i - m_i)$ being respectively normal $(\theta_i, \sigma_i^2/m_i)$ and normal $(\theta_i, \sigma_i^2/(\tilde{n}_i - m_i))$. Hence, for fixed s_i the conditional distribution of $T_1(\tilde{n}_1) - T_2(\tilde{n}_2)$ is normal $(\theta_1 - \theta_2, \sum_i(\sigma_i^2/\tilde{n}_i))$. Hence

$$E\{T_1(\tilde{n}_1) - T_2(\tilde{n}_2)\} = \theta_1 - \theta_2 \quad \text{and} \quad \tilde{V}(A) = E\{(\textstyle\sum_i \sigma_i^2/\tilde{n}_i)\}. \tag{4.2.23}$$

Let $F(u) = P(U(m_1, m_2) \le u)$. Thus from (4.2.21), (4.2.23) we have

$$\begin{aligned}\tilde{V}(A) = {} & \{\sigma_1^2 m_1^{-1} + \sigma_2^2[(A - a_1 m_1)/a_2]^{-1}\}F(a_1 m_1/A) \\ & + \{\sigma_1^2[(A - a_2 m_2)/a_1]^{-1} + \sigma_2^2 m_2^{-1}\}\{1 - F(1 - a_2 m_2/A)\} \\ & + \int_{a_1 m_1/A}^{1 - a_2 m_2/A} \{\sigma_1^2[Au/a_1]^{-1} + \sigma_2^2[A(1-u)/a_2]^{-1}\}\, dF(u).\end{aligned} \tag{4.2.24}$$

Denote by $V^*(A)$ the expression obtained by dropping the square brackets in (4.2.24). Let

$$\rho = \sigma_2/\sigma_1, \qquad c = (a_2/a_1)^{1/2}, \qquad b_1 = (A - a_1 m_1)/(a_2 m_1),$$

$$b_2 = a_1 m_2/(A - a_2 m_2), \qquad r_i = (m_i - 1)/2 \quad (i = 1, 2), \qquad \text{and} \qquad q = r_2/r_1. \tag{4.2.25}$$

Then from (4.2.16) we have $V^0(A) = A^{-1}a_1\sigma_1^2(1 + \rho c)^2$ and we can rewrite the expression for $V^*(A)$ as

$$\begin{aligned}A\{V^*(A) - V^0(A)\} = {} & a_1\sigma_1^2(\rho - cb_1)^2 b_1^{-1} F(a_1 m_1/A) \\ & + a_1\sigma_1^2(\rho - cb_2)^2 b_2^{-1}\{1 - F(1 - a_2 m_2/A)\} \\ & + \int_{a_1 m_1/A}^{1 - a_2 m_2/A} a_1\sigma_1^2\{(1-u)u^{-1} \\ & + \rho^2 c^2 (1-u)^{-1} u - 2\rho c\}\, dF.\end{aligned} \tag{4.2.26}$$

Finally, making the substitution

$$\begin{aligned}W = \rho^2 c^2 u^2/\{\rho^2 c^2 u^2 + q(1-u)^2\} & = (m_1 - 1)(s_1^2/\sigma_1^2)/\textstyle\sum(m_i - 1)(s_i^2/\sigma_i^2) \\ & = \chi^2_{m_1 - 1}/(\chi^2_{m_1 - 1} + \chi^2_{m_2 - 1}),\end{aligned} \tag{4.2.27}$$

so that W has the density function

$$f(w) = \{1/B(r_1, r_2)\} w^{r_1 - 1}(1 - w)^{r_2 - 1} \qquad (0 \le w \le 1) \tag{4.2.28}$$

where $B(r_1, r_2) = \Gamma(r_1)\Gamma(r_2)/\Gamma(r_1 + r_2)$. Thus (4.2.26) reduces to

$$V^*(A)/V^0(A) = 1 + (1 + \rho c)^{-2}\{\rho c I_1 + (\rho - cb_1)^2 b_1^{-1} I_2 + (\rho - cb_2)^2 b_2^{-1} I_3\} \tag{4.2.29}$$

where

$$\begin{aligned} I_1 &= \int_{\beta_1}^{\beta_2} \{(qw)^{1/2}(1 - w)^{-1/2} + (qw)^{-1/2}(1 - w)^{1/2} - 2\} f(w)\, dw, \\ I_2 &= \int_0^{\beta_1} f(w)\, dw, \qquad I_3 = \int_{\beta_2}^{1} f(w)\, dw, \qquad \beta_i = \rho^2/(\rho^2 + c^2 q b_i^2). \end{aligned} \tag{4.2.30}$$

Hence, V^*/V_0 can be computed by means of the tables of the incomplete beta functions. Ghurye and Robbins (1954) have made some calculations of this ratio for $a_1 = a_2 = a$, $A/a = 30$, 50, $m_1 = m_2 = m = (0.2)A/a$, $(0.3)A/a$, $(0.4)A/a$ and various values of ρ. These are presented in Table 4.2.1. When

Table 4.2.1

Comparison of V^* with V^0 and V' for Normal Populations

	ma/A \ ρ	1.00	1.25	1.50	1.75	2.00	2.25	2.50	2.75	3.00
V^*/V^0 for	.2	1.06	1.06	1.05	1.05	1.05	1.04	1.04	1.03	1.03
$A/a = 30$	.3	1.03	1.03	1.03	1.02	1.02	1.02	1.02	1.02	1.02
	.4	1.02	1.01	1.01	1.01	1.02	1.04	1.06	1.08	1.09
V^*/V^0 for	.2	1.03	1.03	1.03	1.03	1.03	1.02	1.02	1.02	1.02
$A/a = 50$	.3	1.02	1.02	1.02	1.01	1.01	1.01	1.01	1.01	1.02
	.4	1.01	1.01	1.01	1.01	1.02	1.04	1.06	1.07	1.09
V'/V^0		1.00	1.01	1.04	1.07	1.11	1.15	1.18	1.22	1.25

$a_1 = a_2$, for estimating $\theta_1 - \theta_2$, one takes $n_1 = n_2 = A/2a$ and the estimator is $T_1(A/2a) - T_2(A/2a)$. Its variance is $V' = 2(\sigma_1{}^2 + \sigma_2{}^2)a/A$. The two-stage procedure seems to provide considerable improvement over the usual one-stage procedure for values of ρ away from 1; and the performance seems to be best for ma/A in the neighborhood of $\sigma_1/(\sigma_1 + \sigma_2)$ if $\sigma_1 < \sigma_2$.

The idea of substituting $s_i(m_i)$ for σ_i in (4.2.15) is based on the belief that as $m_i \to \infty$, the ratio $\tilde{n}_1/\tilde{n}_2$ will approach the optimum value $n_1{}^0/n_2{}^0$ and $\tilde{V}/V^0$ will approach unity. This will be proved in the following theorem.

Theorem 4.2.1 (Ghurye and Robbins, 1954) Let Π_i be normal, and consider $V^*(A)/V^0(A)$ as given in (4.2.29). Let a_1, a_2, and ρ remain fixed while m_1, m_2, and A become infinite in such a way that

$$\begin{aligned} 0 < k \le m_1/m_2 \le k' < \infty \qquad &\text{where} \quad k, k' \text{ are fixed,} \\ m_i/A \to 0 \qquad &(i = 1, 2). \end{aligned} \tag{4.2.31}$$

Then, $V^*(A)/V^0(A) \to 1$.

Proof In (4.2.29)

$$\begin{aligned} I_1 &\le \{q^{1/2}B(r_1 + \tfrac{1}{2}, r_2 - \tfrac{1}{2}) + q^{-1/2}B(r_1 - \tfrac{1}{2}, r_2 + \tfrac{1}{2})\}/B(r_1, r_2) - 2 \\ &= \{q^{1/2}\Gamma(r_1 + \tfrac{1}{2})\Gamma(r_2 - \tfrac{1}{2}) + q^{-1/2}\Gamma(r_1 - \tfrac{1}{2})\Gamma(r_2 + \tfrac{1}{2})\}/\Gamma(r_1)\Gamma(r_2) - 2, \end{aligned}$$

which converges to zero since

$$x^k\Gamma(x - k)\Gamma(x) \to 1 \quad \text{as} \quad x \to \infty. \tag{4.2.32}$$

$$\begin{aligned} I_2 &\le P\{W \le \rho^2/(\rho^2 + c^2qb_1{}^2)\} \\ &\le P\{W^{-1} > c^2q^2b_1{}^2/\rho^2\} \le E(W^{-1})\rho^2/(c^2qb_1{}^2), \end{aligned}$$

and

$$\begin{aligned} I_3 &\le P\{W > \rho^2/(\rho^2 + c^2qb_2{}^2)\} \\ &\le P\{(1 - W)^{-1} > \rho^2/(c^2qb_2{}^2)\} \le \{c^2qb_2{}^2/\rho^2\}E\{(1 - W)^{-1}\} \end{aligned}$$

Now,

$$E(W^{-1}) = \Gamma(r_1 - 1)\Gamma(r_1 + r_2)/\Gamma(r_1)\Gamma(r_1 + r_2 - 1)$$

and

$$E\{(1 - W)^{-1}\} = \Gamma(r_2 - 1)\Gamma(r_1 + r_2)/\Gamma(r_2)\Gamma(r_1 + r_2 - 1),$$

both of which remain bounded on account of (4.2.31) and (4.2.32). Since $b_1 \to \infty$ and $b_2 \to 0$, we have

$$(\rho - cb_1)^2b_1^{-1}I_2 \to 0 \qquad \text{and} \qquad (\rho - cb_2)^2b_2^{-1}I_3 \to 0.$$

Hence $V^*(A)/V^0(A) \to 1$. From the expression (4.2.24) for $\tilde{V}(A)$ it follows that $\tilde{V}(A)/V^0(A) \to 1$. This completes the proof of Theorem 4.2.1.

NONNORMAL POPULATIONS

Next, Ghurye and Robbins (1954) remove the restriction of normality of the Π_i and assume that they have finite variances σ_i^2 and there exist functions g_i such that

$$s_i(n) = g_i(X_{i1}, \ldots, X_{in}; n) \qquad (i = 1, 2) \tag{4.2.33}$$

satisfy conditions (I)–(III) given below.
Let $F_i(s; n) = P(s_i(n) \le s)$ and $C_i(\varepsilon) = [\sigma_i - \varepsilon, \sigma_i + \varepsilon]$. We assume:

(I) There exists a $\gamma > 1$ such that for every fixed $\varepsilon > 0$, $n^\gamma P\{s_i(n) \notin C_i(\varepsilon)\}$ is bounded for all $n > 0$, $i = 1, 2$;

(II) There exists an ε $(0 < \varepsilon < \min(\sigma_1, \sigma_2))$ such that $n\{\int_{C_i(\varepsilon)} s^k \, dF_i(s; n) - \sigma_i^k\}$ is bounded for all $n > 0$ and $k = 2, -2$.

(III) Either $T_i(n)$ and $s_i(n)$ from the same sample are a pair of mutually independent random variables, or Π_i has a finite fourth moment.

Then one will carry out the two-stage procedure given by (4.2.18)–(4.2.22) with $s_i(m_i)$ as given by (4.2.33) instead of (4.2.17). Then, we have

Theorem 4.2.2 (Ghurye and Robbins, 1954) Let a_1, a_2, and ρ remain fixed while m_1, m_2, and A become infinite in such a way that

$$\text{(4.2.31) holds and } A/m^{(\gamma+1)/2} \text{ is bounded.} \tag{4.2.34}$$

Then

$$E\{T_1(\tilde{n}_1) - T_2(\tilde{n}_2)\} \to \theta_1 - \theta_2 \qquad \text{and} \qquad \tilde{V}(A)/V^0(A) \to 1.$$

Proof See Ghurye and Robbins (1954, Theorem 2, pp. 150–153).

Examples (i) If the Π_i are normal, the conditions of Theorem 4.2.2 are satisfied for every $\gamma > 0$, and hence in (4.2.34) A may increase as any power of m_i. From Theorem 4.2.1, it is clear that it is not necessary to restrict A to be of the order of a power of m_i. (ii) If the Π_i are Poisson, $\sigma_i^2 = \theta_i$ and $s_i^2(n) = T_i(n)$. From the fact that $nT_i(n)$ has a Poisson distribution, it can be seen that the conditions of Theorem 4.2.2 are satisfied for every $\gamma > 0$. (iii) If the Π_i are binomial with $P(X_i = 1) = \theta_i$ and $P(X_i = 0) = 1 - \theta_i$, we have $\sigma_i^2 = \theta_i(1 - \theta_i)$ and $s_i^2(n) = T_i(n)(1 - T_i(n))$. Using the fact that $nT_i(n)$ is a binomial random variable, we can show that the conditions of Theorem 4.2.2 hold for every $\gamma > 0$. (iv) If the distribution functions of the Π_i are unknown, we would use the estimates $s_i(m_i)$ of σ_i given by (4.2.17). If Π_i has finite moments of order eight, then one can show that Theorem 4.2.2 is true for the procedure given by (4.2.17)–(4.2.22).

O'Neill and Rohatgi (1973) have given a two-stage procedure for estimating the difference between the mean vectors of two multivariate normal populations.

Problem

Check whether the exponential family of densities satisfy the regularity conditions (I), (II), and (III) (see page 302) so that the two-stage procedure for estimating the difference of the means can be carried out.

4.2.3 Two-Stage Procedure for Estimating the Common Mean

Let Π_1, Π_2 be two populations having the common mean θ and variances σ_1^2, σ_2^2. We wish to estimate θ using a fixed number n of observations. If the population variances are known, the most efficient procedure would be to take all n observations from that population having the smallest variance. When prior information about the variances is lacking or is too vague to be quantified, it is natural to consider taking a preliminary sample of size m from each population, computing estimates of the variances, and then taking the remaining $n - 2m$ observations from the population with the apparently smaller variance. If m is too large or too small, the advantage of the two-stage sampling scheme over the procedure of simply taking $n/2$ observations from each population will be lost. Hence it is of interest to determine for some good estimator an optimal choice of m as a function of n, not dependent on the unknown variances. As an example, suppose that there are two devices for measuring a physical constant, that each measurement is expensive or time consuming so that their total number is limited, and that we wish to estimate the constant as accurately as possible. Richter (1960) has considered this problem and his results will be given below.

Let $X_{i1}, \ldots, X_{i,m}$ be a random sample from $\Pi_i (i = 1, 2)$ and let $\eta = \sigma_2^2/\sigma_1^2$. Also, let

$$R = s_1^2/s_2^2 = \frac{\sum_1^m X_{1i}^2 - (\sum_1^m X_{1i})^2/m}{\sum_1^m X_{2i}^2 - (\sum_1^m X_{2i})^2/m}$$

so that $1/R$ is the usual estimator of η based on $2m$ observations. Then we further take observations $X_{1,m+1}, \ldots, X_{1,n-m}$ if $R < 1$ or take observations $X_{2,m+1}, \ldots, X_{2,n-m}$ otherwise. Write $\overline{X}_{i,N_i} = \sum_{j=1}^{N_i} X_{ij}/N_i$, $i = 1, 2$. We will consider estimators $\hat{\theta}$ of θ which can be written in the form

$$\hat{\theta} = A_1\overline{X}_{1,N_1} + A_2\overline{X}_{2,N_2}$$

where N_1, N_2, A_1, A_2 are random variables such that $N_1 = n - m$ if $R < 1$, $N_1 = m$ if $R \geq 1$, $N_1 + N_2 = n$, $0 \leq A_i \leq 1$ $(i = 1, 2)$, and $A_1 + A_2 = 1$ with probability one; and besides, where A_1 and A_2 are such that

$$E_H \overline{X}_{ik} = E\overline{X}_{ik} \quad (i = 1, 2), \qquad E_H(\overline{X}_{1k}\overline{X}_{2l}) = E_H \overline{X}_{1k} E\overline{X}_{2l},$$
$$E_H X_{ik}^2 = EX_{ik}^2 \qquad (i = 1, 2)$$

for all l and k and where $E_H(\cdot) = E(\cdot \mid H)$ and $H = (A_1, A_2, N_1, N_2)$. If the X_{ij} are assumed to be normally distributed, then the sample mean and sample variance are independent. Hence, the above assumption may be replaced by the assumption that A_1 and A_2 are functions of the sample variances only. Estimators of the form $\hat{\theta}$ seem reasonable since, if observations are available on $X_{ij}, j = 1, \ldots, n_i, i = 1, 2$ the variables are normal, and η is known, $a_1\overline{X}_{1,n_1} + a_2\overline{X}_{2,n_2}$ is the uniformly minimum variance unbiased estimator of θ, where $a_i = n_i\eta/(n_1\eta + n_2)$, $a_1 + a_2 = 1$.

Next, let $V_0 = (1/n)\min(\sigma_1{}^2, \sigma_2{}^2)$, which is the variance of the standard estimator of θ for the case when $\operatorname{sgn}(\sigma_2{}^2 - \sigma_1{}^2)$ is known beforehand, and define $R_n(m; \eta) = V_0^{-1} E(\hat{\theta} - \theta)^2$ to be the risk function associated with the estimator $\hat{\theta}$. The first part of the following theorem implies that $R_n(m; \eta) = V_0^{-1} \operatorname{var} \hat{\theta}$.

Theorem 4.2.3 (Richter, 1960) For any estimator of the form $\hat{\theta}$,

(i) $E\hat{\theta} = \theta$;
(ii) $R_n(m; \eta) = n \max(1, 1/\eta)E\{A_1{}^2/N_1 + \eta A_2{}^2/N_2\}$;
(iii) $R_n(m; \eta) \geq n \max(1, \eta)E\{1/(N_2 + N_1\eta)\} \geq 1$.

Proof Since $E_H\hat{\theta} = A_1E_H\overline{X}_{1,N_1} + A_2E_H\overline{X}_{2,N_2} = \theta$, $E\hat{\theta} = \theta$. Next,

$$E_H(\hat{\theta} - \theta)^2 = E_H(A_1\overline{X}_{1,N_1} + A_2\overline{X}_{2,N_2} - \theta)^2 = A_1{}^2\sigma_1{}^2/N_1 + A_2{}^2\sigma_2{}^2/N_2$$

which proves (ii) since $R_n(m; \eta) = V_0^{-1} E\{E_H(\hat{\theta} - \theta)^2\}$. Finally, $A_1{}^2/N_1 + \eta A_2{}^2/N_2$ has a unique minimum with respect to $A_1 = 1 - A_2$ at $A_1 = N_1\eta/(N_2 + N_1\eta)$ so that $E\{A_1{}^2/N_1 + \eta A_2{}^2/N_2\} \geq \eta E\{1/(N_2 + N_1\eta)\}$ which proves the left-hand inequality of (iii); since $N_2 + N_1\eta \leq n\max(1, \eta)$ the proof is complete.

Let us examine the risk function for the usual one-stage experiment for estimating θ, which would be to observe $n/2$ of the X_{1j} and $n/2$ of the X_{2j}. If we confine ourselves to unbiased estimators θ', and assume the variables normally distributed, then $\operatorname{var} \theta' \geq 2\sigma_2{}^2/n(1 + \eta)$, since $(\eta\overline{X}_1 + \overline{X}_2)/(1 + \eta)$ is the minimum variance unbiased estimator with variance $2\sigma_2{}^2/n(1 + \eta)$ when the variances are known. Then $R(\eta) = \operatorname{var} \theta'/V_0 \geq 2\max(\eta, 1)/(1 + \eta)$

[since $\max(1, \eta) = 1/\min(1, 1/\eta)$] and $2\max(\eta, 1)/(1+\eta) \geq 1$ with equality if and only if $\eta = 1$. Hence, for each fixed $\eta \geq 1$, the risk function is bounded away from unity independent of the sample size. One would hope that the risk function for the two-stage scheme would prove to be smaller, and we shall show that it is so for large samples at least provided m is suitably chosen.

For the two-stage experiment, it is clear that once an estimator is specified, the only variable left at the statistician's disposal is the quantity m. Then given an estimator of the form $\hat{\theta}$, we may say that any real-valued function $m(n)$ such that $4 \leq 2m(n) < n$ for all $n \geq 5$ is a solution to the problem. With respect to an estimator of the form $\hat{\theta}$, $m(n)$ will be called a *uniformly consistent solution* (UCS) if $\sup_\eta R_n(m(n); \eta) \to 1$ as $n \to \infty$. We shall restrict to such solutions if they exist. Further, if $\sup_\eta R_n(m; \eta) < \infty$, a solution which minimizes $\sup_\eta R_n(\eta; m)$ will be called a minimax solution (MMS). If there exists a UCS, then MMS is UC too. Hence the minimax principle affords a means of selecting one solution from the class of UC solutions.

A Simple Estimator

In the following, we shall derive an asymptotic minimax solution for a particular unbiased estimator. For the subsequent considerations, we shall assume that the X_{ij} are normally distributed. Hence $\eta R = s_1{}^2\sigma_2{}^2/s_2{}^2\sigma_1{}^2$ obeys the F-distribution with $m-1$, $m-1$ degrees of freedom, and we may write

$$K(m; \eta) = P(R \geq 1) = B((m-1)/2, (m-1)/2)^{-1} \int_\eta^\infty x^{(m-3)/2}(1+x)^{1-m}\, dx.$$

Now define $\hat{\theta}_1 = A_1\bar{X}_{1,N_1} + A_2\bar{X}_{2,N_2}$, where $A_1 = 1$ or 0 according as $R < 1$ or $R \geq 1$. This estimator has the form of $\hat{\theta}$ and by Theorem 4.2.3, $\hat{\theta}$ is unbiased and

$$\begin{aligned} R_{1n}(m; \eta) &= n \max(1, 1/\eta) E\{A_1{}^2/N_1 + A_2{}^2\eta/N_2\} \\ &= n \max(1, 1/\eta)\left[\frac{1 - K(m; \eta)}{n - m} + \eta \frac{K(m; \eta)}{n - m}\right] \\ &= \max(1, 1/\eta)(1 - m/n)^{-1}[1 + (\eta - 1)K(m; \eta)]. \end{aligned}$$

It is easy to show that $R_{1n}(m; \eta) = R_{1n}(m; 1/\eta)$ by using the fact that $K(m; \eta) = 1 - K(m; 1/\eta)$; thus, when considering $\sup_\eta R_{1n}(m; \eta)$, we may restrict ourselves to the situation $\eta \geq 1$. We need the following lemmas in order to study the risk of the proposed estimator.

Lemma 4.2.1 (i) (Richter, 1960) For $\eta \geq 1$,

$$K(m; \eta) \leq [2\eta^{1/2}/(1+\eta)]^{m-1}.$$

(ii) Let $\eta = 1 + 2\tau m^{-1/2}$. For bounded nonnegative τ, $K(m; \eta) - \Phi(-\tau) = O(m^{-1/2})$ uniformly in τ as $m \to \infty$.

Proof

$$K(m; \eta) = P(\eta R \geq \eta) = P\left(\sum_1^{m-1} U_i \Big/ \sum_1^{m-1} V_i \geq \eta\right) = P(S_{m-1} \geq 0)$$

where $S_{m-1} = Z_1 + \cdots + Z_{m-1}$, $Z_i = U_i - \eta V_i$ and $U_1, \ldots, U_{m-1}, V_1, \ldots, V_{m-1}$ are iid chi-square variables with one degree of freedom. Thus, the Z_i are iid with $EZ_i = 1 - \eta \leq 0$ (since $\eta \geq 1$) and var $Z_i = 2(1 + \eta^2) \geq 4$. The moment-generating function of Z_i is

$$C(t) = E(\exp tZ_i) = [1 + 2(\eta - 1)t - 4\eta t^2]^{-1/2}.$$

Using the Markov inequality,

$$\begin{aligned} K(m; \eta) &= P(S_{m-1} \geq 0) = P(\exp(tS_{m-1}) \geq 1) \\ &\leq E(\exp(tS_{m-1})) = [C(t)]^{m-1}, \qquad t \geq 0. \end{aligned}$$

Since $C(t)$ is a minimum at $t = (\eta - 1)/4\eta$, we obtain (i). Next, let $F_{m-1}(x)$ denote the distribution function of the variable $(S_{m-1} - ES_{m-1})/(\text{var } S_{m-1})^{1/2}$ so that $1 - K(m; \eta) = F_{m-1}(x)$ where $x = (\eta - 1)(m - 1)^{1/2}[2(1 + \eta^2)]^{-1/2}$; if $\eta = 1 + 2\tau m^{-1/2}$, $\tau \geq 0$, and $\tau = O(1)$, then $x = \tau + \varepsilon(m; \tau)$ where $\varepsilon(m; \tau) = O(m^{-1/2})$ uniformly in τ. Noting that τ bounded implies that η is bounded and that

$$|Z_i - EZ_i|^3 \leq 8\{|Z_i|^3 + |EZ_i|^3\} \leq 8\{(U_i + \eta V_i)^3 + (\eta - 1)^3\}.$$

We observe that $E|Z_i - EZ_i|^3$ is bounded. Then the Berry–Esseen theorem (Berry, 1941) states that

$$|F_{m-1}(x) - \Phi(x)| \leq KE\{|Z_i - EZ_i|^3\}/(\text{var } Z_i)^{3/2}(m - 1)$$

for all x where K is a constant. Since $(\text{var } Z_i)^{3/2} \geq 8$ and $(m - 1)^{1/2} > m^{1/2}/2$, we find that $F_{m-1}(x) - \Phi(x) = O(m^{-1/2})$ uniformly in τ. However, $\Phi(x) - \Phi(\tau) = O(m^{-1/2})$ uniformly in τ since $|\Phi(x) - \Phi(\tau)| < |\varepsilon(m; \tau)|$. Hence $F_{m-1}(x) - \Phi(\tau) = O(m^{-1/2})$ uniformly in τ and, since $1 - K(m; \eta) = F_{m-1}(x)$, the proof of Lemma 4.2.1 is complete.

Lemma 4.2.2 (Richter, 1960). For $\eta \geq 1$, $(\eta - 1)K(m; \eta)$ has a unique absolute maximum with respect to η at $\eta = 1 + 2\tau' m^{-1/2} + o(m^{-1/2})$, and $\max_\eta (\eta - 1)K(m; \eta) = cm^{-1/2} + O(m^{-1})$ where τ' is the root of the equation $\Phi(-\tau) - \tau\phi(\tau) = 0$ and $c = 2\tau'\Phi(-\tau')$.

Proof Let $\tilde{K}(m;\eta) = (\eta - 1)K(m;\eta)$ for $\eta \geq 1$ and take $m \geq 4$. Then $\tilde{K}(m;\eta)$ is continuous in η, $\tilde{K}(m;1) = 0$ and $\tilde{K}(m;\eta) > 0$ for $1 < \eta < \infty$. From Lemma 4.2.1(i), $\tilde{K}(m;\eta) < 4\{2\eta^{1/2}/(1+\eta)\}^{m-3}$. Therefore $\tilde{K}(m;\eta) < 4$ and $\tilde{K}(m;\eta) \to 0$ as $n \to \infty$. Hence $\tilde{K}(m;\eta)$ has an absolute maximum with respect to η, $1 \leq \eta < \infty$. Using the integral representation of $K(m;\eta)$ straightforward differentiation yields

$$\frac{\partial^2 \tilde{K}(m;\eta)}{\partial \eta^2} = \frac{(m-3)\eta^{(m-5)/2}\{\eta^2 - 2\eta(m+1)/(m-3) + 1\}}{2(\eta+1)B((m-1)/2, (m-1)/2)}.$$

Hence,

$$\partial^2 K(m;\eta)/\partial\eta^2 \begin{cases} <0 & \text{if } 1 \leq \eta \leq \eta_0, \\ =0 & \text{if } \eta = \eta_0, \\ >0 & \text{if } \eta > \eta_0, \end{cases}$$

where $\eta_0 = 1 + 2^{3/2}(m-1)^{1/2}/(m-3) + 4/(m-3)$. It follows that $\tilde{K}(m;\eta)$ has exactly one maximum wrt η and that the maximizing value of η satisfies the inequality, $1 < \eta \leq \eta_0$. Next, let $\eta = 1 + 2\tau m^{-1/2}$, $\tau \geq 0$, $\tau_0 = (\eta_0 - 1)m^{1/2}2^{-1} = O(1)$, so that $m^{1/2}\max_{\eta<\eta_0} \tilde{K}(m;\eta) = \max_{\tau<\tau_0} 2\tau K(m;\eta) = \max_{\tau<\tau_0} 2\tau\Phi(-\tau) + O(m^{-1/2})$ by Lemma 4.2.1(ii). The root of the equation $\Phi(-\tau) - \tau\phi(\tau) = 0$ gives the τ value that maximizes $\tau\Phi(-\tau)$; call this root τ' (approximately .75) and let $c = 2\tau'\Phi(-\tau')$ (approximately .34). This completes the proof of Lemma 4.2.2.

Lemma 4.2.3 (Richter, 1960). $m(n)$ is uniformly consistent (UC) solution if and only if $m(n)/n \to 0$ and $m(n) \to \infty$ as $n \to \infty$.

Proof Considering the risk of $\hat{\theta}$, we find that $\max_\eta R_{1n}(m;\eta) = \max_{\eta \geq 1} R_{1n}(m;\eta) = (1 - m/n)^{-1}(1 + cm^{-1/2} + O(m^{-1}))$ by Lemma 4.2.2. Suppose now that $\max_\eta R_{1n}(m;\eta) \to 1$ as $n \to \infty$. Then of necessity $(1 - m/n)^{-1} \to 1$ and $cm^{-1/2} \to 0$ as $n \to \infty$, implying respectively that $m/n \to 0$ and $m \to \infty$ as $n \to \infty$. Since the converse is obviously true, the proof is complete.

Theorem 4.2.4 (Richter, 1960) The minimax solution for $\hat{\theta}$ is $m(n) = (cn/2)^{2/3} + O(n^{1/3})$ and $\min_m \max_\eta R_{1n}(m;\eta) = 1 + 3(c/2)^{2/3} n^{-1/3} + O(n^{-2/3})$ where c is defined in Lemma 4.2.2.

Proof For all UC solutions, and hence for the MM solutions $\max_\eta R_{1n}(m; \eta) = 1 + m/n + cm^{-1/2} + \varepsilon$ where $m = m(n)$ and

$$\varepsilon = cm^{1/2}/n + O(m^2/n^2) + O(m^{-1}) = o(m/n) + o(m^{-1/2}).$$

Both m/n and $m^{-1/2}$ converge to zero as $n \to \infty$, one being an increasing function of m, the other a decreasing function. Hence a necessary condition for a minimum is that m/n and $m^{-1/2}$ be of the same order of magnitude. Hence $\max_\eta R_{1n}(m; \eta) \sim 1 + m/n + cm^{-1/2}$ as $n \to \infty$. The expression $m/n + cm^{-1/2}$ has a unique, absolute minimum at $m = (cn/2)^{2/3}$, and finally, taking into account the order of magnitude of ε, we obtain Theorem 4.2.4.

A CLASS OF ESTIMATORS One may ask if better (in the sense of smaller risk) estimators exist and if they do, whether results like Theorem 4.2.4 can be found for such estimators. Richter (1960) provides an affirmative answer to both these questions. Let $\hat{\theta}_2 = (N_1\eta\overline{X}_{1,N_1} + N_2\overline{X}_{2,N_2})/(N_1\eta + N_2)$ whose risk is $R_{2n}(m; \eta) = n \max(1, \eta)E\{1/(N_1\eta + N_2)\}$; then $R_{2n}(m; \eta)$ is a lower bound for the risk of all estimators of form $\hat{\theta}$ by Theorem 4.2.3(iii). However, when η is unknown, one can replace it by $\hat{\eta}$ where $\hat{\eta} \to \eta$ in probability. It is mathematically convenient to use a $\hat{\eta}$ based on the first stage only; we take $\hat{\eta} = 1/R$ and define $\hat{\theta}_3 = (N_1\overline{X}_{1,N_1} + N_2\overline{X}R_{2,N_2})/(N_1 + N_2 R)$. Another motivation for $\hat{\theta}_3$ is as follows. When η is known, for a one-stage experiment, the uniformly minimum variance unbiased (UMVU) estimator is $(n_1\eta\overline{X}_{1,n_1} + n_2\overline{X}_{2,n_2})/(n_1\eta + n_2)$. However, when n_1, n_2, and η are unknown, if one takes $\hat{\eta}$ to be the usual estimator of η based on 2 $\min(n_1, n_2)$ observations, and if one replaces the numbers n_1, n_2 by the random variables N_1, N_2, the result is $\hat{\theta}_3$. Another estimator which might be considered is $\hat{\theta}_4$, the grand mean of all the observations:

$$\hat{\theta}_4 = (N_1\overline{X}_{1,N_1} + N_2\overline{X}_{2,N_2})/n.$$

Then

$$\begin{aligned} R_{4n}(m; \eta) &= (1/n) \max(1, 1/\eta)E\{N_1 + N_2\eta\} \\ &= (1/n) \max(1, 1/\eta)\{(n - m + m\eta)[1 - K(m; \eta)] \\ &\quad + [m + (n - m)\eta]K(m; \eta)\} \\ &= \max(1, 1/\eta)\{1 + (\eta - 1)[m/n + (1 - 2m/n)K(m; \eta)]\} \end{aligned}$$

so that $\sup_\eta R_{4n}(m; \eta) > (m/n) \sup_{\eta > 1}(\eta - 1)$ which tends to ∞ as $\eta \to \infty$. We thus conclude that for $\hat{\theta}_4$ there exist no UC solution and no nontrivial MM solutions. [Note that $\hat{\theta}_4$ would be worthy of consideration if it were known a priori that η was close to unity, since $R_{4n}(m;\ 1) = R_{2n}(m;\ 1) = 1$ and since it can be shown that $R_{1n}(m;\eta) > R_{4n}(m;\eta)$ for $\frac{1}{2} \leq \eta \leq 2$. Consider

$$\begin{aligned} R_{2n}(m;\eta) &= n\max(1,\eta)\{[1 - K(m;\eta)]/[(n-m)\eta + m] \\ &\qquad + K(m;\eta)/(m\eta + n - m)\} \\ &= \max(1, 1/\eta)[1 + r_1(m/n;\eta) + r_2(m/n;\eta)K(m;\eta)], \end{aligned}$$

where

$$\begin{aligned} r_1(m/n;\eta) &= [m(\eta-1)/n\eta]/[1 - m(\eta-1)/n\eta]; \\ r_2(m/n;\eta) &= (1 - 2m/n)(\eta-1)/[1 + (m/n)(1-m/n)(\eta-1)^2/n]. \end{aligned}$$

As for $R_{1n}(\eta;\ m)$ we have $R_{2n}(m;\ \eta) = R_{2n}(m;\ 1/\eta)$. Then, we have the following theorem.

Theorem 4.2.5 (Richter, 1960) For any estimator of the form $\hat{\theta}$ whose risk satisfies $\sup_\eta R_n(m;\ \eta) \leq \max_\eta R_{1n}(m;\ \eta) + o(1)$ as $n \to \infty$, $m(n)$ is UC if and only if the condition in Lemma 4.2.3 is satisfied.

Proof Suppose the estimator is of the form $\hat{\theta}$ such that $\sup_\eta R_n(m;\ \eta) \leq \max_\eta R_{1n}(m;\eta) + o(1)$. If $m/n \to 0$ and $m \to \infty$ as $n \to \infty$, then $\max_\eta R_{1n}(\eta; m)$ and hence $\sup_\eta R_n(m;\eta)$ tend to unity as $n \to \infty$. If on the other hand $\sup_\eta R_n\ (m;\ \eta) \to 1$ as $n \to \infty$, then since $R_{2n}(m;\ \eta) \leq R_n(m;\ \eta)$, $\sup_\eta R_{2n}(m;\eta) \to 1$ as $n \to \infty$. However, $\sup_\eta R_{2n}(m;\eta) = \sup_{\eta \geq 1} R_{2n}(m;\eta)$ and, for $\eta \geq 1$, $R_{2n}(m;\eta) = 1 + r_1 + r_2 K(m;\eta)$ is the sum of three positive terms. Therefore, (i) $\sup_{\eta\geq 1} r_1(m/n;\eta) \to 0$ as $n \to \infty$, and (ii) $\sup_{\eta \geq 1} r_2(m/n;\eta)K(m;\eta) \to 0$ as $n \to \infty$. Since $r_1(m/n;\eta) \geq m(\eta-1)/n\eta$, (i) implies that $m/n \to 0$ as $n \to \infty$; for any fixed $\eta' > 1$, $r_2(m/n;\ \eta')$ is bounded away from zero, so that (ii) implies that $K(m;\eta') \to 0$ as $n \to \infty$ and so $m \to \infty$ as $n \to \infty$.

Richter's (1960) main result is given by the following theorem which we will state without proof.

Theorem 4.2.6 (Richter, 1960) Let the estimator of θ be of the form $\hat{\theta}$ whose risk satisfies $R_n(m;\ \eta) \leq R_{2n}(m;\ \eta) + \varepsilon_n$ where $\varepsilon_n = O(n^{-2/3})$. Then, for the estimator, the class of UC solutions is characterized by the condition in Lemma 4.2.3, and the MM risk and the MMS are respectively given by

$$\begin{aligned} \sup_\eta R_n(m;\eta) &= 1 + c^{2/3}n^{-1/3} + O(n^{-2/3}), \\ m(n) &= (cn)^{2/3} + O(n^{1/3}) \end{aligned}$$

where c is defined in Lemma 4.2.2.

Proof [See Richter (1960, Theorem 4, pp. 1170–1171).]

Richter (1960) obtains

$$R_{3n}(m;\eta) = \max(\eta, 1)\left[\int_0^\eta \frac{(1-m/n)\eta + (m/n)z^2}{[(1-m/n)\eta + (m/n)z]^2}\,dF(z) + \int_\eta^\infty \frac{(m/n)\eta + (1-m/n)z^2}{[(m/n)\eta + (1-m/n)z]^2}\,dF(z)\right]$$

where $dF(z) = [B((m-1)/2,\ (m-1)/2)]^{-1} z^{(m-3)/2}(1+z)^{1-m}\,dz$, and that $R_{3n}(m;\eta) = R_{3n}(m; 1/\eta)$ and $R_{3n}(m;\eta) - R_{2n}(m;1/\eta) < 112/n$ for all η and $6 \le m < n/2$. Thus the solution for the proposed estimator $\hat{\theta}_3$ is specified by Theorem 4.2.6. This implies that the MM risk for $\hat{\theta}_3$ is $\sim 1 + c^{2/3}n^{-1/3}$ and from Theorem 4.2.4 the MM risk for $\hat{\theta}_1$ is $\sim 1 + 3(c/2)^{2/3}n^{-1/3}$. Since $c^{2/3} < 3(c/2)^{2/3}$, we prefer $\hat{\theta}_3$ to $\hat{\theta}_1$; an additional reason for preferring $\hat{\theta}_3$ is as follows: Given any η, $0 < \eta < \infty$, one can find an m_0 such that $R_{3n}(m;\eta) < R_{1n}(m;\eta)$ for $m \ge m_0$. This is easy to see since $R_{1n}(m;\eta) - R_{2n}(m;\eta) = \int_0^\infty (W - \eta T_1)\,dF(z)$ where $W = (1-m/n)^{-1}$ if $z < \eta$, $W = \eta(1-m/n)^{-1}$ if $z \ge \eta$, and $T_1 = [(1-m/n)\eta + m/n]^{-1}$ if $z < \eta$ and $T_1 = [(m/n)\eta + 1 - m/n]^{-1}$ if $z \ge \eta$, and where we assume without loss of generality that $\eta \ge 1$. Noting that $W - \eta T_1 > m/\eta n$ for all z, and the fact $R_{3n}(m;\eta) - R_{2n}(m;\eta) < 112/n$ one obtains the desired result. Hence, if there exists an estimator of the form $\hat{\theta}$, with uniformly smaller risk than that of $\hat{\theta}_3$, it has the same large sample solution.

Using the loss function $r(\hat{\theta};\theta) = n(\hat{\theta} - \theta)^2/\min(\sigma_1{}^2, \sigma_2{}^2)$, Alam (1967) has given a minimax solution of m for moderality large values of n. Yen (1964) has considered the two-sample, two-stage nonparametric estimation of a functional of the population distribution function.

Problem

Is it possible to have a two-stage procedure for estimating the parameter of the exponential density where

(i) $f(x;\theta) = \exp\{-(x-\theta)\}$, $x \ge \theta$
(ii) $f(x;\theta) = \theta^{-1}\exp(-x/\theta)$, $x > 0$?

4.2.4 Two-Stage Procedure for Estimating the Variance of a Normal Population within *d* Units of the True Value

The problem of estimating the variance σ^2 of a normal population arises in many experimental situations. Greenwood and Sandomire (1950) have presented a means of obtaining the sample size required to estimate σ^2

within a given percent of its true value. One may prefer instead to estimate the variance within a given number of units. Birnbaum and Healy (1960) and Graybill and Connell (1964a) have considered this problem. The former set of authors have presented a method which requires the use of Chebyshev's inequality. However the methods of the latter set of authors provide a significantly smaller second sample than the method of Birnbaum and Healy (1960). Hence in the following we shall present the results of Graybill and Connell (1964a).

Let $s_{n_0}^2$ denote the unbiased version of the sample variance based on a preliminary sample of size n_0 taken from a normal density. Let d and $1 - \alpha$ be specified in advance. It is desired to determine n on the basis of the preliminary sample such that

$$P[|s_{n+1}^2 - \sigma^2| < d] > 1 - \alpha \tag{4.2.34a}$$

where $s_{n+1}^2 = (1/n)\sum_{i=1}^{n+1}(X_i - \bar{X}_{n+1})^2$ and $X_1, \ldots, X_{n+1}$ is an additional random sample of size $n + 1$ from a normal population having variance σ^2. Eq. (4.2.34a) can be written as

$$\begin{aligned} P[|s_{n+1}^2 - \sigma^2| < d] &= E[P\{(1 - a) < V < (1 + a)|n\}] \\ &= \int_1^\infty g(n) \int_{1-a}^{1+a} f_1(v|n)\, dv\, dn, \end{aligned} \tag{4.2.34b}$$

where E is expectation with respect to n; $a = d/\sigma^2$, $V = s_{n+1}^2/\sigma^2$; $g(\cdot)$ is the density of n, and $f_1(\cdot\,|n)$ is the density of a chi-square variable divided by n, its degrees of freedom. Assume that $n \geq 1$. By definition

$$f_1(v|n) = \begin{cases} [(n/2)^{n/2}/\Gamma(n/2)]v^{(n/2)-1}e^{-(n/2)v}, & 0 < v < \infty, \\ 0, & -\infty < v \leq 0. \end{cases}$$

It has been shown [see Connell and Graybill (1964)] that

$$\int_{1-a}^{1+a} f_1(v|n)\, dv > \int_{1-a}^{1+a} f_2(v|n)\, dv = 1 - \exp[-(n-1)^{1/2}a/\pi^{1/2}].$$

If a were known, we might set n equal to $1 + [\pi\,(\ln \alpha)^2]/a^2$, since in that case we would have

$$P[|s_{n+1}^2 - \sigma^2 < d] > E\left\{\int_{1-a}^{1+a} f_2(v|n)\, dv\right\} = E(1 - \alpha) = 1 - \alpha.$$

Since a is assumed to be unknown, let

$$n = 1 + [\pi\,(\ln \alpha)^2]ks_{n_0}^4/d^2 \tag{4.2.35}$$

where k is some constant, independent of a, such that

$$E\left\{\int_{1-a}^{1+a} f_2(v\,|\,n)\,dv\right\} = 1 - \alpha. \tag{4.2.36}$$

The density of $s_{n_0}^4$, and consequently that of n, is a known function of σ^2. Also, the expectation of $1 - \exp[-(n-1)^{1/2}a/\pi^{1/2}]$ for n given in (4.2.35) clearly does not involve σ^2. The value of k in (4.2.35) such that (4.2.36) is true is

$$k = (n_0 - 1)[(1/\alpha)^{2/(n_0-1)} - 1]/2\ln(1/\alpha). \tag{4.2.37}$$

Thus, if the sample size

$$n + 1 = (\pi/4)[(1/\alpha)^{2/(n_0-1)} - 1]^2(n_0 - 1)^2 s_{n_0}^4/d^2 + 2 \tag{4.2.38}$$

is used for the second-step sample, the inequality in (4.2.34a) is satisfied. The expected second sample size in (4.2.38) is

$$E(n+1) = (\pi/4)[(1/\alpha)^{2/(n_0-1)} - 1]^2({n_0}^2 - 1)\sigma^4/d^2 + 2. \tag{4.2.39}$$

Table 4.2.3 shows the comparison between the sample size given in (4.2.38)

Table 4.2.3

Value of $h(n_0,\ \alpha) = (n-1)/(n'-1) = E(n-1)/E(n'-1) = (\pi/8)\alpha(n_0-3)(n_0-5) \times [(1/\alpha)^{2/(n_0-1)} - 1]^2$, $n_0 \geq 6$.

n_0	$1-\alpha = .90$	.95	.99
10	.613	.615	.437
100	.820	.704	.344
1000	.832	.705	.334

and the sample size n' obtained by Birnbaum and Healy (1960). It can be shown that

$$h(n_0, \alpha) < h(n_0, \alpha_0) < \lim_{n_0 \to \infty} h(n_0, \alpha_0) = 2\pi e^{-2} = .85$$

where $\alpha_0 = [(n_0 - 5)/(n_0 - 1)]^{(n_0-1)/2}$. With minor modifications, the results of Connell and Graybill (1964) can be used to estimate the mean of the gamma distribution.

4.2.5 Two-Stage Procedure for Estimating the Parameter of the Uniform Density within d Units

Graybill and Connell (1964b) have considered this problem. Let $f(u) = 1/\theta$, for $0 < u < \theta$, and the problem is to determine the sample size n, based on a preliminary sample of size n_0 (given) from $f(u)$ such that

$$P[|\hat{\theta}_n - \theta| < d] > 1 - \alpha \tag{4.2.40}$$

where $\hat{\theta}_n$ is an estimator of θ. The maximum likelihood estimator of θ is the largest observation $Y_{(n)}$ in the sample. Let $d/\theta = a$. Then

$$\begin{aligned} P[|Y_{(n)} - \theta| < d] &= P[(1-a) < Y_{(n)}/\theta < 1] \\ &= E\{P[(1-a) < Y_{(n)}/\theta < 1|n]\} \\ &= E\{1 - (1-a)^n\} = 1 - E(1-a)^n. \end{aligned}$$

Hence, we wish to find an n such that $E(1-a)^n < \alpha$ for all $d < \theta$ $(a < 1)$, that is, for all $0 < a < 1$. If θ were known, then we could let

$$n' = (\ln \alpha)/\ln(1-a), \qquad 0 < a < 1. \tag{4.2.41}$$

Since θ is unknown, a reasonable function of X, the largest observation in the preliminary sample of size n_0, may be obtained by replacing a in (4.2.41) with a constant b times the maximum likelihood estimator of a. We shall use bd/X in place of a where b is some function of n_0 and α $(0 < b < 1)$ and monotonic increasing in n_0. If we specify that $n \geq 1$, then

$$n = \begin{cases} 1, & 0 \leq X \leq bd/(1-\alpha) \\ (\ln \alpha)/\ln(1 - bd/X) & bd/(1-\alpha) \leq X \leq \theta. \end{cases} \tag{4.2.42}$$

Graybill and Connell (1964b) have shown that

$$E\{(1-a)^n\} < b^{n_0}(n_0\alpha + 1)/(n_0 + 1) + \int_b^1 \alpha^{z/b} n_0 z^{n_0-1}\, dz. \tag{4.2.43}$$

By setting the right-hand expression of (4.2.43) equal to α one can solve implicitly for b for each n_0 and α, and one is assured that the use of this b in

(4.2.42) will yield a second sample size n such that (4.2.40) is satisfied. Table 4.2.4 gives the values of b such that (4.2.43) is equal to α. The expected sample

Table 4.2.4

Value of b to Be Used in (4.2.42) for Finding the Sample Size n Such That $P[|Y_{(n)} - \theta| < d] > 1 - \alpha$

$1-\alpha$ \ n_0	2	5	10	15	20	25	30	40	50	75	100
.90	.442	.728	.854	.900	.924	.938	.948	.961	.969	.979	.984
.95	.349	.682	.829	.883	.911	.928	.940	.955	.964	.976	.982
.99	.170	.549	.758	.836	.875	.900	.916	.937	.949	.966	.974

size for the solution available through the use of Chebyshev's inequality [see Birnbaum and Healy (1960)] may be shown to be

$$n'' = [1 + n_0 X^2/(n_0 - 2)\alpha d^2]^{1/2} - 1. \tag{4.2.44}$$

The values in Table 4.2.5 indicate that the method of Graybill and Connell (1964b) yields a significantly smaller expected sample size than the solution given by Eq. (4.2.44). Also the expected sample size is asymptotic to the expected "ideal" sample size, namely n' as n_0 increases.

Table 4.2.5

Showing the Expected Sample Sizes n' [see Eq. (4.2.41)], n [see Eq. (4.2.42)], and n'' [see Eq. (4.2.44)]

$1-\alpha$		.90			.95			.99		
a	n_0	n'	n	n''	n'	n	n''	n'	n	n''
.05	5	44.9	51.5	67.0	58.4	71.7	95.2	89.8	137.4	214.2
.05	20	44.9	46.3	62.5	58.4	61.1	88.8	89.8	97.9	199.8
.05	50	44.9	45.6	62.3	58.4	59.7	88.5	89.8	93.2	199.1
.25	5	8.0	9.3	12.6	10.4	13.1	18.2	16.0	25.6	42.0
.25	20	8.0	8.3	11.7	10.4	11.0	17.0	16.0	17.6	39.2
.25	50	8.0	8.1	11.7	10.4	10.6	16.9	16.0	16.6	39.0
.60	5	2.5	3.1	4.7	3.3	4.4	7.0	5.0	9.1	16.9
.60	20	2.5	2.6	4.3	3.3	3.5	6.5	5.0	5.7	15.7
.60	50	2.5	2.6	4.3	3.3	3.4	6.5	5.0	5.3	15.7

4.2.6 Two-Stage Procedure for Estimating the Binomial and Poisson Parameters

Samuel (1966) has considered the problem of unbiased estimation of the parameter in the binomial and Poisson distributions. The variance of the estimate is required not to exceed a preassigned bound. For the Poisson distribution, this can be achieved only by sequential sampling and in the binomial case, sequential sampling may reduce the expected sample size. The two-stage procedures of Samuel (1966) differ from those proposed by Birnbaum and Healy (1960) in that the preliminary sample is not used to determine the size of the second sample, but is also directly used in estimating the parameter. In the binomial case Samuel's (1966) method always yields a saving of observations, as compared with the methods of Birnbaum and Healy (1960). For the Poisson case, the method is not uniformly better than that of Birnbaum and Healy (1960). Certain numerical comparisons are given. For further details, the reader is referred to Samuel's paper.

4.3 Formulation of the Problem of Sequential Estimation by Intervals or Sets

The general nature of a sequential procedure of estimation by sets may be described as follows. For any positive integer m, we consider a set E_m of samples of size m. These sets must satisfy the condition: if the sample $\mathbf{X}_m$ belongs to E_m and if $\mathbf{X}_{m'}$ $(m' > m)$ belongs to $E_{m'}$, then $\mathbf{X}_m$ must not be equal to the sample consisting of the first m observations in $\mathbf{X}_{m'}$ (that is, the projection of $\mathbf{X}_{m'}$ in R_m must not be $\mathbf{X}_m$). With any sample point $\mathbf{X}_m$ of E_m $(m = 1, 2, \ldots)$ we associate a subset $W(\mathbf{X}_m)$ of the parameter space. We continue to make observations on X until we reach a value n such that $\mathbf{X}_n$ belongs to E_n. At this stage we stop and state that $W(\mathbf{X}_n)$ contains the true parameter point, that is, $W(\mathbf{X}_n)$ is the confidence set resulting from the sequential estimation procedure.

Thus, the sequential estimation procedure is determined by the sets E_1, $E_2, \ldots$ and the set function $W(\mathbf{X})$ defined for all $\mathbf{X}$ in $E_1, E_2, \ldots$. The main problem is a proper choice of E_1, $E_2, \ldots$, and $W(\mathbf{X})$. First we shall impose the following two conditions.

CONDITION I The confidence set $W(\mathbf{X}_n)$ resulting from sequential estimation procedure should satisfy certain requirements regarding its geometric shape.

CONDITION II The confidence set $W(\mathbf{X}_n)$ should satisfy the inequality: $P(\theta \in W(\mathbf{X}_n) | \theta) \geq 1 - \alpha$.

Condition I does not constitute a statistical problem and the geometric shape of a confidence region would be decided on the basis of practical considerations. If the parameter space is one dimensional, then we may require $W(\mathbf{X}_n)$ to be an interval of fixed length, $2d$, or its length to be a function of the midpoint of the interval. The latter criterion is of interest in estimating the mean of a binomial distribution. The specification may be in terms of volume or the diameter (equal to the largest possible distance between two points of the set) of the set.

In general, many sequential procedures satisfy conditions I and II. The criterion for choosing one from among them will be based on the expected number of observations required by the estimation procedure. The problem of finding an optimum estimation procedure (in the sense that it has the smallest expected sample size) is not solved completely.

4.4 Wald's Sequential Estimation Procedures

Let $p_m(X_1, \ldots, X_m; \theta)$ be the joint density of a random sample of size m. For any parameter point θ, let $p_m^*(X_1, \ldots, X_n; \theta)$ be an arbitrary but given probability density of $X_1, \ldots, X_m$. Then according to a lemma of Wald (1947, p. 146), the probability that

$$p_m^*(X_1, \ldots, X_m; \theta)/p_m(X_1, \ldots, X_m; \theta) < A \tag{4.4.1}$$

will hold for all m is greater than or equal to $1 - A^{-1}$ when θ is true. Let $W(\mathbf{X}_n)$ be the set of all θ for which (4.4.1) is fulfilled for all values $m \leq n$. If we set $A = 1/\alpha$, then $W(\mathbf{X}_n)$ will satisfy condition II for each n. Then the estimation procedure is as follows.

Continue taking observations as long as $W(\mathbf{X}_n)$ does not satisfy the requirements of condition I. We stop the process at the smallest n for which $W(\mathbf{X}_n)$ satisfies condition I and then state that the true parameter point θ is included in $W(\mathbf{X}_n)$. This stopping rule ensures the validity of condition I. Wald (1947) has claimed that if $p_m^*(X_1, \ldots, X_m; \theta)$ is chosen so that the probability is unity that the diameter of $W(\mathbf{X}_n)$ will converge to 0 as $m \to \infty$ and condition I is such that any set of sufficiently small diameter satisfies it, the probability is unity that the estimation process will be terminated at a finite stage.

However, there is no proper theory for choosing $p_m^*(X_1, \ldots, X_m; \theta)$ such that the expected sample size is minimized.

4.5 Certain General Concepts of Sequential Estimation

In this section we shall formulate the general loss function involved in sequential estimation and certain stopping rules.

Let $X_1, X_2, \ldots,$ be a sequence of independent random variables having the common density function $f(x; \theta)$. Let $r(\delta(x); \theta)$ denote the loss incurred when we make a terminal decision $\delta(x)$ and θ is the true value of the parameter. We should add to the loss the cost of experimentation namely $C(N)$, the cost of taking N observations (notice that N is random). The statistician's task consists of choosing a stopping rule and a terminal decision rule (estimate for θ). The following are the possible aims he might contemplate:

(i) Limited budget (forcing a bound on expected total cost of observations). That is, subject to $E_\theta\{C(N)\} \le n_0$, he might try to minimize $E_\theta\{r(\delta(X); \theta)\}$.

(ii) Specified accuracy (bound on the expected loss due to bad terminal decisions). That is, he would try to minimize $E_\theta\{C(N)\}$ subject to the condition

$$E_\theta\{r(\delta(X); \theta)\} \le \rho.$$

(iii) Both loses are economically important. That is he would try to minimize $E_\theta\{C(N) + r(\delta(X); \theta)\}$.

In general, there does not seem to be a sequential procedure that satisfies (iii) uniformly in θ unless the criterion is modified. Such modification can be given by a Bayesian criterion of optimality. Regarding (i), the Cramér–Rao inequality implies that an optimal sequential procedure is the one based on a fixed sample of size n_0 whenever the variance of its uniformly minimum variance unbiased estimator attains the Cramér–Rao lower bound. (This is the case, whenever the unbiased estimator in a fixed sample size procedure has a density belonging to the exponential family. If the criterion is (ii), there is a justification for a sequential procedure even in the case of exponential type distributions. This has been shown by DeGroot (1959) and Wasan (1964) in the binomial case. We shall take this up later.

4.6 Cramér–Rao Lower Bound for the Sequential Case

Suppose the statistician is interested in solving the problem (i) posed in Section 4.5 and that the terminal decision is an estimate for θ and let $r(\delta(x); \theta) = [\delta(x) - \theta]^2$. In this case, we might restrict ourselves to unbiased estimates of θ and would then be interested in lower bounds for the variance of such estimates. The Cramér–Rao inequality was extended to the sequential case by Wolfowitz (1947b). Toward this, we need the following lemma pertaining to a random sum of iid random variables.

Lemma 4.6.1 Let $S_N = X_1 + \cdots + X_N$ where the X_i are iid.

(i) If $E(N) < \infty$ and $E(|X|) < \infty$, we have $E(S_N) = E(N)E(X)$.
(ii) If $EX = 0$, $EX^2 < \infty$, and $EN < \infty$, $E(S_N{}^2) = E(N)E(X^2)$.
(iii) If $Eg(X) = Eh(X) = 0$, $E|g(X)h(X)| < \infty$ and $EN^2 < \infty$, then

$$E\left\{\left[\sum_1^N g(X_i)\right]\left[\sum_1^N h(X_i)\right]\right\} = E(N)E[g(X)h(X)]. \tag{4.6.1}$$

Proof For (i) see Theorem 2.4.2 and for (ii) see Theorem 2.4.1. Also (iii) has been established by Lehmann (1950) via a set of general lemmas. Here we shall provide a direct proof of (iii).

$$E\left\{\left[\sum_1^N g(X_i)\right]\left[\sum_1^N h(X_i)\right]\right\} = E\left\{\left[\sum_1^\infty Y_i g(X_i)\right]\left[\sum_1^\infty Y_i h(X_i)\right]\right\}$$

where $Y_i = 1$ if $N \geq i$ and zero otherwise. So, the left-hand expression is equal to

$$\begin{aligned} E\left[\sum_1^\infty Y_i{}^2 g(X_i)h(X_i) + \sum\sum_{i \neq j} Y_i Y_j g(X_i)h(X_j)\right] &= E[(g(X)h(X)]\sum_1^\infty EY_i + 0 \\ &= ENE[g(X)h(X)]. \end{aligned}$$

In order to interchange expectation and infinite summation, the series should be absolutely integrable. That is

$$E|g(X)h(X)| < \infty \qquad \text{and} \qquad \sum\sum_{i \neq j} EY_iY_j < \infty = 2\sum_{i=1}^\infty (j-1)P(N \geq j)$$

which involves EN^2. This completes the proof.

Wolfowitz (1947b) has given a generalized version of Wald's equation [Lemma 4.6.1(i)] which is valid even if the underlying variables are dependent.

Lemma 4.6.2 (Wolfowitz, 1947b) Let $v_i = E(X_i | N \geq i)$ exist for all positive integral i for which $P(N \geq i) \neq 0$. Then, write $v_i' = E(|X_i - v_i| \,|N \geq i)$, $i = 1, 2, \ldots$ and assume that the series $\sum_{i=1}^\infty (\sum_{j=1}^i v_j')P(N = i)$ converges.

Then

$$E\left(S_N - \sum_{i=1}^{N} v_i\right) = 0 \qquad \text{where} \quad S_N = \sum_{i=1}^{N} X_i. \tag{4.6.2}$$

Proof Consider

$$E\left(S_N - \sum_{1}^{N} v_i\right) = E\left[\sum_{i=1}^{\infty} Y_i(X_i - v_i)\right],$$

where $Y_i = 1$ if $N \geq i$ and zero elsewhere. Then

$$\begin{aligned} E\left(S_N - \sum_{1}^{N} v_i\right) &= \sum_{i=1}^{\infty} E[Y_i(X_i - v_i)] \\ &= \sum_{i=1}^{\infty} E\{E[(X_i - v_i)|\, Y_i = 1]\} = 0. \end{aligned}$$

The interchange of integration and infinite summation is justified because

$$\begin{aligned} \sum_{i=1}^{\infty} E[|\, Y_i(X_i - v_i)|\,] &= \sum_{i=1}^{\infty} v_i' P(N \geq i) \\ &= \sum_{j=1}^{\infty} \left(\sum_{i=1}^{j} v_i'\right) P(N = j) < \infty. \end{aligned}$$

Lemma 4.6.3 (Wolfowitz, 1947) Let the X_i be independent rvs having different distributions. Suppose that all v_i and v_i' are equal, except perhaps for those i for which $P(N \geq i) = 0$. Further assume that EN exists. Then (4.6.2) holds.

Notice that Lemma 4.6.3 is a special case of Lemma 4.6.2 and is a generalization of Lemma 4.6.1(i). If the random variables X_i are independent, then $v_i = EX_i$ and $v_i' = E|\, X_i - v_i|$, $i = 1, 2, \ldots$.

In the sequential sample space, let E_1 be the set of values of X_1 for which we stop after taking one observation. Let E_i be the set of values of $X_1, \ldots, X_i$ for which we stop after taking i observations. Notice that $(x_1, \ldots, x_i) \in E_i$ implies that $(x_1, \ldots, x_j) \notin E_j$, $j = 1, \ldots, i-1$ $(i = 1, 2, \ldots)$. The total sample space is $E_1 \cup E_2 \cup \cdots$. Recall that N denotes the random number of observations taken. Then if we define $g(x_1, \ldots, x_N) = g_i(x_1, \ldots, x_i)$ whenever $(x_1, \ldots, x_i) \in E_i (i = 1, 2, \ldots)$ we have

$$E[g(X_1, \ldots, X_N)] = \sum_{i=1}^{\infty} \int_{E_i} g_i(x_1, \ldots, x_i) \prod_{j=1}^{i} f(x_j; \theta)\, dx_j. \tag{4.6.3}$$

We assume that $P_\theta(\bigcup_1^\infty E_i) = 1$.

Theorem 4.6.1 (Wolfowitz, 1947b) Let $T_N = \delta(X_1, \ldots, X_N)$ be an estimate of θ, such that $E_\theta(T_N) = \theta + b(\theta)$. Suppose that differentiation underneath the summation and integral signs is permissible in $E_\theta(1) = 1$ and $E_\theta(T_N) = \theta + b(\theta)$ where $b'(\theta)$ exists. Then

$$\sigma_{T_N}{}^2(\theta) \geq \{E_\theta(N) E_\theta [\partial \ln f(X; \theta)/\partial\theta]^2\}^{-1}[1 + b'(\theta)]^2. \tag{4.6.4}$$

Proof We have

$$\frac{\partial}{\partial\theta} E_\theta(1) = \sum_{j=1}^{\infty} \int_{E_j} \left[\sum_{i=1}^{j} \frac{\partial}{\partial\theta} \ln f(X_i; \theta)\right] \prod_{i=1}^{j} f(x_i; \theta)\, dx_i = 0.$$

That is, $E_\theta(S_N) = 0$ where

$$S_N = \sum_{i=1}^{N} \partial \ln f(X_i; \theta)/\partial\theta, \tag{4.6.5}$$

provided $\int(\partial f/\partial\theta)\, dx = 0$ and $E_\theta|\partial \ln f/\partial\theta| < \infty$. Differentiating $E_\theta(T) = \theta + b(\theta)$ similarly yields

$$E(T_N S_N) = 1 + b'(\theta) \leq \sigma_{T_N}(\theta)\sigma_{S_N}(\theta).$$

However from Lemma 4.6.1(ii) we have

$$\sigma_{S_N}{}^2(\theta) = E(N) E_\theta [\partial \ln f(X; \theta)/\partial\theta]^2,$$

provided $E|\partial \ln f/\partial\theta|^2 < \infty$.

REMARK 4.6.1.1 (Lehmann, 1950) If we are restricting ourselves to sequential estimation procedures for which $E_\theta(N) \leq n_0$ and the regularity conditions of Theorem 4.6.1 hold for all such estimates, then for every unbiased estimate T_N of θ, we have

$$\sigma_{T_N}{}^2(\theta) \geq \{n_0 E_\theta[\partial \ln f(X; \theta)/\partial\theta]^2\}^{-1}. \tag{4.6.6}$$

In the case of a normal distribution with mean θ and variance unity, binomial and Poisson distributions, one can establish the validity of (4.6.6) for all unbiased estimation procedures for which $E_\theta(N) \leq n_0$ provided there is an M such that $P(N \leq M) = 1$. This additional restriction to bounded procedures is somewhat unpleasant from the theoretical point of view, although in practice it is no restriction at all since M could be fairly large.

REMARK 4.6.1.2 If T_N is an unbiased estimator of $h(\theta)$, then one can analogously show [see, for instance Wolfowitz, 1947b)] that

$$\operatorname{var} T_N \geq [h'(\theta)]^2 \Big/ \left\{ E(N) E\left[\frac{\partial}{\partial \theta} \ln f(X;\theta)\right]^2 \right\}. \tag{4.6.7}$$

If $\boldsymbol{\theta} = (\theta_1, \ldots, \theta_k)'$ and if we are interested in unbiased estimation of a single component of $\underset{\sim}{\theta}$ say θ_1, then Wolfowitz has also obtained

$$[\operatorname{var}(T_{N,1})]^{-1} \leq (EN)E[\partial \ln f(X;\boldsymbol{\theta})/\partial\theta_1]^2 \tag{4.6.8}$$

where $T_{N,1}$ denotes any unbiased estimate of θ_1.

Blackwell and Girshick (1947) have shown that the lower bound given by (4.6.4) for the variance of an unbiased estimate of θ is attained only for the sequential process for which $P(N = n) = 1$, if the probability density function $f(x;\theta)$ of X is such that $E(X) = \theta$ and $X_1, + \cdots + X_m$ is a sufficient statistic for all integral values of m, for estimating θ; $X_1, \ldots, X_m$ being independent observations on the random variable X.

Seth (1949) has extended Bhattacharyya's bounds (1946) to the sequential case which in some respects are more general than those of Wolfowitz (1947b). In the following we shall give his result specialized to unbiased estimates of θ on the basis of iid observations.

Theorem 4.6.2 (Seth, 1949) Let $T_N = T(X_1, \ldots, X_N)$ be an unbiased estimate of θ having a finite variance and let θ lie in an open interval I. The derivatives $\partial^i f(x;\theta)/\partial\theta^i)(i = 1, 2, \ldots, k)$ exist for all θ in I and almost all x. Let

$$\lambda_{ij} = E(N)E\left[\frac{\partial^i \ln f(X;\theta)}{\partial\theta^i} \frac{\partial^j \ln f(X;\theta)}{\partial\theta^j}\right], \qquad 1 \leq i, \quad j \leq k \tag{4.6.9}$$

and

$$((\lambda^{ij})) = ((\lambda_{ij}))^{-1}. \tag{4.6.10}$$

Further assume that $E_\theta(1)$ and $E_\theta(T_N)$ are differentiable underneath the infinite summation and the integration at least k times. Then, we have

$$\operatorname{var}(T_N) \geq \lambda^{1,1}. \tag{4.6.11}$$

Proof See Seth (1949, pp. 5–8).

Corollary 4.6.2.1 If $k = 1$, Theorem 4.6.2 reduces to Theorem 4.6.1.

Seth (1949) has also obtained conditions under which the inequality in (4.6.10) [and hence, in (4.6.4)] becomes an equality.

Problem

Show that the exponential family of densities satisfy the regularity conditions of Lemma 4.6.2.

4.7 Sufficiency and Completeness in the Sequential Case

As before, let $X_1, X_2, \ldots$ be a sequence of iid random variables having the common density $f(x; \theta)$. An estimate $\delta(X_1, \ldots, X_N)$ is wanted for θ. In this section we shall assume that the stopping rule is closed [that is, for every θ, $P(N \leq n) \to 1$ as $n \to \infty$, not necessarily uniformly in θ]. This is not of much interest since the choice of a good stopping rule would help us in obtaining a good estimate. Hence, the methods of this section cannot be used for finding optimum stopping rules, the way Cramér–Rao inequality could be used. However, the situation considered here might arise, if the stopping rule was chosen so as to be optimal for testing a hypothesis about θ, and then as an afterthought we decide to estimate θ.

The sample space consists of $E_1 + E_2 + \cdots$ where E_i is contained in R_i and consists of those points $(X_1, \ldots, X_i)$ for which we stop after observing X_i. As before, let N denote the random number of observations taken. Further assume that the conditional probabilities exist as measures.

Assume that for each n the joint density of $X_1, \ldots, X_n$ admits a sufficient statistic. Let $T_n = T(X_1, \ldots, X_n)$ be such a minimal sufficient statistic for the joint density of $X_1, \ldots, X_n$.

Definition 4.7.1 The sequence $(T_1, T_2, \ldots)$ is called a sufficient sequence for the sequential model.

Then we have the following result of E. Fay [see Lehmann (1950)].

Theorem 4.7.1 (E. Fay) If for each n, $T_n = T(X_1, \ldots, X_n)$ is a sufficient statistic for θ in the fixed sample $X_1, \ldots, X_n$, then (T_N, N) is a sufficient statistic for θ in the sequential case.

Proof This theorem was proved by Blackwell (1947) under the assumption that the stopping rule depends only on the Ts. On the other hand, the existence of conditional probabilities is assumed. Let E be any measurable set in $\bigcup_1^\infty E_i$. Then

$$P(E \mid N = n, T_n = t) = P(E, N = n \mid T_n = t)/P(N = n \mid T_n = t)$$
$$= P(E \cap E_n \mid T_n = t)/P(E_n \mid T_n = t). \qquad (4.7.1)$$

Now $E \cap E_n$ and E_n are sets in R_n. Since T_n is sufficient for θ on the basis of $X_1, \ldots, X_n$, both numerator and denominator of the above expression are independent of θ. Hence $P(E \mid N = n, T_n = t)$ is free of θ. Thus (T_N, N) is sufficient for θ in the sequential case.

Let $\{T_N\}$ be a sufficient sequence for the sequence of random variables $X_1, X_2, \ldots,$ Then the SPRT at stage n depends only on the observed value of T_n. Even in sequential estimation, it is reasonable to base our decision on T_n if experimentation is stopped at the nth stage. Then, we are led to the definition of transitivity.

Definition 4.7.2 The sequence of sufficient statistics $\{T_N\}$ is said to be a *transitive sequence* if for every $n > 1$ and for each $\theta \in ,\Theta$ the conditional distribution of T_{n+1} given $(X_1, \ldots, X_n) = (x_1, \ldots, x_n)$ is identical to the conditional distribution of T_{n+1} given that $T_n(X_1, \ldots, X_n) = t_n(x_1, \ldots, x_n)$ [see also Eq. (3.6.5)].

In other words, transitivity of $\{T_N\}$ implies that all the information about T_{n+1} contained in $\mathbf{X}_n = (X_1, \ldots, X_n)$ is carried by the function T_n. Bahadur (1954) has shown that, in the presence of a sufficient and transitive sequence $\{T_n\}$, any closed sequential procedure based on $\{\mathbf{X}_n\}$ (for making a decision about θ) is equivalent to a procedure which at stage n is based only on T_n. $\{T_n\}$ is a transitive sequence for the family of iid random variables if $T_{n+1}(\mathbf{X}_{n+1}) = \psi_n(T_n(\mathbf{X}_n), X_{n+1})$, for every $n > 1$. The exponential family has this property. For a measure-theoretic discussion of this concept and further properties of transitivity, the reader is referred to B. K. Ghosh (1970, Appendix 4, pp. 436–438).

Problem

Verify that the sequence of sufficient statistics for the exponential family of densities is transitive.

Next, we shall consider completeness of (T_N, N). Assume that T_m is complete for every fixed m.

Definition 4.7.3 The family of distributions of (T_N, N) is said to be *complete* if $E_\theta[g(N, T_N)] \equiv 0$ implies that $g(n, T_n) = 0$ almost everywhere.

Definition 4.7.4 The family of distributions of (T_N, N) is said to be *boundedly complete* if for every bounded $g(n, t_n)$, $E_\theta g(N, T_N) \equiv 0$ implies that $g(n, t_n) = 0$ almost everywhere.

Notice that bounded completeness is clearly a weaker property than completeness.

Lehmann and Stein (1950) have found a general necessary condition for completeness of the statistic (N, T_N). It is of interest to explore the stopping rules for which (N, T_N) is complete. Lehmann and Stein (1950) have examined this question where the Xs are normal $(\theta, 1)$, Poisson, or rectangular on $(0, \theta)$. In the case of normal $(\theta, 1)$, (N, T_N) is complete if $N = m$ (here $T_m = \sum_1^m X_i$). In the Poisson case, $T_m = \sum_1^m X_i$. Let S_m be the set of values of T_m for which we stop at the mth observation. A necessary and sufficient condition for (N, T_N) to be complete is that the S_is be disjoint intervals, each lying immediately above the preceding one. For example, if the stopping rule is continue sampling until T_m exceeds c (a given value), then (N, T_N) is not complete. Similarly one can obtain a necessary and sufficient condition in the rectangular case [see Lehmann and Stein (1950, Example 2)].

Example 4.7.1 (Binomial case) Let $P(X_i = 1) = \theta$ and $P(X_i = 0) = 1 - \theta$. Then $T_n = \sum_1^n X_i$ is a sufficient statistic for θ on the basis of the fixed-size sample. Assume that we are given a closed stopping rule which depends only on the T_n. For a given stopping rule, each point (m, T_m) is one of the following types: (i) stopping point, (ii) continuation point, (iii) impossible point. The sample space of points (N, T_N) consists of all stopping points. A continuation point is not a possible value of the sufficient statistic (N, T_N) since we have a closed stopping rule. We are interested in estimating unbiasedly θ. X_1 is unbiased for θ. So, applying the Rao–Blackwell theorem,† we can find

$$Y = E(X_1 | N, T_N) = P(X_1 = 1 | N, T_N)$$

where Y would be unbiased for θ and var $Y \le$ var X_1. Each sample $(X_1, \ldots, X_n)$ can be viewed as a path starting at the origin (0) and ending at a stopping point (n, T_n). The ith step of such a path is either to the point immediately on the right if $X_i = 0$ or to the point immediately above that if $X_i = 1$. Obviously, a path cannot pass through a stopping point before reaching (n, T_n). The probability of any single path from 0 to (m, t) is $\theta^t(1 - \theta)^{m-t}$. Let $\pi(m, t)$ denote the number of paths leading to (m, t) starting at 0, and $\pi^*(m, t)$ be

† The Rao–Blackwell theorem states that if T is sufficient for θ and U is any unbiased estimator for θ, then $V(t) = E(U | T = t)$ is unbiased for θ and var $V \le$ var U.

the number of paths leading to (m, t) starting at $0^* = (1, 1)$. It would help if the reader draws diagrams. Then, we have

$$\begin{aligned} y(m, t) &= P(X_1 = 1 \mid N = m, T_N = t) \\ &= \frac{P(X_1 = 1, (N, T_N) = (m, t))}{P((N, T_N) = (m, t))} = \frac{\theta^t(1-\theta)^{m-t}\pi(m, t)}{\theta^t(1-\theta)^{m-1}\pi^*(m, t)} \\ &= \pi(m, t)/\pi^*(m, t) \end{aligned}$$

That is

$$Y = \pi^*(N, T_N)/\pi(N, T_N). \tag{4.7.2}$$

Let us consider some special stopping rules:

(i) *Sample of fixed size m.* Then

$$\pi(m, t) = \binom{m}{t}, \qquad \pi^*(m, t) = \binom{m-1}{t-1}$$

and consequently $y = t/m$.

(ii) *Continue sampling until c successes are obtained, and then stop.* Then

$$\pi(m, c) = \binom{m-1}{c-1}, \qquad \pi^*(m, c) = \binom{m-2}{c-2}.$$

Hence, $y = (c-1)/(m-1)$.

(iii) *Any stopping rule with* $0^* = (1, 1)$ *as a stopping point.* Then

$$y(m, t) = \begin{cases} 1 & \text{if } (m, t) = 0^*, \\ 0 & \text{otherwise.} \end{cases}$$

That is $Y = X_1$. Notice rule (ii) with $c = 1$ reduces to rule (iii).

(iv) *Curtailed simple sampling.* A lot is to be accepted if fewer than c defectives appear in a sample of size s, and the lot is to be rejected if c or more defectives are observed. Here, one need not take the full sample of size s. We can stop as soon as c defectives or $s - c + 1$ nondefectives are observed. It is customary to inspect all the items in the sample even if the final decision (either to reject or accept the lot) is known before the completion of the inspection of the sample. One reason for this procedure is that an unbiased estimate for θ is not known when the inspection is halted as soon as a decision is reached. Girshick *et al.* (1946) have provided a best unbiased estimate of θ. Here

$$y(x, c) = \binom{c-2+x}{c-2} \Big/ \binom{c+x-1}{c-1} = \left(\frac{c-1}{c+x-1}\right)$$

where x is the number of nondefective items examined, which is the unique unbiased estimate along the horizontal line corresponding to rejection with $c > 1$ defectives. Further, the unique unbiased estimate along the line corresponding to acceptance for $c > 1$ is

$$y(s-c+1, m) = \binom{s-c+m-1}{s-c} \Big/ \binom{s-c+m}{s-c} = \left(\frac{m}{s-c+m}\right)$$

which is equal to the number of defectives observed divided by one less than the number of observations.

Girshick *et al.* (1946) give an example of a general curtailed double sampling plan. They also provide necessary and sufficient conditions for the existence of a unique unbiased estimate of θ. Toward this, we need the following notation.

For every point (n, T_n), n is called the index of the point. The index of a region is the least upper bound of the indices of its accessible (possible) points. A region is said to be *finite* if there exists an m such that the indices of the possible points are less than m. In particular, a region containing a finite number of points is finite. A region is said to be *closed* if the sum of the probabilities of the stopping points in the region is unity. In particular a finite region is closed.

Theorem 4.7.2 (Girshick *et al.*, 1946) A sufficient condition that a region R be closed is that $\liminf_{n\to\infty} D(n)/n^{1/2} = 0$, where $D(n)$ is the number of possible points of index n.

Proof See Girshick *et al.* (1946, pp. 14–15).

Corollary 4.7.2.1 If the number of accessible points of a region R of index n is bounded, the region is closed.

Definition 4.7.5 (Girshick *et al.* 1946). A region R is said to be simple if all the points between any two possible points on the vertical line at n are possible points. That is, no possible points of index n shall be separated on the vertical line at n by impossible points or stopping points. In other words, all possible points of index n form an interval.

Theorem 4.7.3 (Girshick *et al.*, 1946) A necessary and sufficient condition that the estimate Y given by (4.7.2) is the unique unbiased estimate of θ for a closed finite region R is that R be simple.

Proof See Girshick *et al.* (1946, p. 19).

It is of interest to further characterize the unique unbiased estimators of θ and to find their variances.

In the binomial case, a sampling plan is *complete* (*boundedly complete*) if the only estimator g, such that $E_\theta(g) = 0$ for all θ, is the one defined by $g(\gamma) = 0$ for all γ belonging to the boundary (the totality of stopping points).

Theorem 4.7.4 (L. J. Savage, 1947; Wolfowitz, 1946). A necessary and sufficient condition for a closed sampling plan to be boundedly complete is that it be simple.

Theorem 4.7.5 (Lehmann and Stein, 1950) A necessary and sufficient condition for a closed sampling plan to be complete is that it be simple and that the conversion of any boundary point (stopping point) to a continuation point destroy closure.

Sometimes we may be interested in estimating unbounded functions of parameters like $1/\theta$. Hence, completeness is more relevant than bounded completeness.

Problems

1. In the binomial problem, assume that at least two observations are taken. Write 0^{**} for the point (2, 1). Estimate unbiasedly $\theta(1-\theta)$. [Hint: We have $P(0^{**}) = 2\theta(1-\theta)$ and let $Y = \frac{1}{2}P(0^{**} \mid N, T_N) = \pi^{**}(n, T_n)/2\pi(n, T_n)$ where $\pi^{**}(n, t)$ denotes the number of paths from 0^{**} to (m, t). Then Y is an unbiased estimate of $\theta(1-\theta)$.]
2. In the binomial case, check whether removal of (i) 0^*, (ii) (2, 0) destroys closure. Find all unbiased estimates of θ which depend on (N, T_N) only. Among them, there is only one bounded estimate.

Knight (1965) has presented a unified treatment of the inverse sampling applicable to the hypergeometric, binomial and Poisson distributions. He extends the results of Girshick *et al.* (1946) to a more general class of stopping rules called monotone. Estimates of parameters of interest and estimates of the variances of these estimates are obtained.

DeGroot (1959) has considered unbiased estimation of a function of θ, namely $h(\theta)$. From the Cramér–Rao inequality [see Eq. (4.6.7)] we have

$$\operatorname{var} g \geq \theta(1-\theta)[h'(\theta)]^2/E_\theta(N). \tag{4.7.3}$$

Definition 4.7.6 (DeGroot, 1959) A sampling plan S and an estimator g are said to be optimal at $\theta = \theta_0$ if among all procedures with average sample size at θ_0 no larger than that of S, there does not exist an unbiased estimator with smaller variance at θ_0 than that of g.

If a particular estimator for a given sampling plan attains the lower bound in (4.7.3) for its variance at θ_0, then it is concluded that the estimator and the sampling plan are optimal at θ_0. Such an estimator is said to be efficient at θ_0. DeGroot (1959) has shown that the (fixed) single-sample plans and the inverse binomial sampling plans are the only ones that admit an estimator that is efficient at all values of θ. For inverse sampling plan, DeGroot (1959) has given an explicit expression for the unique unbiased estimator of $h(1 - \theta)$. The stopping points of an inverse sampling plan is the totality of $\{\gamma : T(\gamma) = c\}$ where T denotes the number of defectives at the point γ. Then, for each nonnegative integer k there exists a unique stopping point γ_k such that $N(\gamma_k) = c + k$. Since

$$P(N = c + k) = \binom{k + c - 1}{k} \theta^c q^k, \qquad q = 1 - \theta, \tag{4.7.4}$$

then, for any estimator g

$$E(g) = \theta^c \sum_{k=0}^{\infty} g(\gamma_k) \binom{k + c - 1}{k} q^k. \tag{4.7.5}$$

Theorem 4.7.6 (DeGroot, 1959) A function $h(q)$ is estimable unbiasedly if and only if it can be expanded in Taylor's series in the interval $|q| < 1$. If $h(q)$ is estimable, then its unique unbiased estimator is given by

$$g(\gamma_k) = \frac{(c-1)!}{(k+c-1)!} \frac{d^k}{dq^k} \left[\frac{h(q)}{(1-q)^c} \right]_{q=0}, \qquad k = 0, 1 \ldots. \tag{4.7.6}$$

Proof Since $h(q)$ can be expanded in Taylor's series in the given interval, then so also $h(q)/(1 - q)^c$ can be expanded and conversely. Then, suppose that

$$h(q)/(1 - q)^c = \sum_{k=0}^{\infty} b_k q^k.$$

That is

$$h(q) = \theta^c \sum b_k q^k$$

and

$$g(\gamma_k) = b_k \Big/ \binom{k+c-1}{k}$$

yields an estimator g with $E(g) = h(q)$. Suppose now that $h(q)$ is estimable unbiasedly. Then, there exists an estimator g such that

$$h(q) = \theta^c \sum_{k=0}^{\infty} g(\gamma_k) \binom{k+c-1}{k} q^k$$

or

$$\frac{h(q)}{(1-q)^c} = \sum_{k=0}^{\infty} g(\gamma_k) \binom{k+c-1}{k} q^k.$$

Hence

$$\binom{k+c-1}{k} g(\gamma_k) = \frac{1}{k!} \frac{d^k}{dq^k} \left[\frac{h(q)}{(1-q)^c} \right]_{q=0}.$$

The uniqueness of $g(\gamma_k)$ follows from the uniqueness of the Taylor series expansion, which is the basis of the completeness of this sampling plan. This completes the proof.

It is often possible to find the expectation of a given estimator in closed form by using the fact that if the series

$$\psi(x) = \sum_{k=0}^{\infty} b_k x^k \tag{4.7.7}$$

is differentiated m times within its interval of convergence, then

$$\frac{\psi^{(m)}(x)}{m!} = \sum_{k=0}^{\infty} b_{m+k} \binom{k+m}{k} x^k. \tag{4.7.8}$$

As an illustration of the technique involved, the variance of the unbiased estimator of θ and the moment-generating function of N will be determined.

Theorem 4.7.7 (DeGroot, 1959) Let $g(\gamma_k) = (c-1)/(k+c-1)$ which is an unbiased estimator of θ for $c \geq 2$. Then we have

$$E(g^2) = \frac{(c-1)\theta^c}{q^{c-1}} \left[(-1)^{c-1} \ln \theta + \sum_{i=1}^{c-2} \frac{(-1)^{c-i}}{i} \left(\frac{q}{\theta}\right)^i \right] \tag{4.7.9}$$

where $q = 1 - \theta$ and

$$E(e^{tN}) = [\theta e^t/(1 - qe^t)]^c. \tag{4.7.10}$$

Proof

$$E(g^2) = \theta^c(c-1)^2 \sum_{k=0}^{\infty} (k+c-1)^{-2}\binom{k+c-1}{k}q^k$$
$$= \theta^c(c-1)^2 \frac{1}{(c-1)!}\frac{d^{c-1}}{dq^{c-1}}\left[\sum_{k=1}^{\infty} k^{-2}q^k\right],$$

after using (4.7.8). Note that the constant term in the last series on the right-hand side is taken to be zero. Its value can be assigned arbitrarily since it does not appear in the derived series. However,

$$\frac{d}{dq}\left[\sum_1^{\infty} k^{-2}q^k\right] = \sum_1^{\infty} k^{-1}q^{k-1} = q^{-1}\sum_1^{\infty} k^{-1}q^k = [-\ln(1-q)]/q.$$

Consequently,

$$\frac{d^{c-1}}{dq^{c-1}}\left[\sum_1^{\infty} k^{-2}q^k\right] = \frac{-d^{c-2}}{dq^{c-2}}\left[\frac{\ln(1-q)}{q}\right]$$

where

$$\frac{d^m}{dq^m}\left[\frac{\ln(1-q)}{q}\right] = (-1)^m\frac{m!\ln(1-q)}{q^{m+1}} + m!\sum_{i=1}^{\infty}\frac{(-1)^{m-i+1}}{i(1-q)^{m-i+1}}.$$

Using this would complete the proof for (4.7.9). For $t < \ln(1/q)$,

$$E(e^{tN}) = p^c\sum_{k=0}^{\infty} e^{t(c+k)}\binom{k+c-1}{k}q^k$$
$$= (pe^t)^c\sum_{k=0}^{\infty}\binom{k+c-1}{k}(qe^t)^k$$
$$= (pe^t)^c\frac{1}{(c-1)!}\frac{d^{c-1}}{dx^{c-1}}\left[\sum_{k=0}^{\infty} x^k\right]_{x=qe^t}$$
$$= (pe^t)^c\frac{1}{(c-1)!}\frac{d^{c-1}}{dx^{c-1}}[(1-x)^{-1}]_{x=qe^t}$$
$$= (pe^t)^c(1-qe^t)^{-c}.$$

This completes the proof for (4.7.10).

REMARK 4.7.7.1 Haldane (1945) gives $E(g^2)$ in the form of an integral which after repeated integration by parts would yield (4.7.9).

Corollary 4.7.7.1 $E(N) = c/\theta$, and $\operatorname{var} N = c(1-\theta)/\theta^2$ achieving the Cramér–Rao lower bound (4.7.3). Thus, it is an efficient estimator of its expected value. Notice that g is not efficient in this sense. The efficient unbiased estimator of N is $[c!/(k+c-1)!]\ (c-k+2)(c-k+3)\cdots(c+1)$. In the following we shall state DeGroot's (1959) theorem pertaining to the optimality of the single-sample plans and the inverse binomial sampling plans.

Theorem 4.7.8 (DeGroot, 1959) When the stopping points γ are such that $N(\gamma) = n$, any nonconstant function of the form $a + bT$ is an efficient estimator of $a + bn\theta$, and these are the only efficient estimators. When γ is such that $T(\gamma) = c$, any nonconstant function of the form $a + bN$ is an efficient estimator of $a + bc(1/\theta)$, and these are only the efficient estimators where T denotes the number of defective items.

Proof See DeGroot (1959, pp. 86–87).

DeGroot (1959) gives a new characterization of completeness for bounded sampling plans and shows that the dimension of the linear space of unbiased estimators of zero can be determined simply by counting the number of boundary points.

Wasan (1964) also considered unbiased estimation of the binomial parameter. For problem (ii) posed in Section 4.5 [that is, to choose a procedure $\delta(X)$ which minimizes $E(N)$ subject to $E(\delta(X) - \theta)^2 \le \rho$], Wasan (1964) has shown that the fixed-sample procedure is admissible and minimax [see Wasan (1964, Theorem 1, p. 261)].

Consider the following symmetric curtailed sampling procedure $\delta^*(k)$ whose stopping points are $(k+i, k)$ and $(k+i, i)$, $i = 0, 1, \ldots, k-1$. Here

$$P(N = k+i, T_N = k) = \binom{k+i-1}{k-1}\theta^k(1-\theta)^i, \qquad i = 0, 1, \ldots, k-1 \tag{4.7.11}$$

and

$$P(N = k+i, T_N = i) = \binom{k+i-1}{k-1}\theta^i(1-\theta)^k, \qquad i = 0, 1, \ldots, k-1. \tag{4.7.12}$$

Then the unbiased estimate of θ [proposed by Girshick *et al.* (1946)] is given by

$$g(\gamma_k) = (k-1)/(k+i-1), \qquad i = 0, 1, \ldots, k-1,$$
$$= i/(k+i-1), \qquad i = 0, 1, \ldots, k-1 \tag{4.7.13}$$

if $\gamma_k = (k+i, k)$ or $(k+i, i)$ respectively.

Wasan studied the asymptotic optimality of $g(\gamma_k)$ for large k. This is given in the following theorem.

Theorem 4.7.9 (Wasan, 1964) We have

$$\lim_{k\to\infty} k \operatorname{var} g(\gamma_k) = \begin{cases} \theta^2(1-\theta) & \text{if } \theta \geq \frac{1}{2}, \\ (1-\theta)^2\theta & \text{if } \theta \leq \frac{1}{2}, \end{cases} \tag{4.7.14}$$

and

$$\lim k^{-1}E(N(\gamma_k)) = \begin{cases} 1/\theta & \text{if } \theta \geq \frac{1}{2}, \\ 1/(1-\theta) & \text{if } \theta \leq \frac{1}{2}. \end{cases}$$

Proof See Wasan (1964, pp. 264–266). Hence $g(\gamma)$ is asymptotically uniformly better than the fixed-sample size procedure.

Wasan (1964) also demonstrates that the fixed-sample size procedure with $m = 1/2c^{1/2}$ has the smallest total risk $= c^{1/2} = \max_{0<\theta<1} [cm + \theta(1-\theta)/m]$ among all procedures having bounded $E_\theta(N)$. Thus, the fixed-sample procedure with $m = 1/2c^{1/2}$ is minimax and admissible for problem (iii) posed in Section 4.5.

4.8 Minimax Estimation

Let the random variable X have the density function $f(x; \theta)$. We desire to estimate $h(\theta)$. Let $r(\delta(X); \theta)$ be the loss incurred in estimating $h(\theta)$ by $\delta(X)$. The desirability of an estimate is determined by its risk function $R(\delta; \theta) = E_\theta[r(\delta(X); \theta)]$. Two estimates having the same risk will be considered equally good. In Section 4.7 we have restricted ourselves to the class of unbiased estimates. In this section, we require some optimum property of the risk function as a whole, like, the risk function to have as low a supremum as possible.

Definition 4.8.1 An estimate $\delta(X)$ is said to be minimax if it minimizes supremum$_\theta$ of $R(\delta; \theta)$. That is $\delta^*(X)$ is said to be minimax if for any other estimator $\delta(X)$,

$$\sup_\theta R(\delta^*; \theta) \leq \sup_\theta R(\delta; \theta). \tag{4.8.1}$$

Definition 4.8.2 An estimate $\delta(X)$ is said to be inadmissible if there exists another estimate $\delta^*(X)$ having $R(\delta^*;\, \theta) \leq R(\delta;\, \theta)$ for all θ with strict inequality for at least one θ. An estimate $\delta(X)$ is said to be admissible if it is not inadmissible.

Definition 4.8.3 A class $\mathscr{D}$ of estimates is said to be essentially complete if for any given admissible estimate $\delta(X)$ with risk $R(\delta;\, \theta)$ there exists an estimate $\delta^*(X) \in \mathscr{D}$ such that $R(\delta^*;\, \theta) = R(\delta;\, \theta)$, for all θ.

There are two ways of obtaining minimax estimates: (i) a method using Cramér–Rao inequality which was developed by Hodges and Lehmann (1951), (ii) the Bayes method.

METHOD 1: MINIMAX ESTIMATES USING THE CRAMÉR–RAO INEQUALITY

Let $X_1, X_2, \ldots$ be independent each being normal $(\theta,\, 1)$. We want an estimate $\delta(X)$ for θ which minimizes $\sup_\theta R(\delta; \theta) = \sup_\theta E[\delta(X) - \theta]^2$ subject to $E_\theta(N) \leq m$. If $\delta(X)$ is any estimate based on a stopping rule for which $E_\theta(N) \leq m$ and for which there exists an M such that $P(N \leq M) = 1$, we have from the Cramér–Rao inequality

$$\begin{aligned} R(\delta;\, \theta) &= \operatorname{var}(\delta(X)) + b^2(\theta) \\ &\geq b^2(\theta) + [1 + b'(\theta)]^2/E(N) \\ &\geq b^2(\theta) + [1 + b'(\theta)]^2/m. \end{aligned}$$

We know that the estimate $\overline{X}_m = (X_1 + \cdots + X_m)/m$ has risk $R(\overline{X}_m;\, \theta) = 1/m$ and we would like to prove that this estimate is minimax and admissible. Suppose there exists an estimate $\delta(X)$ having $R(\delta;\, \theta) \leq 1/m$ for all θ. Then $R(\delta;\, \theta) \leq 1/m$ implies that

$$b^2(\theta) + [1 + b'(\theta)]^2/m \leq 1/m. \tag{4.8.2}$$

We will show that (4.8.2) implies that $b(\theta) \equiv 0$.

(i) $b(\theta)$ is bounded because $b^2(\theta) \leq 1/m$.

(ii) $b(\theta)$ is nonincreasing because $[1 + b'(\theta)]^2 \leq 1$. That is, $b'(\theta) \leq 0$ for all θ.

(iii) There exists a sequence $\theta_i \to \infty$ for which $b'(\theta_i) \to 0$. Otherwise, $b'(\theta)$ would be bounded away from 0, say $b'(\theta) < -\varepsilon$ as $\theta \to \infty$ and $b(\theta)$ could not be bounded.

(iv) Analogously one can establish that there exists a sequence $\theta_i \to -\infty$ for which $b'(\theta_i) \to 0$. Hence, from (iii) and (iv) we must have $b^2(\theta_i) \to 0$ and so $b(\theta_i) \to 0$.

Now from (ii), (iii), and (iv) it follows that $b(\theta) \equiv 0$. Hence, $R(\delta;\,\theta) \geq 1/m$. That is $R(\delta;\,\theta) \leq 1/m$ implies that $R(\delta;\,\theta) \geq 1/m$. So, $R(\delta,\,\theta) = 1/m$ for all θ. Consequently $\bar{X}_m$ is minimax and admissible. By completeness of $\bar{X}_m$, it follows that $\bar{X}_m$ is the unique estimate that is admissible and minimax.

REMARK 4.8.1 In the above example of the normal distribution, if θ is bounded, then $\bar{X}_m$ is neither admissible nor minimax, for all monotone loss functions. This will be clear from the following result of Lehmann (1950).

Theorem 4.8.1 (Lehmann, 1950) Consider any problem of estimating a parameter θ where $a \leq \theta \leq b$ with loss function $r(\delta(X);\,\theta) = c|\delta - \theta|^k$, $k > 0$ where c is a fixed constant and $\delta(X)$ is any estimate of θ that might take on values outside the interval $(a,\,b)$.

Then δ is neither minimax nor admissible with positive probability.

Proof Consider the estimate δ^* defined by

$$\delta^*(x) = \begin{cases} \delta(x) & \text{if } \ a \leq \delta(x) \leq b, \\ a & \text{if } \ \delta(x) < a, \\ b & \text{if } \ \delta(x) > b. \end{cases}$$

Then, the risk $R(\delta^*;\theta) < R(\delta;\theta)$ for these values of θ and $R(\delta^*;\,\theta) \leq R(\delta;\theta)$ for all θ. This is so because, whatever be θ, δ^* is always on the same side and closer or at least as close to θ as is δ. Hence $\delta(X)$ cannot be admissible. If $\delta(X)$ takes on values outside the interval $(a,\,b)$ with positive probability for all θ, $a \leq \theta \leq b$, then $R(\delta^*;\,\theta) \leq R(\delta;\,\theta)$ for all θ in $[a,\,b]$. If further, there is no sequence of θs for which $R(\delta^*;\,\theta) = R(\delta;\,\theta)$ (continuity of the risk functions is sufficient to ensure this property), we can conclude that $\delta(X)$ cannot be minimax.

In the particular case of the normal population, $\bar{X}_m$ satisfies these stronger conditions. The proof that $\bar{X}_m$ is admissible and minimax when θ is unbounded, now breaks down because conclusion (iii) cannot be made.

Case of Normal Mean Bounded Below

Let $X_1, X_2, \ldots$ be a sequence of independent normal variables each having a normal distribution with mean θ and variance unity where $\theta > \theta_0$. Using a squared loss error, we can show that $\bar{X}_m$ is minimax but inadmissible. Inadmissibility follows in the same way as in the preceding situation. If $\bar{X}_m$ is not minimax, there exists an estimate δ having $R(\delta;\,\theta) \leq m^{-1} - \varepsilon$ for all $\theta > \theta_0$. Since $R(\delta;\,\theta) \geq b^2(\theta) + [1 + b'(\theta)]^2/m$, we have

$$b^2(\theta) + [1 + b'(\theta)]^2/m \leq (1/m) - \varepsilon$$

from which it follows that $b(\theta)$ is bounded and that $b'(\theta) \leq -\varepsilon'$, $\varepsilon' = m\varepsilon$. Hence $b(\theta) \to -\infty$ as $\theta \to \infty$ which contradicts the assumption that $b(\theta)$ is bounded. Hence we cannot have $R(\delta;\, \theta) \leq (1/m) - \varepsilon$ for all $\theta > \theta_0$. So, $\bar{X}_m$ is minimax.

Using the above technique, Wasan (1964) has shown that in the case of the binomial parameter estimation, the fixed-sample procedure $(\bar{X}_m)$ is admissible and minimax.

Problems

1. Let X have a Poisson distribution with parameter θ. An estimate $\delta(X)$ is required for θ. Using the squared loss error show that for every estimate $\delta(X)$, the risk function $R(\delta;\, \theta)$ is unbounded and hence every estimate is minimax. [*Hint*: Use the Cramér–Rao inequality to get a lower bound for $R(\delta;\, \theta)$.]
2. If X has a Poisson distribution and if the loss function is of the form $r(\delta;\, \theta) = (\delta - \theta)^2/\theta$ (which implies that an error of given size is less serious for larger values of θ), an estimate for θ is required. Show that $\bar{X}_m$ is admissible and minimax.
3. If X has normal distribution having mean 0 and variable θ, an estimate $\delta(X)$ for θ is required with $r(\delta;\, \theta) = (\delta - \theta)^2/\theta^2$. Show that $Y = (m+2)^{-1} \sum_1^m X_i^2$ is an admissible minimax estimate for θ.

Method 2: Minimax Estimates Using Bayes Estimation Procedure

In some situations θ may be a random variable with unknown probability distribution, the density of which is $g(\theta)$. We are not even sure whether θ is a nontrivial random variable or not. One way to compute minimax estimates is to compute the Bayes estimates with respect to certain prior density g. Hence, we will be interested in Bayes' estimates. The following definition and theorem relate the Bayes and minimax procedures. For any decision δ and a priori density $g(\theta)$, let

$$W(\delta; g) = ER(\delta; \theta) = \int R(\delta; \theta) g(\theta)\, d\theta. \tag{4.8.3}$$

Definition 4.8.4 $g_0(\theta)$ is said to be a least favorable a priori density function for θ if the Bayes decision function for every other a priori distribution gives at least as small an average risk as does the Bayes decision function for g_0. That is,

$$\min_\delta W(\delta; g_0) \geq \min_\delta W(\delta; g) \tag{4.8.4}$$

where g is any other a priori density function.

Then we have the following theorem which relates the Bayes and minimax estimates.

Theorem 4.8.2 If δ_0 is Bayes for an a priori density g_0 of θ and if

$$W(\delta_0; g_0) = \max_\theta R(\delta_0; \theta), \tag{4.8.5}$$

then δ_0 is a minimax decision function and g_0 gives a least favorable a priori distribution of θ.

Proof δ_0 is Bayes with respect to g_0 means that

$$W(\delta_0; g_0) \le \min_\delta W(\delta; g_0).$$

That is, $W(\delta_0; g_0) \le W(\delta; g_0)$ for all δ. Also

$$W(\delta; g_0) \le \max_\theta R(\delta; \theta)$$

(since if a random variable X is such that $X \le m$, then $EX \le m$). Thus, for any δ

$$\max_\theta R(\delta_0; \theta) = W(\delta_0; g_0) \le W(\delta; g_0) \le \max_\theta R(\delta; \theta).$$

Hence, for every decision function δ

$$\max_\theta R(\delta_0; \theta) \le \max_\theta R(\delta; \theta). \tag{4.8.6}$$

That is, δ_0 is minimax. Further, since δ_0 is a particular decision function, we have for every a priori frequency function g

$$\min_\delta W(\delta; g) \le W(\delta_0; g).$$

Also

$$W(\delta_0; g) \le \max_\theta R(\delta_0; \theta) = W(\delta_0; g_0).$$

Therefore for every a priori g

$$\min_\delta W(\delta; g) \le W(\delta_0; g_0),$$

or

$$\min_{\delta} W(\delta; g) \le \min_{\delta} W(\delta; g_0).$$

That is, g_0 is least favorable a priori density for θ and this completes the proof.

The following theorems provide sufficient conditions under which a nonrandomized estimator is minimax. These theorems can be found in Hodges and Lehmann (1950) or Blyth (1951).

Theorem 4.8.3 Let δ_0 be Bayes with respect to the prior density g_0 and $R(\delta_0; \theta)$ be constant for every $\theta \in \Theta$ (the parameter space), then δ_0 is minimax estimator of θ.

Proof Since $R(\delta_0, \theta)$ is constant for all θ and δ_0 is Bayes for g_0, we have

$$\begin{aligned} R(\delta_0; \theta) &= \inf_{\delta} \int R(\delta; \theta) g_0 \, d\theta \le \sup_{\theta} \inf_{\delta} \int R(\delta; \theta) g(\theta) \, d\theta \\ &\le \inf_{\delta} \sup_{\theta} \int R(\delta; \theta) g \, d\theta \le \inf_{\delta} \sup_{\theta} R(\delta; \theta). \end{aligned}$$

Notice that the second inequality is true in general and is easy to verify. On the other hand,

$$R(\delta_0; \theta) = \sup_{\theta} R(\delta_0; \theta) \ge \inf_{\delta} \sup_{\theta} R(\delta; \theta).$$

Hence,

$$\sup_{\theta} R(\delta_0; \theta) = \inf_{\delta} \sup_{\theta} R(\delta; \theta).$$

By Theorem 4.8.2, δ_0 is minimax.

Theorem 4.8.4 Let $\{g_k, k = 1, 2, \ldots\}$ be a sequence of prior densities on Θ and let $\{\delta_k, k = 1, 2, \ldots\}$ and $W(\delta_k; g_k)$ be the corresponding sequences of Bayes estimators and Bayes risks. If δ_0 is an estimator of θ whose risk function is such that

$$\sup_{\theta} R(\delta_0; \theta) \le \limsup_{k \to \infty} W(\delta_k; g_k), \tag{4.8.7}$$

then δ_0 is minimax.

Proof Suppose that δ_0 is not minimax. Then there exists an estimator δ^* such that $\sup_\theta R(\delta^*; \theta) < \sup_\theta R(\delta_0; \theta)$. Since δ_k are Bayes with respect to g_k $(k \geq 1)$

$$W(\delta_k; g_k) \leq \int R(\delta^*; \theta) g_k \, d\theta \leq \sup_\theta R(\delta^*; \theta) \qquad \text{all} \quad k \geq 1.$$

Hence,

$$\limsup_{k \to \infty} W(\delta_k; g_k) < \sup_\theta R(\delta^*; \theta) < \sup_\theta R(\delta_0; \theta)$$

which contradicts (4.8.7). Hence δ_0 is minimax.

Example 4.8.1 (Wolfowitz, 1950) Let $X_1, X_2, \ldots$ be iid normal with mean θ and variance unity. Here, $T_n = \sum_1^n X_i$ is sufficient for each fixed n. Assume that θ has a prior normal distribution with mean 0 and variance τ^2. Let the loss function be

$$r(\delta; \theta) = \begin{cases} 0 & \text{if } |\delta - \theta| < \varepsilon, \\ 1 & \text{otherwise.} \end{cases}$$

Since the posterior density of θ is normal with mean $T_n/(n + \tau^{-2})$ and variance $(n + \tau^{-2})^{-1}$, the posterior risk is given

$$\begin{aligned} &1 - P[\delta(T_n) - \varepsilon < \theta < \delta(T_n) + \varepsilon | T_n] \\ &= 1 - \left[\Phi\left(\frac{\delta(T_n) + \varepsilon - T_n/(n + \tau^{-2})}{(n + \tau^{-2})^{-1/2}}\right) - \Phi\left(\frac{\delta(T_n) - \varepsilon - T_n/(n + \tau^{-2})}{(n + \tau^{-2})^{-1/2}}\right)\right]. \end{aligned}$$

By minimizing the risk wrt $\delta(T_n)$ we find that $T_n/(n + \tau^{-2})$ is the Bayes estimate of θ. Moreover,

$$\lim_{\tau \to \infty} [T_n/(n + \tau^{-2})] = T_n/n = \bar{X}_n.$$

[In finding the minimum, one should use the following considerations: $\Phi(x + \varepsilon - \mu) - \Phi(x - \varepsilon - \mu)$ is maximum at x_0 if $\phi(x_0 + \varepsilon - \mu) = \phi(x_0 - \varepsilon - \mu)$; $x_0 = \mu$ is a solution since $\phi(\varepsilon) = \phi(-\varepsilon)$. The second-order derivative of the function at x_0 is $-2\varepsilon\phi(\varepsilon) < 0$. Also since $\Phi(x_1 + \varepsilon) - \Phi(x_1 - \varepsilon) = \Phi(x_2 + \varepsilon) - \Phi(x_2 - \varepsilon)$ implies that $x_1 = -x_2$, the solution $x_0 = \mu$ is unique.] Hence, it follows from the earlier theorem that $\bar{X}_n$ is minimax for each fixed $n \geq 1$.

Now, if the other component of the loss function is cN then the risk is $c(EN)$. Also

$$R(\bar{X}_n; \theta) = 1 - [\Phi(\varepsilon n^{1/2}) - \Phi(-\varepsilon n^{1/2})]$$
$$= 2[1 - \Phi(\varepsilon n^{1/2})].$$

Then the minimax procedure can be found as follows. Given n observations, if an additional observation is taken, the minimax risk is reduced from $2[1 - \Phi(\varepsilon n^{1/2})]$ to $2[1 - \Phi(\varepsilon(n+1)^{1/2})]$. On the other hand the cost of samplign will be increased by c. Let $\Delta(n) = 2[\Phi(\varepsilon(n+1)^{1/2}) - \Phi(\varepsilon n^{1/2})]$. It is easy to verify that $\Delta(n)$ is a decreasing function of n, $n = 0, 1, \ldots$. Hence, the minimax stopping rule is to stop sampling at the least integer n ($n \geq 1$) for which $\Delta(n) \leq c$. Since $\Delta(n)$ is independent of the observed values of $X_1, \ldots, X_n$, the sample size can be determined ahead of time. Hence, the minimax estimation procedure is the fixed-sample procedure. If $\Delta(1) > c$, then $\bar{X}_0 = 0$.

REMARK 4.8.2 Wolfowitz (1950) has also solved the problem in which

$$r(\delta; \theta) = \begin{cases} 0 & \text{if } \delta \leq \theta + \varepsilon, \\ 1 & \text{otherwise.} \end{cases}$$

Example 4.8.2 Consider the same problem as in Example 4.8.1, however with $r(\delta; \theta) = (\delta - \theta)^2$ for all δ. In this case one can analogously show that $\bar{X}_n$ is the minimax estimate of θ for fixed n. Hence

$$R(\bar{X}_n; \theta) = cn + 1/n = n^{-1}[(1 - nc^{1/2})^2 + 2nc^{1/2}].$$

Consequently

$$\min_n R(\bar{X}_n; \theta) = 2c^{1/2} \qquad \text{when} \qquad n = c^{-1/2}.$$

Then one can ask, does there exist a sequential procedure δ with $R(\delta; \theta) \leq 2c^{1/2}$ for all θ? The answer is no and a fixed-sample procedure based on size $n = c^{-1/2}$ is minimax. Toward proving this, let N denote the number of observations at which we stop and estimate θ with $\delta = \bar{X}_N = (X_1 + \cdots + X_N)/N$. Then as in Example 4.8.1, taking the limit of the posterior risk as $\tau \to \infty$, we find the risk of δ to be

$$R(\delta; \theta) = cEN + E(N^{-1}) = \sum_{k=1}^{\infty} (ck + k^{-1})p_k$$

where $p_k = P(N = k)$ with $\sum_1^\infty p_k = 1$. Next, since $ck + k^{-1} \geq 2c^{1/2}$ for any k, it follows (either from the convexity argument or the remark that $Y \geq m$ implies that $EY \geq m$), that

$$\sum_1^\infty (ck + k^{-1})p_k \geq 2c^{1/2}.$$

Thus, we could not have done better than a fixed-sample procedure based on $n = c^{-1/2}$. See also Lehmann (1950) for an alternative method of proving this.

Example 4.8.3 Kiefer (1952) considered minimax estimation θ in $f(x; \theta)$ where

$$f(x; \theta) = \begin{cases} 1/\theta, & 0 < x < \theta, \\ 0, & \text{otherwise,} \end{cases}$$

with $r(\delta; \theta) = [(\delta - \theta)/\theta]^2$. We wish to find a sequential procedure that minimizes

$$\sup_\theta [cE_\theta(N) + E_\theta(\delta - \theta)^2/\theta^2].$$

It should be noted that all estimates will be minimax [since the supremum of the risk would be infinite if $r(\delta; \theta) = |\delta - \theta|$, $(\delta - \theta)^2$ or $(\delta - \theta)^2/\theta$]. Another justification for the choice of the loss function is that loss functions which depend only on δ/θ have a structure that essentially simplifies matters when estimating a scale parameter. For each fixed n, $Y_n = \max(X_1, \ldots, X_n)$ is sufficient for θ. Hence, we can restrict to estimation procedures that depend only on Y_n. The density of Y_n is nx^{n-1}/θ^n, for $0 < x < \theta$. Also, let the prior density of θ be given by

$$g_a(\theta) = \begin{cases} -1/(\theta \ln a), & a < \theta < 1, \\ 0 \quad , & \text{otherwise.} \end{cases} \tag{4.8.8}$$

Then, the posterior density of θ given Y_n is

$$h_n(\theta \mid Y_n) = \begin{cases} \dfrac{nz^n}{\theta^{n+1}(1 - z^n)}, & z < \theta < 1, \\ 0 \quad , & \text{otherwise.} \end{cases}$$

where $z = \max(a, Y_n)$. The posterior loss (excluding the cost of experimentation) if one stops after n observations and uses δ to estimate θ is

$$\int_0^1 \left(\frac{\delta - \theta}{\theta}\right)^2 h_n(\theta \mid Y_n)\, d\theta$$

$$= 1 + \frac{n}{z^2(1 - z^n)} \left[z^{n+2}\left(\frac{2\delta}{n+1} - \frac{\delta^2}{n+2}\right) - \left(\frac{2\delta z}{n+1} - \frac{\delta^2}{n+2}\right)\right]$$

and $\delta = (n + 2)(1 - z^{n+1})z/(n + 1)(1 - z^{n+2})$ uniquely minimizes the posterior loss and the corresponding loss function is $1 - n(n + 2)(1 - z^{n+1})^2/(n + 1)^2(1 - z^n)(1 - z^{n+2})$. For $n = 0$ we replace h_n by g_a and with $z = a$, the posterior loss becomes $1 - 2(1 - a)/[-(1 + a) \ln a]$.

The next step is to find the Bayes risk namely the expected value of the posterior loss with respect to the marginal distribution of Y_n. Kiefer (1952) proves that a lower bound for this Bayes risk approaches the function $r(n) = cn + (n + 1)^{-2}$ as $a \to 0$. Moreover there are at most two integers n_0 and $n_0 + 1$ at which $cn + (n + 1)^{-2}$ is minimized for each c. It is worthwhile to point out that although the posterior loss tends to the posterior loss of the minimax, the Bayes estimate does not tend to the minimax estimate as $a \to 0$. Hence Kiefer (1952) has to depart from the traditional way of obtaining the minimax estimate. Thus, the minimax procedure is given by n_0 observations where $r(n_0)$ is the minimum of $r(n)$ (if there are two minima, at n_0 and $n_0 + 1$, one may randomize in any way between the decisions to take n_0 or $n_0 + 1$ observations); and then estimating θ by

$$(n_0 + 2) \max(X_1, \ldots, X_{n0})/(n_0 + 1). \tag{4.8.9}$$

REMARK 4.8.3 If $g_a(\theta)$ is positive for $a < \theta < b$, then one considers the problem of estimating $b\theta$ from the sequence $\{bX_i\}$. Similarly, if the parameter space is given by $\Omega = \{\theta : b < \theta < \infty\}$, one can consider estimating $ba\theta$ from the sequence $\{baX_i\}$. However, the minimax procedure for the two problems would not be admissible if $n_0 > 0$ (or $n_0 \geq 0$ in the second case); for example, a trivially better procedure in the first case when $n_0 > 0$ is to estimate θ by b whenever the expression in (2.1) exceeds b. Finally, if $f(x; \theta) = (2\theta)^{-1}$ for $-\theta < x < \theta$, the problem of estimating θ here would be identical to the one considered, because Y_n is still sufficient for θ even if X_i is replaced by $|X_i| (i = 1, 2, \ldots, n)$. Furthermore, it is of interest to note that the problem may be translated (by considering $Z_i = \exp(-Xi)$, $\mu = e^{-\theta}$) into that of sequential minimax estimation of the parameter μ of the density, $\exp(-(t - \mu))$ for $t > \mu$ and zero elsewhere where $-\infty < \mu < \infty$; the appropriate loss function being $r(\delta; \mu) = [1 - \exp(-(\delta - \mu))]^2$.

Example 4.8.4 So far we have been producing examples in which one cannot do any better than a fixed-sample size procedure if one is interested in minimax estimation. Wald (1950) considered the problem of estimating the mean of a rectangular distribution by means of a minimax sequential procedure and he succeeded in obtaining an honest sequential minimax procedure. Let $f(x; \theta) = 1$ if $\theta - \frac{1}{2} \leq x \leq \theta + \frac{1}{2}$ and zero elsewhere where $-\infty < \theta < \infty$. We wish to find a point estimate for θ. Let the loss function be given by $r(\delta; \theta) = (\delta - \theta)^2$ if δ is designed to be an estimate of θ. Also, let c denote the cost of a single observation. Wald (1950, Section 5.2.5) assumes that θ has a priori density $g(\theta; a, b) = (b - a)^{-1}$ if $a < \theta < b$. Let $u_m = \min(X_1, \ldots, X_m)$ and $v_m = \max(X_1, \ldots, X_m)$. Then the posterior density of θ is again uniform on $I(u_m, v_m)$ which is equal to the common part of the intervals $[a, b]$ and $[v_m - \frac{1}{2}, u_m + \frac{1}{2}]$. For each fixed m, the optimal (Bayes) estimate of θ is $t(u_m, v_m) =$ the midpoint of the interval $I(u_m, v_m)$. Let $l_m = (u_m - v_m + 1)$ be the length of $I(u_m, v_m)$ and $l_0 = b - a$. The posterior risk (or loss) associated with $t(u_m, v_m)$ is $l_m{}^2/12$. Wald (1950) gives

$$E(l_{m+1}^2/12 \mid l_m = l) = \begin{cases} l^2/12 - l^3/24 & \text{if } 0 \leq l \leq 1, \\ 1/12 - 1/24 & \text{if } l > 1. \end{cases} \tag{4.8.10}$$

Then

$$\psi(l) = \frac{l^2}{12} - E(l_{m+1}^2 \mid l_m = l) \begin{cases} \dfrac{l^3}{24} & \text{if } 0 \leq l \leq 1, \\ \dfrac{l^2 - 1}{12} + \dfrac{1}{24l} & \text{if } l > 1. \end{cases} \tag{4.8.11}$$

Notice that $\psi(l)$ is the expected decrease in the posterior risk associated with the terminal decision due to an additional observation when l is the length of $I(u_m, v_m)$. Clearly $\psi(l)$ is monotonic increasing in l in the entire range of l. Thus, the equation $\psi(l) = c$ has exactly one root which will be denoted by $\bar{l}$. Thus the minimax sequential estimation procedure is as follows: For each nonnegative integral m compute l_m. If $l_m < \bar{l}$, stop and estimate θ by $t(u_m, v_m)$. If $l_m > \bar{l}$, take an additional observation. If $l_m = \bar{l}$, randomize between m and $m + 1$ observations.

Blyth (1951) has studied the admissibility of the minimax estimate of the mean of a rectangular population having unit range with $c(n)$ for cost of experimentation and $r(\delta; \theta) = h(|\delta - \theta|)$, nondecreasing in $|\delta - \theta|$. If $c(n)$ is convex, the minimax procedure has sample size fixed or randomized between two consecutive integers and it is admissible.

4.9 A Class of Situations Where Bayes Sequential Estimation Procedure Is Nonsequential

Let $\mathbf{X}_n = (X_1, \ldots, X_n)'$ denote a vector of n observations and let $T(\mathbf{X}_n)$ be designed to estimate the parameter θ that indexes the distribution of X_i. Also, $r(T; \theta)$ denotes the loss incurred in using T for estimating θ. In a decision-theoretic approach, one chooses $T(\mathbf{X}_n)$ to minimize $Er(T; \theta)$. If one is estimating sequentially, then one will choose the estimation rule $T(\mathbf{X}_n)$ and a stopping rule to minimize $E[cn + r(T; \theta)]$ where c denotes the constant cost per observation. However, there are several situations where the optimal stopping turns out to be simply $N = n_0$ where n_0 is a constant. That is, fixed-sample size procedure is optimal. This happens to be the case, if one is estimating the mean θ of a normal distribution of known variance, $r(T; \theta) = (T - \theta)^2$ and the prior distribution of θ is uniform over the real line. Blackwell and Girshick (1954) have obtained a general result characterizing the situation under which the optimal Bayes sequential procedure is nonsequential.

Theorem 4.9.1 (Blackwell and Girshick, 1954) Let $X_1, X_2, \ldots$ be iid sequence of chance variables. If $E[r(T(\mathbf{X}_n); \theta)|\mathbf{X}_n] \leq k(n)$ for all $\mathbf{X}_n$ and n where $k(n) \to 0$ as $n \to \infty$, then the optimal sequential procedure is truncated. If $E[r(T(\mathbf{X}_n); \theta)|\mathbf{X}_n] = k(n)$ (that is, it is a function of n only) the optimal sequential procedure is a fixed-sample size procedure.

Preliminaries for the Proof Let $g(\theta)$ denote the prior density of θ and $g_j(\theta)$ denote the posterior density of θ given the observations $X_1, \ldots, X_j$. That is

$$g_j(\theta) = g(\theta) \prod_{i=1}^{j} f(x_i; \theta) \Big/ \left[\int_\Omega g(\theta) \prod_{i=1}^{j} f(x_i; \theta)\, d\theta\right] \tag{4.9.1}$$

where Ω denotes the parameter space. One can easily verify that

$$g_{j+1}(\theta) = g_j(\theta) f(x_{j+1}; \theta) \Big/ \int_\Omega g_j(\theta) f(x_{j+1}; \theta)\, d\theta. \tag{4.9.2}$$

Thus, the posterior probability density g_j becomes the prior probability density for the next observation (or observations) when the X_i are independent. That is, knowing only g_j and x_{j+1} one can compute the posterior probability density of θ, given all the x_i from 1 to $j + 1$. This no longer holds if the x_i are not independent since in that case additional information, namely, the conditional density of X_{j+1} for given $X_1, \ldots, X_j$ would be required for each

θ. In addition, since the Xs are identically distributed, the functional form of $g_{j+1}(\theta)$ in (4.9.2) is the same for all $j = 0, 1, \ldots$. Let $\mathscr{G}$ denote the space of all densities g. Then one can state that a given sequence of j observations results in a point g_j in $\mathscr{G}$, and an additional observation transforms g_j into a different point in $\mathscr{G}$, the transformation being given by (4.9.2). Denote this transformation by T so that (4.9.2) can be written as $g_{j+1} = Tg_j$.

Let R be a real-valued and bounded function on $\mathscr{G}$ [for example, $R(g) = \inf_a \int_\Omega r(a;\theta)g(\theta)\,d\theta$, where $r(a;\theta)$ denotes the loss incurred in using action a when θ is the true state of nature], and for any g in $\mathscr{G}$, let

$$h(g) = E_0\,[R(Tg)] = E[R(Tg)] \tag{4.9.3}$$

where, by definition

$$E[R(Tg)] = \int_{R_1}\int_\Omega R(Tg)g(\theta)f(x;\theta)\,d\theta\,dx.$$

Now, consider the conditional expectation of g_{j+1} given $X_1, \ldots, X_j$. This can be written as

$$E_j\,[r(Tg_j)] = \int_{R_1}\int_\Omega r(Tg_j)g_j(\theta)f(x;\theta)\,d\theta\,dx. \tag{4.9.4}$$

Since the Xs are identically distributed, the above equation shows that if we set $g' = g_j$,

$$E_j\,[R(Tg_j)] = E[R(Tg')] = h(g')$$

Also, for any bounded function q defined on Ω, the conditional expectation of q for given $X_1, \ldots, X_j$ is given by

$$E_j(q) = \int_\Omega g_j(\theta)q(\theta)\,d\theta. \tag{4.9.5}$$

In particular, the conditional risk and the Bayes risk for given $X_1, \ldots, X_j$ can be expressed as

$$R_j(a;x) = \int_\Omega g_j(\theta)r(a;\theta)\,d\theta \tag{4.9.6}$$

and

$$R_j^*(x) = \inf_a R_j(a;x) = \psi(g_j). \tag{4.9.7}$$

Note that the domain of ψ is the space $\mathscr{G}$, and the functional form of ψ is the same for all j. Let $E_1, E_2, \ldots$ denote the sequence of stopping regions in $R_1, R_2, \ldots$ where R_j denotes the jth dimensional Euclidean space. Notice that E_j can be viewed as a cylindrical set in $R_\infty = R$ and that $\{E_j\}$ are mutually disjoint. Hence, for any truncated sequential sampling plan $\mathscr{E}_n = (E_0, E_1, \ldots, E_n)$ the risk $R(g; \mathscr{E}_n)$ can be written as

$$R(g; \mathscr{E}_n) = \sum_{j=0}^{n} \int_{E_j} [j + \psi(g_j)] p(x_1, \ldots, x_j; g)\, dx_1 \cdots dx_j \tag{4.9.8}$$

where

$$p(x_1, \ldots, x_j; g) = \int_\Omega g(\theta) \prod_{i=1}^{j} f(x_j; \theta)\, d\theta \tag{4.9.9}$$

[see Blackwell and Girshick (1954, Eq. (9.2.11), p. 241)]. Next, we can define functions $h_0, h_1, \ldots$ on $\mathscr{G}$ as follows:

$$h_0(g) = \psi(g) = \inf_a \int_\Omega g(\theta) r(a; \theta)\, d\theta$$

and by induction on j, $j > 0$ one can obtain

$$h_j(g) = \min[\psi(g), 1 + E(h_{j-1}(Tg))]. \tag{4.9.10}$$

Blackwell and Girshick (1954) also show that h_m is in fact the risk function of a sequential procedure truncated at m observations. Define

$$U_j = j + \psi(g_j) \qquad \text{and} \qquad \gamma_{n,n} = U_n. \tag{4.9.11}$$

Then, by backward induction, we have

$$\gamma_{j,n} = \min[U_j, E_j(\gamma_{j+1,n})] \qquad \text{for} \quad j < n. \tag{4.9.12}$$

We note that $\gamma_{n,n} = n + \psi(g_n) = n + \psi(Tg_{n-1}) = n + h_0(Tg_{n-1})$,

$$\begin{aligned}\gamma_{n-1,n} &= \min[n - 1 + \psi(g_{n-1}), n + E_{n-1}(h_0(Tg_{n-1}))]\\ &= n - 1 + \min[\psi(g_{n-1}), 1 + E_{n-1}(h_0(Tg_{n-1}))]\\ &= n - 1 + h_1(g_{n-1}).\end{aligned}$$

The last expression follows from (4.9.10). By a similar argument we have $\gamma_{n-2,n} = n - 2 + h_2(g_{n-2})$. In general we have

$$\gamma_{n-j,n} = n - j + h_j(g_{n-j}). \tag{4.9.13}$$

Lemma 4.9.1.1 $h_j(g)$ is a decreasing function of j for a fixed g.

Proof Replace n by $n+1$ and j by $j+1$ in (4.9.13) and obtain

$$h_j(g) - h_{j+1}(g) = \gamma_{n-j,n} - \gamma_{n-j,n+1}. \tag{4.9.14}$$

Thus it suffices to show that $\gamma_{k,n}$ decreases as n increases. We have, from (4.9.11)

$$\gamma_{n,n+1} = \min[\gamma_{n,n}, E_n(\gamma_{j+1,n+1})] \leq \gamma_{n,n}.$$

Applying an induction backward, we assume that

$$\gamma_{k,n+1} \leq \gamma_{k,n} \qquad \text{for} \quad k > j.$$

Then

$$\begin{aligned}\gamma_{j,n+1} &= \min[U_j, E_j(\gamma_{j+1,n+1})] \\ &\leq \min[U_j, E_j(\gamma_{j+1,n})] = \gamma_{j,n}.\end{aligned}$$

This completes the proof of showing that $h_j(g)$ is decreasing in j.

Lemma 4.9.1.2 (i) There is a monotonically nonincreasing sequence of subsets of $\mathscr{G}_0$ ($\mathscr{G}_0 = \mathscr{G}$), $\mathscr{G}_1, \mathscr{G}_2, \ldots$ such that, for any n and g, the best truncated sequential procedure $\mathscr{E}_n{}^* = (E_{0n}^*, E_{1n}^*, \ldots, E_{nn}^*)$ is described by

$$E_{jn}^* = \{x : g_r \notin \mathscr{G}_{n-r}, \qquad \text{for} \quad r < j,\ g_j \in \mathscr{G}_{n-j}\}.$$

(ii) If $R_n{}^*(g)$ is the Bayes risk from $\mathscr{E}_n{}^*$, then $R_n{}^*(g)$ approaches a limit, say $R^*(g)$ as n tends to infinity.

Proof Blackwell and Girshick (1954, Theorem 9.2.2, p. 243) give the optimal sequential procedure that is characterized by

$$E_{jn}^* = \{x : \gamma_{rn} < U_r \qquad \text{for} \quad r < j,\ \gamma_{jn} = U_j\}. \tag{4.9.15}$$

Hence, writing $r = n - (n - r)$ and employing (4.9.11) and (4.9.13) we have

$$E_{jn}^* = \{x : h_{n-r}(g_r) < \psi(g_r) \qquad \text{for} \quad r < j,\ h_{n-j}(g_j) = \psi(g_j)\}. \tag{4.9.16}$$

Now consider the sets

$$\mathcal{G}_j = \{g : h_j(g) = \psi(g)\}, \qquad j = 0, 1, \ldots. \tag{4.9.17}$$

Then, we have the identity

$$\begin{aligned}\{x : g_r \notin \mathcal{G}_{n-r} \quad &\text{for} \quad r < j, g_j \in \mathcal{G}_{n-r}\} \\ = \{x : h_{n-r}(g_r) \neq \psi(g_r) \quad &\text{for} \quad r < j, h_{n-j}(g_j) = \psi(g_j)\}.\end{aligned} \tag{4.9.18}$$

However from (4.9.10) if $h_j(g) \neq \psi(g)$, then $h_j(g) < \psi(g)$ for all j so that the $\neq$ sign in the right-hand expression of (4.9.18) can be replaced by $<$. This implies that the sequential procedure defined by (4.9.16) can be expressed as

$$E_{jn}^* = \{x : g_r \notin \mathcal{G}_{n-r} \quad \text{for} \quad r < j, g_j \in \mathcal{G}_{n-r}\} \tag{4.9.19}$$

which proves one part of (i).

To show that the sets $\mathcal{G}_j$ are decreasing with j is equivalent to showing that $h_j(g)$ is a decreasing function of j for fixed g. Assume the contrary and let $g \in \mathcal{G}_{j+1}$ and $g \notin \mathcal{G}_j$. Then $h_{j+1}(g) = \psi(g)$ and $h_j(g) \neq \psi(g)$, that is $h_j(g) < \psi(g) = h_{j+1}(g)$ which contradicts Lemma 4.9.1.1. Thus the proof of part (i) of Lemma 4.9.1.2. is complete.

Proof (ii) Since for a given g a possible sequential procedure truncated at $n + 1$ is to discard the $(n + 1)$th observation and employ the procedure $\mathcal{E}_n{}^*$, it must follow that $R_{n+1}^*(g) \le R_n{}^*(g)$. Thus, for fixed g, $R_n{}^*(g)$ is a decreasing function and consequently converges to a limit, $R^*(g)$ (say). Now we are ready to prove Theorem 4.9.1.

Proof of Theorem 4.9.1 From (4.9.13) we have, for all n

$$h_j(g_n) = \min[\psi(g_n), 1 + E_n(h_{i-1}(Tg_n))] \tag{4.9.20}$$

so that for any j, $h_j(g_n) = \psi(g_n)$ whenever $\psi(g_n) < 1$. Since $\psi(g_n) < 1$ for sufficiently large n, say $n \ge n_0$, we have $h_j(g_{n_0}) = \psi(g_{n_0})$ for all j: that is, g_{n_0} is in $\mathcal{G}_j$ for all j where $\mathcal{G}_0 = \mathcal{G}, \mathcal{G}_1, \mathcal{G}_2, \ldots$ is a monotonically nonincreasing sequence of subsets such that for any n and g, the best truncated sequential procedure $\mathcal{E}_n{}^* = (E_{0n}^*, E_{1n}^*, \ldots, E_{nn}^*)$ is described by

$$E_{jn}^* = \{x : g_r \notin \mathcal{G}_{n-r} \quad \text{for} \quad r < j, g_j \in \mathcal{G}_{n-r}\}$$

[see Blackwell and Girshick (1954, Theorem 9.3.1, p. 248)]. Hence $g_{n_0} \in \mathcal{G}^*$ where $\mathcal{G}^* = \bigcap_{i=0}^{\infty} \mathcal{G}_i$. Thus, by Theorem 9.3.1 of Blackwell and Girshick (1954), the optimal procedure does not require more than n_0 observations.

Assume now that $\psi(g_j) = k(j)$; then for any procedure $\mathscr{E}$

$$R(g;\mathscr{E}) = \sum_j \int_{E_j} [j + k(j)]p(x_1, \ldots, x_j; g)\, dx_1 \cdots dx_j$$
$$= \sum_j p(E_j|g_j)[j + k(j)] \geq \min_j[j + k(j)] = n_0 + k(n_0);$$

since the fixed-sample-size procedure with n_0 observations has risk $n_0 + k(n_0)$, it is clearly optimal. Notice that $k(j)$ in general will depend on g.

Example 4.9.1 Let $p(x_1, \ldots, x_n;\ \theta) = \binom{n}{r}\ \theta^r\ (1-\theta)^{n-r}$, $r = \sum_1^n x_i$ and $r(a;\ \theta) = d(a-\theta)^2$, $d > 0$, and $g(\theta) = 1$, for $0 \leq \theta \leq 1$ and zero elsewhere. Then, the Bayes estimate $d(r)$ if sampling terminates with n trials is

$$d(r) = \int_0^1 \theta^{r+1}(1-\theta)^{n-r}\, d\theta \Big/ \int_0^1 \theta^r(1-\theta)^{n-r}\, d\theta$$
$$= (r+1)/(n+2).$$

The Bayes risk

$$\psi(g_n) = \frac{d\int_0^1 \theta^{r+2}(1-\theta)^{n-r}\, d\theta}{\int_0^1 \theta^r(1-\theta)^{n-r}\, d\theta} - d\left(\frac{r+1}{n+2}\right)^2$$
$$= d(r+1)(n-r+1)/(n+2)^2(n+3)$$
$$\leq d(n+1)^2/(n+2)(n+3),$$

since $0 \leq r \leq n$. Thus $\psi(g_j) \to 0$ uniformly in r and by Theorem 4.9.1, the best sequential procedure is truncated.

On the other hand, if the loss is taken as $d(a-\theta)^2/\theta(1-\theta)$, the Bayes estimate is

$$d(r) = \frac{E_n(1/(1-\theta))}{E_n(1/\theta(1-\theta))} = \frac{r}{n}$$

and $\psi(g_j) = d/n$ for all x. Thus the best sequential procedure for this estimation problem is a fixed-sample size procedure. That is, we choose n such that $cn + d/n$ is minimum. That is $n \approx (d/c)^{1/2}$.

Using Theorem 4.9.1(ii) Whittle and Lane (1967) have obtained sufficient conditions under which, the optimal sequential estimate of the parameter (scalar or vector valued) of an exponential family of densities is based on a fixed-sample size. In the following we shall present their results.

Theorem 4.9.2 (Whittle and Lane, 1967) Let X have the probability (density) function $f(x;\,\theta)$ belonging to the exponential family where θ is scalar and let $X_1, X_2, \ldots$ be a sequence of random variables independent and identically distributed as X. If t is designed to estimate θ, let the loss function associated with t be of the form $r(t;\,\theta) = d(t-\theta)^2 I(\theta)$ where d is a constant and $I(\theta)$ is the Fisherian information on θ per observation. That is, $I(\theta) = E[\partial \ln f(X;\,\theta)/\partial\theta]^2$. Further, let θ admit a minimum variance unbiased estimator. Let the prior probability (density) function of θ be the one that is conjugate to $f(x;\,\theta)$ [see Raiffa and Schlaifer (1961, p. 43)]. That is

$$g(\theta) = (\text{const})\exp[Q(\theta)t_0 - \gamma(\theta)\Delta_0] \tag{4.9.21}$$

where t_0, Δ_0 are constants if

$$f(x;\,\theta)\,dx = \exp[Q(\theta)R(x) - \gamma(\theta)]\,dv(x) \tag{4.9.22}$$

where v is some sigma-finite measure. Then if both $\exp[Q(\theta)T - \gamma(\theta)\Delta]$ and $\theta\exp[Q(\theta)T - \gamma(\theta)\Delta]$ are zero at the limits of integration, then a fixed-sample size procedure is optimal.

Proof It is clear from Theorem 4.9.1 that a fixed-sample size procedure will be optimal if $E[r((\mathbf{X}_n);\,\theta)\,|\,\mathbf{X}_n]$ depends on $\mathbf{X}_n$ through n alone and this is consequently all that is necessary to show.

Let $p(\mathbf{X}_n;\,\theta) = \prod_{i=1}^n f(x_i;\,\theta)$. Then

$$\partial \ln p(\mathbf{X}_n;\,\theta)/\partial\theta = Q'(\theta)\sum_{i=1}^{n} R(X_i) - n\gamma'(\theta). \tag{4.9.23}$$

If θ admits a minimum variance unbiased estimate, we can also write (4.9.23) in the form

$$\partial \ln p/\partial\theta = nI(\theta)[u(\mathbf{X}_n) - \theta] \tag{4.9.24}$$

since $E[\partial \ln p/\partial\theta]^2 = nI(\theta)$. By identifying expressions (4.9.23) and (4.9.24) we fix the scale in which θ is measured to within a linear transformation. Let us suppose that $u(\mathbf{X}_n) = \sum_{i=1}^n R(X_i)/n$. Then

$$I(\theta) = Q'(\theta) \qquad \text{and} \qquad \theta Q'(\theta) = \gamma'(\theta). \tag{4.9.25}$$

From (4.9.21) and (4.9.22) we have the posterior density of θ:

$$h(\theta\,|\,\mathbf{X}_n) = (\text{const})[Q(\theta)T - \gamma(\theta)\Delta] \tag{4.9.26}$$

where

$$\Delta = \Delta_0 + n, \qquad T = t_0 + \sum_{i=1}^{n} R(X_i). \tag{4.9.27}$$

Then

$$E[r(t;\theta)|\mathbf{X}_n] = \frac{d\int (t-\theta)^2 \exp[QT - \gamma\Delta]Q'\,d\theta}{\int \exp[QT-\gamma\Delta]\,d\theta}. \tag{4.9.28}$$

The choice of t that minimizes (4.9.28) is

$$t(\mathbf{X}_n) = \frac{\int \exp(QT-\gamma\Delta)\theta Q'\,d\theta}{\int \exp(QT-\gamma\Delta)Q'\,d\theta}. \tag{4.9.29}$$

So that by (4.9.29) and (4.9.25)

$$\begin{aligned} t(\mathbf{X}_n) - \frac{T}{\Delta} &= \frac{\int \exp(QT-\gamma\Delta)\{\gamma' - (T/\Delta)Q'\}\,d\theta}{\int \exp(QT-\gamma\Delta)Q'\,d\theta} \\ &= \frac{-\Delta^{-1}\exp(QT-\gamma\Delta)}{\int \exp(QT-\gamma\Delta)Q'\,d\theta}. \end{aligned}$$

If $\exp\{Q(\theta)T - \gamma(\theta)\Delta\}$ is zero at the limits of integration, the optimal estimate is thus $t(\mathbf{X}_n) = T/\Delta$. Now

$$\begin{aligned} E\{r(t;\theta)|\mathbf{X}_n\} &= \frac{d\int \exp(QT-\gamma\Delta)(T\Delta^{-1}-\theta)^2 Q'\,d\theta}{\int \exp(QT-\gamma\Delta)\,d\theta} \\ &= \frac{d\int \exp(QT-\gamma\Delta)(T\Delta^{-1}-\theta)(Q'T\Delta^{-1}-\gamma')\,d\theta}{\int \exp(QT-\gamma\Delta)\,d\theta}. \end{aligned} \tag{4.9.30}$$

Performing integration by parts once, the numerator becomes

$$\left[\frac{d}{\Delta}\exp(QT-\gamma\Delta)(T\Delta^{-1}-\theta)\right] + \frac{d}{\Delta}\int \exp(QT-\gamma\Delta)\,d\theta$$

so that if both $\exp(QT-\gamma\Delta)$ and $\theta\exp(QT-\gamma\Delta)$ are zero at the limits of integration then

$$E[r(t;\theta)|\mathbf{X}_n] = d/\Delta. \tag{4.9.31}$$

Thus by Theorem 4.9.1 the optimal procedure is based on a fixed-sample of size n_0 which minimizes $cn + d(\Delta_0 + n)^{-1}$ which is approximately equal to $(d/c)^{1/2} - \Delta_0$.

We now extend Theorem 4.9.2 to distributions depending on several parameters. Let

$$f(x; \theta)\, dx = \exp[Q(\theta)'R(x) - \gamma(\theta)]\, dv(x) \tag{4.9.32}$$

where $\theta = (\theta_1, \ldots, \theta_p)'$ and $Q(\theta)$ and $R(x)$ are vectors. Assume that the parameters are all identifiable and that $Q(\theta)$ and $\gamma(\theta)$ are differentiable. We can also choose the parameters so that

$$E\{R(X)\} = \theta \tag{4.9.33}$$

where $Q(\theta) = (Q_1, \ldots, Q_p)'$ and $R(x) = (R_1, \ldots, R_p)'$. Also let

$$r(t; \theta) = d(t - \theta)'I(t - \theta) \tag{4.9.34}$$

where $t = (t_1, \ldots, t_p)'$ is an estimate of θ and I is the information matrix. Further, let

$$g(\theta) = \begin{cases} (\text{const.}) \exp[Q(\theta)'t_0 - \gamma(\theta)\Delta_0], & \theta \in \Theta \\ 0, & \theta \notin \Theta. \end{cases}$$

Then, the posterior density of θ is given by

$$h(\theta \mid \mathbf{X}_n) = \begin{cases} \dfrac{\exp[Q(\theta)'T - \gamma(\theta)\Delta]}{\int \exp[Q(\theta)'T - \gamma\Delta]\, d\theta} & \text{if } \theta \in \Theta \\ 0 & \text{if } \theta \notin \Theta. \end{cases} \tag{4.9.35}$$

where T and Δ are given by (4.9.27) and T is a vector here. Now, we have the following theorem.

Theorem 4.9.3 (Whittle and Lane, 1967). With the above notation and assumptions, if

$$\int_\Theta \frac{\partial}{\partial \theta_j} \{h(\theta \mid \mathbf{X}_n)\}\, d\theta = 0, \qquad j = 1, 2, \ldots, p \tag{4.9.36}$$

and

$$\int_\Theta \frac{\partial}{\partial \theta_j} \{\theta h(\theta \mid \mathbf{X}_n)\}\, d\theta = 0, \qquad j = 1, 2, \ldots, p, \tag{4.9.37}$$

then the optimal estimate of θ based on $\mathbf{X}_n = (X_1, \ldots, X_n)'$ is

$$t = T/\Delta \tag{4.9.38}$$

and the posterior loss is

$$E\{r(t;\theta)|\mathbf{X}_n\} = dp/\Delta. \tag{4.9.39}$$

Proof Since the total probability of (4.9.32) is unity,

$$0 = \frac{\partial}{\partial\theta_j}\int f(x;\theta)\,dx = \sum_{k=1}^{p}\frac{\partial Q_k}{\partial\theta_j}E\{R_k(X)\} - \frac{\partial\gamma}{\partial\theta_j}.$$

That is, using (4.9.33) we have

$$\frac{\partial\gamma}{\partial\theta_j} = \sum_{k=1}^{p}\theta_k\frac{\partial Q_k}{\partial\theta_j}, \qquad j = 1, \ldots, p. \tag{4.9.40}$$

Conversely, if the matrix $\{\partial Q_k/\partial\theta_j\}$ is nonsingular, then (4.9.40) implies (4.9.33). If (4.9.33) holds and the matrix $\{\partial Q_k/\partial\theta_j\}$ is nonsingular then we may define the information matrix $I = ((I_{jk}))$ where

$$I_{jk} = E[-\partial^2 \ln p(\mathbf{X}_n;\theta)/\partial\theta_j\,\partial\theta_k], \qquad p(\mathbf{X}_n;\theta) = \prod_{i=1}^{n} f(X_i;\theta).$$

From (4.9.32) and (4.9.40) we can deduce that

$$\begin{aligned} I_{jk} &= \partial^2\gamma/\partial\theta_j\,\partial\theta_k - \sum_s \theta_s\,\partial^2 Q_s/\partial\theta_j\,\partial\theta_k \\ &= \frac{\partial Q_j}{\partial\theta_k} + \frac{\partial}{\partial\theta_j}\left(\frac{\partial\gamma}{\partial\theta_k} - \sum\theta_s\frac{\partial Q_s}{\partial\theta_k}\right) = \frac{\partial Q_j}{\partial\theta_k} \end{aligned}$$

so that

$$I_{jk} = \partial Q_j/\partial\theta_k = \partial Q_k/\partial\theta_j \tag{4.9.41}$$

or $I(\theta) = \dot{Q}(\theta)$ where the dot denotes differentiation with respect to θ. Also since

$$\begin{aligned} 1 &= \int \exp(Q(\theta)'R(x) - \gamma(\theta))\,d\nu(x) \\ 0 &= \frac{\partial}{\partial\theta}\int = \int [\dot{Q}(\theta)'R(x) - \dot{\gamma}]\exp(Q(\theta) - \gamma(\theta))\,d\nu(x) \\ &= \dot{Q}(\theta)'\theta = \dot{\gamma}, \end{aligned}$$

after using (4.9.33). So,

$$\dot{Q}(\theta)'\theta = \dot{\gamma}. \tag{4.9.42}$$

Now consider

$$E\{r(t;\,\theta)\} = E\{t'I(\theta)t - 2t'I(\theta)\theta - \theta'I(\theta)\theta\}.$$

The t that minimizes $E\{r(t;\,\theta)\}$ is the solution of

$$E\{2I(\theta)t - 2I(\theta)\theta\} = 0.$$

Hence

$$t = \{EI(\theta)\}^{-1}\{EI(\theta)\theta\}. \tag{4.9.43}$$

Now, using (4.9.41) we have

$$\begin{aligned}[EI(\theta)]\left[t - \frac{T}{\Delta}\right] &= \int h(\theta|\mathbf{X}_n)\left[\dot{Q}(\theta)'\theta - \dot{Q}(\theta)'\frac{T}{\Delta}\right]d\theta \\ &= \frac{-1}{\Delta}\int h(\theta|\mathbf{X}_n)[\dot{Q}(\theta)'T - \dot{\gamma}(\theta)]\,d\theta,\end{aligned}$$

after using (4.9.42). Thus the left-hand side equals

$$-\frac{1}{\Delta}\int\frac{\partial}{\partial\theta}h(\theta|\mathbf{X}_n)\,d\theta = 0$$

because of assumption (4.9.36). Consequently $t = T/\Delta$. Next, consider

$$\begin{aligned}E\{r(t;\,\theta)\} &= d\int h(\theta|\mathbf{X}_n)\left(\frac{T}{\Delta} - \theta\right)'\dot{Q}(\theta)'\left(\frac{T}{\Delta}\,\theta\right)d\theta \\ &= d\int h(\theta|\mathbf{X}_n)\left(\frac{T}{\Delta} - \theta\right)'[\dot{Q}(\theta)'\frac{T}{\Delta}\,\dot{\gamma}(\theta)]\,d\theta,\end{aligned}$$

after using (4.9.42). Now using (4.9.36) we obtain

$$\text{L.H.S.} = -\frac{d}{\Delta}\int\theta'\left\{\frac{\partial}{\partial\theta}h(\theta|\mathbf{X}_n)\right\}d\theta. \tag{4.9.44}$$

Since

$$\theta' \frac{\partial h}{\partial \theta} = \sum_i \theta_i \frac{\partial h}{\partial \theta_i} = \sum_i \left[\frac{\partial}{\partial \theta_i} (\theta_i h) - h \right]$$

and using assumption (4.9.37) we obtain

$$E\{r(t; \theta)\} = dp/\Delta. \tag{4.9.45}$$

This completes the proof of Theorem 4.9.3.

REMARK 4.9.1 For a single parameter case, assumptions (4.9.36) and (4.9.37) are equivalent to the vanishing of $h(\theta | \mathbf{X}_n)$ and $\theta h(\theta | \mathbf{X}_n)$ at the boundary values of θ.

REMARK 4.9.2 Whittle and Lane (1967) did not provide the proof for Theorem 4.9.3.

Example 4.9.2 (The binomial distribution) Let $f(x; \theta) = \theta^x (1 - \theta)^{1-x}$, for $x = 0$ or 1 so that $E(X) = \theta$. Then, let $Q(\theta) = \ln[\theta/(1 - \theta)]$, $\gamma(\theta) = -\ln(1 - \theta)$. Also one finds that $I(\theta) = 1/\theta(1 - \theta)$ and the conjugate prior is

$$g(\theta) = (\text{const.}) e^{t_0} (1 - \theta)^{\Delta_0 - t_0}.$$

It would be apparent later that when the proposed loss function is used the integrals involved are properly defined for all samples only if $\Delta_0 > t_0 > 0$. Under these conditions $g(\theta)$ is a proper density. The posterior density of θ given that there are r successes in n trials is

$$h(\theta | \mathbf{X}_n) = h(\theta | r) = \frac{\theta^{r + t_0} (1 - \theta)^{n + \Delta_0 - (r + t_0)}}{\int_0^1 \theta^{r + t_0} (1 - \theta)^{n + \Delta_0 - (r + t_0)} \, d\theta}.$$

In view of Remark 4.9.1, the conditions are obviously satisfied in the present case under the stated conditions. The loss function is $d(t - \theta)^2/\theta(1 - \theta)$, the optimal estimate is $(r + t_0)/(n + \Delta_0)$ and the posterior risk is $d/(n + \Delta_0)$.

This example has already been considered by Blackwell and Girshick (1954) for a uniform prior distribution, but in this case ($\Delta_0 = t_0 = 0$) the results are only formally true by "continuity." For certain samples the integrals in the expressions for the estimate and the posterior risk diverge, and the boundary conditions (4.9.36) and (4.9.37) are not satisfied. As Δ_0 and t_0 play the role of the number of trials and successes respectively in a set of prior experiments the restriction $\Delta_0 > t_0 > 0$ is natural.

Example 4.9.3 (Normal distribution with known variance) Here

$$f(x;\mu) = s(x)\exp\left[\frac{\mu x}{\sigma^2} - \frac{\mu^2}{2\sigma^2}\right].$$

The parameter θ is determined as the expectation of $R(X)$. We choose $R(X) = X$. Here we take $\theta = \mu$ and also find that

$$Q(\theta) = \theta/\sigma^2, \qquad \gamma(\theta) = \theta/2\sigma^2, \qquad I(\theta) = \partial Q/\partial\theta = \sigma^{-2}.$$

Therefore, the loss function as usual is $(t-\theta)^2/\sigma^2$. The boundary conditions (3.9.36) and (3.9.37) require that $\exp(-x^2/2)$ and $x\exp(-x^2/2)$ tend to zero as $|x|$ tends to infinity.

Example 4.9.4 (Normal distribution with unknown mean and variance) Here

$$f(x;\mu,\sigma^2) = \exp\left[\frac{\mu x}{\sigma^2} - \frac{x^2}{2\sigma^2} - \frac{\mu^2}{2\sigma^2} - \log(2\pi)^{1/2})\right].$$

A possible choice of functionals is $R_1(x) = x$, $R_2(x) = x^2$. The parameters must be chosen as

$$\theta_1 = E\{R_1(X)\} = \mu, \qquad \theta_2 = E\{R_2(X)\} = \mu^2 + \sigma^2.$$

Then

$$\begin{aligned} Q_1(\theta) &= \mu/\sigma^2 = \theta_1/(\theta_2 - \theta_1{}^2), \\ Q_2(\theta) &= -1/2\sigma^2 = -1/2(\theta_2 - \theta_1{}^2), \\ \gamma(\theta) &= \frac{\theta_1{}^2}{2(\theta_2 - \theta_1{}^2)} + \frac{1}{2}\ln(\theta_2 - \theta_1{}^2) + \frac{1}{2}\ln(2\pi). \end{aligned}$$

The loss function

$$\begin{aligned} r(t_1, t_2; \theta_1, \theta_2) &= \frac{(\sigma^2 + 2\mu^2)}{\sigma^4}(t_1 - \theta_1)^2 - \frac{2\mu}{\sigma^4}(t_1 - \theta_1)(t_2 - \theta_2) + \frac{1}{2\sigma^4}(t_2 - \theta_2)^2 \\ &= \frac{(t_1 - \mu)^2}{\sigma^2} + \frac{1}{2}\left\{\frac{(t_1 - \mu)^2}{\sigma^2} + \frac{(t_2 - t_1{}^2) - \sigma^2}{\sigma^2}\right\}^2. \end{aligned}$$

The verification of (4.9.36)–(4.9.37) is long but not difficult. From Theorem 4.9.3 it follows that the optimal estimates of θ_1 and θ_2 are

$$\hat{t}_1 = \left(t_{1,0} + \sum_{i=1}^{n} X_i\right)/(\Delta_0 + n)$$

$$\hat{t}_2 = \left(t_{2,0} + \sum_{i=1}^{n} X_i^2\right)/(\Delta_0 + n)$$

where $t_0' = (t_{1,0}, t_{2,0})$. These estimates are biased, of course, as Bayes estimates for loss functions of the type considered here are always biased unless the Bayes risk is zero [see Blackwell and Girshick (1954, Theorem 11.2.3)]. The posterior risk for given $X_1, \ldots, X_n$ is $2/(\Delta_0 + n)$.

Theorem 4.9.4 (Blackwell and Girshick, 1954) Let the loss function be $r(a; \theta) = \lambda(\theta)[a - \theta]^2$, $\lambda(\theta) > 0$. Then, for any given prior density $g(\theta)$ which admits an estimate of finite risk the expectation of the function $\lambda(\theta)$ exists, the Bayes estimate t^* is biased or the Bayes risk is zero.

Proof The existence of $E(\lambda | g) = \int \lambda(\theta) g(\theta)\, d\theta$ implies the existence of $E(\lambda | x, g)$ for all x for which $f(x | g) > 0$ as can be seen from the equation

$$E(\lambda | g) = \int_{-\infty}^{\infty} E(\lambda | x, g) f(x | g)\, dx$$

where $f(x | g)$ denotes the unconditional density of the random variable X. Define

$$g_\lambda(\theta) = \lambda(\theta) g(\theta) / E(\lambda | g). \tag{4.9.46}$$

Then the Bayes estimate with respect to the loss function $r(a; \theta)$ and $g(\theta)$ is the same as the Bayes estimate with respect to the loss function $r^*(a; \theta) = (\theta - a)^2$ and the prior density $g_\lambda(\theta)$. Since g and hence g_λ admits an estimate of finite risk, the Bayes estimate with respect to g_λ exists and is given by $t^*(x) = E(\theta | Z, g_\lambda) = E(\theta | t^*, g_\lambda)$ by Problem 3.11.6 of Blackwell and Girshick (1954, p. 101). Now suppose that t^* is an unbiased estimator of θ for all θ. Then

$$\begin{aligned} E(\theta t^*) &= E[t^* E(\theta | x)] = E t^{*2} \\ &= E[\theta E(t^* | \theta)] = E(\theta^2), \end{aligned}$$

from which it follows that

$$R(t^*; g) = E[t^*(x) - \theta]^2 = 0.$$

REMARK 4.9.3 The assumption of the existence of the second moments of θ and t^* which is implied in the above proof can be removed. If $E(\lambda|g)$ does not exist, then Theorem 4.9.4 does no longer hold, as can be seen from the following example.

Example 4.9.5 Suppose we wish to obtain a Bayes estimate of the binomial probability θ of a success on the basis of n independent trials when the prior density, $g(\theta)$ is uniform on (0, 1) and the loss function is

$$r(a; \theta) = (\theta - a)^2/\theta(1-\theta), \qquad 0 \le a \le 1.$$

Here $\lambda(\theta) = 1/\theta(1-\theta) = 1/\theta + 1/(1-\theta)$, so that $\int_0^1 \lambda(\theta)\, d\theta = \infty$. On the other hand, if r represents the number of successes in the n trials, then the Bayes estimate $t^*(r)$ if $r \neq 0$ or n is given by r/n, which is unbiased. For this example, it is also easily verified that $E[(a-\theta)^2/\theta(1-\theta)|r, g]$ exists for only one value of a, namely $a = 0$, when $r = 0$, and only one value of a, namely $a = 1$ when $r = n$, so that $t^*(0) = 0$ and $t^*(n) = 1$.

Problems

1. The X_i are independent with $P(X_i = 1) = \theta$ and $P(X_i = 0) = 1 - \theta$. Let S denote the following truncated sequential procedure for estimating θ. Stop after the third observation if $X_1 = X_2 = X_3 = 1$ or if $X_1 = X_2 = X_3 = 0$. Stop after the fourth observation if three out of four X_i have the same value. Otherwise take a fifth observation. In any case use as your estimate the usual Bayes estimate corresponding to a uniform prior probability and quadratic loss function. Let N denote the sample size if θ is the true parametric value. Calculate

$$R(S; \theta) = E(S - \theta)^2, \qquad E_\theta(N),$$

and

$$B(S) = \int_0^1 \left[E_\theta(S-\theta)^2 + \frac{1}{200} E_\theta(N)\right] d\theta$$

where $\frac{1}{200}$ is the cost per observation.

2. Same conditions as in (1). Use the fixed-sample size (n) Bayes procedure t_n. Calculate $R(t_n; \theta)$ and then find what integer value n_0 of n minimizes

$$B(t_n) = \int_0^1 R(t_n; \theta)\, d\theta + \frac{n}{200}.$$

Compare the value of $B(t_{n_0})$ to $B(S)$ and also the value of n_0 to $\int_0^1 E_\theta(N)\, d\theta$.

3. Let $f(x; \theta)$ be the normal density with mean 0 and variance θ and assume that θ has the conjugate prior. Taking the loss function to be $r(a; \theta) = (a - \theta)^2/\theta^2$, show that the optimal estimate of θ is the one based on a fixed-sample size procedure.

4.10 Sequential Estimation and Certain Closed Sequential Decision Procedures

In his classical work, Wald (1947, Chapter 10) started a program for dealing with multiple decision problems based on sequential estimation and this program was never completed. Paulson (1964b) obtains a new derivation of his results of 1963 which lends itself to dealing with other situations and sequential confidence intervals are worked out explicitly for two other cases, namely, for the variance and for the ratio of variances of normal distributions. He then applies these results to obtain closed decision procedures for a number of decision problems, including (i) testing a hypothesis regarding the mean of a normal distribution against either one-sided or two-sided alternatives; (ii) comparing the means of k experimental categories with a standard or control; (iii) testing a hypothesis about the variance of a normal distribution; (iv) deciding which of k nonoverlapping intervals contains the variance; (v) testing a hypothesis about the ratio of variances; (vi) comparing the variances of k experimental categories with a standard or control. In addition, a brief discussion of "mixed" problems, where we are concerned with finding a confidence interval for the parameter after a decision has been reached is given. In some of these problems, open sequential solutions (in which there is no upper bound to the number of measurements required to reach a decision) are already known which have some optimum properties. However, for administrative and other reasons it is desirable to restrict to closed sequential procedures. All the sequential procedures proposed by Paulson (1964b) are closed.

Sequential Confidence Limits for the Mean of a Normal Population

Let $X_1, X_2, \ldots$ be a sequence of independent random variables having a common normal distribution with mean θ and variance σ^2. Let

$$\bar{X}_n = \sum_{i=1}^{n} X_i/n,$$

and let

$$f_n(\theta) = [\sigma(2\pi)^{1/2}]^{-n} \exp\left[-\sum_{i=1}^{n} (X_i - \theta)^2/2\sigma^2\right].$$

Lemma 4.10.1 (Wald, 1947, p. 146) Let $X_1, X_2, \ldots$ be a sequence of random variables and let $f_{0,n}(X_1, \ldots, X_n)$ $[f_{1,n}(X_1, \ldots, X_n)]$ denote the joint density of $X_1, \ldots, X_n$ under $H_0\,[H_1]$ for $n = 1, 2, \ldots$ Also, let A be a constant greater than unity. Then

$$P_{H_0}(f_{1,n}/f_{0,n} > A \quad \text{for at least one } n, n = 1, 2, \ldots) \le 1/A. \quad (4.10.1)$$

Proof From the fundamental inequalities associated with Wald's SPRT we have $A = (1 - \beta)/\alpha \leq 1/\alpha$. Hence

$$P_{H_0}(f_{1,n}/f_{0,n} > A, \quad \text{for at least one } n, n = 1, 2, \ldots) = \alpha \leq 1/A.$$

or

$$P_{H_0}(f_{1n}/f_{0n} > 1/\alpha \ \text{ for at least one } n, n = 1, 2, \ldots)$$
$$\leq P_{H_0}(f_{1n}/f_{0n} > A \quad \text{for at least one } n, n = 1, 2, \ldots) = \alpha.$$

Applying this lemma to $f_n(\theta)$, we have

$$P(f_n(\theta + 2d)/f_n(\theta) > 1/\alpha \quad \text{for at least one } n, n = 1, 2, \ldots) \leq \alpha.$$

That is

$$P(\theta < \overline{X}_n - d - (\sigma^2 \ln(1/\alpha))/2dn \quad \text{for at least one } n, n = 1, 2, \ldots) \leq \alpha \tag{4.10.2}$$

which is equivalent to

$$P(\overline{X}_n - d - (\sigma^2 \ln(1/\alpha))/2dn \leq \theta < \infty \quad \text{for every } n; n = 1, 2, \ldots) \geq 1 - \alpha \tag{4.10.3}$$

where all logarithms are to the base e, and d is a positive constant. By starting with $f_n(\theta)/f_n(\theta - 2d)$ we obtain

$$P[-\infty < \theta < \overline{X}_n + d + (\sigma^2 \ln(1/\alpha))/2dn \quad \text{for every } n, n = 1, 2, \ldots] \geq 1 - \alpha. \tag{4.10.4}$$

Now, letting

$$u_n(\alpha) = \min_{1 \leq r \leq n} [\overline{X}_r + d + (\sigma^2 \ln(1/\alpha))/2dr]$$

and

$$v_n(\alpha) = \max_{1 \leq r \leq n} [\overline{X}_r - d - (\sigma^2 \ln(1/\alpha))/2dr],$$

it follows from (4.10.3) and (4.10.4) that

$$P(v_n(\alpha/2) \leq \theta \leq u_n(\alpha/2) \quad \text{for every } n, n = 1, 2, \ldots) \geq 1 - \alpha. \tag{4.10.5}$$

When σ^2 is unknown, suppose that an estimate s^2 of σ^2 is available such that $\nu s^2/\sigma^2$ has the chi-square distribution with ν degrees of freedom and that s^2 is independent of $\bar{X}_{n_0}$. If we let $a(\alpha) = [(1/\alpha)^{2/\nu} - 1](\nu/2)$, then using the results of Paulson (1963) [see also Eq. (3.9.10)] we have

$$P(\bar{X}_n - d - s^2 a(\alpha)/2dn \le \theta < \infty \quad \text{for every } n, n \ge n_0) \ge 1 - \alpha \quad (4.10.6)$$

and

$$P(-\infty < \theta \le \bar{X}_n + d + s^2 a(\alpha)/2dn \quad \text{for every } n \ge n_0) \ge 1 - \alpha. \quad (4.10.7)$$

If we let

$$u_n'(\alpha) = \min_{n_0 \le r \le n} [\bar{X}_r + d + a(\alpha)s^2/2dr],$$

and

$$v_n'(\alpha) = \max_{n_0 \le r \le n} [\bar{X}_r - d - a(\alpha)s^2/2dr],$$

then

$$P[v_n'(\alpha/2) \le \theta \le u_n'(\alpha/2) \quad \text{for every } n, n \ge n_0] \ge 1 - \alpha. \quad (4.10.8)$$

APPLICATIONS TO TESTING HYPOTHESES ABOUT THE MEAN

Let H_0: $\theta \le \theta_0$ and $H_1 : \theta > \theta_0$ and we want a closed sequential procedure for testing H_0 such that $P(\text{rej } H_0 | \theta \le \theta_0) \le \alpha$ and $P(\text{acc } H_0 | \theta > \theta_0 + \Delta) \le \beta$. When σ is known the solution is given as follows: Take observations one at a time, and stop the experiment as soon as either (i) $\bar{X}_n - d - (\sigma^2 \ln(1/\alpha))/2dn > \theta_0$ or (ii) $\bar{X}_n + d + (\sigma^2 \ln(1/\beta))/2dn < \theta_0 + \Delta$. If (i) occurs, reject H_0 and if (ii) occurs, accept H_0. This rule is equivalent to stopping as soon as

$$\sum_{r=1}^{n} (X_r - \theta_0 - \Delta/2) > [\sigma^2 \ln(1/\alpha)/2d] - n(\Delta/2 - d)$$

or

$$\sum_{r=1}^{n} (X_r - \theta_0 - \Delta/2) < -[\sigma^2 \ln(1/\beta)/2d] + n(\Delta/2 - d).$$

For each d with $0 < d < \Delta/2$, the boundaries consist of two intersecting lines and thus a closed sequential procedure is obtained. In order to get some information on what value to choose for d and on the efficiency of the resulting

Table 4.10.1

Average Sample Size for Testing the Hypothesis $\theta \leq 0$ against $0 \geq .25$[a]

			Sequential Procedure		
α	β	θ	$d = \Delta/4$	$d = 3\Delta/8$	SPRT
.05	.05	0, .25	119 (3)	92 (3)	85
		.125	184 (3)	146 (3)	139
.01	.01	0, .25	192 (4)	152 (5)	144
		.125	317 (5)	272 (6)	338

[a] The values $\sigma = 1$ and $\Delta = .25$ were used throughout. The estimated standard error of each empirically determined mean is given in parentheses below the mean. The numbers in the SPRT column are the theoretical values for the sequential probability ratio test of Wald when the excess is neglected.

sequential scheme, a number of Monte Carlo experiments were carried out by Paulson (1964b) and these are summarized in Table 4.10.1.

The data of Table 4.10.1 indicate that Paulson's procedure (with $d = 3\Delta/8$) seems to compare reasonably well with the SPRT of Wald, and in addition to being a closed test has the further advantage that for small values of α and β it substantially reduces the maximum average sample size. In these respects it is somewhat similar to a modification of the Wald SPRT given by Anderson (1960). Although an exact comparison with Anderson's procedure is not possible at present, Paulson's (1964b) procedure seems to be less efficient than Anderson's (1960) for $\alpha = \beta = .05$ but seems to be about as efficient when $\alpha = \beta = .01$. When σ is known, closed sequential procedures were given by Armitage and Schneiderman (1962) for nine different combinations of α and β. Paulson's (1964b) procedure is applicable for any combination of α and β. Another important advantage of Paulson's procedure is that it can easily be modified to solve the problem when σ is unknown, which in the past has been dealt with by means of a sequential t test. Recently an open sequential solution when σ is unknown was given by Hall (1962, Section 3.2) who also gave a closed sequential solution for the special case $\alpha = \beta$, which is similar to but not identical to the following: We start by taking a sample of n_0 observations, compute $s^2 = \sum_{r=1}^{n_0} (X_r - \bar{X}_{n_0})^2/(n_0 - 1)$, take $v = n_0 - 1$ and then take observations one at a time, and stop the experiment and reject H_0 as soon as

$$\sum_{r=1}^{n} (X_r - \theta_0 - \Delta/2) > [s^2 a(\alpha)/2d] - n(\Delta/2 - d),$$

and stop and accept H_0 as soon as

$$\sum_{r=1}^{n} (X_r - \theta_0 - \Delta/2) > [s^2 a(\beta)/2d] + n(\Delta/2 - d).$$

It follows from (4.10.6) and (4.10.7) that we still have $P(\text{rej } H_0 | \theta \le \theta_0) \le \alpha$ and $P(\text{acc } H_0 | \theta > \theta_0 + \Delta) \le \beta$.

When a decision has been reached as to whether H_0 is accepted or rejected, that is, whether $\theta \le \theta_0$ or $\theta > \theta_0$, it may be useful to have a confidence interval for θ. If n denotes the number of observations required to reach a decision about H_0 and if σ is known, it follows from (4.10.5) that $P[v_n(\alpha/2) \le \theta \le u_n(\alpha/2)] \ge 1 - \alpha$. If the resulting interval $[v_n(\alpha/2), u_n(\alpha/2)]$ is too large to be useful, additional observations can be taken. For example, let n' equal the number of additional observations required after H_0 is rejected in order to have $u_{n+n'}(\alpha/2) - v_{n+n'}(\alpha/2) \le l$ (where $l > 2d$). Then, if $n'' = n$ when H_0 is accepted and $n'' = n + n'$ when H_0 is rejected, we can still assert that $P(v_{n''}(\alpha/2) \le \theta \le u_{n''}(\alpha/2)) \ge 1 - \alpha$. When σ is unknown, from (4.10.8) we obtain a similar result, namely $P(v'_{n''}(\alpha/2) \le \theta \le u'_{n''}(\alpha/2)) \ge 1 - \alpha$.

TESTING A HYPOTHESIS ABOUT θ WITH TWO-SIDED ALTERNATIVES

Let H_0: $\theta = \theta_0$ and H_1: $\theta \neq \theta_0$. Let us briefly consider the problem of obtaining a closed sequential procedure for testing the hypothesis H_0 so that

$$P(\text{rej } H_0 | \theta = \theta_0) \le \alpha \quad \text{and} \quad P(\text{acc } H_0 | (\theta - \theta_0) \ge \Delta) \le \beta.$$

This can easily be accomplished by combining two one-sided closed sequential tests considered above. That is, we test $\theta = \theta_0$ against $\theta = \theta_0 - \Delta$ with error probabilities $\alpha/2$ and β, and test $\theta = \theta_0$ against $\theta = \theta_0 + \Delta$ with error probabilities $\alpha/2$ and β. Then, we reject H_0 if either of the two component tests rejects θ_0; otherwise we accept H_0. When explicitly written out it reduces to the following rule.

Accept H_0 as soon as both

$$\sum_{r=1}^{n} (X_r - \theta_0) > [\sigma^2 \ln(1/\beta)]/2d - n(\Delta - d)$$

and

$$\sum_{r=1}^{n} (X_r - \theta_0) < -[\sigma^2 \ln(1/\beta)]/2d + n(\Delta - d).$$

Reject H_0 as soon as either

$$\sum_{r=1}^{n} (X_r - \theta_0) > [\sigma^2 \ln(2/\alpha)]/2d + nd$$

or

$$\sum_{r=1}^{n} (X_r - \theta_0) < -[\sigma^2 \ln(2/\alpha)]/2d - nd.$$

Paulson (1964b) again recommends choosing $d = 3\Delta/8$. If σ^2 is unknown, we start by taking n_0 observations, calculate $s^2 = \sum_{r=1}^{n_0}(X_r - \bar{X}_{n_0})^2/(n_0 - 1)$, take $v = n_0 - 1$ and then taking observations one at a time, we use the above procedure with σ^2 replaced by s^2, $\ln(1/\beta)$ replaced by $a(\beta)$ and $\ln(2/\alpha)$ replaced by $a(\alpha)$, where $a(\alpha) = [(1/\alpha)^{2/v} - 1](v/2)$.

It is reasonable to ask as to how one goes about choosing n_0. A reasonably efficient choice of n_0 is important since the efficiency of the sequential procedure is reduced if n_0 is taken either too small or too large. The following rule of Paulson (1964b) is not optimal. Let $\delta = \min(\alpha, \beta)$ for one-sided testing situation and $\delta = \min(\alpha/2, \beta)$ for the problem of two-sided testing situation. Then select n_0 so that $a(\delta)$ does not exceed its limiting value $\ln(1/\delta)$ by more than 25 percent (say), that is, select the smallest value of n_0 so that $a(\delta) \leq 1.25 \ln(1/\delta)$. When a moderately accurate estimate of $\hat{\sigma}^2$ of σ^2 is available from past experience, it may be desirable to modify the preceding procedure for selecting n_0. Let us illustrate what is involved by an example. Consider the problem of testing one-sided alternative hypothesis and let $\alpha = \beta = .02$. If no prior knowledge of σ^2 is available, then by the above rule we take $n_0 = 20$. Suppose now that a moderately accurate estimate $\hat{\sigma}^2$ of σ^2 is available, let $N(\hat{\sigma})$ denote the number of observations required by the corresponding fixed-sample size procedure when $\sigma = \hat{\sigma}$ and consider the following three cases: (i) $N(\hat{\sigma}) = 120$, (ii) $N(\hat{\sigma}) = 50$, (iii) $N(\hat{\sigma}) = 25$. In case (i) it would seem reasonable to increase n_0, say to $n_0 = 42$, so that $a(.02)$ is within ten percent of its limiting value, in case (ii) n_0 would be left unchanged, while in case (iii) it would seem reasonable to decrease n_0, say to $n_0 = 15$.

SEQUENTIAL CONFIDENCE LIMITS FOR THE VARIANCE We again let $\{X_r\}$ ($r = 1, 2, \ldots$) denote a sequence of independent normal variables having mean θ and variance σ^2. Let $s_n^2 = \sum_{r=1}^{n}(X_r - \bar{X}_n)^2/(n-1)$, and let

$$g_n(s_n^2;\sigma^2) = \frac{1}{\Gamma((n-1)/2)}\left(\frac{n-1}{2\sigma^2}\right)^{(n-1)/2}(s_n^2)^{(n-3)/2}\exp[-(n-1)s_n^2/2\sigma^2]$$

denote the probability density of s_n^2 and let $n^*(\alpha)$ denote the largest integer contained in $1 + \ln(1/\alpha)/\ln\lambda$, where $\lambda > 1$. Then from Lemma 4.10.1 and the transitivity of the invariantly sufficient sequence $\{s_n\}$ [see p. 130 and also the work of Cox (1952a)] we have

$$P[g_n(s_n^2;\lambda^2\sigma^2)/g_n(s_n^2;\sigma^2) > 1/\alpha \quad \text{for at least one } n,\ n = 2, 3, \ldots] \le \alpha. \tag{4.10.9}$$

This simplifies to

$$P\{s_n^2(\lambda^2-1)/2\lambda^2[\ln\lambda + (\ln(1/\alpha))/(n-1)] < \sigma^2 < \infty \quad \text{for all } n,\ n = 2, 3, \ldots\} \ge 1 - \alpha. \tag{4.10.10}$$

In a similar manner we have

$$P\{g_n(s_n^2;\sigma^2)/g_n(s_n^2;\sigma^2/\lambda^2) < \alpha \quad \text{for at least one } n,\ n = 2, 3, \ldots\} \le \alpha.$$

This again simplifies to

$$P\{0 < \sigma^2 \le s_n^2(\lambda^2-1)/2[\ln\lambda - (\ln(1/\alpha))/(n-1)] \quad \text{for all } n,\ n > n^*(\alpha)\} \ge 1 - \alpha. \tag{4.10.11}$$

Now let

$$u_{1n}(\alpha) = \begin{cases} \infty & \text{for } n \le n^*(\alpha) \\ \min\limits_{n^*(\alpha) < r \le n} \{s_r^2(\lambda^2-1)/2[\ln\lambda - (\ln(1/\alpha))/(r-1)], & \text{for } n > n^*(\alpha) \end{cases}$$

$$v_{1n}(\alpha) = \max_{1 < r \le n}\{s_r^2(\lambda^2-1)/2\lambda^2[\ln\lambda + (\ln(1/\alpha))/(r-1)]\}.$$

Then, it follows from (4.10.10) and (4.10.11) that

$$P[v_{1n}(\alpha/2) \le \sigma^2 \le u_{1n}(\alpha/2) \qquad \text{for all } n,\ n = 2, 3, \ldots] \ge 1 - \alpha. \tag{4.10.12}$$

APPLICATION TO TESTING HYPOTHESES ABOUT THE VARIANCE

Let $H_0 : \sigma^2 \geq \sigma_0^2$ and $H_1 : \sigma^2 < \sigma_0^2$. We are interested in finding a closed sequential procedure for testing H_0 satisfying

$$P(\text{rej } H_0 | H_0) \leq \alpha \qquad \text{and} \qquad P(\text{acc } H_0 | \sigma^2 < \sigma_0^2/w^2) \leq \beta,$$

where $w > 1$. Using the preceding results, the following solution is obtained: Take observations one at a time and stop the experiment and (i) accept H_0 if

$$s_n^2(\lambda^2 - 1)/2\lambda^2\{\ln \lambda + [\ln(1/\beta)]/(n-1)\} > \sigma^2/w^2,$$

(ii) reject H_0 if

$$s_n^2(\lambda^2 - 1)/2\{\ln \lambda - [\ln(1/\alpha)]/(n-1)\} < \sigma_0^2 \qquad \text{and} \qquad n > n^*(\alpha).$$

This is equivalent to accepting H_0 as soon as

$$\sum_{r=1}^{n} (X_r - \bar{X}_n)^2 > [2\sigma_0^2\lambda^2/w^2(\lambda^2 - 1)][(n-1)\ln \lambda + \ln(1/\beta)]$$

and rejecting H_0 as soon as

$$\sum_{r=1}^{n} (X_r - \bar{X}_n)^2 < [2\sigma_0^2/(\lambda^2 - 1)][(n-1)\ln \lambda - \ln(1/\alpha)].$$

For each λ with $1 < \lambda < w$ we get a closed sequential procedure. Although the optimal choice of λ is unknown, on the basis of some preliminary calculations Paulson (1964b) recommends the value $\lambda = 1 + .7(w-1)$. Certain Monte Carlo studies carried out by Paulson (1964b) are summarized in Table 4.10.2.

SEQUENTIAL CONFIDENCE LIMITS FOR THE RATIO OF VARIANCES Let $\{X_r\}$ and $\{Y_r\}$ $(r = 1, 2, \ldots)$ be two independent sequences of independent normally distributed random variables with means θ_1 and θ_2 and variances σ_1^2 and σ_2^2 respectively. Let $\rho^2 = \sigma_1^2/\sigma_2^2$ and let $T_n = \sum_{r=1}^{n}(X_r - \bar{X}_n)^2 / \sum_{r=1}^{n}(Y_r - \bar{Y}_n)^2$. Also, let $n^*(\alpha)$ denote the largest integer contained in $1 + \ln(1/\alpha)/\ln \lambda$,

$$u(n, T_n, \alpha) = \lambda T_n[\lambda - \alpha^{1/(n-1)}]/[\lambda\alpha^{1/(n-1)} - 1],$$

$$v(n, T_n, \alpha) = T_n[\lambda - (1/\alpha)^{1/(n-1)}]/\lambda[\lambda(1/\alpha)^{1/(n-1)} - 1],$$

$$u_{2n}(\alpha) = \min_{n^*(\alpha) < r \leq n} [u(r, T_r, \alpha)],$$

Table 4.10.2

Average Sample Size for Testing $\sigma^2 \geq 1$ against $\sigma^2 \leq 2/3$[a]

			Sequential procedure	
α	β	σ^2	$\lambda = 1 + .7(w-1)$	SPRT
.05	.05	2/3	84 (2)	73
		.81	114 (2)	106
		1	56 (2)	56

[a] The estimated standard error of each empirically determined mean is given in parentheses below the mean. The numbers in the SPRT column are the theoretical values of the sequential probability ratio test of Wald, neglecting the excess.

and

$$v_{2n}(\alpha) = \max_{n^*(\alpha) < r \leq n} [v(r, T_r, \alpha)].$$

Further, let

$$h_n(t; \rho^2) = \frac{\Gamma(n-1)}{[\Gamma((n-1)/2)]^2} \frac{t^{(n-3)/2}}{(1 + t/\rho^2)^{n-1}}$$

denote the probability density function of T_n. It follows from the transitivity of the invariantly sufficient sequence $\{T_n\}$ (see p. 130) and Lemma 4.10.1 or the work of Cox (1952a) that

$$P[h_n(T_n; \lambda^2\rho^2)/h_n(T_n; \rho^2) > 1/\alpha \qquad \text{for at least one } n, n = 2, 3, \ldots] \leq \alpha$$

where $\lambda > 1$. This reduces to

$$P[v(n, T_n, \alpha) \leq \rho^2 < \infty \qquad \text{for all } n, n > n^*(\alpha)] \geq 1 - \alpha. \quad (4.10.13)$$

Analogously, we obtain

$$P[h_n(T_n; \rho^2)/h_n(T_n; \rho^2/\lambda^2) < \alpha \qquad \text{for at least one } n, n = 2, 3, \ldots] \leq \alpha$$

which simplifies to

$$P[0 < \rho^2 \le u(n, T_n, \alpha) \qquad \text{for all } n, n > n^*(a)] \ge 1 - \alpha. \quad (4.10.14)$$

Combining (4.10.13) and (4.10.14) we have

$$P[v_{2n}(\alpha/2) \le \rho^2 \le u_{2n}(\alpha/2) \qquad \text{for all } n, n > n^*(\alpha/2)] \ge 1 - \alpha. \quad (4.10.15)$$

APPLICATION TO PROBLEMS INVOLVING THE RATIO OF VARIANCES Let H_0: $\sigma_1^2/\sigma_2^2 \le 1$ and H_1: $\sigma_1^2/\sigma_2^2 > 1$. We now look for a closed sequential procedure for testing this hypothesis subject to the restrictions $P(\text{rej } H_0 | \rho^2 \le 1) \le \alpha$ and $P(\text{acc } H_0 | \rho^2 \ge R^2) \le \beta$, where $R > 1$. This is given by Paulson (1964b) as follows: Let $n^* = \min[n^*(\alpha), n^*(\beta)]$. Start by taking $n^* + 1$ pairs of observations, and then take one pair of observations at a time and stop the experiment and accept H_0 as soon as $n > n^*(\beta)$ and $u(n, T_n, \beta) < R^2$, and stop the experiment and reject H_0 as soon as $n > n^*(\alpha)$ and $v(n, T_n, \alpha) > 1$. For each λ with $1 < \lambda < R$, a closed sequential procedure is obtained. By analogy with the problem of testing for the variance, Paulson (1964b) suggests taking $\lambda = 1 + .7(R - 1)$.

Although taking one observation on each category or population at a time is considered, in practice, sequential sampling by groups may often be preferable. Paulson (1964b) remarks that the above methods can also be used to obtain sequential confidence limits for parameters of certain non-normal distributions like Poisson, exponential, and binomial.

Problems

Let $X_1, X_2, \ldots,$ be a sequence of independent and identically distributed random variables having the density $f(x; \theta)$. Using Paulson's (1964b) technique, set up sequential $(1 - \alpha)$ confidence limits for θ when

(i) $f(x; \theta) = e^{-\theta}\theta^x/x!$, $x = 0, 1, \ldots, \theta > 0$;
(ii) $f(x; \theta) = \theta^x(1 - \theta)^{1-x}$, $x = 0$ or 1, $0 < \theta < 1$; and
(iii) $f(x; \theta) = \theta^{-1}e^{-x/\theta}$, $x, \theta > 0$.

4.11 Estimation by Double Sampling

Suppose that we are interested in the problem of estimation of an unknown parameter θ with assigned accuracy using as few observations as possible. The accuracy could be in terms of standard deviation, that is, the estimate should have variance $a(\theta)$, some given function of θ. Another possibility is to estimate θ by a γ percent confidence interval of predetermined form, for example of given width. In general it is impossible to obtain an estimate with the required properties by taking a sample of fixed size. So one has to resort to some form of sequential sampling. Usually, however, it is extremely difficult to construct a sequential sampling scheme leading to an estimate with the required properties, although Anscombe (1949, 1952) has given a general large-sample theory. Another disadvantage of ordinary sequential procedures is the step by step calculation involved. Cox (1952b) proposes a double-sampling scheme for the above problem. The basic idea is to use the preliminary sample in order to determine how large the total sample should be. Stein's (1945) method for constructing a confidence interval for a normal mean with preassigned length and confidence coefficient is a special case where the exact distribution theory is known. The present double sampling methods of Cox (1952b) are different from those used in industrial inspection, in that in the latter case the second sample, if taken, is of fixed size. The theory of double sampling developed by Cox (1952b) is a large-sample one; however, in practice, it is likely to give reasonable approximations even with quite small samples. In the following we present an estimate of θ with bias $O(n_0^{-2})$ and variance $a(\theta)[1 + O(n_0^{-2})]$, where n_0 is the preliminary sample size and $a(\theta)$ the assigned variance. Also, for any θ the expected sample size for the double-sampling scheme will be greater than the corresponding expected sample size for the "best" sequential procedure. However, the difference is likely to be small except when the sample sizes are very small.

ESTIMATION WITH GIVEN VARIANCE: ONE UNKNOWN PARAMETER Let θ denote the unknown parameter and suppose we wish to estimate it with variance equal to a given function of θ. The main applications are to the estimation of Poisson and binomial means and of a normal mean when the population variance is known. Cox (1952b) provides the large-sample theory in which the assigned variance is small. Consider a sequence of problems $\{S_\lambda\}$ where S_λ denotes the problem of estimating θ with variance $a(\theta)/\lambda$ and the indexing parameter λ is tending to infinity. Assume that in radom samples of any fixed size m one can construct an estimate $T^{(m)}$ of θ such that

(i) $T^{(m)}$ is an unbiased estimate of θ with variance $v(\theta)/m$;

(ii) the skewness coefficient γ_1 of $T^{(m)}$ is asymptotically $\gamma_1(\theta)m^{-1/2}$ and γ_2 (kurtosis) of $T^{(m)}$ is $O(m^{-1})$ as m tends to infinity.

(iii) Asymptotic means and standard errors can be derived for $a(T^{(m)})$, $v(T^{(m)})$, and for combinations of these functions by expansion in series.

The above assumptions are general enough to take care of the present applications. There is no difficulty in considering estimates with nonzero asymptotic skewness or kurtosis. Now, if θ were known, a sample of size

$\lambda n(\theta) = \lambda v(\theta)/a(\theta)$ would give the required accuracy. This suggests the following approximate sampling scheme: Take a random sample of size $n_0 \lambda$ and let T_1 be the estimate of θ from it. Let

$$n(T_1) = v(T_1)/a(T_1). \tag{4.11.1}$$

Take a second random sample of size $\max[0, (n(T_1) - n_0)\lambda]$. (If $(n(T_1) - n_0)\lambda$ is not an integer, take the integer nearest to it.) Let T_2 be the estimate of θ from the second sample. Put

$$T = \begin{cases} [n_0 T_1 + (n(T_1) - n_0)T_2]/n(T_1) & \text{if } n(T_1) \geq n_0, \\ T_1 & \text{if } n(T_1) < n_0. \end{cases} \tag{4.11.2}$$

In particular if $T^{(m)}$ is the sample mean, T is the mean of the pooled sample. We shall say that the above sampling scheme is based on the estimate $T^{(m)}$. The estimate T has the required large-sample properties. We shall outline the method of obtaining closer approximations to the exact theory. The precise statement to be proved is that T has bias $O(\lambda^{-1})$ and variance $a(\theta)\lambda^{-1}[1 + O(\lambda^{-1})]$, Consider

$$E(T) = (1 - \pi)\{n_0 E_1(T_1 m(T_1)) + E_1[(1 - n_0 m(T_1))T_2]\} + \pi E_2(T_1) \tag{4.11.3}$$

where

$$m(T_1) = 1/n(T_1) = a(T_1)/v(T_1), \tag{4.11.4}$$

$\pi = P(n(T_1) < n_0)$ and E_1, E_2 denote the conditional expectations given that $n(T_1) \gtreqless n_0$. Now by (iii), $E[n(T_1)] \sim n(\theta)$ and $\text{var}[n(T_1)] = O(\lambda^{-1})$. Thus if $n_0 < n(\theta)$ and the distribution of $n(T_1)$ dies away exponentially in the lower tail, $\pi = O(\lambda^{-r})$ for all r. If, in addition, $m(T_1)$ is well behaved, the possibility that $n(T_1) < n_0$ may be neglected in (4.11.3). Hereafter we shall assume that

(iv) $n_0 < n(\theta)$ and the distribution of $m(T_1)$ is such that the event $n(T_1) < n_0$ may be ignored.

Thus we write (4.11.3) as

$$E(T) = n_0 E[T_1 m(T_1)] + E[(1 - n_0 m(T_1))T_2].$$

By taking repeated expectations, the second term is $E\{[1 - n_0 m(T_1)]E(T_2 | T_1)\}$ and $E(T_2 | T_1) = \theta$ since the second sample is independent of the first. Thus

$$E(T) = n_0 E[T_1 m(T_1)] + \theta - n_0 \theta E[m(T_1)]. \tag{4.11.5}$$

By (iii), $E[T_1 m(T_1)] = \theta m(\theta) + O(\lambda^{-1})$, and $E(m(T_1)] = m(\theta) + O(\lambda^{-1})$. Hence

$$E(T) = \theta + O(\lambda^{-1}). \tag{4.11.6}$$

Analogously, one can show that

$$\text{var}(T) = a(\theta)\lambda^{-1}[1 + O(\lambda^{-1})]. \tag{4.11.7}$$

Equations (4.11.6) and (4.11.7) show that T has the required properties when the preliminary sample size is large. In the following we shall take a further step in order to obtain closer approximations to the exact sampling theory.

Consider the sampling procedure defined by the above four steps with $n(T_1)$ replaced by $\tilde{n}(T_1)$ defined by

$$\tilde{n}(T_1) = n(T_1)[1 + b(T_1)/\lambda], \tag{4.11.8}$$

where $b(T_1)$ is a function to be determined. Equation (4.11.5) now becomes

$$E(T) = n_0 E\{T_1 m(T_1)[1 - b(T_1)/\lambda]\} + \theta - n_0 E\{m(T_1)[1 - b(T_1)/\lambda]\} + O(\lambda^{-2}). \tag{4.11.9}$$

However by (iii),

$$E[T_1 m(T_1)] = \theta m(\theta) + \frac{1}{2}\frac{d^2}{d\theta^2}[\theta m(\theta)]\frac{v(\theta)}{n_0 \lambda} + O(\lambda^{-2}),$$

and analogous formulas for the other expectations occurring in (4.11.9) hold. After some simplification, we have

$$E(T) = \theta + m'(\theta)v(\theta)\lambda^{-1} + O(\lambda^{-2}). \tag{4.11.10}$$

Set

$$T' = \begin{cases} T - m'(T)v(T)\lambda^{-1} & \text{if } n_0 \le \tilde{n}(T_1), \\ T_1 & \text{if } n_0 > \tilde{n}(T_1). \end{cases} \tag{4.11.11}$$

Then, (4.11.10) shows that T' has bias $O(\lambda^{-2})$. By an analogous computation,

$$\begin{aligned} \text{var}(T') = a(\theta)\lambda^{-1} + \lambda^{-2}\{&2m(\theta)m'(\theta)\gamma_1(\theta)v^{3/2}(\theta) \\ &+ [m'(\theta)v(\theta)]^2 + 2m(\theta)m''(\theta)v^2(\theta) \\ &+ v^2(\theta)m''(\theta)/2n_0 - a(\theta)b(\theta)\} + O(\lambda^{-3}). \end{aligned} \tag{4.11.12}$$

We require T' to have as nearly as possible a variance $a(\theta)/\lambda$. Therefore choose $b(\theta)$ so that the second term in (4.11.12) vanishes. That is, put

$$b(\theta) = n(\theta)v(\theta)[2m(\theta)m'(\theta)\gamma_1(\theta)v^{-1/2}(\theta) + m'^2(\theta) + 2m(\theta)m''(\theta) + m''(\theta)/2n_0]. \qquad (4.11.13)$$

The following summarizes the above result.

(a) Take a preliminary sample of size $n_0\lambda$ and let T_1 be the estimate of θ from it.

(b) Let $\tilde{n}(T_1) = n(T_1)[1 + b(T_1)/\lambda]$ where $n(T_1)$ and $b(T_1)$ are defined by (4.11.1) and (4.11.13).

(c) Take a second sample of size $\max[0, (\tilde{n}(T_1) - n_0)\lambda]$ and let T_2 be the estimate of θ from it.

(d) Define T by (4.11.2) and then T' by (4.11.11). Then under assumptions (i)–(iv), T' has bias $O(\lambda^{-2})$ and variance $a(\theta)\lambda^{-1}[1 + O(\lambda^{-2})]$.

In any particular application, we have one variance function which must be small and one particular preliminary sample size which must be large. So, without loss of generality we can set λ equal to unity.

Example 4.11.1 (Estimation of normal mean θ with given standard error $a^{1/2}$) This is a trivial case when the population variance σ^2 is known and the estimator is the sample mean. We have $n(\theta) = \sigma^2 a^{-1}$ and $b(\theta) = 0$. Thus, the total sample size is constant. The procedure is equivalent to taking a single sample of size the nearest integer to σ^2/a. This is the commonsense solution and will be shown to be optimal.

Example 4.11.2 (Estimation of normal mean θ with given fractional standard error $a^{1/2}$) The population variance is again supposed known. We have

$$a(\theta) = a\theta^2, \qquad n(\theta) = \sigma^2/(a\theta^2), \qquad \text{and} \qquad b(\theta) = 8a + \sigma^2/(n_0\theta^2).$$

Thus the total sample size is

$$\tilde{n}(T_1) = \sigma^2/(aT_1{}^2) + 8\sigma^2/T_1{}^2 + \sigma^4/(n_0 aT_1{}^4), \qquad (4.11.14)$$

where T_1 is the mean of the preliminary sample. The final estimate is $T' = T(1 - 2a)$, where T is the mean of the combined sample. The preliminary sample size n_0 should be chosen as large as possible subject to $n_0 \ll \sigma^2/(a\theta^2)$, that is, it is desirable to know an upper limit to $|\theta|$. The procedure fails if θ is near zero because the function $m(\theta) = \sigma^2/(a\theta^2)$ has a singularity at $\theta = 0$, invalidating the earlier expansions. In practice, if θ is likely to be equal to or near zero, $a(\theta)$ should be modified so that $a(0)$ is finite. This would be roughly equivalent to truncating the procedure (4.11.14).

Example 4.11.3 (Estimation of binomial mean θ with given fractional standard error $a^{1/2}$) If the procedure is based on the sample proportion of defectives, we have

$$a(\theta) = a\theta^2, \qquad v(\theta) = \theta(1-\theta), \qquad \gamma_1(\theta) = (1-2\theta)[\theta(1-\theta)]^{-1/2}.$$

The total sample size is

$$\tilde{n}(T_1) = (1 - T_1)/(aT_1) + 3[T_1(1-T_1)]^{-1} + (an_0 T_1)^{-1}, \tag{4.11.15}$$

and the final estimate is $T' = T - aT(1-T)^{-1}$. We require $n_0 \ll (1-\theta)/a\theta$. The procedure fails if θ is very small for the same reason as in Example 4.11.2. Equation (4.11.15) gives the double-sampling solution corresponding to Haldane's (1945) inverse binomial sampling.

Estimation in the Presence of a Nuisance Parameter We now suppose that the population variance is unknown and in addition to the unknown parameter θ, which is to be estimated with variance $a(\theta)/\lambda$, there is an unknown nuisance parameter ψ. Assume that in samples of any fixed size m, we can find estimates $T^{(m)}$ and $C^{(m)}$ of θ and ψ such that

(i) $T^{(m)}$ and $C^{(m)}$ are unbiased estimates and have variances ψ/m and $\tau\psi^2/m$ where τ is asymptotically constant;

(ii) if m is large, asymptotic means and standard errors can be developed for combinations of $T^{(m)}$, $C^{(m)}$, and $a(T^{(m)})$ by expansion in series.

The following sampling procedure can be justified by an argument similar to the one used in the case of a single unknown parameter. Take a preliminary sample of size $n_0\lambda$ and let T_1, C_1 be the estimates from it. Put

$$\tilde{n}(T_1, C_1) = [C_1/a(T_1)][1 + b(T_1, C_1)/\lambda] \tag{4.11.16}$$

where

$$b(T_1, C_1) = 2a''(T_1) + \frac{[a'(T_1)]^2}{a(T_1)} + \frac{C_1 a''(T_1)}{2n_0 a(T_1)} + \frac{\tau}{n_0}. \tag{4.11.17}$$

Take a second sample of size $\max[0, (\tilde{n}(T_1, C_1) - n_0)\lambda]$ and let T_2 be the estimate of θ from it. Set

$$T = \begin{cases} T_1 & \text{if } n_0 > \tilde{n}(T_1, C_1), \\ [n_0 T_1 + (n(T_1, C_1) - n_0)T_2]/n(T_1, C_1) & \text{if } n_0 \le \tilde{n}(T_1, C_1), \end{cases} \tag{4.11.18}$$

and

$$T' = \begin{cases} T_1 & \text{if } n_0 > \tilde{n}(T_1, C_1), \\ T - a'(T)/\lambda & \text{if } n_0 \leq \tilde{n}(T_1, C_1). \end{cases} \quad (4.11.19)$$

Then, assuming as before that (iv) the possibility that $n_0 > \tilde{n}(T_1, C_1)$ can be ignored, we can show that T' has bias $O(\lambda^{-2})$ and variance $a(\theta)\lambda^{-1}[1 + O(\lambda^{-2})]$.

Example 4.11.4 (Estimation of a normal mean with given standard error $a^{1/2}$) Base the method on the sample mean. Then ψ is the unknown population variance σ^2 and is estimated by the usual sample variance, that is, take $C_1 = s_1{}^2$. Then $\tau = 2$, $a(\theta) = a$, and

$$\tilde{n}(T_1, C_1) = (C_1/a)\left(1 + \frac{2}{n_0}\right) = (s_1{}^2/a)(1 + 2/n_0), \quad (4.11.20)$$

and the final estimate is the pooled sample mean T, which is easily shown to be exactly unbiased. The expected sample size is $\sigma^2(1 + 2/n_0)/a$. Thus, the effect of not knowing σ is to increase the sample size by the ratio $(1 + 2/n_0)$.

Example 4.11.5 (Estimation of the difference between two normal means with given standard error $a^{1/2}$) Let θ denote the difference between the means of two normal populations and σ^2 be twice the separate population variances, assumed equal. Then (4.11.20) of Example 4.11.4 can be applied if we take C_1 to be the sum of the sample estimates of variance. First take a preliminary sample of size n_0 from each population and let C_1 be the sum of the usual sample estimates of variance. Set

$$\tilde{n}(T_1, C_1) = C_1(1 + 1/n_0)/a, \quad (4.11.21)$$

and take a second sample of $\max[0, \tilde{n}(T_1, C_1) - n_0)]$ from each population. The difference between the pooled sample means is the required estimate. If the population variances are not equal, the procedure is still valid if the factor $1 + n_0^{-1}$ in (4.11.21) is replaced by $1 + 2(1 + r^2)(1 + r)^{-2}n_0^{-1}$, where $r = C_1^{(1)}/C_2^{(2)}$ is the ratio of the sample variances in the preliminary samples. However, when the population variances are unequal, the expected sample size is reduced by dividing the second sample between the two populations in a way depending on r^2.

More complicated problems with several unknown parameters can be dealt with in a similar way. Examples of problems requiring the more general analysis are:

(i) to estimate the difference between, or the ratio of, two binomial proportions θ_1, θ_2 with a variance equal to some function of θ_1 and θ_2;

(ii) to estimate the difference $\theta_1 - \theta_2$ between two normal means with a standard error equal to a given percentage of θ_1;

(iii) to estimate the difference between two normal means with unequal population variances with optimum subdivision of the observations between the two populations.

EXPECTED SAMPLE SIZE It will be shown that, except when the preliminary sample size is small, the best double-sampling procedure has an expected sample size only slightly greater than that for the best sequential procedure. For simplicity, consider the single-parameter case. From (4.11.8) the expected sample size is

$$E[\lambda n(T_1)] = \lambda n(\theta) + b(\theta)n(\theta) + v(\theta)n''(\theta)/(2n_0) + o(1). \qquad (4.11.22)$$

To the first approximation the expected sample size is $\lambda n(\theta) = \lambda v(\theta)/a(\theta)$. It follows that the scheme should be based on the estimate $T^{(m)}$ with minimum variance, and that if instead an estimate of efficiency e is used, the mean sample size is increased in the ratio e^{-1}. Now $\lambda n(\theta)$ is the fixed-sample size that gives an estimate with the assigned variance. It is reasonable to expect, at least when $T^{(m)}$ is a sufficient estimate for θ, that no sequential scheme can require fewer than $\lambda n(\theta)$ observations on the average. This can be proved formally as follows. For any sequential estimation procedure giving an unbiased estimate θ^* of θ occurring with density $f(x;\theta)$, we have from Eq. (4.6.4)

$$\operatorname{var} \theta^* \geq [ENE(\partial \ln f/\partial\theta)^2]^{-1}, \qquad (4.11.23)$$

where N is the stopping time for the estimation procedure. Now, for normal, Poisson, and binomial distributions, the variance of the sample mean is $v(\theta)/m$ where $[v(\theta)]^{-1} = E(\partial \ln f/\partial\theta)^2$. Thus, considering estimates for which $\operatorname{var}(\theta^*) \leq a(\theta)/\lambda$, we have for the above distributions

$$EN \geq \lambda v(\theta)/a(\theta) = \lambda n(\theta), \qquad (4.11.24)$$

which is satisfied asymptotically for all distributions. That is, the expected sample size in the best double-sampling scheme is (4.11.22) and in the best sequential scheme it satisfies the inequality (4.11.24). Equation (4.11.23) depends on the assumption that the possibility that $n_0 > \tilde{n}(T_1)$ may be ignored. If this assumption is not valid, the expected sample size will be increased and it is likely that the resulting estimate T' will be more accurate than is required [that is, that (bias of T')^2 + var(T') $\leq a(\theta)/\lambda$]. n_0 should be chosen as large as possible subject to the condition that $P(n_0 > \tilde{n}(T_1))$ is small.

CONFIDENCE INTERVALS We may want to give a confidence interval for θ at the end of one of the sampling procedures already considered. Secondly, the problem may be stated in terms of confidence intervals, that is, we may want to estimate θ by γ percent confidence interval of predetermined form. We shall deal mainly with the first sort of problem. Suppose that we have obtained an estimate T' after a sampling procedure designed to give a variance $a(\theta)/\lambda$. If we want a $(1-2\alpha)100\%$ confidence interval for θ, we let $z_{1-\alpha}$ denote the $(1-\alpha)$th point on the standard normal distribution. Then define θ_-, θ_+ by the equations

$$\theta_- + z_{1-\alpha} a^{1/2}(\theta_-)/\lambda^{1/2} = T', \qquad \theta_+ - z_{1-\alpha} a^{1/2}(\theta_+)/\lambda^{1/2} = T'. \quad (4.11.25)$$

(θ_-, θ_+) will be the required confidence interval. If an explicit solution is impossible, Eq. (4.11.25) is solved by the method of successive approximation which is due to Bartlett (1937).

Example 4.11.6 Suppose that T' is the estimate obtained after the procedure of Example 4.11.2 for estimating a normal mean with given fractional standard error $a^{1/2}$. Then $a(\theta) = a\theta^2$ and (4.11.25) give for, say, the 95% confidence interval

$$\theta_- = T'(1 + 1.96a^{1/2})^{-1}, \qquad \theta_+ = T'(1 - 1.96a^{1/2})^{-1}.$$

Formula (4.11.25) assumes that T' is normally distributed. A refinement of the method depends on evaluating the skewness γ_1, and kurtosis γ_2 of T' and making a correction for these based on the Cornish–Fisher expansion. We shall illustrate the method by a simple example.

Example 4.11.7 (Confidence interval of given breadth for a normal mean, variance unknown) Consider the construction of a confidence interval after the procedure of Example 4.11.4 for estimating a normal mean θ with given standard error $a^{1/2}$. If T' is the final estimate in this case, the sample mean, the $(1-2\alpha)100\%$ confidence interval is, from (4.11.25), $(T' - z_{1-\alpha}a^{1/2}, T' + z_{1-\alpha}a^{1/2})$. Now, it can be shown that for the distribution of T', γ_1 is zero and γ_2 is $6n_0^{-1}$. Thus according to the Cornish–Fisher inversion of the Edgeworth expansion, the normal multiplier should be replaced by $z_{1-\alpha} + (z_{1-\alpha}^3 - 3z_{1-\alpha})/4n_0 = z_{1-\alpha}^*$, say. If we use the normal multiplier, the width of the confidence interval is $2z_{1-\alpha}a^{1/2}$; and if we use the corrected multiplier, the width is $2z_{1-\alpha}^* a^{1/2}$. To solve Stein's problem of arranging that the confidence interval be of width Δ, we take $a^{1/2} = \Delta/(2z_{1-\alpha})$ or $\Delta/(2z_{1-\alpha}^*)$. The corresponding sample size functions are, from (4.11.20),

$$4s_1^{\,2}(1 + 2/n_0)z_{1-\alpha}^2/\Delta^2 \quad (4.11.26)$$

or in the second case

$$\frac{4s_1{}^2}{\Delta^2}\left(1+\frac{2}{n_0}\right)z_{1-\alpha}^{*2} \sim \frac{4s_1{}^2}{\Delta^2}z_{1-\alpha}^2\left(1+\frac{1+z_{1-\alpha}^2}{2n_0}\right). \tag{4.11.27}$$

In Stein's exact solution the corresponding sample size is $4s_1{}^2\Delta^{-2}t_{2\alpha, n_0-1}^2$ where $t_{2\alpha, n_0-1}$ is the two-sided 2α percent point of the t-distribution with $n_0 - 1$ degrees of freedom. From the exact solution we can compute the percentage error of the approximate formulas (4.11.26) and (4.11.27). This has been done in Table 4.11.1. Formula (4.11.26) has a fairly small error even

Table 4.11.1

Estimation of Normal Mean by $(1 - 2\alpha)100\%$ Confidence Interval of Exact and Approximate Solutions

	Preliminary sample size 10		Preliminary sample size 20	
α	Percent error of (4.11.26)	Percent error of (4.11.27)	Percent error of (4.11.26)	Percent error of (4.11.27)
.10	3.1	−2.7	2.5	0.0
.05	−3.4	−4.5	0.0	−1.1
.025	−9.9	−6.8	−3.5	−1.7
.01	−18.4	−10.2	−7.7	−2.6
.005	−24.6	−13.2	−10.8	−3.5

when n_0 is as small as 10, provided that α is less than .025. The correction for kurtosis makes a substantial improvement. These results suggest that the general approximate formulas developed in this section will be reasonably accurate except when n_0 is very small.

If at the end of one of the estimation procedures, we require to test the hypothesis H_0: $\theta = \theta_0$, we proceed in an analogous manner. If H_0 is true T' is distributed with mean θ_0 and variance $a(\theta_0)/\lambda$. Hence, if we assume T' to be normally distributed, the 2α percent significance limits are $\theta_0 \pm z_{1-\alpha}a^{1/2}(\theta_0)\,\lambda^{-1/2}$. Correction for skewness and kurtosis can be introduced if necessary.

DISCUSSION There are two situations in which sequential methods are useful. In the first, observations become available only at infrequent intervals and must be interpreted as soon as they are obtained. An example is the study of accident rates which may only be obtainable at weekly, monthly, etc., intervals. Double-sampling procedures are useless in problems of this sort. The second type of situation is where the number of observations is under the experimenter's control, but observations are expensive, so that the smaller the sample size needed the better. Double sampling is often practical in this sort of problem. Cox (1952b) has also considered the following mixed problem: To construct a scheme for testing $H_0\,(\theta = \theta_0)$ with given power near $\theta = 0$ and for estimating θ with given accuracy when θ is different from zero. [For example, if $|\theta| \gg 0$ we might require our estimate to have given standard error $(a')^{1/2}$.] For an example on this, the reader is referred to Cox (1952b, pp. 225–226).

Problems

1. Using a double-sampling scheme, estimate the Poisson mean θ with given fractional standard error $a^{1/2}$ and also set up a $(1 - 2\alpha)100\%$ confidence interval for θ.
2. Estimate the exponential mean θ by a double-sampling method with given fractional standard error $a^{1/2}$ and also set up a $(1 - 2\alpha)100\%$ confidence interval for θ.

4.12 Large Sample Theory of Sequential Estimation

Anscombe (1949) has provided a large-sample theory of sequential estimation, when there was only one unknown parameter in the distribution of the observations. He has shown using a heuristic argument that an estimation formula valid for fixed-sample size remained valid when the sample size was determined by a sequential stopping rule. Another proof has been given by Cox (1952c) (in the course of deriving a correction term to Anscombe's result) whose work suggests that fixed-sample size formulas might be valid generally for sequential sampling provided the sample size was large. Anscombe (1952) has been able to bypass some of the complexity of his previous work by introducing the condition "uniform continuity in probability" to be satisfied by the statistic used. We shall present his results in the following.

We shall consider random variables which might assume values $\pm\infty$ with positive probabilities. For a one-dimensional random variable X, the distribution function $F(x)$ is equal to the probability that $X \le x$, is nondecreasing, is continuous to the right and satisfies

$$\lim_{x\to-\infty} F(x) \ge 0, \qquad \lim_{x\to\infty} F(x) \le 1. \tag{4.12.1}$$

If equalities hold in (4.12.1), we say that X is a proper random variable and $F(x)$ is a proper distribution function. Let $\{Y_n\}$ $(n = 1, 2, \ldots)$ be an infinite sequence of random variables. Assume that there exists a real number θ, a sequence of positive numbers $\{w_n\}$, and a distribution function $G(x)$ such that the following conditions are satisfied:

(C1) *Convergence of* $\{Y_n\}$ For any x such that $G(x)$ is continuous [a continuity point of $G(x)$], $P(Y_n - \theta \leq xw_n) \to G(x)$ as $n \to \infty$.

(C2) *Uniform continuity in probability of* $\{Y_n\}$ Given any small ε and η, there exists a large v and a small positive c such that, for any $\eta \geq v$,

$$P(|Y_{n'} - Y_n| < \varepsilon w_n \text{ simultaneously for all integers } n' \text{ such that } |n' - n| < cn) > 1 - \eta.$$

Note that, as $n \to \infty$, $Y_n \to \theta$ in probability if $w_n \to 0$ and if $G(x)$ is a proper distribution function. In most applications, $G(x)$ will be both proper and continuous, and in fact will usually be the normal distribution function, w_n is a linear measure of dispersion of Y_n, and might be, for example, the standard deviation or the quartile range. The term "uniform continuity" is used to describe condition (C2) since a property analogous to ordinary uniform continuity is implied. Given any realization of the sequence $\{Y_n\}$, let functions Y_n and w_n be defined for noninteger values of n by linear interpolation between the adjacent integer values. Then, if $\ln w_n$ is uniformly continuous with respect to $\ln n$ for large n, and if (C1) is satisfied with $G(x)$ proper it is easy to see that (C2) implies that, in a probabilistic sense, the uniform continuity of $(Y_n - \theta)/w_n$ holds with respect to $\ln n$.

Theorem 4.12.1 (Anscombe, 1952) Let $\{n_r\}$ be an increasing sequence of positive integers tending to infinity, and let $\{N_r\}$ be a sequence of proper random variables taking positive integer values such that $\{N_r/n_r\} \to 1$ in probability as $r \to \infty$. Then, if the sequence of random variables $\{Y_n\}$ satisfies conditions (C1) and (C2),

$$P\{Y_{N_r} - \theta \leq xw_{n_r}\} \to G(x) \qquad \text{as} \quad r \to \infty. \tag{4.12.2}$$

Proof (Cramér, 1946, §20.6) Given positive ε and η, let v and c satisfy (C2), and let v also be sufficiently large so that for any r for which $n_r > v$

$$P(|N_r - n_r| < cn_r) > 1 - \eta. \tag{4.12.3}$$

For given r such that $n_r > v$, let E denote the event $|N_r - n_r| < cn_r$ and $|Y_{N_r} - Y_{n_r}| < \varepsilon w_{n_r}$. Then

$$P(E) > 1 - 2\eta. \tag{4.12.4}$$

For, if E_1 and E_2 denote the events

$$E_1: \quad |Y_{n'} - Y_n| < \varepsilon w_n \quad \text{simultaneously for all } n' \text{ such that } |n' - n| < cn$$

$$E_2: \quad |N_r - n_r| < cn,$$

we have

$$P(E) \geq P(E_1 E_2) = P(E_1) - P(E_1 \bar{E}_2) \geq P(E_1) - P(\bar{E}_2)$$

and (4.12.4) follows from (4.12.3) and (C2). By considering the event $Y_{N_r} - \theta \leq x w_{n_r}$ as the sum of the exclusive events

$$\text{(i)} \quad Y_{N_r} - \theta \leq x w_{n_r} \quad \text{and} \quad E, \qquad \text{(ii)} \quad Y_{N_r} - \theta \leq x w_{n_r} \quad \text{and} \quad \bar{E},$$

we can easily deduce from (4.12.4) that

$$P\{Y_{n_r} - \theta \leq (x - \varepsilon) w_{n_r}\} - 2\eta < P(Y_{N_r} - \theta \leq x w_{n_r})$$
$$< P\{Y_{n_r} - \theta \leq (x + \varepsilon) w_{n_r}\} + 2\eta,$$

and (4.12.2) follows, if x is a continuity point of $G(x)$, from the fact that the distribution functions are monotonic and almost everywhere continuous. This completes the proof of Theorem 4.12.1.

REMARK 4.12.1 Notice that we have not assumed that the distribution of N_r and Y_n are independent.

APPLICATION 4.12.1 We shall apply this result to the sequential estimation of an unknown parameter θ. In what follows, $X_1, X_2, \ldots$ denote a sequence of observations, Y_n is an estimate of θ calculated from the first n observations, and Z_n is an estimate of the dispersion w_n of Y_n. In order to estimate θ with given small dispersion a, we use the sequential stopping rule: sample until first $Z_n \leq a$ and then calculate Y_n. To show that Y_n as obtained is an estimate of θ with dispersion asymptotically equal to a if a is small, we consider not a single stopping rule, but a sequence of possible stopping rules such that the values of a tend to zero. The situation can be defined in purely probabilistic terms (without the use of interpretative words like *observation*, *unknown*, and *estimate*) as follows.

Let $\{X_n\}$ $(n = 1, 2, \ldots)$ denote a sequence of random variables not necessarily independent. For each n, let Y_n and Z_n be functions of $X_1, X_2, \ldots, X_n$. Assume that $\{Y_n\}$ satisfies (C1) and (C2) above. Let $\{a_r\}$ $(r = 1, 2, \ldots)$ be a decreasing sequence of positive numbers tending to zero. Let $\{N_r\}$ be the sequence of random variables defined by the condition: N_r is the least integer n such that $Z_n \leq a_r$; and let $\{n_r\}$ be the sequence of integers defined by the condition: n_r is the least n such that $w_r \le a_r$. We assume that the following further conditions are satisfied.

(C3) *Convergence of* $\{W_n\}$ $\{w_n\}$ is decreasing, tends to zero, and $w_n/w_{n+1} \to 1$ as $n \to \infty$.

(C4) *Convergence of* $\{N_r\}$ N_r is a proper random variable for all r, and $N_r/n_r \to 1$ in probability as $r \to \infty$.

Then we have the following theorem.

Theorem 4.12.2 (Anscombe 1952) If conditions (C1)–(C4) are satisfied, then

$$P(Y_{N_r} - \theta \le xa_r) \to G(x) \qquad \text{as} \quad r \to \infty, \tag{4.12.5}$$

at all continuity points x of $G(x)$.

Proof (C3) implies that $w_{n_r}/a_r \to 1$ as $r \to \infty$. Now (4.12.5) readily follows from Theorem 4.12.1.

In applying this theorem, it will usually be obvious that (C1) and (C3) are satisfied. The following theorems show that (C2) is satisfied for a wide class of statistics Y_n. (C4) is a weak condition and it is usually easy to discover whether or not it is satisfied. Although these conditions are sufficient for the conclusion of Theorem 4.12.2, it will be shown that they are not necessary.

PARTICULAR FORMS FOR Y_n We shall assume that the observations X_1, $X_2, \ldots$ are independent and identically distributed with the distribution function $F(x)$, and we consider first the statistic $Y_n = \sum_1^n X_i/n$. We shall see that if (C1) and (C3) are satisfied with $G(x)$ the distribution function of a proper random variable V, say, not certainly equal to a constant, then (C_2) is satisfied. From (C1), V has a stable distribution (Lévy, 1937) with characteristic function

$$\int_{-\infty}^{\infty} \exp(itx)\, dG(x) = \exp(c_0|t|^\alpha + ic_1 t|t|^{\alpha-1}),$$

where t, c_0, c_1, α are real, $0 < \alpha \le 2$, $c_0 > 0$, $|c_1| \le c_0|\tan(\pi\alpha/2)|$; $G(x)$ is continuous. If $\alpha = 2$, V is normal and

$$w_n n^{1/2} \to \text{a positive limit or infinity as } n \to \infty \tag{4.12.6}$$

(the former if var $X_r < \infty$, the latter if not). Thus w_n decreases not faster than $n^{1/2}$. This result holds also for $\alpha < 2$ [see Lévy (1937, Theorem 48)]. For $\alpha < 2$, the distribution of X_n has no finite absolute moments of order $>\alpha$

(Lévy, 1937, §60). However, if an absolute moment of order <1 diverges, Y_n cannot tend to a limit in probability as $n \to \infty$ [an easy deduction from Cramér (1937, Theorem 15)] and (C3) implies that $Y_n \to \theta$ in probability. Hence $\alpha \geq 1$ and so from Lévy (1937, see §56) for all finite x, $0 < G(x) < 1$. If $\alpha > 1$ then $EX_n = 0$ and hence $EV = 0$. In all practical applications $G(x)$ will be normal, in which case $0 < G(x) < 1$, for all finite x and consequently mention of (C3) can be omitted from the following statement of Theorem 4.12.3.

Let k be the lower bound of $P(Y_n \geq \theta)$ and of $P(Y_n \leq \theta)$, for $n = 1, 2, \ldots$. Then k is positive. If $k = 0$, it follows from the fact that $0 < G(x) < 1$ that this lower bound is not the lower limit as $n \to \infty$ (for that must be positive), and therefore that it must be attained for some finite $n = n_0$, say. It is then also attained for all integer multiples of n_0 and we have a contradiction. We now show that, given $\varepsilon > 0$, there is a positive c such that for sufficiently large n the probability is as small as we please that

$$|Y_{n'} - Y_n| > \varepsilon w_n \qquad \text{for some } n', \quad n < n' < (1 + c)n. \tag{4.12.7}$$

A similar argument would apply to values of n' where $(1 - c)n < n' < n$. so that (C2) is satisfied. We have

$$Y_n' - Y_n = \frac{1}{n'} \sum_{i=n+1}^{n'} (X_i - \theta) - \frac{n' - n}{n'} (Y_n - \theta).$$

Given $Y_{n'}$, the first term on the right-hand side is a random component of $Y_{n'}$, while the second term represents a nonrandom drift of $Y_{n'}$ toward θ as n' increases. For sufficiently large n, there is a K such that $|Y_n - \theta| < Kw_n$ with probability close to 1. Take $c < \varepsilon/2K$. Then with probability close to 1, $(1 - n/n')|Y_n - \theta| < \varepsilon w_n/2$. The probability that (4.12.7) holds is therefore small if the probability that

$$n^{-1}\left|\sum_{i=n+1}^{n'} (X_i - \theta)\right| > \varepsilon w_n/2 \qquad \text{for some } n', \quad n < n' < (1 + c)n \tag{4.12.8}$$

is small. Given that (4.12.8) holds for some n', and that in fact $n^{-1}\sum_{i=n+1}^{n'} (X_i - \theta) > \varepsilon w_n/2$, the probability is at least k (defined above) that $\sum_{i=n+1}^{(1+c)n} (X_i - \theta) \geq 0$ and therefore that $n^{-1}\sum_{i=n+1}^{(1+c)n}(X_i - \theta) > \varepsilon w_n/2$, and therefore that

$$n^{-1}\left|\sum_{i=n+1}^{(1+c)n} (X_i - \theta)\right| > \varepsilon w_n/2. \tag{4.12.9}$$

The same result follows if (4.12.8) holds with $n^{-1}\sum_{i=n+1}^{n'}(X_i - \theta) < -\varepsilon w_n/2$.

Thus, if p denotes the probability of (4.12.9), P[(4.12.8) holds] $\leq p/k$. Now, p is the probability that $|Y_{cn} - \theta| > \varepsilon w_n/2c$. From (4.12.6) $w_n \geq w_{cn} c^{1/2}$ as $n \to \infty$ with c constant. If therefore we take $c < (\varepsilon/2K)^2$, p is small. Thus we have proved the following theorem.

Theorem 4.12.3 (Anscombe, 1952) Condition (C2) is satisfied if $X_1, X_2, \ldots$ are independent and identically distributed, $Y_n = n^{-1}\sum_1^n X_i$, (C1) and (C3) hold, and $G(x)$ is proper and continuous.

We can also deduce easily that (C2) is satisfied for a wide class of maximum likelihood estimators. Let the distribution of X_n be either discrete or absolutely continuous, and (taking the latter case for the sake of definiteness) let its frequency function be $F'(x) = f(x)$, a known function of x and of the parameter θ to be estimated and perhaps of other unknown parameters $\theta_2, \theta_3, \ldots, \theta_s$. Writing θ_1 instead of θ, we suppose that the derivatives of $f(x)$ with respect to the θ_i satisfy certain regularity conditions (see, for example, Appendix A.5), and that the "information" matrix $((I_{ij}))$ is positive definite, where $I_{ij} = -E(\partial^2 \ln f(X_n)/\partial\theta_i \partial\theta_j)$. Let $((I_{ij}))^{-1} = ((\lambda_{ij}))$ and let random variables $X_n^{(i)}$ and $Y_n^{(i)}$ be defined by

$$X_n^{(i)} = \partial \ln f(X_n)/\partial\theta_i, \qquad Y_n^{(i)} = n^{-1} \sum_{j=1}^{n} X_j^{(i)}.$$

Then, asymptotically for large n the maximum likelihood estimate Y_n of θ_1 is equivalent to Y_n' defined by $Y_n' - \theta_1 = \sum_{i=1}^{s} \lambda_{1i} Y_n^{(i)}$. In other words, for given $\varepsilon > 0$, the probability is as near 1 as we please that

$$|[(Y_n - \theta_1)/(Y_n' - \theta_1)] - 1| < \varepsilon \tag{4.12.10}$$

for sufficiently large n; and if in proving this we use the strong instead of the weak law of large numbers, we see that (4.12.10) holds simultaneously for all sufficiently large n. Therefore if Y_n' satisfies (C2), so does Y_n. Since $X_n^{(i)}$ has mean 0 and variance I_{ii}, $Y_n^{(i)}$ satisfies (C2) for each i by Theorem 4.12.3 and hence so does Y_n'. [It is understood that w_n is chosen to make $G(x)$ normal; we may take $u_n = (\lambda_{11}/n)^{1/2}$, the standard deviation of Y_n'.] We thus are led to the following theorem.

Theorem 4.12.4 (Anscombe, 1952) Under the regularity conditions for the usual maximum likelihood estimation procedure, a maximum likelihood estimate Y_n satisfies (C2).

It is possible to deal in the same way with any other method of estimation which reduces, when n is large, to the averaging of some function of the observations. As a further example, consider a quantile or an interquantile range of the observations. Let $X_{1n} < X_{2n} < \cdots < X_{nn}$ be the ordered X_1, $X_2, \ldots, X_n$. Let X_{rn} be the pth quantile where $r - 1 < np \leq r$, which we will denote by $X_{np,n}$. The (p, q) interquantile range of $X_1, \ldots, X_n$ with $0 < p < q < 1$, so defined to be $X_{nq,n} - X_{np,n}$. Consider $Y_n = X_{np,n}$. First, let the distribution of X_1 be uniform on (0, 1). It is easy to verify that $X_{np,n}$ is asymptotically equivalent to $n^{-1}\sum_{i=1}^{n} X_i'$ where $X_i' = 0$ or 1 according as $X_i < p$ or $> p$. Thus $(Y_n - p)/w_n$ is asymptotically normal if $w_n \sim A/n^{1/2}$ $(A > 0)$, and by Theorem 4.12.3, (C2) is satisfied. This result extends at once to any absolutely continuous distribution of the X_i, provided the frequency function $F'(x)$ is continuous and positive at the pth quantile ξ_p of the distribution, defined by $F(\xi_p) = p$. More generally, we need not assume that $F(x)$ is absolutely continuous everywhere, but only at and near ξ_p. $F(x)$ need not be proper, but ξ_p must be finite. If $Y_n = X_{nq,n} - X_{np,n}$, Y_n again satisfies (C2) because Y_n is asymptotically equivalent to $n^{-1}\sum_{i=1}^{n} X_i''$ where X_i'' is equal to 0 when $X_i < p$ or $X_i > q$ and is equal to -1 when $p < X_i < q$. We assume that ξ_p and ξ_q are distinct, which will certainly be the case if $F(x)$ is proper and continuous since $p \neq q$.

Theorem 4.12.5 (Anscombe, 1952) (C2) is satisfied if $X_1, X_2, \ldots$ are independent and identically distributed, Y_n is a specified quantile or interquantile range of $X_1, X_2, \ldots, X_n$, (C1) holds with $G(x)$ normal, and the frequency function $F'(x)$ of X_n is continuous and positive at the corresponding (finite and distinct) quantiles of the distribution.

A type of statistic to which Theorem 4.12.3 does not apply, but which still satisfies (C2), is given by (i) $Y_n = X_{n,n}$; (ii) $Y_n = X_{1n}$; (iii) $Y_n = X_{nn} - X_{1n}$.

Assume that $F(x)$ is proper. The probability that $Y_{n'}$ differs from Y_n for any n', where $n < n' < (1 + c)n$ is less than c in cases (i) and (ii), and less than $2c$ in case (iii), and is therefore small if c is small; and it is not necessary to inquire how large the difference may be when one occurs. Statistics of this type often consist of one of the three expressions just listed multiplied by a factor not depending on n. In that case, if c is small, the probability is close to 1 that $Y_{n'} - Y_n = (u_{n'}/u_n - 1) Y_n$ and this must with high probability be small compared with w_n, for large n and $|n' - n| < cn$. Thus a condition is imposed on the "continuity" of u_n relative to w_n.

Theorem 4.12.6 (Anscombe, 1952) (C2) is satisfied if $X_1, X_2, \ldots$ have independent identical proper distributions and Y_n is an extremum or the range of $X_1, X_2, \ldots, X_n$ multiplied by a factor u_n (provided that u_n, if not a constant, satisfies the above condition).

Example 4.12.1 Suppose that we wish to estimate, with given small standard error a, the mean θ of a normal distribution, of which the variance

σ^2 is also unknown. $X_1, X_2, \ldots$ denoting the independent observations, we consider the statistic $Y_n = n^{-1}\sum_{i=1}^{n} X_i$ as an estimate of θ; for fixed n this has standard error $w_n = \sigma/n^{1/2}$, estimated (for $n \geq 2$) by $Z_n = [(n(n-1))^{-1} \sum_{i=1}^{n}(X_i - Y_n)^2]^{1/2}$. Conditions (C1) and (C3) are obviously satisfied, and from Theorem 4.12.3 so is (C2). If therefore (C3) holds, Theorem 4.12.2 implies that $(Y_N - \theta)/a$ is asymptotically normal with mean 0 and unit variance if N is the least n for which

$$Z_n \leq a. \tag{4.12.11}$$

Now, (4.12.11) is equivalent to $(n-1)^{-1}\sum_{i=1}^{n-1} \xi_i^2 \leq na$, where the ξ_i are independent normal variables derived from the X_i by a Helmert transformation: $\xi_i = (X_{i+1} - i^{-1}\sum_{j=1}^{i} X_j)[i/(i+1)]^{1/2}$. By the strong law of large numbers, given ε, $\eta > 0$, there is a ν such that

$$P\left(|(n-1)^{-1}\sum_{i=1}^{n-1} \xi_i^2 - \sigma^2| < \varepsilon \quad \text{for all } n > \nu\right) > 1 - \eta. \tag{4.12.12}$$

If a is small enough, the probability exceeds $1 - \eta$ that (4.12.11) is not satisfied for any n in the range $2 \leq n \leq \nu$; and then, given that $N > \nu$, (4.12.12) implies that the probability exceeds $1 - \eta$ that $|N/(\sigma/a)^2 - 1| < \varepsilon/\sigma^2$. Hence (C4) holds as $a \to 0$. To obtain a better approximation to the asymptotic situation when a is not infinitesimally small, it would be advisable to impose a lower limit on the value of N and not consider whether (4.12.11) was satisfied until that limit had been passed. One might for example stipulate that $N \geq (\delta/a)^2$ where δ is a positive quantity believed to be smaller than σ. If in fact $\delta < \sigma$, (C4) and Theorem 4.12.2 would apply just as before.

Example 4.12.2 Suppose that $X_1, X_2, \ldots$ have independent uniform distributions in the interval $(0, \theta)$, and that we desire to estimate θ with given small standard error a. We may take $Y_n = [(n+1)/n] \max_{1 \leq i \leq n} X_i$, $w_n = \theta/n$, $Z_n = Y_n/n$. As $n \to \infty$, $P(Y_n - \theta \leq xw_n) \to G(x) = e^{x-1}$ (for $x \leq 1$), and w_n is asymptotically the standard deviation of Y_n. (C2) is satisfied by Theorem 4.12.6. Hence by Theorem 4.12.2 the required sample size N is the least n for which $\max_{1 \leq i \leq n} X_i \leq [n^2/(n+1)]$. Similar considerations apply regarding (C4) as in the previous example. Notice that in the definition of Y_n we could have omitted the factor $(n+1)/n$. This could have given $G(x) = e^x$ (for $x \leq 0$) and a stopping rule equivalent to the above for large n.

Example 4.12.3 To see what may happen if (C2) is not satisfied, consider independent observations $X_1, X_2, \ldots$ from a normal distribution with unknown mean θ and known variance σ^2, and let us take

$$Y_n = n^{-1/2}X_n + (1 - n^{-1/2})(n-1)^{-1}\sum_{i=1}^{n-1} X_i. \tag{4.12.13}$$

Y_n is normally distributed with mean θ and asymptotic variance $2\sigma^2/n$ for large n. The correlation between Y_n and $Y_{n+1} \to \frac{1}{2}$ and (C2) is not satisfied. Suppose that we wish to estimate θ with given small standard error a. Then we take $N = 2\sigma^2/a^2$ (or the next integer above), a fixed value. The conclusion of Theorem 4.12.2 holds. Now suppose that we wish to estimate θ (assumed to be positive) with given small coefficient of variation a. The coefficient of variation of Y_n and a suitable estimate of it, are $w_n = (\sigma/\theta)(2/n)^{1/2}$, $Z_n = (\sigma/Y_n)(2/n)^{1/2}$. If Theorem 4.12.2 is applied, it would indicate the stopping rule: N is the least n for which

$$nY_n^2 \geq 2\sigma^2/a^2. \tag{4.12.14}$$

When $n = N$ the second member of the right-hand side of (4.12.13) is asymptotically normal with mean $\theta(1 - N^{-1/2})$, but as for the first, as $a \to 0$ the probability tends to 1 that $X_N > \theta + k\sigma$ where k is any positive number. Hence $(Y_N - \theta)/a \to \infty$ in probability, and so does not have the limit distribution $G(x) = \Phi(x/\theta)$ which Theorem 4.12.2 would conclude, where Φ denotes the standard normal distribution function. It is easy to verify that (C4) is satisfied by the rule in (4.12.14).

Thus, when (C2) is not satisfied, the conclusion of Theorem 4.12.2 may hold if N satisfies a stronger condition than (C4), such as being constant. But it seems likely (C2) is necessary if no further condition than (C4) is imposed on N, and in particular if the distribution of N and Y_N are not assumed to be independent.

Problems

1. Let $X_1, X_2, \ldots$ be a sequence of independent Poisson random variables having mean ϑ. Estimate θ (via large-sample theory) with given small standard error a.

2. Let $X_1, X_2, \ldots$ be a sequence of independent random variables having the density $e^{-(x-\theta)}$ for $x \geq \theta$. Estimate θ with given small standard error a.

3. Proceed as in (2) if θ is the parameter in the Bernoulli distribution.

4.13 Sequential Confidence Intervals for the Mean of a Normal Population with Unknown Variance

The general problem of sequential estimation is to postulate a sampling rule that will ensure that the unknown parameter is estimated with given accuracy and with minimum expected sample size. The accuracy could be in the form of the width of the confidence interval, standard error, or coefficient of variation of the estimate. Anscombe (1953, 1954) has studied quite extensively large-sample theory of sequential estimation. He found that for large sample sizes or equivalently for high required accuracy the fixed-sample size formulas for estimation were still approximately applicable in the sequential procedure. Anscombe (1953) has also considered the problem of estimating the mean of a normal population by a confidence interval of prescribed width and coefficient when the population variance is unknown. Ray (1957b) has considered the same problem and certain alternative ones from the exact small-sample approach with somewhat modified stopping rules. Stein and Wald (1947), while considering the sequential confidence intervals for the mean of a normal population with known variance, found that the best sequential procedure (in terms of sample size) was the fixed-sample size procedure.

Let $X_1, X_2, \ldots$ be independent random variables each distributed normally with mean μ and variance σ^2. The problem is to estimate μ by a confidence interval of given width $2d$ and confidence coefficient $1 - \alpha$ $(0 < \alpha < 1)$. For any positive integer n, let

$$Y_n = \sum_{i=1}^{n} X_i/n. \tag{4.13.1}$$

Let u_α be such that the unit normal variable has probability α of lying outside the interval $(-u_\alpha, u_\alpha)$. If σ^2 is known, then the required confidence interval for μ is $(Y_n - d,\ Y_n + d)$ where n is determined by

$$nd^2 = u_\alpha{}^2\sigma^2. \tag{4.13.2}$$

If σ^2 is unknown, the first-order asymptotic procedure is to continue taking observations until

$$s^2 \le d^2 n/u^2 \tag{4.13.3}$$

where $s^2 = \sum_{i=1}^{n}(X_i - Y_n)^2/(n-1)$ and the subscript α in u_α is suppressed. Let N denote the stopping variable. It is convenient to express the stopping rule (4.13.3) in the following form. Make the transformation of $X_1, X_2, \ldots, X_n$ to $U_1, U_2, \ldots, U_{n-1}, Y_n$ by setting

$$U_i = \left(iX_{i+1} - \sum_{j=1}^{i} X_j\right)^2 \Big/ i(i+1) \qquad (i = 1, 2, \ldots, n-1). \tag{4.13.4}$$

Then, the U_i are independent random variables each distributed as $\sigma^2\chi^2$

with one degree of freedom. Also $s^2 = \sum_{i=1}^{n-1} U_i/(n-1)$ and the stopping rule becomes: Stop sampling when

$$\sum_{i=1}^{n-1} U_i \leq d^2 n(n-1)/u^2. \tag{4.13.5}$$

If further observations $X_{n+1}, X_{n+2}, \ldots$ are taken, the sequence U_i is increased by the addition of elements $U_n, U_{n+1}, \ldots$, the earlier elements remaining unchanged.

Anscombe's (1953) procedure when d is small and n is large is given by: Stop sampling when

$$\sum_{i=1}^{n-1} U_i \leq d^2 n(n - 2.676 - \tfrac{1}{2}u^2)/u^2 \qquad (n \geq 4). \tag{4.13.6}$$

The multiplier of nd^2/u^2 on the right-hand side of (4.13.6) may be modified by adding any arbitrary constant or indeed any function of n which is $O(n^{1/2})$ when n is large. For example, if $\alpha = .05$ so that $u = 1.96$ we modify (4.13.6) by the addition of a constant, to $\sum_{i=1}^{n-1} U_i \leq d^2(n-3)(n-1.597)/3.84$. The approximate formula for the expected sample size following the latter rule of procedure (4.13.6) is

$$E(N) = \sigma^2 u^2/d^2 + \tfrac{1}{2}(1 + u^2). \tag{4.13.6a}$$

The first term on the right-hand side of (4.13.6a) is the number of observations that would have been required if σ^2 were known.

If the problem is to estimate μ with given standard error a, then the sampling rule would be to sample sequentially until

$$\sum_{i=1}^{n-1} U_i \leq a^2(n-5)(n+.824) \qquad (n \geq 6). \tag{4.13.7}$$

In this case, the expected sample size is approximately given by

$$E(N) = (\sigma^2/a^2) + 2. \tag{4.13.7a}$$

A related problem is to estimate the difference in means between two normal populations, assuming that each population has a common unknown variance $\sigma^2/2$. Then the sampling is carried out sequentially as before by taking pairs of observations (X_i, X_i'). Corresponding to each sequence of observations from the two populations, corresponding sequence (U_i, U_i')

may be determined. Let $V_i = U_i + U_i'$, then V_i is an estimate of σ^2 distributed as $\sigma^2\chi^2$ having two degrees of freedom. If $Y_n = \bar{X} - \bar{X}'$, the difference between the sample means of the two populations, then the confidence interval for the difference in population means of width $2d$ and coefficient approximately $1 - \alpha$ will be obtained if sampling is terminated when

$$\sum_{i=1}^{n-1} V_i \leq d^2 n(n - 1.505 - \tfrac{1}{4}u^2)/u^2 \qquad (n \geq 3), \tag{4.13.8}$$

and

$$E(N) = (\sigma^2 u^2/d^2) + (1 + u^2)/4. \tag{4.13.8a}$$

Alternatively if an estimate of the difference between the two means is desired with given standard error a, then the sampling rule is: Stop sampling when

$$\sum_{i=1}^{n-1} V_i \leq a^2(n-3)(n + .745) \qquad (n \geq 4), \tag{4.13.9}$$

and

$$E(N) = (\sigma^2/a^2) + 1. \tag{4.13.9a}$$

For proofs of the above results, the reader is referred to Anscombe (1953). In order to develop an exact small sample theory for such estimation problems, Ray (1957b) considered some modified sequential procedures. In the following we shall present his results.

In the case of estimating the normal mean by a confidence interval of width $2d$ and coefficient $1 - \alpha$, Ray (1957b) postulates the sampling rule: Stop sampling when

$$s^2 t_{n-1}^2/n \leq d^2, \tag{4.13.10}$$

where t_{n-1} is such that the random variable T_{n-1} has t distribution with $n - 1$ degrees of freedom having probability α of lying outside the interval $(-t_{n-1}, t_{n-1})$. Rewriting the inequality (4.13.10) in terms of the transformed variables U_i, we have: Stop sampling when

$$\sum_{i=1}^{n-1} U_i \leq d^2 n(n-1)/t_{n-1}^2. \tag{4.13.11}$$

If the procedure is slightly modified to taking first three observations and then subsequently pairs of observations so that there is always an odd number of observations, $2m + 1$ say, then (4.13.11) may be written as: Stop sampling when

$$\sum_{i=1}^{2m} U_i \le d^2(2m + 1)2m/t_{2m}^2. \tag{4.13.12}$$

Let $V_i = U_{2i-1} + U_{2i}$; then V_i is distributed as $\sigma^2\chi^2$ with two degrees of freedom. Now letting $d = c\sigma$ and $D_m = 2m(2m + 1)c^2/t_{2m}^2$, one can write (4.13.12) as $\sum_{i=1}^{m} V_i \le D_m\sigma^2$ or as $\sum_{i=1}^{m} T_i \le D_m$ where T_i is distributed as χ^2 with two degrees of freedom. (One can write it as $T \le 2D_m$ where T has a gamma distribution with parameter m.) Let P_{2m+1} denote the probability that sampling ceases at the $(2m + 1)$th observation. If Z_i denotes the successive partial sums of the sequence T_i, then we have

$$\begin{aligned} P_{2m+1} &= P(Z_1 > D_1, \ldots, Z_{m-1} > D_{m-1}, Z_m \le D_m) \\ &= P(Z_1 > D_1) \cdots P(Z_{m-1} > D_{m-1} | Z_{m-2} > D_{m-2}, \ldots, Z_1 > D_1) \\ &\quad \times P(Z_m \le D_m | Z_{m-1} > D_{m-1}, \ldots, Z_1 > D_1), \end{aligned} \tag{4.13.13}$$

where $Z_j = \sum_{i=1}^{j} T_i$. The overall probability that the interval $(Y_{2m+1} - d, Y_{2m+1} + d)$ includes the unknown mean μ is given by

$$\sum_{m=1}^{\infty} P[|Y_{2m+1} - \mu| \le d]P_{2m+1} = 2\sum_{m=1}^{\infty} \Phi[c(2m + 1)^{1/2}]P_{2m+1} - 1, \tag{4.13.14}$$

where Φ denotes the standard normal distribution function. The P_{2m+1} define the distribution of the sample size from which the expected sample size and its variance can be determined. Thus

$$E(N) = \sum_{m=1}^{\infty} (2m + 1)P_{2m+1}, \tag{4.13.15}$$

and

$$\operatorname{var} N = \sum_{m=1}^{\infty} (2m + 1)^2 P_{2m+1} - (EN)^2, \tag{4.13.16}$$

where N denotes the stopping variable.

DETERMINATION OF P_{2m+1} It is necessary to determine the conditional density functions of the Zs as follows:

$$p(z_1) = p_{\chi_2^2}(z_1) = \tfrac{1}{2}\exp(-z_1/2).$$

Then the truncated distribution $p^-(z_1)$ given that $Z_1 > D_1$ is given by

$$p^-(z_1) = p(z_1)/\int_{D_1}^{\infty} \tfrac{1}{2} e^{-z/2}\, dz = \tfrac{1}{2} \exp[-(z_1 - D_1)/2].$$

Now, $Z_2 = Z_1 + T_2$ and Z_1 and T_2 are independent. Thus the convolution of Z_1 and T_2 or the distribution of Z_2 becomes

$$\begin{aligned} p(z_2 | Z_1 > D_1) &= \int_0^{z_2 - D_1} \{\tfrac{1}{2} \exp[-(z_1 - D_1)/2]\}_{z_1 = z_2 - t} \tfrac{1}{2} e^{-t/2}\, dt \\ &= \int_0^{z_2 - D_1} \tfrac{1}{4} \exp[-(z_2 - D_1)/2]\, dt \\ &= [(z_2 - D_1)/4] \exp^{-(z_2 - D_1)/2}. \end{aligned}$$

The truncated distribution $p(z)$ given that $Z_1 > D_1$, $Z_2 > D_2$ is

$$\begin{aligned} p^-(z) &= p(z | Z_1 > D_1)/\int_{D_2}^{\infty} p(z | Z_1 > D_1)\, dz \\ &= (1/4B_2)(z - D_1) \exp[-(z - D_1)/2] \end{aligned}$$

where $B_2 = \{1 - \Gamma_{(D_2 - D_1)/2}(2)\}$ and $\Gamma_x(k)$ denotes the incomplete gamma function. The convolution of $Z_3 = Z_2 + T_3$ is obtained as

$$\begin{aligned} & p(z | Z_1 > D_1, Z_2 > D_2) \\ &= \int_0^{z - D_2} (1/4\, B_2)(z - t - D_1) \exp[-(z - t - D_1)/2] \tfrac{1}{2}\, e^{-t/2}\, dt \\ &= \int_0^{z - D_2} (1/8\, B_2) \exp[-(z - D_1)/2](z - t - D_1)\, dt \\ &= \frac{\exp[-(z - D_2)/2]}{16 B_2 \exp[(D_2 - D_1)/2]} \left\{ \frac{2(z - D_2)^2}{2!} + 2(D_2 - D_1)(z - D_2) \right\}. \end{aligned}$$

Now one can obtain $p^-(z)$ the truncated density given that $Z_1 > D_1, Z_2 > D_2$, $Z_3 > D_3$. Analogously one can obtain $p(z | Z_1 > D_1, \ldots, Z_{m-1} > D_{m-1})$.

It is now possible to evaluate the probabilities arising in (4.13.13):

$$\begin{aligned}
p_1 &= P(Z_1 \leq D_1) = \Gamma_{D_1/2}(1) \\
p_2 &= \int_{D_1}^{D_2} p(z|Z_1 > D_1)\,dz = \int_{D_1}^{D_2} \tfrac{1}{4}(z - D_1)\exp[-(z - D_1)/2]\,dz \\
&= \Gamma_{(D_2 - D_1)/2}(2). \\
p_3 &= \int_{D_2}^{D_3} p(z|Z_1 > D_1, Z_2 > D_2)\,dz \\
&= \frac{\exp[-(D_2 - D_1)/2]}{2B_2}\left[2\frac{\Gamma_x(3)}{2!} + (D_2 - D_1)\Gamma_x(2)\right]
\end{aligned}$$

where $x = (D_3 - D_2)/2$, $B_2 = \Gamma(2) - \Gamma_{(D_2 - D_1)/2}(2)$.

$$p_4 = \frac{\exp[-(D_3 - D_2)/2]}{2B_3}\left\{2^2\frac{\Gamma_x(4)}{3!} + \frac{2(D_3 - D_1)}{2!}\Gamma_x(3) + \theta_4\Gamma_x(2)\right\}$$

where $x = (D_4 - D_3)/2$ and $B_3 = [\Gamma(3) - \Gamma_y(3) + (D_2 - D_1)][\Gamma(2) - \Gamma_y(2)]$, where $y = (D_3 - D_1)/2$. It is convenient to rewrite these p_i in terms of $I(x, p)$ where $I(x, p) = \Gamma_x(p + 1)/\Gamma(p + 1)$. The expressions for p_i now become

$$p_1 = I(D_1/2, 0), \qquad p_2 = I((D_2 - D_1)/2, 1),$$

$$p_3 = \frac{\exp[-(D_2 - D_1)/2]}{2B_2}[2I(x, 2) + (D_2 - D_1)I(x, 1)]; \qquad x = (D_3 - D_2)/2$$

$$p_4 = \frac{\exp[-(D_3 - D_2)/2]}{2B_3}[2^2 I(x, 3) + 2(D_3 - D_1)I(x, 2) + \theta_4 I(x, 1)];$$
$$x = (D_4 - D_3)/2$$

$$\begin{aligned}
p_5 &= \frac{\exp[-(D_4 - D_3)/2]}{2B_4}[2^3 I(x, 4) + 2^2(D_4 - D_1)I(x, 3) \\
&\quad + 2\theta_5' I(x; 2) + \theta_5 I(x, 1)]
\end{aligned}$$

with $x = (D_5 - D_4)/2$, and so on where

$$B_2 = [1 - I(x, 1)]; \qquad x = (D_2 - D_1)/2$$
$$B_3 = 2[1 - I(x, 2)] + (D_2 - D_1)[1 - I(x, 1)]; \qquad x = (D_3 - D_2)/2$$
$$B_4 = 2^2[1 - I(x, 3)] + 2(D_3 - D_1)[1 - I(x, 2)] + \theta_4[1 - I(x, 1)];$$
$$x = (D_4 - D_3)/2$$
$$\begin{aligned}
B_5 &= 2^3[1 - I(x, 4)] + 2^2(D_4 - D_1)[1 - I(x, 3)] \\
&\quad + 2\theta_5'[1 - I(x, 2)] + \theta_5[1 - I(x, 1)]
\end{aligned}$$

with $x = (D_5 - D_4)/2$ and where

$$\theta_4 = (D_3 - D_2)^2/2! + (D_2 - D_1)(D_3 - D_2);$$
$$\theta_5 = (D_4 - D_3)^3/3! + (D_3 - D_1)(D_4 - D_3)^2/2! + \theta_4(D_4 - D_3);$$
$$\theta_5' = (D_4 - D_3)^2/2! + (D_3 - D_1)(D_4 - D_3) + \theta_4.$$

The necessary expressions for further p_i, B_i, θ_i may be written down almost by inspection (or induction). Rewriting (4.13.13) we now have

$$P_{2m+1} = (1 - p_1)(1 - p_2)\cdots(1 - p_{m-1})p_m;$$

and the confidence coefficient is

$$2\sum_{m=1}^{\infty} \Phi[c(2m+1)^{1/2}]p_m \prod_{i=1}^{m-1}(1 - p_i) - 1 \tag{4.13.17}$$

and

$$E(N) = \sum_{m=1}^{\infty}(2m+1)p_m \prod_{i=1}^{m-1}(1 - p_i). \tag{4.13.18}$$

Application of the New Procedure to Allied Problems If it is required to estimate the mean μ with prescribed standard error, then the following modified procedure can be used: Stop sampling when

$$s^2/(2m+1) \le a^2 \qquad (m \ge 1). \tag{4.13.19}$$

In terms of the transformed variables U_i or V_i (4.13.19) can be written as

$$\sum_{i=1}^{2m} U_i = \sum_{i=1}^{m} V_i \le a^2 2m(2m+1). \tag{4.13.20}$$

This should be compared with the inequality (4.13.8). The determination of P_{2m+1} can be carried out as outlined above, where now, however $a^2 = c^2\sigma^2$ and $D_m = 2m(2m+1)c^2$. Some numerical results for this case for different values of c will be presented later.

In the problem of estimating the difference between the means of two normal populations with common variance $\sigma^2/2$, observations would be taken in pairs, one observation from each population, and sequences U_i, U_i' determined. Then we have

$$E(U_i) = E(U_i') = \sigma^2/2, \qquad EV_i = E(U_i + U_i') = \sigma^2$$

$$s^2 = \left[\sum_{i=1}^{m} (X_i - \bar{X})^2 + \sum_{i=1}^{m} (X_i' - \bar{X}')^2\right] \Big/ 2(m-1)$$

$$= \left(\sum_{i=1}^{m-1} U_i + \sum_{i=1}^{m-1} U_i'\right) \Big/ 2(m-1) = \left(\sum_{i=1}^{m-1} V_i\right) \Big/ 2(m-1) \qquad (m \geq 2)$$

and $E(s^2) = \sigma^2/2$. The sampling rule is: Stop sampling when

$$s(2/m)^{1/2} t_{2(m-1)} \leq d \qquad \text{or} \qquad s^2 \leq d^2 m/2t^2_{2(m-1)}$$

or (4.13.21)

$$\sum_{i=1}^{m-1} V_i \leq d^2 m(m-1)/t^2_{2(m-1)}.$$

To estimate the difference between the means with given standard error a we should sample as above, but stop sampling when

$$2s^2/m \leq a^2 \quad (m \geq 2) \qquad \text{or} \qquad \sum_{i=1}^{m} V_i \leq a^2 m(m+1). \tag{4.13.22}$$

Certain numerical values pertaining to this rule will be given later.

Numerical Results

Ray (1957b) has tabulated the expected sample size and the standard error of the sample size for various values of $d = c\sigma$ for estimating the mean by confidence interval of given width. When σ^2 is known, the required confidence coefficient is always attained provided the sample size is fixed and equal to $u^2\sigma^2/d^2$.

Anscombe (1953, 1954) declares that for small d [or large $(2m + 1)$] the confidence coefficient is close to the nominal confidence coefficient. More precisely his approximate result is

$$P(Y_n - d < \mu < Y_n + d) \sim \{1 - \alpha - 1.176uv^{-1}\phi(u)\} \tag{4.13.23}$$

where $v = \sigma^2 u^2/d^2$, $\phi(u)$ is the standard normal density. For $\alpha = .05$, this becomes $0.95 - .135/v$. The asymptotic confidence coefficients are compared

Table 4.13.1

Giving the Expected Sample Size and Its Standard Error for Estimation of the Mean by a Confidence Interval of Given Width ($\alpha = .05$).

d		2σ	1σ	$.8\sigma$	$.5\sigma$	$.4\sigma$
σ^2 unknown	confidence coefficient $= 1 - \alpha'$ (say)	.99975	.9775	.941	.931	.929
	$E(N)$	4.1	6.6	8.4	16.7	24.9
	σ_N	1.1	2.4	3.1	6.2	8.2
σ^2 known	confidence coefficient $= 1 - \alpha$	.95	.95	.95	.95	.95
	fixed-sample size	.96	3.84	6.0	15.4	24.0

[a] The confidence coefficient has been computed from (3.13.14).

in Table 4.13.2 with the exact coefficient for some values of d expressed in σ units. Table 4.13.1 shows that the actual confidence coefficients obtained by following the modified sequential procedure are very close to the nominal coefficient, namely .95. The true confidence coefficient increases with the width of the confidence interval. The savings in observations obtained when σ^2 is known decreases as the given width of confidence interval decreases. As d decreases, the confidence coefficient appears to tend to a value smaller then .95, contradicting Anscombe's results. This may probably be accounted for by the different starting sample sizes of Anscombe's and Ray's modified procedure.

For estimating the mean with given standard error, we have $\operatorname{var}(Y_N)/a^2 = (\sigma^2/a^2) \times E(1/N)$ which is to be determined for given values of σ^2/a^2. When σ^2 is known, the variance of the mean $\sigma^2/(2m+1)$ is obtained as exactly the required value a^2 if a sample size equal to σ^2/a^2 is taken. These results are given in Table 4.13.3. This table shows that for σ^2/a^2 greater than 5, the necessary accuracy of estimates of the mean is not being achieved. This is related to dispersion of sample size which increases with sample size. There are two simple devices to overcome this deficiency. One of them mentioned by Anscombe (1954) is to take one further observation after sampling has ceased

Table 4.13.2

Comparison of the Exact and Asymptotic Confidence Coefficients

d	2σ	1σ	$.8\sigma$	$.5\sigma$	$.4\sigma$
Modified procedure	.99975	.9775	.941	.931	.929
Asymptotic approach	.810	.915	.9275	.9412	.9444

Table 4.13.3

	σ^2 unknown			σ^2 known	
σ^2/a^2	$E(N)$	σ_N	$\text{var}(Y_N)/a^2$	Fixed-sample size	$\text{var}(Y_N)/a^2$
1	3.1	.44	.327	1	1
5	5.17	2.49	1.178	5	1
10	8.57	4.75	1.680	10	1
12.5	10.46	5.80	1.833	12.5	1

according to a certain rule. Surprisingly this produces a significant improvement in the resultant accuracy of the estimate. Naturally an even greater improvement results if two or more further observations are taken. For example, if rule (4.13.12) is followed and two additional observations are taken after we stop with $d = .4\sigma$, the actual confidence coefficient obtained is increased from .929 to .9427 (see Table 4.13.1). Also, when estimating the mean with given standard error a and $\sigma^2/a^2 = 10$, the variance of the mean Y_N decreases from $1.680a^2$ to $1.176a^2$ (see Table 4.13.3). Another device is to commence the application of the particular sampling rule at higher sampling number. For example, instead of taking first three observations, take five and then proceed with the normal sequential procedure.

ESTIMATING THE DIFFERENCE BETWEEN TWO MEANS WITH GIVEN STANDARD ERROR The sampling rule is given by inequality (4.13.22). The determination of P_m can be effected as described with $a = c\sigma$ and $D_m = m(m+1)c^2$. The variance of $Y_N = \bar{X} - \bar{X}'$ is given by $\text{var}(Y_N) = \sigma^2 E(1/N)$ where N is the number of pairs of observations. $E(N)$, standard deviation of N, and $\text{var}(Y_N)/a^2$ are tabulated in Table 4.13.4 for various values of σ^2/a^2. They can be compared with some exact results of Anscombe (1953) for the same problem but

Table 4.13.4

	σ^2 unknown					σ^2 known	
	Rule (4.13.22)			Rule (4.13.9)			
σ^2/a^2	EN	σ_N	$\text{var}(Y_N)/a^2$	EN	$\text{var}(Y_N)/a^2$	Fixed-sample size	$\text{var}(Y_N)/a^2$
1	2.56	.86	.424	4.14	.243	1	1
5	7.95	4.29	.980	7.04	.760	5	1

with different sampling rule (4.13.9). Since the calculations necessary to obtain P_m become quite lengthy for large values of σ^2/a^2, it has not been possible to make Table 4.13.4 more comprehensive. However, it can be seen that for σ^2/a^2 less than about 5, the accuracy of the estimate is in fact greater than that specified following either of the two sequential procedures.

Ray (1957b) has carried out a Monte Carlo study consisting of 36 samples. On the basis of this study he infers that the results of practical sampling follow closely those predicted from the exact theory.

The Performance of a Sequential Procedure for Estimating the Mean of a Normal Population with Unknown Variance

Starr (1966a) has studied the performance of the sequential procedure which is a modification of the rule (4.13.3). As before let $X_1, X_2, \ldots$ be a sequence of independent random variables distributed normally with mean μ and variance σ^2. For $n \geq 2$, let Y_n and s^2 respectively denote the sample mean and sample variance based on a random sample of size n. Also, let $\psi(x) = \Phi(x) - \Phi(-x)$ and u be a constant for which $\psi(u) = 1 - \alpha$, and

$$\psi_n(x) = [\Gamma((n+1)/2)/(n\pi)^{1/2}\Gamma(n/2)]\int_{-x}^{x}(1+t^2/2)^{-(n+1)/2}\,dt; \quad (4.13.24)$$

$\{u_n\}$ is a sequence of constants such that $\psi_n(u_n) = \alpha$ $(n = 1, 2, \ldots)$; then

$$\lim_{n\to\infty} u_n = u. \quad (4.13.25)$$

Recall that if σ is known, a confidence interval $I_n = [Y_n - d, Y_n + d]$ for μ of width $2d$ and with coverage probability $\geq 1 - \alpha$ is assured provided n is chosen so that

$$n \geq u^2\sigma^2/d^2, \quad (4.13.26)$$

since $P\mu(\in I_n) = \psi(n^{1/2}d/\sigma) \geq 1 - \alpha$. When σ is unknown, there exists no fixed-sample size procedure satisfying the requirement. Hence, one recourse is to a two-stage Stein sampling procedure (1949) which leads to an n which approximately satisfies (4.13.26) with σ^2 estimated from the initial sample of size n_1 (and u increased to u_{n_1} to reflect the limited degrees of freedom). It seems intuitively inefficient not to utilize all of the sample. Hence, consider the following modification of Ray's (1957b) sequential procedure:

(Λ) observe the X_n sequence term by term and stop with X_N where

$$N \text{ is the first integer } n \geq n_0 \text{ such that } s_n^2 \leq nd^2/u_{n-1}^2, \quad (4.13.27)$$

with $n_0 \geq 2$ a fixed integer; then form the interval $I_N = [Y_N - d,\ Y_N + d]$. Heuristically, the interval I_N will be an approximate solution of the problem

$$P(\mu \in I_N) \cong 1 - \alpha. \tag{4.13.28}$$

Furthermore, the expected sample size using Λ will be

$$EN \cong u^2\sigma^2/d^2. \tag{4.13.29}$$

To see why (4.13.28) should hold, set

$$J_n = [Y_n - (u_{n-1}s/n^{1/2}),\ Y_n + (u_{n-1}\,s/n^{1/2})]$$

then, for fixed values of θ, σ and n

$$P(\theta \in J_n) = P[|Y_n - \theta|/s \leq u_{n-1}] = \psi_{n-1}(u_{n-1}) = 1 - \alpha. \tag{4.13.30}$$

Moreover, the N defined by (4.13.27) is such that I_N contains J_N; hence neglecting "the excess," $P(\theta \in I_N) \cong P(\theta \in J_N)$. If we could infer from (4.13.30) that $P(\theta \in J_N) = 1 - \alpha$, then (4.13.28) would follow. Since $\lim_{n\to\infty} s^2 = \sigma^2$ with probability one, it follows from (4.13.27) by neglecting the excess that $\sigma^2 \cong d^2N/u^2$, at least for N large. Thus, EN should satisfy (4.13.29). Finally, a comparison of (4.13.26) and (4.13.29) suggests that not much should be lost due to ignorance of σ if we use Λ instead of a sample of fixed size.

Thus, it is of much interest to establish rigorously the values of the functions

$$C(\lambda) = P(\theta \in I_N), \qquad D(\lambda) = EN \tag{4.13.31}$$

for values $0 < \lambda < \infty$ of the parameter $\lambda = \sigma/d$ and to determine whether (4.13.28) and (4.13.29) are reasonably exact for all σ. Then, let

$$\tau_\lambda = C(\lambda)/(1 - \alpha) \qquad \text{and} \qquad \eta_\lambda = D(\lambda)/a^2\lambda^2 \tag{4.13.32}$$

which are the ratio of the coverage probability using Λ in ignorance of σ to the prescribed coverage probability $1 - \alpha$, and the ratio of the average sample size using Λ in ignorance of σ to (approximately) the smallest value of n satisfying (4.13.26) which would be used if σ were known. τ_λ and η_λ are said to be measures, at the parameter value λ, of the *consistency* and *efficiency* of Λ, respectively. Therefore in order to implement practically a recommendation of Λ we should establish that the procedure is "more or less" consistent ($\tau_\lambda \cong 1$) and efficient ($\eta_\lambda \cong 1$), for all λ.

Λ appears to have been first discussed by Stein (1949), who stated the second-order limiting performance of the procedure. Subsequently, the consistent and efficient (asymptotic) behavior of Λ was proved by Anscombe (1952, 1953) and along somewhat different lines by Gleser *et al.* (1964). The computation of C and D for several moderate values of λ was carried out by Ray (1957b) using a slight modification of (4.13.27) for ease of computation; on the basis of his fragmentary computations, Ray has incorrectly conjectured that this modified version of Λ is asymptotically inconsistent.

PRELIMINARY THEORY At the outset let us show that N is a genuine sampling variable, that is,

Lemma 4.13.1 (Starr, 1966a) $P(N < \infty) = 1$.

Proof By the strong law of large numbers, $\lim_{n\to\infty} s^2 = \sigma^2$ with probability one, which implies

$$P(N = \infty) = P(s^2/n > d^2/u_{n-1}^2 \text{ for all } n \geq n_0) = 0.$$

Define the standard normal variables

$$W_n = \left(\textstyle\sum_{i=1}^n X_i - nX_{n+1}\right)/\sigma[n(n+1)]^{1/2} \qquad (n \geq 1),$$

and observe that $(n-1)s^2/\sigma^2 = \sum_{i=1}^n W_i^2$ $(n \geq 2)$. Therefore (4.13.27) can be rewritten as

$$N \text{ is the first integer } n \geq n_0 \text{ such that } \sum_{i=1}^n W_i^2 \leq n(n-1)/\lambda^2 u_{n-1}^2, \tag{4.13.33}$$

and the probability distribution of N is given for $n \geq n_0$ by

$$\begin{aligned} p_n(\lambda) &= P(N = n) \\ &= P\Bigg\{\sum_{i=1}^k W_i^2 \leq (k+1)k/\lambda^2 u_k^2 \text{ for } k = n-1 \\ &\qquad \text{but not for any } n_0 - 1 \leq k < n-1\Bigg\}. \end{aligned} \tag{4.13.34}$$

The joint probability density of $(W_i^2, i = 1, \ldots, n-1)$ is easily seen to be

$$dF(w_1^2, \ldots w_{n-1}^2)$$
$$= [2^{(n-1)/2}\Gamma^{n-1}(\tfrac{1}{2})]^{-1} \exp\left(-\tfrac{1}{2}\sum_{i=1}^{n-1} w_i^2\right)\left[\prod_{i=1}^{n-1} w_i^2\right]^{-1/2} \prod_{i=1}^{n-1} d(w_i^2)$$

for $0 \le w_i$; $i = 1, \ldots, n-1$. Noting that

$$V_k = \sum_{i=1}^{k} W_i^2 \qquad (k \ge 1) \tag{4.13.35}$$

has the chi-square distribution with k degrees of freedom, the joint probability density of $\{V_k;\, k = 1, \ldots, n-1\}$ is

$$dF(v_1, \ldots, v_k) = [2^{(n-1)/2}\Gamma^{n-1}(\tfrac{1}{2})]^{-1} \exp(-v_{n-1}/2)\left[v_1 \prod_{i=2}^{n-1} (v_i - v_{i-1})^{-1/2} \prod_{i=1}^{n-1} dv_i\right]$$

for $0 \le v_1 \le v_2 \le \cdots \le v_{n-1}$. Hence, subject to Λ, the probability that sampling is terminated at the nth observation is from (4.13.34) for $n \ge n_0$

$$p_n(\lambda) = \int \cdots \int_{[0,\infty]} \int_{B_{n_0-1}} \int_{B_{n_0}} \cdots \int_{B_{n-2}} \int_{\bar{B}_{n-1}} dF(v_1, \cdots, v_{n-1}) \tag{4.13.36}$$

where for $n_0 - 1 \le k \le n-1$, B_k is the interval $((k+1)k/\lambda^2 u_k^2,\ \infty)$, $\bar{B}_k$ is the complement of B_k, and where the multiple integral over $[0, \infty]$ represents an $(n_0 - 2)$-fold integration, while the remainder represents an $(n - n_0 + 1)$-fold integration. Then with $p_n(\lambda)$ defined by (4.13.36), from (4.13.31) we have

$$D(\lambda) = \sum_{n=n_0}^{\infty} n p_n(\lambda), \tag{4.13.37}$$

and, since Y_n is independent of the vector $(W_1, \ldots, W_{n-1})$

$$\begin{aligned} C(\lambda) &= \sum_{n=1}^{\infty} p_n(\lambda) P(\theta \in I_N \mid N = n) \\ &= \sum_{n=n_0}^{\infty} p_n(\lambda) P(n^{1/2}|Y_n - \theta|/\sigma \le n^{1/2} d/\sigma \mid N = n) \qquad (4.13.38) \\ &= \sum_{n=n_0}^{\infty} p_n(\lambda) \psi(n^{1/2}/\lambda). \end{aligned}$$

With this background we are ready to survey briefly the performance of the rule Λ.

CASE 1: SMALL λ

Theorem 4.13.1 With $\lambda = \sigma/d$ we have

$$\lim_{\lambda\to 0} P(N = n_0) = 1 \tag{4.13.39}$$

$$\lim_{\lambda\to 0} C(\lambda) = 1 \tag{4.13.40}$$

$$\lim_{\lambda\to 0} D(\lambda) = n_0 . \tag{4.13.41}$$

Proof Relation (4.13.39) follows readily from (4.13.34). From (4.13.38) and (4.13.39) we have

$$\lim_{\lambda\to 0} C(\lambda) = \lim_{\lambda\to 0} \sum_{n=n_0}^{\infty} p_n(\lambda)\psi(n^{1/2}/\lambda) = \lim_{\lambda\to 0} \psi(n_0^{1/2}/\lambda) = 1,$$

which establishes (4.13.40). In order to establish (4.13.41) we need:

Lemma 4.13.2 (Starr, 1966a) Let $R_\lambda = \sum_{n=n_0-1}^{\infty} P(V_n/n > (n+1)/\lambda^2 u_n{}^2)$, where V_n is defined by (4.13.35). Then

$$\lim_{\lambda\to 0} R_\lambda = 0. \tag{4.13.42}$$

Proof Set $A = 1 + \sup_{n\geq 1} u_n$, and we have for $\lambda < A^{-2}$

$$\begin{aligned} R_\lambda &= \sum_{n=n_0-1}^{\infty} P\{(V_n/n) - 1 > [(n+1)A^2/\lambda u_n{}^2] - 1\} \\ &\leq \sum_{n=n_0-1}^{\infty} P((V_n/n) - 1 > n/\lambda). \end{aligned} \tag{4.13.43}$$

Noting that $E(V_n - n)^2 = 2n$ and applying the Chebyshev's inequality, we obtain from (4.13.43)

$$R_\lambda \leq \sum_{n=n_0-1}^{\infty} [\lambda^2 E((V_n/n) - 1)^2/n^2] = 2\lambda^2 \sum_{n=n_0-1}^{\infty} n^{-3}$$

from which it follows that (4.13.42) holds.

To establish (4.13.41) we remark that

$$\begin{aligned} D(\lambda) &= \sum_{n=1}^{\infty} P(N \geq n) = \sum_{n=1}^{n_0} P(N \geq n) + \sum_{n=n_0-1}^{\infty} P(N > n+1) \\ &= n_0 + \sum_{n=n_0-1}^{\infty} P(N > n+1). \end{aligned}$$

Since, it is easily seen from (4.13.34) that $P(N > n+1) = P(\sum_{i=1}^{k} W_i^2 > (k+1)k/\lambda^2 u_k^2$ for every $n_0 - 1 \leq k \leq n) \leq P(V_n/n > (n+1)/\lambda^2 u_n^2)$, we have

$$\lim_{\lambda \to 0} \sum_{n=n_0-1}^{\infty} P(N > n+1) \leq \lim_{\lambda \to 0} R_\lambda \tag{4.13.44}$$

and the proof is complete after applying Lemma 4.13.2.

CASE 2: LARGE λ Since $p_n(\lambda)$ given by (4.13.36) is relatively intractable, and the computations of C and D are somewhat difficult, it is worthwhile to modify Λ to Λ^* by amending Definition (4.13.27) to

$$N^* \text{ is the first odd integer } n \geq n_0^* \text{ such that } S_n^2 \leq nd^2/u_{n-1}^2, \quad \text{with } n_0^* \text{ a fixed odd integer} \geq 3. \tag{4.13.27*}$$

Then for the modified rule Λ^* in analogy with (4.13.31) and (4.13.32) define

$$C^*(\lambda) = P(\theta \in I_{N^*}) \quad \text{and} \quad D^*(\lambda) = EN^*,$$
$$\tau_\lambda^* = C^*(\lambda)/(1-\alpha) \quad \text{and} \quad \eta_\lambda^* = D^*(\lambda)/u^2\lambda^2.$$

The computations of C^* and D^* given in Table 4.13.1, have been made for $\alpha = .05$ and $\lambda = .5$, 1, 1.25, 2, 2.5 (with $n_0^* = 3$) by Ray (1957b). Then we have the following theorem which was proved by Anscombe (1952, 1953) and Gleser, *et al.* (1964).

Theorem 4.13.2 Let N be the stopping variable defined by (4.13.27) where $\{u_n\}$ is any sequence of positive constants [not necessarily those defined in (4.13.24)] such that $\lim_{n\to\infty} u_n = u$. Then, with $\lambda = \sigma/d$, τ_λ and η_λ defined by (4.13.32)

$$\lim_{\lambda\to\infty} \tau_\lambda = 1 \quad \text{(asymptotic consistency)}, \tag{4.13.45}$$

$$\lim_{\lambda\to\infty} \eta_\lambda = 1 \quad \text{(asymptotic efficiency)}. \tag{4.13.46}$$

Furthermore, (4.13.45) and (4.13.46) are valid when N, τ_λ, and η_λ are replaced by their starred equivalents.

Proof See Gleser *et al.* (1964).

Ray (1957b) apparently being misled by his fragmentary computations, questions the asymptotic consistency of Λ or Λ^*. However, extensive computations carried out by Starr indicate that $C^*(\lambda)$ begins to increase steadily as λ becomes large, apparently tending to $1-\alpha$.

CASE 3: MODERATE λ We have from (4.13.38) that $C(\lambda) \geq \psi(n_0^{1/2}/\lambda)$, $0 < \lambda < \infty$ which with (4.13.40) and (4.13.45), implies that the constant β,

$$\beta = \inf_{0<\lambda<\infty} C(\lambda) \leq \psi(u), \tag{4.13.47}$$

is positive. The value of β depends on the whole sequence $\{u_n\}$ which defines N through (4.13.27) and not just on the limit (4.13.25). Obviously, one can choose the sequence $\{u_n\}$ subject to (4.13.25) so that the value τ,

$$\tau = \inf_{0<\lambda<\infty} \tau_\lambda = \beta/(1-\alpha), \tag{4.13.48}$$

is arbitrarily close to 1, but only at the expense of making the quantity $\eta = \sup_{0<\lambda<\infty} \eta_\lambda$ unreasonably large. For this case, let us restrict ourselves to the sequence $\{u_n\}$ defined by (4.13.24) and Λ^* given by (4.13.27), precisely the rule evaluated by Ray (1957b). Taking $n = 2m + 1$, we have in analogy with (4.13.33)

N^* is the first $n = 2m + 1$ with $m \geq m_0{}^*$ such that

$$\sum_{i=1}^{2m} W_i{}^2 \leq 2m(2m+1)/\lambda^2 u_{2m}^2 = 2b_{m+1} \tag{4.13.49}$$

where

$$m_0{}^* = (n_0{}^* - 1)/2 \geq 1 \qquad \text{and} \qquad b_m = (m-1)(2m-1)/\lambda^2 u_{2m-2}^2. \tag{4.13.50}$$

We observe that $V_2 = W_1{}^2 + W_2{}^2$ has probability density $\frac{1}{2} e^{-v/2}$ $(v > 0)$ and $Z = V_2/2$ has density e^{-z} $(z > 0)$. Then for the rule Λ^* we have in analogy with (4.13.34) for $m \geq m_0{}^*$

$$\begin{aligned} p_m{}^*(\lambda) &= P(N^* = 2m+1) \\ &= P\left\{ \sum_{i=1}^{k} Z_i > b_{2k+1} \quad \text{for every } m_0{}^* \leq k \leq m-1 \right. \\ &\qquad \left. \text{but not for } k = m \right\}, \end{aligned} \tag{4.13.51}$$

and in analogy with (4.13.38) and (4.13.37)

$$\begin{cases} C^*(\lambda) = \displaystyle\sum_{m=m_0^*}^{\infty} p_m{}^*(\lambda)\psi[(2m+1)^{1/2}/\lambda], \\ C^*(\lambda) = \displaystyle\sum_{m=m_0^*}^{\infty} (2m+1)p_m{}^*(\lambda) = 1 + 2 \sum_{m=m_0^*}^{\infty} m p_m{}^*(\lambda). \end{cases} \tag{4.13.52}$$

The $p_m^*(\lambda)$ can be computed by the following method [see Robbins (1959) for the proof]. For the case $n_0^* = 3$, $h_1 \equiv 1$, $c_1 \equiv 1$, and compute recursively

$$h_m(b_n) = \sum_{j=1}^{m-1} [(b_n - b_m)/j!]h_{m-j}(b_m) \qquad (m = 2, 3, \ldots, n = m + 1, m + 2),$$

where b_m is given in (4.13.50), and then compute

$$c_m = \exp(-b_m) \sum_{j=1}^{m-1} h_{m-j}(b_m) \qquad (m = 2, 3, \ldots).$$

Then, for $m = 1, 2, \ldots,$

$$p_m^*(\lambda) = P(N^* = 2m + 1) = c_m - c_{m+1}.$$

(For the case $n_0^* = 5$, take $b_2 = 0$ otherwise compute as above.) Then D^* simplifies to $D^*(\lambda) = 1 + 2 \sum_{m=1}^{\infty} c_m$.

One further modification of the procedure is of interest; namely, that the experimenter using Λ^* be required to take a *fixed* number of additional observations after having decided (nominally) to terminate sampling. Call the amended rule $\Lambda^*(j)$ where

($\Lambda^*(j)$) observe the sequence X_1, $X_2, \ldots$ term by term, stopping with X_{N^*+j} $(j \geq 0)$, where N^* satisfies (4.13.27*). Notice that

$$C_j^*(\lambda) = \sum_{m=m_0^*}^{\infty} p_m^*(\lambda)\psi[(2m + j + 1)^{1/2}/\lambda] \qquad (j \geq 0),$$

$$D_j^*(\lambda) = j + D^*(\lambda) \qquad (j \geq 0),$$

where C_j^* and D_j^* are respectively the coverage probability and expected sample size using $\Lambda^*(j)$. Also notice that taking any *finite* number of additional observations after stopping does not affect the asymptotic theory, nor indeed the theory for small λ, except that (4.13.41) becomes

$$\lim_{\lambda \to 0} D_j^*(\lambda) = n_0^* + j \qquad (j \geq 0) \tag{4.13.53}$$

Starr (1966a) has computed C_j^* and D_j^* for several hundred values of the parameter $0 < \lambda < \infty$ with $\alpha = .95$ ($n_0^* = 3, j = 0, 2, 4$, and $n_0^* = 5, j = 0, 2$) and with $\alpha = .99$ ($n_0^* = 3$, $j = 0$, 2) subject to the requirements of $\Lambda^*(j)$. He also provides a diagrammatic representation of the behavior of $C_j^*(\lambda)$ and in analogy with (4.13.32), of $\eta_\lambda^*(j) = D_j^*(\lambda)/u^2\lambda^2$, for several versions of Λ^*.

Table 4.13.5

Values of $\beta^* = \inf_{0<\lambda<\infty} C^*(\lambda)$ Where $C^*(\lambda) = \sum_{m=m_0^*}^{\infty} p_m^*(\lambda)\psi[(2m+j+1)^{1/2}/\lambda](j \geq 0)$

$\alpha = .95$			$\alpha = .99$		
n_0^*	j	β^*	n_0^*	j	β^*
3	0	.9285	3	0	.9835
5	0	.9379	3	2	.9875
3	2	.9415			
5	2	.9478			
3	4	.9487			

The computed values of Starr sometimes differ from those of Ray (1957b), and the reader is referred to Starr's Table 1. Of special interest are the values [in analogy with (4.13.47)] of $\beta^* = \inf_{0<\lambda<\infty} C^*(\lambda)$ which are given in Table 4.13.5.

Starr (1966a) draws the following conclusions from his extensive computations: τ^* defined in analogy with (4.13.48) is not appreciably less than 1 in any of the versions, so that the cost in coverage probability due to ignorance of σ is "next to nothing." For any of the several versions of Λ^*, $\eta_\lambda{}^*$ appears to decrease monotonically (and rapidly) to 1, being larger only for λ near zero when the expected number of observations $D^*(\lambda)$ is small anyway (and depends largely on n_0^* and j). Hence, the cost of observations due to ignorance of σ is similarly not appreciable. It is obvious that the basic rule Λ will have a minimum coverage probability β not very much less than .928 (when $\alpha = .95$) and efficiency η exceedingly close to 1 for λ bounded somewhat away from zero. Thus, the basic procedure and its several variations are remarkably consistent and efficient, *uniformly* in λ. τ can always be adjusted upward by using a different sequence from that defined by (4.13.24), increasing the starting sample size n_0, or taking j additional observations subsequent to nominally stopping, but only at the expense of increasing η. Alternatively, η can be arbitrarily decreased but only at the expense of concomitantly decreasing τ. There is no unique version of Λ which jointly reduces the loss in observations and minimum coverage probability due to ignorance of σ. So, the choice of the procedure should be dictated by the particular requirements of the applications.

Comparison of Λ with Stein's Two-Stage Sampling Procedure

Seelbinder (1953) and Moshman (1958) have considered the problem of obtaining an appropriate determination of n_1, the size of the initial sample for Stein's (1945) two-stage sampling procedure. Let N_s be the total size using two-stage sampling, and $E_\lambda(N_s)$ = expectation of N_s at the parameter point $0 < \lambda < \infty$. Seelbinder (1953) proves a somewhat stronger limiting theorem than did Stein (1945); namely

$$\lim_{\lambda \to \infty} [E_\lambda(N_s) - u_{n-1}^2 \lambda^2] = 0. \tag{4.13.54}$$

Noting that the sequence $\{u_n\}$ defined by (2.13.24) has the property: $u_1 > u_2 > u_3 > \cdots$, it follows from (2.13.25) that Stein's procedure is asymptotically less efficient than Λ; in fact (4.13.54) implies that

$$\lim_{\lambda \to \infty} E_\lambda(N_s)/u^2\lambda^2 = u_{n-1}^2/u^2 > 1 \qquad (n_1 \text{ fixed})$$

compared to the asymptotic result (4.13.45) proved for Λ. Further, it is not difficult to show that

$$\lim_{\lambda \to 0} E_\lambda(N_s) = n_1. \tag{4.13.55}$$

Seelbinder (1953, p. 647) suggests in using Stein's two-stage sampling procedure that if one is uncertain about the value of σ but believes it is such that $\lambda < 10$, "then the first part of the sample (n_1) should be taken to be 250 or thereabouts," so as to decrease the possibility of having $E_\lambda(N_s)$ inordinately large for λ bounded somewhat away from zero. Therefore comparing (4.13.55) with (4.13.53) (for which in practice none of the values n_0, n_0^*, or j will be large) we see that for λ near zero Λ and its several variants are distinctly more efficient than the Stein procedure. Seelbinder has computed $E_\lambda(N_s)$ for $\alpha = .01, .02, .05, .10$, and $\lambda^{-1} = .01(.01).1(.1)1.0$, for a number of values of n_1 (the initial sample size).

Now, for $0 < \lambda < \infty$, $E_\lambda(N_s)$ depends on the choice of the first-stage sample size n_1. Indeed if one had *precise* knowledge of σ, the value $\xi_\lambda(n_1)$ [$\xi_\lambda(n_1) = E_\lambda(N_s | n_1)$] could be minimized at λ by an inspection of Seelbinder's table. Starr (1966a) infers on the basis of his computations and comparison with Seelbinder's table, that

$$D^*(\lambda) < \min_{2 \le n_1 \le \infty} \xi_\lambda(n_1) \tag{4.13.56}$$

for all λ for which comparisons are possible. Starr also conjectures that in fact (4.13.56) holds for all λ. If we had precise knowledge of σ, we would not rely on any sequential procedure, but simply preassign n the smallest integer value satisfying (4.13.26). Therefore the only case of interest to sequential analysis is when σ is unknown. In this event, an inappropriate choice of n_1

can have costly results. We conclude that Λ^* is more efficient than a two-stage sampling procedure for λ very small and very large; that it *appears* that the former procedure is always more efficient than the latter for $0 < \lambda < \infty$; and that Λ is always reasonably efficient irrespective of the value of λ and what we "think" this value may be. On the other hand, two-stage sampling is somewhat more consistent than Λ in assuring a coverage probability $\geq 1 - \alpha$, whereas Λ and its variants assure only $\beta < 1 - \alpha$. However, in normal experimentations it would seem prudent to settle for a minimum coverage probability (irrespective of the magnitude of σ) which is not conspicuously less than $1 - \alpha$, say .9415 [using $\Lambda^*(4)$ with $n_0^* = 3$] when $1 - \alpha = .95$ or .9875 [using $\Lambda^*(2)$ with $n_0^* = 3$] when $1 - \alpha = .99$, with the certainty that (at worst) our expected sample size does not sensibly exceed that of two-stage sampling and that we do not run the risk, in our ignorance of σ, of requiring extra needless observations.

Starr (1966b) has also considered the general loss function in estimating μ by $Y_n = \sum_1^n X_i/n$. Suppose, that for fixed, r, $t > 0$ the loss incurred in estimating μ by Y_n from a sample of fixed size n is

$$r(Y_n; \mu) = A|Y_n - \mu|^r + n^t = r_n \quad \text{(say)} \qquad (A > 0) \qquad (4.13.57)$$

with risk

$$R_n(\sigma) = E_\sigma r_n = AE_\sigma|Y_n - \mu|^r + n^t. \qquad (4.13.58)$$

When σ is known the problem of finding the value of n say n^0, for which the risk (4.13.58) is minimum is straightforward. Let

$$R(\sigma) = R_{n^0}(\sigma) = \min_{n>0} R_n(\sigma).$$

On the other hand, in ignorance of σ no fixed sample size procedure will minimize (4.13.58) simultaneously for all $0 < \sigma < \infty$. Let $\mathscr{R}$ denote a sequential rule that specifies the stopping variable N. In analogy with (4.13.57) the loss in using $\mathscr{R}$ is for fixed r, $t > 0$ and N

$$r_N = A|Y_N - \mu|^r + N^t \qquad (A > 0) \qquad (4.13.59)$$

with risk

$$\bar{R}(\sigma) = E_\sigma r_N = AE_\sigma|Y_N - \mu|^r + E_\sigma N^t.$$

It would be of much interest and importance to compare the values of $R(\sigma)$ and $\bar{R}(\sigma)$ for values of $0 < \sigma < \infty$. Robbins (1959) has computed $\bar{R}$ for a number of values of σ in the special case $r = t = 1$, verifying that the difference between $\bar{R}$ and R is negligible at the values of σ considered. Starr (1966b) has studied the conditions under which the risk efficiency

$$\eta(\sigma) = \bar{R}(\sigma)/R(\sigma) \tag{4.13.60}$$

converges to 1 as σ becomes infinite. The sole condition turns out to be simply

$$n_0 > [r^2/(r + 2t)] + 1 \tag{4.13.61}$$

where n_0 is the initial sample size of the sequential procedure. Thus, for example, in the special case $r = 2$, $t = 1$ (that is, squared-error and unit cost of sampling), $\mathscr{R}$ is asymptotically risk efficient if and only if it requires that $N \geq 3$.

σ Is Known We wish to find the value n^0 (a function of σ) for which the risk (4.13.58) is a minimum. Notice that (4.13.57) can be rewritten as

$$r_n = (A\sigma^r/n^{r/2})|Z|^r + n^t, \qquad 0 < \sigma < \infty, \tag{4.13.62}$$

where for fixed n, $Z = n^{1/2}(Y_n - \mu)/\sigma$ is a standard normal variable and the risk (4.13.58) becomes

$$\begin{aligned} R_n(\sigma) &= (A\sigma^r/n^{r/2})(2/\pi)^{1/2}\int_0^\infty z^r \exp(-z^2/2)\, dz + n^t \\ &= (2/r)K\sigma^r/n^{r/2} + n^t, \end{aligned} \tag{4.13.63}$$

where

$$K = K(A, r) = (r/2)A\Gamma((r + 1)/2)2^{r/2}/\Gamma(\tfrac{1}{2}) \tag{4.13.64}$$

depends on A and r only. Treating $n \geq 0$ as a continuous variable, we have

$$\partial R_n(\sigma)/\partial\sigma = -K\sigma^r/n^{1+(r/2)} + tn^{t-1} \tag{4.13.65}$$

which when set equal to zero, has the unique positive root

$$n^0 = (K\sigma^r/t)^{2/(r+2t)}; \tag{4.13.66}$$

since from (4.13.63)

$$\lim_{n\to 0} R_n(\sigma) = \lim_{n\to\infty} R_n(\sigma) = \infty,$$

n^0 is the value of $n \geq 0$ which minimizes (4.13.58). After simple algebra, we find that the minimum of risk (4.13.58) is

$$R(\sigma) = [(2/r)t + 1](n^0)^t. \tag{4.13.67}$$

σ IS UNKNOWN We will consider the following sequential procedure. Let K be defined by (4.13.64), and let $\{k_n, n \geq 1\}$ be any sequence of positive constants such that

$$\lim_{n\to\infty} k_n = K; \tag{4.13.68}$$

then in analogy with (4.13.66) we define the sequential rule $\mathcal{R}$: observe the sequence of Xs term by term, stopping with X_N, where

$$N \text{ is the first integer } n \geq n_0 \text{ such that } n \geq (k_{n-1}s^r/t)^{2/(r+2t)} \tag{4.13.69}$$

with the starting sample size $n_0 \geq 2$ a fixed integer. As in Robbins (1959) and Starr (1966a) define the standard normal variables W_n and $V_k = \sum_{j=1}^{k} W_j^2$ $(k \geq 2)$. Observing that $V_n = (n-1)s^2/\sigma^2$, it is easily seen that (4.13.69) may be rewritten as

$$N \text{ is the first integer } n \geq n_0 \text{ such that } V_n \leq l(n; \sigma), \tag{4.13.70}$$

where for $n \geq 2$, $0 < \sigma < \infty$,

$$l(n; \sigma) = (t/k_{n-1})^{2/r}(n-1)n^{(r+2t)/r}/\sigma^2; \tag{4.13.71}$$

then the probability distribution of N is given for $n \geq n_0$ by

$$p_n(\sigma) = P(N = n) = P(V_{k+1} \leq l(k+1; \sigma) \quad \text{for} \quad k = n-1$$
$$\text{but not for any } m - 1 \leq k \leq n-1). \tag{4.13.72}$$

We shall evaluate the performance of $\mathcal{R}$ in terms of the risk $\bar{R}$ defined by (4.13.59). Since for fixed $n \geq 2$, Y_n is independent of the vector $(W_1, \ldots, W_{n-1})$, r_n and the event $\{N = n\}$ are independent for all $n \geq n_0$, and consequently

$$\begin{aligned}\bar{R}(\sigma) &= \sum_{n=n_0}^{\infty} p_n(\sigma)E_\sigma(r_N|N=n) = \sum_{n=n_0}^{\infty} p_n(\sigma)R_n(\sigma)\\ &= \sum_{n=n_0}^{\infty} p_n(\sigma)\{(2/r)K\sigma^r/n^{r/2} + n^t\}\\ &= (2/r)K\sigma^r E(N^{-r/2}) + E(N^t), \qquad 0 < \sigma < \infty, \end{aligned} \tag{4.13.73}$$

and

$$\eta(\sigma) = \bar{R}(\sigma)/R(\sigma) = (2/r)K\sigma^r E(N^{-r/2})/\{(2/r)t + 1\}(n^0)^t \\ + EN^t/[(2/r)t + 1)](n^0)^t, \tag{4.13.74}$$

which, after some algebra, simplifies to

$$\eta(\sigma) = [(2/r)t + 1]^{-1}[(2/r)t(n^0)^{r/2}E(N^{-r/2}) + E(N/n^0)^t]. \tag{4.13.75}$$

To show that $\mathscr{R}$ is asymptotically risk efficient, that is, that

$$\lim_{\sigma \to \infty} \eta(\sigma) = 1, \tag{4.13.76}$$

we could verify that for arbitrary fixed ω

$$E(N^\omega) \sim (n^0)^\omega \qquad \text{as} \quad \sigma \to \infty, \tag{4.13.77}$$

since (4.13.76) follows immediately from (4.13.75). However, it is apparent that the validity of (4.13.76) depends in some way on the size n_0 of the starting sample of $\mathscr{R}$. For suppose that the Xs have a very large variance. Then, if r is "large" relative to t in a specialization of the loss (4.13.57), the portion of loss assignable to error of estimate will effectively "dominate" that associated with the cost of sampling. This suggests that the risk may blow up "out of proportion" if the possibility of early termination of the sampling is permitted. In the following we shall present a necessary and sufficient criterion of Starr (1966b) on the starting sample size n_0 for which (4.13.76) holds; this condition amounts to the verification of (4.13.77) for $\omega = t$ and, when (4.13.61) is satisfied, for $\omega = -r/2$.

Theorem 4.13.3 (Starr, 1966b). With n^0 defined by (4.13.66) and with N defined by (4.13.69) or (4.13.70), we have

$$p\lim_{\sigma \to \infty} (N/n^0) = 1. \tag{4.13.78}$$

Proof It is a corollary to Lemma 1 of Chow and Robbins (1965) and it may be proved directly by extending the methods of Gleser *et al.* (1964).

Theorem 4.13.4 (Starr, 1966b). For $\omega > 0$ fixed,

$$\lim_{\sigma \to \infty} E\{(N/n^0)^\omega\} = 1. \tag{4.13.79}$$

Proof See Starr (1966b, Theorem 2).

Theorem 4.13.5 (Starr, 1966b) For $\omega > 0$ fixed,

$$\lim_{\sigma \to \infty} (n^0)^\omega E(N^{-\omega}) = 1, \quad 1 + Q(n_0, \omega), \quad \text{or} \quad \infty \tag{4.13.80}$$

according as m is greater than, equal to, or less than $[2r\omega/(r + 2t)] + 1$, respectively, where

$$Q(m, \omega) = \{2^{-(m-3)/2} m^{-\omega}/(m - 1)\Gamma((m - 1)/2)\} \times (K/t)^{2\omega/(r+2t)}[\sigma^2 l(m; \sigma)]^{(m-1)/2}. \tag{4.13.81}$$

Proof See Starr (1966b, Theorem 3).

Corollary 4.13.5.1 (Starr, 1966b) For r, $t > 0$ fixed

$$\lim_{\sigma \to \infty} \eta(\sigma) = 1, \quad 1 + (2/r)tQ(n_0, r/2)/\{(2/r)t + 1\} \quad \text{or} \quad \infty \tag{4.13.82}$$

according as n_0 is greater than, equal to, or less than $[r^2/(r + 2t)] + 1$, respectively.

Proof The corollary follows from Theorem 4.13.3 (with $\omega = t$) and Theorem 4.13.5 (with $\omega = r/2$) and (4.13.75).

REMARKS Starr (1966b) modifies the cost of sampling in (4.13.57) to

$$r_n{}^* = A|Y_n - \mu|^r + \ln n \qquad (A > 0, \quad r > 0) \tag{4.13.83}$$

and extends Theorems 4.13.3, 4.13.4, and 4.13.5 to this loss function. The reader is referred to Starr (1966b, Theorems 1*, 2*, and 3* of Section 3). He also obtains the asymptotic consistency and asymptotic efficiency of the rule Λ [see (4.13.27)] when $\sigma \to \infty$ where u_n is any sequence of positive constants converging to u which denotes the $(1 - \alpha/2)$th fractile of the standard normal distribution. The reader is referred to Starr (1966b, Theorems 1_*, 2_*, and 3_* of Section 4).

Starr and Woodroofe (1969) have studied in more detail the case when $t = 1$ in the loss (4.13.62). In particular they show that the regret, equal to the difference in the expected loss with optional stopping and the expected loss which would be incurred if σ were known, is a bounded function of σ. Let the regret be defined as

$$\omega(\sigma) = \bar{R}(\sigma) - R(\sigma). \tag{4.13.84}$$

Then, Starr and Woodroofe (1969) have obtained the following results.

Theorem 4.13.6 (Starr and Woodroofe, 1969). As $\sigma \to \infty$

$$\omega(\sigma) = 0(1) \tag{4.13.85}$$

if and only if $n_0 \geq r + 1$.

Proof See Starr and Woodroofe (1969, pp. 287–288).

Before concluding this section, from the historical point of view, it is worthwhile to give Stein's (1949) fixed width ($2d$) sequential procedure for estimating the normal mean with unknown variance. Let $C_1 = C_2 = 0$ and

$$C_n = \max[(d/u)^2(n^2 - \tfrac{1}{2}(3 + u^2)n), 0], \qquad \text{for} \quad n \geq 3 \tag{4.13.86}$$

where u denotes the $(1 - \alpha/2)$th fractile of the standard normal distribution. Continue sampling term by term and stop with the Nth observation where

$$N \text{ is the first integer } n \geq 3 \text{ such that } (n - 1)s_n{}^2 \leq C_{n-1}, \tag{4.13.87}$$

then form the confidence interval $I_N = (Y_N - d,\ Y_N + d)$ where as before Y_N denotes the sample mean of N observations. Stein (1949) asserts that it can be shown that

$$P(\mu \in I_N) = 1 - \alpha + (2\pi)^{-1/2}(d/u\sigma)^4$$
$$\times u \exp(-u^2/2)E(C_{N-1} - (N - 1)s_N{}^2) + O(\sigma^{-3}) \tag{4.13.88}$$

as $\sigma \to \infty$ where

$$0 \leq (d/u\sigma)^2 E(C_{N-1} - (N - 1)s_N{}^2) \leq 1.5 + O(1) \tag{4.13.89}$$

and

$$E(N) = \left(\frac{u\sigma}{d}\right)^2 + \frac{1 + u^2}{2} + \left(\frac{d}{u\sigma}\right)^2 E(C_{N-1} - (N - 1)s_N{}^2) + O(\sigma^{-1}). \tag{4.13.90}$$

If $(d/u\sigma)^2 E(C_{N-1} - (N - 1)s_N{}^2)$ were known to a first approximation, say $\rho + O(\sigma^{-1})$ it would be possible to choose

$$C_n = \max\left\{\left(\frac{d}{u}\right)^2\left[n^2 - \left(\frac{3 + u^2}{2} - \rho\right)n\right], 0\right\} \qquad \text{for} \quad n \geq 3$$

and obtain

$$P(\mu \in I_N) = 1 - \alpha + O(\sigma^{-3})$$

and

$$E(N) = (u\sigma/d)^2 + \tfrac{1}{2}(1 + u^2) + O(\sigma^{-1}).$$

Stein (1949) also points out that although these results are asymptotic in nature, they seem to err on the side of excessive caution for small or moderate σ^2. We also note that $(1 + u^2)/2$ is roughly the number of additional observations which must be taken due to ignorance of σ. For Stein's (1945) two-stage procedure, the average number of added observations is nearly a fixed multiple of σ^2 as $\sigma^2 \to \infty$.

The above results were proved using Taylor's expansion and the first two derivatives of Wald's fundamental identity.

Problems

1. Let 89, 102, 108, 92, 98, 110, 88, 96, 94, 105, 107, 87, 112, 95, 99, ... constitute a sequence of independent observations from a normal population having an unknown mean μ and variance σ^2. Estimate μ by a confidence interval of given width 6 and confidence coefficient 0.90. Also find the expected sample size assuming that $\sigma^2 = 10$.

2. For the above data estimate μ with prescribed standard error $a = 1$.

4.14 The Asymptotic Theory of Fixed-Width Sequential Confidence Intervals for the Mean

In this section we shall present the large-sample fixed-width sequential confidence intervals for the population mean. The main results are based on certain convergence theorems, the understanding of which requires some knowledge of measure theory. For the sake of completeness, the convergence theorems will be stated as lemmas without proofs and the basic tools of Chow and Robbins (1965) will be given as lemmas while their main results and Nàdas' (1969) results will be stated as theorems.

Lemma 4.14.1 (Dominated convergence theorem) Let $\{f_n\}$ be a sequence of measurable functions with respect to some measure ν. If $f_n \to f$ as $n \to \infty$ almost everywhere (a.e.) or merely in measure on Ω and for every n: $|f_n| \le g$ a.e. on Ω, with $\int_\Omega g\, d\nu < \infty$, then

$$\lim_{n\to\infty} \int_\Omega f_n\, d\nu = \int_\Omega f\, d\nu = \int_\Omega \lim f_n\, d\nu. \qquad (4.14.1)$$

Lemma 4.14.2 (Bounded convergence theorem) If $f_n \to f$ as $n \to \infty$ a.e. or merely in measure on Ω and there exists a constant M such that for every n: $|f_n| \leq M$ a.e. on Ω, then (4.14.1) is true.

Lemma 4.14.3 (Monotone convergence theorem) If $f_n \geq 0$ and $f_n \uparrow f$ a.e. on Ω, then (4.14.1) is again true provided that $+\infty$ is allowed as a value for either member.

Lemma 4.14.4 (Fatou's lemma) If $f_n \geq 0$ a.e. on Ω, then

$$\int_\Omega \left(\liminf_{n\to\infty} f_n\right) dv \leq \liminf_{n\to\infty} \int_\Omega f_n \, dv. \tag{4.14.2}$$

Lemma 4.14.5 (Chow and Robbins, 1965) Let Y_n $(n = 1, 2, \ldots)$ be any sequence of random variables such that $Y_n > 0$ a.s. (almost surely) $\lim_{n\to\infty} Y_n = 1$ a.s., let $f(n)$ be any sequence of constants such that $f(n) > 0$, $\lim_{n\to\infty} f(n) = \infty$, $\lim_{n\to\infty} f(n)/f(n-1) = 1$, and for each $t > 0$ define

$$N = N(t) = \text{smallest } k \geq 1 \text{ such that } Y_k \leq f(k)/t. \tag{4.14.3}$$

Then N is well defined and nondecreasing as a function of t,

$$\lim_{t\to\infty} N = \infty \quad \text{a.s.,} \qquad \lim_{t\to\infty} EN = \infty, \tag{4.14.4}$$

and

$$\lim_{t\to\infty} f(N)/t = 1 \quad \text{a.s.} \tag{4.14.5}$$

Proof (4.14.4) can easily be verified. In order to prove (4.14.5) we observe that for $N > 1$, $Y_N < f(N)/t < [f(N)/f(N-1)]Y_{N-1}$, whence (4.14.5) follows as $t \to \infty$.

Lemma 4.14.6 (Chow and Robbins, 1965) If the assumptions of Lemma 4.14.5 hold and if $E(\sup_n Y_n) < \infty$, then

$$\lim_{t\to\infty} Ef(N)/t = 1. \tag{4.14.6}$$

Proof Let $Z = \sup_n Y_n$; then $EZ < \infty$. Choose m such that $f(n)/f(n-1) \leq 2$ $(n > m)$. Then for $N > m$

$$f(N)/t = [f(N)/f(N-1)][f(N-1)/t] < 2Y_{N-1} < 2Z.$$

Hence, for all $t \geq 1$

$$f(N)/t \leq 2Z + f(1) + \cdots + f(m). \tag{4.14.7}$$

Now, (4.14.6) follows from (4.14.5), (4.14.7), and Lemma 4.14.1 (dominated convergence theorem).

Let $X_1, X_2, \ldots$ be a sequence of independent observations from some population. We wish to set up a confidence interval of prescribed width $2d$ and prescribed coverage probability $1 - \alpha$ for the unknown mean μ of the population. If the variance of the population σ^2 is known, and if d is small compared to σ^2, this can be done as follows. For any $n \geq 1$ define

$$\overline{X}_n = n^{-1} \sum_{1}^{n} X_i, \qquad I_n = [\overline{X}_n - d, \overline{X}_n + d]$$

and let u_α (to be written as u hereafter) denote the $(1 - \alpha/2)$th fractile of the standard normal distribution. Then, for a sample size n determined by

$$n = \text{smallest integer} \geq (u^2\sigma^2)/d^2 \tag{4.14.8}$$

the interval I_n has coverage probability

$$P(\mu \in I_n) = P(n^{1/2} |\overline{X}_n - \mu|/\sigma \leq dn^{1/2}/\sigma).$$

Since (4.14.8) implies that $\lim_{d \to 0} (d^2 n)/(u^2\sigma^2) = 1$, it follows from the central limit theorem that $\lim_{d \to 0} P(\mu \in I_n) = 1 - \alpha$.

We shall be concerned with the situation where the nature of the population and hence σ^2 is unknown, so that no fixed-sample size method is available. Let

$$V_n^2 = n^{-1} \sum_{1}^{n} (X_i - \overline{X}_n)^2 + n^{-1} \quad (n \geq 1), \tag{4.14.9}$$

$u_1, u_2, \ldots$ be any sequence of positive constants such that $\lim_{n \to \infty} u_n = u$, and define

$$N = \text{smallest } k \geq 1 \text{ such that } V_k^2 \leq (d^2 k)/u_k^2. \tag{4.14.10}$$

Then, we have the following theorem of Chow and Robbins (1965).

Theorem 4.14.1 (Chow and Robbins, 1965) Under the sole assumption that $0 < \sigma^2 < \infty$ we have

$$\lim_{d\to 0} (d^2N)/(u^2\sigma^2) = 1 \qquad \text{a.s.,} \tag{4.14.11}$$

$$\lim_{d\to 0} P(\mu \in I_N) = 1 - \alpha \qquad \text{(asymptotic consistency),} \tag{4.14.12}$$

$$\lim_{d\to 0} (d^2EN)/(u^2\sigma^2) = 1 \qquad \text{(asymptotic efficiency).} \tag{4.14.13}$$

Proof In Lemma 4.14.5 set $Y_n = V_n^2/\sigma^2 = (1/n\sigma^2)\{\sum_1^n (X_i - \overline{X}_n)^2 + 1\}$,

$$f(n) = (nu^2)/u_n^2, \qquad t = (u^2\sigma^2)/d^2; \tag{4.14.14}$$

then (4.14.10) can be written as $N = N(t) =$ smallest $k \geq 1$ such that $Y_k \leq f(k)/t$. By Lemma 4.14.5,

$$1 = \lim_{t\to\infty} f(N)/t = \lim_{d\to 0} (d^2N)/(u^2\sigma^2) \quad \text{a.s.,} \tag{4.14.15}$$

which proves (4.14.11). Now,

$$P(\mu \in I_N) = P(|X_1 + \cdots = X_N - N\mu|/\sigma N^{1/2} \leq dN^{1/2}/\sigma).$$

By (4.14.15) $dN^{1/2}/\sigma \to u$ and $N/t \to 1$ in probability as $t \to \infty$; it follows from a result of Anscombe (1952, see, for instance, Theorem 4.12.1) that as $t \to \infty$ $(X_1 + \cdots + X_N - N\mu)/\sigma N^{1/2}$ behaves like a standard normal variable. Hence $\lim_{t\to\infty} P(\mu \in I_N) = 1 - \alpha$ which proves (4.14.12). Now (4.14.13) is an immediate consequence of Lemma 4.14.6 whenever the distribution of the X_i is such that

$$E\left\{\sup_n \left(n^{-1} \sum_1^n (X_i - \overline{X}_n)^2\right)\right\} < \infty, \tag{4.14.16}$$

for then

$$\lim_{t\to\infty} [Ef(N)/t] = 1, \tag{4.14.17}$$

and from the fact that the function $f(n)$ defined by (4.14.14) is $n + o(n)$ it follows from (4.14.17) that

$$1 = \lim_{t\to\infty} [EN/t] = \lim_{d\to 0} [(d^2EN)/(u^2\sigma^2)].$$

For (4.14.16) to hold, finiteness of the fourth moment of the X_i would suffice; however, the following lemma shows that (4.14.13) is valid without such a restriction.

Lemma 4.14.7 (Chow and Robbins, 1965) If the conditions of Lemma 4.14.5 hold, if $\lim_{n\to\infty} [f(n)/n] = 1$, if for N defined by (4.14.3)

$$EN < \infty \quad (\text{all } t > 0), \qquad \limsup_{t\to\infty}[E(NY_N)/EN] \leq 1, \tag{4.14.18}$$

and if there exists a sequence of constants $g(n)$ such that

$$g(n) > 0, \qquad \lim_{n\to\infty} g(n) = 1, \qquad Y_n \geq g(n)Y_{n-1},$$

then

$$\lim_{t\to\infty} EN/t = 1. \tag{4.14.19}$$

Proof For any $0 < \varepsilon < 1$, choose m so that $f(n-1) \geq (1-\varepsilon)f(n)$, $f(n-1) \geq (1-\varepsilon)n$ and $g(n) \geq 1-\varepsilon$ for $n \geq m$, and $E(NY_N) \leq (1+\varepsilon)EN$ for $t \geq m$. On the set $A = \{N \geq m\}$ it follows that

$$\begin{aligned}[(1-\varepsilon)^2/t]N^2 &= (1-\varepsilon)N(1-\varepsilon)N/t \\ &\leq g(N)Nf(N-1)/t \leq g(N)NY_{N-1} \leq NY_N.\end{aligned}$$

Hence

$$[(1-\varepsilon)^2/t]\left(\int_A N\right)^2 \leq [(1-\varepsilon)^2/t]\int_A N^2 \leq \int_A NY_N \leq E(NY_N),$$

$$[(1-\varepsilon)^2/t]\int_A N \leq E(NY_N)\Big/\left(\int_A N\right),$$

$$[(1-\varepsilon)^2/t](EN-m) \leq E(NY_N)/(EN-m).$$

From (4.14.4) and (4.14.18) it follows

$$(1-\varepsilon)^2 \limsup_{t\to\infty} [EN/t] \leq \limsup_{t\to\infty} E(NY_N)/(EN) \leq 1,$$

so that

$$\limsup_{t\to\infty} \{EN/t\} \leq 1. \tag{4.14.20}$$

Now, let $Y_n' = \min(1, Y_n)$. Then

$$0 \le Y_n' \le 1, \qquad Y_n' \le Y_n, \qquad \lim_{n\to\infty} Y_n' = 1 \quad \text{a.s.}$$

Let $N' = N'(t) =$ smallest $k \ge 1$ such that $Y_k' \le f(k)/t$. From Lemma 4.14.6, since $\sup_n(Y_n') \le 1$,

$$1 = \lim_{t\to\infty} [Ef(N)]/t = \lim_{t\to\infty} (EN')/t.$$

However, since $Y_n' \le Y_n$, $N' \le N$ and hence $EN' \le EN$. Thus

$$\liminf_{t\to\infty} (EN)/t \ge \liminf_{t\to\infty} (EN')/t = 1,$$

which with (4.14.20) proves (4.14.19).

In order to prove (4.14.13), fix $t > 0$, choose m such that $f(n)/t \ge 1$ $(n \ge m)$, choose $\delta > 0$ such that $(n-1)f(n-1) > \delta n^2$ $(n \ge 2)$, and let, for any $r \ge m$, $M = \min(N, r)$. By Wald's theorem on cumulative sums

$$E\left(\sum_1^M (X_i - \mu)^2\right) = (EM)E(X_i - \mu)^2 = \sigma^2(EM).$$

Hence, by the definition of Y_n in the proof Theorem 4.14.1

$$\begin{aligned} E(MY_M) &= \sigma^{-2}E\left(\sum_1^M (X_i - \bar{X}_M)^2 + 1\right) \\ &\le \sigma^{-2}E\left(\sum_1^M (X_i - \mu)^2 + 1\right) = EM + \sigma^{-2}. \qquad (4.14.21) \end{aligned}$$

Set $g(n) = (n-1)/n$, $(n \ge 2)$; then

$$Y_n \ge (1/n\sigma^2)\sum_1^{n-1} (X_i - \bar{X}_{n-1})^2 + (1/n\sigma^2) = [(n-1)/n]Y_{n-1} = g(n)Y_{n-1}$$

Hence

$$\begin{aligned} E(MY_M) &\ge \int_{\{N>r\}} rY_r + \int_{\{N\le r\}} NY_N \ge [rf(r)/t]P(N>r) + \int_{\{2\le N\le r\}} NY_N \\ &\ge rP(N>r) + \int_{\{2\le N\le r\}} [Ng(N)f(N-1)]/t \\ &\ge rP(N>r) + (\delta/t)\int_{\{2\le N\le r\}} N^2. \end{aligned}$$

Consequently by (4.14.21)

$$\int_{\{N\leq r\}} N \geq (\delta/t)\int_{\{2\leq N\leq r\}} N^2 - (1/\sigma^2) \geq (\delta/t)\left(\int_{\{2\leq N\leq r\}} N\right)^2 - \sigma^{-2},$$

and letting $r \to \infty$ it follows that

$$EN = \lim_{r\to\infty}\int_{\{N\leq r\}} N < \infty$$

which is the first part of (4.14.18). Again by Wald's theorem, $E(NY_N) \leq EN + \sigma^{-2}$. So by (4.14.4)

$$\limsup_{t\to\infty}[E(NY_N)]/(EN) \leq 1$$

which is the second part of (4.14.18). Thus the conditions of Lemma 4.14.7 hold and hence

$$1 = \lim_{t\to\infty} EN/t = \lim_{d\to 0}(d^2EN)/(u^2\sigma^2) \leq 1,$$

which is (4.14.13).

REMARK 4.14.1.1 If the random variables X_i are continuous, definition of V_n^2 in (4.14.9) can be modified to $V_n^2 = n^{-1}\sum_1^n (X_i - \bar{X}_n)^2$. The only purpose of the term n^{-1} is to ensure that $Y_n = V_n/\sigma^2 > 0$ a.s., this fact having been used in the proof of Lemma 4.14.5 to guarantee that $N \to \infty$ a.s. as $t \to \infty$. Also it is evident from the proofs, N in (4.14.10) could be defined as the smallest (or the smallest odd, etc.) integer $\geq n_0$ such that the indicated inequality holds, where n_0 is any fixed positive integer.

REMARK 4.14.1.2 The method of proof of Theorem 4.14.1 is a modification of that used by Doob (1948) to prove the elementary renewal theorem. Also Theorem 4.14.1 has been proved when the X_i are normal with mean μ and variance σ^2 by Stein (1949), Anscombe (1952, 1953), Gleser, *et al.* (1964). Ray (1957b), who was apparently misled by having considered too few values of d in his numerical computations, doubts the validity of (4.14.12). Extensive numerical computations of Starr (1966a) for the normal case indicate, for example, that for $1 - \alpha = .95$ the lower bound for all $d > 0$ of $P(\bar{X}_N - d \leq \mu \leq \bar{X}_N + d)$, where N is the smallest odd integer $k \geq 3$ such that

$$(k-1)^{-1}\sum_1^k (X_i - \bar{X}_k)^2 \leq (d^2k)/u_k^2,$$

is about 0.929 if the values u_k are taken from the t-distribution with $k - 1$ degrees of freedom.

Nádas (1969) has extended the theorem of Chow and Robbins (1965) in order to take care of other prescribed "accuracies." We speak of "absolute accuracy" when estimating μ by

$$I_n = \{\mu : |\overline{X}_n - \mu| \le d\} \qquad (d > 0); \tag{4.14.22}$$

and if $\mu \neq 0$, we speak of "proportional accuracy" when estimating μ by

$$J_n = \{\mu : |\overline{X}_n - \mu| \le p|\mu|\}, \qquad (0 < p < 1). \tag{4.14.23}$$

Denote by ρ the coefficient of variation $\sigma/|\mu|$ and define

$$n(d) = \min_{n \ge 1} \{n : \sigma^2 \le n(d/u)^2\}, \tag{4.14.24}$$

$$m(p) = \min_{n \ge 1} \{n : \rho^2 \le n(p/u)^2\} \tag{4.14.25}$$

where as before u denotes the $(1 - \alpha/2)$th fractile of the standard normal variable. Then (4.14.24) and (4.14.25) increase without bound as the arguments tend to zero. Hence, for small arguments one can (at least approximately) achieve the required probability of coverage $1 - \alpha$ by taking the "sample size" n no smaller than $n(d)$ (for absolute accuracy) or $m(p)$ (for proportional accuracy). If, however, σ^2 (or ρ^2) is unknown, then (4.14.24) or [(4.14.25)] is not available. On the other hand, if we let V_n^2 be given by (4.14.9) then the stopping rules

$$N = \min_{n \ge 1} \{n : V_n^2 \le n(d/u)^2\} \tag{4.14.26}$$

and

$$M = \min \{n : (V_n/\overline{X}_n)^2 \le n(p/u)^2\} \tag{4.14.27}$$

are well defined. In the event that ρ^2 is known (but not σ^2) and one insists on absolute accuracy, or if σ^2 is known (but not ρ^2) in the proportional case, then one has

$$N^* = \min_{n \ge 1} \{n : \overline{X}_n^2 + n^{-2} \le n(d/\rho u)^2\} \tag{4.14.28}$$

and

$$M^* = \min_{n \ge 1} \{n : \overline{X}_n^2 \ge n^{-1}(u\rho/p)^2\} \tag{4.14.29}$$

as the sequential analogues of (4.14.24) and (4.14.25). Denote by K any one of (4.14.26)–(4.14.29), let k be the corresponding fixed "sample size" (4.14.24) or (4.14.25), and let H_k be the corresponding interval estimate (4.14.22) or (4.14.23). Then we have the following theorem of Nádas (1969).

Theorem 4.14.2 (Nádas, 1969) With the above notation, we have

$$\lim K/k = 1 \qquad \text{a.s.,} \tag{4.14.30}$$

$$\lim P(\mu \in H_k) = 1 - \alpha \qquad \text{(asymptotic consistency),} \tag{4.14.31}$$

$$\lim EK/k = 1 \qquad \text{(asymptotic efficiency),} \tag{4.14.32}$$

as d or p tends to zero.

Proof Chow and Robbins (1965) have proved Theorem 4.14.2 (see Theorem 4.14.1) in the case $K = N$. Nádas (1969) has improved on (4.13.31) by giving a uniform upper bound on $EN - n(d)$. He also shows that $EM^* - m(p)$ is bounded above. Under the additional assumption that the distribution function F of X_1 is continuous with a finite fourth moment, he shows that $EM - m(p)$ is also bounded above.

For every definition (4.14.26)–(4.14.29) of K, (4.14.30) and (4.14.31) follow from (a trivial modification of) Lemma 4.14.5 and a result of Anscombe (1952), just as in the work of Chow and Robbins (1965). We have from (4.14.30) and Fatou's lemma (see Lemma 4.14.4) that

$$\liminf (EK/k) \geq 1. \tag{4.14.33}$$

In order to prove that

$$\limsup (EK/k) \leq 1 \tag{4.14.34}$$

and to find upper bounds. We consider separate cases.

THE CASE $K = N$. Let $t = (u\sigma/d)^2$ and for $n = 1, 2, \ldots$, let

$$\sigma^2 S_n = 1 + \sum_{k=1}^{n} (X_k - \mu)^2. \tag{4.14.35}$$

Define

$$N(t) = \min_{n \geq 1} \{n : S_n \leq n^2/t\}.$$

Then $N(t)$ is well defined, it is no smaller than N, and it can be shown [e.g., by truncating $N(t)$ and proceeding as below] that $EN(t) < \infty$. By Wald's (1947) theorem on cumulative sums,

$$\begin{aligned}\sigma^{-2} + EN(t) &= ES_{N(t)} \\ &\geq \int_{\{N(t)=1\}} S_1 + \int_{\{N(t)>1\}} S_{N(t)-1} \\ &\geq t^{-1} \int_{\{N(t)>1\}} (N(t)-1)^2 = t^{-1} E\{(N(t)-1)^2\} \\ &\geq t^{-1} E^2(N-1). \qquad (4.14.36)\end{aligned}$$

Thus,

$$EN(t) \in \{x : x^2 - (t+2)x + (1 - t\sigma^{-2}) \leq 0\}. \qquad (4.14.37)$$

Hence, for all $t > 0$ we have $EN(t) - t \leq 2 + \sigma^{-2}$. If F is continuous, then the n^{-1} term in the definition of V_n^2 can be dropped and we obtain in the same way that for all $t > 0$ and all $\sigma > 0$

$$EN(t) - t \leq 2. \qquad (4.14.38)$$

THE CASE $K = N^*$ AND $K = M^*$ In these cases (4.14.34) is a consequence of the following lemma.

Lemma 4.14.8 (Nádas, 1969) Let $Z_1, Z_2, \ldots$ be a sequence of independent and identically distributed random variables such that $0 < EZ_1 < \infty$. Denote by S_n $(n = 1, 2, \ldots)$ their nth cumulative sum. Let c be a positive number.

(i) Assume that $Z_1 \geq 0$ a.s. Let $\gamma > 0$ and $t = (EZ_1/c)^{1/\gamma}$. Define

$$T = \min_{n \geq 1} \{n : S_n \leq cn^{1+\gamma}\}. \qquad (4.14.39)$$

Then, T is well defined, $ET^{1+\gamma} < \infty$, and the family $\{T/t : 0 < c < 1\}$ is uniformly integrable.

(ii) Let $0 < \gamma < 1$ and $t = (c/EZ_1)^{1/(1-\gamma)}$. Define

$$T = \min_{n \geq 1} \{n : |S_n| \geq cn^{\gamma}\}. \qquad (4.14.40)$$

Then, T is well defined, $ET < \infty$, and $\lim_{c \to \infty}(ET/t) = 1$.

Proof The lemma can be proved by truncating both the stopping rule T (to T') and the Z_n in each case (i) and (ii) and then approximating the roots of a polynomial in the variable ET'. The lemma is related to a result of Chow (1966a) and this has been generalized by Siegmund (1967).

In the case $K = N^*$, (4.14.34) is established by using (i) above to prove the uniform integrability of the family $\{N^*/n(d) : 0 < d < 1\}$. In the case $K = M^*$, (4.14.34) follows by a change of notation in (ii) above. The boundedness (from above) of $EM^* - m(p)$ can be shown by the method used when $K = M$ as shown below. The same is true for $EN^* - n(d)$ provided one makes the (highly unsatisfactory) assumption that the X_n are positive.

THE CASE $K = M$ Without loss of generality assume that $\mu > 0$. Let $t = (u\rho/p)^2$ and put, for $n = 1, 2, \ldots,$

$$S_n = \sum_{k=1}^{n} X_i \qquad \text{and} \qquad Q_n = 1 + \sum_{k=1}^{n} (X_i - \mu)^2. \tag{4.14.41}$$

The random variable

$$M(t) = \min_{n \geq 1} \{n : S_n \geq \rho^{-1}(tQ_n)^{1/2}\} \tag{4.14.42}$$

is well defined and it is no smaller than M. For $r = 1, 2, \ldots,$ define

$$R = \min(M(t), r) \qquad \text{and} \qquad B = (1 < M(t) \leq r).$$

We apply Wald's theorem for cumulative sums to each of S_R, $Q_R - 1$ and $X_1{}^2 + \cdots + X_R{}^2$ to obtain

$$\begin{aligned}
\mu(ER) = ES_R &= \int_{\{R=1\}} X_1 + \int_B S_{R-1} + \int_{\{M(t)>r\}} S_r + \int_B X_R \\
&\leq E|X_1| + \rho^{-1}\int_B (tQ_{R-1})^{1/2} + \rho^{-1}\int_{\{M(t)>r\}} (tQ_r)^{1/2} + \int_B |X_R| \\
&\leq E^{1/2}(X_1{}^2) + \rho^{-1}E(tQ_R)^{1/2} + E^{1/2}\left(\sum_{k=1}^{R} X_k{}^2\right) \\
&\leq (\sigma^2 + \mu^2)^{1/2} + \rho^{-1}(t + t\sigma^2 ER)^{1/2} + (\sigma^2 + \mu^2)^{1/2}E^{1/2}R \\
&\leq (\sigma^2 + \mu^2)^{1/2} + \rho^{-1}t^{1/2} + [\mu t^{1/2} + (\sigma^2 + \mu^2)^{1/2}]E^{1/2}R
\end{aligned}$$

since $(a + b)^{1/2} \leq a^{1/2} + b^{1/2}$ for $a, b > 0$. Thus

$$E^{1/2}R \in \{x : x^2 - [t^{1/2} + (\rho^2 + 1)^{1/2}]x - [\sigma^{-1}t^{1/2} + (\rho^2 + 1)^{1/2}] \leq 0\}. \tag{4.14.43}$$

Consequently

$$ER \leq \{t^{1/2} + o(t^{1/2})\}^2. \tag{4.14.44}$$

It follows from the a.s. monotone convergence $R \uparrow M(t)$ as $r \to \infty$ that $EM(t) < \infty$ and that (4.14.44) holds with R replaced by $M(t)$. Hence

$$\limsup_{t \to \infty} EM(t) \leq 1. \tag{4.14.45}$$

In order to show that $EM - m(p)$ is bounded above, assume that $EX_1^4 < \infty$ and that F is continuous so that the constant term 1 in the definition of nV_n^2 and Q_n can be omitted. Consider the line

$$L(x;t) = \mu(x+t)/2, \qquad x > 0, \quad t > 0, \tag{4.14.46}$$

which is the tangent to the curve $\mu(tx)^{1/2}$, in an x-y plane, at the point of inter-section of the curve with the line μx. Then

$$\begin{aligned} M(t) &= \min_{n \geq 1} \{n : S_n \geq \mu(tQ_n\sigma^{-2})^{1/2}\} \\ &\leq \min_{n \geq 1} \{n : S_n \geq L(Q_n\sigma^{-2}; t)\} \\ &= \min_{n \geq 1} \left\{n : \sum_{k=1}^{n} Z_k \geq t\right\} \end{aligned}$$

where

$$\mu Z_k = 2X_k - \mu(X_k - \mu)^2/\sigma^2 \qquad (k = 1, 2, \ldots). \tag{4.14.47}$$

Let ψ_A denote the indicator of the set A. If we denote by

$$U(t) = E\sum_{n=0}^{\infty} \psi\left(\sum_{k=1}^{\infty} Z_k < t\right) \tag{4.14.48}$$

the "renewal measure" of the transient ($EZ_1 > 0$) random walk $\sum_{k=1}^{n} Z_k$, then it follows [see Feller, (1970, p. 372)] that

$$\limsup_{t\to\infty} EM - t \leq \lim_{t\to\infty} U(t) - t = EZ_1^2/2E^2Z_1 < \infty.$$

CERTAIN REMARKS (i) So far we are unaware of reasonably general conditions on F that would ensure that $EK - k$ is bounded below.

(ii) In practice one may wish to replace the constant u in (4.14.26)–(4.14.29) by a positive convergent sequence $\{u_n\}$ in the hope of improving the coverage probability for noninfinitesimal d or p. It may be done without invalidating the asymptotic results. The same is true if the range of K is restricted to certain infinite subsets of the positive integers.

(iii) Anscombe (1953) made asymptotic expansions for the distribution of (essentially) $N - n(d)$ on the assumption that F is not purely discrete, that F has finite eighth moment, and that the tails of the distribution of $N - n(d)$ are of certain orders of magnitude. The latter are not in terms of F.

(iv) Anscombe (1953) and Siegmund (1967) have used the idea of approximating a convex stopping boundary for a random walk by the appropriate tangent.

(v) One can get a minor insight into the nature of (4.14.32) by noting that when $K = N$, (4.14.32) is equivalent to

$$\lim_{d \to 0} EV_N{}^2 = \sigma^2.$$

(vi) The stopping rules $U = \max(N, M)$ and $L = \min(N, M)$ are appropriate for some "mixed" accuracy criteria that arise in practice, that is, for estimating the mean by the intervals

$$I_U \cap J_U = \{\mu : |\overline{X}_U - \mu| \leq \min(d, p|\mu|)\},$$
$$I_L \cup J_L = \{\mu : |\overline{X}_L - \mu| \leq \max(d, p|\mu|)\}.$$

Let $u(d, p)$ and $l(d, p)$ be the corresponding fixed "sample sizes," and let the ratio d/p be fixed. It is fairly easy to see that Theorem 4.14.2 is valid for L, and that it is valid for U is a straightforward consequence of the uniform integrability of $N/n(d)$ and $M/m(p)$.

Simons (1968) has shown that the mean of a normal distribution with unknown variance σ^2 may be estimated to lie within an interval of given fixed width at a prescribed confidence level using a procedure which overcomes ignorance about σ^2 with no more than a finite number of observations. That is, the expected sample size exceeds the (fixed) sample size one would use if σ^2 were known by a finite amount, the difference depending on the confidence level $1 - \alpha$ but not depending on the values of the mean μ, the variance σ^2, and the interval width $2d$. In the following we shall present his results. Hereafter, let

$$u_n \equiv u, \tag{4.14.49}$$

where u denotes the $(1-\alpha/2)$th fractile of the standard normal population. Also, let

$$C = u^2\sigma^2/d^2. \tag{4.14.50}$$

Let the stopping variable N be given by

$$N \equiv \text{smallest index} \quad n \geq n_0 \geq 2 \quad \text{for which} \quad n \geq u_n{}^2 s_n{}^2/d^2 \tag{4.14.51}$$

where $s_n{}^2$ denotes the sample variance. Then, in an unpublished manuscript Chernoff and Simons have shown a stronger efficiency result than that of Chow and Robbins (1965) namely, when

$$u_n \equiv u, \qquad EN \leq C + n_0 + 1 \tag{4.14.52}$$

(independent of d, u, and the distribution of X). Thus, one is tempted to claim that the "cost of ignorance" in not knowing σ^2 is at most $n_0 + 1$ observations. However, the objective is not to achieve asymptotic consistency but rather to achieve

$$P(|Y_N - \mu| < d) > 1 - \alpha \text{ for all values of } \mu \text{ and } \sigma^2. \tag{4.14.53}$$

One must be able to satisfy (4.14.51) before one can properly assess the true cost of ignorance. Starr (1966a) conducted a numerical study for normally distributed X using a particular sequence of u_n [of the form $u + O(n^{-1})$]. It seems that he rarely achieves (4.14.53). Simons (1968) remarks that one can show for $u_n = u + O(n^{-1})$ that

$$EN = C + O(1) \qquad \text{as} \quad d \to 0. \tag{4.14.54}$$

Thus, it seems likely (and we shall verify it for normally distributed X) that there exist stopping rules N for which (4.14.53) holds and for which the cost of ignorance, $EN - C$ is uniformly bounded for all μ and σ^2 and $d > 0$. Specifically, for some integer k, we can achieve these objectives by taking k more observations after rule (4.14.51). We shall need $n_0 \geq 3$ (not 2) and $u_n \equiv u$.

Lemma 4.14.9 (Simons, 1968) With $u_n \equiv u$ and N defined by (4.14.51), for an arbitrary continuous X, we have

$$EN \leq C + n_0. \tag{4.14.55}$$

Proof See Simons (1968, Theorem 1, p. 1948).

Lemma 4.14.10 (Simons, 1968) With $u_n \equiv u$ and N defined by (4.14.51), for a normally distributed X we have

$$EN^j \leq C^j + 0(C^{j-1}) \qquad \text{as} \quad C \to \infty \text{ for } j = 1, 2, \ldots, \tag{4.14.56}$$

and

$$EN^j \geq C^j + 0(C^{j-1}), \qquad \text{as} \quad C \to \infty \text{ for any } n_0 \geq 3 \text{ for } j = 1, 2, \ldots. \tag{4.14.57}$$

Proof See Simons (1968, Theorem 2, p. 1949).

The proofs of Lemmas 4.14.9 and 4.14.10 are based on the definition of a *reverse stopping variable* and the properties of a *reverse martingale.*

Theorem 4.14.3 (Simons, 1968) Let $\lambda = \sigma/d$ so $C = u^2\lambda^2$. If the X_i are normal with mean μ and variance σ^2, and if the value of any stopping variable N is determined by $s_2{}^2, s_3{}^2, \ldots$, then

$$P(|Y_N - \mu| < d) = 2E\Phi(N^{1/2}/\lambda) - 1 \qquad \text{for all} \quad \mu, \sigma^2. \tag{4.14.58}$$

For N defined by (4.14.51) with $u_n \equiv u$ and $n_0 \geq 3$, we have for some finite integer $k \geq 0$,

$$E\Phi((N + k)^{1/2}/\lambda) \geq \Phi(u) = 1 - (\alpha/2) \qquad \text{for all } \mu, \sigma^2, \text{ and } d. \tag{4.14.59}$$

Then

$$P(|Y_N - \mu| < d) \geq 1 - \alpha \qquad \text{for all } \mu, \sigma^2, \text{ and } d, \tag{4.14.60}$$

and

$$E(N + k) \leq C + n_0 + k \qquad \text{for all } \mu, \sigma^2 \text{ and } d. \tag{4.14.61}$$

Proof When X is normal, the random variables Y_n and the vector $(s_2{}^2, s_3{}^2, \ldots, s_n{}^2)$ are independent for $n = 2, 3, \ldots$. Thus the events $\{N = n\}$ and $|Y_N - \mu| < d$ are mutually independent. Hence

$$\begin{aligned} P(|Y_N - \mu| < d) &= \sum_{n=n_0}^{\infty} P(|Y_n - \mu| < d \text{ and } N = n) \\ &= \sum_{n=n_0}^{\infty} P(|Y_n - \mu| < d)P(N = n) = 2E\Phi(N^{1/2}/\lambda) - 1. \end{aligned}$$

For $g(x) = \Phi(x^{1/2}/\lambda)$, we have $g'(x) = \phi(x^{1/2}/\lambda)/\lambda(2x^{1/2})$ and

$$g''(x) = -\{(x\lambda^{-2} + 1)/\lambda\}\phi(x^{1/2}/\lambda)/(4x^{3/2})$$

where $\phi(y)$ denotes the standard normal density function. Expand $g(x)$ in Taylor series about $x = C$ with a second-degree remainder term. We find for arbitrary θ, $0 < \theta < 1$, that

$$\begin{aligned} E\Phi((N+k)^{1/2}/\lambda) \geq {} & E\Phi((N+k)^{1/2}/\lambda)I_{\{N+k\geq\theta^2C\}} \\ & - u(u^2\theta^2+1)\phi(u\theta)(8\theta^3C^2)^{-1}E(N+k-C)^2 \\ & \times I_{\{N+k\geq\theta^2C\}}, \\ \geq {} & \Phi(u) + u\phi(u)(2C)^{-1}E(N+k-C) \\ & - u(u^2\theta^2+1)\phi(u\theta)(8\theta^3C^2)^{-1}E(N+K-C)^2 \\ & + \{-\Phi(u) + u\phi(u)(1-\theta^2)/2 + u(u^2\theta^2+1) \\ & \times \phi(u\theta)(1-\theta^2)^2/(8\theta^3)\}P\{N+k\leq\theta^2C\}. \end{aligned}$$

For small $\theta > 0$ the coefficient of $P(N + k \leq \theta^2 C)$ is positive and for such θ,

$$\begin{aligned} E\Phi((N+k)^{1/2}/\lambda) \geq {} & \Phi(u) + u\phi(u)(2C)^{-1}[k + E(N-C)] \\ & - u(u^2\theta^2+1)\phi(u\theta)(8\theta^3C^2)^{-1} \\ & \times [k^2 + 2kE(N-C) + E(N-C)^2]. \end{aligned}$$

By (4.14.56) and (4.14.57) we have

$$\begin{aligned} E\Phi((N+k)^{1/2}/\lambda) - \Phi(u) & \geq O(C^{-2})k^2 + \{u\phi(u)(2C)^{-1} + O(C^{-2})\}k + O(C^{-1}) \\ & = \{u\phi(u)(2C)^{-1} + O(C^{-2})\}\{O(C^{-1})k^2 + k + O(1)\}. \end{aligned}$$

Thus, for large k and for large C (say $C \geq C_0$), (4.14.59) holds and clearly for some large k (4.14.59) holds for all $C < C_0$. Thus (4.14.59) holds for some integer $k \geq 0$. (4,14.60) follows from (4.14.58) and (4.14.59), and (4.14.61) from (4.14.55). This completes the proof.

Problem

Let X and Y have a bivariate normal distribution with unknown mean vector and variance–covariance matrix. Also, let (X_i, Y_i), $i = 1, 2, \ldots$ denote a sequence of independent observations from this population. Set up a fixed-width confidence interval for $p = P(X < Y)$ and study its asymptotic properties when the width is small.

4.15 Asymptotic Theory of Fixed-Size Sequential Confidence Bounds for Linear Regression Parameters

Gleser (1965) has extended Chow and Robbins' (1965) results to the linear regression problem. Let $y_1, y_2, \ldots$ be a sequence of independent observations with

$$y_i = \beta' x^{(i)} + \varepsilon_i, \tag{4.15.1}$$

where β' is an unknown $1 \times p$ vector, $x^{(i)}$ a known $p \times 1$ column vector, and ε_i a random observation obeying a (possibly) unknown distribution function F with mean 0 and finite, but unknown variance σ^2. We wish to find a region R in p-dimensional Euclidean space such that $P(\beta \in R) = 1 - \alpha$ and such that the length of the interval cut off on the β_i axis by R has width $\leq 2d, i = 1, \ldots, p$. As has already been noted for $p = 1$, no fixed-sample procedure meeting the requirements exists. Hence, we are led to consider sequential procedures.

σ Is Known. Since the least-squares (Gauss–Markov) estimate of β has componentwise uniformly minimum variance among all linear unbiased estimates of β, has good asymptotic properties [such as consistency, see Eicker (1963b)] and performs reasonably well against nonlinear unbiased estimates [see Anderson (1962)], the usual practice would be to use the least squares estimate of β in the construction of our confidence region. It is well known that the least squares estimate of β in our problem is

$$\hat{\beta}(n) = (X_n X_n')^{-1} X_n Y_n \tag{4.15.2}$$

where $Y_n = (y_1, \ldots, y_n)'$, $X_n = (x^{(1)}, \ldots, x^{(n)})$: a $p \times n$ $(p \leq n)$ matrix and where we assume that X_p is of full rank. [This is usually possible to achieve in practice—if not sample until p independent $x^{(i)}$ are found, start with the p corresponding y_i and save the remainder for future use in the sequential procedure. Such a procedure does not bias the results and is equivalent to starting after a fixed number of observations n_0.]

Since the covariance matrix of the $\hat{\beta}(n)$ is $\sigma^2(X_n X_n')^{-1}$, the usual practice would be to construct the confidence region

$$\{\hat{\beta}(n) - \beta\}'(X_n X_n')^{-1}\{\hat{\beta}(n) - \beta\} \leq d^2$$

which, if F were the standard normal distribution function, would then have probability of coverage equal to $P\{\sigma^2 \chi_p^{\,2} \leq d^2\}$ (and hopefully, would have this property asymptotically for any F). Unless such a region has a probability of coverage equal to $1 - \alpha$, it will not be of any use to us. To find a confidence interval of width $2d$ for any one of the β_i, we could [in analogy with Chow and Robbins (1965)] use the interval $\hat{\beta}_i(n) \pm d$. Also, for any normal linear combination $a'\beta$, $a' = 1 \times p$, $a'a = 1$, of the β_i, $i = 1, \ldots, p$, we could use the confidence interval $a'\hat{\beta}(n) \pm d$. Now, a region R_n that would be contained in all of these confidence intervals is

$$R_n = \{z: (z - \hat{\beta}(n))'(z - \hat{\beta}(n)) \le d^2\}, \tag{4.15.3}$$

since for any a such that $a'a = 1$, and any $z \in R_n$

$$\{a'(z - \hat{\beta}(n))\}^2 \le \max_{a'a=1} \{a'(z - \hat{\beta}(n))\}^2 = \{z - \hat{\beta}(n)\}'\{z - \hat{\beta}(n)\} \le d^2.$$

This region has been adopted by Gleser (1965) for the confidence procedure.

When σ^2 is unknown, the least squares estimate of σ^2 is

$$\hat{\sigma}^2(n) = n^{-1} Y_n'(I_n - X_n'(X_n X_n')^{-1} X_n) Y_n. \tag{4.15.4}$$

Before presenting the class of sequential procedures $\mathscr{C}$, we shall consider some asymptotic properties of $\hat{\beta}(n)$ and $\hat{\sigma}^2(n)$, which will be relevant to the discussion of the asymptotic properties of the class $\mathscr{C}$. The asymptotic distribution theory for $\hat{\beta}(n)$ is merely a corollary to the following well-known theorem.

Theorem 4.15.1 If $z_1, z_2, \ldots$ are independent and identically distributed random variables, each with mean 0 and variance 1, and cumulative distribution fuction G, if b_{ni}, $i = 1, \ldots, n$, $n = 1, 2, \ldots$ is a fixed array of constants with $\sum_{i=1}^{n} b_{ni}^2 = 1 (n = 1, 2, \ldots)$, and if

$$\max_{1 \le i \le n} |b_{ni}| \to 0 \qquad (n \to \infty), \tag{4.15.5}$$

then we have

$$\lim_{n\to\infty} P\left(\sum_{i=1}^{n} b_{ni} z_i \le t\right) = \Phi(t). \tag{4.15.6}$$

Proof This theorem is an immediate consequence of the "particular case" of Gnedenko and Kolmogorov (1954, Theorem 3, p. 103). Of interest in connection with Theorem 4.15.1 is the work of Eicker (1963a).

For our problem, let $U_n = (X_n X_n')^{-1/2} X_n = ((u_{nij}))$.

Corollary 4.15.1.1 If

$$\max_{i,j} |u_{nij}| \to 0 \qquad \text{as} \quad n \to \infty \tag{4.15.7}$$

then

$$(\hat{\beta}(n) - \beta)'(X_n X_n')^{1/2} \sim \text{normal}\,(0, \sigma^2 I_p) \qquad (n \to \infty). \tag{4.15.8}$$

Proof For any $1 \times p$ vector a' such that $a'a = 1$, consider

$$t_n(a) = \sigma^{-1} a'(X_n X_n')^{1/2}(\hat{\beta}(n) - \beta) = \sigma^{-1} a' U_n(Y_n - EY_n).$$

Then, $t_n(a)$ is a linear combination of the elements $Z_n = \sigma^{-1}(Y_n - EY_n)$, the elements of Z_n being independent, identically distributed with zero means, variances 1, and cumulative distribution functions $F(x/\sigma)$. Further, $a'U_n U_n'a = 1$ so that we may apply Theorem 4.15.1 provided the maximum element of the vector $a'U_n$ tends to zero as $n \to \infty$. This follows from the fact that [letting $U_n = (U_n^{(1)}, \ldots, U_n^{(n)})$, $U_n^{(i)}$, $p \times 1$]

$$|a'U_n^{(j)}| = \left|\sum_{i=1}^{n} a_i u_{nij}\right| \le \sum_{i=1}^{p} |a_i|\,|u_{nij}| \le p^{1/2} \max_{ij} |u_{nij}| \to 0 \quad \text{all } j.$$

Thus, by Theorem 4.15.1, $t_n(a)$ is asymptotically distributed as the standard normal variable. Since this is true for all a, $a'a = 1$, it follows from well-known theorems in multivariate analysis and large-sample theory that (4.15.8) holds.

A sufficient condition for (4.15.7) to hold is that assumptions 4.15.1 and 4.15.2 hold, where:

Assumption 4.15.1 There exists a $p \times p$ positive definite matrix Σ such that

$$\lim_{n\to\infty} n^{-1}(X_n X_n') = \Sigma. \tag{4.15.9}$$

Assumption 4.15.2 $\lim_{n\to\infty} X_n/n^{1/2} = \mathbf{0}$.

Under these assumptions we can find the asymptotic probability of coverage of the region R_n.

Corollary 4.15.1.2 Under assumptions 4.15.1 and 4.15.2, we have

$$P\{n(\hat{\beta}(n) - \beta)'(\hat{\beta}(n) - \beta) \le d^2\} = P\{T(\lambda_1, \ldots, \lambda_p) \le d^2/\sigma^2\} \tag{4.15.10}$$

where $\lambda_1, \ldots, \lambda_p$ are the characteristic roots of Σ^{-1} and $T(\lambda_1, \ldots, \lambda_p)$ has the distribution of a weighted sum of p independent chi-square variables with one degree of freedom, the λ_i being the weights.

Proof By Corollary 4.15.1.1, we have that $(\hat{\beta}(n) - \beta)'(X_n X_n')^{1/2} \sim$ normal $(0, \sigma^2 I_p)$ $(n \to \infty)$. Further,

$$(\hat{\beta}(n) - \beta)'(\hat{\beta}(n) - \beta) = (\hat{\beta}(n) - \beta)'(X_n X_n')^{1/2}(X_n X_n')^{-1}(X_n X_n')^{1/2}(\hat{\beta}(n) - \beta).$$

Since $n^{-1}X_n X_n'$ converges to Σ, it follows that

$$n(\hat{\beta}(n) - \beta)'(\hat{\beta}(n) - \beta) \sim w'\Sigma^{-1}w \qquad \text{(in distribution)} \qquad (4.15.11)$$

where w is normal $(0, \sigma^2 I_p)$. An application of a well-known theorem in multivariate analysis completes the proof.

Finally, we need to consider the consistency of $\hat{\sigma}^2(n)$.

Theorem 4.15.2 (Gleser, 1965) If $\max_{ij} |u_{nij}| \to 0$ as $n \to \infty$, then

$$\lim \hat{\sigma}^2(n) = \sigma^2, \quad \text{a.s.} \qquad (4.15.12)$$

Proof Since $Z_n = Y_n - EY_n = Y_n - \beta' X_n$ and the introduction of $\beta' X_n$ does not alter the value of $\hat{\sigma}^2(n)$ [see, for instance, Rao (1965, p. 186)]

$$\begin{aligned}\hat{\sigma}^2(n) &= n^{-1}Z_n'(I_n - U_n'U_n)Z_n, \\ &= n^{-1}Z_n'Z_n(1 + o(1))\end{aligned}$$

which by the strong law of large numbers converges almost surely to σ^2.

REMARK 4.15.1 Notice that Theorem 4.15.1 is a special case of a more general theorem of Eicker (1963a). Generalizations of the results of Gleser (1965) in directions suggested by the theorem of Eicker (1963a) are possible.

ASYMPTOTIC PROPERTIES OF THE CLASS $\mathscr{C}$ Given d and α and for a fixed sequence of x vectors, $x^{(1)}, x^{(2)}, \ldots$ arranged so that X_p is nonsingular and so that assumptions 4.15.1 and 4.15.2 are satisfied, let $\{u_n\}$ be any sequence of constants converging to the number u satisfying

$$P(T(\lambda_1, \ldots, \lambda_p) \le u) = 1 - \alpha. \qquad (4.15.13)$$

Then, this sequence $\{u_n\}$ determines a member of the class $\mathscr{C}$ of sequential procedures as follows:

Start by taking $n_0 \geq p$ observations $y_1, \ldots, y_{n_0}$. Then, sample term by term, stopping according to the stopping variable N defined by

$$N = \text{smallest } k \geq n_0 \quad \text{such that} \quad k^{-1}(\hat{\sigma}^2(k) + k^{-1}) \leq d^2/u_k; \tag{4.15.14}$$

when sampling is stopped at $N = n$, construct the region R_n described in (4.15.3).

Then the procedures in the class $\mathscr{C}$ are asymptotically consistent and efficient, as $d \to 0$.

Theorem 4.15.3 (Gleser, 1965) Under the assumption that $0 < \sigma^2 < \infty$

$$\lim_{d \to 0} (d^2 N)/(u\sigma^2) = 1 \quad \text{a.s.}, \tag{4.15.15}$$

$$\lim_{d \to 0} P(\beta \in R_N) = 1 - \alpha, \tag{4.15.16}$$

and

$$\lim_{d \to 0} (d^2 EN)/(u\sigma^2) = 1. \tag{4.15.17}$$

Proof Since $\hat{\sigma}^2(n)$ converges to σ^2 a.s., and since $\hat{\sigma}^2(n) + n^{-1}$ is a.s. positive, Lemma 4.14.5 [which is Chow and Robbins (1965, Lemma 1)] implies (4.15.15). Further, Lemma 4.14.7 [see Lemma 3 and the discussion thereof in Chow and Robbins (1965)] applies to $\hat{\sigma}^2(n) + n^{-1}$ as well, and thus (4.15.17) follows. It only remains to prove (4.15.16). Since

$$\begin{aligned} P(\beta \in R_N) &= P((\hat{\beta}(N) - \beta)'(\hat{\beta}(N) - \beta) \leq d^2) \\ &= P\{N(\hat{\beta}(N) - \beta)'(\hat{\beta}(N) - \beta)/\sigma^2 < Nd^2/\sigma^2\}, \end{aligned}$$

and since $Nd^2/\sigma^2 \to u$ a.s. and since by (4.15.11)

$$\lim_{n \to \infty} P(n(\hat{\beta}(n) - \beta)'(\hat{\beta}(n) - \beta)/\sigma^2 \leq u) = 1 - \alpha,$$

it follows as a trivial extension (to the distribution of $T(\lambda_1, \ldots, \lambda_p)$) of a result of Anscombe (1952) that (4.15.16) holds.

REMARK 4.15.2 The addition of n^{-1} to $\hat{\sigma}^2(n)$ in (4.15.14) is unnecessary if F is continuous. N could be defined as the smallest odd, even, etc., integer $\geq n_0$ such that (4.15.14) holds and Theorem 4.15.3 will be still valid. Very little is known about the properties of any member of the class $\mathscr{C}$ for moderate values of σ^2/d^2.

Gleser (1966) points out a number of corrections to his earlier paper and provides further justification for his assertions: For example, it was not clear why $n^{-1}Z_n'U_n'U_nZ_n = o(1)$, since U_n does not tend to zero. In order to show that we need the following lemma of Chow (1966b).

Lemma 4.15.1 (Chow, 1966b). Let $z_1, z_2, \ldots$ be iid random variables with means $Ez_i = 0$ and variance $Ez_i^2 = \sigma^2$ $(0 < \sigma^2 < \infty)$. Let b_{nm} be any array of real numbers $m \le n$, $n = 1, 2, \ldots$ satisfying $\lim_{n\to\infty} \sum_{m=1}^{n} b_{nm}^2 = 1$. Then

$$n^{-1/2} \sum_{m=1}^{n} b_{nm} z_m \to 0 \quad \text{a.s.} \qquad \text{as} \quad n \to \infty.$$

Proof of Theorem 4.15.2 (continued) Let $U_n = ((u_{nij}))$, $Z_n = (z_1, \ldots, z_n)'$; we have

$$n^{-1}Z_n'U_n'U_nZ_n = \sum_{i=1}^{p} \left(n^{-1/2} \sum_{j=1}^{n} u_{nij} z_j\right)^2 .$$

Now since $U_n = (X_n X_n')^{1/2} X_n$, it follows that $U_n U_n' = I_p$ or $\sum_{j=1}^{n} u_{nij}^2 = 1$ for all i and all n. Applying Lemma 4.15.1 we obtain

$$n^{-1/2} \sum_{j=1}^{n} u_{nij} z_j \to 0 \quad \text{a.s.} \qquad (i = 1, \ldots, p)$$

which completes the proof of Theorem 4.15.2.

The proof of the assertion (4.15.16) is still incomplete. Toward this consider the following lemma.

Lemma 4.15.2 Let $z_1, z_2, \ldots$ be iid with zero means and unit variances. Let $a_1, a_2, \ldots$ be any sequence of numbers such that $n^{-1}\sum_{i=1}^{n} a_i^2 \to 1$ and $n^{-1} \max_{1\le i\le n} a_i^2 \to 0$ as $n \to \infty$. Define $v_n = n^{-1/2} \sum_{i=1}^{n} a_i z_i$ and $N = N(t)$ be a positive integer valued random variable such that $N(t)/t$ tends to 1 in probability as $t \to \infty$. Then v_N has a limiting standard normal distribution as $t \to \infty$.

Proof This lemma is an immediate corollary to Theorem 2 of Mogyoródi (1961).

Proof of Theorem 4.15.3 (continued) From assumptions (4.15.1) and (4.15.2) and Lemma 4.15.2 we infer that $\lim_{d\to 0} \{N^{-1/2}c'X_N Z_N\}$ tends in distribution to a normal variable having zero mean and $c'\Sigma c$ for the variance, for all $1 \times p$ vectors c', $c'c = 1$. Thus $\lim_{d\to 0}\{N^{-1/2}X_N Z_N\}$ is multivariate normal with mean 0 and Σ for the covariance matrix, which in turn implies that (4.15.16) holds.

Gleser's assumptions (4.15.1) and (4.15.2) have been found to be too strong and they have been weakened by Albert (1966) and Srivastava (1967b, 1971). Also, the latter authors obtain spherical confidence regions for the regression parameters. Consider the following example.

Example 4.15.1 Let $y_i = \alpha + \beta i + \varepsilon_i$, $i = 1, 2, \ldots, n$, where ε_i are iid having mean 0 and variance σ^2, so $p = 2$ and $X_n = \begin{pmatrix} 1 & 1 & \cdots & 1 \\ 1 & 2 & \cdots & n \end{pmatrix}$. Hence

$$X_n X_n' = \begin{pmatrix} n & n(n+1)/2 \\ n(n+1)/2 & n(n+1)(2n+1)/6 \end{pmatrix}.$$

It is clear that $n^{-1}(X_n X_n') \neq$ a positive definite matrix as $n \to \infty$. Thus assumption (4.15.1) is not satisfied. The characteristic roots of $X_n X_n'$ are given by

$$\lambda = \{n(2n^2 + 3n + 7) \pm n[(2n^2 + 3n + 7)^2 - 12(n+1)(n-1)]^{1/2}\}/12.$$

Hence $\lambda_{\max}(X_n X_n')/\lambda_{\min}(X_n X_n') \to \infty$. Hence one of the assumptions of Albert (1966) is not satisfied (the limit should go to unity). Srivastava (1971) has given alternative sufficient conditions that are weaker than Albert's (1966) and Gleser's (1965). It is known [see, for example, Roy (1957)] that X_n can be written as $X_n = T_n L_n$ where T_n is a $p \times p$ triangular matrix with positive diagonal elements (hence unique); and L_n is a $p \times n$ semiorthogonal matrix; $L_n L_n' = I_p$ and I_p is a $p \times p$ identity matrix. Hence

$$T_n^{-1} X_n = L_n = (l_n^{(1)}, \ldots, l_n^{(n)}) = ((l_{ij}(n))).$$

Let $\lambda_n = \lambda_{\min}(X_n X_n')$. Srivastava (1971) has shown that the basic result of Gleser (namely Theorem 4.15.3) holds under the weaker conditions:

(i) $\lim_{n\to\infty} \max_{1 \le i \le n} l_n^{(i)\prime} l_n^{(i)} = 0$;

(ii) $\lim_{c\to 0} \sup_n \| \sum_{i=n+1}^{n^*} l_{n^*}^{(i)} l_{n^*}^{(i)\prime} \| = 0$,

where $n^* = [n(1+c)] + 1$ where $[\cdot]$ denotes here the largest integer in $(\cdot)$, and $\|B\| = [\lambda_{\max}(B)]^{1/2}$;

(iii) For some $c \ge 1$, $\lambda_{c(n+2)} - \lambda_{c(n+1)} \ge \lambda_{c(n+1)} - \lambda_{cn}$.

Proof See Srivastava (1971).

Example Let

$$X_n = \begin{pmatrix} 0 & 1 & 0 & 2 & 0 & 3 & 0 & 4, \ldots \\ 1 & 0 & 2^{3/2} & 0 & 3^{3/2} & 0 & 4^{3/2} & 0, \ldots \end{pmatrix}.$$

This also satisfies (i)–(iii).

Srivastava (1971) gives the following sequential rule in order to construct confidence regions whose limiting coverage probability is $1 - \alpha$ and the maximum diameter of the region is almost $2d$.

PROCEDURE Start taking $n_0 \geq p$ observations $y_1, y_2, \ldots, y_{n_0}$. Then sample observations one at a time and stop when

$$N = \text{smallest } k \geq n_0 \quad \text{such that} \quad (\hat{\sigma}^2(k) + k^{-1}) \leq d^2\lambda_k/u_k$$

where λ_n is the smallest characteristic root of $(X_n X_n')$, $P(\chi_p{}^2 < u) = 1 - \alpha$, $u_n \to u$. When sampling is stopped at $N = n$, construct the region R_n defined by

$$R_n = \{z : (z - \hat{\beta}(n))'(X_n X_n')(z - \hat{\beta}(n)) \leq d^2\lambda_n\}.$$

The spheroid region R_n' can be similarly obtained as in the work of Srivastava (1967b).

Callahan (1969b) has considered sequential procedures for finding confidence regions for regression parameters $\beta_1, \ldots, \beta_p$, such regions having a specified "size" d and average probability at least $1 - \alpha$. By "size" we mean that the confidence regions are required to intercept each β_i axis in an interval of width no greater than $2d$. Four sequential procedures are studied, each consisting of a stopping variable N and a corresponding terminal confidence region, the regions being (i) elliptical, (ii) spherical, and (iii) rectangular. A general stopping rule, of which the above four rules are special cases, is formulated, and the properties of such a rule are discussed.

4.16 Confidence Sequences for Other Parameters

4.16.1 Confidence Sequences for Mean, Variance, and Median

Let $X_1, X_2, \ldots$ be independent and identically distributed (iid) random variables and put $C(t) = E(e^{tX})$, $\overline{X}_n = \sum_{i=1}^{n} X_i/n$. In earlier papers Darling and Robbins (1967b, c) have proved the following lemma (which is also given as Theorem 3.13.3).

Lemma 4.16.1 (Darling and Robbins, 1967a) Assume that $C(t) \leq \exp(t^2/2)$ for all $t \geq 0$. Let $\gamma > 1$, $c > 1$ be arbitrary constants and set (with $(\ln)^2 n = \ln \ln n$)

$$g_n(A) = \left(\frac{1 + \gamma}{2\gamma^{1/2}}\right)\{(2c(\ln)^2 n - 2c(\ln)^2\gamma + 2 \ln A)/n\}^{1/2},$$

and

$$Q_m = 1\Big/\left\{(c-1)\left(\frac{\ln m}{\ln \gamma} - \frac{1}{2}\right)^{c-1}\right\}.$$

Then for any $m \geq \gamma$, $A \geq 1$

$$P(\overline{X}_n \geq g_n(A) \text{ for some } n \geq m) < Q_m/A. \tag{4.16.1}$$

Example 4.16.1 (Confidence interval for mean) Let $Y_1, Y_2, \ldots$ be independent normal with mean μ and variance σ^2. Applying Lemma 4.16.1 to $X_i = \pm (Y_i - \mu)/\sigma$ we obtain

$$P(|\overline{Y}_n - \mu| \geq \sigma g_n(A) \text{ for some } n \geq m) < 2Q_m/A. \tag{4.16.2}$$

For any $n \geq m$, define the intervals

$$I_n = (\overline{Y}_n - \sigma g_n(A), \overline{Y}_n + \sigma g_n(A)), \; J_n = \bigcap_{i=m}^{n} I_i.$$

Then from (4.16.2) we have

$$P(\mu \in J_n \text{ for every } n \geq m) > 1 - 2Q_m/A. \tag{4.16.3}$$

The nested sequence of intervals $\{J_n\}$, $n \geq m$, is such that as $n \to \infty$ the length of $J_n \to 0$, and the probability that μ belongs to the intersection of the J_n exceeds $1 - 2Q_m/A$ which for any fixed $m \geq \gamma$ can be made arbitrarily near 1 by choosing A sufficiently large. We may call $\{J_n\}$ a "confidence sequence" for an unknown normal μ when σ is known.

Example 4.16.2 (Confidence interval for the mean and variance) For the Y_i defined in Example 4.16.1 let $s_n{}^2$ denote the sample variance and $\overline{V}_n = n^{-1}\sum_i^n V_i$. Then $s_{n+1}^2 = \sigma^2 \overline{V}_n$ where the V_i are squares of standard normal variables. Setting $X_i = (1 - V_i)/2^{1/2}$, we have for all $t \geq 0$ $C(t) = \exp(t/2^{1/2})$. $(1 + t2^{1/2})^{-1/2} \leq \exp(t^2/2)$. Hence, by Lemma 4.16.1

$$\begin{aligned} &P(s_n{}^2 \leq \sigma^2(1 - 2^{1/2} g_{n-1}(A)) \text{ for some } n \geq m+1) \\ &\quad = P(\overline{X}_n \geq g_n(A) \text{ for some } n \geq m) < Q_m/A. \end{aligned} \tag{4.16.4}$$

Define for $n > m$

$$w_n = \min_{m < i \leq n} s_i(1 - 2^{1/2}\, g_{i-1}(A))^{-1/2},$$

assuming that m is so large that $g_m(A) < 2^{-1/2}$. Then from (4.16.4)

$$P(\sigma < w_n \text{ for every } n \geq m + 1) > 1 - Q_m/A, \tag{4.16.5}$$

which provides a one-sided confidence sequence for an unknown normal σ when μ is unknown. Finally if we put for $n > m$

$$J_n' = \sum_{i=m+1}^{n} (\bar{Y}_i - w_n g_i(A),\ \bar{Y}_i + w_n g_i(A))$$

then from (4.16.2) and (4.16.5) we have in analogy with (4.16.3)

$$P(\mu \in J_n' \text{ for every } n > m) > 1 - 3Q_m/A, \tag{4.16.6}$$

a confidence sequence for an unknown normal μ when σ is also unknown.

Example 4.16.3 (Confidence interval for the median) Let $Z_1, Z_2, \ldots$ be iid with $P(Z_i \leq \xi) = \frac{1}{2}$. Put $X_i = 1\ (-1)$ according as $Z_i \leq \xi\ (>\xi)$. Then

$$C(t) = E(\exp(tX_i)) = (e^t + e^{-t})/2 \leq \exp(t^2/2) \qquad \text{for all } t. \tag{4.16.7}$$

Hence, by Lemma 4.16.1 we have

$$P(|\bar{X}_n| \geq g_n(A) \text{ for some } n \geq m) < 2Q_m/A. \tag{4.16.8}$$

Define

$$r = r(n) = \text{largest integer} \leq n(1 - g_n(A))/2$$
$$s = s(n) = \text{smallest integer} \geq n(1 + g_n(A))/2,$$

and let $Z_{1n} \leq Z_{2n} \leq \cdots \leq Z_{nn}$ be the ordered values $Z_1, \ldots, Z_n$. Then by (4.16.8)

$$P(Z_{rn} \leq \xi < Z_{sn} \text{ for all } n \geq m) = P[(2r/n) - 1 \leq \bar{X}_n \leq (2s/n) - 1 \text{ for all } n \geq m]$$
$$\geq P(|\bar{X}_n| < g_n(A) \text{ for all } n \geq m) > 1 - 2Q_m/A.$$

For example, if $\gamma = c = 2$, $m = 32$, $A = 80/9$ we have

$$g_n(A) = 3\{[(\ln)^2 n + 1.457]/2n\}^{1/2}, \qquad 1 - 2Q_m(A) = 0.95.$$

The intervals $I_n = [Z_{rn}, Z_{sn}]$ for $n \geq 32$ form *a* (more than) 95 percent confidence sequence for an unknown ξ as do the nested intervals $J_n = \bigcup_{i=32}^{n} I_i$.

The above method can readily be generalized to the estimation of any quantile.

REMARK 4.16.1 There is a discussion of how to choose the constants γ and c optimally. Perhaps it is simpler to set them each equal to 2. The choice of m is dictated by the problem. Then the confidence coefficient $1-\alpha$ determines the constant A. Notice that there is no stopping variable here. The confidence coefficient of the above confidence sequences can be made arbitrarily close to 1 by choosing A arbitrarily large. For extended ideas on this, see Robbins (1970).

4.16.2 Fixed Width Confidence Interval for the Difference of Means of Two Normal Populations

Let $X_1, X_2, \ldots$ and $Y_1, Y_2, \ldots$ be two independent sequences of rvs the Xs being iid normal with mean μ_1 and variance $\sigma_1{}^2$ and the Ys being normal with mean μ_2 and variance $\sigma_2{}^2$. The Xs and Ys are mutually independent. The four parameters $\mu_1, \mu_2, \sigma_1, \sigma_2$ are assumed to be unknown. We wish to find a confidence interval I of width $2d$ and with coverage probability $\geq 1-\alpha$ for the parameter $\Delta = \mu_1 - \mu_2$ where $0 < d < \infty$ and $0 < \alpha < 1$ are preassigned constants.

If σ_1 and σ_2 were known, we could proceed as follows. Take r observations on X and t observations on Y, and let

$$\overline{X}_r = \sum_{i=1}^{r} X_i/r, \qquad \overline{Y}_r = \sum_{j=1}^{t} Y_j/t$$

be the respective means. If

$$I = [\overline{X}_r - \overline{Y}_t - d, \overline{X}_r - \overline{Y}_t + d] \tag{4.16.9}$$

is the interval of width $2d$ centered at $\overline{X}_r - \overline{Y}_t$, then

$$P(\Delta \in I) = 2\Phi(d/(\sigma_1{}^2/r + \sigma_2{}^2/t)^{1/2}) - 1.$$

Hence, defining the constants u and v by $2\Phi(u) - 1 = 1 - \alpha$, $v = (u/d)^2$, we have $P(\Delta \in I) \geq 1 - \alpha$ provided that r and t satisfy the inequality

$$\sigma_1{}^2/r + \sigma_2{}^2/t \leq 1/v. \tag{4.16.10}$$

Regarding r and t as continuous variables, the pair (r^*, t^*) which satisfies (4.16.10) and for which $n = r + t$ is a minimum is given by

$$r^* = v\sigma_1(\sigma_1 + \sigma_2), \qquad t^* = v\sigma_2(\sigma_1 + \sigma_2). \tag{4.16.11}$$

For this pair,

$$r^*/t^* = \sigma_1/\sigma_2 \tag{4.16.12}$$

and the total sample size is

$$n^* = r^* + t^* = v(\sigma_1 + \sigma_2)^2. \tag{4.16.13}$$

THE SEQUENTIAL PROCEDURE WHEN σ_1, σ_2 ARE UNKNOWN

Robbins *et al.* (1967) have given a sequential procedure for determining r and t as random variables in such a way that (4.16.11) will hold approximately with high probability. The procedure consists of (i) a *sampling scheme* which tells us at each stage whether to take the next observation on X or Y, and (ii) a *stopping rule* which determines r and t and therefore I by (4.16.9). Let

$$s_{1i}^2 = (i-1)^{-1}\sum_{k=1}^{i}(X_k - \overline{X}_i)^2, \qquad s_{2j}^2 = (j-1)^{-1}\sum_{k=1}^{j}(Y_k - \overline{Y}_j)^2$$

be the usual estimates of ${\sigma_1}^2$ and ${\sigma_2}^2$ for which $s_{1i} \to \sigma_1$, $s_{2j} \to \sigma_2$ almost surely (a.s.) as $i, j \to \infty$. We take $n_0 \geq 2$ observations on X and on Y to begin with. Then if at any stage we have taken i observations on X and j on Y, with $n = i + j \geq 2n_0$, we take the next observation on X or Y according as $i/j \leq s_{1i}/s_{2j}$ or $i/j > s_{1i}/s_{2j}$. This procedure generates an infinite sequence of observations and does not depend on the value of α or d. Later on it will be shown that $i/j \to \sigma_1/\sigma_2$ a.s. as $n \to \infty$.

We now give three more or less equivalent stopping rules. Let $\{u_n\}$ be a given sequence of positive constants such that $u_n \to u$ as $n \to \infty$ and set $v_n = (u_n/d)^2$.

$\mathcal{R}_1$: Stop with the first $n \geq 2n_0$ such that, if r observations on X and t observations on Y have been taken, with $r + t = n$

$$n \geq v_n(s_{1r} + s_{2t})^2 \qquad \text{[cf (4.16.13)]}. \tag{4.16.14}$$

$\mathcal{R}_2$: The same with (4.16.14) replaced by

$$s_{1r}^2/r + s_{2t}^2/t \leq 1/v_n \qquad \text{[cf (4.16.10)]}. \tag{4.16.15}$$

$\mathcal{R}_3$: The same, with (4.16.14) replaced by

$$r \geq v_n s_{1r}(s_{1r} + s_{2t}) \quad \text{and} \quad s \geq v_n s_{2t}(s_{1r} + s_{2t}) \qquad \text{[cf (4.16.11)]}. \tag{4.16.16}$$

It is easy to check that the sample sizes n_k determined by these three rules $\mathcal{R}_k$ are such that $n_1 \leq n_2 \leq n_3$.

Let $\mu_1, \mu_2, \sigma_1, \sigma_2, \alpha, n_0, \{u_n\}$ be fixed and let $d \to 0$, so that $n^* \to \infty$ where n^* is the optimal fixed-sample size determined by (4.16.13). Denote by $n = r + t$ the sample size determined by any one of the stopping rules $\mathscr{R}_k$, $k = 1, 2, 3$, and let I be the interval (4.16.9) for r, t. We need the following nonstochastic lemma in order to prove the asymptotic optimality of the sequential procedure.

Lemma 4.16.2 (Robbins, *et al.*, 1967) Given constants c_{ij} $(i, j = 1, 2, \ldots)$ such that $0 < c_{ij} \to c > 0$ as $i, j \to \infty$ and any integer $n_0 \geq 1$, define $i(2n_0) = j(2n_0) = n_0$ and for $n \geq 2n_0$ let

(I) $i(n+1) = i(n) + 1, j(n+1) = j(n)$ if $i(n)/j(n) \leq c_{i(n), j(n)}$;

(II) $i(n+1) = i(n), j(n+1) = j(n) + 1$ if $i(n)/j(n) > c_{i(n), j(n)}$.

Then $i(n)/j(n) \to c$ as $n \to \infty$.

Proof Clearly $i(n) \to \infty$ and $j(n) \to \infty$ as $n \to \infty$. Call an integer $n \geq 2n_0$ of type I or II according as the right-hand side of I or II holds for n. Then, for all sufficiently large n there exists a largest integer n', $2n_0 \leq n' < n$ of type I, and a largest integer n'', $2n_0 \leq n'' < n$, of type II, and $n', n'' \to \infty$ as $n \to \infty$. Then

$$i(n)/j(n) \leq (i(n') + 1)/j(n') \leq c_{i(n'), j(n')} + 1/j(n') \to c,$$

$$\begin{aligned} i(n)/j(n) \geq i(n'')/(j(n'') + 1) &= \{i(n'')/j(n'')\}\{1 - 1/(j(n'') + 1)\}^{-1} \\ &\geq c_{i(n''), j(n'')}\{1 - 1/(j(n'') + 1)\} \to c, \end{aligned}$$

and this completes the proof of the lemma.

In particular, we can put $c_{ij} = u_i/v_j$ and obtain that $i(n)/j(n) \to \sigma_1/\sigma_2$ a.s. as $n \to \infty$. For any of the stopping rules, since $n \to \infty$, a.s. as $d \to \infty$ it follows that as $d \to 0$

$$r/t \to \sigma_1/\sigma_2 \quad \text{a.s.} \tag{4.16.17}$$

Theorem 4.16.1 (Robbins *et al.*, 1967) As $d \to 0$

$$n/v(\sigma_1 + \sigma_2)^2 \to 1 \quad \text{a.s.} \qquad En/v(\sigma_1 + \sigma_2)^2 \to 1, \tag{4.16.18}$$

and

$$P(\Delta \in I) \to 1 - \alpha. \tag{4.16.19}$$

Proof We shall prove (4.16.18) for $\mathscr{R} = \mathscr{R}_3$ where we stop sampling with the first $n = r + t \geq 2n_0$ such that

$$r \geq v_n s_{1r}(s_{1r} + s_{2t}) \qquad \text{and} \qquad t \geq v_n s_{2t}(s_{1r} + s_{2t}) \tag{4.16.20}$$

where $v_n = (u_n/d)^2$, $v = (u/d)^2$. Suppose that $r > n_0$ and that just before the rth observation on X there were $r - 1$ observations on X and j observations on Y. Then by the sampling described in the sequential procedure,

$$(r-1)/j \leq s_{1,r-1}/s_{2j}$$

and this implies that $(r-1) \leq v_{r-1+j} s_{1,r-1}(s_{1,r-1} + s_{2j})$ for otherwise we would have $j \geq (r-1)s_{2j}/s_{1,r-1} > v_{r-1+j} s_{2j}(s_{1,r-1} + s_{2j})$, and sampling would have stopped at the $(r-1, j)$th stage. Hence even if $r = n_0$

$$r \leq v_{r-1+j} s_{1,r-1}(s_{1,r-1} + s_{2j}) + n_0 . \tag{4.16.21}$$

As $d \to 0$, $r \to \infty$, $r/j \to \sigma_1/\sigma_2$ so $j \to \infty$ a.s., and hence $\limsup r/v \leq \sigma_1(\sigma_1 + \sigma_2)$ a.s. The reverse inequality for the lim inf is obvious from (4.16.20), so

$$\lim_{d\to 0} r/v = \sigma_1(\sigma_1 + \sigma_2), \qquad \lim_{d\to 0} t/v = \sigma_2(\sigma_1 + \sigma_2) \quad \text{a.s.}, \tag{4.16.22}$$

which implies the first part of (4.16.18). The second part of (4.16.18) will follow from the dominated convergence theorem if we can bound the random variable r/v by an integrable random variable which is independent of d. Let $s_1 = \sup_{i \geq 2} s_{1i}$, $s_2 = \sup_{j \geq 2} s_{2j}$; then $s_1^{\,2} \leq 2 \sup_{i \geq 2}\{(1/i)\sum_{k=1}^{i}(X_k - \mu_1)^2\}$ and the right-hand side is integrable [by Wiener's (1939) dominated ergodic theorem] since the fourth moment of X is finite. Thus, $Es_1^{\,2} < \infty$ and similarly $Es_2^{\,2} < \infty$. However, from (4.16.21) for $d \leq u, r/v \leq Cs_1(s_1 + s_2) + n_0$ so we can apply the dominated convergence theorem to conclude from (4.16.22) that $Er < \infty$, $Et < \infty$ and

$$\lim_{d\to 0} Er/v = \sigma_1(\sigma_1 + \sigma_2), \qquad \lim_{d\to 0} Et/v = \sigma_2(\sigma_1 + \sigma_2) \tag{4.16.23}$$

and consequently the second part of (4.16.18) follows. Now (4.16.19) follows from (4.16.22) by a simple extension of the proof of Rényi's (1957) theorem to the case of two populations. This completes the proof of the theorem for $\mathscr{R} = \mathscr{R}_3$. Because $n_1 \leq n_2 \leq n_3$, the other cases also follow.

REMARK 4.16.2 It should be noted that Theorem 4.16.1 remains valid for nonnormal populations having finite fourth moments. The latter condition may also be relaxed as will be shown later.

SMALL SAMPLE BEHAVIOR

Robbins *et al.* (1967) have carried out Monte Carlo studies in order to find approximately the values of $P(\Delta \in I)$ and En for finite values of n^* using

$\alpha = .05$, $u = 1,.96$, $n_0 = 5$, $u_n{}^2 = [(n+4)/(n-4)](1.96)^2$ for which (4.16.13) becomes

$$n^* = [1.96\lambda(\Lambda + 1)]^2$$

where $\Lambda = \sigma_1/\sigma_2$, $\lambda = \sigma_2/d$. Values $\Lambda = \frac{1}{2}, \frac{1}{4}, \frac{1}{8}, \frac{1}{16}$ and $n^* = 10, 20, \ldots, 200$ were used. 2000 sequences of X and Y were generated for the desired value of En and $P(\Delta \in I)$ using rule $\mathscr{R}_k$ $(k = 1, 2, 3)$. They draw the following conclusions.

(i) Although (4.16.18) merely asserts that $E(n) = n^* + O(n)$, the Monte Carlo studies suggest that the difference $En - n^*$ may be bounded for all $d > 0$. That this is so will be shown in the following.

(ii) As is to be expected, for values of $\Lambda < \frac{1}{8}$, the rule $\mathscr{R}_3$ is somewhat more successful in keeping the coverage frequency $\geq .95$ for small values of n^*.

(iii) The authors conjecture that the minimum coverage probability for all n^* is attained in the computed range $n^* = 10, \ldots, 200$.

(iv) To test $H_0\colon \mu_1 = \mu_2$ so as to generate a type I error $\leq \alpha$ and type II error $\leq \alpha/2$ for alternatives such that $|\mu_1 - \mu_2| \geq 2d$, irrespective of σ_1 and σ_2, we may reject H_0 if and only if $|\bar{X}_r - \bar{Y}_t| > d$ [see Robbins and Starr (1965) for the corresponding one-population test of $H_0\colon \mu = 0$].

The Cost of Ignorance

Theorem 4.16.2 (Robbins *et al.*, 1967) Let $u_n = u + O(n^{-1})$ as $n \to \infty$. Then for $\mathscr{R}_3$ (and a fortiori for $\mathscr{R}_1$ and $\mathscr{R}_2$), $En - n^*$ is less than some finite constant for all $d > 0$.

Proof For some $0 \leq M < \infty$, we have

$$v_n = (u_n/d)^2 \leq v(1 + Mn^{-1}) \qquad \text{where} \quad v = (u/d)^2.$$

Define $S_{1i}^2 = (i-1)s_{1i}^2$, $S_{2j}^2 = (j-1)s_{2j}^2$. We have $S_{1,i+1}^2 \geq S_{1i}^2$, $S_{2,j+1}^2 \geq S_{2j}^2$, and

$$ES_{1r}^2 \leq E\{\textstyle\sum_{k=1}^{r} (X_k - \mu_1)^2\} = \sigma_1{}^2 Er, \qquad ES_{2t}^2 \leq \sigma_2{}^2 Et.$$

Finally, the function

$$g(r, n) = (r - n_0)^2 r^2 (r + M)^{-1} n^{-1}$$

is convex in r and n for $r \geq n_0$, $n > 0$, $M \geq 0$. As in the proof of (4.16.18) suppose that $r > n_0$ and that just before the rth observation on X, there were j observations on Y. Then (as before)

$$(r-1)/j \leq s_{1,r-1}/s_{2j}, \qquad (r-1) \leq v_{r-1+j} s_{1,r-1}(s_{1,r-1} + s_{2j})$$

and hence

$$r - 1 \leq v(1 + Mn^{-1})s_{1,r-1}\{s_{1,r+1} + j(r-1)^{-1}s_{1,r-1}\}$$
$$\leq v(1 + Mr^{-1})S_{1r}^2(n-1)(r-1)^{-1}(r-2)^{-1}.$$

Since $n_0 \geq 2$ this implies that

$$(r - n_0)^2 r^2/(r + M)n \leq vS_{1r}^2$$

and this holds even if $r = n_0$. By Jensen's inequality

$$E^2(r - n_0)^2 E^2 r/(Er + M)En \leq vES_{1r}^2 \leq v\sigma_1{}^2 Er.$$

Hence from (4.16.18) and (4.16.23)

$$E^2(r - n_0) \leq v\sigma_1{}^2\{En + MEn/Er\} \leq v\sigma_1{}^2\{En + O(1)\}. \quad (4.16.24)$$

Similarly,

$$E^2(t - n_0) \leq v\sigma_2{}^2\{En + O(1)\}. \quad (4.16.25)$$

Since $n - 2n_0 = (r - n_0) + (t - n_0)$ it follows that

$$E^2(n - 2n_0) \leq v(\sigma_1 + \sigma_2)^2\{En + O(1)\}$$

and hence $En \leq v(\sigma_1 + \sigma_2)^2 + O(1) = n^* + O(1)$, which was to be proved. From this and (4.16.24) and (4.16.25) we also obtain that for all $d > 0$

$$Er \leq r^* + O(1), \qquad Et \leq t^* + O(1). \quad (4.16.26)$$

REMARK 4.16.3 These results hold whenever the distributions of X and Y are such that $\sigma_1{}^2 < \infty$, $\sigma_2{}^2 < \infty$, and $En < \infty$ for all $d > 0$; as noted, finite fourth moments for X and Y will guarantee this. In fact by truncating n and using a modification of the preceding argument one can prove that the second part of (4.16.18) holds, and hence $En < \infty$, under the sole condition that $\sigma_1, \sigma_2 < \infty$. Consequently Theorems 4.16.1 and 4.16.2 are valid when $\sigma_1, \sigma_2 < \infty$.

Quite small upper bounds for the constants in (4.16.26) can be obtained in the case $u_n = u$ if they depend only on n_0. If $u_n = u + O(n^{-1})$, useful estimates of these constants depending on n_0, μ, σ_1, σ_2 could perhaps be found by improving the above argument.

Paulson (1969) also has considered sequential interval estimation for the difference of two normal means. In the following we shall present his results.

Lemma 4.16.3 (Paulson, 1969) Let $X_1, X_2, \ldots$ be a sequence of independent normal variables having mean θ and variance σ^2. Let $z(p)$ denote the $(1-p)$th fractile of the standard normal distribution. Let t_1 be the smallest integer $\geq \sigma^2[z(\alpha/4)]^2/d^2$, and let $A_1 = dt_1$. Then

$$P\left(-A_1 \leq \sum_{i=1}^{n}(X_i - \theta) \leq A_1 \text{ for all } n,\ 1 \leq n \leq t_1\right) \geq 1 - \alpha. \quad (4.16.27)$$

Proof By considering the complement of the event in (4.16.27) we have

L.H.S. of (4.16.27)

$$\geq 1 - 2P\left(\sum_{i=1}^{n}(X_i - \theta) > A_1 \text{ for at least one } n,\ 1 \leq n \leq t_1\right).$$

Now, if $X(t)$ denotes a Wiener process such that $X(0) = 0$ and $X(t)$ has zero mean and variance $= \sigma^2 t$, it is obvious that

$$\begin{aligned} P&\left(\sum_{i=1}^{n}(X_i - \theta) > A_1 \text{ for at least one } n,\ 1 \leq n \leq t_1\right) \\ &\leq P(X(t) > A_1 \text{ for at least one } t,\ 0 < t < t_1) \qquad (4.16.28) \\ &= 2[1 - \Phi(A_1/\sigma t_1^{1/2})] \end{aligned}$$

by a standard result [see, for instance, Cox and Miller (1965, p. 221)]. Hence, the left-hand side expression in (4.15.27) is

$$\geq 1 - 4[1 - \Phi(A_1/\sigma t_1^{1/2})] \geq 1 - 4[1 - \Phi(z(\alpha/4))] = 1 - \alpha,$$

This completes the proof of the lemma. If we now define $\mathscr{U}_n^{(1)}$ and $\mathscr{L}_n^{(1)}$ by

$$\mathscr{U}_n^{(1)} = \min_{1 \leq r \leq n} [\bar{X}_r + A_1/r] \quad \text{and} \quad \mathscr{L}_n^{(1)} = \max_{1 \leq r \leq n} [\bar{X}_r - A_1/r]$$

then it follows from (4.16.27) and the definitions of A_1 and t_1 that

$$\mathscr{U}_{t_1}^{(1)} - \mathscr{L}_{t_1}^{(1)} \leq 2d$$

and

$$P(\mathscr{L}_n^{(1)} \leq \theta \leq \mathscr{U}_n^{(1)} \text{ for all } n,\ 1 \leq n \leq t_1) \geq 1 - \alpha,$$

Now, if σ is unknown, let s^2 be an estimate of σ^2 with v degrees of freedom such that s is independent of $\sum_{i=1}^{n_0} X_i$ and vs^2/σ^2 has the chi-square distribution with v degrees of freedom. Let $t(p, v)$ denote the $(1-p)$th fractile of the Student's distribution with v degrees of freedom. Let t_2 denote the smallest integer $\geq \max\{n_0, s^2[t(\alpha/4, v)]^2/d^2\}$, and let $A_2 = dt_2$. Then, we have the following lemma.

Lemma 4.16.4 (Paulson, 1969) With the above notation we have

$$P\left(-A_2 \leq \sum_{i=1}^{n} (X_i - \theta) \leq A_2 \text{ for all } n,\ n_0 \leq n \leq t_2\right) \geq 1 - \alpha. \tag{4.16.29}$$

Proof Since $\sum_{i=1}^{n} X_i$ was assumed to be independent of s^2, it follows from (4.16.28) that

$$P\left(-A_2 \leq \sum_{i=1}^{n} (X_i - \theta) \leq A_2 \text{ for all } n,\ n_0 \leq n \leq t_2 \,|\, s^2\right)$$
$$\geq 1 - 4\{1 - \Phi(s/\sigma)t(\alpha/4, v))\}.$$

Taking expectations, we obtain for the unconditional probability that the left-hand side of (4.16.29) exceeds

$$1 - 4\{1 - E\Phi[(s/\sigma)t(\alpha/4, v)]\} = 1 - 4P[T_v > t(\alpha/4, v)] = 1 - \alpha,$$

where the random variable T_v is distributed as Student's t with v degrees of freedom. [For values of α or v for which $t(\alpha/4, v)$ is not tabulated, the approximation $t(p, v) = z(p)\{2v/(2v - 1 - z^2(p))\}^{1/2}$ given by Elfving (1961) should ordinarily be accurate enough for most applications.]

If $\mathscr{L}_n^{(2)}$ and $\mathscr{U}_n^{(2)}$ for $n = n_0, n_0 + 1, \ldots, t_2$ are defined as

$$\mathscr{U}_n^{(2)} = \min_{n_0 \leq r \leq n} [\bar{X}_r + A_2/r] \quad \text{and} \quad \mathscr{L}_n^{(2)} = \max_{n_0 \leq r \leq n} [\bar{X}_r - A_2/r],$$

then $\mathscr{U}_{t_2}^{(2)} - \mathscr{L}_{t_2}^{(2)} \leq 2d$ and from (4.16.29) we have

$$P(\mathscr{L}_n^{(2)} \leq \theta \leq \mathscr{U}_n^{(2)} \text{ for all } n,\ n_0 \leq n \leq t_2) \geq 1 - \alpha. \tag{4.16.30}$$

PAULSON'S SEQUENTIAL PROCEDURE FOR INTERVAL ESTIMATION FOR THE DIFFERENCE OF TWO MEANS Let X_{1r} and X_{2r} denote the measurements taken at the rth stage of the experiment ($r = 1, 2, \ldots$) with categories Π_1 and Π_2. Assume that X_{ir} is normal with mean μ_i and variance σ_i^2 ($i = 1, 2$). Let $\bar{X}_{in} = \sum_{r=1}^{n} X_{ir}/n$, and let $X_r = X_{1r} - X_{2r}$ so that the variables $X_r (r = 1, 2, \ldots)$

are normally distributed with mean $\theta = \mu_1 - \mu_2$ and variance $\sigma^{*2} = \sigma_1^2 + \sigma_2^2$. In finding a confidence interval for θ, it seems reasonable to insist that the width of the confidence interval should not exceed a preassigned value $2d$ as long as $\theta = 0$ is included in the confidence interval, but to progressively relax this requirement as the distance between zero and the confidence interval increases. Suppose that $\{(\mathscr{L}_n, \mathscr{U}_n)\}$ denote a confidence sequence for $\theta = \mu_1 - \mu_2$, that is $P(\mathscr{L}_n \le \theta \le \mathscr{U}_n$, for all $n) \ge 1 - \alpha$. Let ω denote a positive constant which together with $2d$ and α is selected in advance of the experiment on the basis of practical considerations. Consider the following stopping rules; $\mathscr{R}_1$: Stop the experiment and decide that $\mathscr{L}_n \le \theta \le \mathscr{U}_n$ as soon as one of the inequalities

$$\text{(a)}\quad \mathscr{U}_n - \mathscr{L}_n \le 2d \qquad \text{(b)}\quad \mathscr{L}_n > 0 \quad \text{and} \quad \mathscr{U}_n - \mathscr{L}_n \le 2d + \omega\mathscr{U}_n$$

$$\text{(c)}\quad \mathscr{U}_n < 0 \quad \text{and} \quad \mathscr{U}_n - \mathscr{L}_n \le 2d - \omega\mathscr{U}_n$$

is satisfied. When Π_1 represents an experimental category and Π_2 represents a standard or control, then Π_1 may not be of practical value unless $\mu_1 - \mu_2 \ge \delta$ where δ is a nonnegative constant also specified in advance of the experiment. Then consider the following alternative stopping rule. $\mathscr{R}_2$: Stop the experiment and decide that $\mathscr{L}_n \le \mu_1 - \mu_2 \le \mathscr{U}_n$ as soon as one of the inequalities

$$(\text{a}_1)\quad \mathscr{U}_n - \mathscr{L}_n \le 2d, \qquad (b_1)\quad \mathscr{L}_n \ge \delta \quad \text{and} \quad \mathscr{U}_n - \mathscr{L}_n \le 2d + \omega(\mathscr{L}_n - \delta),$$

$$(\text{c}_1)\quad \mathscr{U}_n < \delta,$$

is satisfied.

When $\sigma^{*2} = \sigma_1^2 + \sigma_2^2$ is assumed known, we take

$$\mathscr{U}_n = \min_{1 \le r \le n} [\bar{X}_{1r} - \bar{X}_{2r} + A_1/r] \quad \text{and} \quad \mathscr{L}_n = \max_{1 \le r \le n} [\bar{X}_{1r} - \bar{X}_{2r} - A_1/r]$$

where $A_1 = dt_1$ and t_1 is the smallest integer $\ge \sigma^{*2}[z(\alpha/4)]^2/d^2$. With this choice of $\mathscr{L}_n$ and $\mathscr{U}_n$, it follows directly from Lemma 4.16.3 that the sequential procedure based on stopping rule $\mathscr{R}_1$ or $\mathscr{R}_2$ yields $P(\mathscr{L}_n \le \mu_1 - \mu_2 \le \mathscr{U}_n]$ $\ge 1 - \alpha$. Also notice that the rules $\mathscr{R}_1$ and $\mathscr{R}_2$ are closed since $n \le t_1$. There is a small possibility that at the termination of the experiment we might have $\mathscr{L}_n > \mathscr{U}_n$ in which case the confidence interval is empty. However, if it occurs, Paulson (1969) recommends repeating the experiment. Paulson (1969) has carried out certain numerical studies with the stopping rule $\mathscr{R}_2$ (see his Table 1). If σ^* is unknown, take a sample of n_0 pairs of measurements

$(X_{11}, X_{21}), (X_{12}, X_{22}), \ldots, (X_{1n_0}, X_{2n_0})$. If it is known a priori that $\sigma_1 = \sigma_2$, estimate $\sigma^{*2} = \sigma_1{}^2 + \sigma_2{}^2$ by

$$s^2 = \left[\sum_{r=1}^{n_0} (X_{1r} - \overline{X}_{1n_0})^2 + \sum_{r=1}^{n_0} (X_{2r} - \overline{X}_{2n_0})^2\right] / (n_0 - 1)$$

with $v = 2(n_0 - 1)$ degrees of freedom, otherwise estimate σ^{*2} by

$$s^2 = \sum_{r=0}^{n_0} [(X_{1r} - X_{2r}) - (\overline{X}_{1n_0} - \overline{X}_{2n_0})]^2/(n_0 - 1)$$

with $v = (n_0 - 1)$ degrees of freedom. Next take

$$\mathscr{U}_n = \min_{n_0 \le r \le n} [\overline{X}_{1r} - \overline{X}_{2r} + A_2/r] \qquad \text{and} \qquad \mathscr{L}_n = \max_{n_0 \le r \le n} [\overline{X}_{1r} - \overline{X}_{2r} - A_2/r]$$

where $A_2 = dt_2$ and t_2 is the smallest integer $\ge \max\{n_0, s^2[t(\alpha/4, v)]^2/d^2\}$. After the sample of n_0 pairs is taken, one pair of measurements is to be taken at a time until one of the inequalities in rule $\mathscr{R}_1$ (or $\mathscr{R}_2$) is satisfied. The resulting sequential procedure is closed since it needs at most $t_2 - n_0$ additional pairs in order to terminate the experiment. Regarding the choice of n_0 when there is a little or no prior information available about the value of σ^*, we might select n_0 as the smallest integer so that $t(\alpha/4, v)$ is within a specified percent (perhaps ten percent) of its limiting value $z(\alpha/4)$.

REMARK 4.16.4 One drawback of Paulson's (1969) procedure is that it is not completely sequential and the latter data are not utilized in estimating σ^*. The sequential procedures here are preferable to the earlier ones of Paulson (1964a,b) since the former come fairly close to minimizing the average sample size in the least favorable situation when the population means are all equal while still allowing a substantial saving when the parameter configuration is favorable. Paulson (1969) also gives simultaneous confidence intervals for the means of k populations. The reader is referred to Paulson (1969, Section 3.2) for the same.

Estimating the Mean of the Exponential Population

Starr and Woodroofe (1971) have considered this problem and we briefly describe the procedure. Let $f(x) = (1/\mu) \exp(-x/\mu)$ for $x > 0$ where $\mu > 0$ is unknown. Given a sample $X_1, \ldots, X_k$ of size $k \ge 1$, estimate μ by $\overline{X}_k = (X_1 + \cdots + X_k)/k$ incurring the loss

$$r_k = A(\overline{X}_k - \mu)^2 + k, \qquad A > 0. \tag{4.16.31}$$

The expected loss $E(r_k) = (A\mu^2/k) + k$ is minimized by taking a sample of size $k = c$, where $c = A^{1/2}\mu$ (which is obtained by treating k as a continuous variable). The minimum expected loss then is $R_c = A^{1/2}\mu + A^{1/2}\mu = 2c$.

SEQUENTIAL PROCEDURE Define the random sample size N to be the least integer $k \geq m$ for which $k \geq A^{1/2}\ \overline{X}_k$ where $m \geq 2$ denotes the starting sample size. (The choice of m plays a crucial role in determining the efficiency of the procedure.) When sampling is terminated, estimate μ by $\overline{X}_N$ incurring the loss (4.16.31). The expected loss is

$$\overline{R}_c = AE[(\overline{X}_N - \mu)^2 + E(N)$$

which is a function of c. The authors show that $\omega(c) = \overline{R}_c - R_c$ is bounded as $c \to \infty$. For further details, the reader is referred to Starr and Woodroofe (1971).

Problems

1. A batch of twelve rats was fed a particular diet from birth until age three months and the following weight gains in grams were observed: 55, 62, 54, 57, 65, 64, 60, 63, 58, 67, 63, and 61. Another batch of twelve rats was fed a different diet from birth until age three months and the following weight gains in grams were observed: 70, 66, 68, 72, 65, 75, 59, 63, 69, 76, 80, 74. Set up a fixed-width (equal to 6) confidence interval with coverage probability exceeding 0.9 for the parameter $\Delta = \mu_2 - \mu_1$, the difference between the means of the populations which can be assumed to be normal.

2. Let 90, 95, 87, 78, 69, 102, 110, 97, 88, 75, denote the lives of light bulbs (in hours) before failure. Assume that the time of failure of a light bulb is negative exponential with mean μ. Set up a sequential procedure for estimating μ assuming the loss function to be $r(T_n; \mu) = (T_n - \mu)^2 + n$ where T_n is an estimate of μ that is based on n observations.

SEQUENTIAL ESTIMATION OF TAIL PROBABILITIES IN THE EXPONENTIAL DISTRIBUTION

Let X, X_1, ... be a sequence of iid random variables having a negative exponential distribution with mean θ $(0 < \theta < \infty)$. The tail probability of X is $\rho(\theta) = P_\theta[X \geq \tau] = e^{-\tau/\theta}$ where $0 < \tau < \infty$ is specified. Simons and Zacks (1967) have considered the problem of estimating $\rho(\theta)$ on the basis of N (possible random number of) observations so that for given $0 < \delta < 1$, $0 < \gamma < 1$

$$P_\theta[|\hat{\rho}_N - \rho(\theta)| < \delta\rho(\theta)] \geq \gamma \qquad \text{for all} \quad 0 < \theta < \infty. \tag{4.16.32}$$

The probability in (4.16.32) is called the proportional closeness of $\hat{\rho}_N$ to ρ. The proportional closeness criterion is an appropriate requirement in many estimation problems, especially when there is a functional relationship between the parameter estimated and the variance of the estimators. Furthermore, it is a common practice in the engineering and physical sciences to measure the precision of an estimator proportional to the value estimated.

The maximum likelihood estimate (mle) of ρ is $\hat{\rho}_N = \exp(-\tau/\overline{X}_n)$, where $\overline{X}_n$ denotes the mean of a (fixed) sample of size n. It is simple to verify that the Fisher information for $\rho(\theta)$, contained in observation X is $I_{\rho\rho} = \theta^2\rho^{-2}(\theta)/\tau^2$. Thus the mle $\hat{\rho}_n$, as $n \to \infty$, is distributed as a normal variable having mean $\rho(\theta)$ and variance $n^{-1}(\tau/\theta)^2 \rho^2(\theta)$. This property implies that if the fixed sample size n is not smaller than

$$n_0(\delta) = c\tau^2/\theta^2\delta^2, \tag{4.16.33}$$

then the mle $\hat{\rho}_n$ satisfies asymptotically as $\delta \to 0$ the proportional closeness criterion (4.16.32) where c denotes the γth fractile of chi-square with one degree of freedom. However, θ is unknown and Simons and Zacks (1967) present a family of sequential procedures based on the properties of the mle of $\rho(\theta)$, which asymptotically as $\delta \to 0$ guarantees the prescribed proportional closeness. Further, they have shown that for all values of δ, the expected sample size of the sequential procedure does not exceed $n_0(\delta)$ by more than a uniformly bounded number of observations.

The estimation of tail probabilities of a negative exponential distribution is a common reliability problem. The mle $\hat{\rho}_n$ is widely used. A study of the proportional closeness of the mle of $\rho(\theta)$ has been presented by Zacks and Even (1966) for fixed-sample procedures. In reliability studies, one is interested in a lower confidence interval; that is we should impose

$$P_\theta[\rho(\theta) > \hat{\rho}_N - \delta\rho(\theta)] \geq \gamma \qquad \text{for all} \quad 0 < \theta < \infty. \tag{4.16.34}$$

The sample size to use when θ is known is approximately $c^*\tau^2/\theta^2\delta^2$ where c^* (for $\gamma > \frac{1}{2}$) is the $(2\gamma - 1)$th fractile of chi-square with one degree of freedom. The sequential procedure presented below, while designed with goal (4.16.32), is reasonable for goal (4.16.34) when c is replaced by c^*. Lemmas 4.16.5 and 4.16.6 and Theorem 4.16.2 have obvious counterparts.

The Sequential Procedure Based on the mle

Without loss of generality we can set $\tau = 1$ since when τ is specified, the transformation $x \to x/\tau$ transforms $\theta \to \theta/\tau$. The sequential procedure is based on the mle after each observation (replacing θ^2 by its mle $\overline{X}_n{}^2$) and terminates sampling according to the stopping rule:

$$N(\delta) = \text{least } n \geq 1, \text{ such that } n \geq c_n/\overline{X}_n{}^2\delta^2, \tag{4.16.35}$$

where $\{c_n\}$ is a sequence of positive constants approaching c ($c_n = \infty$ allowed). The stopping variable $N(\delta)$ can be written in terms of the cumulative sums $S_n = \sum_{i=1}^{n} X_i$, as

$$N(\delta) = \text{least integer } n \geq 1, \text{ such that } S_n \geq (c_n^{1/2}/\delta)\, n^{1/2}.$$

As an immediate consequence of Lemma 1 of Chow and Robbins (1965) (see Lemma 4.14.5) we obtain

Lemma 4.16.5 (Simons and Zacks, 1967) The sample size $N(\delta)$ is a well-defined stopping variable, nondecreasing function of δ, satisfying:

$$\lim_{\delta \to 0} N(\delta) = \infty \quad \text{a.s.} \qquad \lim_{\delta \to 0} E_\theta\{N(\delta)\} = \infty, \qquad 0 < \theta < \infty,$$

and (4.16.36)

$$\lim_{\delta \to 0} \{N(\delta)/n_0(\delta)\} = 1, \quad \text{a.s.}$$

Furthermore, for each $0 < \delta < 1$,

$$E_\theta\{N(\delta)\} < \infty \qquad \text{for all} \quad 0 < \theta < \infty.$$

The asymptotic consistency of the sequential procedure follows from Anscombe's theorem (1952) (see Theorem 4.12.1). We thus have:

Lemma 4.16.6 (Simons and Zacks, 1967)

$$\lim_{\delta \to 0} P_\theta[|\hat{\rho}_N - \rho| < \delta\rho] \geq \gamma \qquad \text{for all} \quad 0 < \theta < \infty. \tag{4.16.37}$$

Now we are ready to state the main result of Simons and Zacks (1967) pertaining to the asymptotic efficiency of the sequential procedure.

Theorem 4.16.2 (Simons and Zacks, 1967) For each $0 < \theta < \infty$,

$$\lim_{\delta \to 0} [E_\theta\{N(\delta)\}/n_0(\delta)] = 1. \tag{4.16.38}$$

Moreover, if $c_n \leq c(1 + O(n^{-1}))$, then

$$E_\theta\{N(\delta)\} \leq n_0(\delta) + O(1) \qquad \text{as} \quad \delta \to 0. \tag{4.16.39}$$

Proof Since $N(\delta)/n_0(\delta) \geq 0$ and $N(\delta)/n_0(\delta) \to 1$ a.s., we obtain from Fatou's lemma that $\liminf_{\delta \to 0} [E_\theta\{N(\delta)\}/n_0(\delta)] \geq 1$. Hence, in order to establish (4.16.38) it suffices to show that $\limsup_{\delta\to 0}[E_\theta\{N(\delta)\}/n_0(\delta)] \le 1$. For any given $\varepsilon > 0$, there exists an M such that, $n \ge M$ implies $c_n^{1/2} \le c^{1/2}(1+\varepsilon)$. Define the auxiliary stopping variable

$$N^*(\delta) = \text{least integer } n \ge M \text{ such that } S_n \ge c^{1/2}(1+\varepsilon)\, n^{1/2}/\delta. \quad (4.16.40)$$

Define

$$W(\delta) = \begin{cases} 0 & \text{if } N^*(\delta) = M, \\ S_{N^*(\delta)} - c^{1/2}(1+\varepsilon)(N^*(\delta))^{1/2}/\delta & \text{if } N^*(\delta) > M. \end{cases} \quad (4.16.41)$$

Notice that $N(\delta) \le N^*(\delta)$ and that

$$S_{N^*(\delta)} \le c^{1/2}(1+\varepsilon)(N^*(\delta))^{1/2}/\delta + W(\delta) + S_M. \quad (4.16.42)$$

Applying Wald's identity [see Eq. (2.5.6)] and Jensen's inequality, we have

$$\begin{aligned} \theta E_\theta\{N^*(\delta)\} &\le [c^{1/2}(1+\varepsilon)/\delta]E_\theta\{N^{*1/2}(\delta)\} + E_\theta\{W(\delta)\} + \theta M \\ &\le [c^{1/2}(1+\varepsilon)/\delta]E_\theta^{1/2}\{N^*(\delta)\} + E_\theta\{W(\delta)\} + \theta M. \end{aligned} \quad (4.16.43)$$

By using an argument similar to the one used by Wald (1947, p. 172), we conclude that $E_\theta\{W(\delta) \mid N^*(\delta) > M\} \le \theta$. Employing this in (4.16.43) we deduce

$$[E_\theta\{N^*(\delta)\}/n_0(\delta)]^{1/2} \le 1 + \varepsilon + (1+M)/\{n_0(\delta)E_\theta N^*(\delta)\}^{1/2}. \quad (4.16.44)$$

Hence

$$\limsup_{\delta\to 0}[E_\theta\{N(\delta)\}/n_0(\delta)\} \le \limsup_{\delta\to 0}[E_\theta N^*(\delta)/n_0(\delta)] \le (1+\varepsilon)^2 \quad (4.16.45)$$

which proves (4.16.38).

If $c_n \le c(1 + O(n^{-1}))$ as $n \to \infty$, there exists finite positive constants M_0 and K (free of δ) such that if $n \ge M_0$, then $c_n^{1/2} \le c^{1/2}(1 + Kn^{-1})$. Define then another auxiliary stopping variable:

$$\begin{aligned} \tilde{N}(\delta) &= \text{least integer } n \ge \tilde{M} = \max(M_0, 3K), \quad \text{such that} \\ &S_n \ge c^{1/2}[1 + Kn^{-1}]n^{1/2}/\delta. \end{aligned} \quad (4.16.46)$$

Clearly $N(\delta) \le \tilde{N}(\delta)$. Defining $\tilde{w}(\delta)$ analogous to $w(\delta)$ as in (4.16.41) one can obtain an inequality for $S_{\tilde{N}(\delta)}$ from which we deduce

$$\theta E_\theta\{\tilde{N}(\delta)\} \le c^{1/2}E\{(1 + K(\tilde{N}(\delta))^{-1})(\tilde{N}(\delta))^{1/2}/\delta\} + (\tilde{M}+1)\theta. \quad (4.16.47)$$

Since $(1 + Kn^{-1})n^{1/2}$ is a concave function of n, for $n \geq 3K$,

$$E\{(1 + K(\tilde{N}(\delta))^{-1})(\tilde{N}(\delta))^{1/2}\} \leq (1 + K(E(\tilde{N}(\delta)))^{-1})(E\tilde{N}(\delta))^{1/2}. \quad (4.16.48)$$

Hence

$$E\{\tilde{N}(\delta)\} \leq [n_0(\delta)E(\tilde{N}(\delta))]^{1/2}[1 + K(E(\tilde{N}(\delta))^{-1}] + 1 + \tilde{M}. \quad (4.16.49)$$

Now (4.16.38) and (4.16.49) imply (4.11.39).

REMARK 4.16.5 If one is interested in sequential estimation of the tail probabilities in a Weibull distribution with a prescribed proportional closeness, one can use the above results by transforming the Weibull data into exponential data. [If Y is Weibull with shape parameter m (known) and scale parameter θ, then $X = Y^m$ is negative exponential with scale parameter θ].

Problem

Let $X_1, X_2, \ldots$ be a sequence of independent random variables each having a Weibull distribution $F(x; \theta) = 1 - \exp(-x^m/\theta)$ where m is a known constant. We wish to estimate the tail probability $\rho(\theta) = P_\theta(X \geq \tau) = \exp(-\tau^m/\theta)$ where $0 < \tau < \infty$ is specified such that for given $0 < \delta < 1, 0 < \gamma < 1$

$$P_\theta[\,|\hat{\rho}_N - \rho(\theta)| < \delta\rho(\theta)] \geq \gamma \qquad \text{for all } 0 < \theta < \infty.$$

Set up a sequential procedure and give its properties.

4.17 Sequential Estimation of the Mean Vector of a Multivariate Normal Distribtuion

Khan (1968) has considered the sequential estimation of the mean vector of a p-variate normal distribution $\mathscr{N}(\mu, \Sigma)$ when the dispersion matrix Σ is a diagonal matrix of unknown elements. In the following we shall present his results.

Consider a p-variate normal distribution having unknown $\mu = (\mu_1, \ldots, \mu_p)'$ and unknown variance–covariance matrix $\Sigma \equiv \operatorname{diag}(\sigma_1{}^2, \ldots, \sigma_p{}^2)$ $(\sigma_i{}^2 < \infty)$. Consider the loss function

$$r(n) = (\overline{X}_n - \mu)' D(\overline{X}_n - \mu) + n \quad (4.17.1)$$

where $\bar{X}_n = (\bar{X}_{1n}, \ldots, \bar{X}_{pn})'$ and $D = \text{diag}(\lambda_1, \ldots, \lambda_p)$, $\lambda_i \geq 0$ and the suffix n indicates that the sample means are based on the samples of size n. Now,

$$r(n) = \sum_{i=1}^{p} \lambda_i(\bar{X}_{in} - \mu_i)^2 + n = \sum_{i=1}^{p} \lambda_i \left[\frac{\bar{X}_{in} - \mu_i}{\sigma_i/n^{1/2}}\right]^2 \frac{\sigma_i^2}{\text{n}} + n.$$

Hence

$$\psi(n) = E(r_n) = \sum_{i=1}^{p} \lambda_i(\sigma_i^2/n) + n. \tag{4.17.2}$$

Minimizing $\psi(n)$, we obtain the minimizing value of n as

$$n^0 = \left(\sum_{i=1}^{p} \lambda_i \sigma_i^2\right)^{1/2}. \tag{4.17.3}$$

Let $\bar{X}_{in} = \sum_{j=1}^{n} X_{ij}/n$, $s_{in}^2 = (n-1)^{-1} \sum_{j=1}^{n}(X_{ij} - \bar{X}_{in})^2$, $n \geq 2$. If the σ_i are known, then $n = n^0$ is the optimum sample size that minimizes the expected loss (4.17.2). If σ_i are unknown, the following stopping rule in analogy with (4.17.3) is adopted.

$\mathscr{R}_1$: Stop at the first $n \geq p$ for which

$$n \geq \left(\sum_{i=1}^{p} \lambda_i s_{in}^2\right)^{1/2} \quad \text{or} \quad \Sigma\, \lambda_i s_{in}^2 \leq n^2. \tag{4.17.4}$$

Let N denote the stopping variable for rule $\mathscr{R}_1$ that is, $\{N = n\} \Leftrightarrow$ (4.17.4).

Lemma 4.17.1 (Khan, 1968) The sequential procedure based on rule $\mathscr{R}_1$ terminates finitely with probability one, that is, $P(N = \infty) = 0$.

Proof $s_{in} \to \sigma_i$ with probability one, implying that the right-hand member of (4.17.4) tends to n^0 with probability one, which in turn implies that $P(N = \infty) = 0$.

Now since the events $\{N = n\}$ and $\bar{X}_{in}$ are mutually independent [since $\{N = n\}$ depends on s_{in} $(i = 1, 2, \ldots p)$ only] $E(r(N) | N = n) = E(r(n))$. Thus

$$\begin{aligned} E(r(N)) &= \sum_{n=p}^{\infty} P(N = n) E(r(n) | N = n) \\ &= \sum_{n=p}^{\infty} P(N = n) E(r(n)) = \sum_{n=p}^{\infty} P(N = n)\left(\sum_{i=1}^{p} (\lambda_i \sigma_i^2/n) + n\right) \\ &= (n^0)^2 E(N^{-1}) + E(N). \end{aligned}$$

Let $\sigma = (\sigma_1, \ldots, \sigma_p)'$ and $\sigma \to \infty$ means $\sigma_i \to \infty$. Towards the limiting behavior of N as $\sigma \to \infty$, we have the following result.

Theorem 4.17.1 (Khan, 1968) With the above notation, we have (i) $\lim_{\sigma \to \infty} (N/n^0) = 1$ a.s. (ii) $\lim_{\sigma \to \infty} E(N)/n^0 = 1$.

Proof (i) follows from Chow and Robbins (1965, Lemma 4.14.5) by setting

$$Y_n = \left(\sum_{i=1}^{p} \lambda_i s_{in}^2 / \sum_{i=1}^{p} \lambda_i \sigma_i^2\right)^{1/2}, \qquad f(n) = n,$$

$$t = \left(\sum_{i=1}^{p} \lambda_i \sigma_i^2\right) = n^0, \qquad \text{where} \quad t \to \infty \quad \text{as} \quad \sigma \to \infty.$$

Also, (ii) follows from Lemma 4.14.6 [see Chow and Robbins (1965)] as the conditions of the lemma are satisfied via the existence of the fourth moment of the normal distribution.

REMARK 4.17.1 Rohatgi and O'Neill (1973) have shown that the risk $E(r_N)$ is bounded.

SEQUENTIAL CONFIDENCE REGION FOR μ

We are interested in a confidence region for μ with given coverage probability $1 - \alpha$ such that the length of the μ_i axis of the ellipsoid is bounded by $2l_i$ $(i = 1, \ldots, p)$. If σ_i^2 are unknown, a fixed-sample size procedure ensuring bounds on the lengths of the axes of the ellipsoid for a given coverage probability is not available. Thus, Khan (1968) provides a sequential procedure for the problem which approximately does well and which is asymptotically an exact solution.

Let us start with finding the optimum fixed-sample size which ensures bounds on the lengths of the axes of the ellipsoid with a given coverage probability $1 - \alpha$ when the σ_i are known. Let a be a positive number such that $P(\chi_p^2 > a) = \alpha$. Then

$$P(\mu \in S_p^n) = 1 - \alpha, \qquad S_p^n = \{\mu : n(\bar{X}_n - \mu)'\Sigma^{-1}(\bar{X}_n - \mu) \le a\}. \quad (4.17.5)$$

Consider the ellipsoid,

$$\sum_{i=1}^{p} n(\bar{X}_{in} - \mu_i)^2 / a\sigma_i^2 = 1. \quad (4.17.6)$$

Suppose we require that the lengths of the μ_i axis be less than or equal to $2l_i$ $(i = 1, 2, \ldots, p)$; we must have

$$a\sigma_i^2/n \le l_i^2, \qquad \text{that is,} \qquad n \ge a\sigma_i^2/l_i^2. \quad (4.17.7)$$

Let

$$n^0 = \max_{1 \le i \le p} (a\sigma_i^2/l_i^2). \tag{4.17.8}$$

Thus, if σ_i are known, n^0 given by (4.17.8) ensures bounds on the lengths of the μ_i axis of the ellipsoid and coverage probability $1 - \alpha$. When σ_i are unknown, we adopt the following sequential rule.

$\mathscr{R}_2$: Stop at the first $n \ge p$ for which

$$n \ge as_{in}^2/l_i^2 \qquad \text{for every} \qquad i = 1, \ldots, p. \tag{4.17.9}$$

Let N denote the stopping variable for the rule $\mathscr{R}_2$, that is, $\{N = n\} \Leftrightarrow$(4.17.9).

Lemma 4.17.2 (Khan, 1968) The sequential procedure based on rule $\mathscr{R}_2$ terminates finitely with probability one.

Proof Similar to that of Lemma 4.17.1.

Let $L = \text{diag}(l_1^2, \ldots, l_p^2)$ and let the confidence region for μ be defined as

$$S_p^{*N} = \{\mu : (\bar{X}_N - \mu)' L^{-1} (\bar{X}_N - \mu) \le 1\}. \tag{4.17.10}$$

The region S_p^{*N} is the required confidence region for μ which ensures the coverage probability $1 - \alpha$ and the lengths of the axes of the ellipsoid bounded by $2l_i$. Now we have the following result pertaining to the asymptotic properties of the region.

Theorem 4.17.2 (Khan, 1968) We have

$$\text{(i)} \quad P(\mu \in S_p^{*N}) \simeq 1 - \alpha, \qquad \text{(ii)} \quad E(N) \simeq n^0.$$

Proof To prove (i) let $R_p^N = \{\mu : n(\bar{X}_n - \mu) S_n^{-1} (\bar{X}_n - \mu) \le a\}$ where $S_n = \text{diag}(s_{1n}^2, \ldots, s_{pn}^2)$. Using (4.17.9) it is easily seen that $R_p^n \subset S_p^{*n}$ which implies $R_p^N \subset S_p^{*N}$. Now

$$\lim_{n \to \infty} P(\mu \in R_p^n) = \lim_{n \to \infty} P\{n(\bar{X}_n - \mu) S_n^{-1} (\bar{X}_n - \mu) \le a\} = P(\chi_p^2 \le a) = 1 - \alpha.$$

Therefore, at least asymptotically, $P(\mu \in S_p^{*N}) \simeq 1 - \alpha$. To prove (ii) from (4.17.9) it follows that at least for large N, $N \simeq \max_{1 \le i \le p} (a\sigma_i^2/l_i^2) = n^0$, which implies that $E(N) \simeq n^0$.

Let $l = (l_1, \ldots, l_p)'$ and $l \to 0$ means $l_i \to 0$.

Theorem 4.17.3 (Khan, 1968) We have

(i) $\lim_{l_i \to 0}(Nl_i/a\sigma_i^2) = 1$ a.s. for each i.
(ii) $\lim_{l \to 0} (N/n^0) = 1$ a.s.
(iii) $\lim_{l \to 0} P(\mu \in S_p^{*N}) = 1 - \alpha$ (asymptotic consistency).
(iv) $\lim_{l \to 0} E(N)/n^0 = 1$ (asymptotic efficiency).

Proof (i) follows from Lemma 4.14.5 by setting $Y_n = s_{in}^2/\sigma_i^2$, $f(n) = n$, $t = a\sigma_i^2/l_i^2$ where $t \to \infty$ as $l_i \to 0$. (ii) follows from (i) and the definition of n^0.

To prove (iii) we note that $N/t \to 1$ as $t \to \infty$ where $t = a\sigma_i^2/l_i^2 \to \infty$ as $l_i \to 0$. Therefore from a result of Anscombe (1952) (see Theorem 4.12.1) it follows that as $t \to \infty$, $N^{1/2}(\overline{X}_{iN} - \mu_i)/\sigma_i \sim$ standard normal variable. Furthermore, from (i) it follows that $\sigma_i^2/Nl_i^2 \to 1/a$ in probability as $t \to \infty$ $(l_i \to 0)$. Therefore,

$$\begin{aligned}\lim_{l \to 0} P(\mu \in S_p^{*N}) &= \lim_{l \to 0} P\left\{\sum_{i=1}^{p} (\overline{X}_{iN} - \mu_i)^2/l_i^2 \le 1\right\} \\ &= \lim_{t \to \infty} P\left\{\sum_{i=1}^{p} \frac{N(\overline{X}_{iN} - \mu_i)^2}{\sigma_i^2} \frac{\sigma_i^2}{Nl_i^2} \le 1\right\} \\ &= P(\chi_p^2/a \le 1) = 1 - \alpha,\end{aligned}$$

and (iv) follows from Lemma 4.14.6 as the conditions of Lemma 4.14.5 are satisfied and $E(\sup_n Y_n) < \infty$ by the existence of the fourth moment in a normal distribution.

Rohatgi and Rastogi (1973) have considered sequential estimation of the parameter

$$\theta = \sum_{i=1}^{p} b_i \mu_i \tag{4.17.11}$$

where b_i $(i = 1, 2, \ldots, p)$ are given real numbers. Without loss of generality assume that $b_i \neq 0$ $i = 1, \ldots, p$. Let

$$Y_n = \sum_{i=1}^{p} b_i \overline{X}_{in}, \qquad n \ge 2 \tag{4.17.12}$$

which is an unbiased estimate of θ and measure the loss incurred by

$$r_n = |Y_n - \theta|^r + n \tag{4.17.13}$$

where $r > 0$ is a given real number. Clearly

$$\psi(n) = Er_n = a(r)\left(\sum_{1}^{p} b_i^2 \sigma_i^2\right)^{r/2} n^{-r/2} + n \tag{4.17.14}$$

where $a(r) = (2\pi)^{-1/2}2^{(r+1)/2}\,\Gamma((s+1)/2)$. The risk $\psi(n)$ is minimized for $n = n^0$ given by

$$n^0 = \beta^{2/(r+2)}(\Sigma\, b_i{}^2\sigma_i{}^2)^{r/(r+2)} \tag{4.17.15}$$

where $\beta = ra(r)/2$. If the σ_i are known, we take n^0 observations and estimate θ by Y_n. The risk in doing so is given by

$$R(\sigma) = \psi(n^0) = (1 + 2/r)n^0. \tag{4.17.16}$$

Since $\sigma = (\sigma_1, \ldots, \sigma_p)'$ is not known, Rohatgi and Rastogi (1973) propose the following sequential procedure. Let

$$N = \text{smallest integer } n \geq m \text{ for which}$$
$$n > \left(\beta^{2/r} \sum_1^p b_i{}^2 s_{in}^2\right)^{r/(r+2)}, \tag{4.17.17}$$

where m is the initial sample size. Let C denote a generic positive constant. Write

$$\sigma_* = \min_{1\leq i\leq p} \sigma_i, \qquad \sigma^* = \max_{1\leq i\leq p} \sigma_i, \tag{4.17.18}$$

$$b_*{}^2 = \min_{1\leq i\leq p} b_i{}^2, \qquad b^{*2} = \max_{1\leq i\leq p} b_i{}^2. \tag{4.17.19}$$

and assume that

$$\sigma^*/\sigma_* \to 1 \qquad \text{as} \quad \sigma \to \infty \qquad (\text{that is } \sigma_i \to \infty,\ i = 1, \ldots, p). \tag{4.17.20}$$

Note that

$$\beta^{2/(2+r)}(p\, b_*{}^2\sigma_*{}^2)^{r/(r+2)} \leq n^0 \leq \beta^{2/(r+2)}(p\, b^{*2}\sigma^{*2})^{r/(r+2)}.$$

Along the lines of Khan (1968), the following lemma and theorem readily follow.

Lemma 4.17.3 (Rohatgi and Rastogi, 1973) $P(N < \infty) = 1$.

Theorem 4.17.4 (Rohatgi and Rastogi, 1973)

(i) $\lim_{\sigma\to\infty} (n^0)^{-1}N = 1$ a.s, (ii) $\lim_{\sigma\to\infty} E(N/n^0) = 1$.

Notice that Lemma 4.17.3 and Theorem 4.17.4 are valid if the loss function r_n is modified to

$$r_n{}^* = |Y_n - \theta|^r + \ln n. \tag{4.17.21}$$

Lemma 4.17.4 (Rohatgi and Rastogi, 1973) $P(N = m) = O(\sigma_*^{-p(m-1)})$ as $\sigma \to \infty$ and (4.17.20) holds.

Lemma 4.17.5 (Rohatgi and Rastogi, 1973) For fixed ξ, $0 < \xi < 1$ $P(N \le \xi n^0) = O(\sigma_*^{-p(m-1)})$ as $\sigma \to \infty$ and (4.17.20) holds.

The proofs of Lemmas 4.17.4 and 4.17.5 can easily be constructed by minor modification of the methods of Simons (1968).

Theorem 4.17.5 (Rohatgi and Rastogi, 1973) As $\sigma \to \infty$ and (4.17.20) holds

$$\omega(\sigma) = \bar{R}(\sigma) - R(\sigma) = O(1) \tag{4.17.22}$$

if and only if $m \ge 1 + (r/p)$ where

$$\bar{R}(\sigma) = Er_N = (2/r)(n^0)^{(r+2)/2}\, E(N^{-r/2}) + EN, \tag{4.17.23}$$

and $R(\sigma)$ is given by (4.17.16).

Proof This result follows from a modification of the proof of the theorem in the work of Starr and Woodroofe (1969).

Problems

1. Let X denote the toughness of a specimen of steel and Y denote the percentage of nickel in the specimen. Assume that (X, Y) follow a bivariate normal distribution. Let the following constitute independent observations from this population.

X	47	50	52	52	54	56	58	59	60	60	62	64	65	66
Y	2.5	2.7	2.8	2.8	2.9	3.2	3.2	3.3	3.4	3.5	3.5	3.6	3.7	3.8

Set up a sequential estimation procedure for the mean vector of the bivariate population, the loss function being $r_n = (\bar{X}_n - \mu_1)^2 + (\bar{Y}_n - \mu_2)^2 + n$.

2. For the above data set up a confidence region having coverage probability .90 such that the lengths of the μ_1, μ_2 axes of the ellipse are bounded by 8 and 2 respectively.

3. Carry out a sequential procedure for estimating $\theta = \mu_1 - 20\mu_2$ using loss function $r_n = (Y_n - \theta)^2 + n$.

Callahan (1969a) also has considered the problem of estimating sequentially the mean of a random normal vector when the covariance matrix Σ is unknown. This estimation is studied not only from the decision-theoretic point of view but also from a confidence region viewpoint. For the point estimation problem, two loss functions are formulated; these take into account both the error of estimation and the cost of sampling. The risk efficiency, defined as the ratio of the risk which results from using the sequential procedure, to the minimum risk which results from using the sample of fixed size that we would use if Σ were known, is studied for various values of the trace of Σ. It is found that as $\operatorname{tr} \Sigma$ goes to zero, the risk efficiency goes to one, and that the risk efficiency goes to one as $\operatorname{tr} \Sigma$ goes to infinity if and only if the initial sample size n_0 is of a certain stipulated size.

Sequential Estimation of the Common Mean of Two Normal Populations

Suppose (X, Y) has a bivariate normal distribution having mean vector $\mu = (\mu, \mu)'$ and variance–covariance matrix $\left(\begin{smallmatrix} \sigma^2 & 0 \\ 0 & 1 \end{smallmatrix}\right)$ where μ and $\sigma^2 (\neq 1)$ are unknown. We wish to estimate μ sequentially when $\eta = \sigma^{-2}$ is unknown and we pay c dollars per observation. We assume a squared error loss function

$$r(\hat{\mu}; \mu) = (\hat{\mu} - \mu)^2 + cn \tag{4.17.24}$$

when $\hat{\mu}$ is an estimator of μ. Recall that this problem without cost function has been solved by Richter (1960, Section 4.2) by a two-stage procedure. If we know that $\eta > 1$ $[\eta < 1]$, we should take all our observations from the X $[Y]$ population and use an estimate of μ based on observations on X $[Y]$. Since we do not know η, a reasonable procedure will start by taking observations from the X population, then by estimating η as well as μ. After each observation, we have to decide whether $\eta > 1$ or $\eta < 1$. As long as we suspect that $\eta > 1$, we take another observation from X; however, once we are convinced that $\eta < 1$, we switch to the Y population. Once we switch to Y, no new information on the variance of X is obtained, and there seems to be no point in switching back to X. The appropriate number of observations to be taken on Y can be determined at the time of switching. Further it is possible never to switch to Y in the first place. So we have to find a sequential rule that tells us when to switch to the Y population and a stopping rule that tells us when to stop taking observations from the X population in case we decide not to switch to the Y population. If σ were known and it were less than unity, the appropriate (fixed) number of observations n would be $\sigma/c^{1/2}$ which minimizes $cn + \sigma^2/n$. If σ were known and was greater than unity, we would take 0 observations from X but $c^{-1/2}$ observations from Y population. Mallik (1971) has given a solution to the sequential aspect of this problem, and his method combines the solutions of two related problems, namely the sequential estimation of the mean of a normal population when the variance is unknown [considered by Robbins (1959) and Starr (1966b); see Section 4.13] and the one-armed bandit problem [considered by Chernoff (1968)]. In the following we shall briefly describe the one-armed bandit problem.

One-Armed Bandit Problem

Let $X_1, \ldots, X_n$ be iid normal random variables having unknown mean μ and known variance σ^2. An integer n such that $n \le n_0$ (n_0 is a specified integer) is to be selected sequentially so as to maximize

$$E(X_1 + \cdots + X_n) = E(n\mu). \tag{4.17.25}$$

Viewing $X_1, \ldots, X_n$ as the winnings from a "one-armed bandit" by a player who can stay almost long enough to play n_0 games, we may call this a one-armed bandit problem.

For the normal continuous time version of this problem, let $X(t)$, representing the gambler's cumulated win at time t, be a Wiener process (see Appendix 4) with mean drift μ and known variance σ^2 per unit time. Then the continuous time one-armed bandit problem consists of finding a sequential procedure for selecting a stopping time $t \le n_0$ so as to maximize

$$E\{X(t)\} = E(t\mu). \tag{4.17.26}$$

If μ has a normal prior distribution with mean μ_0 and variance $\sigma_0{}^2$, the posterior distribution of μ, in the discrete time version, for given n observations $X_1, \ldots, X_n$ is normal with mean Y_n and variance $(\sigma_0^{-2} + n\sigma^{-2})^{-1}$ where

$$Y_n = \{\mu_0 \sigma_0^{-2} + (X_1 + \cdots + X_n)\sigma^{-2}\}/(\sigma_0^{-2} + n\sigma^{-2}). \tag{4.17.27}$$

In the continuous time version, the posterior distribution of μ given $X(t)$ is normal with mean $Y(s)$ and variance s where

$$Y(s) = \{\mu_0 \sigma_0^{-2} + X(t)\sigma^{-2}\}/(\sigma_0^{-2} + t\sigma^{-2}) \tag{4.17.28}$$

and

$$s^{-1} = \sigma_0^{-2} + t\sigma^{-2}. \tag{4.17.29}$$

Chernoff (1968) has shown that if μ is random, Y is a Wiener process in the $-s$ scale starting from $Y = \mu_0$ and $s = \sigma_0{}^2$. The payoff with discrete time case is $X_1 + \cdots + X_n$ and in the continuous time case it is

$$X(t) = \{Y(s)/s - \mu_0 \sigma_0^{-2}\}\sigma^2.$$

Thus the problem of selecting a stopping variable that maximizes $E(X(t))$ subject to $t \le n_0$ is equivalent to maximizing $E\{Y(s)/s\}$ subject to $s \ge s_0 = (\sigma_0^{-2} + n_0\sigma^{-2})^{-1}$. We can standardize this problem to the case $s_0 = 1$ by the following transformation. Let

$$T = \sigma_0^{-2} + n_0 \sigma^{-2} = s_0^{-1}, \qquad Y^*(s^*) = T^{1/2} Y(s), \qquad s^* = Ts. \quad (4.17.30)$$

It is easy to see that Y^* is a Wiener process in the $-s^*$ scale with mean drift 0 and variance 1 per unit decrease in s^*. Since the payoff is proportional to $Y/s = T^{1/2}(Y^*/s^*)$ and $s^* = 1$ when $s = s_0 = T^{-1}$, it follows that the optimal procedure for the one-armed bandit problem can be expressed in terms of the solutions of the standardized version of the problem involving Y^*, s^*, where s^* is restricted by $s^* \geq 1$.

The optimal procedure for the standardized problem has been obtained by Chernoff and Ray (1965) and is given by: Stop when

$$Y^*(s^*) \leq -(s^*)^{1/2}\, \gamma(s^{*-1}) \qquad \text{or when} \qquad s^* = 1 \qquad (4.17.31)$$

whichever comes first. An approximation to $\gamma(\cdot)$ is given by Chernoff (1967). Transforming to the Y, s scale by use of (4.17.30) one should stop when

$$Y(s) \leq -s^{1/2}\, \gamma(1/Ts) \qquad \text{or when} \qquad s = T^{-1} \qquad (4.17.32)$$

whichever comes first. Suppose that the prior information is negligible, that is, $\sigma_0^2 \to \infty$. Then applying (4.17.31), and $1/Ts = t/n_0$, the procedure calls for stopping when

$$X(t)/\sigma t^{1/2} \leq -\gamma(t^*), \qquad t^* = t/n_0, \qquad (4.17.33)$$

where t^* represents the proportion of the total potential information attained at time t. Notice that under the hypothesis $\mu = 0$, $X(t)/\sigma t^{1/2}$ denotes the number of standard deviations of $X(t)$ away from its mean. Thus, the procedure can be interpreted as a one-sided significance test of $\mu \geq 0$ with a nominal significance level

$$\alpha(t^*) = \Phi(-\gamma(t^*)) \qquad (4.17.34)$$

where Φ denotes the standard normal distribution function. Notice that α varies as t^* increases, going from 0 (strictly) to $\frac{1}{2}$ as t^* increases from 0 to 1 (that is, as t increases from 0 to n_0). The decision to reject $\mu \geq 0$ corresponds to the decision to stop playing. Chernoff and Ray (1965) have obtained the following asymptotic expressions for $\gamma(t^*)$ and $\alpha(t^*)$:

$$\alpha(t^*) \simeq 2t^*, \qquad \gamma(t^*) \simeq \{-2 \ln t^* - \ln(-16\pi \ln t^*)\}^{1/2} \qquad \text{as} \quad t^* \to 0. \qquad (4.17.35)$$

$$\alpha(t^*) \simeq 1/2, \qquad \gamma(t^*) \simeq 0.639(1 - t^*)^{1/2} \qquad \text{as} \quad t^* \to 1. \qquad (4.17.36)$$

Table 4.17.1

Nominal Significance Level $\alpha(t^*)$

t^*	$\alpha(t^*)$	t^*	$\alpha(t^*)$
1.0	.5	.034	.049
.404	.25	.013	.021
.102	.108	.010	.017

Applying a refined version of the backward induction used by Chernoff (1967), Mallik (1971) has tabulated the values of $\alpha(t^*)$, $0 \le t^* \le 1$ and a selection of values is given in Table 4.17.1.

The solution of the continuous time one-armed bandit problem will serve as an approximate solution to the discrete time case. Further, in the discrete time case, the expected loss that may be attributed to our ignorance of μ when we take n observations is

$$\begin{cases} -n\mu & \text{when} \quad \mu < 0, \\ (n_0 - n)\mu & \text{when} \quad \mu > 0. \end{cases} \tag{4.17.37}$$

DESCRIPTION OF THE SEQUENTIAL PROCEDURE FOR ESTIMATING THE COMMON MEAN

Mallik (1971) has given the following sequential procedure for estimating the common mean. Suppose we have already taken $n \ge n^*$ (where n^* is a fixed integer) observations on X, with sample mean $\bar{X}_n = \sum_1^n X_i/n$ and sum of squared deviations $S_n = \sum_1^n (X_i - \bar{X}_n)^2$. Stop sampling [which is analogous to Robbins' procedure (1959)] and estimate μ by $\hat{\mu} = \bar{X}_n$ if

$$n \ge c^{-1/2}\{S_n/(n-1)\}^{1/2},$$

i.e., if

$$S_n \le cn^2(n-1) \equiv a_n. \tag{4.17.38}$$

After each observation on X, if we decide not to stop, we switch to Y if $S_n \ge d_n$ where d_n is chosen so that

$$P(\chi_{n-1}^2 \ge d_n) \le \alpha_n = \alpha(t^*) \tag{4.17.39}$$

where $t^* = n/n_0$, $n_0 = c^{-1/2}$, α is the nominal significance level given in Table 4.17.1, and χ_{n-1}^2 has the chi-square distribution with $n-1$ degrees of freedom (that is, the data tend to reject the hypothesis $\sigma \leq 1$ at a suitable significance level). If we switch to Y after n observations on X, we take

$$m = c^{-1/2} - n\hat{\eta} \tag{4.17.40}$$

observations on Y and estimate μ by

$$\hat{\mu} = (n\hat{\eta}\bar{X}_n + m\bar{Y}_m)/(n\hat{\eta} + m) \tag{4.17.41}$$

where

$$\hat{\eta} = n/S_n. \tag{4.17.42}$$

If $a_n < S_n < d_n$, we continue sampling from the X population. Note that $a_{n_0} = n_0 - 1$ while d_{n_0} is approximately but slightly less than $n_0 - 1$, thus guaranteeing that there will be no more than $n = c^{-1/2}$ observations on X. This is not surprising, since continuing with X corresponds to a belief in $\sigma < 1$, in which case the appropriate number of observations, $\sigma c^{-1/2}$, would be less than $n_0 = c^{-1/2}$.

The asymptotic properties of Robbins' (1959) procedure are applicable to the above procedure. Mallik (1971) draws the following conclusions: (i) the attainment of $S_n \leq cn^2(n-1)$ should be regarded as a sign that sampling should terminate; (ii) when $\eta > 1$ ($\sigma^2 < 1$), this stopping rule would lead to a relatively small cost of ignorance, i.e. $O_p(c)$; and (iii) the sample size when this rule is applied (that is, n that satisfies: $n \geq c^{-1/2}\{S_n/(n-1)\}^{1/2}$) will tend to be relatively close to $\sigma c^{-1/2}$.

Selection of m, the Number of Observations on Y

Let us consider a hypothetical problem. Assume that we switch to Y after taking n observations on X and that σ^2 (i.e., η) is known. Let m be the number of observations one takes on Y and estimate μ by

$$\hat{\mu} = (n\eta\bar{X}_n + m\bar{Y}_m)/(n\eta + m)$$

where $\bar{X}_n = \sum_1^n X_i/n$ and $\bar{Y}_m = \sum_1^m Y_i/m$. Then the risk is

$$R_0(n, m, \eta) = c(n + m) + (n\eta + m)^{-1}.$$

Treating m as a continuous variable, the value of m that minimizes $R_0(n, m, \eta)$ is $\max(c^{-1/2} - n\eta, 0)$. Now if η is unknown, we replace η by $\hat{\eta}$ and choose m to be

$$m = m(n; \hat{\eta}) = \max(c^{-1/2} - n\hat{\eta}, 0) \tag{4.17.43}$$

where $\hat{\eta}$ given by (4.17.42) is the maximum likelihood estimate of η. If we estimate μ by $\hat{\mu}$ given by (4.17.41), then

$$\operatorname{var}(\hat{\mu} \mid n, \hat{\eta}) = [(n\hat{\eta}^2/\eta) + m]/(m + n\hat{\eta})^2$$

and the conditional risk, given n and $\hat{\eta}$ is

$$R(n, \eta, \hat{\eta}) = c(n + m) + \{(n\hat{\eta}^2/\eta) + m\}/(m + n\hat{\eta})^2. \tag{4.17.44}$$

We have the following propositions regarding the conditional risk.

Proposition 4.17.1 (Mallik, 1971) If $n\hat{\eta} > c^{-1/2}$ and $n = \sigma c^{-1/2} + O(\sigma^{1/2}c^{-1/4})$, then $R(n, \eta, \hat{\eta}) = 2\sigma c^{1/2} + O(c)$.

Proof According to (4.17.43), $m = 0$ and

$$\begin{aligned} R(n, \eta, \hat{\eta}) &= cn + 1/n\eta \\ &= 2\sigma c^{1/2} + cn - 2\sigma c^{1/2} + \sigma^2/n \\ &= 2\sigma c^{1/2} + cn^{-1}(n - \sigma c^{-1/2})^2 \\ &= 2\sigma c^{1/2} + O(c). \end{aligned}$$

Proposition 4.17.2 (Mallik, 1971) If $n\hat{\eta} < c^{-1/2}$, then

$$R(n, \eta, \hat{\eta}) = 2c^{1/2} + cn(1 - \eta) + O_p(-c \ln c).$$

Proof See Mallik (1971, pp. 20–23).

Mallik (1971) has carried out extensive Monte Carlo simulations with $n^* = 4$, in order to measure the efficiency of the procedure. The simulation studies indicate that when $\sigma^2 > 1$, the cost of ignorance of σ^2 is equal to that of wasting some observations; the number of observations wasted is 5, 7, and 10 when $c = 10^{-3}$, 10^{-4}, and 10^{-5} respectively.

4.18 A General Method of Determining Fixed-Width Confidence Intervals

Khan (1969) has given a method for determining stopping rules in order to obtain a fixed-width confidence interval of prescribed coverage probability for an unknown parameter of a distribution involving possibly some unknown nuisance parameters. The results are asymptotic in nature and lean heavily on the asymptotic theory of Chow and Robbins (1965). In the following we shall present Khan's (1969) results.

Let $p(x; \theta_1, \theta_2)$ denote the probability density function of a random variable X (for convenience with respect to Lebesgue measure) with real-valued parameters θ_1 and θ_2 where θ_2 is regarded as a nuisance parameter. For the sake of simplicity, we assume that there is a single nuisance parameter since the case of several nuisance parameters would be analogous. We wish to determine a confidence interval of fixed width $2d$ $(d > 0)$ for θ_1 when both θ_1 and θ_2 are unknown, with preassigned coverage probability $1 - \alpha$ $(0 < \alpha < 1)$.

Assumption We assume that all the regularity conditions of maximum likelihood estimation are satisfied. (See for instance, Appendix 5).

Let N denote a bona fide stopping variable (that is, N is a positive integer-valued random variable such that the stopping set $\{N = n\} \in \mathscr{F}_n$ where $\mathscr{F}_n \uparrow$ sub σ-algebras, are the σ-algebras of subsets generated by $X^{(n)} = (X_1, \ldots, X_n)'$ and $P(N < \infty) = 1$). Let n denote a fixed value assumed by N. Also, let Fisher's information matrix [see for instance, Rao (1965, p. 270)] be denoted by $I(\theta) = ((I_{ij}))$, $i, j = 1, 2$ where $\theta = (\theta_1, \theta_2)'$ and

$$I_{ij} = E[\{\partial \ln p(X; \theta_1, \theta_2)/\partial \theta_1\}\{\partial \ln p(X; \theta_1, \theta_2)/\partial \theta_2\}], \qquad 1 = i, j \leq 2.$$

We assume that $((I_{ij}))$ is positive definite and let $((I_{ij}))^{-1} = ((\lambda_{ij})) = \Lambda$, that is $I^{-1}(\theta) = \Lambda$. $\hat{\theta}_1(n)$ and $\hat{\theta}_2(n)$ will denote the maximum likelihood estimators (mle) of θ_1, θ_2 respectively (possibly unbiased which can be done in most cases by trivial modification of the mle, though unbiasedness is not essential for the foregoing discussion) based on a random sample of size n. It should be noted that $\hat{\theta}_1(n)$ is asymptotically normal with mean θ_1 and variance λ_{11}/n where $\lambda_{11} = \lambda_{11}(\theta_1, \theta_2)$ since, in general, the I_{ij} are functions of θ_1 and θ_2.

Let $\{u_n, n \geq 1\}$ be a sequence of positive constants converging to a constant u where $\Phi(u) = 1 - (\alpha/2)$. Let $J_n = [\hat{\theta}_1(n) - d, \hat{\theta}_1(n) + d]$ and

$$n_d = \text{smallest integer } n \geq u^2 \lambda_{11}(\theta_1, \theta_2)/d^2 = n^0. \tag{4.18.1}$$

From (4.18.1) it follows that $\lim_{d \to 0} n_d = \infty$ and

$$\lim_{d \to 0} [d^2 n_d / u^2 \lambda_{11}(\theta_1, \theta_2)] \geq 1.$$

Hence,

$$\begin{aligned}\lim_{d \to 0} P(\theta_1 \in J_{n_d}) &= \lim_{d \to 0} P\{n_d^{1/2} |\hat{\theta}_1(n_d) - \theta_1| / \lambda_{11}^{1/2} \leq d(n_d/\lambda_{11})^{1/2}\} \\ &= 2\Phi(u') - 1 \geq 1 - \alpha,\end{aligned}$$

since $u' > u$. We will treat n^0 as the optimum sample size if θ_1 and θ_2 were known, which is justified in the strict sense but will serve as a standard for comparison with the stopped random variable in the sequential procedure to be adopted. In some cases n^0 will turn out to be optimum if only θ_2 were known and $\lambda_{11}(\theta_1, \theta_2)$ depends only on θ_2, for example, in the case of a normal distribution with θ_1 as the mean and θ_2 as the variance.

When θ_1 and θ_2 are unknown, a fixed n such as that determined by (4.18.1) will not be available to guarantee fixed-width $2d$ and coverage probability $1 - \alpha$. So, in analogy to (4.18.1) we adopt the following sequential rule. Let m be a fixed positive integer: Starting with $n \geq m$, stop whenever for the first time

$$n \geq u_n{}^2 \lambda_{11}(\hat{\theta}_1(n), \hat{\theta}_2(n))/d^2, \tag{4.18.2}$$

that is $N = \inf\{n \geq m : n \geq u_n{}^2 \hat{\lambda}_{11}(n)/d^2\}$ where $\hat{\lambda}_{11}(n) = \lambda_{11}(\hat{\theta}_1(n), \hat{\theta}_2(n))$.

Lemma 4.18.1 (Khan, 1969) Under the sole assumption that $\lambda_{11}(\theta_1, \theta_2) < \infty$, the sequential procedure terminates finitely with probability 1.

Proof Under the regularity assumptions, $\hat{\lambda}_{11}(n) \to \lambda_{11}(\theta_1, \theta_2)$ with probability 1 [see for instance, Rao (1965, pp. 300–301)]. Hence the lemma follows from the fact that the right-hand member of (4.18.2) tends to n^0 with probability one which in turn implies $P(N = \infty) = 0$.

Then we have the following asymptotic result in the sense of Chow and Robbins (1965).

Theorem 4.18.1 (Khan, 1969) If

$$E(\sup_n \hat{\lambda}_{11}(n)) < \infty \tag{4.18.3}$$

then we have (i) $\lim_{d\to 0} N/n^0 = 1$ a.s. (ii) $\lim_{d\to 0} P(\theta_1 \in J_N) = 1 - \alpha$ (asymptotic consistency); (iii) $\lim_{d\to 0} E(N)/n^0 = 1$ (asymptotic efficiency).

Proof To prove (i), let $Y_n = \hat{\lambda}_{11}(n)/\lambda_{11}$, $f(n) = nu^2/u_n{}^2$ and $t = u^2 \times \lambda_{11}(\theta_1, \theta_2)/d^2 = n^0 \to \infty$ as $d \to 0$. Then conditions of Lemma 4.14.5 (Chow and Robbins, 1965, Lemma 1) are satisfied and hence

$$\lim_{t\to\infty} N/t = \lim_{d\to 0} N/n^0 = 1 \quad \text{a.s.}$$

To prove (ii), we observe that $N(t)/t \to 1$ a.s. as $t \to \infty$ and hence $N(t)/n_t \to 1$ a.s. as $t \to \infty$ where $n_t = [t]$, here the largest integer $\leq t$. We have already noted that $\hat{\theta}_1(n)$ satisfies the conditions C_1 and C_2 of Theorem 4.12.1 (Anscombe, 1952, Theorem 1) and hence it follows that

$$(n_t/\lambda_{11}(\theta_1, \theta_2))^{1/2}\{\hat{\theta}_1(N(t)) - \theta_1\}$$

tends to the standard normal variable in distribution as $t \to \infty$. Further,

$$\{N(t)/\lambda_{11}(\theta_1, \theta_2)\}^{1/2}\{\hat{\theta}_1(N(t)) - \theta_1\}$$
$$= \{N(t)/n_t\}^{1/2}(n_t/\lambda_{11})^{1/2}\{\hat{\theta}_1(N(t)) - \theta_1\}.$$

Now, since $N(t)/n_t \to 1$ a.s. as $t \to \infty$, therefore a well-known theorem of Cramér (1946) implies that $\{N(t)/\lambda_{11}(\theta_1, \theta_2)\}^{1/2}\{\hat{\theta}_1(N(t)) - \theta_1\}$ tends to the standard normal variable in distribution as $t \to \infty$. Also from (ii) it follows that $d(N/\lambda_{11})^{1/2} \to u$ a.s. as $d \to 0$. Hence

$$\lim_{d\to 0} P\{\theta_1 \in J_N\} = \lim_{t\to\infty} P\{(N(t)/\lambda_{11})^{1/2}|\hat{\theta}_1(N(t)) - \theta_1| \le d(N(t)/\lambda_{11})^{1/2}\}$$
$$= \Phi(u) - \Phi(-u) = 1 - \alpha.$$

Finally (iii) follows from Lemma 4.14.6 (Chow and Robbins, 1965, Lemma 2).

REMARK 4.18.1 It should be noted that (i) and (ii) of Theorem 4.18.1 are universally valid and assumption (4.18.3) is required only for the validity of (iii). However, in some cases it might be possible to establish (iii) without (4.18.3) by using Lemma 4.14.7 (Chow and Robbins, 1965, Lemma 3).

Examples (a) Consider the normal population having mean μ and variance σ^2 $(0 < \sigma^2 < \infty)$. Take $\theta_1 = \mu$, $\theta_2 = \sigma^2$. Then

$$((I_{ij})) = \begin{pmatrix} 1/\theta_2 & 0 \\ 0 & 1/2\theta_2^2 \end{pmatrix} \quad \text{and} \quad \lambda_{11}(\theta_1, \theta_2) = \theta_2 = \sigma^2.$$

The mle of θ_1 and θ_2 are $\hat{\theta}_1(n) = \bar{X}_n = n^{-1} \sum_{i=1}^n X_i$ and $\hat{\theta}_2(n) = n^{-1} \sum_{i=1}^n (X_i - \bar{X}_n)^2$. Instead of $\hat{\theta}_2(n)$, we can use $\hat{\theta}_2(n) = n\hat{\theta}_2(n)/(n-1) = s_n^2$ which is an unbiased and consistent estimator for θ_2. Hence, the following stopping rule is obtained:

$$N = \inf\{n \ge 2 : n \ge u_n^2 s_n^2/d^2\} \quad \text{and} \quad n^0 = (u^2\sigma^2)/d^2.$$

(b) For the normal population of example (a), let $\theta_1 = \sigma^2$, $\theta_2 = \mu$. Then

$$((I_{ij})) = \begin{pmatrix} 1/2\theta_1^2 & 0 \\ 2 & 1/\theta_1 \end{pmatrix} \quad \text{and} \quad \lambda_{11}(\theta_1, \theta_2) = 2\theta_1^2 = 2\sigma^4.$$

Thus, $N = \inf\{n \ge 2 : n \ge (2u_n^2 s_n^4)/d^2\}$ and $n^0 = (2u^2\sigma^4)/d^2$. Graybill and

Connell (1964a) have given a two-stage procedure for estimating σ^2 (see Section 4.2.4).

(c) Let $p(x; \theta) = \theta\, e^{(-\theta x)}$, $x \geq 0$, $0 < \theta < \infty$. Then

$$i(\theta) = E\{\partial \ln p(X; \theta)/\partial\theta\}^2 = \theta^{-2}$$

and

$$N = \inf\{n \geq 1 : n \geq u_n^2/(\overline{X}_n^2 d^2)\} \quad \text{and} \quad n^0 = u^2\theta^2/d^2.$$

The validity of assumption (4.18.3) in (a) and (b) follows from the following lemma which is proved from Wiener's (1939) theorem. However, it is not true in (c) and hence (iii) cannot be concluded from Lemma 4.14.6. We first state without proof Wiener's (1939) theorem.

Theorem 4.18.2 (Wiener, 1939) Let $\{X_n, n \geq 1\}$ be a sequence of iid random variables with $E|X_n|^r < \infty$ or $E|X_n|^r \ln^+|X_n| < \infty$ according as $r > 1$ or $r = 1$. Then

$$E\left(\sup_{n \geq 1} n^{-r}\left|\sum_{i=1}^{n} X_i\right|^r\right) < \infty$$

and conversely.

Lemma 4.18.2 (Khan, 1969) Under the assumption $0 < \sigma^2 < \infty$,

$$E\left(\sup_{n \geq 2} s_n^q\right) < \infty \qquad \text{for} \quad q \geq 2$$

where s_n^2 is the sample variance (unbiased version).

Proof For $q = 2$,

$$s_n^2 = (n-1)^{-1}\sum_{i=1}^{n}(X_i - \overline{X}_n)^2 = (n-1)^{-1}\sum_{i=1}^{n} X_i^2 - n(n-1)^{-1}\overline{X}_n^2.$$

Since

$$s_n^2 \leq (n-1)^{-1}\sum_{i=1}^{n} X_i^2 - n(n-1)^{-1}\overline{X}_n^2 + (n-1)^{-1}\overline{X}_n^2$$

$$+ 2\overline{X}_n^2 \leq X_1^2 + (n-1)^{-1}\sum_{i=2}^{n} X_i^2 + \overline{X}_n^2,$$

we have

$$\sup_{n\geq 2} s_n^2 \leq X_1^2 + \sup_{n\geq 2}(n-1)^{-1}\sum_{i=2}^{n} X_i^2 + \sup_{n\geq 2} n^{-2}\left(\sum_{i=1}^{n} X_i\right)^2.$$

Consequently using Theorem 4.18.2

$$E\left(\sup_{n\geq 2} s_n^2\right) < \infty \qquad \text{if} \quad EX^2 \ln^+|X|^2 < \infty \quad \text{and} \quad EX^2 < \infty.$$

However, $EX^2 \ln^+|X|^2 \leq EX^4 < \infty$ and $EX^2 < \infty$ are true for normal distribution with finite variance. Now let $q > 2$. Then

$$\begin{aligned} s_n^q = (s_n^2)^{q/2} &\leq \left[(n-1)^{-1}\sum_{i=1}^{n} X_i^2 + \overline{X}_n^2\right]^{q/2} \\ &\leq 2^{q/2}\left[(n-1)^{-q/2}\left(\sum_{i=1}^{n} X_i^2\right)^{q/2} + |\overline{X}_n|^q\right], \end{aligned}$$

after using the inequality $|a+b|^p \leq 2^p(|a|^p + |b|^p)$, $p \geq 1$. That is

$$s_n^q \leq 2^{q/2}\left[2^{q/2}\left\{|X_1^q| + (n-1)^{q/2}\left(\sum_{i=2}^{n} X_i^2\right)^{q/2}\right\} + |\overline{X}_n|^q\right].$$

we have

$$\sup_{n\geq 2} s_n^q \leq 2^{q/2}\left[2^{q/2}\left\{|X_1^q| + \sup_{n\geq 2}(n-1)^{-q/2}\left(\sum_{i=2}^{n} X_i^2\right)^{q/2}\right\} + \sup_{n\geq 1} n^{-q}\left|\sum_{i=1}^{n} X_i\right|^q\right].$$

Hence, $E(\sup_{n\geq 2} s_n^q) < \infty$ if $E|X_1|^q < \infty$ which is true for a normal distribution with finite variance. This completes the proof of the lemma.

REMARK 4.18.2 In the case of a single parameter family of distributions, the stopping variable N takes the form: $N = \inf\{n \geq m : n \geq u_n^2/d^2 i(\hat{\theta}_n)\}$ where $i(\theta) = E\{\partial \ln p(X;\theta)/\partial\theta\}^2$ and $\hat{\theta}_n$ is the mle of θ. However, if $i(\theta)$ is independent of θ, no sequential procedure is required since the bounded length confidence intervals of given coverage probability can be based on normal theory. More generally, no sequential procedure is required when $\lambda_{11}(\theta_1, \theta_2)$ depends only on θ_2 which is known. As an example, consider a normal distribution with unknown mean and known variance.

Problems

1. Set up a fixed-width confidence interval for (i) θ_1 when θ_2 is unknown and for (ii) θ_2 when θ_1 is unknown where $f(x; \theta_1, \theta_2) = \theta_2^{-1} \exp[-(x - \theta_1)/\theta_2]$, $x \geq \theta_1$.
2. Assume that the underlying population has the distribution function $F(x; \theta_1, \theta_2) = 1 - \exp\{-[(x - \theta_1)/\theta_2]^m\}$ for $x \geq \theta_1$ where m is a given positive constant. Set up a fixed-width confidence interval for (i) θ_1 when θ_2 is unknown and for (ii) θ_2 when θ_1 is unknown.

4.19 Sequential Estimation of the Size of a Finite Population

Consider the following model. An urn contains an unknown number N of white balls, and no others. An estimate of N is desired based on the following sampling scheme. Balls are drawn at random, one at a time, from the urn. A white ball is colored black before it is returned, a black ball is returned unchanged. The ball is always returned before the next ball is drawn. We are interested in two problems: (i) to find a stopping rule t, and (ii) to estimate N after we stop sampling. This problem has applications in estimating wildlife of a certain species or the number of fish in a pond, etc. This problem (also in a more general setup) has been considered by several authors, notably Goodman (1953), Chapman (1954), Darroch (1958), Darling and Robbins (1967d) and Samuel (1968). We shall present their results with more emphasis on Samuel's (1968) results.

Let w_i, b_i denote (the random) number of white balls and black balls respectively, observed in the first i draws ($w_i + b_i = i$). Then, consider the following stopping rules.

(A) Let $A > 0$ be a fixed integer. $t_A = A$.
(B) Let $B > 0$ be a fixed integer $t_B = \inf\{i : b_i = B\}$.
(C) Let $C > 0$ be fixed. $t_C = \inf\{i : b_i \geq Cw_i\} = \inf\{i : i \geq (C + 1)w_i\}$.
(D) Let $-\infty < D < \infty$ be fixed.

$$\begin{aligned} t_D &= \inf\{i : b_i \geq \max(1, w_i \ln w_i + w_i D)\} \\ &= \inf\{i : i \geq \max(w_i + 1, w_i \ln w_i + w_i(D + 1)\}. \end{aligned}$$

(E) Let $\{D_j\}$ be such that $\lim D_j = \infty$.

$$t_E = \inf\{i : b_i \geq \max(1, w_i \ln w_i + w_i D_{w_i})\}.$$

Since $w_i \leq N$, each of these rules is closed and thus clearly terminates finitely with probability one. Of these rules, rule B has been studied most [see Goodman (1953), Chapman (1954), and Darroch (1958)]. Rule D has been investigated by Darling and Robbins (1967d) who show that for any $0 < \alpha < 1$ and a suitable choice of D one can have $P_N(W_D = N) \geq 1 - \alpha$ uniformly in N, where W_D is the total of white balls observed before stopping.

We shall state the limiting distribution of w_i as $N \to \infty$, in a sample of fixed size i for various relationships between i and N. Different parts of the theorem have been proved by various authors [see, for example, Rényi (1962) where proper references are given]. Let $u_i - N - w_i$ = number of unobserved white balls in the sample of size i (= number of white balls in the urn after i draws). Since there are linear relationships between w_i, b_i, u_i, we shall express the limiting distribution for one of these variables. Let Φ denote the standard normal distribution function.

Theorem 4.19.1 Let $N \to \infty$.

(a) If $i = (2N\lambda_N)^{1/2}$ where $\lambda_N \to 0$, then $P_N(b_i = 0) \to 1$.
(b) If $i = (2N\lambda_N)^{1/2}$ where $\lambda_N \to \lambda$ and $0 < \lambda < \infty$, then

$$P_N(b_i = k) = e^{-\lambda}\, \lambda^k/k!, \qquad k = 0, 1, \ldots .$$

(c) If $i = Na_N$ where $\xi_N N^{-1/2} < a_N < \ln N - \mu_N$ and $\xi_N \to \infty$, $\mu_N \to \infty$, then

$$P_N((w_i - Ew_i)/(\text{var } w_i)^{1/2} \leq x) \to \Phi(x), \qquad -\infty < x < \infty.$$

(d) If $i = N \ln N + Na_N$ where $a_N \to a$ and $-\infty < a < \infty$, then

$$P_N(u_i = k) \to e^{-\lambda}\lambda^k/k!, \qquad k = 0, 1, \ldots, \quad \text{where} \quad \lambda = e^{-a}.$$

(e) If $i = N \ln N + Na_N$ where $a_N \to \infty$, then $P_N(w_i = N) \to 1$.

Maximum Likelihood Estimation of N

Samuel (1968) has considered mainly maximum likelihood estimate (mle) of N, denoted by $\hat{N}$. It turns out that if when we stop we have observed w white and b black balls, then $\hat{N} = \hat{N}(w, b)$ and does not depend on the stopping rule used, although the distribution of $\hat{N}$ clearly will depend on the stopping rule. Next, we shall derive the value of $\hat{N}(w, b)$. For any stopping rule t, the probability of having observed exactly w white balls and b black balls when we stop clearly depends on t as well as on N. Let $P_N(w, b\,|\,t)$ denote this probability. It can be shown that

$$P_N(w, b\,|\,t) = (N)_w\, h(w, b)/N^{w+b}, \quad w = 1, 2, \ldots, \quad b = 0, 1, \ldots, \qquad (4.19.1)$$

where $(N)_i = N(N-1)\cdots(N-i+1)$, and where $h(w, b)$ depends on t but not on N. For example, for rule A,

$$h(w, b) = \sum_{j=0}^{w} (-1)^{w-j} j^A \{j!(w-j)!\}^{-1} = S_A^{(w)}, \qquad A = w + b, \quad (4.19.2)$$

where $S_A^{(w)}$ is a Stirling number of the second kind and is defined by the identity in (4.19.2).

For rule B,

$$P_N(t_B = k) = (N)_{k-B}(k-B)S_{k-1}^{k-B}/N^k, \qquad k = B+1, \ldots, B+N, \quad (4.19.3)$$

and for rule C,

$$P_N(W_C = k) = P_N(t_C = [\gamma k] + 1) = (N)_k h_C(k)/N^{[\gamma k]+1}, \qquad k = 1, \ldots, N, \quad (4.19.4)$$

where $C + 1 = \gamma$, $[\gamma k]$ is the largest integer contained in γk, and $h_C(k)$ are constants given by

$$h_C(1) = 1, \quad h_C(k) = \left\{1 - \sum_{i=1}^{k-1} (k)_i \, h_C(i) k^{-[\gamma i]-1}\right\} k^{[\gamma k]+1} (k!)^{-1}, \qquad k \geq 2.$$

For proofs of the above, the reader is referred to Samuel (1968). Thus, for any w, b such that $h(w, b) \neq 0$, $\hat{N}(w, b)$ is the positive integer which maximizes $(N)_w/N^{w+b}$. Maximum likelihood estimation for the above problem has been considered by Darroch (1958), and for a mathematically equivalent problem by Lewontin and Prout (1956). Samuel (1968) points out that the claim of asymptotic normality of the mle in the work of Lewontin and Prout (1956, p. 22), is generally false in view of Theorem 4.19.1. Direct inspection yields

$$\hat{N}(w, 0) = \infty \quad \text{for} \quad w \geq 2, \qquad \hat{N}(1, b) = 1 \quad \text{for} \quad b \geq 1. \quad (4.19.5)$$

(The case $w = 1$, $b = 0$ is of no interest, since the first ball drawn is always white, and thus more than one draw must take place in order to obtain information about N.) Treating $(N)_w/N^{w+b}$ as a function of a positive real variable $N(N > w - 1)$, we have

$$d\{(N)_w/N^{w+b}\}/dN = \left\{\sum_{j=0}^{w-1} (N-j)^{-1} - (w+b)N^{-1}\right\}(N)_w N^{-(w+b)}. \quad (4.19.6)$$

Equating the right-hand side of (4.19.6) to zero yields that the mle $\tilde{N} = \tilde{N}(w,b)$ satisfies

$$(w+b)/\tilde{N} = \sum_{j=0}^{w-1} (\tilde{N}-j)^{-1}. \tag{4.19.7}$$

Clearly (4.19.7) has a unique finite solution, and

$$\tilde{N}(w,b) = [\tilde{N}(w,b)] \qquad \text{or} \qquad [\tilde{N}(w,b)] + 1 \qquad \text{or both,}$$

where $[x]$ is here the largest integer contained in x. The right-hand side of (4.19.7) is less than $\ln\{\tilde{N}/(\tilde{N}-w)\}$ and greater than $\ln\{(\tilde{N}+1)/(\tilde{N}-w+1)\}$. The solution of

$$(w+b)/x = \ln\{x/(x-w)\} \tag{4.19.8}$$

is given by $x = (w+b)/m(s)$, where $s = w/(w+b)$ and

$$m(s) \text{ is the solution of } s = (1-e^{-m})/m, \tag{4.19.9}$$

and can be obtained from existing tables. Thus, approximately

$$\tilde{N}(w,b) \approx (w+b)/m(s) \qquad \text{when} \quad s = w/(w+b). \tag{4.19.10}$$

The interpretation of (4.19.10) is that the mle is approximately proportional to the sample size, with the proportionality factor a function only of the proportion of white balls in the sample drawn. More accurate information about $\hat{N}$ can be obtained by considering the ratio of $(N)_w/N^{w+b}$ to $(N-1)_w/(N-1)^{w+b}$. In the following we shall give the results of Samuel (1968) without proofs.

Theorem 4.19.2 (Samuel, 1968) (i) For every fixed A and rule A

$$P_N(\hat{N} = \infty) > 0 \quad \text{for all } N \geq A, \qquad \text{and} \qquad \lim_{N\to\infty} P_N(\hat{N} = \infty) = 1. \tag{4.19.11}$$

Also, for every fixed N

$$\lim_{A\to\infty} P(\hat{N} = N) = 1. \tag{4.19.12}$$

(ii) For every fixed B and rule B

$$\lim_{N\to\infty} P_N(2B\hat{N}/N \leq x) = F_{2B}(x), \qquad -\infty < x < \infty$$

when F_{2B} is the distribution function of a χ^2_{2B} variable.

Proof See Samuel, (1968, proofs of Assertions 1 and 2, pp. 1060–1061).

For rule B, Goodman (1953) has shown that there exists a unique uniformly minimum variance unbiased estimator (UMVUE). Let $W_B = t_B - B =$ number of white balls seen until stopping. The UMVUE is given by

$$W_B{}^2/2B + (\tfrac{2}{3} - 1/6B)W_B + P_1/P_2 \tag{4.19.13}$$

where P_1 and P_2 are polynomials of degree $2B - 1$ in W_B. The mle and the UMVUE thus have the same asymptotic distribution. Darroch (1958, p. 351) has shown that the UMVUE equals $S_{t_B}^{(W_B)}/S_{t_B-1}^{(W_B)}$. For rule C also, it is easily seen that there exists a unique UMVUE and it is a function of W_C which we denote by $a_C(W_C)$ where $a_C(k)$ is the (unique) solution of the equations

$$\sum_{i=1}^{k} a_C(i)(k)_i\, h_C(i)k^{-[\gamma i]-1} = k, \qquad k = 1, 2, \ldots. \tag{4.19.14}$$

The solution is given by the inductive formula

$$a_C(1) = 1, \qquad a_C(k) = \left\{k - \sum_{i=1}^{k-1} a_C(i)(k)_i h_C(i)k^{-1-[\gamma i]}\right\}k^{-1-[\gamma k]}/h_C(k)k!,$$

$$k \geq 2. \tag{4.19.15}$$

Theorem 4.19.3 (Samuel, 1968) For rule C, let s be fixed, let $m(s)$ be as in (4.19.9) and let

$$H(s) = \{1 - sm(s)\}/(s - 1 + sm(s)). \tag{4.19.16}$$

Then,

$$[(w + b)/m(s) - H(s)] \leq \hat{N}(w, b) < (w + b)/m(s)$$

$$\text{for all } w, b \text{ with } w/(w + b) = s. \tag{4.19.17}$$

Proof See Samuel (1968, Assertion 3).

Theorem 4.19.4 (Samuel, 1968) Let $C + 1 = \gamma$,

$$\sigma^2 = m(\gamma^{-1})(\gamma - m(\gamma^{-1}))/(m(\gamma^{-1}) - C) \quad \text{and} \quad \sigma^{*2} = \sigma^2/(m(\gamma^{-1}))^2. \tag{4.19.18}$$

Then

$$\liminf_{N\to\infty} P_N((\hat{N} - N)N^{-1/2}\sigma^{*-1} \leq x) \geq \Phi(x), \qquad -\infty < x < \infty. \quad (4.19.19)$$

Proof See Samuel, (1968, Assertion 4). She did not succeed in obtaining the exact asymptotic distribution of $\hat{N}$.

The definition of t_D as given, rather than $\inf\{i : b_i \geq w_i \ln w_i + w_i D\}$ is necessary in order to prevent us from stopping with one observation only, that is, with no black balls observed, which we would have to whenever $D \leq 0$, according to the latter definition. For $D > 0$, the modification is redundant. A similar remark holds for rule E. t_D can take on the values $k + 1$ for $k = 1, \ldots, [e^{-D}]$, and the values $[k \ln k + k(D + 1)] + 1$ for $k = [e^{-D}] + 1, \ldots, N$. The exact distribution of t_D or t_E can be obtained along the same lines as the distribution of t_C and similarly the UMVUE can be obtained. Toward the asymptotic distribution of U_D = the number of unobserved white balls when we stop and $\hat{N}$ we have the following result.

Theorem 4.19.5 (Samuel, 1968) For rule D and every fixed D,

$$-\infty < D < \infty,$$

(i) $P_N(U_D = j) \to e^{-\lambda}\lambda^j/j!$, $j = 0, 1, \ldots$ where $\lambda = e^{-(D+1)}$;
(ii) $\lim_{N\to\infty} P_N(\hat{N} = N - j + [\lambda]) = e^{-\lambda}\lambda^j/j!$, $\quad j = 0, 1, \ldots$;
and for rule E, $\lim_{N\to\infty} P_N(\hat{N} = N) = 1$.

Proof See Samuel, (1968, Assertions 6 and 8).

Problem

Let a lot contain N, an unknown number, of good items and no others. An estimate of N is desired based on the following sampling scheme. Items are drawn at random one at a time from the urn. A good item is marked defective before it is returned to the lot; a defective item is returned as it is. Notice that the item is always returned before the next item is drawn. Suppose we have the following sequence of data: GGGDGGDDGGGDGD... where G denotes a good item and D denotes a defective item. Find a stopping rule t and propose an estimate for N after you stop sampling. Also give some of the properties of your estimate.

4.20 Consistency of Certain Sequential Estimators

Given an unbiased estimate and the sufficient statistic for a certain parameter of interest, one can follow the Rao–Blackwell procedure for obtaining another (perhaps better) estimator by taking the expectation of the given unbiased estimator conditional on the sufficient statistic. Wolfowitz (1947a) has considered the consistency of such estimators for the binomial parameter in the sequential case. Loynes (1969) has studied the consistency of the Rao–Blackwell type of sequential estimators for the general case. Berk (1969) has considered the strong consistency of such estimators.

Let $X_1, X_2, \ldots$ be a sequence of independent and identically distributed (iid) random variables, taking their values in a space $\mathscr{X}$. Let N denote a stopping time defined on the sequence $X_1, X_2, \ldots,$ taking positive integral values, with the property that for each $n \geq 1$, the event $\{N = n\}$ is determined by conditions on $X_1, X_2, \ldots, X_n$ only. We assume that all stopping times are finite with probability one. For brevity, we denote the n-tuple $(X_1, \ldots, X_n)$ by $X^{(n)}$. By Z_n we mean the order statistic calculated from $X^{(n)}$. If the X_i are real valued, this is just the usual order statistic; otherwise we can regard it as the function from $\mathscr{X}$ to the integers which describes how many of $X_1, X_2, \ldots, X_n$ are equal to a given x in $\mathscr{X}$. It is worthwhile to note that Z_{n+1} can be calculated from a knowledge of Z_n and X_{n+1}. We make the following basic assumptions.

(A1) For each n, a statistic V_n is given which is sufficient for $X^{(n)}$; V_n is not necessarily real valued.

(A2) For each n, V_n is a function of Z_n. (Which will certainly be satisfied if $V_n = Z_n$ or if V_n is a minimal sufficient statistic.)

When a stopping rule N is given we shall write V_N for the random variable which equals V_n when $N = n$; without causing any confusion we shall write V in the place of V_N. If we are dealing with a sequence of stopping rules N_i, we shall write $V^{(i)}$ in the place of V_{N_i}. Assume that for some parameter θ we have constructed an unbiased estimator T (defined on the sequence $X_1, X_2, \ldots$) which will remain fixed throughout the discussion.

(A3) $E_\theta T = \theta$; in particular $E_\theta T$ exists.

The estimator T will be assumed to depend on a finite (but not necessarily bounded) number of observations only; there is a stopping rule M for T such that if $M = n$, T is a function of $X_1, X_2, \ldots, X_n$ only. Any such estimator we call a sequential estimator; a fixed sample size estimator, depending on a fixed number of X_i, will be a special case of this obtained by setting $M = m$ with probability one for an appropriate integer m.

(A4) Any stopping rule N or N_i considered satisfies $N \geq M$ or $N_i \geq M$ with probability one.

The role of (A4) is, in conjunction with (A1), to ensure that the quantities $U(N)$ and $U(N_i)$ (to be defined later) are functions of the observations only.

If A is an event, $I(A)$ will denote the indicator function of A, which is unity on A and zero elsewhere.

CONSTRUCTION OF RAO–BLACKWELL ESTIMATORS Given the estimator T and a stopping rule N we define the new estimator $U = U(N)$ as the conditional expectation of T given N and V. That is,

$$U(N) = E(T \mid N, V_N). \tag{4.20.1}$$

Equivalently, $U = U_n(v)$ when $N = n$ and $V_n = v$ where

$$U_n(v) = E(T \mid N = n, V_n = v) = E(TI(N = n) \mid V_n = v)\{E[I(N = n) \mid V_n = v]\}^{-1}. \tag{4.20.2}$$

Assumptions (A1) and (A4) clearly imply that $U_n(v)$ is a function of $X_1, X_2, \ldots, X_n$ only, so that U is indeed a statistic with stopping rule N. Then U is also unbiased for θ, and has at least as small a risk as T for any convex function (indeed unbiasedness is not necessary for this property).

As a particular case we may take $V_n = Z_n$ which would lead in the case of fixed M and N to U statistics [see, for example, Fraser (1957)], and we may therefore regard such estimators as generalized U statistics. We note that in any case, because of (A2), U is a symmetric function of $X_1, \ldots, X_N$. Also, it is of interest to note that U is the unique minimum variance unbiased estimator when the pair (N, V) is complete.

CONSISTENCY OF $U_N(V)$. Wolfowitz (1947a) has shown that $\{U_N(V)\}$, when estimating the binomial parameter, is a consistent sequence provided the sequence of stopping rules $\{N_i\}$ satisfy a certain condition. His condition is that $n_{0,i} \to \infty$ as $i \to \infty$ where $n_{0,i}$ is the smallest value of n for which $P(N_i = n) > 0$. Loynes (1969) has shown that even in the general case consistency follows from weaker and more appealing conditions on the N_i. We shall write

$$U^{(i)} = U(N_i) = E(T \mid N_i, V^{(i)}). \tag{4.20.3}$$

Theorem 4.20.1 (Loynes, 1969) Suppose that assumptions (A1)–(A4) and in addition the following conditions (C1) and (C2) are satisfied: (C1). For each fixed k, $P(N_i \le k) \to 0$ as $i \to \infty$. (C2) There exists an integer $\lambda(i, k)$, which is monotone in k and which tends to ∞ if both i and k tend to ∞, such that for each k, $N_i = k$ if and only if $N_i \ge k$ and the set of random variables $Z_{\lambda(i,k)}, X_{\lambda(i,k)+1}, \ldots, X_k$ satisfies some condition (depending of course on i and k). Then $U^{(i)}$ converges to θ in probability and in the mean of order 1 as $i \to \infty$.

REMARKS (i) (C1) states that $N_i \to \infty$ in probability. Wolfowitz's (1947a) condition implies that $N_1 \to \infty$ with probability one. (ii) (C2) is not necessary, since for example, it obviously need only be required to hold for large i, but some condition obviously is. Its role is (effectively) to ensure that T and N_i are nearly independent for large i. (We do not need to make this precise, but the intention is to exclude stopping rules such as $N_i = [iX_i]$.) The form chosen here is a convenient one which is reasonable for most (though not all) applications. A somewhat simpler condition which implies (C2) is

(C2′) For each k, $N_i = k$ if and only if $N_i \geq k$ and Z_k satisfies some condition.

This is very similar to the extra condition imposed by Blackwell (1947). The stronger condition (C2′) is satisfied by the stopping rules (a) stop when $\sum_{j=1}^{n} X_j$ first crosses a barrier, (b) stop when the estimated variance of the sample mean, s^2/n first becomes smaller than some prescribed value, and (in the binomial case) (c) stop when the point whose coordinates are number of successes and number of failures first enters a region R. An example in which (C2) is satisfied but not (C2′) is given by defining $N_i = n$ if X_n is the first X_j for which $X_j > \max(X_1, X_2, \ldots, X_i)$.

Proof of Theorem 4.20.1 It would be straightforward to prove the theorem if Wolfowitz's (1947a) condition were satisfied, and the main difficulty is in taking advantage of the fact that it is almost satisfied. By (C1) there exists an increasing sequence of integers m_i tending to infinity with i, with the property that $P(N_i < m_i) \to 0$. For convenience, let $A_i = \{N_i \geq m_i\}$ and $k_i = \lambda(i, m_i)$ so that $k_i \to \infty$ and $P(A_i) \to 1$. Define

$$W_i = E[E[T \mid I(A_i), Z_{k_i}, X_{k_i+1}, X_{k_i+2}, \ldots] \mid N_i, V^{(i)}]; \qquad (4.20.4)$$

then when $N_i = j$ and $V_j = v$

$$W_i = E[E[T \mid I(A_i), Z_{k_i}, X_{k_i+1}, \ldots] I(N_i = j) \mid V_j = v]\{E[I(N_i = j) \mid V_j = v]\}^{-1} \qquad (4.20.5)$$

Now, if $j \geq m_i$ it follows from (C2) that $N_i = j$ if and only if $N_i \geq m_i$ and the set of random variables $Z_{k_i}, X_{k_i+1}, \ldots, X_j$ satisfy some condition mentioned in the preceding remark; consequently if $j \geq m_i$

$$\begin{aligned} W_i &= E[E[TI(N_i = j) \mid I(A_i), Z_{k_i}, X_{k_i+1}, \ldots] \mid V_j = v]\{E[I(N_i = j) \mid V_j = v]\}^{-1} \\ &= E[TI(N_i = j) \mid V_j = v]\{E[I(N_i = j) \mid V_j = v]\}^{-1} \end{aligned} \qquad (4.20.6)$$

because of (A2). Hence $W_i = U^{(i)}$ on $\{N_i \geq m_i\}$ and since $P(A_i) \to 1$

$$W_i - U^{(i)} \to 0 \qquad \text{in probability.} \qquad (4.20.7)$$

Now write

$$Y_i = E[T \mid I(A_i), Z_{k_i}, X_{k_i+1}, \ldots] \tag{4.20.8}$$

so that $W_i = E[Y_i \mid N_i, V^{(i)}]$. Then when $N_i \geq m_i$ or $I(A_i) = 1$,

$$Y_i = E[TI(A_i) \mid Z_{k_i}, X_{k_i+1}, \ldots]\{E[I(A_i) \mid Z_{k_i}, X_{k_i+1}, \ldots]\}^{-1}. \tag{4.20.9}$$

We have

$$\begin{aligned} &E[\,|1 - E[I(A_i) \mid Z_{k_i}, X_{k_i+1}, \ldots]|\,] \\ &\quad = E[\,|E[1 - I(A_i) \mid Z_{k_i}, X_{k_i+1}, \ldots]|\,] \leq E[1 - I(A_i)] \to 0 \end{aligned} \tag{4.20.10}$$

so that the denominator of (4.12) tends in the mean of order 1 and a fortiori in probability, to 1. If now we write

$$S_i = E[T \mid Z_{k_i}, X_{k_i+1}, \ldots] \tag{4.20.11}$$

it follows similarly that the difference between the numerator of (4.20.9) and S_i tends in probability to zero, and hence finally, recalling (C1)

$$Y_i - S_i \to 0 \qquad \text{in probability.} \tag{4.20.12}$$

Now, the S_i form a backward martingale, so that by Theorem 4.2 in Chapter VII of Doob (1953), S_i converges with probability 1 and in the mean of order 1 to a limiting random variable having expectation $ET = \theta$. From the zero–one law of Hewitt and Savage (1955) it follows that the limit is in fact constant, and therefore equal to θ with probability 1. Thus $S_i \to \theta$. Hence by (4.20.12) we have $Y_i \to \theta$ in probability. Now if $\varepsilon > 0$, writing $Y_i' = Y_i - \theta$ we have

$$\begin{aligned} \mathrm{E}[\,|Y_i - \theta|\,] &> E[\,|Y_i - \theta| I(|Y_i'| > \varepsilon)] + E[\,|Y_i - \theta| I(|Y_i'| \leq \varepsilon)] \\ &\leq E\{E|T - \theta|\,|I(A_i), Z_{k_i}, X_{k_i+1}, \ldots][I(|Y_i'| > \varepsilon)]\} + \varepsilon \\ &= E\{E[\,|T - \theta| I(|Y_i'| > \varepsilon) \mid I(A_i), Z_{k_i}, X_{k_i+1}, \ldots]\} + \varepsilon \\ &= E[\,|T - \theta| I(|Y_i'| > \varepsilon)] + \varepsilon \end{aligned} \tag{4.20.13}$$

which tends to zero as $i \to \infty$, and then $\varepsilon \to 0$ because $Y_i \to \theta$ in probability.

Consequently from the definition of W_i

$$\begin{aligned}E(|W_i-\theta|) &= E\{|E|Y_i-\theta||N_i,V^{(i)}]|\}\le E\{E[|Y_i-\theta|N_i,V^{(i)}]\}\\ &= E[|Y_i-\theta|]\to 0,\end{aligned}\tag{4.20.14}$$

and a fortiori W_i converges to θ in probability, so that by (4.20.7) $U^{(i)}\to\theta$ in probability. That $U^{(i)}$ also converges to θ in the mean of order 1 follows by applying an argument similar to that used in (4.20.13). This completes the proof of the theorem.

Examples Loynes (1969) has given five examples, the first two of which are entirely concerned with fixed-sample sizes. All except the fifth have T depending only on a fixed-sample size. All except the fourth satisfy (C2′).

(i) Suppose that X_j are normally and independently distributed with mean μ and variance 1, that $T=X_1{}^2-1$ so that $\theta=\mu^2$, that $N_i=i$, and that we use the minimal sufficient statistic $V_n=\sum_1^n X_j$. Then we find

$$U^{(i)}=\left(\sum_1^i X_j\right)^2 i^{-2}-i^{-1},\tag{4.20.15}$$

which converges to θ in probability and in the mean of order 1 according to Theorem 4.20.1 and with probability 1 according to the strong law. As a matter of interest the variance of $U^{(i)}$ is

$$2i^{-2}+4\mu^2 i^{-1}\tag{4.20.16}$$

whereas that of the U statistic associated with T is $2i^{-1}+4\mu^2 i^{-1}$.

(ii) Suppose that the X_j are independently distributed, uniformly on $(0,2\theta)$, that $T=X_1$, that $N_i=i$, and that $V_n=M_n$ where

$$M_n=\max(X_1,X_2,\ldots,X_n)\tag{4.20.17}$$

the minimal sufficient statistic. Then $U^{(i)}=(i+1)(2i)^{-1}M_i$, whose behavior is in general terms the same as that in example (i).

(iii) Let the (real) X_j be independent and identically distributed with continuous distribution extending to $+\infty$, let $V_n=Z_n$, and let N_i be the first n for which $M_n>i$; again suppose $T=X_1$. Then if $N_i=1$, $U^{(i)}=X_1$; otherwise, if $N_i=n\ge 2$ and $Z_n=(z_1,z_2,\ldots,z_n)'$ where $X_n=z_n>i$ and $z_1\le z_2\le\cdots\le z_{n-1}$, we have $X_1=z_j$ with probability $(n-1)^{-1}$ for each j

$(1 \le j \le n-1)$. So

$$U^{(i)} = \begin{cases} z_1 = X_1 & \text{if} \quad N_i = 1, \\ \sum_{j=1}^{n-1} z_j(n-1)^{-1} = \sum_{j=1}^{n-1} X_j(n-1)^{-1} & \text{if} \quad N_i = n \ge 2. \end{cases} \tag{4.20.18}$$

By Theorem (4.20.1) $U^{(i)}$ converges to $\theta = EX$ in probability and in the mean of order 1. In both this example and the next, we can also show that convergence with probability 1 occurs, because Theorem 1 of Richter (1965) applies.

(iv) Suppose T, X_j, and V_n are as in example (iii), and let N_i be the first $n > i$ for which $M_n > M_i$. Then when $N_i = n$ and $Z_n = (z_1, z_2, \ldots, z_n)$ we have $X_n = z_n$ and z_{n-1} occurs among $X_1, X_2, \ldots, X_i$; thus $T = z_{n-1}$ with probability i^{-1}, and $T = z_j$ with probability $(i-1)/(n-2)i$ for $1 \le j \le n-2$. It follows that

$$U^{(i)} = z_{n-1}i^{-1} + (i-1)i^{-1}(z_1 + z_2 + \cdots + z_{n-2})(n-2)^{-1}. \tag{4.20.19}$$

The convergence behavior is as in example (iii).

(v) Suppose the situation is as in example (iii), except that $T = X_{N_1}$, and consider as before the case when $N_i = n$ and $Z_n = (z_1, z_2, \ldots, z_n)'$. Then clearly, if $n = 1$, $N_1 = 1$; if $n > 1$ and $z_n \le 1$, $N_1 = n$; and if $n > 1$ and $z_1 < z_2 < \cdots < z_r \le 1 < z_{r+1} < \cdots < z_n$, $X_{N_1} = z_j$ with probability

$$(n-r-1)^{-1} \quad \text{for} \quad r+1 \le j \le n-1.$$

Thus

$$U^{(i)} = \begin{cases} z_1 = X_1 & \text{if} \quad n = 1, \\ z_n = X_n & \text{if} \quad n > 1 \quad \text{and} \quad z_{n-1} \le 1, \\ (z_{r+1} + \cdots + z_{n-1})(n-r-1)^{-1} & \text{if} \quad n > 1 \quad \text{and} \quad z_r \le 1 < z_{r+1} \end{cases} \tag{4.20.20}$$

and convergence occurs in probability and in the mean of order 1 to $\theta = ET$.

Intuition suggests that almost sure convergence of $U^{(i)}$ to θ ought to hold, provided the requirements on N_i are strengthened. Possible conditions which suggest themselves are $N_i \to \infty$ with probability 1 or $N_{i+1} \ge N_i$ for all i. Loynes (1969) could not prove this in general except in a certain nonsequential situation. Berk (1969) has shown that the sequence of Rao–Blackwell type of estimators in the sequential case are strongly consistent provided the sequence of stopping times $\{N_i\}$ is such that $N_i \to \infty$ a.s. and the N_i are C-ordered [see Berk (1969, Definition (2.3))]. The reader is referred to Berk's (1969) paper for further details on the assertion and more specific considerations.

4.21 Asymptotically Optimal Bayes Sequential Estimation

Every sequential procedure S prescribes a stopping rule N and an estimator d. Moreover, the posterior distribution of the parameter θ, given $\{N = n\}$ is independent of the stopping rule; consequently the estimation rule d is independent of the stopping rule N for any specified prior distribution $G(\theta)$ and the loss function $r(d; \theta)$. Let $r_1(S; \theta)$ denote the loss due to erroneous estimation and $r_2(S; \theta)$ the loss (cost) due to stopping according to the rule N. The prior risk functions (or the expected risk) for a prior probability measure G on $(\Theta, \mathscr{C})$ will be denoted by $R_1(S; G)$ and $R_2(S; G)$ where

$$R_1(S; G) = \int dG(\theta) \sum_{n=0}^{\infty} \int_{\{N=n\}} r_1(d(x_1, \ldots, x_n); \theta) \prod_{i=1}^{n} dF(x_i; \theta) \tag{4.21.1}$$

and

$$R_2(S; G) = \int dG(\theta) \sum_{n=0}^{\infty} \int_{\{N=n\}} r_2(S(n); \theta) \prod_{i=1}^{n} dF(x_i; \theta). \tag{4.21.2}$$

If $r_2(S; \theta) = cN(S; G)$ then $R_2(S; G)$ will assume a somewhat simpler form. In general, it is somewhat difficult to determine the optimal Bayes sequential stopping rule. However, one can determine stopping rules that are asymptotically as good as the Bayes stopping rules. If $r_2(S; \theta)$ is a function $cK(N(S, G))$, where $N(S; G)$ is the random sample size obtained by S under G, then if $R_2(S; G) < \infty$, $R_2(S; G) \to 0$ as $c \to 0$. Thus, as $c \to 0$, the optimal stopping rule will generally give very large samples. Let T denote the class of all stopping rules.

Definition 4.21.1 (Kiefer and Sacks, 1963) A stopping rule N^0 in T is *asymptotically optimal* (AO) if

$$\limsup_{c \to 0} \left\{ R(N^0; G) \left[\inf_{N \in T} R(N; G) \right]^{-1} \right\} \le 1 \tag{4.21.3}$$

where

$$R(N^0; G) = R_1((N^0; \hat{\theta}_G); G) + R_2(N^0; G), \tag{4.21.4}$$

and $\hat{\sigma}_G$ is the Bayes estimate which minimizes $R_1(S; G)$.

Bickel and Yahav (1967, 1969) and Yahav and Bickel (1968) have investigated stopping rules that are asymptotically pointwise optimal in the Bayesian sense. Let $R(X_n; G)$ denote the Bayes (posterior) risk (that is, the expected loss with respect to the posterior distribution of θ) associated with the Bayes estimator when the sample size is n, and $r_2(S; \theta) = cK(N(S, G))$.

Definition 4.21.2 (Bickel and Yahav, 1967) A stopping rule N is said to be *asymptotically pointwise optimal* (APO) if

$$\limsup_{c \to 0} \left\{ [R(X_N; G) + cK(N)] \left[\inf_{N \in T} \{R(X_N; G) + cK(N)\} \right]^{-1} \right\} \leq 1 \quad \text{a.s.} \tag{4.21.5}$$

Notice that under a very weak condition for constant cost of observation the property of APO implies the property of asymptotic optimality (AO), given by (4.21.3) [see Bickel and Yahav (1967, Theorem 3.1)]. For the following discussion, let Y_n denote the Bayes posterior risk when the sample is of size n. Obviously $Y_n > 0$ a.s. Assume also that $Y_n \to 0$ a.s. The conditions for the validity of this assumption will be discussed later. Let $K(n)$ be a strictly increasing function of n, with $\lim_{n \to \infty} K(n) = \infty$. Define, for each c $(0 < c < \infty)$

$$X(n, c) = Y_n + cK(n). \tag{4.21.6}$$

If $\mathscr{B}_n$ denotes the Borel sigma-subfield generated by the first n observations, $Y_n \in \mathscr{B}_n$, all $n \geq 1$. Let $\tilde{T}$ be the class of all stopping rules $N(c)$ defined on the sequence $\{\mathscr{B}_n, n \geq 1\}$. We have $\{N(c) = n\} \in \mathscr{B}_n$ for all $n \geq 1$ and all c. Let $\beta > 0$ be real and suppose the following conditions are satisfied:

(i) $n^\beta Y_n \to V$ a.s. where $P(0 < V < \infty) = 1$.

(ii) For any c $(0 < c < \infty)$ and x $(0 < x < \infty)$ let $N(x, c)$ be the least positive integer n that minimizes the function $h(x, c, n) = xn^{-\beta} + cK(n)$. We assume that $N(x, c) < \infty$.

(iii) Let $\Delta_n K(n) = K(n + 1) - K(n)$. Then $n^{-(1+\beta)} = o(\Delta_n K(n))$, as $n \to \infty$.

Assumption (iii) guarantees that it would be optimal to stop sampling. According to (i) the Bayes risk is asymptotically of the order $O(n^{-\beta})$. Hence if $\Delta_n K(n) = O(n^{-(1+\beta)})$ it is worthwhile to continue sampling, provided the sample is sufficiently large. Consider the following stopping rule:

$$\bar{N}(c) = \text{least positive integer } n \text{, such that}$$
$$Y_n\{1 - (n/(n + 1))^\beta\} \leq c\, \Delta_n K(n). \tag{4.21.7}$$

Then we have the following theorem.

Theorem 4.21.1 (Yahav and Bickel, 1968) If assumptions (i)–(iii) hold, the stopping rule $\overline{N}(c)$ is APO, that is,

$$\lim_{c\to 0}\left\{X(\overline{N}(c), c)\left[\inf_{N\in T} X(N, c)\right]^{-1}\right\} \le 1 \quad \text{a.s.} \tag{4.21.8}$$

Proof From (i) $Y_n\{1-(n/(n+1))^{\beta}\}$ is asymptotically equivalent to $Vn^{-(\beta+1)}$ and by (iii) we have $P(\overline{N}(c)<\infty)=1$ for all $c>0$. By (ii) we have

$$X(\overline{N}(c), c) = \min_n h(\overline{N}^{\beta}(c)Y_{\overline{N}(c)}, c, n). \tag{4.21.9}$$

Let $n_0(c)$ be the first n for which

$$X(n_0(c), c) = \min_n X(n, c). \tag{4.21.10}$$

The a.s. existence of $n_0(c)$ follows from the fact that $Y_n>0$ a.s. and $Y_n\to 0$ a.s. and $K(n)\uparrow\infty$ strictly as $n\to\infty$.

From (4.21.9) and (4.21.10) we obtain

$$X(n_0(c), c) \le X(\overline{N}(c), c) \le \overline{N}^{\beta}(c)Y_{\overline{N}(c)}/n_0{}^{\beta}(c) + cK(n_0(c)). \tag{4.21.11}$$

The above facts and (4.21.10) also imply that as $c\to 0$, $n_0(c)\uparrow\infty$ a.s. The facts about Y_n and (4.21.9) imply that $\overline{N}(c)\uparrow\infty$. Hence $\overline{N}^{\beta}(c)Y_{\overline{N}(c)}\to V$ a.s.; and since $n_0(c)\uparrow\infty$ a.s., we get $n_0{}^{\beta}(c)Y_{n_0(c)}\to V$ a.s. as $c\to 0$. We remark that for any three sequences of real numbers x_n, x_n', a_n such that $x_n\to x$, $x_n'\to x$ with $x\neq 0$, we have $(x_n+a_n)/(x_n'+a_n)\to 1$. Upon dividing the left-hand side expression of (4.21.11) by the right-hand side expression and using the preceding remark, we have

$$\limsup_{c\to 0}\{X(\overline{N}(c), c)/X(n_0(c), c)\} = 1 \quad \text{a.s.} \tag{4.21.12}$$

which implies that $\overline{N}(c)$ is APO.

Corollary 4.21.1.1 Let $h(x, c, n) = xn^{-\beta} + cK(n)$. If (i)–(iii) hold, then

$$\limsup_{c\to 0}\left\{X(\overline{N}(c), c)\left[\inf_n h(V, c, n)\right]^{-1}\right\} = 1. \tag{4.21.13}$$

Proof See Yahav and Bickel (1968, Corollary 2.2).

A variant of $\bar{N}(c)$, which is also APO is the following stopping rule:

$$N'(c) = \text{least positive integer } n \text{ such that } Y_n \beta(n+1)^{-1} \leq c. \quad (4.21.14)$$

Then, we have the following results of Bickel and Yahav (1967) and Yahav and Bickel (1968).

Theorem 4.21.2 (Yahav and Bickel, 1968) If a sequence $\{Y_n\}$ of Bayes posterior risks satisfy the conditions (i)–(iii) with $K(n) = n$, and

$$\sup_{n \geq 1} n^{\beta} E(Y_n) < \infty, \quad (4.21.15)$$

then the stopping rules $\bar{N}(c)$ and $N'(c)$ are AO.

Proof See Yahav and Bickel (1968, Theorem 3.1).

If $r_1(d; \theta) = (d - \theta)^2$, θ real, and the distribution of X is regular in the sense of Cramér–Rao, then the posterior risk of the Bayes estimator is asymptotically equivalent to the variance of the maximum likelihood estimator (mle).

Theorem 4.21.3 (Bickel and Yahav, 1967) If the density function of X is $f(x; \theta)$, $\theta \in \Theta$, which is an interval on the real line, and the following conditions hold:

(i) $(\partial^2/\partial\theta^2) \ln f(X; \theta)$ is continuous in θ a.s. $[P_\theta]$;

(ii) $E_\theta\{\sup_{|\psi-\theta| \geq \varepsilon}[\ln f(X; \psi) - \ln f(X; \theta)]\} < 0$ for almost all θ and all $\varepsilon > 0$;

(iii) $E_\theta\{\sup_{|\psi-\theta| < \varepsilon}[(\partial^2/\partial\theta^2) \ln f(X; \psi)]^+\} < \infty$ for some $\varepsilon > 0$ and almost all θ;

(iv) the Fisher information function $I(\theta)$ exists and is positive;

(v) the mle $\hat{\theta}_n$ of θ exists and $\hat{\theta}_n \to \theta$ a.s. $[P_\theta]$;

(vi) the prior distribution $G(\theta)$ has a continuous and bounded density and has a finite second moment.

Then, $nY_n \to I^{-1}(\theta)$ a.s. $[P_\theta]$ where $Y_n = \text{var}\{\theta \mid X_1, \ldots, X_n\}$ and P^θ denotes the underlying probability measure.

Proof See Bickel and Yahav (1967, Theorem 3.2).

Notice that if $\beta = 1$, Theorem 4.21.2 holds and the stopping rules $\bar{N}(c)$ and $N'(c)$ are APO. Yahav and Bickel (1968), also give conditions for the APO of $\bar{N}(c)$ and $N'(c)$ when the loss function is $r_1(d; \theta) = W(|d - \theta|)$, θ is vector valued, the cost of observations is $cK(n)$ and β is positive and real.

In summary, under certain regularity conditions, Bickel and Yahav (1967), and Yahav and Bickel (1968) have shown that

$$X(\overline{N}(c), c) = 2c^{1/2}I^{-1/2}(c) + o(c^{1/2}) \quad \text{a.s.}, \tag{4.21.16}$$

$$X(\overline{N}(c), c) - X(c) = o(c^{1/2}) \quad \text{a.s.}, \tag{4.21.17}$$

$$E(X(c)) = 2c^{1/2}E(I^{-1}(\theta)) + o(c^{1/2}), \tag{4.21.18}$$

and

$$E(X(\overline{N}(c), c)) - E(X(c)) = o(c^{1/2}) \tag{4.21.19}$$

where $X(c) = \inf_n X(n, c)$. Bickel and Yahav (1969) show that the term $o(c^{1/2})$ in (4.21.16) and (4.21.17) can be refined to be $o(c^{3/4-\varepsilon})$ for all $\varepsilon > 0$ (see their Theorem 2.1). They also refine the $o(c^{1/2})$ term in (4.21.18) and (4.21.19) and show

$$E[X(c) - 2[I^{-1}(\theta)c]^{1/2}] = \max[o(c^{1/2+\delta(\lambda,b)-\varepsilon}), o(c)] \tag{4.21.20}$$

where (see their Theorems 3.1 and 3.2)

$$\delta(\lambda, b) = \tfrac{1}{2}(\lambda - 1)b(b + (\lambda - 1)^{-1}) \tag{4.21.21}$$

and b and λ depend on the problem (typically $\lambda = \frac{3}{2}$). Also,

$$E(X(\overline{N}(c), c) - 2[I^{-1}(\theta)c]^{1/2})^{+} = \max(O(c^{\lambda/2}), O(c)) \tag{4.21.22}$$

where again typically $\lambda = 3/2$. (See their Theorem 4.1.)

Example 4.21.1 (Estimation of the normal mean) We wish to estimate μ with quadratic loss on the basis of $X_1, \ldots, X_n, \ldots$ where the X_i are iid normal having mean μ and variance 1. Also μ has a normal prior having mean μ_0 and variance σ^2. In this case it is easy to compute

$$Y_n = (n + \sigma^{-2})^{-1} \tag{4.21.23}$$

and a direct computation yields that the Bayes rule is taking a fixed sample of size $N(c)$ observations where $N(c)$ is the positive integer closest to $(c^{1/2}\sigma^{-1} - \sigma^{-2})$. Similarly, $\overline{N}(c)$ takes $\frac{1}{2}\{-(1 + \sigma^{-2}) + ((1 - \sigma^{-2})^2 + 4c^{-1})^{1/2}\}$ observations and $|N(c) - \overline{N}(c)| = O(c)$, which is the best possible bound.

Bickel and Yahav (1969) also deal with estimation of the binomial parameter when it has a beta prior. The reader is referred to their paper for further details.

4.22 Sequential Nonparametric Confidence Intervals for Certain Parameters of the Population

4.22.1 p-Point of a Distribution Function

Farrell (1966a, b) has considered this problem and we shall present his results. For $0 < p < 1$, a number $\gamma(p, F)$ is a p-point of the distribution function F if $F(\gamma) \geq p$ while $F(\gamma-) \leq p$. Given $d > 0$ and $0 < \alpha < 1$, a (sequential) confidence interval procedure is a $d - \alpha$ bounded length confidence interval procedure if when sampling stops an interval of length not exceeding $2d$ is given which covers $\gamma(p, F)$ with probability at least $1 - \alpha$. Farrell (1966a) defines a measure of flatness by

$$\varepsilon_F = \sup_{0<\rho<1} \min[F(\gamma(p, F) + 2\rho d) - p, p - F(\gamma(p, F) + 2(\rho - 1) d]. \tag{4.22.1}$$

We shall be interested in confidence interval procedures that may be applied to observations on $F \in \mathscr{F}$, where $F \in \mathscr{F}$ implies that $\varepsilon_F > 0$. Then Farrell (1966a) has obtained the following negative result.

Theorem 4.22.1 (Farrell, 1966a) If $m \geq 1$ and if $\mathscr{F}$ contains all distribution functions F having bimodal density functions, then there does not exist an m-stage $d - \alpha$ bounded length confidence interval procedure that works for all $F \in \mathscr{F}$.

Proof See Farrell (1966a).

Farrell (1966a) points out that a similar (unpublished) result has been proved by Kiefer and Weiss (1964). $F \in \{\mathscr{F} : \varepsilon_F > 0\}$ implies that F is unimodal. Weiss (1960) has shown the existence of two-stage $d - \alpha$ bounded length confidence interval procedures for quantiles of unimodal populations. Examples of sequential procedures have also been constructed by Kiefer and Weiss (1964), however, these results have not been published. Farrell (1966b) proposes two sequential procedures for constructing bounded-width confidence intervals for the p-point of a distribution function. The procedures constructed are measurable functions of the iid random variables $\{X_n, n \geq 1\}$. Two sets of functions $\{u_{i,n}, n \geq 1\}$ and $\{v_{i,n}, n \geq 1\}$ $i = 1, 2$ are constructed. Properties of the sets of functions are as follows.

(P1) For $i = 1, 2$, and $n \geq 1$, the $u_{i,n}$ and $v_{i,n}$ are real-valued Borel-measurable functions defined on Euclidean n-space, and

$$u_{i,n}(x_1, \ldots, x_n) \leq v_{i,n}(x_1, \ldots, x_n) \quad \text{for all} \quad (x_1, \ldots, x_n) \text{ in } n\text{-space.}$$

(P2) For $i = 1, 2$ and $n \geq 1$ let $U_{i,n} = u_{i,n}(X_1, \ldots, X_n)$ and $V_{i,n} = v_{i,n}(X_1, \ldots, X_n)$. There exist real numbers α and β such that $0 < \alpha < 1$ and $0 < \beta < 1$ and such that if $F \in \mathscr{F}$ then

$$P_F(\text{all } n \geq 1, U_{i,n} \leq \gamma(p, F)) \geq 1 - \beta$$

and (4.22.2)

$$P_F(\text{all } n \geq 1, V_{i,n} \geq \gamma(p, F)) \geq 1 - \alpha.$$

(P3) If F is a distribution function and $\gamma(p, F)$ is the unique p-point of F then for $i = 1, 2$,

$$P_F\left(\lim_{n\to\infty} (V_{i,n} - U_{i,n}) = 0\right) = 1. \tag{4.22.3}$$

(P4) If g is a strictly increasing function mapping the real line into the real line, then for $i = 1, 2$, $n \geq 1$ and all $x_1, \ldots, x_n$,

$$\begin{aligned} u_{i,n}(g(x_1), \ldots, g(x_n)) &= g(u_{i,n}(x_1, \ldots, x_n)) \\ v_{i,n}(g(x_1), \ldots, g(x_n)) &= g(v_{i,n}(x_1, \ldots, x_n)) \end{aligned} \tag{4.22.4}$$

(P5) For $i = 2$, and $n \geq 1$, $u_{i,n} \leq u_{i,n+1} \leq v_{i,n+1} \leq v_{i,n}$.

The stopping variable $N \geq 1$ (relative to $\{X_n, n \geq 1\}$) is such that the event $\{N \geq j\}$ is independent of $\{X_n, n \geq j\}$, $j \geq 1$. Then the basic result of Farrell (1966b) is Theorem 4.22.2.

Theorem 4.22.2 (Farrell, 1966b) Let N be a stopping variable. If F is a distribution function, then $P_F(U_{i,N} \leq \gamma(p, F) \leq V_{i,N}) \geq 1 - \alpha - \beta$. If F is a continuous distribution function, then the numbers P_F [all $n \geq 1$, $V_{i,n} \leq \gamma(p, F)$] and P_F [all $n \geq 1$, $V_{i,n} \geq \gamma(p, F)$] are free of F. If N is a stopping variable and if G is obtained from F by translation of the random variables, then

$$P_G(U_{i,N} \leq \gamma(p, F) \leq V_{i,N}) = P_F(U_{i,N} \leq \gamma(p, F) \leq V_{i,N}).$$

Proof

$$\begin{aligned} 1 - P_F(U_{i,N} &\leq \gamma(p, F) \leq V_{i,N}) \\ &\leq P_F(\text{some } n \geq 1, U_{i,n} > \gamma(p, F)) \\ &\quad + P_F(\text{some } n \geq 1, V_{i,n} < \gamma(p, F)) < \alpha + \beta. \end{aligned}$$

To prove that various probabilities are free of F, if F is continuous, we may choose a strictly increasing function g such that if Y_1 is uniform on $[0, 1]$, then $g(Y_1)$ has F for its distribution function. Then if $\{Y_n, n \geq 1\}$ is a sequence of iid uniform variables, for $i = 1, 2$ we have

$$P(\text{all } n \geq 1, u_{i,n}(Y_1, \ldots, Y_n) \leq p) = P(\text{all } n \geq 1, u_{i,n}(g(Y_1), \ldots, g(Y_n)) \leq g(p))$$
$$= P_F(\text{all } n \geq 1, U_{i,n} \leq \gamma(p, F)).$$

Analogous proofs can be constructed for the other parts of the theorem.

We shall usually choose N to be the least integer n such that $V_{i,n} - U_{i,n} \leq L$, $i = 1, 2$. By virtue of (4.22.3), $P_F(N < \infty) = 1$ if N is defined in this manner. The functions $u_{i,n}$ and $v_{i,n}$ for $i = 1, 2$ and $n \geq 1$ are constructed from the order statistics, namely $X_{n,1} \leq X_{n,2} \leq \cdots \leq X_{n,n}$. We will use the integer sequences $\{a_n, n \geq 1\}$ and $\{b_n, n \geq 1\}$ satisfying

$$\text{If } n \geq 1, \text{ then } a(n) > b(n); \lim_{n\to\infty} a(n)/n = \lim_{n\to\infty} b(n)/n = p. \tag{4.22.5}$$

As an indication of the method, we define $u_{1,n}$ and $v_{1,n}$, $n \geq 1$ as

$$U_{1,n} = X_{n,b(n)+1}, \qquad V_{1,n} = X_{n,a(n)}. \tag{4.22.6}$$

The functions so defined clearly possess properties (P1) and (P4). Relation (4.22.3) follows from Theorem 4.22.3.

Theorem 4.22.3 (Farrell, 1966b) Let $\gamma^+(p, F) = \inf\{\gamma : F(\gamma) > p\}$ and $\gamma^-(p, F) = \sup\{\gamma : F(\gamma) < p\}$. Then

$$1 = P_F\left(\liminf_{n\to\infty} X_{n,b(n)+1} \geq \gamma^-(p, F)\right) = P_F\left(\limsup_{n\to\infty} X_{n,a(n)} \leq \gamma^+(p, F)\right).$$

Proof For real t, and $n \geq 1$ let $S_n(t) = \sum_{i=1}^n I(X_i \leq t)$ where $I(B)$ denotes the indicator function of the event B. Then $\{X_{n,b(n)+1} \leq t\} = \{S_n(t) \geq b(n)\}$ and $\{X_{n,a(n)} > t\} = \{S_n(t) < a(n)\}$. Hence

$$P_F\left\{\liminf_{n\to\infty} S_{n,b(n)+1} \geq t\right\} \geq P(\text{all but a finite number of } n, S_n(t) \leq b(n)).$$

However, $S_n(t)$ is the sum of n iid Bernoulli random variables such that $P_F(S_1(t) = 1) = F(t)$. Hence, $t < \gamma^-(p, F)$ implies that $F(t) < p$, and since $\lim_{n\to\infty} b(n)/n = p$, it follows from the law of large numbers that P_F (all but a finite number of n, $S_n(t) \leq b(n)) = 1$. Since this holds for all $t < \gamma^-(p, F)$, the first half of Theorem 4.22.3 follows. Similar argument holds for the other half.

Now $P_F(N < \infty) = 1$ if (4.22.5) holds and if $\gamma^+(p, F) - \gamma^-(p, F) < L$. Now, it remains to verify (4.22.2) for the above procedure. This has to do with the proper choice of the integer sequences $\{a(n),\ n \geq 1\}$ and $\{b(n),\ n \geq 1\}$. The basic discussion is contained in the works of Farrell (1962, 1964a). It has been shown there that it is possible to choose sequences satisfying (4.22.5) such that for all F,

$$\begin{aligned} F(t) \leq p \quad &\text{implies} \quad P_F(\text{some } n \geq 1,\ S_n(t) \geq a(n)) \leq \alpha, \\ F(t) \geq p \quad &\text{implies} \quad P_F(\text{some } n \geq 1,\ S_n(t) \leq b(n)) \leq \beta. \end{aligned} \tag{4.22.7}$$

So, all that is needed in the sequel is (4.22.7). The reader is referred to Farrell (1962, 1964a) for the method. For the above sequential procedure,

$$\begin{aligned} P\ (\text{some } n \geq 1,\ U_{1,n} > \gamma(p, F)) &= P_F(\text{some } n \geq 1,\ X_{n,\,b(n)+1} > \gamma(p, F)) \\ &= P_F[\text{some } n \geq 1,\ S_n(\gamma(p, F)) \leq b(n)] \leq \beta \end{aligned}$$

Analogously

$$P_F(\text{some } n \geq 1,\ V_{1,n} < \gamma(p, F)) \leq \alpha.$$

Hence (4.22.2) follows. So, the sequential procedure will always give an interval covering $\gamma(p, F)$ with probability not less than $1 - (\alpha + \beta)$. However the closure of the stopping variable and more generally the expected sample size, depend on the flatness of F about its p-points. Farrell (1966b) proves the following result.

Theorem 4.22.4 (Farrell, 1966b) There exist choices of the sequences $\{a(n),\ n \geq 1\}$ and $\{b(n), n \geq 1\}$ such that if $\mathscr{F} = \{F\colon \varepsilon_F > 0$ and F is continuous$\}$, then

$$\limsup_{\varepsilon_F \to 0^+} (\varepsilon_F)^2 \{\ln(\ln \varepsilon_F)^{-1})\}^{-1} E_F N \leq 4(1 - \alpha)p(1 - p). \tag{4.22.8}$$

Proof See Farrell (1966b, Theorem 3.1).

For the second procedure Farrell (1966b) defines the random variables

$$\begin{aligned} U_{2,n} &= \sup\{t\colon N_t \leq n,\ S_{N_t}(t) \leq b(N_t)\}, \\ V_{2,n} &= \inf\{t\colon N_t \leq n,\ S_{N_t}(t) \geq a(N_t)\}. \end{aligned} \tag{4.22.9}$$

For details on this procedure the reader is referred to Farrell (1966b, Section 2.)

4.22.2 Other Population Parameters

Geertsema (1970a) has applied the methods of Chow and Robbins (1965) in order to construct a sequential nonparametric confidence interval procedure for certain population parameters. In the following we shall present his results.

A General Method for Constructing Bounded Length Confidence Intervals

Let $X_1, X_2, \ldots, X_n$ be a fixed random sample of size n from a population having F for its cumulative distribution function (cdf) and let θ be a parameter of the population. We are interested in constructing for θ a confidence interval of length not larger than $2d$. For each positive integer n, consider two statistics L_n and U_n (not depending on d) based on the first n observations, such that $L_n < U_n$ a.s. and $\lim_{n\to\infty} P(L_n \le \theta \le U_n) = 1 - \alpha$ [so that, for n large, (L_n, U_n) is a confidence interval for θ with coverage probability approximately $1 - \alpha$]. Define a stopping variable N to be the first integer $n \ge n_0$ such that $U_n - L_n \le 2d$, where n_0 is a positive integer. Take (L_N, U_N) as the confidence interval. Then one could ask: (i) what is the coverage probability of the procedure? (ii) what is the expected sample size? These questions can, under the following regularity assumptions, be answered asymptotically as $d \to 0$.

(A1) $L_n < U_n$ a.s. (L_n and U_n independent of d);

(A2) $n^{1/2}(U_n - L_n) \to 2K_\alpha/A$ a.s. as $n \to \infty$ where $A > 0$ and $\Phi(K_\alpha) = 1 - \alpha/2$ and where Φ denotes the standard normal cdf;

(A3) $n^{1/2}(L_n - \theta) = Z_n/A - K_\alpha/A + o(1)$ a.s. as $n \to \infty$ where Z_n is a standardized average of iid random variables with finite second moment;

(A4) the set $\{Nd^2\}_{d>0}$ is uniformly integrable.

Theorem 4.22.5 (Geertsema, 1970a) Under the assumptions (A1)–(A4),

(i) N is well defined, $EN < \infty$ for all $d > 0$, N $(=N(d))$ is a function of d which is nondecreasing as d decreases, $\lim_{d\to 0} N = \infty$ a.s., and $\lim_{d\to 0} EN = \infty$.

(ii) $\lim_{d\to 0} Nd^2 = K_\alpha^2/A^2$ a.s. and $\lim_{d\to 0} ENd^2 = K_\alpha^2/A^2$.

(iii) $\lim_{d\to 0} P(L_N \le \theta \le U_N) = 1 - \alpha$.

Proof (i) That N is well defined, that is, finite a.s. for all $d > 0$ follows from assumption (A2), while $EN < \infty$ for all $d > 0$ follows from (A4). The definition of N and the monotone convergence theorem (see Lemma 4.14.3) imply the remaining statements.

(ii) By definition of N, one gets $U_N - L_N \le 2d$ and $U_{N-1} - L_{N-1} > 2d$.

Using the fact that $N \to \infty$ a.s. as $d \to 0$ together with (A2), we obtain

$$\liminf_{d\to 0} 2dN^{1/2} \geq \liminf_{d\to 0} N^{1/2}(U_N - L_N) = 2K_\alpha/A \quad \text{a.s.}$$

$$\limsup_{d\to 0} 2d(N-1)^{1/2} \leq \limsup_{d\to 0} (N-1)^{1/2}(U_{N-1} - L_{N-1}) = 2K_\alpha/A \quad \text{a.s.}$$

These together yield the first statement. The second statement readily follows from (A4) [see Loéve (1962, p. 163)].

(iii) A theorem of Anscombe (1952) together with the first statement of (ii) of the present theorem implies that Z_N is distributed asymptotically as a standard normal variable. From (A3) it follows that the asymptotic distribution of $N^{1/2}(L_N - \theta)$ is asymptotically normal with mean $-K_\alpha/A$ and variance $1/A^2$ as $d \to 0$, so that

$$\lim_{d\to 0} P(0 \leq N^{1/2}(\theta - L_N) \leq 2K_\alpha/A) = 1 - \alpha.$$

However, $N^{1/2}(U_N - L_N) = 2K_\alpha/A + o_p(1)$ as $d \to 0$ by (A2) and the first part of (ii) so that (iii) follows.

A Procedure Based on the Sign Test Let $X_1, X_2, \ldots, X_n$ be iid random variables having a unique median γ. For testing the hypothesis $\gamma = 0$, one uses the sign test based on the statistic $\sum_{i=1}^n I(X_i > 0)$ where $I(B)$ denotes the indicator function of the set B. In the case of a fixed-sample of size n, a confidence interval for γ can be derived from the sign test in a standard way. This confidence interval is of the form $(X_{n,b(n)}, X_{n,a(n)})$ where $X_{n,1} \leq X_{n,2} \leq \cdots \leq X_{n,n}$ are the ordered Xs and where $a(n)$ and $b(n)$ are integers depending on n. The limiting coverage probability as $n \to \infty$ of such confidence interval is $1 - \alpha$ if

$$a(n) \sim n/2 + K_\alpha n^{1/2}/2 \qquad \text{and} \qquad b(n) \sim n/2 - K_\alpha n^{1/2}/2. \tag{4.22.10}$$

From this confidence interval one can thus obtain a sequential procedure as follows: Let N be the first integer $n > n_0$ for which $X_{n,a(n)} - X_{n,b(n)} \leq 2d$ and choose as resulting confidence interval $(X_{N,b(N)}, X_{N,a(N)})$ where $\{a(n)\}$ and $\{b(n)\}$ are sequences of positive integers satisfying assumption (A5) which is given below and n_0 is some integer. This procedure is similar to Farrell's (1966b) procedure discussed in Section 4.22.1. The following assumption will be needed.

(A5) Let $X_1, X_2, \ldots$ be a sequence of iid random variables with common cdf $F(x - \gamma)$ where $F(x)$ is symmetric about zero. F has two derivatives in a neighborhood of zero and the second derivative is bounded in this neighborhood so that γ is the unique median of the Xs. The sequences $\{a_n\}$ and $\{b_n\}$ are defined by

$$b_n = \max\{1, [n/2 - K_\alpha n^{1/2}/2]\}, \qquad a_n = n - b_n + 1$$

where $[x]$ here denotes the largest integer contained in x. We shall now show that the above procedure satisfies (A1)–(A4). Without loss of generality we can assume that $\gamma = 0$. A strong use is made of the following representation of sample quantiles by Bahadur (1966): Under assumption (A5),

$$X_{n,k(n)} - \xi = [k(n)/n - F_n(\xi)]/f(\xi) + O(n^{-3/4} \ln n) \quad \text{a.s.} \quad (4.22.11)$$

where $\{k(n)\}$ is a sequence of positive integers satisfying $k(n) = np + o(n^{1/2} \ln n)$, $0 < p < 1$, $F(\xi) = p$, $F'(\xi) = f(\xi)$, and F_n is the empirical cdf of the Xs.

Lemma 4.22.1 (Geertsema, 1970a) We have

$$n^{1/2}(X_{n,a(n)} - X_{n,b(n)}) \to K_\alpha/f(0) \quad \text{a.s.} \qquad \text{as} \quad n \to \infty.$$

Proof It follows immediately from (4.22.11) that

$$n^{1/2}(X_{n,a(n)} - X_{n,b(n)}) = \frac{a(n) - b(n)}{n^{1/2}} \frac{1}{f(0)} + O(n^{-1/4} \ln n) \quad \text{a.s.}$$
$$\to K(\alpha)/f(0) \qquad \text{as} \quad n \to \infty \quad \text{a.s.}$$

Lemma 4.22.2 (Geertsema, 1970a) We have

$$n^{1/2} X_{n,b(n)} = Z_n/A - K_\alpha/A + o(1) \quad \text{a.s.} \quad \text{as} \quad n \to \infty,$$

where $A = -2f(0)$, $Z_n = 2n^{1/2}(F_n(0) - \frac{1}{2})$.

Proof This readily follows from (4.22.11).

Lemma 4.22.3 (Yahav and Bickel, 1968) Let $\{Y_n\}$ be a sequence of random variables on some probability space $(\Omega, \mathscr{B}, P)$ tending in law to some random variable Y. Let $a_{mn} = P(Y_n > m)$: and $\sup_n a_{mn} = a_m$. Then

$$\sum a_m < \infty \qquad (4.22.12)$$

implies that $E(Y_n)$ and $E(Y)$ are finite and $E(Y_n) \to E(Y)$.

Proof

$$\int_{\{Y_n > m\}} |Y_n| \, dP \leq \sum_{k=m}^{\infty} (k+1)P(k < |Y_n| \leq k+1)$$
$$= (m+1)P(|Y_n| > m) + \sum_{k=m}^{\infty} P(|Y_n| > k)$$
$$\leq (m+1)a_m + \sum_{k=m}^{\infty} a_k .$$

If (4.22.12) holds, since $a_m \downarrow$, by a classical lemma of Abel $ma_m \to 0$. Thus (4.22.12) implies uniform integrability of the $\{Y_n\}$.

Lemma 4.22.4 (Geertsema, 1970a) The set $\{Nd^2\}_{d>0}$ is uniformly integrable.

Proof According to Lemma 4.22.3, it suffices to show that

$$\sum_{m=1}^{\infty} \sup_{0<d<d_0} P[N(d)d^2 > m] < \infty \quad \text{for some} \quad d_0 .$$

Now

$$P[N(d)\, d^2 > m] = P[N(d) > m(d)] \quad \text{where} \quad m(d) = [m/d^2]$$
$$\leq P[X_{m(d),\, a(m(d))} - X_{m(d),\, b(m(d))} > 2d]$$
$$\leq P[S_{m(d)}(d) < a(m(d))] + P[S_{m(d)}(-d) \geq b(m(d))]$$
$$= 2P[B(m(d), F(-d)) \geq b(m(d))]$$

where $S_n(t) = \sum_{i=1}^{n} I(X_i \leq t)$ and $B(n, p)$ denotes a binomial random variable with parameters n and p. Then from Hoeffding (1963, Theorem 1) we have

$$P[N(d)d^2 > m] \leq 2 \exp\{-2m(d)t_{m(d)}^2\}$$

where $t_{m(d)} = \{b(m(d)) - m(d)F(-d)\}/m(d) \geq d\delta$ for some $\delta > 0$ for $d \leq d_0$ and $m \geq M_0$ where $d_0 > 0$ and M is sufficiently large. Hence, for $d \leq d_0$ and $m \geq M_0$ and a constant A

$$P[N(d)\, d^2 > m] < A \exp(-2m\delta^2)$$

which proves the uniform integrability since $\sum_{m=M_0}^{\infty} \exp(-2m\delta^2) < \infty$. The following theorem is a direct consequence of Theorem 4.22.5 and the preceding lemmas.

Theorem 4.22.6 (Geertsema, 1970a) The confidence interval procedure based on the sign test has asymptotic coverage probability $1 - \alpha$ as $d \to 0$. This stopping variable N satisfies $\lim_{d \to 0} ENd^2 = K_\alpha^{\,2}/4f^2(0)$.

REMARK 4.22.6.1 Unlike in the work of Chow and Robbins (1965), no assumption of the finiteness of the second moment of F is made in Theorems 4.22.5 and 4.22.6.

A PROCEDURE BASED ON THE WILCOXON ONE-SAMPLE TEST

The Wilcoxon one-sample test is based on the statistic

$$\textstyle\sum_{i=1}^{n} \sum_{j=1}^{n} I((X_i + X_j)/2 > 0)$$

where $X_1, \ldots, X_n$ is a sample from a distribution symmetric about γ. The test is used to test the hypothesis $\gamma = 0$ against shift alternatives. A confidence interval for γ based on a fixed-sample of size n is of the form $(Z_{n,\,b(n)}, Z_{n,\,a(n)})$ where $Z_{n,1} \leq Z_{n,\,2} \leq \cdots \leq Z_{n,\,n(n+1)/2}$ are ordered averages $(X_i + X_j)/2$, for $i, j = 1, \ldots, n$ and $i \leq j$. The limiting coverage probability of such an interval is $1 - \alpha$ if

$$\begin{aligned} a(n) &\sim [n(n+1)/4] + K_\alpha[n(n+1)(2n+1)/24]^{1/2}, \\ b(n) &\sim [n(n+1)/4] - K_\alpha[n(n+1)(2n+1)/24]^{1/2}. \end{aligned} \qquad (4.22.12')$$

THE DERIVED SEQUENTIAL PROCEDURE Let N be the first integer $n \geq n_0$ for which $Z_{n,\,a(n)} - Z_{n,\,b(n)} \leq 2d$ and choose $(Z_{N,\,b(N)}, Z_{N,\,a(N)})$ as the resulting confidence interval. $\{a(n)\}$ and $\{b(n)\}$ are sequences of positive integers satisfying (4.22.12′) and n_0 is some positive integer.

The asymptotic analysis of this procedure is somewhat complicated because it is based on ordered dependent random variables, namely the $Z_{n,\,k}$, $k = 1, 2, \ldots, n(n+1)/2$. Fortunately the theory of U statistics [see Hoeffding (1948, 1963)] can be applied. The statistic

$$\binom{n}{2}^{-1} \sum_{i=1}^{n} \sum_{j=i+1}^{n} I((X_i + X_j)/2 > 0) \qquad (4.22.13)$$

is a one-sample U statistic and the test based on it is asymptotically equivalent to the Wilcoxon one-sample test.

THE MODIFIED SEQUENTIAL PROCEDURE Let N be the first integer $n \geq n_0$ for which $W_{n,\,a(n)} - W_{n,\,b(n)} \leq 2d$ and choose $(W_{N,\,b(N)}, W_{N,\,a(N)})$ as the resulting confidence interval. $\{a(n)\}$ and $\{b(n)\}$ are sequences of integers satisfying (4.22.12), n_0 is some positive integer, and $W_{n,\,1} \leq W_{n,\,2} \leq \cdots \leq W_{n,\,n(n-1)/2}$ are the ordered averages $(X_i + X_j)/2$ for $i < j$ and $i, j = 1, \ldots, n$. So, let us confine ourselves to the modified sequential procedure. We need the following assumption.

(A6) $X_1, X_2, \ldots$ is a sequence of iid variables having common cdf $F(x - \gamma)$ where F is symmetric about 0. F has density f which satisfies $\int f^2(x)dx < \infty$. $G(x - \gamma)$ denotes the cdf of $(X_1 + X_2)/2$ and G has a second derivative in some neighborhood of 0 with G'' bounded in the neighborhood. G' is denoted by g where it exists. $\{a(n)\}$ and $\{b(n)\}$ are sequences of positive integers defined by

$$b(n) = \max\{1, [n(n+1)/4 - K_\alpha\{n(n+1)(2n+1)/24\}^{1/2}]\}$$
$$a(n) = n(n+1)/2 - b(n) + 1.$$

The following facts can easily be established.

(i) The assumptions on F guarantee the existence of a derivative for G.
(ii) If f has a Radon–Nikodym derivative f' where $\int |f'| < \infty$ and $\int (f')^2 < \infty$, then the assumptions on G are satisfied.
(iii) Assumption (A6) implies that $G'(0) > 0$ since $G'(0) = 2\int f^2(x)\, dx$.

Without loss of generality we can set $\gamma = 0$. We need the following lemmas.

Lemma 4.22.5 (Hoeffding, 1963) Let $U = \binom{n}{r}^{-1}\Sigma_c \psi(X_1, \ldots, X_r)$ where $X_1, X_2, \ldots, X_n$ are iid random variables, ψ is a function symmetric in its r variables, and c denotes summation over all $\binom{n}{r}$ combinations of r Xs. Then

$$P(U - EU > t) \le \exp(-2kt^2) \tag{4.22.14}$$

if $0 \le \psi(x_1, \ldots, x_r) \le 1$, $t \ge 0$ and $k = [n/r]$. Also

$$P(|U - EU| \ge t) \le 2e^{-h}, \quad h = kt^2/2\{\sigma^2 + \tfrac{1}{3}t \max(z, 1, -z)\} \tag{4.22.15}$$

if $|\psi(x_1, \ldots, x_r)| \le 1$, $t \ge 0$, $z = E\psi(X_1, \ldots, X_r)$, $\sigma^2 = \operatorname{var} \psi(X_1, \ldots, X_r)$, and $k = [n/r]$.

Lemma 4.22.6 (Geertsema, 1970a) If $\{k(n)\}$ is any sequence of integers satisfying $k(n)/\binom{n}{2} = \frac{1}{2} + o(n^{-1/2} \ln n)$ as $n \to \infty$, then

$$W_{n,k(n)} = \left\{k(n)\Big/\binom{n}{2} - G_n(0)\right\}\Big/g(0) + O(n^{-3/4}\ln n) \quad \text{a.s.,}$$

$$G_n(x) = \binom{n}{2}^{-1} \sum_{i<j} I((X_i + X_j)/2 \le x). \tag{4.22.16}$$

Proof The proof consists in extending Bahadur's (1966) result so as to apply to the ordered averages of pairs of observations that are being considered. The proof follows closely the method of Bahadur's proof [for details see Geertsema (1970a, Lemma 4.2)].

Corollary 4.22.6.1

$$n^{1/2}(W_{n,a(n)} - W_{n,b(n)}) \to K_\alpha[3^{1/2} \int f^2(x)\,dx] \quad \text{a.s.}$$

as $n \to \infty$.

Proof This follows from Lemma 2.22.6 in conjunction with (4.22.12) and the fact that $G'(0) = 2\int f^2(x)\,dx$.

Lemma 4.22.7 $n^{1/2}W_{n,b(n)} = Z_n/A - K\ /A + o(1)$ a.s. as $n \to \infty$, where

$$A = 3^{1/2}g(0), \qquad Z_n = (12n)^{1/2}\left[n^{-1}\sum_{i=1}^{n} F(X_i) - \tfrac{1}{2}\right].$$

Proof This lemma follows from Lemma 4.22.6 together with Lemma 4.4 and Corollary 4.2 of Geertsema (1970a) upon noting that $G_n(0)$ is a U statistic.

Lemma 4.22.8 $\{Nd^2\}_{d>0}$ is uniformly integrable.

Proof Analogous to that of Lemma 4.22.4.

REMARK From the fact that $\tilde{G}_n(x) = G_n(x) + O(1/n)$ where

$$\tilde{G}_n(x) = [2/n(n+1)] \sum_{i \le j} I((X_i + X_j)/2 \le x)$$

one can prove all the above results for the sequential procedure based on the one-sample Wilcoxon test.

The following theorem now follows from Theorem 4.22.5 and the above lemmas.

Theorem 4.22.7 (Geertsema, 1970a) Both the confidence interval procedures based on the Wilcoxon one-sample test have asymptotic coverage probability $1 - \alpha$ as $d \to 0$. The stopping variable N satisfies

$$\lim_{d \to 0} ENd^2 = K_\alpha^2/12 \left\{\int f^2(x)\,dx\right\}^2.$$

ASYMPTOTIC EFFICIENCIES OF THE PROCEDURES Consider two bounded-length confidence interval procedures T and S, for estimating the mean of symmetric population by means of an interval of prescribed length $2d$. Denote by N_T and N_S the stopping variables and by p_T and p_S the coverage probabilities associated with the procedures T and S respectively.

Definition 4.22.1 The asymptotic efficiency as $d \to 0$ of procedure T relative to S is $\rho(T, S) = \lim_{d \to 0} EN_S/EN_T$ provided $\lim_{d \to 0} p_T = \lim_{d \to 0} p_S$ and that all the limits exist. Denote by M the procedure of Chow and Robbins (1965) and by S and W the procedures based on sign test and Wilcoxon test respectively. Then it follows from Theorems 4.14.1, 4.22.6 and 4.22.7 (under the assumptions (A5), (A6), and $\sigma^2 < \infty$)

$$e(S, M) = 4\sigma^2 f^2(0), \qquad E(W, M) = 12\sigma^2\left(\int f^2(x)\,dx\right)^2$$
$$e(S, W) = f^2(0)/3\left(\int f^2(x)\,dx\right)^2. \tag{4.22.17}$$

If one regards the procedures M, S, and W as based on the t-test, sign test and the Wilcoxon one-sample test respectively, one sees that the above efficiencies are the same as the Pitman efficiencies of the respective tests relative to each other.

Monte Carlo Studies Geertsema (1970b) has performed a series of Monte Carlo studies for different values of d and for a few symmetric populations (normal, contaminated normal, uniform, and double exponential) to compare the behavior of the procedures with the asymptotic results. Geertsema (1970a) points out that the actual coverage probability is quite close to the asymptotic coverage probability and that the actual coverage probability for the two procedures that are based on rank tests is higher than that of the procedures based on the t-test. The results also suggest that $EN \le \{K_\alpha^2/3[G(d) - \frac{1}{2}]^2\} + C$ in the case of the W procedure where C is a constant. The results illustrate the upper bound $EN \le K_\alpha^2\sigma^2/d^2 + n_0$ for the procedure based on the t-test [see Simons (1968)].

4.22.3 Bounded Length Sequential Confidence Intervals Based on One-Sample Rank Order Statistics

Sproule (1969) has extended the results of Chow and Robbins (1965) to the class of Hoeffding's (1948) U statistics and in the particular cases of the signed-rank and sign statistics (which are both U statistics). Sen and Ghosh (1971) have considered the problem of providing a sequential confidence interval for the median of a symmetric (but otherwise unknown) distribution based on a general class of one-sample rank order statistics. Of particular interest is the procedure based on the absolute normal scores (or one-sample normal scores) statistic. This procedure is shown to be asymptotically (that is, as the prescribed bound on the width of the confidence interval is made to converge to zero) at least as efficient as the Chow–Robbins procedure for a broad class of parent distributions.

Let $X_1, X_2, \ldots$ be a sequence of iid random variables having an absolutely continuous distribution function $F_\theta(x)$ having location parameter θ (unknown). It is desired to determine (sequentially) a confidence interval $I_N = \{\theta : \hat{\theta}_{L,N} \leq \theta \leq \hat{\theta}_{U,N}\}$ of width $\leq 2d$ where for each positive integer n, $\hat{\theta}_{L,n}$ and $\hat{\theta}_{U,n}$ are two statistics (not depending on d) based on the first n observations such that $\lim_{n\to\infty} P(\theta \in I_n) = 1 - \alpha$ (the desired confidence coefficient) while N is the stopping variable defined to be the first integer $n \geq n_0$ (some positive integer) such that $\hat{\theta}_{U,N} - \hat{\theta}_{L,N} \leq 2d$. The procedure of Sen and Ghosh (1971) rests on the following class of one-sample rank order statistics. Let $c(u) = 0$ or 1 according as $u < 0$ or not. Then, let

$$R_{n\alpha} = \sum_{\beta=1}^{n} c(|X_\alpha| - |X_\beta|), \qquad \alpha = 1, \ldots, n, \quad \mathbf{X}_n = (X_1, \ldots, X_n)', \tag{4.22.18}$$

and

$$T_n = T_n(\mathbf{X}_n) = n^{-1} \sum_{\alpha=1}^{n} c(X_\alpha) J_n(R_{n\alpha}/(n+1)) \tag{4.22.19}$$

where $\{J_n(u) : 0 < u < 1\}$ is generated by a score function $\{J(u), 0 < u < 1\}$ in either of the following two ways:

(a) $J_n(u) = J(i/(n+1))$, $(i-1)/n < u \leq i/n$, for $i = 1, \ldots, n$;
(b) $J_n(u) = EJ(U_{ni})$, $(i-1)/n < u \leq i/n$, $1 \leq i \leq n$ where U_{ni} $(i = 1, \ldots, n)$ are the standard uniform order statistics.

Also assume that $J(u) = \Psi^{-1}((1+u)/2)$, $0 \leq u < 1$ where $\Psi(u)$ is an absolutely continuous cdf defined on $(-\infty, \infty)$ satisfying the conditions

$$\text{(a)} \quad \Psi(-x) + \Psi(x) = 1, \qquad \Psi'(x) = \psi(x) = \psi(-x) \quad \text{for all } x, \tag{4.22.20}$$

$$\text{(b)} \quad \psi(x)/\{1 - \Psi(x)\} \quad \text{is nondecreasing for all} \quad x \geq x_0 \geq 0. \tag{4.22.21}$$

The statistic T_n when Ψ is the standard normal cdf is termed the normal scores statistic; and when Ψ is uniform over $(-1, 1)$, it is called the signed-rank statistic. Note that by definition $J(0) = 0$, $J(u)$ is increasing in u $(0 < u < 1)$. Also,

$$\begin{aligned} \psi(x)/[1 - \Psi(x)] &= [2J'(2\Psi(x) - 1)\{1 - \Psi(x)\}]^{-1} \\ &= [J'(u)(1-u)]^{-1}, \qquad u = 2\Psi(x) - 1. \end{aligned}$$

Hence by (4.22.21)

$$J'(u) \le K(1-u)^{-1}, \qquad 0 < u < 1, \qquad \text{where} \quad 0 < K < \infty \tag{4.22.22}$$

and by integration $J(u) \le K[-\ln(1-u)]$, $0 < u < 1$. Hence

$$M(t) = \int_0^1 \exp[tJ(u)]\, du \le \int_0^1 (1-u)^{-Kt} du < \infty \qquad \text{for all} \quad t \le t_0 \le 1/K. \tag{4.22.23}$$

Notice that (4.22.22) and (4.22.23) hold for the normal, the logistic, double exponential, and many other dfs. Also, let

$$\bar{J}_n = n^{-1} \sum_{i=1}^{n} J_n(i/(n+1)), \qquad A_n{}^2 = n^{-1} \sum_{i=1}^{n} J_n{}^2(i/(n+1)) \tag{4.22.24}$$

$$\mu = \int_0^1 J(u)\, du \qquad \text{and} \qquad A^2 = \int_0^1 J^2(u)\, du. \tag{4.22.25}$$

Notice that if $\theta = 0$, T_n has a distribution, free of F, and symmetric about $\bar{J}_n/2$. Hence there exist two constants $T_{n,\alpha}^{(1)}$ and $T_{n,\alpha}^{(2)} = \bar{J}_n - T_{n,\alpha}^{(1)}$ and a known α_n (close to the specified α) such that

$$P\{T_{n,\alpha}^{(2)} \le T_n(X_n) \le T_{n,\alpha}^{(1)} \mid \theta = 0\} = 1 - \alpha_n \to 1 - \alpha \qquad \text{as} \quad n \to \infty. \tag{4.22.26}$$

For large n, it is known [see Hájek and Šidak (1967, p. 166)] that

$$\lim_{n\to\infty} (T_{n,\alpha}^{(2)} - \tfrac{1}{2}\bar{J}_n) = -\tfrac{1}{2}AK_\alpha, \qquad \lim_{n\to\infty} n^{1/2}(T_{n,\alpha}^{(1)} - \tfrac{1}{2}\bar{J}_n) = \tfrac{1}{2}AK_\alpha \tag{4.22.27}$$

where $\Phi(K_\alpha) = 1 - \alpha/2$. It is known [either from Chernoff and Savage (1958) with further assumption on $J''(u)$ or from Govindarajulu *et al.* (1967) under much weaker condition on $J'(u)$ than (4.22.22) or from Pyke and Shorack (1968, Corollary 5.1) under (4.22.22)] that

$$n^{-1} \sum_{i=1}^{n} |J_n(i/(n+1)) - J(i/(n+1))| = o(n^{-1/2}) \tag{4.22.28}$$

which implies that the use of either of the scores will lead to the same asymptotic results.

THE PROCEDURE FOR OBTAINING I_n Since $J(u)$ is increasing in u $(0 < u < 1)$ $T_n(\mathbf{X}_n - a\mathbf{1}_n)$ (where $\mathbf{1}_n = (1, \ldots, 1))'$ is decreasing in $a, -\infty < a < \infty$. Define

$$\begin{aligned} \hat{\theta}_{L,n} &= \sup\{a : T_n(\mathbf{X}_n - a\mathbf{1}_n) > T_{n,\alpha}^{(1)}\} \\ \hat{\theta}_{U,n} &= \inf\{a : T_n(\mathbf{X}_n - a\mathbf{1}_n) < T_{n,\alpha}^{(2)}\}. \end{aligned} \tag{4.22.29}$$

It follows from (4.22.26) and (4.22.29) that $P_\theta\{\hat{\theta}_{L,n} \le \theta \le \hat{\theta}_{U,n}) = 1 - \alpha_n \to 1 - \alpha$ as $n \to \infty$. Upper and lower limits for θ given by (4.22.29) are known to be robust for outliers or gross errors. We shall state the basic result of Sen and Ghosh (1971) pertaining to the asymptotic properties of the above procedure.

Theorem 4.22.8 (Sen and Ghosh, 1971) Under the above assumptions $N = N(d)$ is a nonincreasing function of d, $N(d)$ is finite a.s. and $EN(d) < \infty$ for all $d > 0$; $\lim_{d\to 0} N(d) = \infty$ a.s., and $\lim_{d\to 0} EN(d) = \infty$;

$$\begin{gathered} \lim_{d\to 0} N(d)/v(d) = 1 \quad \text{a.s.} \qquad \lim_{d\to 0} P_\theta(\theta \in I_{N(d)}) = 1 - \alpha; \\ \lim_{d\to 0} EN(d)/v(d) = 1 \end{gathered} \tag{4.22.30}$$

where $v(d) = A^2 K_\alpha{}^2/\{4d^2 B^2(F)\}$ and

$$B(F) = \int_0^1 (d/dx)\{J(2F(x) - 1)\}\, dF(x) = \int_0^1 J'(H(x)) f(x)\, dH(x);$$

$$H(x) = F(x) - F(-x).$$

Proof See Sen and Ghosh (1971, Section 5).

When $\sigma^2 = \text{var}\, X_1 < \infty$, the asymptotic efficiency of the above procedure G, relative to the Chow and Robbins' (1965) procedure M, is $e(G, M) = 4\sigma^2 B^2(F)/A^2$ which is equal to the Pitman efficiency of a general rank order test relative to Student's t-test. Hence, from the variational argument of Govindarajulu (1961, Theorem 6C.1) we infer that $e(G, M) \ge 1$ for all cdf F having a density f and a finite second moment, the equality sign being achieved if and only if F is normal with zero mean and variance σ^2. When $J(u) = u$, then $e(G, M) = e(W, M)$ obtained in Section 4.22.2.

4.22.4 Fixed-Width Confidence Intervals for $P(X < Y)$

Let X and Y denote two independent random variables having either purely discrete or absolutely continuous F and G for their distribution functions, respectively. There are some practical situations [see, for instance, Birnbaum (1956)] where one is interested in estimating $p = P(X < Y)$ on the basis of random samples $X_1, \ldots, X_n$ and $Y_1, \ldots, Y_n$ from $F(x)$ and $G(y)$ respectively. It is well known (due to the completeness of the order statistic) that the minimum variance unbiased estimate of p is the Mann–Whitney statistic given by $\hat{p}_n = \int_{-\infty}^{\infty} F_n(x)\, dG_n(x)$ where F_n and G_n denote the empirical distribution functions based on the random samples of size n of X and Y. Suppose we wish to set up a confidence interval of prescribed width $2d$ and prescribed coverage probability $1 - \alpha$ for the unknown parameter p. Let

$$\sigma^2 = 2 \iint_{x<y} F(x)[1 - F(y)]\, dG(x)\, dG(y) + 2 \iint_{x<y} G(x)[1 - G(y)]\, dF(x)\, dF(y). \tag{4.22.31}$$

If σ^2 is known, and if d is small compared to σ^2, the fixed-sample size procedure is given as follows: For any $n \geq 1$ define $I_n = (\hat{p}_n - d,\ \hat{p}_n + d)$ and let u_α denote the $(1 - \alpha/2)$th fractile of the standard normal distribution. Then, for a sample of size n determined by

$$n = \text{smallest integer} \geq (u_\alpha{}^2 \sigma^2)/d^2 \tag{4.22.32}$$

the interval I_n has coverage probability

$$P(p \in I_n) = P(n^{1/2} |\hat{p}_n - p| \leq dn^{1/2}/\sigma)$$

and (4.22.32) implies that $\lim_{d\to 0} (d^2 n)/(u_\alpha{}^2\sigma^2) = 1$. It follows from the asymptotic normality of $\hat{p}_n$ [see Govindarajulu (1968b, Theorem 2.1) which could be easily extended to the situation where F and G are purely discrete since F and G can be made continuous by the continuization process described by Govindarajulu *et al.* (1967)] that $\lim_{d\to 0} P(p \in I_n) = 1 - \alpha$. However, in real situations F and G and consequently σ^2 are unknown, so that no fixed-sample size method is available.

SEQUENTIAL FIXED-WIDTH PROCEDURE Let

$$V_n = \hat{\sigma}^2 = 2 \iint_{x<y} F_n(x)[1 - F_n(y)]\, dG_n(x)\, dG_n(y) + 2 \iint_{x<y} G_n(x)[1 - G_n(y)]\, dF_n(x)\, dF_n(y) \tag{4.22.33}$$

and let $\{u_n\}$ be any sequence of positive constants tending to u_α as $n \to \infty$. Then define

$$N = \text{smallest } k \geq 2 \text{ such that } V_k < (d^2 k)/u_k^{\,2}. \tag{4.22.34}$$

Thus, we have the following theorem establishing the asymptotic properties of the above procedure.

Theorem 4.22.9 Under the sole assumption that $0 < \sigma < \infty$ we have

$$\lim_{d \to 0} (d^2 N)/(u_\alpha^{\,2}\sigma^2) = 1 \quad \text{a.s.} \qquad \text{(asymptotic optimality)}, \tag{4.22.35}$$

$$\lim_{d \to 0} P(p \in I_N) = 1 - \alpha \qquad \text{(asymptotic consistency)}, \tag{4.22.36}$$

and

$$\lim_{d \to 0} (d^2 EN)/(u_\alpha^{\,2}\sigma^2) = 1 \qquad \text{(asymptotic efficiency)}. \tag{4.22.37}$$

Proof In Lemma 4.14.5 set $Y_n = V_n/\sigma^2, f(n) = nu_\alpha^{\,2}/u_n^{\,2}$, and $t = u_\alpha^{\,2}\sigma^2/d^2$. Then (4.22.34) can be rewritten as $N = N(t) = \text{smallest } k \geq 1$ such that $Y_k \leq f(k)/t$. By Lemma 4.14.5

$$1 = \lim_{t \to \infty} f(N)/t = \lim_{d \to 0} (d^2 N)/(u_\alpha^{\,2}\sigma^2) \quad \text{a.s.} \tag{4.22.38}$$

which proves (4.22.35). Now

$$P(p \in I_N) = P(N^{1/2} \,|\, \hat{p}_N - p \,|\, /\sigma \leq d\, N^{1/2}/\sigma).$$

By (4.22.38) $d\, N^{1/2}/\sigma \to u_\alpha$ and $N/t \to 1$ in probability as $t \to \infty$; it follows from Theorem 4.12.1 which is due to Anscombe (1952) that as $t \to \infty$, $N^{1/2}(\hat{p}_N - p)/\sigma$ behaves like a standard normal variable. Hence

$$\lim_{t \to \infty} P(p \in I_N) = 1 - \alpha$$

which establishes (4.22.36). Now the property of asymptotic efficiency (4.22.37) follows immediately from Lemma 4.14.6 whenever the distributions of the X_i and Y_i are such that

$$E\left(\sup_n V_n/\sigma^2\right) < \infty.$$

However, this is trivially true since $V_n \leq 2$ and $\sigma^2 > 0$. This completes the proof of Theorem 4.22.9.

It should be remarked that the above considerations can easily be extended to the case when F and G are dependent and have an unknown joint distribution $H(x, y)$ and we observe pairs $(X_1, Y_1), (X_2, Y_2), \ldots$.

4.23 Certain Problems of Optimal Stopping

We shall give a host of problems that enjoy the identical optimal stopping rule. It is impossible to pin down the originators of these problems. For a historical discussion of this, the reader is referred to Gilbert and Mosteller (1966).

BEAUTY CONTEST Suppose that a boy is to have a date with his choice of one of n unseen and unknown girls, and suppose he wishes to choose the prettiest. The girls are presented for him to see one at a time in a random order, and he must choose or reject a girl when she appears. Once he chooses his date, he sees the rest and he is disappointed if his date is not the prettiest. Problem is how to maximize his probability of choosing the prettiest of the lot.

SECRETARY PROBLEM An executive wishes to hire a secretary from a group of n girls. He can interview the girls one at a time, after each interview accepting or rejecting that particular girl. If he rejects a girl, he may not recall her. At any stage he knows how the girl he is now interviewing ranks among her predecessors, but does not know how she compares with the girls he has not yet seen. We assume that he sees the girls in a random order. Problem is how he should go about deciding whom to hire. In particular, we might ask for a rule (i) that maximizes his probability of choosing the best of the n girls, or (ii) that maximizes the expected absolute rank of the girl chosen n being the rank of the best girl, $n - 1$ the rank of the next best, ..., 1 the rank of the worst).

THE DOWRY PROBLEM Each of n slips of paper has a different dowry written on it, and a lucky bachelor has one chance to choose, the girl with the largest dowry. Slips are drawn at random without replacement. If he chooses the slip with the largest dowry, he gets the girl and dowry, otherwise nothing.

The above three problems are mathematically equivalent. Problems of optimal stopping are obviously sequential decision problems and a basic reference is Arrow *et al.* (1949). Problems of optimal stopping require delicate theoretical points in order to establish the existence and optimality of a strategy or rule. Chow *et al.* (1971) have established the optimal rule for (ii) using the method of backward induction [see Chow *et al.* (1971, Sections 3.3)] and abstract measure-theoretic concepts. Dynkin and Yushkevich (1967) have given an elementary proof of the optimality of the strategy for (i). We shall present the latter in the following.†

We shall slightly change the mathematical formulation of the problem. Assume that there are n objects that are ordered according to their quality. Imagine that they represent points on the line, where better objects are denoted by points more to the right. Let a_1 be the object we see first, which could be any of the n points with the same probability. Similarly, a_2 may be any of the remaining $n - 1$ points with the same probability. Numbering the objects in the order we see them, we arrive at an ordering $a_{i_1}, a_{i_2}, \ldots, a_{i_n}$ where each of the $n!$ permutations has the same probability. We are observing this permutation step by step. After the kth trial we know only the relative position of $a_1, \ldots, a_k$. The problem is to recognize the rightmost point when it appears. That is, we have to devise a method of guessing which maximizes the probability of correct guess. The latter probability does not tend to zero as $n \to \infty$ [see the illustration of Dynkin and Yushkevich (1967)]. Suppose now that the position of points $a_1, \ldots, a_k$ on the line is already known. Then, the probability with which the point a_{k+1} falls into any of the intervals induced by the points $a_1, \ldots, a_k$ is $(k + 1)^{-1}$, regardless of the order of the points $a_1, \ldots, a_k$ [for a proof, see Dynkin and Yushkevich (1967)].

We should never stop at a point which is not maximal among those observed up to the given moment. If we continue sampling after the first maximal point, we will have a second maximal point which is to the right of all preceding points. Thus, we will have a succession of maximal points. Thus, if a_k is maximal, we have to decide whether to stop or continue sampling. We know, in such a situation, the positions of $a_1, \ldots, a_k$ among which a_k is the rightmost.

Since we can choose among $a_k, \ldots, a_n$ only, the decision should depend on the prediction of the mutual positions of $a_k, \ldots, a_n$. Nothing but the conditional probabilities of permutations of $a_k, \ldots, a_n$, given that $a_1, \ldots, a_k$ are known, can influence our decision. It is fairly simple to show (see Dynkin and Yushkevich, 1967) that these probabilities depend only on k, and not on the mutual position of $a_1, \ldots, a_{k-1}$. This would imply that when a maximal point a_k appears, the decision whether to stop or continue will depend only on k. Thus, to make the optimal decision, it is sufficient to remember only the numbers of successive maximal points. Denote these numbers by $X(0)$, $X(1), \ldots$. Clearly $X(0) = 1$, while $X(1), X(2), \ldots$ are random; none of them

† I am grateful to Professor Robert Bartoszyński of the Institute of Mathematics (Polish Academy of Science, Warszawa) for his help in translating into English the relevant parts of Dynkin and Yushkevich (1967, Chapter III).

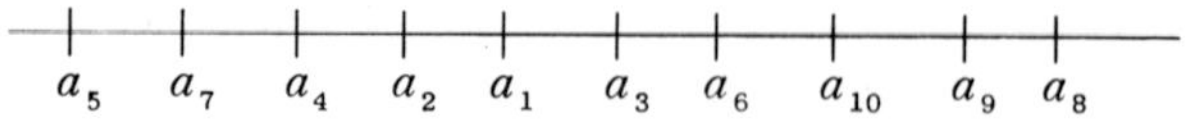

Fig. 4.23.1

exceed n; the number of these numbers is also random. The last maximal among $X(i)$ is the number of the rightmost point a_k, and it is to be guessed with maximal probability. Thus, if the order of points is (for $n = 10$) as shown in Fig. 4.23.1, then $X(0) = 1$. The successive maximals are: $X(1) = 3, X(2) = 6$, $X(3) = 8$ and here the process stops. The sequence $\{X(i)\}$ form a Markov chain.

Lemma 4.23.1 (Dynkin and Yushkevich, 1967) The $\{X(i)\}$ form a Markov chain, that is, the probability that $X(i + 1) = l$ given the values of $X(0), \ldots, X(i)$ depends only on the value assumed by $X(i)$.

Proof Suppose that $X(0) = 1$, $X(1) = b, \ldots, X(i) = k$. This means that a_k is maximal and we know the mutual positions of $a_1, \ldots, a_{k-1}$. The event $X(i + 1) = l$ means that the points $a_{k+1}, \ldots, a_{l-1}$ are to the left of a_k, and a_l is to the right of a_k. However, as noted earlier, given that a_k is maximal, the conditional probability of events expressed by $a_k, a_{k+1}, \ldots, a_n$ depends only on k, and not on the mutual positions of $a_1, \ldots, a_{k-1}$. Thus

$$P(X(i + 1) = l \mid X(0) = 1, X(1) = b, \ldots, X(i) = k)$$

depends only on k and l (and possibly on n, the total number of points). Denote this probability by $p(l, k)$. Here $\Sigma_l p(l, k)$ may be less than 1; in terms of a random walk, we interpret $1 - \Sigma_l p(l, k)$ as the probability of the particle vanishing—there will be no more maximal points. We have

$$p(l, k) = P\{X(i) = k, X(i + 1) = l\}/P(X(i) = k), \qquad k, l = 1, \ldots, n. \quad (4.23.1)$$

Clearly, $P(l, k) = 0$ if $l \leq k$ (there are no jumps to the left). The event $X(i) = k$, $X(i + 1) = l$ $(k < l)$ means that among $a_1, \ldots, a_l$ the points a_k and a_l are to the right of all points, and a_l to the right of a_k. Consequently, this probability is $1/l(l - 1)$. Consequently,

$$p(l, k) = k/l(l - 1), \qquad 1 \leq k < l \leq n. \quad (4.23.2)$$

This completes the proof of Lemma 4.23.1.

Next, let us turn to the construction of optimal rule for stopping. For each maximal point which occurs at the kth place, we have to decide whether to stop or not to stop. Let Γ denote the set of these numbers where we ought to stop. There are 2^n subsets of the n points, each subset denoting a strategy, and we ought to pick out the optimal one. There are, of course, other strategies. For instance, let ξ be the first of the values $X(0)$, $X(1)$, ... that exceed k, i.e. $\xi = X(i)$ with $X(0) < \cdots < X(i-1) < k$, $X(i) > k$. We can consider the strategy: stop at next number after ξ, i.e. on $X(i+1)$. Such a strategy is not optimal, but it will be used in the subsequent evaluations.

Let $q(k)$ be the probability of winning if we stop at $X(i) = k$,

$$q(k) = 1 - \sum_{l=k+1}^{n} p(l, k) = 1 - \sum_{l=k+1}^{n} \{k/l(l-1)\}$$
$$= 1 - k \sum_{l=k+1}^{n} ((l-1)^{-1} - l^{-1}) = k/n. \tag{4.23.3}$$

To compare, let us find the conditional probability of winning $q'(k)$, if in the same situation we decide to continue exactly one more step, i.e. to stop at $X(i+1)$. We have, by the law of absolute probability:

$$q'(k) = \sum_{l=k+1}^{n} p(l, k)q(l) = \sum_{l=k+1}^{n} \frac{k}{l(l-1)} \frac{l}{n} = \frac{k}{n}\left(\frac{1}{k} + \cdots + \frac{1}{n-1}\right)$$
$$= q(k)\left(\frac{1}{k} + \cdots + \frac{1}{n-1}\right) \tag{4.23.4}$$

for $k < n$ and $q'(k) = 0$ if $k = n$. Since the sum $k^{-1} + \cdots + (n-1)^{-1}$ decreases with the increase in k, the ratio $q'(k)/q(k)$ decreases, becoming 0 for $k = n$. Thus, $q'(k) \leq q(k)$ for some interval $r, r+1, \ldots, n$ of the sequence $1, \ldots, n$. We shall show that the set $\Gamma = \{r, r+1, \ldots, n\}$ is the optimal rule: i.e. one should continue as long as $X(i) < r$ and stop when for the first time $X(i) \geq r$ (r will depend on n which we shall suppress for the sake of clarity).

In the sequel, we put $n \geq 3$ (for $n = 1$ the situation is trivial, for $n = 2$ we can stop wherever we want). For $n \geq 3$,

$$q'(1) = q(1)\left(1 + \frac{1}{2} + \cdots + \frac{1}{n-1}\right) > q(1)$$

hence $r > 1$. It is *not* optimal to stop at $X(0) = 1$ as it leads to success with probability $q(1)$, while if we take one more observation, we get $q'(1)$. Now since $p(1, k) > 0$ for $2 \leq k \leq n$, with positive probability, we can expect that any number will occur. Suppose that the strategy S tells us to stop at $k < r$. Then the strategy S', identical to S if the particle does not fall in k, and which tells us to stop after one more observation if the particle falls in k, is better than S. Indeed for S' the probability of success $p_{S'}(k)$ will exceed the probability of success $p_S(k)$ by $p_S(k)[q'(k) - q(k)]$. Thus, optimal strategy excludes stopping at $1, 2, \ldots, r - 1$.

By backward induction we shall show that at points $r, r + 1, \ldots, n$, the optimal strategy tells us to stop immediately. Clearly, for all those points we have $q'(k) < q(k)$. If the strategy S required continuing on at the point n, then S' telling us to stop at n, and behaving in the same way otherwise, would be superior to S. (It would increase the probability of success by $p_S(n)$, hence S would not be optimal.) Assume that the assertion holds for $k + 1, k + 2, \ldots, n$. If S tells us to go on for k, then S' requiring us to stop at k, and otherwise identical, would be superior to S. Indeed if we fall to k, S' tells us to stop, while S tells us to stop exactly after one more observation. Hence the difference in probability of success would be $p_S(k)[q(k) - q'(k)]$, and S is not optimal. It follows that optimal strategy requires stopping also at k. Thus, at $1, \ldots, r - 1$ the optimal strategy tells us to go on; at $r + 1, \ldots, n$ it tells us to stop. If for $k = r$ we have $q'(r) \le q(r)$, we can extend the induction to $k = r$, and see that the optimal rule requires us to stop also at r. If for some n we have $q'(r) = q(r)$, we may stop or continue (let us agree to stop). In fact $q'(k) = q(k)$ is possible only for $n = 2$, $k = 1$. Instead, among $k, k + 1, \ldots, n$ exactly one is divisible by maximal power of 2 not exceeding $n - 1$. Thus, after reducing the sum $k^{-1} + (k + 1)^{-1} + \cdots + (n - 1)^{-1}$ to a common denominator, the numerator will be odd. For $n > 2$, the denominator will be even, hence $k^{-1} + \cdots + (n - 1)^{-1} \neq 1$.

Thus, the optimal rule is: Let the first $r - 1$ objects go, and stop the first time you observe an object better than all the preceding ones. r is the least integer for which $q'(k) \le q(k)$, i.e. for which

$$k^{-1} + (k + 1)^{-1} + \cdots + (n - 1)^{-1} \le 1,$$

or equivalently r is the least integer such that

$$r^{-1} + \cdots + (n - 1)^{-1} \le 1 < (r - 1)^{-1} + r^{-1} + \cdots + (n - 1)^{-1}. \quad (4.23.5)$$

Let us illustrate the above with an example when $n = 4$ for the dowry problem. In Table 4.23.1, (1, 2, 3, 4) denote the ranks of the dowries. Let "candidate" denote any draw whose value exceeds those of earlier draws. We consider strategies that pass the first r draws and choose the first candidate thereafter, if any. Thus * denotes the strategy with $r = 2$, and † denotes the strategy with $r = 3$. From the table we see that there are six ways to win by choosing the first candidate, 11 ways to win by passing the first number, 10 ways by passing the first two numbers and six ways by passing the first three. Among these four strategies, the strategy with $r = 2$ is optimal and it improves considerably

Table 4.23.1

Wins for the Dowry Problem When $n = 4$[a]

1234	2134	*†3124	4123
†1243	*†2143	*†3142	4132
†1324	†2314	*†3214	4213
†1342	†2341	*†3241	4231
*1423	*2413	*3412	4312
*1432	*2431	*3421	4321

[a] The asterisk indicates, wins by passing first number; the †, wins by passing first two numbers

upon the strategy of a random choice, which gives $\frac{1}{4}$ as the probability of winning.

Next let us compute the probability of winning with strategy of the optimum form.

Theorem 4.13.1 The probability of picking the true maximum of a sequence of n elements when we pass the first $r - 1$ draws and choose the first later candidate is given by

$$\pi(r, n) = \frac{r-1}{n} \sum_{k=r}^{n} (k-1)^{-1} = \frac{1}{\binom{n}{r-1}} \sum_{i=2}^{n-r+2} \frac{\binom{n-i}{r-2}}{(i-1)}. \tag{4.23.6}$$

REMARK 4.13.1 The first equality has been obtained by many, notably by Chow *et al.* (1971), Gilbert and Mosteller (1966) and the second equality by Bartoszyński (1974).

Proof Clearly for $r = 1$, the procedure leads to picking the first candidate and $\pi(1, n) = 1/n$. Hence assume that $2 \le r \le n$. Let $y_1, \ldots, y_n$ be a permutation of the ranks $1, \ldots, n$. Now, in order that the procedure be successful, the value n must appear in the permutation $y_1, \ldots, y_n$ at place k with $r \le k \le n$ and if n appears at place k (probability $1/n$), then the maximum among $y_1, \ldots, y_{k-1}$ must occur at one of the places $1, \ldots, r - 1$ [probability $(r - 1)/(k - 1)$]. The product $(r - 1)/n(k - 1)$ gives the probability of a win at draw k, $r \le k \le n$. Summing over k we obtain the first equality in (4.23.4). For the proof of the second equality, see Bartoszyński (1974).

Gilbert and Mosteller (1966) have tabulated the optimum values of r and $\pi(r, n)$ accurate to five decimals for $n = 1(1)50(10)100,1000$ and a selection of

Table 4.23.2

Optimum Values of r Together With $\pi(r, n)$

n	r	$\pi(r, n)$	n	r	$\pi(r, n)$
1	1	1.000	15	6	.389
2	1, 2	.500	20	8	.384
3	2	.500	25	10	.381
4	2	.458	30	12	.379
5	3	.433	35	14	.377
6	3	.428	40	16	.376
7	3	.414	45	17	.375
8	4	.409	50	19	.374
9	4	.406	100	38	.371
10	4	.399	1000	369	.368
			∞	n/e	.368

these are given in Table 4.23.2. The following inequality for r has also been given by Gilbert and Mosteller (1966). Since

$$1 \leq \sum_{j=r-1}^{n-1} (1/j) < \int_{r-(3/2)}^{n-(1/2)} dx/x = \ln[(n - \tfrac{1}{2})/(r - \tfrac{3}{2})].$$

Also

$$1 \geq \sum_{j=r}^{n-1} (1/j) > \int_r^n dx/x + \tfrac{1}{2}(1/r - 1/n).$$

The first inequality implies $r < (n - \frac{1}{2})/e + \frac{3}{2}$, the second that

$$\begin{aligned} e &> (n/r)e^{(1/2)(1/r-1/n)} > (n/r)[1 + \tfrac{1}{2}(1/r - 1/n)] \\ &> (n/r)[1 + (\tfrac{1}{2})(1/\{(n - \tfrac{1}{2})/e + \tfrac{3}{2}\} - 1/n)]. \end{aligned}$$

Consequently, we have

$$\frac{n - 1/2}{e} + \frac{1}{2} - \frac{3e - 1}{2(2n + 3e - 1)} \leq r \leq \frac{n - 1/2}{e} + \frac{3}{2} \tag{4.23.7}$$

which pins down r to an interval of length $1 + (1.79)(n + 1.79)^{-1}$. We also have the following asymptotic expressions after noting $\sum_{j=1}^{n} 1/j \simeq \gamma + \ln n$ where γ is Euler's constant,

$$\pi(r, n) \simeq \frac{r-1}{n} \ln\left(\frac{n-1}{r-1}\right) \simeq \frac{r}{n} \ln\left(\frac{n}{r}\right). \tag{4.23.8}$$

Similarly approximating the left- or right-hand sum in (4.23.5) shows that $\ln(n/r) \simeq 1$ and so $r \approx n/e$. Substituting these into (4.23.8) gives us the well-known asymptotic result

$$\lim_{n\to\infty} \pi(r, n) = 1/e = .368, \qquad r \approx n/e. \tag{4.23.9}$$

[see also Chow *et al.* (1971, pp. 51–52)]. To sum up, the optimum asymptotic strategy is to pass the fraction $1/e$ of the objects and choose the first candidate thereafter, and then the probability of winning is asymptotically $1/e$.

Gilbert and Mosteller (1966) have considered the dowry problem with two or more choices and provide exact and asymptotic results.

APPENDIX 1

Solution to Wald's Equation

While deriving the OC function, we assumed that there exists an $h \neq 0$ such that $E_\theta\{f(X; \theta_1)/f(X; \theta_0)\}^h = 1$. In the following lemma of Wald we shall give the sufficient conditions under which such a solution exists.

Lemma A-1.1 (Wald, 1947) Let Z be a random variable such that the following conditions are satisfied:

(i) The expected value of $Z = \ln\{f(X; \theta_1)/f(X; \theta_0)\}$ exists and is not equal to zero.

(ii) For any real value h, the expected value $E(e^{hZ}) = g(h)$ exists.

Then there exists one and only one real value $h_0 \neq 0$ such that $E[\exp(h_0 Z)] = 1$.

Proof For any positive h, we obtain

$$g(h) > P(e^Z > 1 + \delta)(1 + \delta)^h.$$

Hence, because of (ii),

$$\lim_{h \to \infty} g(h) = \infty. \tag{A.1.1.}$$

Similarly for $h < 0$, we have $g(h) > P(e^Z < 1 - \delta)(1 - \delta)^h$. Hence

$$\lim_{h \to -\infty} g(h) = \infty. \tag{A.1.2.}$$

Since $g''(h) = E(Z^2 e^{hZ})$, it follows from (ii) that

$$g''(h) > 0, \text{ all real } h. \tag{A.1.3.}$$

Then (A.1.1), (A.1.2), and (A.1.3) imply that there exists exactly one real value h^* such that $g(h)$ takes its minimum value for $h = h^*$. Since $g'(0) = E(Z) \neq 0$, we notice that $h^* \neq 0$, and $g(h^*) < g(0) = 1$. Further, it is clear that the function $g(h)$ is strictly monotonically decreasing for h in $(-\infty, h^*)$ and is strictly monotonically increasing for h in (h^*, ∞). Since $g(0) = 1$ and $g(h^*) < 1$, there exists exactly one real value $h_0 \neq 0$ such that $g(h_0) = 1$. This completes the proof of lemma 2.3.1.

REMARKS A-1.1 It follows that $h^* > 0$ ($h^* < 0$) implies that $h_0 > 0$ ($h_0 < 0$). Further note, if $h^* > 0$, then $E(Z) = g'(0) < 0$; and if $h^* < 0$, then $E(Z) = g'(0) > 0$. That is, h_0 and $E(Z)$ are opposite in sign.

APPENDIX 2

Differentiation of an Expectation

Wald (1946) has established the following theorem for differentiating under the expectation sign of the fundamental identity of the SPRT.

Theorem A.2 Let $Z = \ln\{f(X; \theta_1)/f(X; \theta_0)\}$ and $C(t) = E(e^{Zt})$. If $C(t)$ exists for all real values of t, then the identity, namely $E(\exp(S_N t)\{C(t)\}^{-N}) = 1$ can be differentiated under the expectation sign any number of times with respect to t at any value t such that $C(t) \geq 1$.

Proof First we shall obtain an upper bound for $E(\exp(tS_N) \mid N = m)$ for any fixed integer m. Let us consider the case $t > 0$. Let $\ln A = a$, $\ln B = b$.

$$E(\exp(tS_N) \mid N = m) \leq E(\exp(tS_n) \mid S_N \geq a,\ N = m) \qquad (t > 0). \quad \text{(A.2.1)}$$

Hence

$$E(\exp(tS_N)|\,S_N \geq a,\ N = m,\ \exp(tS_{N-1}) = \rho e^{at}) = e^{at}\,\rho E(e^{Zt}|\,e^{Zt} \geq \rho^{-1}). \tag{A.2.2}$$

Let $l(t)$ denote the least upper bound of the expression

$$\rho E(e^{Zt}|\,e^{Zt} \geq \rho^{-1}) \tag{A.2.3}$$

with respect to ρ over the interval $(e^{-(a-b)|t|}, 1)$. The existence of $C(t)$ implies the finiteness of $l(t)$. Thus, it follows from (A.2.1) and (A.2.2) that

$$E(\exp(S_N t)|\,N = m) \leq e^{at} l(t) \qquad (t > 0). \tag{A.2.4}$$

Hence

$$E(\exp(S_N t)) \leq e^{at} l(t) \qquad (t > 0). \tag{A.2.5}$$

Analogously, for $t < 0$, one can obtain

$$E(\exp(S_N t)) \leq e^{bt} l(t). \qquad (t < 0). \tag{A.2.6}$$

To prove Theorem A.2, it suffices to show that the derivatives of all orders of $\exp(S_N t)[C(t)]^{-N}$ with respect to t exist for all t such that $C(t) \geq 1$ and that the rth derivative is bounded uniformly for all t belonging to an interval I by an integrable function. The existence of all derivatives follows from the existence of $C(t)$ for all t since $d^r \exp(S_N t)[C(t)]^{-N}/dt^r$ is equal to the sum of a finite number of terms of the type $S_N^{r_1} N^{r_2} \exp(S_N t)[C(t)]^{-N}$. Consider

$$|S_N^{r_1} N^{r_2} \exp(S_N t)[C(t)]^{-N}| \leq |S_N^{r_1}|\, N^{r_2} \exp(|S_N|\, t_0) \tag{A.2.7}$$

where t_0 is an upper bound of $|t|$ in I. Let t_1 be a value $> t_0$. Then for a suitable constant c

$$|S_N|^{r_1} \exp(|S_N|\, t_0) < c \exp[|S_N|(t_1 - t_0)] \exp[|S_N|\, t_0] = C \exp[|S_N|\, t_1]. \tag{A.2.8}$$

Hence

$$\begin{aligned} |S_N^{r_1} N^{r_2} \exp(S_N t)[C(t)]^{-N}| &\leq cN^{r_2}[\exp(S_N t_1) + \exp(-S_N t_1)] \\ &= D_{r_1, r_2}(S_N, N) \quad \text{(say)} \end{aligned} \tag{A.2.9}$$

for all t in I and

$$E[D_{r_1, r_2}(S_N, N)] = c \sum_{m=1}^{\infty} p_m m^{r_2} E\{(\exp(S_N t_1) | N = m) + E(\exp(-S_N t_1) | N = m)\} \tag{A.2.10}$$

where $p_m = P(N = m)$. Now using (A.2.4) and (A.2.6) in (A.2.10) we obtain

$$\begin{aligned} E[D_{r_1, r_2}(S_N, N)] &\le c\{\exp(at_1)l(t_1) + \exp(-bt_1)l(-t)\} \sum_1^{\infty} p_m m^{r_2} \\ &= c\{\exp(at_1)l(t_1) + \exp(-bt_1)l(-t)\}E(N^{r_2}) < \infty. \end{aligned}$$

This completes the proof of the assertion.

APPENDIX 3

On the Moments of a Random Variable

Lemma A.3.1 Let X be a random variable having $F(x)$ for its distribution function and assume that $E|X|^k < \infty$. Then

$$E(X^k) = k\int_0^\infty x^{k-1}[1 - F(x) + (-1)^k F(-x)]\,dx, \; k = 1, 2, \ldots. \quad \text{(A.3.1)}$$

Proof

$$\begin{aligned} E(X^k) &= \int_{-\infty}^0 x^k\,dF(x) - \int_0^\infty x^k\,d[1 - F(x)] \\ &= -k\int_{-\infty}^0 F(x)x^{k-1}\,dx + k\int_0^\infty x^{k-1}[1 - F(x)]\,dx, \end{aligned}$$

after performing integration by parts once in each integral. Now, the desired result follows after setting $x = -y$ in the first integral.

Corollary A.3.1 If X is symmetric about zero, and $E(X^{2k}) < \infty$, then

$$E(X^{2k}) = 4k \int_0^\infty x^{2k-1}[1 - F(x)]\, dx, \qquad k = 1, 2, \ldots. \tag{A.3.2}$$

Corollary A.3.2 If X takes on nonnegative integers only, then

$$E(X) = \sum_{i=0}^{\infty} [1 - F(i)] = \sum_{i=0}^{\infty} P(X > i). \tag{A.3.3}$$

For integer-valued random variables and $k \geq 1$, analogously to (A.3.3) one gets factorial moments of X. Thus, for example

$$E(X(X+1)) = 2 \sum_{1}^{\infty} iP(X \geq i), \tag{A.3.4}$$

so that $EX^2 = \sum_1^\infty (2i - 1)P(X \geq i)$.

APPENDIX 4

The Normal Diffusion Process

The Wiener Process

A stochastic process $\{X(t), t \in I\}$ is defined as a family of random variables $X(t)$ indexed by the time parameter t whose domain is a set of real numbers I. Let the distribution function of $X(t)$ for any $t \in I$ be denoted by $F_t(x) = P(X(t) \leq x)$. If $X(t)$ is discrete (continuous) its probability function (probability density function) can be denoted by $f_t(x)$. The process is said to have independent increments if for all choices of the indices $t_0 < t_1 < \cdots < t_n$, the n random variables $X(t_1) - X(t_0), \ldots, X(t_n) - X(t_{n-1})$ are mutually independent. Further, the process is said to have *stationary independent increments* if, in addition $X(t_2 + h) - X(t_1 + h)$ has the same distributions as $X(t_2) - X(t_1)$ for all choices of t_1 and t_2 and every $h > 0$.

In the theory and applications of stochastic processes, Wiener process plays a fundamental role. For instance, the Wiener process provides a model for the Brownian motion and thermal noise in electric circuits. The motion of a particle immersed in a liquid or gas exhibits ceaseless and irregular motions. Such a motion of a particle is called Brownian motion. Let $X(t)$ denote the displacement (from its starting point) after time t of a particle in Brownian motion, where $X(0) = 0$. The particle undergoes perpetual motion due to continual "impacts" upon it due to the force field of its surrounding medium.

The displacement of the particle over a time interval (s, t) which is long compared to the time between impacts can be regarded as the sum of a large number of small displacements. Hence, by the central limit theorem, it is reasonable to assume that $X(t) - X(s)$ is normally distributed. Also, it is reasonable to assume that the probability distribution of the displacement $X(t) - X(s)$ should be the same as that of $X(t + h) - X(s + h)$ for any $h > 0$, since the medium is supposed to be in equilibrium and the probability law of a particle's displacement over an interval should depend only on the length $t - s$ of the interval, and not on the time at which we begin observation. Next, we also assume that the displacements over nonoverlapping intervals are independent since the number and size of impacts in nonoverlapping intervals are independent. Thus we are led to define formally the Wiener or Wiener–Lévy process. A stochastic process $\{X(t), t > 0\}$ is said to be a Wiener process if

(i) $\{X(t), t > 0\}$ has stationary independent increments;
(ii) for every $t > 0$, $X(t)$ is normally distributed;
(iii) for all $t > 0$, $E(X(t)) = 0$; and
(iv) $X(0) = 0$.

Because of (i) and (iv), it suffices to state the probability law of $X(t) - X(s)$ for any $s < t$. Since its probability law is normal, it is determined by its mean and variance. We have $E\{X(s) - X(t)\} = 0$ and var $[X(t) - X(s)] = \sigma^2 |t - s|$. Thus the probability law of Wiener process is determined by axioms (i)–(iv) up to a parameter σ^2 which is an empirical characteristic of the process and which is to be determined from observations. In Brownian motion, σ^2 is the mean square displacement of the particle per unit of time.

The Normal Diffusion Process

If, in the Brownian motion average displacement per unit of time, μ is not zero, then the process is called a normal diffusion process. In other words, if the movement of the particle is symmetric (i.e. $\mu = 0$), then the normal diffusion process would become the Wiener process. In the following we shall study the normal diffusion process with *drift* (average displacement per unit of time) μ and *diffusion constant* $\sigma/2^{1/2}$ (notice σ^2 is the variance of the displacement per unit of time).

Let $X(t)$ denote the process in which the increment ΔX to X in a time interval Δt has an independent normal probability density having mean $\mu\,\Delta t$ and variance $\sigma^2\,\Delta t$, denoted by $\phi((\Delta x - \mu\,\Delta t)/\sigma\,(\Delta t)^{1/2})$ [where $\phi(z)$ denotes the standard normal density]. Let $f(x;t)\,\Delta x$ denote the probability that at time t the process [namely $X(t)$] lies between $x + \Delta x$ and x. Then we have

$$f(y; t + \Delta t) = \int_{-\infty}^{\infty} f(x; t)\phi((y - x - \mu\,\Delta t)/\sigma\,(\Delta t)^{1/2})\,dx. \quad \text{(A.4.1)}$$

This is the process studied by Armitage (1957) and Anderson (1960) (see Section 2.11). If $\Delta t \to 0$, then we get the normal diffusion process, which is more tractable from the mathematical point of view than the process with discrete Δt. Since $\phi((y - x - \mu\,\Delta t/\sigma\,(\Delta t)^{1/2})$, is concentrated near $y - x = \mu\,\Delta t = \Delta x$ (say) when Δx is small, we have, approximately,

$$f(x; t) = f(y; t) - \Delta x \left.\frac{\partial f}{\partial x}\right|_{x=y} + \frac{(\Delta x)^2}{2} \left.\left(\frac{\partial^2 f}{\partial x^2}\right)\right|_{x=y} + O((\Delta x)^3). \quad \text{(A.4.2)}$$

Using (A.4.2) in (A.4.1) we obtain

$$f(y; t + \Delta t) = f(y; t) - \mu\Delta t \left.\left(\frac{\partial f}{\partial x}\right)\right|_{x=y} + \frac{\sigma^2(\Delta t)}{2} \left.\left(\frac{\partial^2 f}{\partial x^2}\right)\right|_{x=y} + O((\Delta t)^2). \quad \text{(A.4.3.)}$$

Now, letting $\Delta t \to 0$ we have

$$\frac{\partial f(x, t)}{\partial t} + \mu \frac{\partial f(x, t)}{\partial x} = \frac{\sigma^2}{2} \frac{\partial^2 f(x, t)}{\partial x^2}. \quad \text{(A.4.4)}$$

An unrestricted solution of (A.4.4) (which can be verified by substitution) is given by

$$f(x; t) = (2\pi\sigma^2 t)^{-1/2} \exp[-(x - \mu t)^2/2\sigma^2 t]. \quad \text{(A.4.5)}$$

If $X(0) = x_0$, the solution would be the right-hand side expression of (A.4.5) with x replaced by $x - x_0$. [For a proof of (A.4.5), see for instance Cox and Miller (1965, pp. 209–210).] That is

$$X(t) = x_0 + \mu t + \sigma Y(t) \quad \text{(A.4.6)}$$

where $Y(t)$ is normal with mean 0 and variance t. Let us assume that $x_0 = 0$ and that there is an absorbing barrier at $x = a$. Then, we have the further restriction on the density function:

$$f(a; t) = 0, \qquad t > 0.$$

The solution is a weighted linear combination of $f(x; t)$ and $f(x - 2a: t)$ where $f(x; t)$ is given by (A.4.5). [The solution is obtained by the principle of images used in solving problems of heat conduction; for details, see, for instance, Cox and Miller (1965, pp. 220–221).] Then the density is given by

$$g(x; t) = \begin{cases} f(x; t) - \exp(2a\mu/\sigma^2) f_1(x; t), & x \leq a, \\ 0, & x > a, \end{cases} \tag{A.4.7}$$

where $f_1(x; t) = (2\pi\sigma^2 t)^{-1/2} \exp[-(x - 2a - \mu t)^2/2t\sigma^2]$.

Notice that $g(x; t)$ is zero along $x = -a$ and the solution is equivalent to subtracting from the standard diffusion density g_0 that due to an image source at $x = 2a$, say g_{2a}. If two absorbing barriers exist at $x = a$ and $x = -b$ $(b > 0)$ we shall analogously obtain a solution in terms of the infinite series of images at $2a$, $2a + 2b$, $4a + 2b$, $\ldots$, $-2b$, $-2a - 2b$, $-2a - 4b$, $\ldots$. That is,

$$f = g_0 - g_{2a} - g_{-2b} + g_{2a+2b} + g_{-2a-2b} - g_{4a+2b} - g_{-2a-4b} \cdots \tag{A.4.8}$$

for $-b \leq x \leq a$ and zero elsewhere.

Let $H(t)$ denote the distribution function of the absorption time. That is

$$H(t) = P(\text{the particle is absorbed before time } t). \tag{A.4.9}$$

Then we have

$$H(t) = 1 - \int_{-b}^{a} f(x; t)\, dx, \tag{A.4.10}$$

$f(x; t)$ denoting the density of the process, and $h(t) = H'(t) = (-d/dt) \int_{-b}^{a} f(x; t)\, dx$ provided $f(x; 0)$ is normalized. Further, we have

$$H(\infty) = \left[1 - \int_{-b}^{a} f(x; t)\, dx\right]_{t=0}^{t=\infty} = 1 \tag{A.4.11}$$

provided that the upper limit $t = \infty$ vanishes.

MOMENT-GENERATING FUNCTION OF $h(t)$. These are obtained by Bartlett (1946) and we present below the main results.

(i) *Case* $b = -\infty$. Then we have

$$H(t) = 1 - \Phi\left(\frac{a - \mu t}{\sigma t^{1/2}}\right) + \exp(2a\mu/\sigma^2)\Phi\left(\frac{-a - \mu t}{\sigma t^{1/2}}\right), \qquad \text{(A.4.12)}$$

which was used by Armitage (1957), where Φ denotes the standard normal distribution function. Thus

$$h(t) = a(2\pi\sigma^2 t^3)^{-1/2} \exp[-(a - \mu t)^2/2\sigma^2 t]. \qquad \text{(A.4.13)}$$

Theorem A.4.1 (Bartlett, 1946) Let $C(\xi) = E(e^{\xi T})$ where T is the random time of absorption of the particle at $x = a$. Then

$$C(\xi) = \exp[\gamma^2 - (\gamma^4 - 2\gamma^2 a\xi/\mu)^{1/2}], \qquad \text{(A.4.14)}$$

where $\gamma^2 = a\mu/\sigma^2$.

Proof Let $u = \sigma^2 t/a^2$. Then

$$h(t)\,dt = (2\pi u^3)^{-1/2} \exp[-(\gamma^4 u + u^{-1} - 2\gamma^2)/2]\,du. \qquad \text{(A.4.15)}$$

So, using (A.4.11) we have

$$\int_0^\infty (2\pi u^3)^{-1/2} \exp[-(\gamma^4 u + u^{-1})/2]\,du = \exp(-\gamma^2). \qquad \text{(A.4.16)}$$

Consequently (since $T = a^2 U/\sigma^2$), we have

$$\begin{aligned} C(\xi) &= E(\exp(\xi a^2 U/\sigma^2) \\ &= \exp(\gamma^2)\int_0^\infty (2\pi\mu^3)^{-1/2} \exp[-\tfrac{1}{2}(u^{-1} + u(\gamma^4 - 2\gamma^2 a\xi/\mu))]\,du \\ &= \exp[\gamma^2 - (\gamma^4 - 2\gamma^2 a\xi/\mu)^{1/2}], \end{aligned}$$

using (A.4.16). This completes the proof.

Corollary A.4.1.1 It follows from Theorem A.4.1, that $E(T) = a/\mu$ and $\operatorname{var} T = a\sigma^2/\mu^3$.

(ii) *Case of symmetric barriers*, (that is $b = +a$). From (A.4.8) one can obtain [see Bartlett (1955, p. 48) or Wetherill (1966, p. 125)] the density of the process as

$$f(x;t) = \frac{1}{\sigma(2\pi t)^{1/2}} \sum_{r=-\infty}^{\infty} \left[\exp\left\{\frac{\mu x_r'}{\sigma^2} - \frac{(x - x_r' - \mu t)^2}{2\sigma^2 t}\right\} - \exp\left\{\frac{\mu x_r''}{\sigma^2} - \frac{(x - x_r'' - \mu t)^2}{2\sigma^2 t}\right\}\right] \tag{A.4.17}$$

for $-a < x \leq a$ where the images are $x_r' = 4ra$ and $x_r'' = 2a - 4ra$, $r = 0, \pm 1, \pm 2, \ldots$. By integrating the above with respect to x in the interval $(-a, a)$ and then differentiating with respect to t, the probability density of the time t at termination is obtained as

$$h(t) = \frac{a}{\sigma(2\pi)^{1/2} t^{3/2}} \sum_{-\infty}^{\infty} (4r - 1) \exp\left[-\frac{(4r-1)a^2}{2\sigma^2 t}\right] \tag{A.4.18}$$

when $\mu = 0$. The probability density function of $1/T$ (T denoting the time for termination) for $\mu \neq 0$ is obtained as a sum of terms of the form

$$(2\pi)^{-1/2} \exp\left[-\frac{(4ra - a + \mu t)^2}{2\sigma^2 t}\right] \cdot \frac{(4ra - a + \mu t)}{\sigma t^{3/2}} \tag{A.4.19}$$

which is zero at $1/t = 0$.

(iii) *Case when b is finite.* The above solution provides an approximation for the double barrier problem provided $b\mu/\sigma^2 \gg 1$. Bartlett (1946) has derived the exact expressions for f, h, and the moment-generating function of T. In this case

$$C(\xi) = E(e^{\xi T}) = \frac{\exp(\gamma^2 \xi)(1 - \exp[2\delta^2(\xi - 1)])}{1 - \exp[2(\gamma^2 + \delta^2)(\xi - 1)]} + \frac{\exp[\delta^2(\xi - 2)]\{1 - \exp[2\gamma^2(\xi - 1)]\}}{1 - \exp[2(\gamma^2 + \delta^2)(\xi - 1)]} \tag{A.4.20}$$

where $\delta^2 = b\mu/\sigma^2$.

By setting $\xi = 1$, the probabilities of absorption at the barrier are given by

$$p_a = \frac{1 - \exp(-2\delta^2)}{1 - \exp[-2(\gamma^2 + \delta^2)]}, \qquad p_b = \frac{[1 - \exp(-2\gamma^2)]\exp(-2\delta^2)}{1 - \exp[-2(\gamma^2 + \delta^2)]}. \tag{A.4.21}$$

We also have

$$\frac{\mu^2(ET)_a}{\sigma^2} = \gamma^2 - \frac{2\delta^2 \exp(-2\delta^2)}{1 - \exp(-2\delta^2)} + \frac{2(\gamma^2 + \delta^2)\exp[-2(\gamma^2 + \delta^2)]}{1 - \exp[-2(\gamma^2 + \delta^2)]} \tag{A.4.22}$$

and

$$\frac{\mu^4(\text{var } T)_a}{\sigma^4} = \frac{\mu^2(ET)_a}{\sigma^2} - \frac{4\delta^4 \exp(-4\delta^2)}{[1 - \exp(-2\delta^2]^2} + \frac{4(\gamma^2 + \delta^2)^2 \exp[-4(\gamma^2 + \delta^2)]}{\{1 - \exp[-2(\gamma^2 + \delta^2)]\}^2} \tag{A.4.23}$$

where $(ET)_a$ and $(\text{var } T)_a$ respectively denote the expectation and the variance of time for absorption at the barrier a. Similar expressions can be obtained for $(ET)_b$ and $(\text{var } T)_b$ by interchanging the roles of γ and δ.

REMARK A.4.1.1 Bartlett (1946) justifies the approximate boundary values for the SPRT in terms of error probabilities proposed by Wald (1947), by viewing the SPRT as the normal diffusion process with absorbing barriers.

APPENDIX 5

Large Sample Properties of Maximum Likelihood Estimates

As pointed out in Section 3.7, when one is faced with testing composite hypotheses, one resorts to maximum likelihood procedures (either in the non-sequential or the sequential case) when every other method fails. Nuisance parameters can be eliminated by replacing them by the maximum likelihood estimators, since these have desirable large-sample properties, like consistency and asymptotic normality. These properties have been studied by Cramér (1946) and the regularity conditions have been relaxed to a large extent by LeCam (1970). Let $f(x; \theta)$ be the probability density of a sequence of independent and identically distributed random variables, where $\theta \in \Theta \subset R_k$. Let $\Psi(x; \theta) = \ln f(x; \theta)$.

Assumption A.5.1 There is a vector-valued function $\psi(x; \theta)$ and matrix-valued functions $B(x; \theta)$ and $B(x; \theta, t)$ such that

(i) $\Psi(x; \theta + t) - \Psi(x; \theta) = t'\psi(x; \theta) - \frac{1}{2}t'B(x; \theta, t)t$;

(ii) $E_\theta \psi(X;\theta) = 0$, also $M(\theta) = E_\theta[\psi(X;\theta)\{\psi(X;\theta)\}']$ exists;
(iii) $\lim_{\varepsilon\to 0} \sup_{|t|<\varepsilon} E_\theta \|B(x;\theta,t) - B(x;\theta)\| = 0$
(iv) $\tilde{C}(\theta) = E_\theta\{B(X;\theta)\}$ exists.

We shall also need a stronger assumption as follows.

Assumption A.5.2 All the conditions of Assumption A.5.1 are satisfied, but part (iii) of Assumption A.5.1 is replaced by

$$\lim_{\varepsilon\to 0} E_\theta \sup_{|t|\le\varepsilon} \|B(X;\theta,t) - B(X;\theta)\| = 0.$$

where $\|\cdot\|$ denotes the norm of the matrix $(\cdot)$ (namely, the square root of the sum of the squares of the elements). Then, we have the following results of LeCam (1970).

Lemma A.5.1 (LeCam, 1970) Assume that Assumption A.5.1 is satisfied; then, for any sequence t_n such that $\theta + t_n/n^{1/2} \in \Theta$, and $\sup|t| < \infty$, the difference

$$Z_n(\theta, \theta + (t_n/n^{1/2})) - t_n{}'\Delta_n + \tfrac{1}{2}t_n'\tilde{C}(\theta)t_n$$

tends to zero in probability, where

$$Z_n(\theta, t) = \sum_{i=1}^{n} [\Psi(X_i; t) - \Psi(X_i; \theta)]$$

and

$$\Delta_n = n^{-1/2} \sum_{i=1}^{n} \psi(X_i; \theta).$$

Proof See LeCam (1970, p. 817).

Theorem A.5.1 (LeCam, 1970) If Assumption A.5.1 is satisfied and $t_n \to t$, then $Z_n(\theta, \theta + (t_n/n^{1/2}))$ is asymptotically normally distributed with mean $-\tfrac{1}{2}t'\tilde{C}(\theta)t$ and variance $t'M(\theta)t$. The difference $\tilde{C}(\theta) - M(\theta)$ is positive semidefinite (note that prime denotes transpose).

Proof See LeCam (1970, p. 187).

Theorem A.5.2 (LeCam, 1970) Suppose that Assumption A.5.2 is satisfied, that the matrix $\tilde{C}(\theta)$ is nonsingular, and that Θ has a convex tangent set at θ. Then there is a $\delta > 0$ with the following property. Let $\hat{\theta}_n$ be any point $\hat{\theta}_n \in \Theta$ such that

$$\sum_{i=1}^{n} \Psi(X_j; \hat{\theta}_n) \geq \sup_{t \in \Theta} \left\{ \sum_{i=1}^{n} \Psi(X_j; t) : |t - \theta| < \delta \right\} - n^{-1}.$$

Let T_n be any point $T_n \in \Theta$ such that

$$\begin{aligned} n^{-1} + n^{1/2}(T_n - \theta)'\Delta_n - \tfrac{1}{2}n(T_n - \theta)'\tilde{C}(\theta)(T_n - \theta) \\ \geq \sup\{n^{1/2}t'\,\Delta_n - \tfrac{1}{2}nt'\tilde{C}(\theta)t : \theta + t \in \Theta\}. \end{aligned}$$

Then $n^{1/2}(T_n - \theta)$ tends to zero in probability (with respect to the underlying probability measure).

Proof See LeCam (1970, p. 819).

APPENDIX 6

A Table of Standard Distributions

Distribution	Probability or probability density function
Bernoulli	$f(x) = p^x(1-p)^{1-x}$, $x = 0, 1$ and $0 < p < 1$.
Discrete uniform	$f(x) = \frac{1}{N}$, $x = 1, 2, \ldots, N$.
Binomial	$f(x) = \binom{n}{x} p^x(1-p)^{n-x}$, $x = 0, 1, \ldots, n$, $0 < p < 1$.
Geometric	$f(x) = p(1-p)^x$, $x = 0, 1, \ldots$.
Hypergeometric	$f(x) = \binom{M}{x}\binom{N-M}{n-x} \Big/ \binom{N}{n}$, $x = 0, 1, \ldots, n$.
Poisson	$f(x) = e^{-\lambda}\lambda^x/x!$, $x = 0, 1, \ldots$.
Negative binomial	$f(x) = \binom{r+x-1}{x} p^r(1-p)^x$, $x = 0, 1, \ldots$, $0 < p < 1$.
Uniform	$f(x) = \begin{cases} (b-a)^{-1}, & a < x < b, \\ 0, & x < a \text{ or } x > b. \end{cases}$
Negative exponential	$f(x) = \begin{cases} \lambda e^{-\lambda x}, & x > 0, \\ 0, & x < 0. \end{cases}$
Cauchy	$f(x) = [\pi(1+x^2)]^{-1}$, $-\infty < x < \infty$.
Normal	$f(x) = (2\pi\sigma^2)^{-1/2}\exp[-(x-\mu)^2/2\sigma^2]$,
Weibull	$f(x) = mx^{m-1}\exp(-x^m)$, $x, m > 0$.
Gamma	$f(x) = \begin{cases} e^{-x}x^{p-1}/\Gamma(p), & x, p > 0, \\ 0, & \text{otherwise.} \end{cases}$
Beta	$f(x) = \frac{\Gamma(r+s)}{\Gamma(r)\Gamma(s)} x^{r-1}(1-x)^{s-1}$, $0 < x < 1$, $r, s > 0$
Exponential family	$f(x) = B(\theta)e^{Q(\theta)R(x)}h(x)$.

Mean	Variance
p	$p(1-p)$
$(N+1)/2$	$(N^2-1)/12$
np	$np(1-p)$
$(1-p)/p$	$(1-p)/p^2$
$\frac{nM}{N}$	$n\frac{M}{N}\frac{N-M}{N}\frac{N-n}{N-1}$
λ	λ
$r(1-p)/p$	$r(1-p)/p^2$
$(a+b)/2$	$(b-a)^2/12$
λ^{-1}	λ^{-2}
does not exist	does not exist
μ	σ^2
$\Gamma(m^{-1})$	$\Gamma\left(\frac{2}{m}\right)-\Gamma^2\left(\frac{1}{m}\right)$
p	p
$\frac{r}{r+s}$	$\frac{rs}{(r+s)(r+s+1)}$
$E[R(x)]=-\frac{d\ln B(\theta)}{dQ(\theta)}$	$\operatorname{var} R(x)=-\frac{d^2\ln B(\theta)}{d[Q(\theta)]^2}$

APPENDIX 7

A Useful Convergence Theorem of Cramér

Here, we shall present a useful convergence theorem of Cramér (1946) and provide an application of the same.

Definition A.7.1 $a_n = O(n^\alpha)$ $[o(n^\alpha)]$ if a_n/n^α tends to a constant [zero] as n becomes large.

Theorem A.7.1 (Cramér, 1946) Let $X_1, X_2, \ldots$ be a sequence of independent and identically distributed random variables. Let $U_n = U_n(X_1, \ldots, X_n)$ and $V_n = V_n(X_1, \ldots, X_n)$ be two sequences of statistics such that

(i) $E(U_n) = \mu + O(n^{-1})$;
(ii) $E(V_n) = \nu + O(n^{-1})$;
(iii) $\text{var}(U_n) = a/n + o(n^{-1})$;
(iv) $\text{var}\,(V_n) = b/n + o(n^{-1})$; and
(v) $\text{Cov}(U_n, V_n) = c/n + o(n^{-1})$,

where μ, v, a $(a > 0)$, b $(b > 0)$, and c are free of n. Let $\psi(x, y)$ be a function that is continuous in some neighborhood of $x = \mu$ and $y = v$ with continuous partial derivatives of first and second orders. Then $T_n = n^{1/2}[\psi(U_n, V_n) - \psi(\mu, v)]$ is asymptotically normal with zero mean and variance A where

$$A = a\left[\frac{\partial\psi(\mu, v)}{\partial\mu}\right]^2 + 2c\left[\frac{\partial\psi(\mu, v)}{\partial\mu}\frac{\partial\psi(\mu, v)}{\partial v}\right] + b\left[\frac{\partial\psi(\mu, v)}{\partial v}\right]^2.$$

APPLICATION Let $X_1, X_2, \ldots$ be independent normal with mean θ and variance σ^2. Let $U_n = \bar{X}_n$ and $V_n = s_n$ where $\bar{X}_n$ and s_n respectively denote the sample mean and sample standard deviation. Then an application of Stirling's formula and Theorem A.7.1 yields that $T_n = n^{1/2}(\bar{X}_n s_n - \theta/\sigma)$ is asymptotically normal with mean 0 and variance $1 + \theta^2/2\sigma^2$, since $\mu = \theta$, $v = \sigma$, $a = \sigma^2$, $b = \sigma^2/2$, and $c = 0$ and the function $\psi(x, y) = x/y$ satisfies the conditions of Theorem A.7.1. Further, if Theorem A.7.1 is valid, *Chebyshev's* inequality shows that $\psi(U_n, V_n$ converges in probability to $\psi(\mu, v)$.

APPENDIX 8

Backward Induction

The backward induction or "dynamic programming" plays an important role in sequential decision theory. Reference to such a tool has been made in Sections 3.12, 4.18, and 4.23. Sakaguchi (1961) has also dealt with this technique for simple payoff functions. Chow *et al.* (1971) have given a general treatment of the same. In the following we shall give the result of Chow *et al.* (1971) and refer the reader to their book for the proof.

Let $X_1, X_2, \ldots$ denote a sequence of observable random variables. In some cases, for example, the secretary problem of Section 4.23, there are only a finite number n of random variables to be observed. Even in the infinite case it is of interest to study stopping rules that are bounded by a fixed number n. In all such cases a complete solution is provided in principle by the method of backward induction that will be described as follows: Let $g_i(X_1, \ldots, X_i)$ (to be abbreviated as g_i) be the payoff after observing $X_1, \ldots, X_i$ ($i = 1, 2, \ldots, n$). For example, $g_i(X_1, \ldots, X_i) = X_i - (n - i)c$ [which is considered by

Sakaguchi (1961)] or $g_i = X_1 + \cdots + X_i$, where c denotes the cost per observation. Assume that g_i is integrable. Let C^n denote the class of all stopping variables t such that $t \leq n$ and define

$$V^n = \sup_{t \in C^n} Eg_t.$$

Notice that $E|g_t| \leq \sum_{i=1}^n E|g_i| < \infty$ and hence Eg_t exists for all $t \in C^n$. Also $t^*(n)$ (to be abbreviated as t^*) is said to be optimal in C^n if $t^* \in C^n$ and $Eg_{t^*} = V^n$. It is convenient to consider the more general problem of which the above problem is a special case. Let $C_{n^*}^n$ denote the set of all stopping variables t such that $n^* \leq t \leq n$ and define $V_{n^*}^n = \sup_{t \in C_{n^*}^n} Eg_t$. (Note that $C^n = C_1{}^n$ and $V^n = V_1{}^n$.) We shall show how to find a stopping variable which is optimal in $C_{n^*}^n$. If $n^* = n$, the problem is trivial since $t = n$ is the only stopping rule in $C_{n^*}^n$, and hence $V_{n^*}^n = Eg_n$. For $n^* = n - 1$, it is intuitively obvious that we should compare g_{n-1} with $E(g_n | X_1, \ldots, X_{n-1})$ and use the rule

$$t = \begin{cases} n-1 & \text{if} \quad g_{n-1} \geq E(g_n | X_1, \ldots, X_{n-1}), \\ n & \text{if} \quad g_{n-1} < E(g_n | X_1, \ldots, X_{n-1}). \end{cases}$$

These considerations motivate the following "dynamic programming" theorem of Chow *et al.* (1971) which formalizes the principle of backward induction.

Theorem A.8.1 (Chow *et al.*, 1971) Let n be a fixed positive integer. Define successively, $\gamma_n{}^n, \gamma_{n-1}^n, \ldots, \gamma_1{}^n$ by setting

$$\gamma_n{}^n = g_n, \qquad \gamma_{n^*}^n = \max[g_{n^*}, E(\gamma_{n^*+1}^n | X_1, \ldots, X_{n^*}),], \qquad n^* = n-1, \ldots, 1.$$

For each $n^* = 1, \ldots, n$, let

$$t^*(n, n^*) = \text{first } i \geq n^* \text{ such that } g_i = \gamma_i{}^n.$$

Then $t^*(n, n^*) \in C_{n^*}^n$ and

$$E(g_{t^*} | X_1, \ldots, X_{n^*}) = \gamma_{n^*}^n \geq E(g_t | X_1, \ldots, X_{n^*}), \ t \in C_{n^*}^n;$$

so that

$$Eg_{t^*} = E\gamma_{n^*}^n \geq Eg_t, \qquad t \in C_{n^*}^n, \qquad \text{and} \qquad V_{n^*}^n = E\gamma_{n^*}^n.$$

Proof See Chow *et al.* (1971, Theorem 3.2, p. 50).

Chow *et al.* (1971) apply this to the secretary problem where $g_i = P$ (rank of ith girl is $n | X_1, \ldots, X_i$). For details see Chow *et al.* (1971, pp. 51–52). Sakaguchi (1961) has considered the case of $g_i = X_i - (n - i)c$ where c denotes the cost per observation and obtains the optimal rule: Stop with the first $X_m > \gamma$ where γ is the root of the equation

$$\int_{\gamma}^{\infty} (x - \gamma)\, dF(x) = c$$

and $F(x)$ is the common distribution function of the X_i. Further, if we are allowed to make infinitely many successive draws from the population, the optimal stopping rule is as follows:

If $0 > \gamma(c)$, do not enter the game;
if $0 \leq \gamma(c)$, take the first observation;

and then, continuing the successive observations, stop with the first $X_m > \gamma(c)$, where $\gamma(c)$ is the root of the above equation. The expected payoff is then $\max(\gamma(c), 0)$.

References and Author Index

The following is a list of the references that are cited in the text. The numbers of the pages on which the references are cited are given in brackets at the end of each reference. A few uncited references, denoted by asterisks, that are relevant to sequential analysis are also included, and the chapter or section to which each pertains is indicated. No claim is made that the reference list is complete.

The references are arranged alphabetically according to the authors' names. Multiple-authored articles are listed according to the names of the primary authors only. The following abbreviations (as in *Mathematical Reviews*) of the Journals are used.

Acta Math. Acad. Sci. Hungar.	*Acta Mathematica Academiae Scientiarum Hungaricae*
Ann. Math. Statist.	*The Annals of Mathematical Statistics*
Ann. Statist.	*The Annals of Statistics*
Bull. Amer. Math. Soc.	*Bulletin of the American Mathematical Society*
Bell Syst. Tech. J.	*Bell System Technical Journal*

Calcutta Statist. Assoc. Bull.	*Calcutta Statistical Association Bulletin* (Calcutta)
Comput. J.	*The Computer Journal* (London)
Dokl. Akad. Nauk SSSR	*Doklady Akademii Nauk SSSR* (Moscow)
Duke Math. J.	*Duke Mathematical Journal*
J. Amer. Statist. Assoc.	*Journal of the American Statistical Association*
J. Roy. Statist. Soc. Ser. A (*or B*)	*Journal of the Royal Statistical Society, Series A* (*or B*).
J. Roy. Statist. Soc. Suppl.	*Journal of the Royal Statistical Society Supplement*
J. Austral. Math. Soc.	*The Journal of the Australian Mathematical Society* (Sydney)
Kodai Math. Sem. Rep.	*Kodai Mathematical Seminar Reports* (Tokyo)
Nat. Bur. Standards Appl. Math. Ser.	*National Bureau of Standards, Applied Mathematics Series* (Washington, D.C.)
Naval Res. Logist. Quart.	*Naval Research Logistics Quarterly*
Pacific J. Math.	*Pacific Journal of Mathematics*
Philos. Trans. Roy. Soc. London Ser. A	*Philosophical Transactions of the Royal Society of London, Series A*
Proc. Amer. Math. Soc.	*Proceedings of the American Mathematical Society*
Proc. Cambridge Philos. Soc.	*Proceedings of the Cambridge Philosophical Society* (Cambridge, England)
Proc. Berkeley Symp. Math. Stat. Prob., *n*th ($n = 1, \ldots, 6$)	*Proceedings of the Berkeley Symposium on Mathematical Statistics and Probability*, *n*th ($n = 1, \ldots, 6$) (Berkeley, California)
Proc. Nat. Acad. Sci. U.S.A.	*Proceedings of the National Academy of Sciences of the United States of America*
Publ. Math. Inst. Hungar. Acad. Sci.	*Publications of the Mathematical Institute, Hungarian Academy of Sciences*
Rep. Statist. Appl. Res. Un. Japan Sci. Engrs.	*Reports of Statistical Application Research, Union of Japanese Scientists and Engineers* (Tokyo)
Rev. Inst. Internat. Statist.	*Review of the International Statistical Institute* (Hague)
Sankhyā	*Sankhyā, The Indian Journal of Statistics*
Skand. Aktuarietidskr.	*Skandinavisk Aktuarietidskrift*
South African Statist. J.	*South African Statistical Journal*
Theor. Probability Appl.	*Theory of Probability and Its Applications*
Trabajos Estadist.	*Trabajos de Estadistica Y de Investigation Operation*
Trans. Amer. Math. Soc.	*Transactions of the American Mathematical Society*
Uspehi Mat. Nauk. Akademija Nauk SSSR	*Moskovskoe Mathematičeskoe Obščestvo, Uspehi Mathematičeskih Nauk* (Moscow–Leningrad)
Z. Wahrscheinlichkietstheorie und Verw. Gebiete	*Zeitschrift für Wahrscheinlichkeitstheorie und Verwandte* Gebiete (Berlin)

ABRAHAM, J. K. (1969). *The Local Power of Sequential Tests Subject to an Expected Sample Size Restriction.* Unpublished Tech. Rep. Dept. of Statist., Stanford Univ., Stanford California. [285, 286].

AIVAZIAN, S. A. (1959). A comparison of the optimal properties of the Neyman–Pearson and the Wald sequential probability ratio tests. *Theor. Probability Appl.* **4**, 86–93. [216, 218]

ALAM, K. (1967). A two-sample estimate of a common mean. *Ann. Inst. Statist. Math.* **19**, 261–270. [310]

ALBERT, A. (1961). The sequential design of experiments for infinitely many states of nature. *Ann. Math. Statist.* **32**, 774–799. [238]

ALBERT, A. (1966). Fixed size confidence ellipsoids for linear regression parameters. *Ann. Math. Statist.* **37**, 1602–1630. [434]

ALBERT, G. E. (1947). A note on the fundamental identity of sequential analysis. *Ann. Math. Statist.* **18**, 593–596. [47]

ALBERT, G. E. (1948). Correction to "A note on the fundamental identity of sequential analysis." *Ann. Math. Statist.* **19**, 426. [47]

*ALBERT, G. E. (1954). On the computation of the sampling characteristics of a general class of sequential decision problems. *Ann. Math. Statist.* **25**, 340–356. [Chapter 3].

*ALBERT, G. E. (1956). Accurate sequential tests on the mean of an exponential distribution. *Ann. Math. Statist.* **27**, 460–470. [Chapter 3].

ANDERSON, R. L., and BANCROFT, T. A. (1952). *Statistical Theory in Research*, p. 192. McGraw-Hill, New York. [208]

ANDERSON, T. W. (1955). The integral of a symmetric unimodal function. *Proc. Amer. Math. Soc.* **6**, 170–176. [63]

ANDERSON, T. W. (1960). A modification of the sequential probability ratio test to reduce the sample size. *Ann. Math. Statist.* **31**, 165–197. [61, 62, 89, 90, 91, 117, 194, 219, 361, 522]

ANDERSON, T. W. (1962). Least squares and best unbiased estimates. *Ann. Math. Statist.* **33**, 266–272. [428]

ANDERSON, T. W., and FRIEDMAN, M. (1960). A limitation of the optimum property of the sequential probability ratio test. In *Contributions to Probability and Statistics* (*Essays in Honor of Harold Hotelling*) (I. Olkin, *et al.*, eds.), No. 6, pp. 57–69. Stanford Univ. Press, Stanford California. [5, 16, 18, 19, 81]

ANSCOMBE, F. J. (1949). Large sample theory of sequential estimation. *Biometrika* **36**, 455–458. [368, 377]

ANSCOMBE, F. J. (1952). Large sample theory of sequential estimation. *Proc. Cambridge Philos. Soc.* **48**, 600–607. [271, 368, 377, 378, 380, 382, 383, 398, 401, 415, 418, 420, 432, 450, 456, 466, 492, 502]

ANSCOMBE, F. J. (1953). Sequential estimation. *J. Roy. Statist. Soc. Ser. B* **15**, 1–29. [386, 387, 388, 393, 395, 398, 401, 418, 424]

ANSCOMBE, F. J. (1954). Fixed-sample size analysis of sequential observations. *Biometrics* **10**, 89–100. [386, 393, 394]

ARMITAGE, P. (1947). Some sequential tests of Student's hypothesis. *J. Roy. Statist. Soc. Suppl.* **9**, 250–263. [163, 174, 259,]

ARMITAGE, P. (1950). Sequential analysis with more than two alternative hypotheses, and its relation to discriminant function analysis. *J. Roy. Statist. Soc. Ser. B* **12**, 137–144. [163, 175, 199]

ARMITAGE, P. (1957). Restricted sequential procedures. *Biometrika* **44**, 9–26. [89, 91, 93, 522, 524]

*ARMITAGE, P. (1960). *Sequential Medical Trials.* Blackwell, Oxford.

ARMITAGE, P., and SCHNEIDERMAN, M. A. (1962). A family of closed sequential procedures. *Biometrika* **49**, 41–56. [361]

ARNOLD, K. J. (1951). Tables to facilitate sequential t-tests. *Nat. Bur. Stand.* (*U.S.*) *Appl. Math. Ser.* **7**, 82. [107]

ARROW, K. J., BLACKWELL, D., and GIRSHICK, M. A. (1949). Bayes and minimax solutions of sequential decision problems. *Econometrica* **17**, 213–244. [504]

BAHADUR, R. R. (1954). Sufficiency and statistical decision functions. *Ann. Math. Statist.* **25**, 423–462. [323]

BAHADUR, R. R. (1958). A note on the fundamental identity of sequential analysis. *Ann. Math. Statist.* **29**, 534–543. [43, 44, 45]

BAHADUR, R. R, (1966). A note on quantiles in large samples *Ann. Math. Statist.* **37**, 577–580. [493, 497]

BAKER, A. G. (1950). Properties of some tests in sequential analysis. *Biometrika* **37**, 334–346. [109, 111]

*BARNARD, G. A. (1947). Review of A. Wald's sequential analysis. *J. Amer. Statist. Assoc.* **42**, 658–664 [Chapter 2]

BARNARD, G. A. (1952). The frequency justification of certain sequential tests. *Biometrika* **39**, 144–150. [133]

BARNARD, G. A. (1954). Sampling inspection and statistical decisions. *J. Roy. Statist. Soc. Ser. B* **16**, 151–174. [227, 228]

BARTLETT, M. S. (1937). Some examples of statistical methods of research in agriculture and applied biology. *J. Roy. Statist. Soc. Suppl.* **4**, 137–170. [375]

BARTLETT, M. S. (1946). The large-sample theory of sequential tests. *Proc. Cambridge Philos. Soc.* **42**, 239–244. [64, 91, 149, 151, 524, 525, 526]

BARTLETT, M. S. (1955). *An Introduction to Stochastic Processes.* Cambridge Univ. Press, London and New York. [524]

BARTOSZYŃSKI, R. (1974). A combinatorial identity. To appear in Colloquium Mathematicum **29**. [509]

BECHHOFER, R. (1960). A note on the limiting relative efficiency of the Wald sequential probability test. *J. Amer. Statist. Assoc.* **55**, 660–663. [216, 219, 226]

BECHHOFER, R. E., and SOBEL, M. (1958). *A Sequential Multiple Decision Procedure for Ranking Parameters of Koopman-Darmois Populations with Special Reference to Ranking Means of Normal Populations with a Common Known Variance.* Part II of a two-part-monograph in the *Statist. Res. Monographs Ser.* co-sponsored by the Inst. of Math. Statist. and the Univ. of Chicago. [219]

*BECHHOFER, R., KIEFER, J., and SOBEL, M. (1968). *Sequential Identification and Ranking Problems*, with special reference to Koopman-Darmois populations. *Statist. Res. Monographs Ser.*, Vol. III., Univ. of Chicago Press, Chicago, Illinois. [Chapter 3]

BERK, R. H. (1969). Strong consistency of certain sequential estimators. *Ann. Math. Statist.* **40**, 1492–1495; see also correction: *Ann. Math. Statist.* **42**, 1135–1137 (1971). [476, 481]

BERK, R. H. (1970). Stopping time of SPRTS based on exchangeable models. *Ann. Math. Statist.* **41**, 979–990. [131, 132, 133, 139]

BERK, R. H. (1972). Locally most powerful sequential tests. Unpublished manuscript. [284, 285, 287, 289, 291, 292]

BERK, R. H. (1973). Some asymptotic aspects of sequential analysis. *Ann. Statist.* **1**, 1126–1138. [97, 98, 99, 100, 144, 145, 146, 220, 224, 289, 290]

BERK, R. H., and BICKEL, P. J. (1968). On invariance and almost invariance. *Ann. Math. Statist.* **39**, 1573–1576. [145]

BERK, R. H., and SAVAGE, I. R. (1968). The information in a rank-order and stopping time of some associated SPRTS. *Ann. Math. Statist.* **39**, 1661–1674. [271, 273, 277, 282]

BERRY, A. C. (1941). The accuracy of the Gaussian approximation to the sum of independent variates. *Trans. Amer. Math. Soc.* **49**, 122–136. [306]

BESSLER, S. A. (1960). *Theory and Application of Sequential Design of Experiments, k-actions and Infinitely Many Experiments.* Tech. Rep. No. 55. Dept. of Statist., Stanford Univ. Stanford, California. [238]

*BESSLER, S. A., CHERNOFF, H., and MARSHALL, A. W. (1962). An optimal sequential accelerated life test. *Technometrics* **4**, 367–379. [Chapter 3]

BHATE, D. H. (1955). *Some Properties of Sequential Tests,* Doctoral Dissertation, Univ. of London, London. [146, 147]

BHATE, D. H. (1959). Approximations to the distribution of sample size for sequential tests. I. Tests of simple hypothesis. *Biometrika* **46**, 130–138. [63]

BHATTACHARYYA, A. (1946). On some analogues of the amount of information and their use in statistical estimation. *Sankhyā* **8**, 1–14. [321]

BICKEL, P. J., and YAHAV, J. A. (1967). Asymptotically pointwise optimal procedures in sequential analysis. *Proc. Symp. Math. Statist. Probability, 5th, Berkeley,* 1965, **1,** pp. 401–413. Univ. of California Press, Berkeley. [482, 483, 485, 486]

BICKEL, P. J., and YAHAV, J. A. (1969). On an A.P.O. rule in sequential estimation with quadratic loss. *Ann. Math. Statist.* **40**, 417–426. [238, 482, 486]

BILLARD, L., and VAGHOLKAR, M. K. (1969). A sequential procedure for testing a null hypothesis against a two-sided alternative hypothesis. *J. Roy. Statist. Soc. Ser. B.* **31**, 285–294. [173, 199, 200, 206, 208]

BIRNBAUM, A., and HEALY, W. C., JR. (1960). Estimates with prescribed variance based on two-stage sampling. *Ann. Math. Statist.* **31**, 662–676. [311, 312, 314, 315]

BIRNBAUM, Z. W. (1956). On a use of the Mann-Whitney Statistics. *Proc. Symp. Math. Statist. Probability, 3rd, Berkeley,* 1956, pp. 13–17. Univ. of California Press, Berkeley. [502]

BIRNBAUM, Z. W., and TINGEY, F. H. (1951). One-sided confidence contours for probability distribution functions. *Ann. Math. Statist.* **22**, 592–596. [256]

BLACKWELL, D. (1946). On an equation of Wald. *Ann. Math. Statist.* **17**, 84–87. [56]

BLACKWELL, D. (1947). Conditional expectation and unbiased sequential estimation. *Ann. Math. Statist.* **18**, 105–110. [322, 478]

BLACKWELL, D. (1953). Extension of a renewal theorem. *Pacific J. Math.* **3**, 315–320. [155]

BLACKWELL, D., and GIRSHICK, M. A. (1947). A lower bound for the variance of some unbiased sequential estimates. *Ann. Math. Statist.* **18**, 277–280. [321]

BLACKWELL, D., and GIRSHICK, M. A. (1954). *Theory of Games and Statistical Decisions* pp. 254–255. Wiley, New York. [343, 345, 346, 347, 354, 356]

BLYTH, C. R. (1951). On minimax statistical decision procedures and their admissibility. *Ann. Math. Statist.* **22**, 22–42. [337, 342]

BRADLEY, R. A. (1967). Topics in rank order statistics. *Proc. Symp. Math. Statist. Probability, 5th, Berkeley,* 1965, **1**, pp. 609–638. Univ. of California Press, Berkeley. [259]

BRADLEY, R. A., MARTIN, D. C., and WILCOXON, F. (1965). Sequential rank tests I. Monte Carlo studies of the two-sample procedure. *Technometrics* **7**, 463–483. [260]

BRADLEY, R. A., MERCHANT, S. D., and WILCOXON, F. (1966). Sequential rank tests II. Modified two-sample procedure. *Technometrics* **8**, 615–623. [260, 267]

BURKHOLDER, D. L., and WIJSMAN, R. A. (1963). Optimum properties and admissibility of sequential tests. *Ann. Math. Statist.* **34**, 1–18. [79]

BURMAN, J. P. (1946). Sequential sampling formulae for a binomial population. *J. Roy. Statist. Soc. Suppl.* **8**, 98–103. [230]

CALLAHAN, J. D. (1969a). *On Sequential Estimation Procedures for the Mean Vector of a Multi-variate Normal Distribution.* Tech. Rep. No. 132. Dept. of Statist., Johns Hopkins Univ., Baltimore, Maryland. [458]

CALLAHAN, J. D. (1969b). *Fixed-size Confidence Regions for Regression Parameters*. Tech. Rep. No. 129. Dept. of Statist., Johns Hopkins Univ., Baltimore, Maryland. [435]

CHANDA, K. C. (1971). Asymptotic distribution of the sample size for a sequential probability ratio test. *J. Amer. Statist. Assoc.* **66**, 178–183. [64, 65, 66]

CHAPMAN, D. G. (1954). The estimation of biological populations. *Ann. Math. Statist.* **25**, 1–15. [470]

CHERNOFF, H. (1952). A measure of asymptotic efficiency for tests of a hypothesis based on the sum of observations. *Ann. Math. Statist.* **23**, 493–507. [135, 219, 225, 226, 262]

CHERNOFF, H. (1956). Large-sample theory: Parametric case. *Ann. Math. Statist.* **27**, 1–22. [216, 218]

CHERNOFF, H. (1959). Sequential design of experiments. *Ann. Math. Statist.* **30**, 755–770. [238, 243]

CHERNOFF, H. (1960). Sequential tests for the mean of a normal distribution. *Proc. Symp. Math. Statist. Probability, 4th, Berkeley*, 1960, **1**, pp. 79–92. Univ. of California Press, Berkeley. [235]

CHERNOFF, H. (1967). Sequential models for clinical trials. *Proc. Symp. Math. Statist. Probability, 5th, Berkeley*, 1965, **4**, pp. 805–812. Univ. of California Press, Berkeley. [461, 462]

CHERNOFF, H. (1968). Optimal stochastic control. *Sankhyā Ser. A* **30**, 221–252. [236, 237, 459, 460]

*CHERNOFF, H. (1971). *The Efficient Estimation of a Parameter Measurable by Two Instruments of Unknown Precision*. Tech. Rep. No. 39. Stanford Univ., Stanford, California. [Section 4.17].

CHERNOFF, H., and RAY, S. N. (1965). A Bayes sequential sampling inspection plan. *Ann. Math. Statist.* **36**, 1387–1407. [461]

CHERNOFF, H., and SAVAGE, I. R. (1958). Asymptotic normality and efficiency of certain nonparametric test statistics. *Ann. Math. Statist.* **29**, 972–994. [500]

CHOW, Y. S. (1966a). On the moments of some one-sided stopping rules. *Ann. Math. Statist.* **37**, 382–387. [422]

CHOW, Y. S. (1966b). *Some convergence theorems for independent random variables*. Mimeo. Ser. No. 63. Dept. of Statist., Purdue Univ., Lafayette, Indiana. [433]

CHOW, Y. S., and ROBBINS, H. (1965). On the asymptotic theory of fixed-width sequential confidence intervals for the mean. *Ann. Math. Statist.* **36**, 457–462. [409, 412, 413, 414, 415, 416, 419, 420, 425, 428, 429, 432, 450, 454, 464, 466, 467, 491, 495, 498, 501]

CHOW, Y. S., ROBBINS, H., and TEICHER, H. (1965). Moments of randomly stopped sums. *Ann. Math. Statist.* **36**, 789–799. [31, 32]

CHOW, Y. S., ROBBINS, H., and SIEGMUND, D. (1971). *Great Expectations: The Theory of Optimal Stopping*, Houghton, Boston. [504, 505, 509, 511, 534, 535, 536]

CHUNG, K. L. (1949). An estimate concerning the Kolmogorov limit distribution. *Trans. Amer. Math. Soc.* **67**, 36–50. [258]

CONNELL, T. L., and GRAYBILL, F. A. (1964). A Tchebycheff type inequality for chi-square. Unpublished manuscript. [311, 313]

COX, C. P., and ROSEBERRY, T. D. (1966). A note on the variance of the distribution of sample number in sequential probability ratio tests. *Technometrics* **8**, 700–704. [64]

COX, D. R. (1952a). Sequential tests for composite hypotheses. *Proc. Cambridge Philos. Soc.* **48**, 290–299. [119, 121, 364, 366]

COX, D. R. (1952b). Estimation by double sampling. *Biometrika* **39**, 217–227. [368, 377]

COX, D. R. (1952c). A note on the sequential estimation of means. *Proc. Cambridge Philos. Soc.* **48**, 447–450. [377]

COX, D. R. (1963). Large-sample sequential tests for composite hypotheses. *Sankhyā Ser. A* **25**, 5–12. [149, 150, 152]

Cox, D. R., and Miller, H. D. (1965). *The Theory of Stochastic Processes.* Wiley, New York. [201, 204, 444, 522, 523]

Cramér, H. (1937). *Random Variables and Probability Distributions* (Cambridge Tracts in Math., No. 36). Cambridge Univ. Press, London and New York. [381]

Cramér, H. (1946). *Mathematical Methods of Statistics.* Princeton Univ. Press, Princeton, New Jersey. [148, 280, 378, 467, 527, 532]

Dantzig, G. B. (1940). On the nonexistence of tests of "Students" hypothesis having power functions independent of σ. *Ann. Math. Statist.* **11**, 186–192. [7, 8, 11, 293]

Darling, D. A., and Robbins, H. (1967a). Confidence sequences for mean, variance, and median. *Proc. Nat. Acad. Sci. U.S.A.* **58**, 66–68. [246, 249, 435]

Darling, D. A., and Robbins, H. (1967b). Iterated logarithm inequalities. *Proc. Nat. Acad. Sci. U.S.A.* **57**, 1188–1192. [244, 246, 248, 435]

Darling, D. A., and Robbins, H. (1967c). Inequalities for the sequence of sample means. *Proc. Nat. Acad. Sci. U.S.A.* **57**, 1577–1580. [244, 249, 250, 435]

Darling, D. A., and Robbins, H. (1967d). Finding the size of a finite population. *Ann. Math. Statist.* **38**, 1392–1398. [470]

Darling, D. A., and Robbins, H. (1968a). Some further remarks on inequalities for sample sums. *Proc. Nat. Acad. Sci. U.S.A.* **60**, 1175–1182. [250, 251, 252, 285]

Darling, D. A., and Robbins, H. (1968b). Some nonparametric sequential tests with power one. *Proc. Nat. Acad. Sci. U.S.A.* **61**, 804–809. [253, 254, 256, 258]

Darroch, J. N. (1958). The multiple recapture census. I. Estimation of a closed population. *Biometrika* **45**, 343–359. [470, 472, 474]

David, H. T., and Kruskal, W. H. (1956). The WAGR sequential t-test reaches a decision with probability one. *Ann. Math. Statist.* **27**, 797–805. [104]

Degroot, M. H. (1959). Unbiased binomial sequential estimation. *Ann. Math. Statist.* **30**, 80–101. [317, 327, 328, 329, 331]

Dodge, H. F., and Romig, H. G. (1929). A method of sampling inspection. *Bell System Tech. J.* **8**, 613–631. [3]

Donnelly, T. G. (1957). *A Family of Truncated Sequential Tests.* Doctoral Dissertation. Univ. of North Carolina, Chapel Hill. [89]

*Doob, J. L. (1948). Renewal theory from the point of view of the theory of probability. *Trans. Amer. Math. Soc.* **63**, 422–438. [Section 2.4]. [418]

Doob, J. L. (1953). *Stochastic Processes.* Wiley, New York. [479]

Dvoretsky, A., Kiefer, J., and Wolfowitz, J. (1953). Sequential decision problems for processes with continuous time parameter: Testing hypotheses. *Ann. Math. Statist.* **24**, 254–264. [217]

Dynkin, E. B., and Yushkevich, A. A. (1967). *Theorems and Problems in Markov Processes*, in Russian. Izd. Nauka, Moscow. [505, 506]

Eicker, F. (1963a). Central limit theorems for families of sequences of random variables. *Ann. Math. Statist.* **34**, 439–447. [429, 431]

Eicker, F. (1963b). Asymptotic normality and consistency of the least squares estimators for families of linear regression. *Ann. Math. Statist.* **34**, 447–457. [428]

Elfving, G. (1961). An expansion principle for distribution functions with applications to Student's statistic and the one-dimensional classification statistic. In *Studies in Item Analysis and Prediction* (H. Solomon, ed.). Stanford Univ. Press., Stanford, California. [445]

Fabian, V. (1974). Note on Anderson's sequential procedures with triangular boundary. *Ann. Statist.* **2**, 170–176. [92, 93]

Farrell, R. H. (1962). Bounded length confidence intervals for the zero of a regression function. *Ann. Math. Statist.* **33**, 237–247. [490]

FARRELL, R. H. (1964a) Asymptotic behavior of expected sample size in certain one-sided tests. *Ann. Math. Statist.* **35**, 36–72. [154, 250, 490]

*FARRELL, R. H. (1964b). Limit theorems for stopped random walks. *Ann. Math. Statist.* **35**, 1332–1343. [Section 3.14].

FARRELL, R. H. (1966a). Bounded length confidence intervals for the p-point of a distribution function, II. *Ann. Math. Statist.* **37**, 581–585. [487]

FARRELL, R. H. (1966b). Bounded length confidence intervals for the p-point of a distribution function, III. *Ann. Math. Statist.* **37**, 586–592. [487, 488, 489, 490, 492]

FELLER, W. (1966). *An Introduction to Probability Theory and Its Applications*, 2nd ed. Wiley, New York. [154, 291]

FELLER, W. (1970). *An Introduction to Probability Theory and Its Applications*, 2nd ed., Vol. II, pp. 209–210. Wiley, New York. [423]

FERGUSON, T. S. (1967). *Mathematical Statistics*: A Decision Theoretic Approach, pp. 378–383. Academic Press, New York. [49, 52]

FISZ, M. (1963). *Probability Theory and Mathematical Statistics.* Wiley, New York. [280]

FRASER, D. A. S. (1956). Sufficient statistics with nuisance parameters. *Ann. Math. Statist.* **27**, 838–842. [123]

FRASER, D. A. S. (1957). *Nonparametric Methods in Statistics.* Wiley, New York. [477]

GEERTSEMA, J. C. (1970a). Sequential confidence intervals based on rank tests. *Ann. Math. Statist.* **41**, 1016–1026. [491, 493, 494, 495, 496, 497, 498]

GEERTSEMA, J. C. (1970b). A Monte Carlo study of sequential confidence intervals based on rank tests. *South African Statist. J.* **4**, [498]

GHOSH, B. K. (1969). Moments of the distribution of sample size in a SPRT. *J. Amer. Statist. Assoc.* **64**, 1560–1574. [63]

GHOSH, B. K. (1970). *Sequential Tests of Statistical Hypotheses.* Addison-Wesley, Reading, Massachusetts. [79, 87, 91, 93, 94, 123, 125, 126, 137, 141, 142, 143, 151, 164, 170, 172 173, 175, 195, 196, 197, 198, 199, 210, 211, 213, 214, 215, 216, 323]

GHOSH, J. K. (1960a). On some properties of sequential t-test. *Calcutta Statist. Assoc. Bull.* **9**, 77–86. [106, 107]

GHOSH, J. K. (1960b). On the monotonicity of the OC of a class of sequential probablity ratio tests. *Calcutta Statist. Assoc. Bull.* **9**, 139–144. [87, 96]

GHOSH, M. N. (1960). Bounds for the expected sample size in a sequential probability ratio test. *J. Roy. Statist. Soc. Ser. B* **22**, 360–367. [54, 56, 63]

GHURYE, S. G., and ROBBINS, H. (1954). Two-stage procedures for estimating the difference between means. *Biometrika* **41**, 146–152. [297, 300, 301, 302]

GILBERT, J. P., and MOSTELLER, F. (1966). Recognizing the maximum of a sequence. *J. Amer. Statist. Assoc.* **61**, 35–73. [504, 509, 510, 511]

*GIRSHICK, M. A. (1946a). Contributions to the theory of sequential analysis. I. *Ann. Math. Statist.* **17**, 123–143. [Chapter 2]

GIRSHICK, M. A. (1946b). Contributions to the theory of sequential analysis. II, III. *Ann. Math. Statist.* **17**, 282–298. [49]

GIRSHICK, M. A., MOSTELLER, F., and SAVAGE, L. J. (1946). Unbiased estimates for certain binomial sampling problems with applications. *Ann. Math. Statist.* **17**, 13–23. [326, 327, 332]

GLESER, L. J. (1965). On the asymptotic theory of fixed-size sequential confidence bounds for linear regression parameters. *Ann. Math. Statist.* **36**, 463–467. [428, 429, 431, 432, 434]

GLESER, L. J. (1966). Correction note. *Ann. Math. Statist.* **37**, 1053–1055. [433]

GLESER, L. J., ROBBINS, H., and STARR, N. (1964). *Some Asymptotic Properties of Fixed-width Sequential Confidence Intervals for the Mean of a Normal Population with Unknown Variance.* Tech. Rep. Dept. of Math. Statist., Columbia Univ., New York. [398, 401, 409, 418]

GNEDENKO, B. V., and KOLMOGOROV, A. N. (1954). *Limit Distributions for Sums of Independent Random Variables*, translated from the Russian by K. L. Chung. Addison-Wesley, Reading, Massachusetts. [429]

GNEDENKO, B. V. and KOROLYUK, V. S. (1951). On the maximal deviation between two empirical distibutions. *Dokl. Akad. Nauk. SSSR* **80**, 525–528. [254]

GOODMAN, L. A. (1953). Sequential sampling tagging for population size problems. *Ann. Math. Statist.* **24**, 56–69. [470, 474]

GOVINDARAJULU, Z. (1961). *Central Limit Theorems and Asymptotic Efficiency for One-Sample Nonparametric Procedures.* Doctoral Dissertation. Univ. of Minnesota, Minneapolis. [501]

GOVINDARAJULU, Z. (1968a). Asymptotic normality of two-sample rank order sequential probability ratio test based on Lehmann alternatives. Unpublished manuscript. [98, 268, 269, 270]

GOVINDARAJULU, Z. (1968b). Distribution-free confidence bounds for $P(X < Y)$. *Ann. Inst. Statist. Math.* **20**, 229–238. [502]

GOVINDARAJULU, Z. (1969). Stopping time for a c-sample rank order SPRT. Unpublished manuscript. [281, 282]

GOVINDARAJULU, Z., LECAM, L., and RAGHAVACHARI, M. (1967). Generalizations of the theorems of Chernoff and Savage on the asymptotic normality of test statistics. *Proc. Symp. Math. Statist. Probability, 5th, Berkeley, 1965*, **1**, pp. 603–638. Univ. of California Press, Berkeley. [268, 269, 271, 500, 502]

GRAYBILL, F. A., and CONNELL, T. L. (1964a). Sample size for estimating the variance width in d units of the true value. *Ann. Math. Statist.* **35**, 438–440. [311, 467]

GRAYBILL, F. A., and CONNELL, T. L. (1964b). Sample size required to estimate the parameter in the uniform density within d units. *J. Amer. Statist. Assoc.* **59**, 550–556. [313, 314]

GREENWOOD, J. A., and SANDOMIRE, M. M. (1950). Sample size required for estimating the standard deviation as a percent of its true value. *J. Amer. Statist. Assoc.* **45**, 257–260. [310]

GROENVELD, R. A. (1971). A note on the sequential sign test. *Amer. Statist.* **25**, 2, 15–16. [279, 280]

HÁJEK, J. and ŠIDAK, Z. (1967). *Theory of Rank Tests.* Academic Press, New York. [280, 500]

HALDANE, J. B. S. (1945). On a method of estimating frequencies. *Biometrika* **33**, 222–225. [331, 372]

HALL, W. J. (1962). Some sequential analogs of Stein's two-stage test. *Biometrika* **49**, 367–378. [109, 111, 115, 118, 361]

HALL, W. J. (1965). *Methods of Sequentially Testing Composite Hypotheses with special Reference to the Two-Sample Problem.* Mimeo, series. No. 441, pp. 40–41. Inst. of Statist., Univ. of North Carolina, Chapel Hill. [259]

HALL, W. J., WIJSMAN, R. A., and GHOSH, J. K. (1965). The relationship between sufficiency and invariance with applications in sequential analysis. *Ann. Math. Statist.* **36**, 575–614. [130, 132, 141, 142, 196, 259, 260]

HARDY, G. H., LITTLEWOOD, J. E., and POLYA, G. (1959). *Inequalities.* Cambridge Univ. Press, London and New York. [165]

HEWITT, E., and SAVAGE, L. J. (1955). Symmetric measures on cartesian products. *Trans. Amer. Math. Soc.* **80**, 470–501. [132, 479]

HODGES, J. L., JR., and LEHMANN, E. L. (1950). Some problems in minimax estimation. *Ann. Math. Statist.* **21**, 182–197. [337]

HODGES, JR., J. L., and LEHMANN, E. L. (1951). Some applications of the Cramér-Rao inequality. *Proc. Symp. Math. Statist. Probability, 2nd, Berkeley*, 1950, **1**, pp. 13–22. Univ. of California Press, Berkeley. [333]

HOEFFDING, W. (1948). A class of statistics with asymptotically normal distribution. *Ann. Math. Statist.* **19**, 293–325. [495, 498]

HOEFFDING, W. (1953). A lower bound for the average sample number of a sequential test. *Ann. Math. Statist.* **24**, 127–130. [56, 57, 58]

HOEFFDING, W. (1960). Lower bounds for the expected sample size and the average risk of a sequential procedure. *Ann. Math. Statist.* **31**, 352–368. [58, 59, 62, 64, 117, 118]

HOEFFDING, W. (1963). Probability inequalities for sums of bounded random variables *J. Amer. Statist. Assoc.* **58**, 13–30. [494, 495, 496]

HOEL, D. G. (1970). On the monotonicity of the OC of a SPRT. *Ann. Math. Statist.* **41**, 310–314. [96]

HOEL, P. G. (1954). On a property of the sequential t-test. *Skand. Aktuarietidskr.* **37**, 19–22. [107]

HOEL, P. G. (1955). On a sequential test for the general linear hypothesis. *Ann. Math. Statist.* **26**, 136–139. [119]

IFRAM, A. F. (1965). On the asymptotic behavior of densities with applications to sequential analysis. *Ann. Math. Statist.* **36**, 615–637. [93, 131]

*JACKSON, J. E. (1960). Bibliography on sequential analysis. *J. Amer. Statist. Assoc.* **55**, 561–580.

JOHNSON, N. L. (1953). Some notes on the application of sequential test for the general linear hypothesis. *Ann. Math. Statist.* **24**, 614–623. [119]

JOHNSON, N. L. (1959). A proof of Wald's theorem on cumulative sums. *Ann. Math. Statist.* **30**, 1245–1247. [29]

*JOHNSON, N. L. (1961). Sequential analysis: A survey. *J. Roy. Statist. Soc.Ser. A* **124**, 372–411.

*JOINER, B. L., LAUBSCHER, N. F., BROWN, E. S., and LEVY, B. (1970). *An author and permitted title index to selected statistical journals. Nat. Bur. Stand. Spec. (U.S.) Publ.* **No. 321**, 279–281.

KARLIN, S. (1957). Polya type distributions II. *Ann. Math. Statist.* **28**, 281–308. [85]

KEMP, K. W. (1958). Characteristic and the average sample number of some sequential tests/AE for calculating the operating. *J. Roy. Statist. Soc. Ser. B* **20**, 379–386. [67, 68, 71, 76]

KEMPERMAN, J. H. B. (1961). *The Passage Problem for a Stationary Markov Chain.* Univ. of Chicago Press, Chicago, Illinois.[49, 50, 52, 53]

KHAN, R. A. (1968). Sequential estimation of the mean vector of a multivariate normal distribution. *Sankhyā Ser. A* **30**, 331–334. [452, 453, 454, 455, 456, 457]

KHAN, R. A. (1969). A general method of determining fixed-width confidence intervals. *Ann. Math. Statist.* **40**, 704–709. [464, 465, 466, 468]

KIEFER, J. (1952). Sequential minimal estimation for the rectangular distribution with known range. *Ann. Math. Statist.* **23**, 586–593. [340, 341]

KIEFER, J., and SACKS, J. (1963). Asymptotically optimum sequential inference and design. *Ann. Math. Statist.* **34**, 705–750. [238, 241, 243, 244, 482]

KIEFER, J., and WEISS, L. (1957). Some properties of generalized sequential probability ratio tests. *Ann. Math. Statist.* **28**, 57–74. [88, 219]

KIEFER, J., and WEISS, L. (1964). Unpublished work. [487]

KINGMAN, J. F. C. (1962). Some inequalities for the queue-GI/G/1. *Biometrika* **48**, 315–324. [189, 191]

KNIGHT, W. A. (1965). A method of sequential estimation applicable to the hypergeometric, binomial, Poisson, and exponential distributions. *Ann. Math. Statist.* **36**, 1494–1503. [327]

KULLBACK, S. (1959). *Information Theory and Statistics.* Wiley, New York. [220]

KURTZ, T. E. (1963). *Basic Statistics.* Prentice-Hall, Englewood Cliffs, New Jersey. [152]

LECAM, L. (1970). On the assumptions used to prove asymptotic normality of maximum likelihood estimates. *Ann. Math. Statist.* **41**, 802–828. [148, 527, 528, 529]

LECHNER, J. A. (1964). Optimality and the OC curve for the Wald SPRT. *J. Amer. Statist. Assoc.* **59**, 464–468. [289]

LEHMANN, E. L. (1950). *Notes on the Theory of Estimation* (Lecture notes recorded by C. Blyth). Associated Student's Store, Univ. of California, Berkeley. [318, 320, 322, 334, 340]

LEHMANN, E. L. (1953). The power of rank tests. *Ann. Math. Statist.* **24**, 23–43. [259]

LEHMANN, E. L. (1959). *Testing Statistical Hypotheses*, pp. 104–110. Wiley, New York. [78, 79, 86, 129, 287]

LEHMANN, E. L., and STEIN, C. (1950). Completeness in the sequential case. *Ann. Math. Statist.* **21**, 376–385. [324, 327]

LÉVY, P. (1937). *Théorie de l'addition des variables aleatoires.* Gauthier-Villars, Paris. [380, 381]

LEWONTIN, R. C., and PROUT, T. (1956). Estimation of the number of different classes in a population. *Biometrics* **12**, 211–223. [472]

*LINDGREN, B. W. (1966). *Statistical Theory*, 2nd ed. Macmillan, New York. [Chapter 2].

LINDLEY, D. V., and BARNETT, B. N. (1965). Sequential sampling, two decision problems with linear losses for binomial and normal random variables. *Biometrika* **52**, 507–532. [232, 234, 235, 236]

LOÉVE, M. (1962). *Probability Theory*, 3rd ed. Van Nostrand-Reinhold, Princeton, New Jersey. [492]

LORDEN, G. (1967). Integrated risk of asymptotically Bayes sequential tests. *Ann. Math. Statist.* **37**, 1399–1422. [187, 191]

LORDEN, G. (1970). On excess over the boundary. *Ann. Math. Statist.* **41**, 520–527. [153, 154, 157]

LORDEN, G. (1972). Likelihood ratio tests for sequential k-decision problems. *Ann. Math. Statist.* **43**, 1412–1427. [189, 190, 191, 192, 194]

LORDEN, G. (1973). Open-ended tests for Koopman–Darmois families. *Ann. Statist.* **1**, 633–643. [157, 159, 162, 185, 192, 194]

LOTKA, A. J. (1939). Théorie analytique des associations biologiques II. *Actualités Sci. Ind.* **780**, 123–136. [31]

LOYNES, R. M. (1969). The consistency of certain sequential estimators. *Ann. Math. Statist.* **40**, 568–574. [476, 477, 480, 481]

*MCCABE, G. P., JR. (1970). *Sequential Estimation of a Poisson Integer Mean.* Mimeo. Ser. No. 232. Dept. of Statist., Purdue Univ., Lafayette, Indiana. [Chapter 4]

*MCCABE, G. P., JR. (1974). *Sequential Estimation of a Restricted Mean Parameter of an Exponential Family. Ann. Inst. Statist. Math.* **26**, 103–116. [Chapter 4]

MALLIK, A. K. (1971). *Sequential Estimation of the Common Mean of Two Normal Populations.* Tech. Rep. No. 42. Dept. of Statist., Stanford Univ., Stanford, California. [459, 462, 463, 464]

MATTHES, T. K. (1963). On the optimality of sequential probability ratio tests. *Ann. Math. Statist.* **34**, 18–21. [79]

MIKHALEVICH, V. S. (1956). Sequential Bayes solutions and optimal methods of statistical acceptance control. *Theor. Probability Appl.* **1**, 395–421. [236]

MILLER, R. G. (1970). Sequential signed rank test. *J. Amer. Statist. Assoc.* **65**, 1554–1561. [277, 278, 279]

MOGYORÓDI, J. (1961). On limiting distributions for sums of a random number of independent random variables. *Publ. Math. Inst. Hungar. Acad. Sci.* **6**, 365–371. [433]

MORIGUTI, S., and ROBBINS, H. (1962). A Bayes test of $p \leq 1/2$ against $p > 1/2$. *Rep. Statist. Appl. Res. Uu. Japan Sci. Engrs.* **9**, 39–60. [235]

MOSHMAN, J. (1958). A method for selecting the size of the initial sample in Stein's two-sample procedure. *Ann. Math. Statist.* **29**, 1271–1275. [11, 13, 404]

NÁDAS, A. (1969). An extension of a theorem of Chow and Robbins on sequential confidence intervals for the mean. *Ann. Math. Statist.* **40**, 667–671. [412, 419, 420, 421]

NELDER, J. A., and MEAD, R. (1965). A simple method for function minimization. *Comput. J.* **7**, 308–313. [206]

*NEYMAN, J. (1937). Outline of a theory of statistical estimation based on the classical theory of probability. *Philos. Trans. Roy. Soc. London Ser. A* **236**, 333–380. [Chapter 4]

NEYMAN, J., and PEARSON, E. S. (1933). On the problem of the most efficient tests of statistical hypotheses. *Philos. Trans. Roy. Soc. London. Ser. A.* **231**, 289–337. [14]

*NOETHER, G. (1954). A sequential test of randomness against linear trend. *Ann. Math. Statist.* **25**, 176 (abstract). [Chapter 3]

O'NEILL, R. T., and ROHATGI, V. K. (1973). A two-stage procedure for estimating the difference between the mean vectors of two multi-variate normal populations. *Trabajos Estadist.* **24**, 123–130. [303]

PAGE, E. S. (1954). An improvement to Wald's approximation for some properties of sequential tests. *J. Roy. Statist. Soc. Ser. B* **16**, 136–139. [67, 68, 71, 76]

PARENT, JR., E. A. (1965). *Sequential Ranking Procedures.* Tech. Rep. No. 80. Dept. of Statist., Stanford Univ., Stanford, California. [274, 275]

PAULSON, E. (1947). A note on the efficiency of the Wald sequential test. *Ann. Math. Statist.* **18**, 447–450. [216, 218, 223]

PAULSON, E. (1963). A sequential procedure for choosing one of k hypotheses concerning the unknown mean of a normal distribution. *Ann. Math. Statist.* **34**, 549–554. [175, 177, 180, 360]

PAULSON, E. (1964a). A sequential procedure for selecting the population with the largest mean from k normal populations. *Ann. Math. Statist.* **35**, 174–180. [93, 447]

PAULSON, E. (1964b). Sequential estimation and closed sequential decision procedures. *Ann. Math. Statist.* **35**, 1048–1058. [182, 183, 358, 361, 363, 365, 367, 447]

PAULSON, E. (1969). Sequential interval estimation for the means of normal populations. *Ann. Math. Statist.* **40**, 509–516. [444, 445, 446]

PERNG, S. K. (1969). A comparison of the asymptotic expected sample sizes of two sequential procedures for ranking problems. *Ann. Math. Statist.* **40**, 2198–2202. [92]

PHATARFOD, R. M. (1965). Sequential analysis of dependent observations I. *Biometrika* **52**, 157–165. [93]

PUTTER, J. (1951). Sur une methode de double e'chantillonnage, pour estimer la moyenne d'une population Laplacienne stratifice. *Rev. Inst. Internat. Statist.* **19**, 231–238. [297]

PYKE, R., and SHORACK, G. R. (1968). Weak convergence of a two-sample empirical process and a new approach to Chernoff-Savage theorems. *Ann. Math. Statist.* **39**, 755–771. [500]

RAIFFA, H., and SCHLAIFER, R. (1961). *Applied Statistical Decision Theory.* Harvard Univ. Press, Cambridge, Massachusetts. [235, 349]

RAO, C. R. (1965). *Linear Statistical Inference and Its Applications.* Wiley, New York. [431, 465, 466]

RAY, W. D. A. (1957a). A proof that the sequential probability ratio test (SPRT) of the general linear hypothesis terminates with probability unity. *Ann. Math. Statist.* **28**, 521–522. [119, 120]

RAY, W. D. A. (1957b). Sequential confidence intervals for the mean of a normal population with unknown variance. *J. Roy. Statist. Soc. Ser. B* **19**, 133–143. [386, 388, 393, 396, 398, 401, 402, 404, 418]

RÉNYI, A. (1957). On the asymptotic distribution of the sum of random number of random variables. *Acta Math. Acad. Sci. Hungar.* **8**, 193–199. [441]

RÉNYI, A. (1962). Three new proofs and a generalization of a theorem of Irving Weiss. *Publ. Math. Inst. Hungar. Acad.Sci. Ser. A* **7**, 203–214. [471]

RICHTER, D. (1960). Two-stage experiments for estimating a common mean. *Ann. Math. Statist.* **31**, 1164–1173. [303, 304, 306, 307, 308, 309, 310, 459]

RICHTER, W. (1965). Limit theorems for sequences of random variables with sequences of random indices. *Theor. Probability Appl.* **10**, 74–84. [481]

ROBBINS, H. (1952). Some aspects of the sequential design of experiments. *Bull. Amer. Math. Soc.* **58**, 527–535. [297]

ROBBINS, H. (1959). Sequential estimation of the mean of a normal population. *Probability and Statistics* (The Herald Cramér Volume), pp. 235–245. Almqvist & Wiksell, Stockholm. [402, 406, 408, 459, 462, 463]

ROBBINS, H. (1970). Statistical methods related to the law of the iterated logarithm. *Ann. Math. Statist.* **41**, 1397–1409. [258, 259, 285, 438]

ROBBINS, H., and STARR, N. (1965). *Remarks on Sequential Hypothesis Testing.* Tech. Rep. No. 68. Dept. of Statist., Univ. of Minnesota, Minneapolis. [13, 442]

ROBBINS, H., SIMONS, G., and STARR, N. (1967). A sequential analogue of the Behrens-Fisher problem. *Ann. Math. Statist.* **38**, 1384–1391. [439, 440, 441, 442]

*ROBBINS, H., SOBEL, M., and STARR, N. (1968). A sequential procedure for selecting the largest of k means. *Ann. Math. Statist.* **39**, 88–92. [Chapter 3]

ROHATGI, V. K., and O'NEILL, R. T. (1973). On the sequential estimation of the mean vector of a multinormal population. *Trabajos Estadist.* **24**, part 3, 123–130. [454]

ROHATGI, V. K., and RASTOGI, S. C. (1973). On sequential estimation of a certain estimable function of the mean vector of a multivariate normal distribution. *J. Austral. Math. Soc.* **15**, 291–295. [456, 457, 458]

*ROMANI, J. (1956). Tests no parametricos en forma secuencial. *Trabajos Estadist.* **7**, 1–96. [Section 3.14]

ROY, S. N. (1957). *Some Aspects of Multivariate Analysis.* Wiley, New York. [434]

RUSHTON, S. (1950). On a sequential t-test. *Biometrika* **37**, 326–333. [104, 151]

SACKS, J. (1965). A note on the sequential t-test. *Ann. Math. Statist.* **36**, 1867–1869. [108, 109]

SAKAGUCHI, M. (1961). Dynamic programming of some sequential sampling designs. *J. Math. Anal. Appl.* **2**, 446–466. [534, 535, 536]

SAKAGUCHI, M. (1967). Relative efficiency of the Wald SPRT and the Chernoff information number. *Kōdai Math. Sem. Rep.* **19**, 138–145. [219]

SAMUEL, E. (1966). Estimators with prescribed bound on the variance for the parameters in the binomial and Poisson distributions, based on two-stage sampling. *J. Amer. Statist. Assoc.* **61**, 220–227. [315]

SAMUEL, E. (1968). Sequential maximum likelihood estimation of the size of a population. *Ann. Math. Statist.* **39**, 1057–1068. [470, 471, 472, 473, 474, 475]

SAVAGE, I. R. (1956). Contributions to the theory of rank order statistics. *Ann. Math. Statistics.* **27**, 590–615. [261]

SAVAGE, I. R. (1957). Contributions to the theory of rank order statistics–the "trend" case. *Ann. Math. Statist.* **28**, 968–977. [283]

SAVAGE, I. R. (1959). Contributions to the theory of rank order statistics–the one-sample case. *Ann. Math. Statist.* **30**, 1018–1023. [276]

SAVAGE, I. R., and SAVAGE, L. J. (1965). Finite stopping time and finite expected time. *J. Roy. Statist. Soc. Ser. B* **27**, 284–289. [94, 95, 96]

SAVAGE, I. R. and SETHURAMAN, J. (1966). Stopping time of a rank order sequential probability ratio test based on Lehmann alternatives. *Ann. Math. Statist.* **37**, 1154–1160; [260,

261, 263, 265, 266, 267, 273, 276, 281]; see also correction. *Ann. Math. Statist.* **38**, 1309 (1967). [263]

SAVAGE, L. J. (1947). A uniqueness theorem for unbiased sequential estimation. *Ann. Math. Statist.* **18**, 295–297. [327]

SCHWARZ, G. (1962). Asymptotic shapes of Bayes sequential testing regions. *Ann. Math. Statist.* **33**, 224–236. [157, 159, 184, 188, 238, 239, 240, 241]

SEELBINDER, B. M. (1953). On Stein's two-stage procedure. *Ann. Math. Statist.* **24**, 640–649. [404, 405]

SEN, P. K. and GHOSH, M. (1971). On bounded length sequential confidence intervals based on one-sample rank-order statistics. *Ann. Math. Statist.* **42**, 189–203. [498, 499, 501]

SETH, G. R. (1949). On the variance of estimates. *Ann. Math. Statist.* **20**, 1–27. [321, 322]

SETHURAMAN, J. (1964). On the probability of large deviations of families of sample means. *Ann. Math. Statist.* **35**, 1304–1316. [262]

SETHURAMAN, J. (1970). Stopping time of a rank order sequential probability ratio test based on Lehmann alternatives-II. *Ann. Math. Statist.* **41**, 1322–1333. [267, 268, 270, 282]

SETHURAMAN, J. (1970). Correction note. *Ann. Math. Statist.* **41**, 1376–1380. [262]

SIEGMUND, D. O. (1967). Some one-sided stopping rules. *Ann. Math. Statist.* **38**, 1641–1646. [98, 422, 424]

SIMONS, G. (1967a). Lower bounds for average sample number of sequential multi-hypothesis tests. *Ann. Math. Statist.* **38**, 1343–1364. [164, 172]

SIMONS, G. (1967b). A sequential three hypothesis test for determining the mean of a normal population with known variance. *Ann. Math. Statist.* **38**, 1365–1375. [173, 176]

SIMONS, G. (1967c). A class of sequential procedures for choosing one of k hypotheses concerning the unknown drift parameters of the Wiener process. *Ann. Math. Statist.* **38**, 1376–1383.[175]

SIMONS, G. (1968). On the cost of not knowing the variance when making a fixed-width confidence interval for the mean. *Ann. Math. Statist.* **39**, 1946–1952. [424, 425, 426, 458, 498]

SIMONS, G., and ZACKS, S. (1967). *A Sequential Estimation of Tail Probabilities in Exponential Distribution with a Prescribed Proportional Closeness.* Tech. Rep. No. 21. Dept. of Statist., Stanford Univ., Stanford, California. [448, 449, 450]

SMIRNOV, N. V. (1944). Approximation of random variables by empirical data. *Uspehi Mat. Nauk.* **10**, 179–206. [256]

SOBEL, M., and WALD, A. (1949). A sequential decision procedure for choosing one of three hypotheses concerning the unknown mean of a normal distribution. *Ann. Math. Statist.* **20**, 502–522. [118, 163, 164, 177, 208]

SPROULE, R. N. (1969). *A Sequential Fixed-Width Confidence Interval for the Mean of a U-statistic.* Doctoral Dissertation. Univ. of North Carolina, Chapel Hill. [498]

*SRIVASTAVA, M. S. (1966). Some asymptotically efficient sequential procedures for ranking and slippage problems. *J. Roy. Statist. Soc. Ser. B* **28**, 370–380.

*SRIVASTAVA, M. S. (1967a). The performance of a sequential procedure for a slippage problem. Unpublished manuscript. [Chapter 3]

SRIVASTAVA, M. S. (1967b). On fixed-width confidence bounds for regression parameters and mean vector. *J. Roy. Statist. Soc. Ser. B* **29**, 132–140. [434, 435]

SRIVASTAVA, M. S. (1971). On fixed-width confidence bounds for regression parameters. *Ann. Math. Statist.* **42**, 1403–1411. [434, 435]

*SRIVASTAVA, M. S., and OGILVIE, J. (1968). The performance of some sequential procedures for a ranking problem. *Ann. Math. Statist.* **39**, 1040–1047. [Chapter 3].

STARR, N. (1966a). The performance of a sequential procedure for the fixed-width interval estimate. *Ann. Math. Statist.* **36**, 36–50. [396, 398, 400, 403, 404, 405, 408, 418, 425]

STARR, N. (1966b). On the asymptotic efficiency of a sequential procedure for estimating the mean. *Ann. Math. Statist.* **37**, 1173–1185. [406, 407, 409, 410, 459]

*STARR, N., and WOODROOFE, M. B. (1968). Remarks on stopping time. *Proc. Nat. Acad. Sci. U.S.A.* **61,** 1215–1218 [Chapter 4]

STARR, N., and WOODROOFE, M. B. (1969). Remarks on sequential point estimation. *Proc. Nat. Acad. Sci. U.S.A.* **63**, 285–288. [410, 411, 458]

STARR, N., and WOODROOFE, M. B. (1971). *Further Remarks on Sequential Estimation: The Exponential Case*. Tech. Rep. No. 7. Dept. of Statist, Univ. of Michigan, Ann Arbor. [447, 448]

STEIN, C. (1945). A two-sample test for a linear hypothesis whose power is independent of the variance. *Ann. Math. Statist.* **16**, 243–258. [8, 10, 13, 115, 293, 294, 295, 368, 404, 405, 412]

STEIN, C. (1946). A note on cumulative sums. *Ann. Math. Statist.* **17**, 498–499. [25, 56, 133, 134, 135, 267]

STEIN, C. (1949). Some problems in sequential estimation. *Econometrica* **17**, 77–78. [396, 398, 411, 412, 418]

STEIN, C., and WALD, A. (1947). Sequential confidence intervals for the mean of a normal distribution with known variance. *Ann. Math. Statist.* **18**, 427–433. [386]

STONE, C. J. (1966). Absolutely continuous components and renewal theory. *Ann. Math. Statist.* **37**, 271–275. [154]

TALLIS, G. M., and VAGHOLKAR, M. K. (1965). Formulas to improve Wald's approximation for some prperties of sequential tests. *J. Roy. Statist. Soc. Ser. B* **27**, 74–81. [71, 72, 76, 77]

*TEICHER, H. (1967). A dominated ergodic type theorem. *Z. Wahrscheinlichkeitstheorie und Verw. Gebiete* **8**, 113–116. [Chapter 4]

VAGHOLKAR, M. K. (1955). *Application of Statistical Decision Theory to Sampling Inspection Schemes*. Doctoral Dissertation. Univ. of London, London. [231]

VAGHOLKAR, M. K., and WETHERILL, G. B. (1960). The most economical binomial sequential probability ratio test. *Biometrika* **47**, 103–109. [227, 231, 232]

WALD, A. (1945). Sequential tests of statistical hypotheses. *Ann. Math. Statist.* **16**, 117–186. [54, 56, 57]

WALD, A. (1946). Differentiation under the expectation sign in the fundamental identity of sequential analysis. *Ann. Math. Statist.* **17**, 493–497. [41, 515]

WALD, A. (1947). *Sequential Analysis*. Wiley, New York. [5, 18, 21, 32, 40, 41, 43, 47, 49, 54, 56, 57, 60, 61, 63, 67, 68, 71, 76, 77, 80, 97, 98, 99, 102, 103, 104, 107, 110, 121, 133, 147, 150, 153, 156, 162, 188, 193, 195, 250, 258, 259, 260, 279, 316, 358, 421, 451, 513, 526]

WALD, A. (1950). *Statistical Decision Functions*. Wiley, New York. [342]

WALD, A., and WOLFOWITZ, J. (1948). Optimum character of the sequential probability ratio test. *Ann. Math. Statist.* **19**, 326–339. [78, 79, 208, 232]

WASAN, M. T. (1964). Sequential optimum procedures for unbiased estimation of a binomial parameter. *Technometrics* **6**, 259–272. [317, 331, 332, 335]

WEED, J. H. D., Jr. (1968). *Sequential One-sample Grouped Rank Tests for Symmetry*. Ph.D. Dissertation, Florida State University Library, Tallahassee, Florida. [275]

WEED, JR., H. D., and BRADLEY, R. A. (1971). Sequential one-sample grouped signed rank tests for symmetry. *J. Amer. Statist. Assoc.* **66**, 321–326. [274]

WEED, JR., H. D., BRADLEY, R. A., and GOVINDARAJULU, Z. (1974). Stopping times of two rank order sequential probability ratio tests for symmetry based on Lehmann alternatives. *Ann. Statist.* **2**, 1314–1322. [276, 277]

WEISS, L. (1953). Testing one simple hypothesis against another. *Ann. Math. Statist.* **24**, 273–281. [87, 88, 89]

WEISS, L. (1956). On the uniqueness of Wald sequential tests. *Ann. Math. Statist.* **27**, 1178–1181. [19, 81]

WEISS, L. (1960). Confidence intervals of preassigned length for quantiles of unimodal populations. *Naval Res. Logist. Quart.* **7**, 251–256. [487]

WETHERILL, B. G. (1957). *Application of the Theory* of *Decision Functions to Sampling Inspection with Special Reference to Cost of Inspection.* Doctoral Dissertation. Univ. of London, London. [232]

WETHERILL, B. G. (1966). *Sequential Methods in Statistics* (Methuen Monographs on Appl. Probability and Statist.). Methuen, London. [121, 146, 293, 524]

WHITTAKER, E. T., and WATSON, G. S. (1947). *Modern Analysis.* Cambridge Univ. Press, London and New York. [254]

WHITTLE, P., and LANE, R O.D. (1967). A class of situations in which a sequential estimation procedure is nonsequential. *Biometrika* **54**, 229–234. [348, 349, 351, 354]

WIENER, N. (1939). The ergodic theorem. *Duke Math. J.* **5**, 1–18. [441, 468]

WIJSMAN, R. A. (1958). On the existence of Wald's sequential test. *Ann. Math. Statist.* **29**, 938–939 (abstract). [19]

WIJSMAN, R. A. (1960). A monotonicity property of the sequential probability ratio test. *Ann. Math. Statist.* **31**, 677–684. [19, 80, 81, 82, 84]

WIJSMAN, R. A. (1963). Existence, uniqueness and monotonocity of sequential probability ratio tests. *Ann. Math. Statist.* **34** 1541–1548. [85]

WIJSMAN, R. A. (1967a). Cross sections of orbits and their application to densities of maximal invariants. *Proc. Symp. Math. Statist. Probability, 5th Berkeley*, 1965, **1**, pp. 389–400 Univ. of California Press, Berkeley. [93, 133, 142]

WIJSMAN, R. A. (1967b). General proof of termination with probability one of invariant sequential probability ratio tests based on multivariate normal observations. *Ann. Math. Statist.* **38**, 8–24. [131]

WIJSMAN, R. A. (1968). Bounds on the sample size distribution for a class of invariant sequential probability ratio tests. *Ann. Math. Statist.* **39**, 1048–1056. [135]

WIJSMAN, R. A. (1971). Exponentially bounded stopping time of invariant sequential probability ratio tests. *Proc. Symp. on Decision Theory and Related Topics, Purdue Univ. Lafayette, Indiana*, 1970, pp. 217–224. Academic Press, New York. [133, 134, 139]

WIJSMAN, R. A. (1972). Examples of exponentially bounded stopping time of invariant sequential probability radio tests when the model may be false. *Proc. Symp. Math. Statist. Probability, 6th, Berkeley*, 1970, **1**, 109–128. Univ. of California Press, Berkeley. [133, 134, 135, 137, 138, 196]

WILCOXON, F., RHODES, L. J., and BRADLEY, R. A. (1963). Two sequential two-sample grouped rank tests with applications to screening experiments. *Biometrics* **19**, 58–84. [259]

WIRJOSUDIRDJO, S. (1961). *Limiting Behavior of a Sequence of Density Ratios.* Doctoral Dissertation. Univ. of Illinois, Urbana; See also *Ann. Math. Statist.* **33**, 296–297 (1962) (abstract). [260]

*WOINSKY, M. N., and KURTZ, L. (1969). Sequential nonparametric two-way classification with prescribed maximum asymptotic error probability. *Ann. Math. Statist.* **40**, 445–455. [Sections 3.14 and 4.22].

WOLFOWITZ, J. (1946). On sequential binomial estimation. *Ann. Math. Statist.* **17**, 489–493. [327]

WOLFOWITZ, J. (1947a). Consistency of sequential binomial estimates. *Ann. Math. Statist.* **18**, 131–135. [475, 477, 478]

WOLFOWITZ, J. (1947b). Efficiency of sequential estimates and Wald's equation for sequential processes. *Ann. Math. Statist.* **18**, 215–230. [318, 319, 320, 321]

WOLFOWITZ, J. (1950). Minimax estimates of the mean of a normal distribution with known variance. *Ann. Math. Statist.* **21**, 218–230. [338, 339]

WONG, S. P. (1968). Asymptotically optimum properties of certain sequential tests. *Ann. Math. Statist.* **39**, 1244–1263. [158, 185, 188]

*WOODROOFE, M. (1971). *The Bounded Modification of the SPRT.* Tech. Rep. No. 8. Dept. of Statist. Univ. of Michigan, Ann Arbor. [Chapter 2].

YAHAV, J. A., and BICKEL, P. J. (1968). Asymptotically optimal Bayes and minimax procedures in sequential estimation. *Ann. Math. Statist.* **39**, 442–456. [482, 484, 485, 486, 493]

YEN, E. H. (1964). On two-stage nonparametric estimation. *Ann. Math. Statist.* **35**, 1099–1114. [310]

ZACKS, S. (1971). *The Theory of Statistical Inference.* Wiley, New York. [239, 240]

ZACKS, S., and EVEN, M. (1966). The efficiencies in small samples of the maximum likelihood and best unbiased estimators of reliability functions. *J. Amer. Statist. Assoc.* **61**, 1033–1051. [449]

Subject Index

D

E

P

R

Z

Probability and Mathematical Statistics

A Series of Monographs and Textbooks

---

1. Thomas Ferguson. Mathematical Statistics: A Decision Theoretic Approach. 1967
2. Howard Tucker. A Graduate Course in Probability. 1967
3. K. R. Parthasarathy. Probability Measures on Metric Spaces. 1967
4. P. Révész. The Laws of Large Numbers. 1968
5. H. P. McKean, Jr. Stochastic Integrals. 1969
6. B. V. Gnedenko, Yu. K. Belyayev, and A. D. Solovyev. Mathematical Methods of Reliability Theory. 1969
7. Demetrios A. Kappos. Probability Algebras and Stochastic Spaces. 1969
8. Ivan N. Pesin. Classical and Modern Integration Theories. 1970
9. S. Vajda. Probabilistic Programming. 1972
10. Sheldon M. Ross. Introduction to Probability Models. 1972
11. Robert B. Ash. Real Analysis and Probability. 1972
12. V. V. Fedorov. Theory of Optimal Experiments. 1972
13. K. V. Mardia. Statistics of Directional Data. 1972
14. H. Dym and H. P. McKean. Fourier Series and Integrals. 1972
15. Tatsuo Kawata. Fourier Analysis in Probability Theory. 1972
16. Fritz Oberhettinger. Fourier Transforms of Distributions and Their Inverses: A Collection of Tables. 1973
17. Paul Erdös and Joel Spencer. Probabilistic Methods in Combinatorics. 1973
18. K. S•rkadi and I. Vincze. Mathematical Methods of Statistical Quality Control. 1973
19. Michael R. Anderberg. Cluster Analysis for Applications. 1973
20. W. Hengartner and R. Theodorescu. Concentration Functions. 1973
21. Kai Lai Chung. A Course in Probability Theory, Second Edition. 1974
22. L. H. Koopmans. The Spectral Analysis of Time Series. 1974
23. L. E. Maistrov. Probability Theory: A Historical Sketch. 1974
24. William F. Stout. Almost Sure Convergence. 1974

25. E. J. McShane. Stochastic Calculus and Stochastic Models. 1974

26. Z. Govindarajulu. Sequential Statistical Procedures. 1975

In Preparation

Roger Cuppens. Decomposition of Multivariate Probabilities

Robert B. Ash and Melvin F. Gardner. Topics in Stochastic Processes

Avner Friedman. Stochastic Differential Equations and Applications, Volumes 1 and 2

Harry Dym and Henry P. McKean. Gaussian Processes: Complex Function Theory and the Inverse Spectral Method

A
B 5
C 6
D 7
E 8
F 9
G 0
H 1
I 2
J 3